Biotechnology

Second Edition

Volume 11c

Environmental Processes III

Biotechnology

Second Edition

Fundamentals

Volume 1
Biological Fundamentals

Volume 2
Genetic Fundamentals and Genetic Engineering

Volume 3
Bioprocessing

Volume 4
Measuring, Modelling and Control

Products

Volume 5a
Recombinant Proteins, Monoclonal Antibodies and Therapeutic Genes

Volume 5b
Genomics and Bioinformatics

Volume 6
Products of Primary Metabolism

Volume 7
Products of Secondary Metabolism

Volumes 8a and b
Biotransformations I and II

Special Topics

Volume 9
Enzymes, Biomass, Food and Feed

Volume 10
Special Processes

Volumes 11a–c
Environmental Processes I–III

Volume 12
Legal, Economic and Ethical Dimensions

All volumes are also displayed on our Biotech Website: http://www.wiley-vch.de/books/biotech

A Multi-Volume Comprehensive Treatise

Biotechnology

Second, Completely Revised Edition

Edited by
H.-J. Rehm and G. Reed
in cooperation with
A. Pühler and P. Stadler

Volume 11c

Environmental Processes III

Solid Waste and Waste Gas Treatment, Preparation of Drinking Water

Edited by
J. Klein and J. Winter

Weinheim · New York · Chichester · Brisbane · Singapore · Toronto

Series Editors:
Prof. Dr. H.-J. Rehm
Institut für Mikrobiologie
Universität Münster
Corrensstraße 3
D-48149 Münster
FRG

Prof. Dr. A. Pühler
Biologie VI (Genetik)
Universität Bielefeld
Postfach 100131
D-33501 Bielefeld
FRG

Dr. G. Reed
1029 N. Jackson St. #501-A
Milwaukee, WI 53202-3226
USA

Prof. Dr. P. J. W. Stadler
Artemis Pharmaceuticals
Geschäftsführung
Pharmazentrum Köln
Neurather Ring 1
D-51063 Köln
FRG

Volume Editors:
Prof. Dr. J. Klein
DMT-Gesellschaft für
Forschung und Prüfung GmbH
Am Technologiepark 1
D-45307 Essen
FRG

Prof. Dr. J. Winter
Universität Karlsruhe (TH)
Institut für Ingenieurbiologie und
Biotechnologie des Abwassers
Am Fasanengarten
Postfach 6980
D-76128 Karlsruhe
FRG

Library of Congress Card No.: applied for

British Library Cataloguing-in-Publication Data:
A catalogue record for this book is available from the British Library

Die Deutsche Bibliothek – CIP-Einheitsaufnahme

A catalogue record for this book
Is available from Der Deutschen Bibliothek
ISBN 3-527-28336-6

Printed on acid-free and chlorine-free paper.

Composition and Printing: Zechner Datenservice und Druck, D-67346 Speyer.
Bookbinding: J. Schäffer, D-67269 Grünstadt.
Printed in the Federal Republic of Germany

Preface

In recognition of the enormous advances in biotechnology in recent years, we are pleased to present this Second Edition of "Biotechnology" relatively soon after the introduction of the First Edition of this multi-volume comprehensive treatise. Since this series was extremely well accepted by the scientific community, we have maintained the overall goal of creating a number of volumes, each devoted to a certain topic, which provide scientists in academia, industry, and public institutions with a well-balanced and comprehensive overview of this growing field. We have fully revised the Second Edition and expanded it from ten to twelve volumes in order to take all recent developments into account.

These twelve volumes are organized into three sections. The first four volumes consider the fundamentals of biotechnology from biological, biochemical, molecular biological, and chemical engineering perspectives. The next four volumes are devoted to products of industrial relevance. Special attention is given here to products derived from genetically engineered microorganisms and mammalian cells. The last four volumes are dedicated to the description of special topics.

The new "Biotechnology" is a reference work, a comprehensive description of the state-of-the-art, and a guide to the original literature. It is specifically directed to microbiologists, biochemists, molecular biologists, bioengineers, chemical engineers, and food and pharmaceutical chemists working in industry, at universities or at public institutions.

A carefully selected and distinguished Scientific Advisory Board stands behind the series. Its members come from key institutions representing scientific input from about twenty countries.

The volume editors and the authors of the individual chapters have been chosen for their recognized expertise and their contributions to the various fields of biotechnology. Their willingness to impart this knowledge to their colleagues forms the basis of "Biotechnology" and is gratefully acknowledged. Moreover, this work could not have been brought to fruition without the foresight and the constant and diligent support of the publisher. We are grateful to VCH for publishing "Biotechnology" with their customary excellence. Special thanks are due to Dr. Hans-Joachim Kraus and Karin Dembowsky, without whose constant efforts the series could not be published. Finally, the editors wish to thank the members of the Scientific Advisory Board for their encouragement, their helpful suggestions, and their constructive criticism.

H.-J. Rehm
G. Reed
A. Pühler
P. Stadler

Scientific Advisory Board

Contributors

Prof. Reinhard Böhm
Institut für Umwelt- und Tierhygiene
Universität Hohenheim – 460
D-70593 Stuttgart
Germany
Chapter 9

Dr. Derek E. Chitwood
University of Southern California
Los Angeles, CA 90098-1450
USA
Chapter 17

Prof. Dr. Horst Chmiel
Institut für Umweltkompatible
Prozesstechnik UPT
Am Stadtwald 47
D-66123 Saarbrücken
Germany
Chapter 13

Prof. Dr. Joseph S. Devinny
University of Southern California
Los Angeles, CA 90098-1450
USA
Chapter 17

Prof. Dr. Karl-Heinrich Engesser
Institut für Siedlungswasserbau, Wassergüte
und Abfallwirtschaft
Abt. Biologische Abluftreinigung
Universität Stuttgart
Bandtäle 2
D-70569 Stuttgart
Germany
Chapter 12

Dr.-Ing. Klaus Fischer
Institut für Siedlungswasserbau, Wassergüte
und Abfallwirtschaft
Abt. Biologische Abluftreinigung
Universität Stuttgart
Bandtäle 2
D-70569 Stuttgart
Germany
Chapter 14

Prof. Dr. Hans-Curt Flemming
Universität Duisburg
Fachbereich 6
Aquatische Mikrobiologie
Geibelstrasse 41
D-47057 Duisburg
Germany
Chapter 21

Dr.-Ing. Ina Körner
TU Hamburg-Harburg
Arbeitsbereich Abfallwirtschaft
Harburger Schloßstraße 37
D-21071 Hamburg
Germany
Chapter 4

Dr. Claudia Gallert
Universität Karlsruhe (TH)
Institut für Ingenieurbiologie und
Biotechnologie des Abwassers
Am Fasanengarten
Postfach 6980
D-76128 Karlsruhe
Germany
Chapters 1, 23

Dr. Uta Krogmann
Department of Environmental Sciences
Rutgers University
P.O. Box 231
New Brunswick, NJ 08903
USA
Chapter 4

Prof. Dr. Rolf Gimbel
Universität Duisburg
IWW Rheinisch-Westfälisches Institut für
Wasserforschung
Moritzstr. 26
D-45476 Mülheim/Ruhr
Germany
Chapter 18

Prof. Dr. Hans Jürgen Kutzner
Dresdener Straße 16
D-64372 Ober-Ramstadt
Germany
Chapter 2

Dr. Veerle Herrygers
University of Gent
Coupure L 653
B-9000 Gent
Belgium
Chapter 11

Dipl.-Ing. Oliver Lämmerzahl
Rütgers Kunststofftechnik GmbH
Research and Development
Nürtinger Str. 25
D-73275 Köngen
Germany
Chapter 15

Dipl.-Ing. Kai-Uwe Heyer
Ingenieurbüro für Abfallwirtschaft
Bleicherweg 6
D-21073 Hamburg
Germany
Chapter 6

Dr.-Ing. Hans-Joachim Mälzer
Universität Duisburg
IWW Rheinisch-Westfälisches Institut für
Wasselforschung
Moritzstr. 26
D-45476 Mülheim/Ruhr
Germany
Chapter 18

Prof. Dr. Ralf Otterpohl
TU Hamburg-Harburg
Arbeitsbereich Abwasserwirtschaft
Eißendorfer Str. 42
D-21073 Hamburg
Germany
Chapter 10

Dr. Norbert Rilling
TU Hamburg-Harburg
Arbeitsbereich Abfallwirtschaft und Stadttechnik
Harburger Schloßstr. 37
D-21079 Hamburg
Germany
Chapter 5

Dr. Werner Philipp
Institut für Umwelt- und Tierhygiene
Universität Hohenheim – 460
D-70593 Stuttgart
Germany
Chapter 9

Dr. Peter Schalk
Wehrle Werk AG
Energie- und Umwelttechnik
Bismarckstr. 1-11
D-79312 Emmendingen
Germany
Chapter 8

Dipl.-Biol. Thorsten Plaggemeier
Universität Stuttgart
Institut für Siedlungswasserbau, Wassergüte und Abfallwirtschaft
Abt. Biologische Abluftreinigung
Bandtäle 2
70569 Stuttgart
Germany
Chapters 12, 15

Dr.-Ing. Egbert Schippert
Institut für Umweltkompatible Prozesstechnik UPT
Am Stadtwald 47
D-66123 Saarbrücken
Germany
Chapter 13

Dr. Gudrun Preuß
Dortmunder Energie- und Wasserversorgung und Institut für Wasserforschung
Zum Kellerbach 46
D-58239 Schwerte
Germany
Chapter 20

Prof. Dr. Dirk Schoenen
Hygiene Institut
Universität Bonn
Sigmund-Freud-Straße 25
D-53105 Bonn
Germany
Chapter 19

Dr.-Ing. Martin Reiser
Institut für Siedlungswasserbau
Arbeitsbereich Technik und Analytik der Luftreinhaltung
Universität Stuttgart
Bandtäle 2
D-70569 Stuttgart
Germany
Chapter 16

Dr. Frank Schuchardt
Bundesforschungsanstalt für Landwirtschaft Braunschweig-Völkenrode (FAL)
Institut für Technologie
Bundesallee 50
D-38116 Braunschweig
Germany
Chapter 3

Dipl.-Geol. Ulrich Schulte-Ebbert
Dortmunder Energie- und Wasserversorgung
und Institut für Wasserforschung
Zum Kellerbach 46
D-58239 Schwerte
Germany
Chapter 20

Prof. Dr.-Ing. Rainer Stegmann
TU Hamburg-Harburg
Arbeitsbereich Abfallwirtschaft
Harburger Schloßstraße 37
D-21071 Hamburg
Germany
Chapter 6

Prof. Dr. Dr. h.c. Dieter Strauch
Institut für Umwelt- und Tierhygiene
Universität Hohenheim – 460
D-70593 Stuttgart
Germany
Chapter 9

Prof. Dr. Michael S. Switzenbaum
Environmental Engineering Program
Department of Civil and Environmental
Engineering
University of Massachusetts
Amherst, MA 01003-5205
USA
Chapter 7

Dipl.-Ing. Wolfgang Uhl
Universität Duisburg
IWW Rheinisch-Westfälisches Institut für
Wasserforschung
D-45476 Mülhein/Ruhr
Germany
Chapter 22

Dr. Herman Van Langenhove
University of Gent
Coupure L 653
B-9000 Gent
Belgium
Chapter 11

Prof. Dr. Willy Verstraete
University of Gent
Coupure L 653
B-9000 Gent
Belgium
Chapter 11

Dr. Muthumbi Waweru
University of Gent
Coupure L 653
B-9000 Gent
Belgium
Chapter 11

Prof. Dr. Josef Winter
Universität Karlsruhe (TH)
Institut für Ingenieurbiologie und
Biotechnologie des Abwassers
Am Fasanengarten
Postfach 6980
D-76128 Karlsruhe
Germany
Chapters 1, 23

Contents

III Drinking Water Preparation

Introduction

JOSEF WINTER
Karlsruhe, Germany

Although humans or the human society have always reacted early and in a very sensitive manner towards environmental pollution, counteractions by industrial enterprises or by governmental policy were slow and time-consuming and seemed to be influenced by economic considerations of the respective company or the policy of the country. Except for single production accidents with a spatially restricted and exceptionally high level of pollution, permanent environmental pollution was usually caused by gaseous, liquid, or solid production residues of many industries which had to be treated by so-called end-of-pipe technologies. Treatment of these wastes primarily was costly and reduced the profit. As a consequence of the strict (legal) requirements to reduce or even totally avoid certain wastes in developed countries and because of cheap labor elsewhere, many companies moved production into less developed countries with less stringent environmental regulations. Alternatively, they had to develop new, environmentally sound production processes. Nevertheless, in industrialized countries a high standard for the treatment of liquid and solid domestic and industrial wastes and of accidential spillage soil has been achieved within the last centuries. In contrast, in most developing countries the range of environmental protection activities reached from no treatment of residues at all over low-tech partial treatment to truly high-tech waste or wastewater treatment processes.

This was the situation when planning Volumes 11 a–c of the Second Edition of *Biotechnology*. The Editors had to cope with a wealth of highly advanced knowledge and a much broader field of environmental biotechnology compared to the First Edition. Numerous new processes or procedures, sometimes leading to unexpected drawbacks of the original expectations, are meanwhile widely encountered. The three volumes on environmental biotechnology cover the present technological status and the microbiological basis of industrial and domestic wastewater treatment (Volume 11a), of soil (bio-)remediation processes (Volume 11b), and of solid waste handling, off-gas purification techniques, and drinking water preparation (Volume 11c).

Wastewater led to visible damage of lakes, rivers, and the sea and for this reason, the development of treatment systems already began early in the 19th century. The comparably small amounts of mainly bioorganic solid wastes were used together with cattle dung as a fertilzer in agricultural areas or they were dumped within restricted landfill sites at the borders of big cities. Hence, they were almost "invisible" for most of the population harming only the direct neighbors of dump sites. Air pollution by sanitary landfill gas or groundwater contamination by leachates was either not known, not expected, or not thought to be really relevant at that time. Later, air pollution from the fast development of industrial production processes was well recognized, but for a long time was thought to be not really relevant because of the immense dilution capacity of the atmosphere. Similarly, the drinking water supply from unpolluted mountain regions or from high-quality groundwater seemed to be unlimited.

At the time of prosperous industrial development and increased consumption after World War II in Europe the per capita amount of solid waste increased drastically to more than 250 kg per year. This was in particular due to the increase of package wastes. Sanitary landfills grew to huge mountains, causing atmospheric and groundwater pollution and, as a consequence, they endangered the groundwater reservoir in general and especially in the vicinity of settlement areas. Therefore, solid waste treatment – apart from just deposition in one way or another – was obligately required, either by incineration or mechanical/biological treatment to "inertisize" the material. However, both treatment modes are relatively young technologies and due to the inhomogeneities of the material the treatment steps must be arranged in a rather sophisticated manner in order to maintain the daily throughput. Many different ways of solid waste handling are meanwhile practised by city authorities or industry.

This book gives a general overview over the presently applied solid waste treatment procedures, off-gas purification processes, and techniques of drinking water preparation. Except for the development of solid waste inertization procedures, soil remediation arose as a new problem of not just local significance during the last 30 years, due to the infiltration of harmful pollutants into soil and groundwater and the dilution and distribution of pollutants with the groundwater. For centuries, non-clean industrial production processes made whole industries intolerable or even abundant. Industrial sites highly contaminated with oil, chemicals, or inorganic material had to be decontaminated for the use of other industries or settlement areas or to prevent long-term groundwater contamination. The methods suitable to cope with contaminated soil or groundwater are summarized in Volume 11b.

Since experience with environmental protection or remediation techniques in the field meanwhile are numerous, an adequate overview covering all important facts or at least the general tendencies required a division into related fields. Volume 11c is one of three volumes on environmental biotechnology and should, therefore, mainly cover biotechnological processes of solid waste, off-gas, and drinking water treatment, but most practically applied procedures are combinations of biological and/or chemical and physical reactions embedded in a technologically engineered process. The treatment procedures for, e.g., solid waste inertization or off-gas purification sometimes are chemical or physico-chemical processes without a biological component. For this reason, we felt that we at least should mention, e.g., the principles of classical waste or sludge incineration and of newly developed systems, since these are the processes by which most of the total solid waste mass is presently conditioned or inertisized before final disposal of the slags. Some of these processes might become the methods of choice in the near future. Whether one or the other process is favored is more or less an individual decision of the responsible authority and must be discussed in conneciton with the available infrastructure, including waste collection and waste separation. There may be situations when even the biowaste fraction is better incinerated than composted and other situations, where the non-recyclable waste fraction should be treated mechanically and biologically before disposal instead of incineration.

In rich European countries waste inertization by incineration may be the most intriguing process, since it leads to the highest reduction of volume and mass. Deposition of the slags is supposed to cause less aftercare than deposition of otherwise stabilized wastes. To be competitive on the world market, however, especially in developing countries it is necessary not only to provide solutions of the most sophisticated technological standard, but also to provide or supply less costly solutions with a high, but maybe not the highest standard of environmental protection. It would, e.g., not be realistic to expect a slaughterhouse in a megacity of a developing country to operate a waste and wastewater treatment system which requires more investment and operating costs than the slaughterhouse itself. Thus, it is important for developed countries to improve the techniques for application in the own country, but also to supply experience with less advanced technologies. Taking out 80% of the pollutants from wastewater or wastes countrywide is a better option for the environment in developing countries than taking out 100% in a few places at the same total costs. The decision must, however, be made for each individual case.

The editors of this volume hope to provide the reader with an overview of the present state of solid waste treatment, off-gas purification techniques, and the principles of drinking water preparation. For more detailed insight into single processes the reader is referred to special references given in the respective chapters. Together with Volume 11a on wastewater purifcation systems and Volume 11b on soil remediation this volume rounds out the actual status of environmental biotechnology in different fields und presents the state of the art and of future trends.

Karlsruhe, March 2000 Josef Winter

I Solid Waste Treatment

1 Bio- and Pyrotechnology of Solid Waste Treatment

CLAUDIA GALLERT
JOSEF WINTER
Karlsruhe, Germany

List of Abbreviations

AOX	adsorbable halogenated organic compounds to charcoal
COD	chemical oxygen demand
CSTR	continuously stirred tank reactor
DOC	dissolved organic carbon
PCDD	polychlorinated dibenzodioxines
PCDF	polychlorinated dibenzofurans
PCB	polychlorinated biphenyls
TOC	total organic carbon
dm	dry matter

1 Introduction

As a consequence of improving life standards in industrialized as well as in developing countries increasing masses of municipal and industrial solid and liquid wastes are generated and have to be processed for recycling or handled for disposal. In developing countries a regular waste collection and an appropriate waste handling (e.g., by deposition in orderly and controlled sanitary landfills) is still missing. In the developed countries wastes are collected by the municipalities at a regular time schedule. In Germany municipal wastes are either disposed in sanitary landfills, incinerated or separated for recycling (Fig. 1). The greatest part of the municipal wastes (47.6%) is still directly deposited in controlled, "orderly" sanitary landfills. Those modern sanitary landfills are supposed not to create pollution of the groundwater by leachates at all or to cause only the minimum pollution that can be achieved by the present state of the art if the technical standards for construction and waste processing are applied. In the past sanitary landfills often were more or less "wild" deposits of refuse, without bottom seal, control of the input material and control of the gaseous and liquid emissions. In the near future in the European Community sanitary landfills will be exclusively reserved for inert, "earthlike" materials, that can be mixed and deposited without adverse environmental impact (REINHART and TOWNSEND, 1998). For European countries possible input materials for sanitary landfills were defined by the European parliament, and solid waste deposition techniques are regulated by laws of the European community (superior authority). These recently defined requirements must subsequently be transferred into state laws of the partnership countries and put into force at the latest in the year 2005.

According to the German "technical directive on recycling, handling and disposal of municipal waste" (TA Siedlungsabfall, 1993) for sanitary landfills a multiple layer barrier system is required as a bottom seal that should prevent leachate water to trickle into soil and groundwater and a special finishing layer on top should keep most of the rainwater from seeping through the body of the sanitary landfill. Yet control and guidance of chemical and

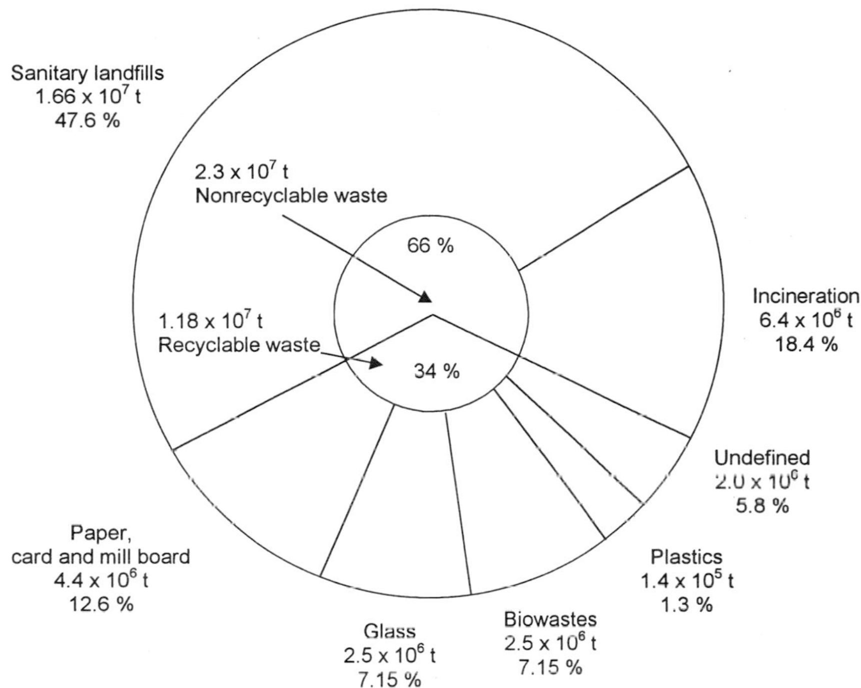

Fig. 1. Municipal waste generation and waste exploitation in 1999 in Germany (according to data of UBA, 1999).

biological processes in the disposed mixed wastes are still difficult, if not impossible. Efforts to handle sanitary landfills as a controlled bioreactor system by laboratory, pilot and full scale studies have been undertaken worldwide and results have been summarized, e.g., by REINHART and TOWNSEND (1998). The situation concerning sanitary landfills is totally unsatisfactory for older waste dumping places not only in Europe, but almost everywhere and for recent dumping places in developing countries, that often resemble our former "wild waste deposits". In those landfills highly concentrated and toxic substances containing leachates are generated which trickle unhindered into the groundwater, due to the lack of a suitable bottom seal and of a leachate collection system. For this reason, in the near future most of these sanitary landfills presumably will have to be completely restructured or reclamed by "landfill mining" to prevent a permanent pollution of the groundwater with toxicants and to achieve remediation.

To reduce to a minimum or to prevent chemical and biological reactivity in sanitary landfills in the future three classes were defined in the German "technical directive on recycling, handling and disposal of municipal waste", liable from the year 2005 on: Class I and class II sanitary landfills are supposed to obtain "mixed municipal wastes" that have been pretreated to contain $\leqslant 3\%$ (class I) or $\leqslant 5\%$ (class II) organic dry matter content, alternatively $\leqslant 1\%$ or $\leqslant 3\%$ total organic carbon, respectively. Class III dumping places will be monodeposits for, e.g., overlay soil, industrial dried solid residues or rubble from road and the construction industry.

In order to reduce the chemical and biological reactivity of solid wastes for a deposition in class I or class II sanitary landfills, the reactive organic and inorganic matter of the wastes must be quantitatively converted to non-reactive residues, e.g., by incineration. An only partial reduction of the organic content to reduce the reactivity upon deposition by a combined mechanical-biological treatment or even by pyrolysis (degassing of wastes at approximately 650 °C) was either not considered or not considered sufficient, when defining the landfill classes. For this reason quantitative oxidation of solid wastes by modern pyro- or incineration technologies was included in this overview. Mechanical separation of bulky waste and biological treatment of the fine particulate organic fraction or alternatively pyrolysis of municipal waste would reduce those chemical and microbial reactions in the sanitary landfills, that lead to self-heating, highly contaminated leachate and methane gas emissions to a minimum. Only little biological activity must be expected upon deposition of such pretreated wastes so that short-term mass losses and sagging would be minimized. Economically, a partial inertization of wastes by a mechanical-biological treatment or even pyrolysis would be much cheaper than incineration.

The recently defined new classes of sanitary landfills are supposed to behave neutral, without disturbance of the structural arrangement of the dumped material through volume reduction by microbial degradation of organic matter and subsequent sagging. Whether such landfills will really react neutral or whether by the action of heterotrophic and lithotrophic microorganisms some of the residual DOC (dissolved organic carbon) will be converted to methane or whether heavy metal oxides will be resolubilized by chemical reactions or microbial reduction will have to await long-term future experience.

2 Waste Generation and Handling

In 1993 in Germany $3.48 \cdot 10^7$ t a^{-1} of solid household wastes were collected by the municipalities (Fig. 1). According to Umweltbundesamt (UBA, 1992, http://www.umweltbundesamt.de) this amount includes the wastes of private households and small business enterprises, but not mining residues ($6.78 \cdot 10^7$ t), production wastes of the industry ($7.77 \cdot 10^7$ t), rubble, soil and road material ($1.431 \cdot 10^8$ t) and a non-definable portion of different origin ($6.4 \cdot 10^6$ t). Of the solid wastes collected by the municipalities $2.3 \cdot 10^7$ t (66%) were non-recyclable material and $1.18 \cdot 10^7$ t were recyclable (Fig. 1). The non-recyclable material was either deposited ($1.66 \cdot 10^7$ t) or incinerated ($6.4 \cdot 10^6$ t), the recyclable fraction separated into paper, card and millboard, glass, plastics, biowaste and an undefined residue containing mixed materials (Fig. 1). In addition to the wastes mentioned above, about $9 \cdot 10^6$ t of wastes, that were highly toxic for man or dangerous for soil, surface and groundwater as well as for the atmosphere, were generated in the producing industry and in hospitals (http://www.umweltbundesamt.de).

The average total solid waste generation per person in 1993 was 428 kg (http://www.umweltbundesamt.de). The amount of household wastes per person and annum varied between 130 and 375 kg, depending on the settling structure and the size of the single households (UBA, 1992; TABASARAN, 1994) and included metals, paper, glass, part of the package waste and the biowaste fraction. Package waste, marked with a green round symbol ("the green point". "Der Grüne Punkt") in Germany, must be recycled by "Duales System Deutschland", an organization taking care of recyclable package material for business enterprises. Recycling rates for different materials have to follow a predefined, year-by-year increasing ratio. The costs for recycling of package wastes are payed by the trading enterprises to "Duales System Deutschland" and are already included in the retail prices for food and household goods.

Whereas recycling processes for metals, paper and glass are available and recycling rates for, e.g., paper ware or glass have reached almost 100%, other materials such as plastic polymers cannot yet be recycled to a similar extent, due to the impurities of different plastic fractions. The difficulties in an accurate source-sorting of plastic waste material lead to a secondary raw material of low quality, mainly because of the inhomogeneous melting behavior. For ecological and economical reasons material recycling of plastics would only be favorable if a clean, strictly source-sorted homogeneous material is available. Since on the other hand plastic material is almost as energy-efficient as fossil fuels, mixed plastic material could replace part of the fossil fuels in industry, leaving the crude oil for processing of new plastic material. At present a recycling of waste material to products of the initial quality is only achievable for heavy and light metals and for glass. In 1997, e.g., about $2.7 \cdot 10^6$ t glass were recycled in Germany, covering 78% of the market volume for container glass (http://www.umweltbundesamt.de). High recycling quota were also obtained for paper, but paper recycling is better described as downcycling, since every recycling round leads to products of a lower quality. If the fiber slurry is not supplemented with new, long fibers containing cellulosic material the tensile strength of the recycled paper decreases, so that, e.g., it cannot be used in fast rotating printing machines.

The separation of different waste materials in municipal garbage ideally has to start in each household. The city authorities may choose between different collection systems for the household wastes, including so-called "bring systems" and "fetch systems".

The "bring-by-the-consumer system" has a relatively long tradition for glass ware. Green, brown and white glass bottles must be brought by the consumers to collection containers for either glass color, stationed on public areas all over the city and in the suburbs. The collected glass is recycled by the glass processing industry.

A "fetch-by-the-trader system" for newspaper with periodical collection was introduced already decades ago by charitable church organizations. Upon announcement they organized a house-to-house collection of paper bundels and sold the paper ware to paper factories as a secondary raw material for recycling. At present, in many communities the paper should be thrown into the container for recyclable waste ("Wertstofftonne"). Then it is source-sorted from other material by the staff of "Duales System Deutschland" for paper recycling.

3 Waste Collection Systems

Several collection systems for household wastes are practiced by the municipalities, distinguishing between different waste materials or waste fractions (Tab. 1). The selection of a certain collection system by the municipality depends on direct reutilization possibilities, reutilization after pretreatment or the locally available possibilities for final disposal. The most common collection systems in use are:

(1) Glass bottles must be disposed by the consumer in public glass containers, distinguishing between brown, green or white glass. All other wastes must be separated into recyclable material, including paper (called "Wertstoffe") and non-recyclable material, including wet kitchen wastes or biowastes (called "Restmüll").
(2) Glass has to be disposed of in public glass containers, distinguishing between brown, green or white glass. All other wastes must be sorted into recyclable material, non-recyclable material and biowastes (e.g., wet kitchen wastes, fruits, leaves, etc.).
(3) Old batteries and hazardous wastes, like paint, household cleaners, solvents, etc., have to be kept away from the garbage. They are collected once or twice a year or can be brought to special municipal deposits for interim storage (bring system). There they are separated into fractions for underground deposition, chemical inertization or physical inertization in a hazardous waste incineration plant.
(4) Green garden residues such as grass, leaves and branches have to be brought by the owners of the house or the garden to collection containers that are placed outside, but close to the living areas of the city. The containers are transported to and emptied at the composting plant by the municipality or the private operator of the composting plant, whenever it is full. The plant

Tab. 1. Waste Types or Fractions and their Reutilization or Intermediary and Final Disposal

Waste Type	Collection Mode	(Pre-)Treatment	Reutilization/Final Disposal
Glass ware	"bring system"	color sorting, cleaning	high quality of new glasses, (→ real recycling)
Paper	"fetch system"	pulping, de-inking	Paper production with a reduced quality (→ down cycling)
Metals	together with "Restmüll"	separation ferrous/aluminum	melting to high quality raw material
"Valuables" (recyclable fraction)	"fetch system"	separation into several fractions	recycling and downcycling of fractions, disposal and incineration of non-recyclable residues
Biowastes	"fetch system"	separation of non-bio-material, composting or fermentation/composting	soil conditioner, solid biofertilizer, liquid fertilizer and biogas
Non-recyclable material	"fetch-system"	homogenization	energetic recycling (incineration) and deposition of ashes
		homogenization	direct deposition
		mechanical-biological pretreatment	deposition of waste compost and energetic utilization of biogas from leachates

wastes are shredded to a suitable size for composting.

(5) Bulky wastes are collected once or twice a year upon announcement by the city authorities.

The containers for recyclable material and non-recyclable material are emptied every week or every second week. The waste in the container for recyclable material is separated for recycling into fractions as outlined, e.g., in Fig. 1. The non-recyclable wastes must either be processed for deposition by a mechanical and biological treatment or by pyrolysis, or must be incinerated. The biowaste from the bio-dustbins can be either composted together with more bulky green garden waste or is fermented anaerobically in a biogas reactor to yield biogas.

4 Material Recycling from Garbage and Waste Treatment for Final Disposal

In the German waste recycling law ("Kreislaufwirtschaftsgesetz" KrW/AbfG) of 1996 the highest priority was given to the prevention of wastes, whereever it is possible. Only if waste generation cannot be prevented, re-utilization of compounds from municipal waste and garbage, either by a direct material recycling (first priority) or by an indirect recycling through utilization of its energy content by incineration (second priority), is the next option, followed by direct deposition of the waste. If this "ranking of options" is obeyed all organic wastes that cannot be avoided or recycled as a secondary raw material have to be incinerated for energy recycling. Only the residing slags and ashes may then be deposited. Due to these legally defined priorities to handle waste material, municipal wastes must be separated into different recyclable and non-recyclable fractions (Tab. 1), either before or after collection. Until the year 2005 the non-recyclable waste fraction may still be directly deposited in sanitary landfills. Alternatively it may be subjected to a mechanical separation into a fraction of bulky material and a fraction of fine particulate matter, rich in organic material for biological treatment. The fine particulate fraction can either be composted or fermented anaerobically to reduce the biological reactivity to a minimum before deposition. Due to regulations of the German technical directive on recycling, handling or disposal of municipal wastes (TA Siedlungsabfall, 1993) deposition of municipal waste and litter after the year 2005 requires a reduction of the organic content to less than 3% or 5% dry matter, depending on whether a class I or class II sanitary landfill is available. If this directive in its present wording will be strictly prosecuted, alternatives to direct disposal, such as a combination of mechanical separation of bulky material and biological treatment of the fine particulate fraction before disposal or even pyrolysis of the waste and disposal of the pyrolysis coke will not suffice for deposition of the residues in sanitary landfills. The only means to reach the low organic carbon content of wastes, demanded by the TA Siedlungsabfall (1993) for deposition in sanitary landfills, will be waste incineration in a modern waste incineration plant or pyrolysis and waste coke conversion to syngas (e.g., by the procedure of Thermoselect), leaving only cinders and slags with $\leqslant 5\%$ or $\leqslant 3\%$ organic dry matter. In older waste incineration plants with wide-slit roasts the slags often contain $>5\%$ organic dry matter. Presently, a high proportion of the incineration slags is processed by a sequence of conditioning (storage for approximately 3 month to reach a steady state moisture content and to reduce the hydrophobicity), crunching and sieving for utilization as a construction material in a mixture with sand. However, due to the high content of toxic heavy metals and other toxicants, this mode of recycling of incineration residues should be strictly controlled and critically reevaluated.

5 Uncontrolled Long-Term Bioconversion of Garbage by Deposition in Sanitary Landfills

For deposition of the non-recyclable fraction of municipal wastes the modern sanitary landfills class I or class II have to be equipped with a multiple layer barrier protection system (Fig. 2) to prevent the underground and the groundwater from pollution by soluble toxic organic or inorganic compounds of the leachate. For this purpose the leachates must be collected quantitatively and treated biologically or subjected to ultrafiltration, reverse osmosis, evaporation, etc., to concentrate and/or to dry the pollutants. The concentrated leachates must be distilled to dryness to make possible underground disposal of the solids in abandoned mining tunnels.

After deposition of municipal wastes that contain a high proportion of organic material, in the sanitary landfills chemical reactions and biological conversions of the waste constituents proceed for several decades. The sanitary landfills are mutating from a short aerobic phase via an anoxic and anaerobic acidification phase to finally a sulfate reducing and methanogenic phase (Fig. 3a). Biological degradation of organic material and leaching of solubles with the drainage water lead to volume and mass reduction (Fig. 3b). In the final stage the landfills consist of inorganic insoluble material and undegradable organic waste material, that still contains a high proportion of toxic substances. These xenobiotics are permanently leached out by rain water, which trickles through the landfill body.

In a relatively short period after deposition, when air is still filling the pores and niches between the waste particles, aerobic degradation of organic waste compounds is the predominant process in the dump heap. Respiration of organic waste material to carbon dioxide and water is accompanied by heat production. Self-heating of spots of intensive biological reactions in parallel with heat-generating chemical reactions may raise the temperature into the thermophilic range if the heat is not

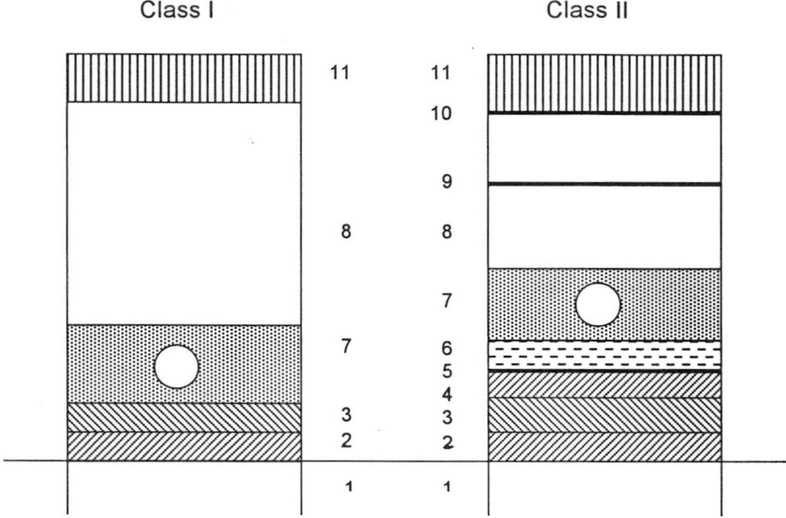

Fig. 2. Section across sanitary landfills class I and II: Multi-layer bottom leak protection systems, class I: 1 underground; 2, 3 mineral layers, each ⩾50 cm (e.g., clay, bentonite, etc.: $k = 10^{-8}$–10^{-10} m s^{-1}); 7 gravel layer with drainage system, ⩾30 cm; 8 waste, 11 recultivation layer ~1 m; class II: 1 geological barrier (>3 m), e.g., shid rocks; 2, 3, 4 mineral layers, each ⩾75 cm ($k > 10^{-10}$ m s^{-1}); 5 HDPE (high density polyethylene) foil, >2.5 mm thick; 6 protection layer, e.g., sand; 7 gravel layer with drainage system, ⩾30 cm; 8 waste; 9 intermediary planum; 10 surface protection layer; 11 recultivation layer ~1 m.

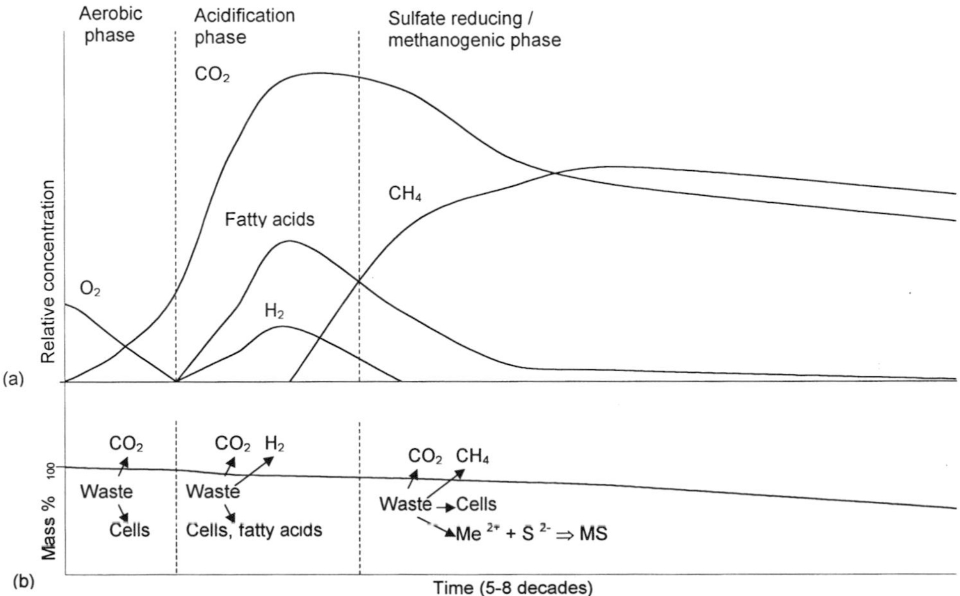

Fig. 3a, b. Assessment of relative metabolite concentration (**a**) and of mass loss (**b**) during aging of sanitary landfills.

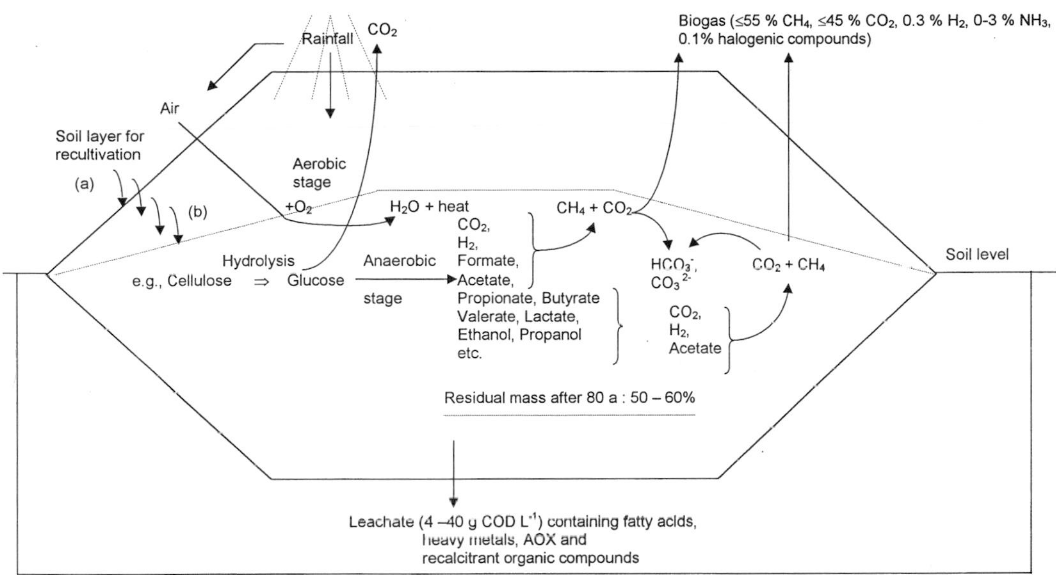

Fig. 4a, b. Overall mass flow and general microbial reactions in a sanitary landfill (**a**) after closure, (**b**) after 5–8 decades aging.

ventilated and may even lead to self-ignition. When the oxygen supply by diffusion becomes more difficult or finally is no longer possible due to compactation and self-compression of the waste material with length of deposition time, the biodegradable material is fermented in the subsequent acid phase of the landfill. Microbial fermentation of complex organic matter abundantly yields organic acids, that lower the pH below 5 and presumably initiate solubilization processes of primarily immobile minerals, e.g., of aluminum salts. Later on, sulfidogenesis and methanogenesis start from anaerobic niches in the center of waste particles or bacterial agglomerates, that are surrounded by aerobic or aerotolerant microorganisms on the surface. The pH of the dump heap slowly increases to the neutral range as soon as the microbes convert the volatile organic acids to biogas. Degradation of the organic material of municipal waste to biogas proceeds with decreasing intensity for several decades. A water film on the surface of particles, that contains suspended microorganisms or that moistens the bacterial colonies growing on the surface of particles, is periodically renewed by trickling rainwater, carrying suspended microorganisms through the waste. Due to an unrestricted time span for chemical and biological reactions, microorganisms, that possess or aquire the capability to degrade or bioconvert recalcitrant compounds may be enriched after depletion of the readily degradable organic compounds. Under these substrate limiting conditions humic substances presumably serve as an electron reservoir for biochemical reactions, as reported for other ecosystems (Benz et al., 1998).

A generalized scheme of the main subsequent biological reactions during the aging of a sanitary landfill is shown in Fig. 4. In a few decades time about 50% of the initial mass of the compacted organic waste are converted by microbial degradation to gaseous compounds or lost by washout of salts and soluble organic material from sanitary landfills by leachates. Sprinkling of "young" sanitary landfills with the leachate increases the moisture content and intensifies the microbial activity in the waste body through recycling of readily biodegradable volatile solids. By leachate recycling the dump heap is operated as a more or less controlled bioreactor system. The biogas that is released from sanitary landfills during 40–80 years of deposition must be collected and should be used as an energy source. Since it contains up to 5 vol% impurities (e.g., hydrogen, ammonia, volatile halogenic compounds, etc.), it must be purified before utilization as a fuel for gas engines to prevent corrosion and atmospheric pollution by the off-gas (see Chapter 16, this volume). A detailed list of components, including trace contaminations, found in the gas of sanitary landfills and in leachates was presented by WESTLAKE (1995).

Only in the late phases of deposition, when the gas production is decreasing, incineration to CO_2 by the use of gas funnels may be a more economic means of methane conversion to carbon dioxide. An almost quantitative landfill gas recovery can be achieved by gas sucking wells. These must be arranged in regular arrays over the ground area after preparation of the bottom seal and installation must be continued piece by piece during deposition of garbage until the maximal landfill height is reached (BILITEWSKI et al., 1994).

Due to rain water, that cannot completely be kept from trickling through the top soil layer of sanitary landfills and due to some water, that is produced during chemical and biological reactions, leachate is permanently trickling to the bottom of sanitary landfills. The leachate, containing soluble and toxic compounds, must be collected, e.g., via a perforated pipe system, and prestored for treatment or treated immediately. Whereas the amount of leachate may be relatively constant over the whole landfill age its pollutants concentration and its pollutants composition change drastically with time, making a biological treatment difficult. The leachate of the acid phase of a young dump heap, e.g., contains high concentrations of biologically degradable volatile fatty acids together with non-degradable, toxic compounds (chemical oxygen demand, COD, around 40 g L^{-1}). Later, only low concentrations of mainly xenobiotic pollutants are dominant, due to their resistance to microbial conversion or degradation or because they represent dead end bioconversion products. The collected leachate must be purified, e.g., by aerobic or anaerobic biological treatment, followed by adsorption of xenobiotic substances to charcoal. Alternatively, the leachate may be concentrated, e.g., by reversed osmosis through ultrafiltration membranes. The concentrated residues are subjected to distillation to obtain a dry material for deposition.

Leachate from young sanitary landfills may be recycled by sprinkling onto the top soil layer, if the landfill is operated as a bioreactor system for maximal methane generation (REINHART and TOWNSEND, 1998). With this procedure biological reaction rates in the waste body can be improved and the gassing activity of the landfill concentrated to a shorter time span.

The biogas production of the sanitary landfills of Germany was estimated between 180,000 and 315,000 t a^{-1} methane (SCHÖN and WALZ, 1993). If the landfill gas is not recovered and handled properly there is a danger of explosions, or the gas may prevent growth of the vegetation in the vicinity of the landfill or it may contribute to climatic changes. If the landfill gas is released into the atmosphere its methane content causes photochemical reactions in the troposphere (HEINZ and REINHARDT, 1993). For this reason sanitary landfills have to be actively degassed and, as mentioned above, the gas must be purified and should be used as an energy source or otherwise be converted to carbon dioxide.

If only a low yield of landfill gas is to be expected, as, e.g., in the late phases of older sanitary landfills or in sanitary landfills that obtained thermally pretreated, almost "inert" waste material, the little expected amount of landfill gas could be oxidized in the top soil layer biologically by methylotrophic bacteria (Eq. 1).

$$CH_4 + 2\,O_2 \xrightarrow[\textit{Methylocystis}]{\text{e.g., }\textit{Methylosinus,}} CO_2 + 2\,H_2O \tag{1}$$

Biological pretreatment of the wastes before disposal may reduce the maximal gassing activity to about 20% of those dump heaps, that obtained untreated waste (FIGUEROA, 1998). The dissimilation of methane proceeds via methanol, formaldehyde and formate to carbon dioxide and requires 2 mol of oxygen, whereas for assimilation of methane during cell growth 1.1–1.7 mol of oxygen are required (FIGUEROA, 1998).

To avoid emission of methane from passively degassing sanitary landfills into the atmosphere biological methane oxidation in the surface layer of the sanitary landfills must be actively improved. A minimum oxygen concentration of 9 vol% was found to be necessary for maximal methane oxidation (FIGUEROA, 1998). For this reason the top soil layer of the sanitary landfill must have a loose, porous structure to facilitate gas exchange and in addition the soil water content must be high enough for optimal bacterial metabolism, e.g., around 50% as for composting. The temperatures must be $>10\,^{\circ}C$. Ammonia in the landfill gas may competitively inhibit methane oxidation. The observed methane oxidation rates in soil, covering waste deposits vary from 0.01–20 L methane oxidized m^{-2} h^{-1}. Commonly 0.1–2 L m^{-2} h^{-1} of methane were oxidized (KRÜMPELBECK et al., 1998).

Methane oxidation by methylotrophic bacteria is a suitable means for landfill gas bioconversion of older sanitary landfills, if the gas amount is no longer sufficient to economically operate a gas engine and the gas composition has changed towards too much CO_2 for combustion. For this reason in older sanitary landfills the gas engines should be replaced by biofilters containing wet humus or composted plant material as a substratum for immobilization of bacteria that oxidize the methane. HUMER and LECHNER (1997) and KUSSMAUL and GEBERT (1998) obtained maximal methane oxidation rates of 0.24–0.50 kg m^{-3} d^{-1} CH_4 (equivalent to 336–700 L m^{-3} d^{-1}) in a biofilter with composted municipal wastes as filter material and with gas from a sanitary landfill as source of methane.

6 Controlled Aerobic or Anaerobic Bioconversion of Wastes

Waste material can be stabilized before deposition by aerobic or anaerobic degradation processes to reduce the biodegradable organic material. A general scheme of the mass flow for aerobic respiration of organic material during composting or biogas formation during anaerobic degradation is shown in Fig. 5. It indicates, that even for the same degradation efficiency much more solids reside in the compost than after anaerobic fermentation. This is due to the high biochemical energy amount conserved during respiration, allowing an extensive growth of aerobic bacteria. For this reason compost contains more bacterial biomass than concentrated anaerobic digester effluent.

6.1 Composting of Wastes

Composting of municipal wastes may follow two main purposes:

(1) decreasing the reactivity of municipal wastes or waste fractions before deposition in sanitary landfills or
(2) preparation of a soil conditioner or a natural fertilizer from plant materials and biowaste fractions.

Composting of mainly plant residues is a semi-dry aerobic stabilization process of organic material requiring about 50% moisture (for details, see Chapter 3, this volume). Compared to the starting material, the ripe compost contains a higher proportion of nitrogenous compounds, due to respiration of carbonaceous material to CO_2 and water. The composting process may be technically simple like in windrow or heap composting, or technically sophisticated if performed under strictly controlled conditions, maintained in actively aerated and thermostated rotting boxes or rotting drums. In two-stage composting processes the first stage normally is an intensive rotting phase in closed rotting systems for raw compost preparation, followed by a compost ripening phase in windrows. Some commercially operated composting systems with an intensive rotting phase followed by windrows for compost ripening are compared in Tab. 2. Depending on the total composting time, a raw compost (grade I or II after 2–4 weeks of intensive composting) or a ripe compost (grade 3–5 after up to 5 months of subsequent windrow composting) is obtained.

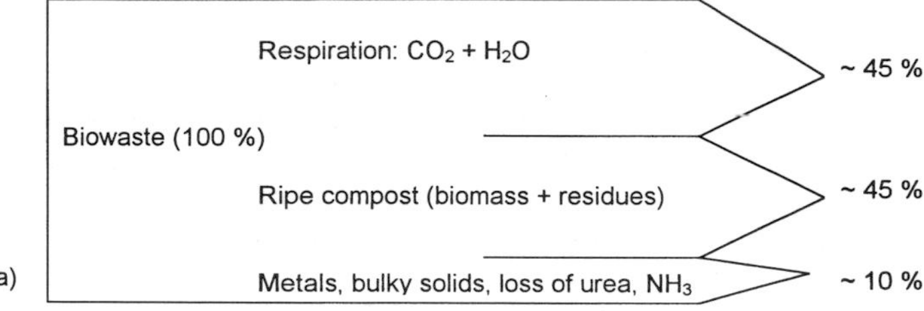

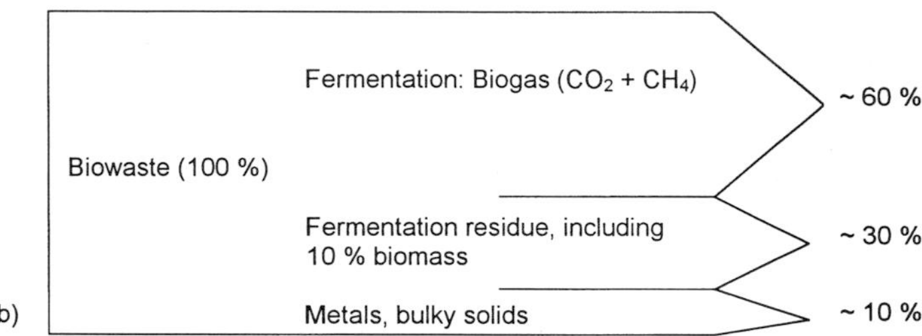

Fig. 5a, b. Mass flow during (**a**) aerobic and (**b**) anaerobic treatment of wastes.

Tab. 2. Different Composting Processes (modified according to BIEHLER and NUDING, 1996)

Intensive Rotting Phase Varying Systems	Compost Ripening Phase Windrows	Capacity $[t\ a^{-1}]$	Energy Input $[kWh \cdot t^{-1}]$	System
Rotting drums 1–10 d	2–4 months	1,300–50,000	<20	Lescha Envital Altvater
Rotting boxes and container systems 7–14 d	2–3 months	1,300–30,000	20–25	Herhof Mab-Lentjes
Windrow roofed 2–3 month	2.5 months	5,000–50,000	30–36	Bühler Thyssen Engineering
Tunnel composting 3–11 weeks	3–5 months	1,000–15,000	30–45	Passavant- Schönmackers Babcock
Brikollare 5–6 weeks	2–2.5 months	>5,000	30	Rethmann

Composting requires a structured material that allows unhindered oxygen diffusion into the center of rotting boxes or windrows for some weeks. Even after sagging and compactation of the waste material its overall structure must be loose enough to maintain the minimum oxygen transfer required for respiration. Whereas rotting boxes or container systems must be actively aerated, aeration of rotting tunnels or of windrows is achieved by a periodical loosening and turning of the material with stack turning machines. The energy requirement for composting ranges from <20–45 kWh t^{-1} input material (Tab. 2). In the intensive rotting phase heat evolution leads to self-heating up to more than 70 °C. In parallel, the compost population changes from mesophilic microorganisms via thermotolerant to thermophilic species. Later in the ripening phase at ambient temperatures a mesophilic population is developing again (see Chapter 2, this volume). In order to obtain a hygienically safe compost, the composting process must be guided in a manner that 70 °C are maintained in all parts of the compost heap for at least 2 weeks. Infectious plant and animal pathogenic microorganisms apparently do not survive the elevated thermophilic temperatures and are inactivated during the self-heating phase of composting. This is in particular essential for plant-pathogenic microorganisms to avoid spreading of plant diseases upon utilization of compost as a natural fertilizer (for details, see Chapter 9, this volume).

6.2 Composting of Green Plant Material

Many municipalities have placed containers for collection of green plant material from private gardens close to the living areas. The garden owners bring the green plant material to the containers, free of charge. The containers are transported to a central composting plant, when full with grass, leaves, rotten fruits, stems of bushes, etc. The plant material is shredded and composted by the staff of municipalities or by private operators of composting plants. The ripe compost is sieved and the different fractions are marketed as soil conditioners for horticulture, ornamental flower breeding or as organic fertilizer to farmers or for landscape gardening of city planners. The quality of such composts normally is excellent and the state quality criteria for compost application in Baden-Württemberg (Komposterlass, 1994) or of the working sheet M10 of the federal working group on compost handling (LAGA M10, 1995) are met. Quality criteria for compost of different state ordinances of European states include the C–N ratio, the salt content and the content of toxic compounds, e.g., heavy metal ions, PCB, PCDD, PCDF, and AOX (Tab. 3).

6.3 Composting of Biowastes

Whereas green plant residues are loosely structured and are dry enough or even have to be moistened for composting, biowastes (kitchen or restaurant residues, etc.) normally are too wet for composting. Furthermore, they contain only little structural material, which are both unfavorable conditions for composting. For this reason, dry structure giving material such as straw, wood chips, leaves, branches, etc., have to be mixed with biowastes to reduce the water content and to improve the loose structure of the material for composting. If biowastes are intented to be used for composting quality control requires additional analyses, such as an extended heavy metal spectrum, content of pathogenic bacteria, toxicity tests, etc. This is especially important in view of the German soil protection law ("Bodenschutzgesetz"), that was put into force in March 1998 (BBodSchG, 1998). If compost is used as a soil conditioner its quality has to meet the requirements defined, e.g., by the German ordinance on the utilization of sewage sludge ("Klärschlammverordnung") or LAGA M10 (Bilitewski and Rotter, 1998). Some of the maximally allowed parameters of compost, as required by state laws or recommendations of expert working groups are summarized in Tab. 3. The main criteria concern heavy metal concentrations, salt content, organic dry matter and ash content, AOX, PCB, PCDD/PCDF, etc. These quality measures are considered necessary, since more than $6.3 \cdot 10^6$ t of biowastes and green plant material from gardens or parks were composted in more than 517 composting plants (BGK, 1997) and are used as an organic fertilizer for soil.

6.4 Anaerobic Digestion of Biowastes

An alternative to composting of biowastes is anaerobic fermentation. More than 25 anaerobic reactors for methanation of the 70–100 kg biowastes per inhabitant and year, including kitchen wastes, plant and animal residues, have been installed or are under construction (SCHERER, 1995). At present the total capacity of biowaste reactors in Germany is 380,000 t (KORZ, 1999) with an increasing trend.

Anaerobic conversion of the carbohydrates, protein and lipid fractions of biowastes to biogas generally proceeds as outlined in Fig. 6. The overall degradation scheme is similar during dry or conventional wet fermentation in the different types of reactors that are in use. Only in 2-stage fermentation processes the hydrolytic and fermentative phase may partially or completely be separated from acetogenesis and methanogenesis (Fig. 6). A selection of different anaerobic one and two-stage wet and 1-stage dry systems, that are operated at full scale for biowaste treatment in European countries, is presented in Tab. 4. An overview over biowaste reactor systems for methanation of solid residues was given by SCHERER (1995). If the volumetric biogas production rates from sewage sludge, animal manures and municipal

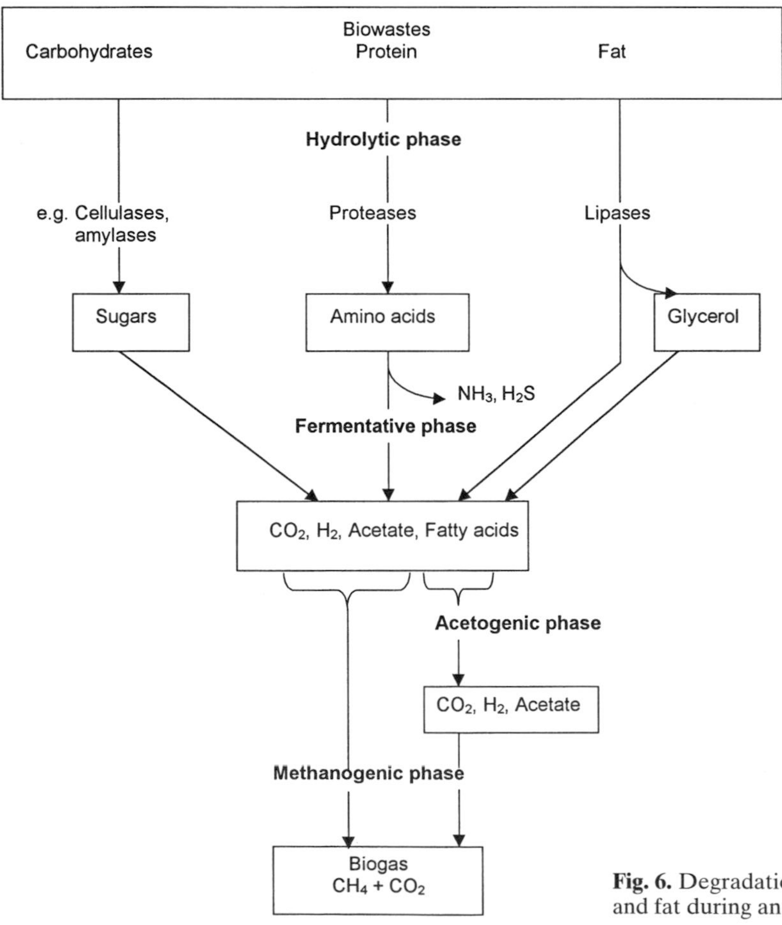

Fig. 6. Degradation of carbohydrates, protein and fat during anaerobic waste stabilization.

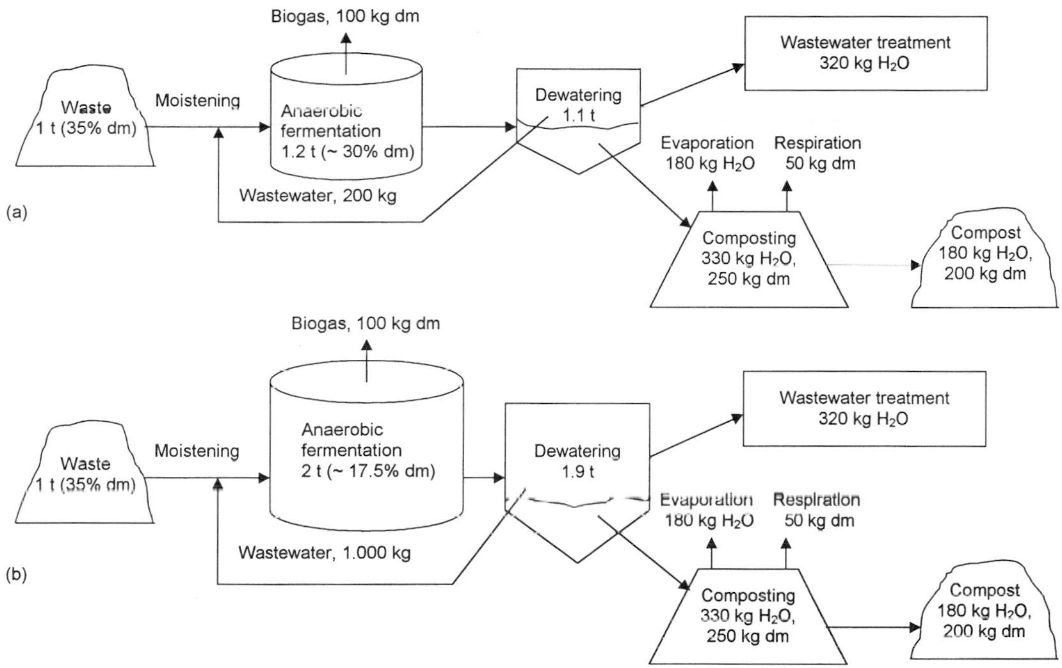

Fig. 7a, b. Comparison of mass balances of (**a**) dry anaerobic fermentation and (**b**) wet anaerobic fermentation, dm: dry matter.

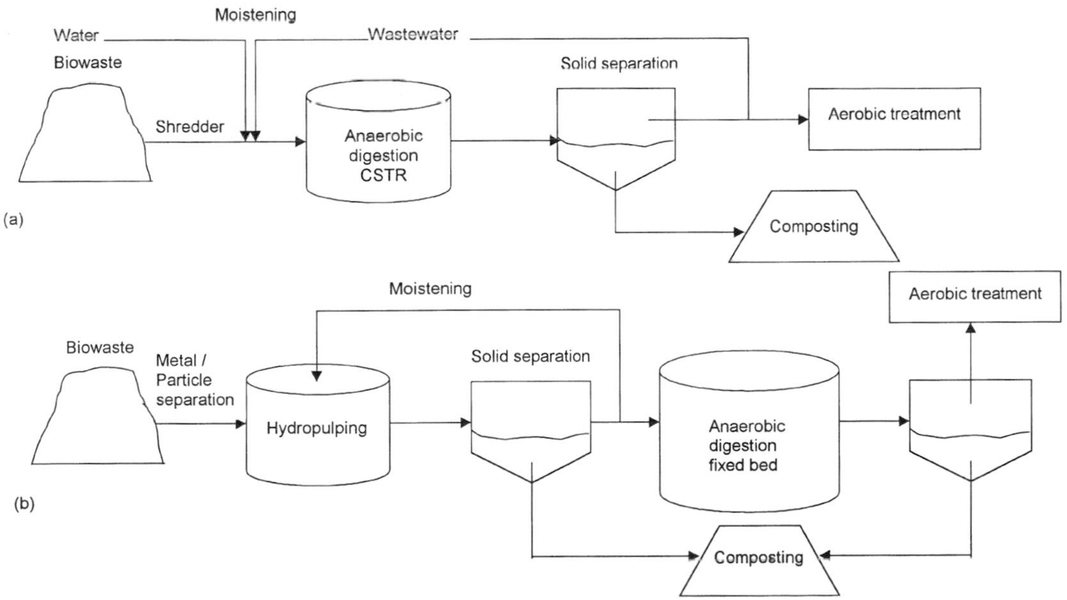

Fig. 8. Different processes for wet anaerobic digestion of biowaste.

Tab. 3. Maximally Allowed Values of Toxic Compounds in Compost Used as a Fertilizer in Agriculture

Recommended by		Heavy Metals [mg kg^{-1}]							PCB [mg kg^{-1}]	PCDD/F [ng kg^{-1}]	AOX [mg kg^{-1}]
		Pb	Cd	Cr	Cu	Ni	Zn	Hg			
RAL environmental standard 45 (Blue Angel, 1996)		100	1	100	75	50	300	1			
Federal compost association[a] [RAL 251 (BGK, 1994)]		150	1.5	100	100	50	400	1			
Ordinance of Baden-Württemberg for biowaste composting[a] (Komposterlass, 1994)		100	1	100	75	50	300	1	0.2	17	
Bavarian compost recommendation[a] (Komposthinweis Bayern, 1994)		100	1	100	75	50	300	1	0.12	17	100
Austrian standard 2200 for composting[a] (Österreichische Norm 2200 für Kompost		100	1	100	100	60	300	1			
Swiss material ordinance (StoV, 1986)		150	3	150	150	50	500	2			
German fertilizer ordinance[a] (DüngeV, 1996)		200	4		200	30	750	4			
Instruction sheet M 10 of LAGA[a]	category I	150	1.5	100	100	50	400	1			
(LAGA M 10, 1995)	category II	250	2.5	200	200	100	750	2			
Ordinance of biowaste (Bioabfallverordnung; BioAbfV, 1998)		100	1	70	70	35	300	0.7			
Federal ordinance of sewage sludge	soil I	900	10	900	800	200	2,500	8	0.2^{b}	100	500
(Klärschlammverordnung; AbfKlärV, 1992)	soil II	900	5	900	800	200	2,000	8	0.2^{b}	100	500

AOX: adsorbable halogenated organic compounds to charcoal, category I: agriculture, vegetable raising; category II: horticulture, landscape gardening, PCB: polychlorinated biphenyls ($\Sigma 6$ congeners), PCDD: polychlorinated dibenzodioxines (ng international toxicity equivalents), PCDF: polychlorinated dibenzofurans (ng international toxicity equivalents) soil I: normal soil; soil II: soil with pH 5–6, clay content $<5\%$

[a] Standardized to 30% organic dry matter content of the compost

[b] Single compounds.

Tab. 4. Full Scale Anaerobic Processes for the Treatment of Biowastes (modified according to KRULL, 1995 and SCHÖN, 1994)

Process Name	Mode	1st Stage				2nd Stage				Location/Capacity
		Purpose	Reactor	Temperature Range	HRT [d]	Purpose	Reactor	Temperature Range	HRT [d]	
Biocel	wet	H, M	PC	m	>20		no 2nd stage			t'Zand (NL) 12,000 t a^{-1}
										Flevoland (NL) 35,000 t a^{-1}
Bio-Stab	wet	M	CSTR	t	~20		no 2nd stage			Kaufbeuren (D) 3,000 t a^{-1}
Italba	wet	M	CSTR	m	n.c.		no 2nd stage			Bellaria (I) 30,000 t a^{-1}
Solidigest										
Linde/KCA	wet	M	CSTR	t	12–15		no 2nd stage			Himmelgarten (D) 25,000 m^3 a^{-1}
WAASA,	wet	M	CSTR	m	15–20		no 2nd stage			Vasa (SF) 26,000 t a^{-1}
Wabio										Bottrop (D) 6,500 t a^{-1}
Paques/BFI	wet	H	CSTR	m	1–5	M	SB/UASB	m	0.5–1	Breda (NL) 25,000 t a^{-1}
Prethane/Rudad										Leiden (NL) 100,000 t a^{-1}
BTA	wet	H	CSTR	m	1.5–2	M	FB	m	1–2	Helsingör (DK) 20,000 t a^{-1}
										Karlsruhe (D) 8,000 t a^{-1}
Plauener Verfahren	wet	H	CSTR	a	3–5	M	CSTR	m	8–12	Zobes (D) 15,000 t a^{-1}
DSD/CTA										
Uhde-Schwarting	wet	H	CSTR	m	n.c.		CSTR	t	n.c.	Finsterwalde (D) 90,000 t a^{-1}
AN/Biothane	wet	H	PC	m	4–10	M	SB	m	0.5–1	Ganderkesee (D) 3,000 t a^{-1}
ATF	dry	M	PFR	m	20–25		no 2nd stage			TU Hamburg (D) pilot scale
Dranco	dry	M	CSTR	t	10–20		no 2nd stage			Brecht (B) 10,500 t a^{-1}
										Salzburg (A) 20,000 t a^{-1}
Valorga	dry	M	CSTR	t	12–18		no 2nd stage			Amiens (F) 72,000 t a^{-1}
Kompogas	dry	M	PFR	t	15–20		no 2nd stage			Bachenbülach (CH) 10,000 t a^{-1}
										Kempten (D) 10,000 t a^{-1}

[a] ambient, CSTR: continuous stirred tank reactor, FB: fixed bed reactor, H: hydrolysis, M: methanation, m: mesophilic, n.c.: no comment, PC: percolator, PFR: plug flow reactor, SB: sludge bed reactor, t: termophilic temperature, UASB: upflow anaerobic sludge blanket.

Tab. 5. Comparison of Volumetric Biogas Production Rates from Different Substrates for Stable Methane Fermentation

Substrate	Biogas [$m^3\,m^{-3}\,d^{-1}$]	HRT [d]
Primary sewage sludge	0.9–3.0	22–5[a]
Secondary sewage sludge (surplus sludge)	0.7–2.4	22–5
Municipal biowaste fractions	2.4–3.6	30–19
Non recyclable municipal waste ("Restmüll")	0.3–0.6	ca. 20
Primary and secondary sewage sludge	1	10[b]
Cattle manure	1	10[c]
Swine manure	1	10[d]

Equivalent to a space loading of 1.5–8[a], 2.9[b], 5.8[c] and 1.8[d] kg COD $m^{-3}\,d^{-1}$; data compiled from Temper et al. (1983), Wildenauer and Winter (1984), Gallert and Winter (1997) and Gallert et al. (1998).

wastes, obtained in laboratory studies or in pilot plant or full scale experiments, are compared, the highest gas production resulted from biowaste methanation (Tab. 5). This was, however, not surprising, since biowastes are original material, whereas the others are fermentation feed and food residues. The volume related maximal biogas production rates were obtained as long as the space loading of the reactor was low enough or the hydraulic retention time was long enough to allow the maximal degree of degradation. The highest loading (shortest retention time) was possible with biowaste fractions, followed by primary sewage sludge, swine manure, secondary sewage sludge and cattle manure.

The principle differences of dry and wet anaerobic biowaste digestion processes are outlined in Fig. 7. Under ideal conditions both processes lead to the same amount of biogas (although for dry anaerobic processes methane generation is not as fast as for wet anaerobic fermentation), wastewater and compost (Fig. 7). However, due to the higher recycling rate of wastewater during wet anaerobic digestion the salt concentration in the anaerobic digester is increasing faster and may lead to inhibiting concentrations of ammonia and other salts. Concerning ammonia tolerance, it was shown, that a thermophilic flora in biowaste digesters was almost twice as ammonia (NH_3) tolerant than a mesophilic flora (Gallert and Winter, 1997; Gallert et al., 1998).

About 70% of the full scale biowaste reactors are operated in the wet mode (Korz, 1999). For wet anaerobic digestion two principally different processes are established on the market:

(1) Moistening of presorted and shredded biowaste with water (or anaerobically digested effluent after sludge separation), anaerobic digestion of the waste suspension in a continuously stirred tank reactor followed by separation of residual solids and composting of the solids together with structure giving dry plant material (Fig. 8a). If wastewater is utilized for moistening purposes, the ammonia concentration in the reactor and also the salt concentration increase quite significantly. Methanogenesis may then be negatively affected (Langhans, 1999).

(2) Moistening of presorted biowaste, hydropulping, solids separation and methanation of the liquid fraction in a fixed bed reactor (e.g., BTA-process, Fig. 8b). Sludge from gravity sedimentation of the pulped material and surplus sludge from anaerobic digestion of the liquid waste suspension are then mixed with dry plant material and composted.

Disruption of fibrous substrates such as vegetable residues, potato peels, grass, leaves etc. by hydropulping improves the bioavailability by increasing the specific surface of the material. Hydrolysis of particulate waste matter, which is normally the rate limiting degradation step is improved and may then no longer be the rate limiting step. To obtain the same degradation efficiency the hydraulic retention time of the wastewater fraction in the anaerobic reactor can be shorter than in a reactor, operated with a non-pulped waste suspension. Even for treatment of suspended waste fractions as described under (1) digestion of evenly disrupted material yields a higher biogas productivity and less solid residues are found in the effluent (PALMOWSKI und MÜLLER, 1999).

6.5 Cofermentation of Biowastes with Animal Manures or Sewage Sludge

An alternative to the fermentation of solely biowastes is cofermentation with manure of swine, chicken or cattle or with municipal sewage sludge. However, a cofermentation of biowastes and sludges of different origin requires a pasteurization of the biowastes in order to prevent spreading of infectious diseases (see also Chapter 9, this volume). A pre- or post-pasteurization may be applied. Prepasteurization bears the danger of reinfinfection of the material with pathogenic organisms during anaerobic stabilization, whereas pasteurization of the digester effluent leads to a partial hydrolysis of residual volatile solids, and thus causes a destabilization of the effluent.

An advantage of anaerobic cofermentation of animal manures with biowastes is the possibility of a direct utilization of the liquid as a fertilizer on farm land, not requiring solid–liquid separation and solids composting, compost marketing and wastewater treatment (LANGHANS, 1998). Moreover, the farmer obtains a higher yield of biogas from manure–biowaste mixtures and more electricity by gas combustion in gas engines, coupled with electricity generators (LANGHANS, 1997). A cofermentation of animal manures with municipal biowastes, food residues, rumen content or floating fat increased the volume-specific amount of biogas by 14% at stable process conditions and the effluent did not contain increased values for AOX, heavy metals or salts (HOPPENHEIDT et al., 1999). Quality requirements of the wastes to be met for fermentation (e.g., pasteurization) and by the final products are defined in the "ordinance for the application of biowastes on soils for agriculture, forestry and horticulture" (BioAbfV, 1998).

In waste suspensions sufficient water is available for a free motility of bacteria and the solubilization of organic and inorganic material to non-inhibiting concentrations. But in dry fermentation systems only a thin water film around the particles is available that contains much higher concentrations of microorganisms and of solubilized growth promoting or growth inhibiting material. Nevertheless, even if process inhibiting concentrations of, e.g., sulfide or heavy metals would be reached in the water film, due to the unrestricted fermentation time finally a complete degradation of the material would be obtained as long as the toxicants cause only a partial inhibition. An advantage of the dry fermentation system is that in water films the diffusion distance of, e.g., exoenzymes for polymer solubilization is much shorter and the enzyme concentrations are much higher, leading presumably to increased hydrolysis rates. Another advantage of semi-dry or dry fermentation is facilitated colony or biofilm formation on the particles covered by a water film. This may lead also to an enrichment of those microorgansims that are growing more slowly than others and that have or may aquire the capability of degrading recalcitrant material. The degradation of biopolymers follows the general scheme of Fig. 6. In suspensions containing a high proportion of solids, hydrolysis of insoluble biopolymers may be a rate limiting factor (WINTER and COONEY, 1980), whereas in wastewater with mainly solubilized components it is more likely that acetogenesis and methanogenesis are rate limiting steps (WINTER, 1984). Methanogenesis may have to compete with sulfate reduction, if the wastewater contains sulfate. Sulfidogenesis on the other hand abolishes the toxic effect of soluble heavy metal salts by sulfide precipitation. If the biowaste material is free of toxic compounds the digester effluent should be utilized as a fertilizer in agriculture, provided that the hygienic requirements are met (see Chapter 9, this volume).

6.6 Direct Anaerobic Fermentation of Non-Recyclable Municipal Solid Waste ("Restmüll") of which Packing Material and the Biowaste Fraction have been Separated

The 40–60 mm sieving fraction of non-recyclable municipal waste without packing material and biowastes has an energy content of about 5,000 kJ kg^{-1} and thus contains not enough energy for incineration (>9,000 kJ kg^{-1}). Since incineration of such material requires energy support by fossil fuels, other criteria to reach waste inertization for disposal on sanitary landfills are under discussion, such as respiratory rest activity after aerobic stabilization processes or residual biogas production after anaerobic digestion. Although neither aerobic nor anaerobic biological treatment processes are effective enough to reduce the organic content even close to the demanded 5% to meet future legal disposal criteria, the remaining biological activity would be low enough to allow a safe disposal without substantial sagging. Another biochemical parameter for a rough estimation of the residual biological activity could be a low proportion of cellulose to lignin. To meet the mentioned, however not yet legally approved inertization criteria, SCHERER and VOLMER (1999) suggested a thermophilic two-stage anaerobic wet digestion. They calculated a biogas production of 797 NL kg^{-1} organic dry matter, whereas LOOCK et al. (1999) experimented with a dry anaerobic digestion and obtained ca. 400 NL biogas kg^{-1} organic dry matter.

6.7 Combined Mechanical and Biological Processing of Non-Recyclable Municipal Waste

For a partial inertization of municipal wastes before deposition in sanitary landfills a combined mechanical and biological treatment may be performed to reduce the biologically degradable carbon (Fig. 9). For this purpose metals and bulky solids must be separated from the wastes. After removal of metals and the bulky material the remaining fraction may be disrupted (optional) and put into a percolator. The soluble and fine particulate material is swept out by water that is sprinkled on top and that percolates through the disrupted wastes. With increasing percolation time metabolites of biopolymer hydrolysis are also released and leached out. The overall percolation time should be not less than 2–5 d. By

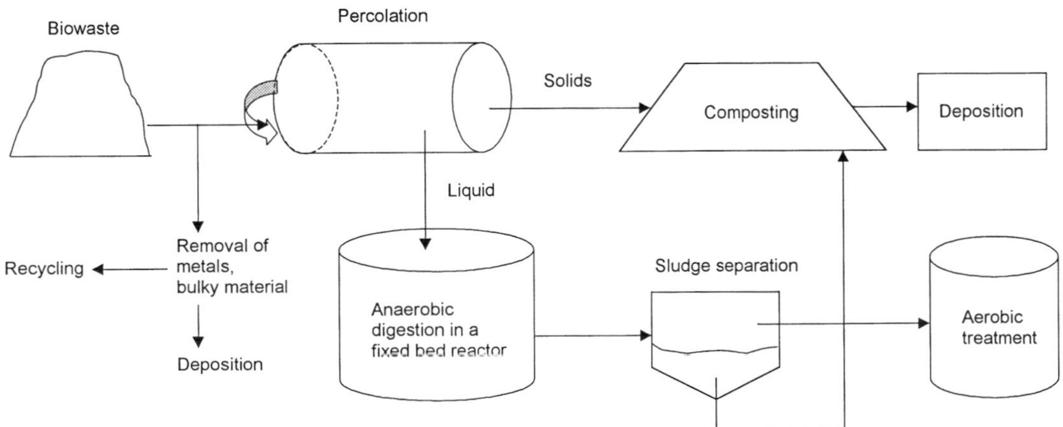

Fig. 9. Mechanical separation of non-recyclable municipal wastes and biological treatment of solids and suspended or solubilized material.

percolation up to 30% of the waste material of non-recyclable municipal waste (containing no biowaste), may be leached out, either in solution or suspended in the trickling water. After sand sedimentation and separation, the percolation liquid is subjected to anaerobic treatment, preferably in a fixed bed digester to yield biogas. The sludge of the digester effluent is sedimented by gravity and the supernatant guided into a municipal wastewater treatment plant for final stabilization. The solid residues from the percolator and the surplus sludge of anaerobic digestion are composted together with structure giving material before the stabilized waste compost is disposed in a sanitary landfill. By this combination of mechanical and biological pretreatment of non-recyclable wastes ("Restmüll") a great part of the organic fraction, that is bioavailable for aerobic or anaerobic degradation, is removed (see Chapter 8, this volume). The residual waste compost, that normally does not meet compost standards for use as a fertilizer (e.g., Tab. 3) can be disposed without intensive gassing reactions after deposition. Some emissions, however, still have to be expected: Although a high density of the solid residues of mechanically and biologically pretreated waste is reached during deposition with k_{liquid} values $<10^{-9}$ m s^{-1} and k_{gas} values $<10^{-15}$ m^{-2} little leachate still has to be expected. The leachate resembles that of older depositories, that are in the stable methane phase. In addition, a constant emission of small amounts of biogas indicates a slowly proceeding anaerobic digestion of organic material (DACH, 1999).

7 Thermic Treatment of Wastes

The recyling and solid waste law of Germany (Kreislaufwirtschaftsgesetz; KrW-/AbfG, 1996), that was put into force in October 1996, demands a pretreatment of solid wastes to obtain a chemically and biologically inert material before disposal. According to this law it will not be allowed to deposit wastes that contain more than 1 or 3% total organic carbon (TOC), equivalent to 3 or 5% organic dry matter content in sanitary landfills class I or class II after a transition period until the year 2005. The low carbon content can only be achieved by thermic treatment of the waste, favoring or even exclusively demanding waste incineration, since, e.g., degassing by pyrolysis would not meet this requirement. Only one plant for pyrolysis of garbage is in operation in Germany at Burgau (Bavaria) since 1983 with an annual maximal capacity of 25,000 t (information sheet of local authorities). The degassed pyrolysis coke, which contains still more than 5% organic dry matter content is then disposed in a sanitary landfill.

At present and in future all unavoidable wastes have to be either separated for a direct recycling of the material or utilized as an energy source for energy recovery, leaving behind only the slags and ashes for final deposition. The criteria on whether to recycle certain wastes or to use them as a replacement of fossil energy sources, e.g., waste plastics, are still a matter of controversy and of commercial interest. Whereas most of the existing 58 incineration plants with a total capacity of around $1.3 \cdot 10^7$ t of wastes (http://www.umweltbundesamt.de; HAASE, 1999) cannot be operated at their full capacity due to waste shortage, the operators of sanitary landfills intend to deposit as much waste as possible until the year 2005 to exploit the capacity of the depositories and to accumulate money for landfill maintenance for centuries after closure. The preference of the waste producers for deposition over incineration has its reason in the lower costs for deposition, even if the wastes have to be transported over long distances to the dumping places.

Thermic treatment can be used for domestic refuse and toxic industrial solid and liquid wastes as well. Simultaneous incineration of solid and liquid industrial wastes requires different feeding mechanisms and suitable oven systems for liquid and/or solids combustion. The focus must lie on a quantitative oxidation of all organic components to CO_2. The inorganic matter is oxidized to insoluble metal oxides and forms the majority of the slag mass.

The off-gas of waste incineration plants consists mainly of carbon dioxide and water vapor, but contains chlorine and sulfur dioxide as major contaminants as well as traces of organic and inorganic volatile substances, such as PAK, AOX, dioxins, mercury, etc. The contaminants must be separated from the bulk mass of the gas by off-gas purification procedures (see respective chapters in this volume). With the different incineration techniques, both, the waste volume and waste mass are highly reduced in lieu with utilization of the energy ("energetic recycling"), either directly as heat (classical incineration plants, Siemens KWU process) or indirectly as syngas. Syngas, a mixture of carbon monoxide and hydrogen, is produced during the Noell thermal conversion process or the Thermoselect process (for explanation, see below). Fly ashes of waste incineration plants are highly toxic and must be filtered out of the off-gasses. The fly ashes are then filled into big bags and are stored in abandoned mining cavernes, whereas the residing slags of waste incineration are partially deposited and partially reconditioned to replace gravel products in the construction industry. For this purpose it is a major advantage to melt the residing minerals at temperatures >2,000 °C in a melting oven and to immobilize the melted oxides by shock cooling, as it is intended in the Thermoselect process or the Siemens KWU process. Although full scale plants have been built for the latter two processes, the waste and waste coke degassing technology is still in an improving experimental phase.

The 51 waste incineration plants that are in operation in Germany and the 7 incineration plants that are under construction or planned (HAASE, 1999) have a total capacity of $1.3–1.45 \cdot 10^7$ t a^{-1} (ATV Information, 1999). Thermal conversion of organic non-recyclable waste fractions to harmless products is a future task of high importance and can be achieved either by classical waste incineration or by the above mentioned different processes of pyrolysis and pyrolysis coke conversion:

(1) One-step classic waste incineration plant for mineralization of municipal wastes, e.g., in tunnel ovens, to CO_2 and water in the presence of excessive oxygen. Incineration includes drying, complete degassing (pyrolysis) and waste coke and syngas oxidation (Fig. 10). Except for solid waste incineration in classical roast ovens, fluidized bed ovens for solid wastes or revolving grate ovens for either solid, pasty and liquid wastes may be used.

(2) Two-step thermal waste treatment processes with a separated drying–pyrolysis process and a subsequent waste coke conversion process with restricted oxygen supply. In the Noell thermal conversion process pyrolysis of slightly compressed wastes and waste coke degassing are completely separated in two oven systems (Fig. 11), whereas in the Thermoselect process, pyrolysis of highly compressed wastes and degassing of the waste coke are spacially connected in a combined oven system (Fig. 12).

(3) The two-step waste incineration plant consisting of a pyrolysis reactor and a pyrolysis coke and syngas incineration unit. The pyrolysis coke and the syngas are thermally oxidized with excess oxygen in a process developed by Siemens KWU (Fig. 13). Whereas in the Thermoselect and Noell thermal conversion process syngas production is intended during pyrolysis and waste coke conversion to power a gas engine, in the Siemens KWU process the syngas, which is produced during pyrolysis is burned together with the waste coke with excess oxygen to yield thermal energy for steam generation.

For pyrolysis the waste material can either be applied in a more or less loose form (Noell thermal conversion process) or highly compressed to reduce nitrogen introduction into the pyrolysis reactor (Thermoselect process). The advantage of pyrolysis of compressed waste is that due to the low nitrogen content the NO_x content of the syngas can be kept very low, abolishing catalytic NO_x removal. A disadvantage of pyrolysis of compressed waste is, however, the low energy conductivity through insulating materials (e.g., sport shoes, bundles of paper, bulky waste, etc.), that prevent a complete degassing. Since pyrolysis and waste coke degassing are directly coupled in the Thermoselect process, there is no possibility to remove metals or non-carbonized material before final oxidation of the pyrolysis coke in the

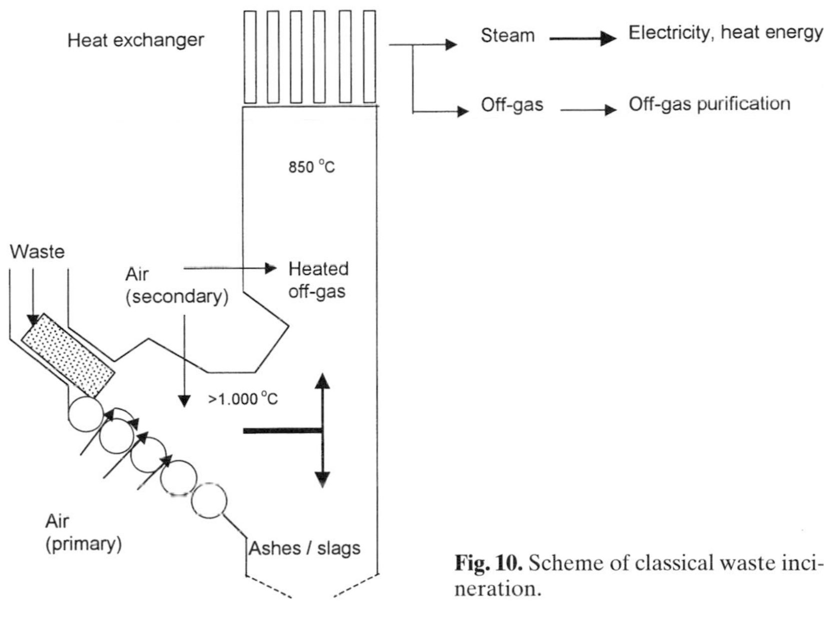

Fig. 10. Scheme of classical waste incineration.

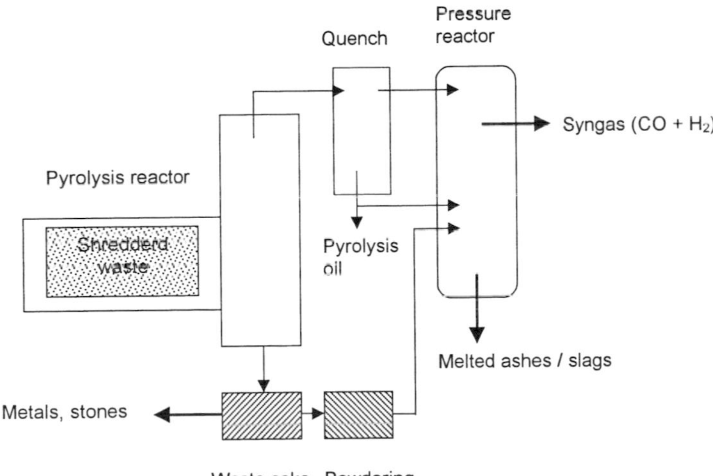

Fig. 11. Thermal conversion process of Noell Co. (Würzburg, Germany).

melting oven. The difficulty to dose the exact amount of oxygen for waste coke oxidation may lead to explosions in the melting oven. A separation of a pure metal fraction after pyrolysis, before the waste coke is ground to yield a powder for injection into a fly–stream degasser, is part of the Noell thermal conversion process.

The carbonizing-burning process ("Schwel-Brenn-Verfahren") of Siemens KWU combines pyrolysis and waste coke gasification with an intermittant cooling for metals and stones removal. This process works with surplus oxygen for a complete oxidation of the waste coke to carbon dioxide in the melting chamber. In addition the syngas of pyrolysis is

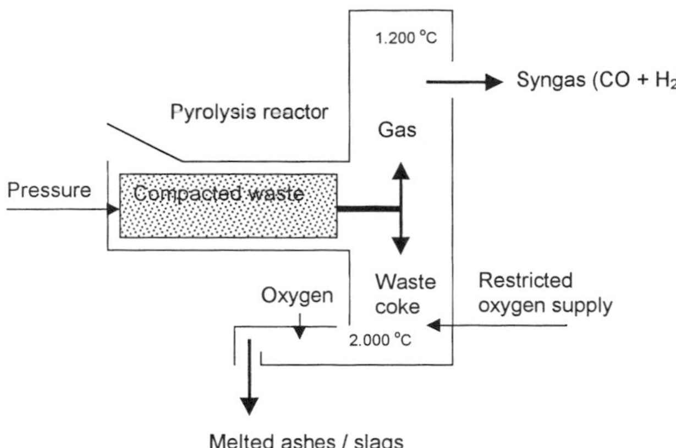

Fig. 12. Two-step thermal waste conversion by the Thermoselect process (EnBW, Karlsruhe, Germany).

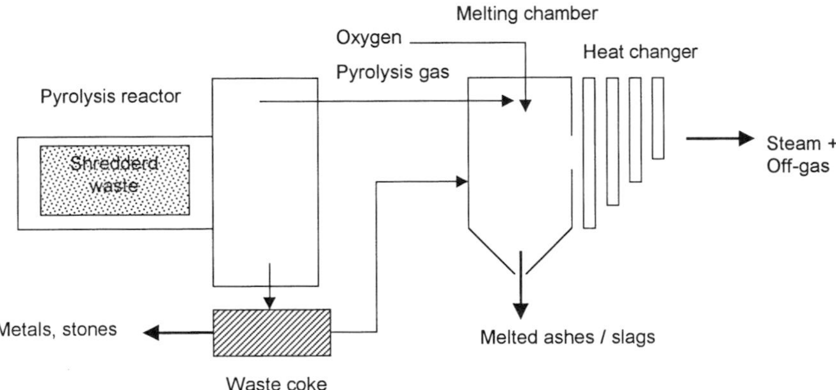

Fig. 13. Two-step thermal waste conversion by the carbonization burning process of Siemens KWU (Offenbach, Germany).

also burned in the melting chamber to reach the high temperatures for the slags to melt. Heat energy is recovered by heat exchangers and electricity is generated in steam turbines.

Very important for all waste incineration processes is the purification of the off-gases by wet, semi-wet or dry gas purification processes (see, e.g., Bilitewski et al., 1994; Chapter 16, this volume). All gas purification procedures in general start with a particle separation in a cyclone, by gas filtration or by electrofiltration, followed by acidic and alkaline washing. Nitrogen compounds are removed catalytically thereafter and poisonous trace gasses must be adsorbed onto charcoal.

The organic residues of slags and ashes from waste incineration plants still serve some microorganisms as a carbon source after deposition. Gimmler et al. (1998) investigated the microbial activity in incineration slags that have been disposed at least for 1 year. The slags contained 17.3 g TOC kg^{-1} material and had a nitrogen content of 0.43 g N kg^{-1}. The biodegradable portion of the total organic carbon was 3.7% with 1.2% dissolved organic carbon, 0.9% lipophilic substances and 1.6% starch (Gimmler et al., 1998).

Microbial colonization of the incineration slags was apparently highly dependent on growth of plants on the top soil layer. Exudates of plant roots were apparently favoring fermentation of the biodegradable portion of the slag material. A high pH, a low content of phosphate and nitrogen and the presence of heavy metals were unfavorable conditions for microorganisms to grow. If during long-term deposition the metabolizable portion of slags would be degraded to biogas, then about 440 mL biogas kg^{-1} of slags would be produced, compared to 150–350 L biogas kg^{-1} of municipal waste that was deposited directly. Since the population density of microorganisms in slag deposits is very low, the gas production rate was estimated below 1.2 mL kg^{-1} d^{-1} (GIMMLER et al., 1998).

Except for a significant reduction of toxic waste components by incineration and an immobilization of toxic residues in the ashes, energy generation is a major aim. Waste incineration with the different processes mentioned above yields different products (Fig. 14) and a different amount of energy. According to BORN (1995) the Siemens KWU process was the most energy-efficient thermal conversion process, while HEUSSER et al. (1998) reported, that the classical waste incineration plants and the Siemens KWU process operate more energy efficient than the Thermoselect process and the Noell thermal conversion process (Tab. 6). Practical experience for all waste incineration processes at full scale is not yet available and must prove the theoretical calculations.

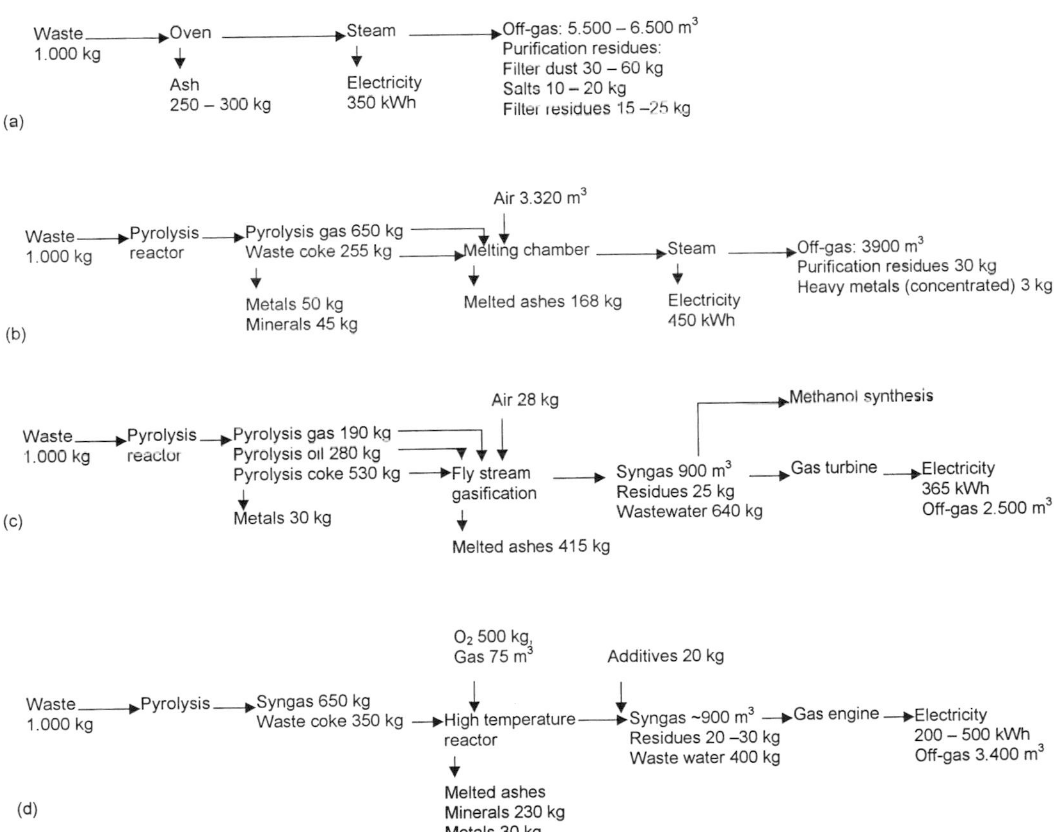

Fig. 14a–d. Comparison of input and output data of different processes for thermal waste treatment (BORN, 1995, modified), (**a**) classical waste incineration, (**b**) carbonizing-burning process of Siemens KWU, (**c**) thermal conversion process of NOELL, (**d**) Thermoselect process.

Tab. 6. Energy Balance for Different Waste Conversion and Incineration Processes (according to HEUSSER et al., 1998)

Process	Input [kWh]		Process Demand	Output [kWh] Grit Delivery
	Waste Energy	Primary Energy		
Classical waste incineration (roast oven)				
Without slag melting	2,800	280	160	540
With slag melting	2,800	730	160	540
Noell thermal conversion process	2,800	–	410	340
Siemens KWU carbonizing-burning process	2,800	280	210	560
Thermoselect process	2,800	220	300	360

8 Co-Incineration of Wastes as a Fuel Supplement in Power Plants or for Industrial Processes with a High Energy Demand

In principle it is possible to coincinerate wastes in power plants. Problems arise with the off-gas since for power plants less stringent pollution standards are valid (13. BImSchV, 1983; BImSchV: Bundesimmissionsschutzverordnung) than for municipal waste and sewage sludge incineration plants (17. BImSchV, 1990). The differences concern the emission of dust, CO, HF, HCl, carbon, SO_2 and NO_2, as well as the emissions of heavy metals (e.g., mercury), dioxines, and furanes. If a coal power plant is fueled with 25% sewage sludge and 75% coal for co-incineration more SO_2, HCl, mercury, and more ashes are emitted (HOFFMAN 1997).

Except for the use as a supplementary fuel in power plants, waste material could also be used as a fuel in the cement industry. At present the proportion of fuels covered by "secondary fuels" such as tires, used oils, etc., in the cement industry is about 13% of the total demand. The maximal proportion of secondary fuels that could substitute primary energy sources in energy intensive industries was estimated around 30%, equal to 2–3 $\cdot 10^6$ t of wastes per year if energy rich production-specific wastes from industry could also be used for this purpose (MASELLI, 1999).

9 Waste Processing Costs and Management

The total costs for waste processing in different regions of Germany vary within a broad range. According to HAASE (1999) the costs include 35–55% of the total amount for waste collection, 10–30% for transport and transshipment or turnover, 15–50% for pretreatment and final disposal and 5–7% for administrational requirements. The fees for the waste producer include the effort for collection, treatment, disposal, and landfill maintenance and may be calculated according to the local waste management concept. Waste management concepts have to follow "the state of the art" within the frame of ecological requirements and economical feasability. In 1998 treatment and disposal of domestic and industrial, domestic-like solid wastes in European countries had a market volume of $2.957 \cdot 10^{10}$ US \$. These costs that were calculated by the authorities included costs for the collection mode, for treatment (e.g., separation of valuables and recycling or disposal: direct disposal after biological treatment or disposal of slags after incineration) and for transport to the final deposition.

In 1998 the costs for incineration of non-recyclable solid waste varied from 80–440 US \$ per ton (BILITEWSKI and HEILMANN, 1998). A significant cost factor was the size of the incineration plants: In decentralized plants with one oven and capacities of $25 \cdot 10^4$–$100 \cdot 10^4$ t a^{-1} incineration caused costs within a narrow range of 133–207 US \$ per ton (RIEMANN and SONNENSCHEIN, 1998) whereas bigger plants were more economic if run at full capacity. Many incineration plants with more than one oven had, however, a shortage of waste and thus caused much higher operation costs.

As an alternative to waste incineration only recently waste stabilization by a sequence of mechanical and biological treatment steps has proceeded. The specific costs for this combined procedure depend on the kind of manual-mechanical separation of the wastes, grinding of non-recyclable residues for direct composting or, alternatively, for percolation. The solid residues of percolation are composted and the leachates are treated in a biogas reactor. Depending on whether the treatment residues can be disposed on sanitary landfills or must be incinerated the total costs for mechanical and biological treatment were between 38 and 100 US \$, respectively (BILITEWSKI and HEILMANN, 1998).

Costs for waste composting in Germany vary between 95 US \$ for windrow composting and 150 US \$ for container composting for plants with a capacity of less than 30,000 t a^{-1}. The operation costs for anaerobic digesters of $<15{,}000$ t a^{-1} annual capacity vary between 100 and 130 US \$, of $<25{,}000$ t a^{-1} between 60 and 106 US \$ and of $<50{,}000$ t a^{-1} between 44 and 75 US \$ (SCHÖN, 1994). The deposition costs for solid residues of composting or digestion vary within a broad range. Since due to the TA Siedlungsabfall (1993) it will no longer be permitted to deposit untreated wastes after the year 2005, many operators of sanitary landfills offer dumping prices to fill up their deposition capacity. It is thus difficult to specify the real costs, that would include the costs for maintenance after closure.

10 References

AbfKlärV (1992), *Abfallklärschlammverordnung* vom 15. 04. 1992, Bundesgesetzblatt Jahrgang 1992, Part I, p. 912.

ATV Information (1999), *Thermische Behandlung von Abfällen*, ATV Abwasser Abfall Gewässerschutz (Ed.). Theodor-Heuss-Allee 17, D-53773 Hennef.

BENZ, M., SCHINK, B., BRUNE, A. (1998), Humic acid reduction by *Propionibacterium freudenreichii* and other fermenting bacteria, *Gen. Microb. Ecol.* **11**, 4507–4512.

BGK (1994), Bundesgütegemeinschaft Kompost e.V. *RAL 251*. Schönhauserstr. 3, D-50968 Köln.

BGK (1997), Humuswirtschaft und Kompost 2/97. Informationsdienst der Bundesgütegemeinschaft Kompost e.V. Köln: BGK.

BIEHLER, M. J., NUDING, R. (1996), Vergleich verschiedener Verfahren zur Verwertung von Bio- und Grünabfällen auf kommunaler Ebene, *Abfallwirtschafts.J.* **9**, 28–38.

BILITEWSKI, B., HEILMANN, A. (1998), Kosten der mechanisch-biologischen Abfallbehandlung im Vergleich zur thermischen Behandlung, *Entsorgungspraxis* **10**, 26–31.

BILITEWSKI, B., HÄRDTLE, G., MAREK, K. (1994), *Abfallwirtschaft* (BILITEWSKI, B., HÄRDTLE, G., MAREK, K., Eds.). Berlin: Springer-Verlag.

BILITEWSKI, B., ROTTER, S. (1998), Übersicht über die behördlichen Anforderungen an Anlagen zur biologischen Abfallbehandlung, *Wasser Boden* **50**, 13–16.

13. BImSchV (1983), *Dreizehnte Verordnung zur Durchführung des Bundes-Immissionsschutzgesetzes* – Verordnung über Großfeuerungsanlagen – 13 BImSchV vom 22. 06. 1983, Bundesgesetzblatt p. 719ff.

17. BImSchV (1990), *Siebzehnte Verordnung zur Durchführung des Bundesimmissionsschutzgesetzes* – Verordnung über Verbrennungsanlagen für Abfälle und ähnliche brennbare Stoffe – 17. BImSchV vom 23. 11. 1990, Bundesgesetzblatt p. 2545 (17th ordinance to the act on protection from the harmful effects on air pollution, noise, vibrations and other nuisances – ordinance on in-

cineration plants for wastes and other combustible materials).

BioAbfV (1998), *Bioabfallverordnung* – Verordnung über die Verwertung von Bioabfällen auf landwirtschaftlich, forstwirtschaftlich und gärtnerisch genutzten Böden, Bundesgesetzblatt Jahrgang 1998 Part I No. 65, pp. 2955–2981 (Ausgegeben zu Bonn am 28. September 1998).

Blue Angel (1996), *Blue Angel RAL UZ 45 1996*. Deutsches Institut für Gütesicherung und Kennzeichnung e.V. Bonn.

BBodSchG (1998), Bundes-Bodenschutzgesetz – Gesetz zum Schutz vor schädlichen Bodenveränderungen und zur Sanierung von Altlasten, BBodSchG vom 17. 03. 1999, Bundesgesetzblatt Part I, p. 502.

BORN, M. (1995), Zur Bewertung thermischer Verfahren in der Abfallwirtschaft, *Energieanwendung* **4**, 18–25.

DACH, J. (1999), Emissionen aus Deponien nach mechanisch-biologischer Abfallbehandlung, *Entsorgungspraxis* **6**, 26–30.

DüngeV (1996), *Düngeverordnung* – Verordnung der guten fachlichen Praxis beim Düngen, Bundesgesetzblatt 26. Januar 1996, pp. 118ff.

FIGUEROA, R. A. (1998), Untersuchungen zur mikrobiellen Methanoxidation in Rekultivierungsschichten von Abfalldeponien, *Müll Abfall* **5**, 324–342.

GALLERT, C., WINTER, J. (1997), Mesophilic and thermophilic anaerobic digestion of source-sorted organic wastes: effect of ammonia on glucose degradation and methane production, *Appl. Microbiol. Biotechnol.* **48**, 405–410.

GALLERT, C., BAUER, S., WINTER, J. (1998), Effect of ammonia on the anaerobic degradation of protein by a mesophilic and thermophilic biowaste population, *Appl. Microbiol. Biotechnol.* **50**, 495–450.

GIMMLER, H., TRACK, C., DRESCH, H. (1998), Müllverbrennungsschlacke als Baumaterial, *Umweltwiss. Schadstoffforsch.* **10**, 90–98.

HAASE, H. (1999), Logistische Aufgabenstellung im Spannungsfeld der Kreislaufwirtschaft, *Wasser Luft Boden* **7–8**, 63–65.

HEINZ, A., REINHARDT, G. A. (1993), *Chemie und Umwelt*, 3rd Edn. Braunschweig/Wiesbaden: Vieweg.

HEUSSER, P., LAUBBACHER, R., BORN, M. (1998), Verfahrens- und Kostenvergleich für die thermische Abfallbehandlung, *Chem. Techn.* **6**, 281–286.

HOFFMANN, V. (1997), Abfallmitverbrennung in Kraftwerken – rechtliche und ökologische Aspekte am Beispiel Klärschlamm, *Wasser Boden* **9**, 49–54.

HOPPENHEIDT, K., HIRSCH, P., KOTTMAIR, A., NORDSIECK, H., SWEREV, M., MÜCKE, W. (1999), Gemeinsame Vergärung von Bio- und Gewerbeabfall, *EntsorgungsPraxis* **6**, 13–20.

HUMER, M., LECHNER, P. (1997), Deponiegasentsorgung von Altlasten mit Hilfe von Mikroorganismen, *Österr. Wasser- und Abfallwirtschaft* **7–8**, 164–171.

KompostErlass (1994), Kompostiererlass des Landes Baden Württemberg vom 30. Juni 1994. Umweltministerium Baden-Württemberg, Stuttgart.

Komposthinweis Bayern (1994), Bayerischer Staatsanzeiger vom 20. 9. 1994.

KORZ, D. J. (1999), Naßvergärungsanlagen in Deutschland, *Entsorgungspraxis* **3**, 39–41.

KrW-/AbfG (1996), Kreislaufwirtschafts- und Abfallgesetz – Gesetz zur Förderung der Kreislaufwirtschaft und Sicherung der umweltverträglichen Beseitigung von Abfällen (Kreislaufwirtschafts- und Abfallgesetz – KrW-/AbfG) vom 27. 09. 1994. Bundesgesetzblatt BGBL. I, 2705 pp.

KRULL, R. (1995), Bioabfälle im Verbund vergären und kompostieren, *Umwelt* **3**, 86–90.

KRÜMPELBECK, I., HÖRING, K., EHRIG, H.-J. (1998), Zukünftige Entwicklung der Deponiegasmengen, *AbfallwirtschaftsJ.* **7–8**, 30–35.

KUSSMAUL, M., GEBERT, J. (1998), Ein neues Verfahren zum biologischen Methan- und Geruchsabbau von Gasen aus Abfalldeponien mit passiver Entgasung, *Müll Abfall* **8**, 512–518.

LAGA M10 (1995), Qualitätskriterien und Anwendungsempfehlungen für Kompost, *LAGA Merkblatt M10*. Länderarbeitsgemeinschaft Abfall (LAGA) (Ed.). Berlin: Erich Schmidt Verlag.

LANGHANS, G. (1997), Bemessung und Bilanzierung der Biogasausbeuten in der Abfallvergärung, *AbfallwirtschaftsJ.* **1–2**, 35–38.

LANGHANS, G. (1998), Bioabfallvergärung oder Co-Fermentation – Eine schwere Entscheidung, *Entsorgungspraxis* **3**, 26–32.

LANGHANS, G. (1999), Stoffströme in die Umwelt – der Output von Vergärungsanlagen, *Entsorgungspraxis* **1–2**, 27–34.

LOOCK, R., FETSCH, S., FOELLMER, T., NÖLDEKE, P. (1999), Trockenfermentation von Restmüll – Einstufige anaerobe Trockenfermentation von Restmüll, Verfahrensentwicklung und Ergebnisse, *Müll Abfall* **2**, 86–90.

MASELLI, J. (1999), Status quo und mittelfristige Entwicklung von thermischer Behandlung und energetischer Verwertung von Siedlungsabfällen, *Müll Abfall* **7**, 435–440.

PALMOWSKI, L., MÜLLER, J. (1999), Einfluß der Zerkleinerung biogener Stoffe auf deren Bioverfügbarkeit, *Müll Abfall* **6**, 368–372.

REINHARDT, D. R., TOWNSEND, T. G. (1998), *Landfill Bioreactor Design and Operation* (REINHARDT, D. R., TOWNSEND, T. G., Eds.). Boca Raton, FL: Lewis Publ., New York: CRC Press.

RIEMANN, K.-A., SONNENSCHEIN, H. (1998), Kosten einer thermischen Restabfallbehandlung in dezentralen Kleinanlagen, *Entsorgungspraxis* **10**, 32–35.

SCHERER, P. (1995), Aktuelle Marktübersicht zu Vergärungsanlagen für feste Abfälle – Vorteile gegenüber Kompostieranlagen, *Müll Abfall* **12**, 845–856.

SCHERER, P. A., VOLLMER, G.-R. (1999), Entwicklung eines einfachen Hochleistungsvergärungsverfahrens zur Behandlung von Restmüll, *Müll Abfall* **3**, 150–158.

SCHÖN, M. (1994), Verfahren zur Vergärung organischer Rückstände in der Abfallwirtschaft, *Abfallwirtschaft in Forschung und Praxis* Vol. 66. Berlin: Erich Schmidt Verlag.

SCHÖN, M., WALZ, R. (1993), Emissionen der Treibhausgase Distickstoffoxid und Methan in Deutschland, in: Forschungsbericht des Bundensministers für Umwelt, Naturschutz und Reaktorsicherheit, *UBA-Forschungsbericht 93–121*. Berlin: Erich Schmidt Verlag.

StoV (1986), Schweizerische Stoffverordnung vom 9. Juni 1996 mit Novellierung vom 16. September 1992.

TA Siedlungsabfall (1993), *Dritte Allgemeine Verwaltungsvorschrift zum Abfallgesetz* (TA Siedlungsabfall). Technische Anleitung zur Verwertung, Behandlung und sonstigen Entsorgung von Siedlungsabfällen, 14. Mai 1993. (Technical directive on recycling, handling and disposal of municipal waste), Bundesministerium der Justiz (Ed.) Bundesanzeiger 45, No. 99a, 29. 5. 1993.

TABASARAN, O. (1994), *Abfallwirtschaft Abfalltechnik – Siedlungsabfälle* (TABASARAN, O., Ed.). Berlin: Ernst & Sohn.

TEMPER, U., WINTER, J., KANDLER, O. (1983), Methane fermentation of wastes at mesophilic and thermophilic temperatures, in: *Energy from Biomass, 2nd E. C. Conference.* (STRUB, A., CHARTIER, P., SCHLESER, G., Eds.), pp. 521–525. London, New York: Elsevier Applied Science Publishers.

UBA (1992), Daten zur Umwelt 1990/91 Umweltbundesamt Fachgebiet I 1.2 "Umweltforschung, Umweltstatistik" (Ed.) Bismarckplatz 1, 1000 Berlin 33. Berlin: Erich Schmidt Verlag.

WESTLAKE, K. (1995), Landfill, in: *Waste Treatment and Disposal, Issues in Environmental Science and Technology* (HESTER, R. E., HARRISON, R. M., Eds.), pp. 43–67. The Royal Society of Chemistry, Thomas Graham House, Science Park, Cambridge CB4 4 WF.

WILDENAUER, F., WINTER, J. (1984), Efficiency of anaerobic digestion of sewage sludge, cattle manure and piggery waste, in: *Bioenergy 84*, Vol. III *Biomass Conversion* (EGNÉUS, H., ELLEGÁRD, A., Eds.), pp. 431–439. London, New York: Elsevier Applied Science Publishers.

WINTER, J. (1984), Anaerobic waste stabilization, *Biotechnol. Adv.* **2**, 75–99.

WINTER, J. U., COONEY, C. L. (1980), Fermentation of cellulose and fatty acids with enrichments from sewage sludge, *Eur. J. Appl. Microbiol. Biotechnol.* **11**, 60–66.

2 Microbiology of Composting

HANS JÜRGEN KUTZNER

Ober-Ramstadt, Germany

List of Abbreviations

APPL	acid precipitable, polymeric lignin
CFU	colony forming unit
GC	gas–liquid chromatography
HDMF	3-hydroxy-4,5-dimethyl-2(5H)-furanone
MS	mass spectroscopy
PVC	polyvinyl chloride
SAR	systemic acquired resistance
TMV	tobacco mosaic virus
TOC	total organic compound
v.s.	volatile solids

1 Introduction

In his comprehensive monographs, HAUG (1980, 1993) defines composting as *"the biological decomposition and stabilization of organic substrates under conditions which allow development of thermophilic temperatures as a result of biologically produced heat, with a final product sufficiently stable for storage and application to land without adverse environmental effects"*. This definition differentiates composting from the mineralization of dead organic matter taking place in nature above the soil or in its upper layers leading to a more or less complete decomposition – besides the formation of humic substances; it thus describes the compost pile as a man made microbial ecosystem. Composting has been carried out for centuries, originally as an agricultural and horticultural practice to recycle plant nutrients and to increase soil fertility (HOWARD, 1948); nowadays it has become also part of the management of waste disposal to get rid of the huge amounts of diverse organic waste produced by our civilized urban life. In most cases, the product *compost* has to be regarded as a by-product which hardly finances its production now often being carried out in highly mechanized plants (FINSTEIN et al., 1986; FINSTEIN, 1992; JACKSON et al., 1992; STEGMANN, 1996).

Composting has frequently been regarded as *more an art than a science*; this view, however, ignores the fact that its scientific base is well understood; of course, successful application of the principles requires experience as is more or less true for all applied sciences. In fact, the basic rules of composting have been known for decades as can be seen from numerous reviews and monographs of the last 25 years, beginning with UPDEGRAFF (1972) and ending with DE BERTOLDI et al. (1996). These surveys also indicate the broad interest of scientists of various disciplines in this process, disciplines such as agriculture, horticulture, mushroom science, soil science, microbiology and sanitary engineering. The literature on composting is vast, comprising numerous broad reviews and minireviews of which only few can be cited in addition to those mentioned above: GASSER (1985), BIDDLESTONE et al. (1987), MATHUR (1991), MILLER (1991, 1993), HOITINK and KEENER (1993) and SMITH (1993); in addition, there exist also specific journals devoted solely or primarily to the subject, e.g., *"Compost Science"*, *"Agricultural Wastes"*, *"Müll & Abfall"*. Being well aware of the literature covered in these reviews, the author has tried to avoid repetition as much as possible; thus only selected papers will be considered, in addition to paying regard to some older work not reviewed until now because of its "hidden" publication.

This review is primarily concerned with the microbiology of composting. However, since composting touches many related disciplines, even the restriction to this selected field has to take various aspects into consideration which may seem at first glance rather remote from the composting process *per se*:

(1) The microbiology of self-heating of moist, damp organic matter has first been extensively studied in the case of agricultural products, e.g., hay, grain and wool. This phenomenon very early led to the concept of heat generation as part of microbial (and organismic in general) metabolism.
(2) The microbiology of composting is somehow related to soil microbiology and litter decomposition, i.e., soil fertility, turnover of organic matter in nature and formation of humic substances.
(3) The control of pathogenic agents in wastes to be composted, and the emission of pathogenic agents from compost plants are of concern to medical microbiologists. This aspect has to be extended to agents causing plant diseases and to the effect of compost on plant pathogens.
(4) Mushroom cultivation includes the preparation of a compost substrate, a special process whose study contributed much to the general understanding of composting.

The main focus of this chapter will be *the compost pile as a microbial ecosystem*, and a more proper title for it would be *"A Microbiologist's View of Composting"*. Most of the reviews cited above also deal with the microbiology of composting, and there are several which specifically discuss this aspect, e.g., FINSTEIN and MORRIS (1975) and LACEY (1980). Many papers mentioned there will not be cited in this review, and it is hoped that their authors will have some understanding for this approach: a reviewer has to make a selection of topics and of the literature to be cited, which inevitably leads to a somewhat personal view, not entirely free of bias.

2 Heat Production by Microorganisms

Any metabolism – from microbes to man – leads inevitably to the production of heat (Fig. 1, Tab. 1). This is actually a consequence of the 2nd law of thermodynamics, i.e., only part of the energy consumed can be transformed into *useful work*, e.g., biosynthesis, while the rest is liberated as heat to increase the entropy of the surroundings. Very often, mostly just for simplification, the degradation of a carbohydrate (e.g., glucose) serves as an example to demonstrate this context: Tab. 2 gives an energy balance for the aerobic metabolism of 2 M glucose, assuming that 1 of them enters the energy metabolism producing 38 ATP M^{-1} glucose, whereas the other supplies the precursors for the biosynthesis of new biomass which consumes the 38 ATP: According to this calculation, which follows the reasoning of DIEKERT (1997), the catabolism has a physiological efficiency of 61–69%, whereas the anabolism of only 40%. A very similar balance has been found by TERROINE and WURMSER (1922) for the mold *Aspergillus niger* as discussed in detail by BATTLEY (1987, pp. 108 ff): 59% of the energy (not weight!) of the glucose consumed were incorporated into new biomass (mycelium), whereas 41% were liberated as heat.

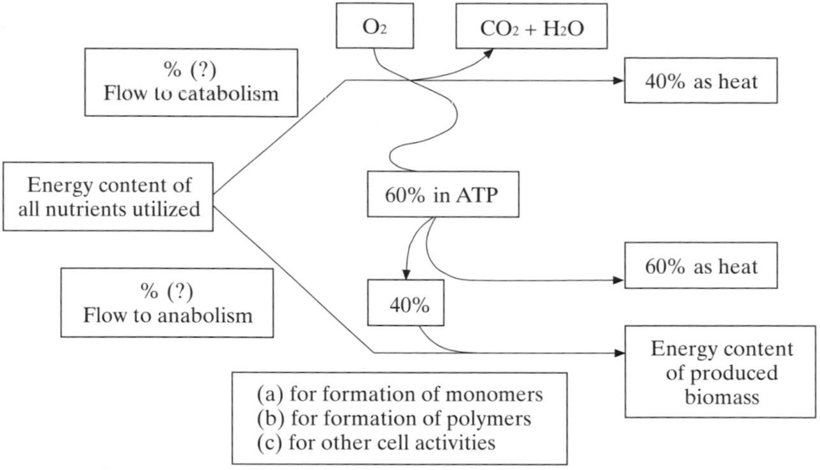

Fig. 1. Energy flow in aerobic metabolism of bacteria (for further explanation see text and Tab. 1).

Tab. 1. Energy flow in Microorganisms with Glucose as Substrate: Proportioning of the Substrate Energy to New Biomass and Liberated Heat as well as especially the Y_{ATP} Value Depend on the Number of ATP per Mol Glucose

	% Glucose Utilized for		% Substrate Energy in				
	Catabolism (Energy Production)	Anabolism (Biosynthesis)	New Biomass	Liberated Heat	ATP Glucose^{-1}	Y_s	Y_{ATP}
A	25	75	81.1	18.9	38	0.565	10.7
B_1	33.33	66.66	74.8	25.2	38	0.502	7.137
B_2	33.33	66.66	72.2	27.8	26	0.502	10.43
C_1	50	50	62.2	37.8	38	0.376	3.568
C_2	50	50	58.3	41.7	26	0.376	5.215

Tab. 2. Energy Balance of the Aerobic Metabolism of Glucose by Bacteria (Free Energy of Hydrolysis $ATP + H_2O \rightarrow ADP + P_i$: A = 52 kJ, B = 46 kJ)

Metabolism	A	B
Catabolic metabolism		
$C_6H_{12}O_6 + 6\ O_2 \rightarrow 6\ CO_2 + 6\ H_2O$	$\Delta G^{0\prime} = -2{,}872$ kJ	
Invested into 38 ATP (38 · 52 or 46 kJ)	1,976 kJ = 69%	1,748 kJ = 61%
Liberated as heat	896 kJ = 31%	1,124 kJ = 39%
Anabolic metabolism		
Free energy of hydrolysis of 38 ATP	$\Delta G^{0\prime} = -1{,}976$ kJ	
Invested in biosynthesis, transport, movement	790 kJ = 40%	699 kJ = 40%
Liberated as heat	1,186 kJ = 60%	1,049 kJ = 60%
Total balance		
2 M glucose (2 · 2,872)	$\Delta G^{0\prime} = -5{,}744$ kJ	
Liberated as heat	2,082 kJ = 36%	2,173 kJ = 38%
Fixed in new biomass	3,662 kJ = 64%	3,571 kJ = 62%

Note that the heat of combustion of 1 M glucose amounts to $\Delta H_c = -2{,}816$ kJ.

The percentages of the substrate (1) employed for energy formation (catabolism) and (2) utilized for biosynthesis depend on the energy source and the kind of metabolism, (e.g., amount of ATP M^{-1} substrate).

For *E. coli* (26 ATP M^{-1} glucose) DIEKERT (1997) proposed the following balance: One third of the substrate (glucose) is used for the production of ATP, whereas two thirds [more correctly 4 of the 6 carbon atoms (Eq. 1)] appear in the biomass; this results in an Y_s of about 0.5 and an Y_{ATP} of about 10 (Tab. 3).

The heat produced in the metabolism of microbes cultivated on a small scale is rapidly dissipated to the environment and hardly noticed in laboratory experiments. Therefore, this phenomenon, although of great theoretical importance, is surprisingly not discussed in most textbooks of microbiology, a rare exception being the one by LAMANNA and MALLETTE (1959, pp. 586–589). Of course, heat production is of great practical significance in the mass culture of microorganisms and, therefore, treated in books on biochemical engineering, e.g., BAILY and OLLIS (1977, pp. 473–482) and CRUEGER and CRUEGER (1984, pp. 58–59); it has been extensively discussed by LUONG and

Tab. 3. Equations of Microbial Growth Calculated for Various Growth Efficiencies

	Equation	Yield
(1)	$C_6H_{12}O_6 + 0.8\ NH_3 + 2.0\ O_2 \rightarrow 0.8\ [C_5H_7O_2N] + 2.0\ CO_2 + 4.4\ H_2O$	$Y_s = 90.4/180 = 0.502$
(2)	$C_6H_{12}O_6 + 0.7\ NH_3 + 2.5\ O_2 \rightarrow 0.7\ [C_5H_7O_2N] + 2.5\ CO_2 + 4.6\ H_2O$	$Y_s = 79.1/180 = 0.430$
(3)	$C_6H_{12}O_6 + 0.6\ NH_3 + 3.0\ O_2 \rightarrow 0.6\ [C_5H_7O_2N] + 3.0\ CO_2 + 4.8\ H_2O$	$Y_s = 67.8/180 = 0.376$
(4)	$C_6H_{12}O_6 + 0.5\ NH_3 + 3.5\ O_2 \rightarrow 0.5\ [C_5H_7O_2N] + 3.5\ CO_2 + 5.0\ H_2O$	$Y_s = 56.5/180 = 0.314$
(5)	$C_6H_{12}O_6 + 0.4\ NH_3 + 4.0\ O_2 \rightarrow 0.4\ [C_5H_7O_2N] + 4.0\ CO_2 + 5.2\ H_2O$	$Y_s = 45.2/180 = 0.251$
(6)	$C_6H_{12}O_6 + 0.3\ NH_3 + 4.5\ O_2 \rightarrow 0.3\ [C_5H_7O_2N] + 4.5\ CO_2 + 5.4\ H_2O$	$Y_s = 33.9/180 = 0.188$
(7)	$C_6H_{12}O_6 + 0.2\ NH_3 + 5.0\ O_2 \rightarrow 0.2\ [C_5H_7O_2N] + 5.0\ CO_2 + 5.6\ H_2O$	$Y_s = 22.6/180 = 0.125$

HAUG (1993, p. 248) considered $Y_s = 0.1–0.2$ as a typical growth yield in composting; for $Y_s = 0.1$ he presented the following balance (here reduced to one mole of glucose).

(8) $C_6H_{12}O_6 + 0.16\ NH_3 + 5.2\ O_2 \rightarrow 0.16\ [C_5H_7O_2N] + 5.2\ CO_2 + 5.7\ H_2O$ $\quad Y_s = 18.8/180 = 0.104$

1. Note: Calculation in g (NB = New Biomass)

(a) Complete oxidation of glucose without production of biomass
100 g glucose + 106,7 g $O_2 \rightarrow$ 146.7 g CO_2 + 60 g H_2O

(b) Eqs. (1) and (6) from above in g:
$Y_s = 0.502$: [100 + 7.5] substrate + 35.6 $O_2 \rightarrow$ 50.2 NB + 48.9 CO_2 + 44.0 H_2O
$Y_s = 0.188$: [100 + 2.8] substrate + 80.0 $O_2 \rightarrow$ 18.8 NB + 110.0 CO_2 + 54.0 H_2O

(c) The following equation has been used for the hypothetical composting process discussed in Sect. 9 (Fig. 15 and Tab. 15, Equ. 8a
$Y_s = 0.2008$: [100 + 3.02] substrate + 78.22 $\rightarrow$ 20.08 NB + 107.56 CO_2 + 53.6 H_2O

2. Note Calculation of oxygen consumption in relation to loss of volatile solids, Δ v.s., in g:
$Y_s = 0.502$: Δ v.s. = 107.5 − 50.2 = 57.3 + 35.6 $O_2 \rightarrow$ 48.9 CO_2 + 44.0 H_2O
Δ v.s. : 100 + 62.13 $O_2 \rightarrow$ 85.34 CO_2 + 76.79 H_2O
$Y_s = 0.188$: Δ v.s. = 102.8 − 18.8 = 84.0 + 80.0 $O_2 \rightarrow$ 110.0 CO_2 + 54.0 H_2O
Δ v.s. : 100 + 95.23 $O_2 \rightarrow$ 130.95 CO_2 + 64.28 H_2O
$Y_s = 0.167$: Δ v.s. = 102.5 − 16.7 = 85.8 + 83.0 $O_2 \rightarrow$ 114.1 CO_2 + 54.7 H_2O
Δ v.s. : 100 + 96.73 $O_2 \rightarrow$ 132.98 CO_2 + 63.75 H_2O

VOLESKY (1983), in the monograph by BATTLEY (1987) and in Vol. 1 of the Second Edition of *Biotechnology* by POSTEN and COONEY (1993, pp. 141–143).

3 The Phases of the Composting Process

If the heat produced by the metabolism of microorganisms is prevented by some kind of insulation from being dissipated to the environment, the temperature of the habitat increases. This is the case when damp organic matter is collected in bulky heaps or kept in tight containers, as it is done when organic waste is composted either in large piles (windrows) or in boxes of various kinds. If the composting process is carried out as a batch culture – as opposed to a continuous operation – it proceeds in various more or less distinct phases which are recognized superficially by the stages of temperature rise and decline (Fig. 2). These temperature phases are, of course, only the reflection of the activities of successive microbial populations performing the degradation of increasingly more recalcitrant organic matter.

As shown in Fig. 2, the time–temperature course of the composting process can be divided into 4 phases:

(1) During the first phase a diverse population of mesophilic bacteria and fungi proliferates, degrading primarily the readily available nutrients and thereby raising the temperature to about 45 °C. At this point their activities cease, the vegetative cells and hyphae die and eventually lyse, and only heat resistant spores survive.

(2) After a short lag period (not always discernible) there occurs a second more or less steep rise of temperature. This second phase is characterized by the development of a thermophilic microbial population comprising some bacterial species, actinomycetes and fungi. The temperature optimum of these microorganisms is between 50 and 65 °C, their activities terminate at 70–80 °C.

(3) The third phase can be regarded as a stationary period without significant changes of temperature because microbial heat production and heat dissipation balance each other. The microbial population continues to consist of thermophilic bacteria, actinomycetes, and fungi.

(4) The fourth phase is characterized by a gradual temperature decline; it is best described as the maturation phase of the composting process. Mesophilic microorganisms having survived the high temperature phase or invading the cooling down material from the outside succeed the thermophilic ones and extend the degradation process as far as it is intended.

Fig. 2 presents just one of numerous examples of the temperature course that can be found in the literature, very typical ones having been published by CARLYLE and NORMAN (1941), WALKER and HARRISON (1960), NIESE (1959). In all cases the 4 phases mentioned have been observed more or less distinctly leaving no doubt that they characterize very closely the composting process.

Since the optimum temperature for composting is regarded to be about 50–60 °C, measures are being taken to prevent further self-heating except for a rather short period up to 70 °C to guarantee the elimination of pathogens (see Sect. 7.1). However, 70 °C appears to be not the limit of microbial heat production which can easily reach 80 °C as practised in the Beltsville process (see Sect. 4.3). Under certain conditions even much higher temperatures leading to ignition can be reached, but neither the exact requirements for such an event nor the mechanism of ignition appear to be well understood (BOWES, 1984). Whereas there are only rare cases of self-ignition of manure piles or compost heaps (JAMES et al., 1928), this phenomenon is not uncommon in the storage of damp hay (GLATHE, 1959, 1960; CURRIE and FESTENSTEIN, 1971; HUSSAIN, 1972) and fat contaminated pie wool (WALKER and WILLIAMSON, 1957).

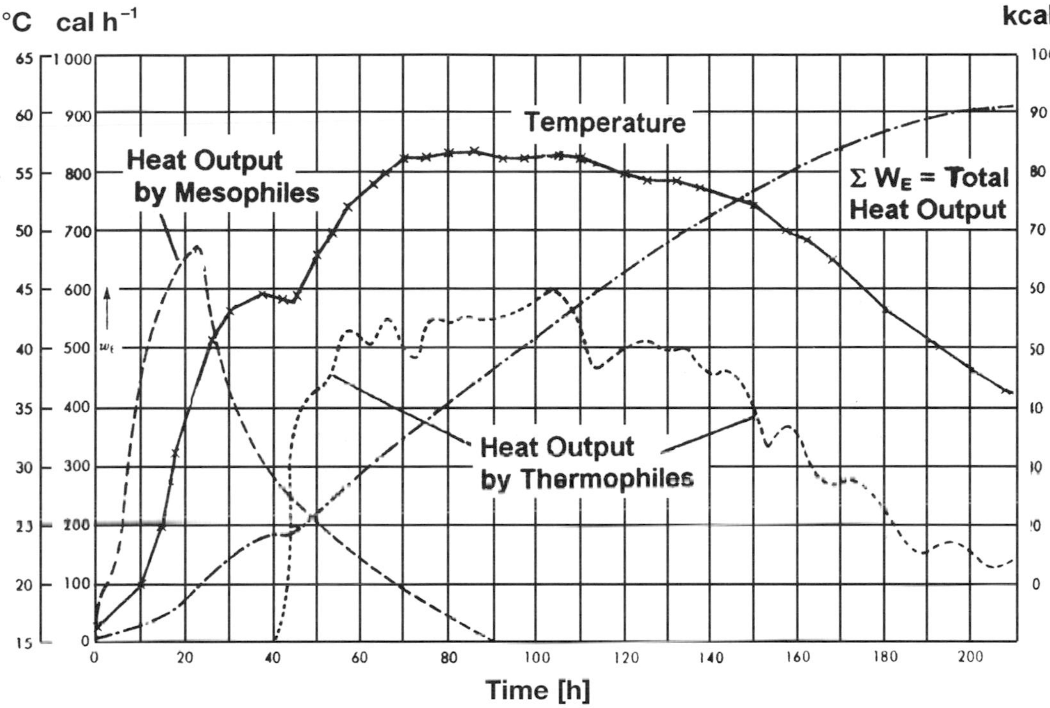

Fig. 2. Temperature course during the composting of urban garbage: four phases, *mesophilic, thermophilic, stationary*, and *maturation*, can easily be recognized (from PÖPEL, 1971).

As mentioned above the temperature phases are just a reflection of the activities of successive microbial populations. This has been demonstrated by various means – besides by a detailed analysis of the bacterial, actinomycete and fungal population:

(1) Fig. 3, taken from NIESE (1969), shows that the microbial community of fresh refuse plus sewage sludge exhibits a respiratory activity only at 28 and 38 °C, i.e., it consists primarily of mesophiles. On the contrary, the samples taken from the self-heated material started instantaneously to take up oxygen when incubated at 58 and 48 °C; the relatively high respiration rate at 38 °C is probably due to the broad temperature range of several thermophiles (Sect. 6, Fig. 8, Tab. 9).

(2) Fig. 4, taken from FERTIG (1981), illustrates the O_2 uptake and CO_2 production during the temperature course of composting: 4 maxima of microbial activity can be observed, surprisingly within the very short time of 54 h. Two or three maxima of CO_2 evolution during composting have been observed by numerous authors, e.g., SIKORA et al. (1983) who discussed also earlier observations of this kind; VIEL et al. (1987) reported three maxima of oxygen consumption.

(3) Finally, a detailed analysis of adaptation and succession of microbial populations in composting of sewage sludge has been undertaken by MCKINLEY and VESTAL (1984, 1985a,b), the main aim of their study being to ascertain the optimal temperature for the composting process: The microbial communities from hotter samples were better adapted to higher temperatures than those from cooler samples and *vice versa*, as

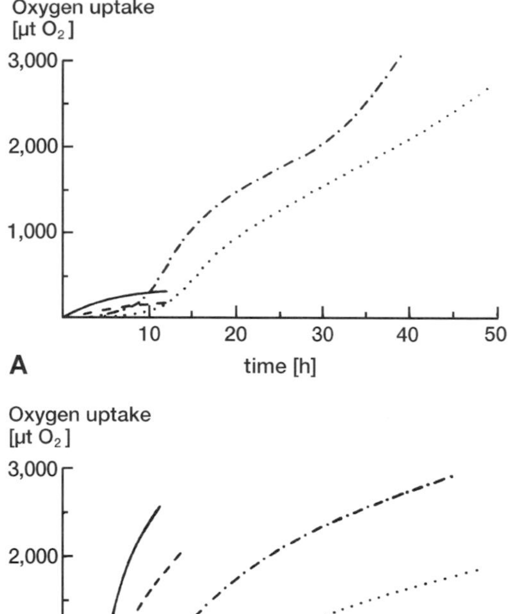

Fig. 3a,b. Oxygen uptake of microbial communities in Warburg flasks at different temperatures: **A** fresh garbage plus sewage sludge, **B** composting material removed from the pile during the high temperature phase, ·········· 28 °C — · — 38 °C - - - - - 48 °C —— 58 °C (according to NIESE, 1969).

shown by the determination of the rate of [^{14}C]-acetate incorporation into cellular lipids and calculation of its apparent energies of activation and inactivation. Lipid phosphate was used as indicator of viable bacterial biomass. The authors came to the conclusion, that the composting temperature should not be allowed to exceed 55 °C – in agreement with numerous other investigators.

4 The Compost Pile as a Microbial Habitat

In order to secure fast stabilization of the waste material, the microorganisms performing this task have to be provided with *nutrients, water* and *oxygen*. Of course, the demand for nutrients appears to be contradictory since material without nutrients does not need to be stabilized. However, because organic waste material in any case lends itself to decomposition the nutritional state of the starting material deserves consideration.

A fourth parameter of composting is the *temperature*, which plays actually a dual role in this habitat: It is the result of microbial activity – without necessity of being taken care of at the commencement of the process – and at the same time it is a selective agent determining the microbial population at any stage of the composting process, eventually demanding its regulation by technical measures.

Finally, the *pH* of the habitat can be considered as environmental factor.

It is obvious that the various parameters are intimately related; this should be kept in mind when in Sects. 4.1–4.5 they are necessarily treated separately.

4.1 Organic Wastes as Nutrients

Waste suitable for composting comes from very diverse sources: grass clippings, leaves, hedge cuttings, food remains, fruit and vegetables waste from the food industry, residues from the fermentation industry, solid and liquid manure from animal houses, wastes from the forest, wood and paper industries, rumen contents from slaughtered cattle and sewage sludge from wastewater treatment plants. Thus, the starting material of composting varies tremendously in its coarse composition, and in addition there is often a seasonal variation of the material arriving at the compost plant. Since many of the materials listed above cannot be easily composted if supplied by themselves alone because of nutritional and/or structural reasons (water content), they have

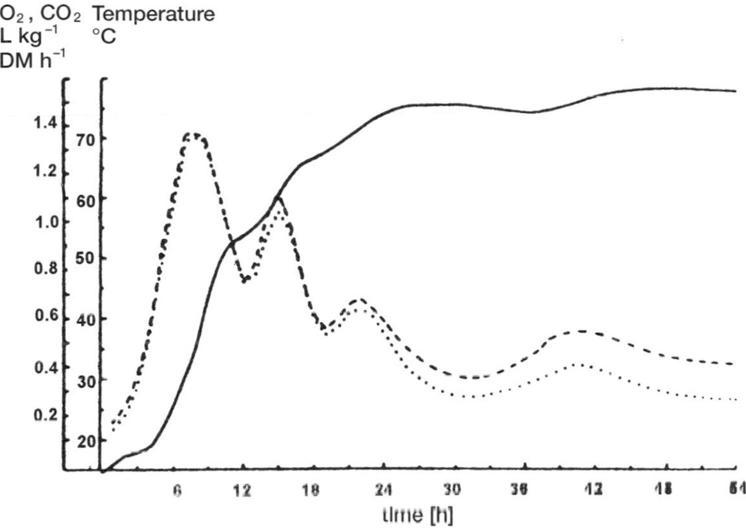

Fig. 4. Oxygen uptake and CO_2 production during laboratory composting: four maxima occurring within the first 2 d (!) are easily recognized (according to FERTIG, 1981).

to be mixed purposely if they are not delivered as a mixture in the first place.

Tables listing the chemical composition of the materials mentioned, e.g., contents of carbohydrates, proteins, fat, hydrocarbons, lignin and ash are given by BIDLINGMAIER (1983), and KROGMANN (1994), and can be found in various reviews cited above. Unfortunately, the data of most of the ingredients are rather incomplete making a strict comparison difficult. These tables sometimes contain empirical formulae of the substrates involved, e.g., for sewage sludge [$C_{10}H_{19}O_3$], for the organic fraction of domestic garbage [$C_{64}H_{104}O_{37}N$] for residues from vegetables [$C_{16}H_{27}O_8N$], and for grass [$C_{23}H_{38}O_{17}N$]. However, these figures are almost meaningless, except that they indicate the carbon–nitrogen ratio (see also Sect. 9.1, Tab. 16).

Of greater relevance is the *biochemical composition* of the various waste materials because this determines their susceptibility to microbial degradation. Those wastes containing carbohydrates, lipids and proteins, would be the most suitable carbon and energy sources for microbes, whereas materials with a high lignocellulose fraction and a shortage of nitrogenous compounds will be only slowly degraded. In fact, the biodegradability of organic matter in composting may be related to the lignin content (HAUG, 1993, pp. 312–314) employing a formula which has been derived originally for anaerobic digestion by CHANDLER et al. (1980):

biogradable fraction of volatile solids (v.s.) = 0.830–0.028 x lignin content in % of v.s. (9)

According to this formula a substrate containing no lignin would only achieve a maximum degradability of 83% because the decomposition of the substrate organics is coupled with production of bacterial by-products, some of which themselves are not readily degradable. However, since the waste material has to support the growth of several successive microbial populations, which have different nutritional requirements and different capabilities to attack macromolecules of organismic origin, the waste material need not (and, in fact, should not) consist solely of easily degradable materials.

It can be more or less safely assumed that the starting materials – at least mixtures of those listed above – contain the essential nutrients or elements for microbial growth. Whereas carbon compounds for energy metabolism and biosynthesis are in most cases in excess, the nitrogen supply is usually rather limited. In fact, the carbon–nitrogen ratio is considered a significant criterion of the starting material as well as of the product compost. A rule of thumb says that the C–N ratio at the beginning of composting should be about 30:1 and will be reduced to about 10:1 in the course of the process. Of course, there is a theory behind this empirical recommendation which has been seldom considered: The decrease of the C–N ratio can only be understood if we assume that there are several microbial populations, each deteriorating at the end of its growth phase and supplying its nitrogen to the next population. The factor by which this process advances depends on three parameters:

(1) the C–N ratio of the new biomass,
(2) the yield coefficient Y_s, and
(3) the rate of turnover of the biomass; the latter is, however, a matter of conjecture.

In Fig. 5, Tab. 4 a bacterial biomass of $[C_5H_7O_2N]$ is assumed (C–N ratio 4.28), and a C turnover rate of 75%; the calculation has been carried out for seven yield coefficients, using the conception depicted in Tab. 5.

As mentioned above, the rate of biomass turnover is open for discussion. However, whichever reasonable value will be employed, the result will correspond to Fig. 5, Tab. 3, only the slope of the straight lines varying. From this figure it can be deduced that at $Y_s = 0.5$ four succeeding populations reduce the C–N ratio from 25.7–12.8; at the same time, they decompose 50% of the organic matter. The same calculation can be done with fungal biomass $[C_{10}H_{18}O_5N]$, C–N ratio = 8.57: In this case,

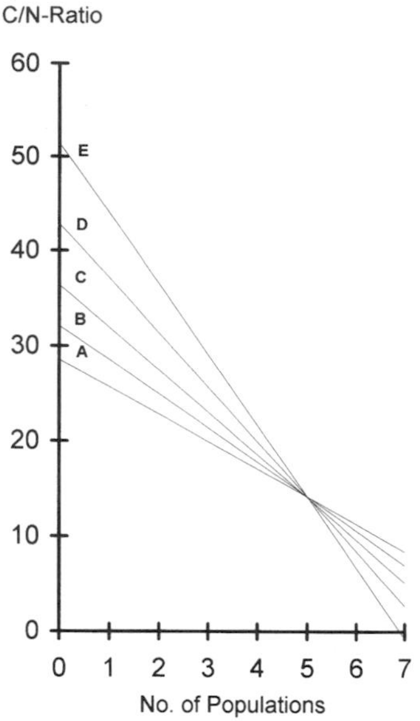

Fig. 5. The stepwise decrease of the C–N ratio by succeeding populations of bacteria (carbon turnover rate of the cell biomass=75%), values see Tab. 4.

Tab. 4. Stepwise Decrease of the C–N Ratio by Succeeding Populations of Bacteria (Fig. 5)

	Y_s	Decrease per Population Δ Volatile Solids (as Glucose)	 Δ C–N	Narrowing the C–N Ratio by A–F: 4 Populations G: 3 Populations	Concomitant Degradation of Volatile Solids in % (as Glucose)
A	0.565	78.75	2.50	23.5 → 13.6	43
B	0.502	90.00	3.21	25.7 → 12.8	50
C	0.439	101.25	4.13	28.5 → 11.9	58
D	0.376	112.50	5.36	32.1 → 10.7	67
E	0.313	123.75	7.07	37.3 → 9.0	76
F	0.251	135.00	9.64	45.0 → 6.4	86
G	0.188	146.25	13.93	57.8 → 16.1	72

Tab. 5. Calculation of the Decrease of the C–N Ratio of the Nutrient Supply by the Growth of One Bacterial Population at a 75% Carbon Turnover Rate and $Y_s = 0.502$ (see Eq. 1 in Tab. 3)

Start	$(C_6H_{12}O_6)_4 + 0.8\ NH_3$	C–N = 288 $(11.2)^{-1}$ = 25.71
Growth	$(C_6H_{12}O_6)_1 + 0.8\ NH_3 + 2\ O_2 \rightarrow 0.8\ [C_5H_7O_2N] + 2\ CO_2 + 4.4\ H_2O$	
Lysis/turnover	$0.8\ [C_5H_7O_2N] + 1\ O_2 + 1.4\ H_2O \rightarrow 0.5$ glucose $+ 0.8\ NH_3 + 1\ CO_2$	
Balance	$(C_6H_{12}O_6)_4 + 0.8\ NH_3 + 3\ O_2 \rightarrow (C_6H_{12}O_6)_{3.5} + 0.8\ NH_3 + 3\ CO_2 + 3\ H_2O$	
Rest for next population	$(C_6H_{12}O_6)_{3.5} + 0.8\ NH_3$	C–N = 252 $(11.2)^{-1}$ = 22.5

Δ M glucose = 4 − 3.5 = 0.5 = 90 g "volatile solids"; Δ C–N = 3.21.

three populations ($Y_s = 0.52$) diminish the C–N ratio from 32.1–12.8, degrading concomitantly 60% of the volatile solids.

Since the carbon–nitrogen ratio of the various types of the waste material deviates from the ratio considered optimum, they have to be mixed to arrive at a value which is required to lead – at least theoretically – to the fixation of the nitrogen in new biomass and in humic substances, or as ammonium adsorbed by inorganic and organic particles. Otherwise, nitrogen in excess will be lost as NH_3 to the air. If, on the other hand, nitrogen is deficient, the compost when applied as fertilizer will lead to the so-called nitrogen depression well known to farmers, i.e., soil nitrogen instead of being available for plant growth will be used for the further degradation of surplus carbon and thereby temporarily incorporated into microbial biomass.

4.2 Water Availability

General experience shows that organic matter can be stored without any risk of deterioration if kept dry, e.g., containing less than about 12% of moisture. In fact, drying is the most ancient method to preserve foodstuffs and animal feed. Less thorough drying (or inadvertent wetting) leads to instantaneous growth of microorganisms inherent in any organic matter (if not intentionally sterilized). Thus, water is certainly the initiator of microbial development on dead organic matter.

The water–microbe relationships in a compost pile are manifold. One would expect that there is an optimum moisture content on a mere weight basis, but this is not the case. This is because water exists in different states which are unequally available to microbes: water films covering the solid particles, capillary water, and matrix water. The various materials to be composted differ widely in their water holding capacity; i.e., the same moisture content in % of dry matter can result in a very different water availability. Thus, some materials require for optimum composting a water content of 75–90% (saw dust, straw), whereas others (grass clippings, food remains) need only a water content of 50–60%. Therefore, two other criteria are more suitable to characterize the water status:

(1) the *water activity*, expressed by the so-called a_w value (a_w: vapor pressure of water in a solution/vapor pressure of pure water.
(2) the *water potential* Ψ (more exactly "potential energy of water") which is related to the a_w value by Eq. (10) (TEMPLE, 1981):

$$\Psi = RT\ V_w^{-1} \cdot \ln a_w \quad \text{(dimension kg m}^{-2}\text{)} \tag{10}$$

V_w: partial molal volume of water.

Water activity is always less than 1.0, and water potential is always negative in real systems, since they express the availability of water in the real system contrasted to the availability of pure water under the same conditions.

The use of *water activity* to characterize the water status of a system has now been widely replaced by the measurements of the *water potential*, as outlined by PAPENDICK and MULLA (1986). This is, because water activity is much too insensitive in systems with a high amount of readily available water; instead, the water potential is regarded as the only approach for investigating water limitations caused by dryness, as discussed in detail by MILLER (1989, 1991) and DIX and WEBSTER (1995, pp. 59–66). Water potential is made up of several components, i.e., osmotic, matric, pressure and gravimetric. In composting systems, the matrix water potential (as measured with a tensiometer and expressed as a negative pressure in units of pascals [Pa]) is the most important one; it determines the extent of filling of the capillaries with water: a potential of −20 to −50 kPa is regarded as optimal, whereas −5 kPa stand for a too wet matrix, and −100 kPa for a too dry matrix. At −300 kPa pores of just ≤1 μm are water saturated, and −1,380 kPa correspond to the wilting point of vascular plants; this latter value is equivalent to a water activity of 0.990.

Apart from being essential for microbial growth, the moisture content of the starting material influences the course of the composting process: Too high a content hinders aeration and thus reduces the supply of oxygen for aerobic microbial growth, thereby favoring the establishment of anaerobic niches with the consequence of anaerobic metabolism leading to the formation of acid fermentation products. Even more important, however, is the fact that a high water content delays the self-heating because of the relatively high heat capacity of water. On the other hand, too little water, which of course can be easily corrected, retards the composting process.

The water content of the starting material is only one aspect of water and composting. In fact, the dynamics of water changes within a compost pile are rather complex. First, water is produced by aerobic microbial metabolism, about 0.45 kg of water per 1.0 kg of decomposed organic material. MILLER (1991) collected five values from the literature which are somewhat higher: WILEY and PEARCE (1957): 0.63 g H_2O g^{-1} garbage decomposed; GRIFFIN (1977): 0.55 g H_2O g^{-1} cellulose, HAUG (1979): 0.72 g H_2O g^{-1} sewage sludge; HOGAN et al. (1989): 0.5–0.53 g H_2O g^{-1} rice hulls | rice flour; HARPER et al. (1992) 0.5–0.6 g H_2O g^{-1} straw and poultry manure. (Although this is not of great influence in composting, this effect has to be considered when storing foods and feeds just at the threshold of the moisture requirements of microbial growth.) – Second, water is continually removed from the compost by the air supplied to meet the oxygen demand of the microorganisms and to remove heat from the compost pile to avoid temperatures above 60–70 °C. This withdrawal of water, which is actually desirable, must, of course, not proceed faster than the composting process, i.e., before the material is "stabilized"; otherwise, water has to be added for optimum completion of the process. At any rate, the water content of the compost pile decreases during the process, let's say from 50–70% to about 30%. This leads to a reduction of the microbial acitivities in general, but to an encouragement of microbes adapted to rather dry conditions, e.g., xerophilic fungi (DIX and WEBSTER, 1995, pp. 332–340).

4.3 Structure, Oxygen Supply and Aeration

The aerobic decomposition of organic matter requires oxygen in a definite stoichiometric relation. According to an equation, which will be used in Sect. 9.1, Tab. 15 for balancing the process, about 80 g of oxygen are used up for the degradation of 100 g of organic matter. It can be easily imagined that this amount is not initially contained in the compost pile and that it hardly reaches its interior just by passive diffusion. Thus, a very active aeration is necessary for an effective composting process. However, aeration has to fulfill another purpose which, as it turns out, is quantitatively of even greater importance: In a well isolated compost pile, the temperature can soon reach 80 °C and even higher. This is not compatible with microbial life, thus leading to *microbial suicide* (FINSTEIN, 1989). This heat has to be removed by ventilative cooling. As will be shown in Sect. 9.2 about 5 times as much air are needed for the removal of the heat as for the supply of the oxygen necessary for microbial metabolism.

Before further discussing aeration, another aspect has to be dealt with, i.e., the structure of the compost pile. This topic has been studied extensively by SCHUCHARDT (1977): The compost pile is a 3-phasic system comprising solid matter, water and gas. For optimum performance of the process, the *free air space* should amount to 20–30% of the total volume. Since, in the course of composting, this value tends to decrease, it has to be kept that way by a repeated turning over of the pile. By this means also larger air channels, which are often built-up, are destroyed. The relationship between *pore volume* (water and gas volume) and *volume of solid matter* does not describe the system completely because the variance of particle size, of diameters of air channels, and of capillaries penetrating the individual particles finally determine the provision of the microbial population with oxygen. Oxygen reaches the microbial cells by a succession of various mechanisms: convection and diffusion within the free air space, and dissolution in the liquid phase. Even if thoroughly aerated, it appears that anaerobic microniches are left, allowing anaerobic microbial metabolism; thus, in practice composting appears to be not an entirely aerobic process (DERIKX et al., 1989; ATKINSON et al., 1996). This can be deduced from the formation of organic acids, leading to a drop of pH (especially during the first phase), and the appearance of traces of gases from anaerobic metabolism in the exhaust air, e.g., methane and N_2O (denitrification) (HELLMANN et al., 1997; LEINEMANN, 1998).

The amount of air to be supplied to a compost pile, usually measured in m^3 air kg^{-1} dry organic matter h^{-1}, is certainly a matter of practical experience. Of course, the uptake of oxygen by microbial metabolism can now be analyzed on-line, giving information about the degradative activity; alternatively, and possibly more conveniently, the CO_2 content of the exhaust air can be determined. According to Strom et al. (1980) the O_2 content of the exhaust air should not drop below 5%, and BIDLINGMAIER (1983) regards 10% as tolerable. DE BERTOLDI et al. (1983) recommend even 18% O_2, whereas SUHLER and FINSTEIN (1977) found no difference in composting efficiency between 10 and 18% O_2 in the exhaust air.

Aeration based on oxygen consumption has been one of the possible strategies for controlling the composting process, i.e., the Beltsville process. This approach, however, has been strongly opposed by FINSTEIN et al. (1986), who convincingly showed that aeration is quantitatively more important for regulating the temperature of the compost pile by *ventilative cooling* (Rutgers strategy) than for supplying oxygen to the microbes (see also Sect. 9.2). Since much more air is necessary to meet this requirement, any considerations regarding the amount of oxygen needed for the decomposition of organic matter are secondary. In addition, only part of the organic matter is degraded during a certain stage (and this is not known in advance, unless O_2–CO_2 analysis of the exhaust air is carried out on-line). Therefore, the oxygen requirement to be met cannot be calculated exactly to arrive at an optimum aeration. The wide variation of organic waste in its composition and, thus, in its degradability adds further uncertainty. And finally, since the efficiency of the air supply to carry out its tasks depends also on the structure of the waste material, the wide range of values for "optimum" composting to be found in the literature is not surprising. Of course, the engineer planning a composting plant must have some guidance to calculate the aeration devices, but these calculations can hardly be based on pure microbiological or thermodynamical data.

4.4 Temperature

Since production of heat and its preservation within the compost pile – at least for a certain period of time – is an outstanding characteristic of the composting ecosystem as compared with other terrestrial habitats, the parameter *temperature* has found the special interest of compost microbiologists and composting practitioners. The temperature relationship of microorganisms are dealt with in numerous treatises (e.g., INGRAHAM, 1962; SCHLEGEL and JANNASCH, 1992) and monographs (e.g., DIX and WEBSTER, 1995, pp. 53–54, pp. 322–332). Elementary information on this subject can be found in any textbook of general microbiology (e.g., LAMANNA and MALLETTE, 1959, pp. 422–444). Thus, there is no need here for a further discussion of this topic.

Temperature plays a dual role in composting:

(1) As discussed in Sects. 2 and 3 heat is produced in the course of microbial metabolism and is maintained in the system leading to a temperature rise. By this means the indigenous microorganisms produce a favorable environment for themselves, first for a mesophilic population, then for a thermophilic one. Since ambient temperatures in our climate (temperate zone) are for long periods of the year below optimum even for mesophilic microorganisms, the warm compost pile offers excellent conditions for microbial growth and concomitant decomposition of organic matter.
(2) At the same time, the temperature reached at any stage of the process acts as selective agent,
   - eventually eliminating just that population whose activity raised the temperature to a certain level, i.e., its growth maximum,
   - smoothing the path for a new population favored by the elevated temperature; this then lifts the temperature further until its maximum is reached.

Thus, the temperature course induces a succession of various microbial populations. This topic will be discussed in detail in Sect. 6.2.

There has been an extensive and also controversial discussion as to the optimal temperature of composting (FINSTEIN et al., 1986; FINSTEIN, 1989). In contrast to earlier views which regarded temperatures of 60 °C (NAKASAKI et al., 1985d) and even higher (up to 80 °C) as optimum and necessary (the latter because of sanitary reasons), the general view now considers temperatures in the mid to upper 50s as most appropriate (MILLER, 1991). It may well be, however, that there exists no optimum temperature for all materials to be composted, i.e., various substrates, e.g., municipal garbage, agricultural wastes, sewage sludge, may need, in fact, different composting strategies as far as the temperature control is concerned. The influence of temperature on the composting process has been discussed in the numerous reviews listed above; observations made in bench scale composting are given in Tab. 8.

4.5 Hydrogen Ion Concentration, pH

Although the hydrogen ion concentration, the pH, is an important growth parameter in many microbial habitats as well as in pure culture studies, it appears to play only a minor role in composting. Numerous observations point to the fact, that the *initial pH* of the materials to be composted may vary in a wide range without significant effects on the commencement of the self-heating process. Thus, an adjustment of the pH is rarely necessary. As discussed by MATHUR (1991) in extreme cases sulfur may be used to create and lime to neutralize acidity. Addition of ammonia releasing compounds (blood, urine, urea) to materials of a wide C–N ratio (e.g., forest residues) also helps to correct an initial low pH.

Otherwise, the pH is not a constant environmental factor in composting, but changes due to microbial activities. Of course, the development of the pH value in either direction depends on the composition of the substrates and also other circumstances of the process, e.g., aeration. Nevertheless, in many instances one can observe a certain regularity: At the outset of the process the pH may slightly decrease due to the formation of organic acids – even in well aerated composts. This may be followed by an increase of pH caused by the decomposition of proteins and thereby liberation of ammonium; this development depends very much on the C–N ratio of the starting material. During the maturation phase the degradation of organic acids as well as nitrification finally leads to a more or less neutral reaction. This is, of course, only a very rough characterization of the pH situation in composting. For reports dealing with the – rather small – effects of pH on composting the reader is referred to MILLER (1991). An extensive discussion of the pH changes in self-heated hay as well as the pH relationships of fungi can be found in the review of LACEY (1980).

5 Laboratory Composting

It is obvious that composting as it is carried out on a large scale, i.e., in windrows, boxes or drums, is not a very suitable system for an exacting, basic study of the composting process *per se* as well as of the individual parameters mentioned above: the conditions (composting material, performance) vary considerably from plant to plant as well as within a large composting mass. This situation has induced numerous investigators to study composting on a laboratory scale, imitating as far as possible the situation within a compost pile. It should be mentioned, however, that even before compost scientists elaborated devices for this study, microbiologists being concerned about the phenomenon of self-heating of damp hay, grain or wool carried out respective experiments on a laboratory scale (see Sect. 3).

Tab. 6 lists the individual parameters to be considered when carrying out laboratory composting, the most critical ones being the strategies for *insulation* and *heat exchange*: Their importance is obvious because composting of small quantities of biomass lacks the insulating properties characteristic of large compost piles, and these circumstances – e.g., to minimize *conductive heat transfer* to the surrounding – received greatest attention by investigators in this field. The other parameters with their alternatives listed in Tab. 6 are rather self-understood or will become evident in the following discussion.

An often employed method to prevent *conductive* heat loss from the laboratory composter to the environment consists in heating the surrounding fluid (air or water) in such a way that its temperature is just 1 °C or less below that within the reactor. Such an apparatus is shown in Fig. 6. To minimize heat loss by *ventilative cooling* the supplied air is brought to the temperature of the reactor and saturated with water. Such a strategy is sometimes called *adiabatic*. It should be stressed, however, that

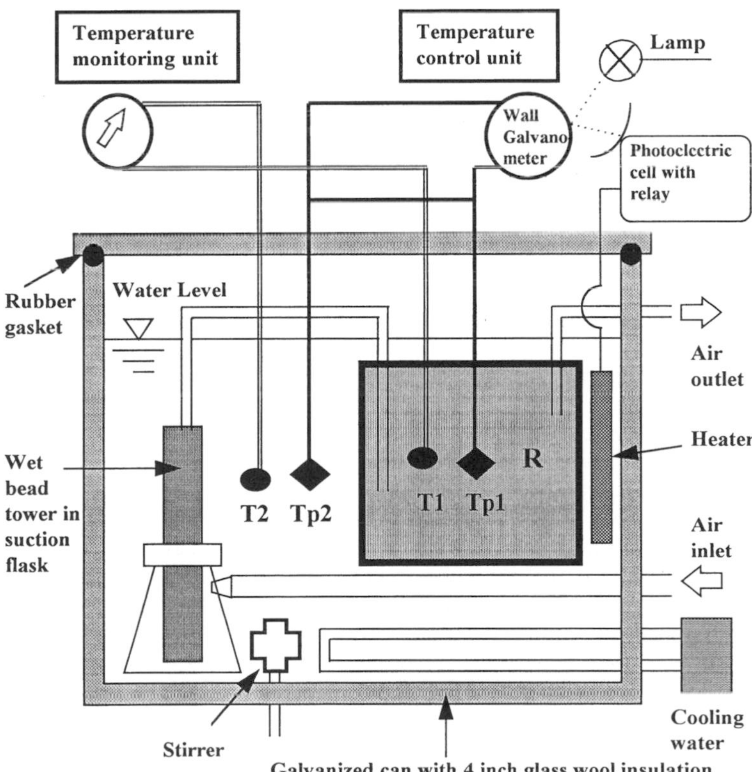

Fig. 6. Diagram of an adiabatic apparatus for the study of thermogenesis by plant materials (redrawn from NORMAN et al., 1941), R: reactor, glass quart fruit jar, flanged brass cover with rubber gasket (clamps not shown), T1, T2: thermocouples for temperature monitoring, Tp1, Tp2: thermopiles for temperature control.

Tab. 6. Strategies of Laboratory Composting

Parameter	Alternatives		
Temperature regime	spontaneous self-heating	programmed warming	isothermal
Heat exchange	extensive avoidance of heat losses, e.g., by dynamic heating of the environment, not quite correctly termed *adiabatic*		heat release by controlled conduction and ventilative cooling: *physical model* of HOGAN (1989)
Aeration	ambient air	prewarmed air	water saturated air
Insulation	Dewar flask	cover of insulating material	incubator or water bath
Performance	static	dynamic: stirrer rotating drum	
	batch	semi-continuous (only thermophilic)	continuous (only thermophilic)

even these precautions do not lead to an adiabatic system (or better process) in the strict thermodynamic sense, since there is still some heat flow to the environment, especially after the temperature maximum has been attained and the composting mass cools down. In addition, there is a mass transfer between the reactor and the environment by O_2 uptake and CO_2 release which should not be allowed to take place in a strict adiabatic system. Therefore, in Tab. 7 the term "so-called adiabatic" is used.

Combination of the various alternatives listed in Tab. 6 allows the construction of a large variety of laboratory units for composting. However, these all can be reduced to 3 main types as listed in Tab. 7 which has been taken from HOGAN et al. (1989). Although it has to be admitted that the use of the so-called adiabatic apparatus as well as the constant temperature apparatus yielded important contributions to our knowledge of the composting process, they mimic rather incompletely the real situation within a large compost pile; e.g., they consider the composting mass more or less as a homogenous batch culture in regard to temperature, moisture content and air supply, neglecting the fact, that these 3 parameters occur in gradients within a true composting mass. An indication of the departure from the actual compost pile is the water balance of these 2 systems: Because water is produced in the course of decomposition of organic matter and no water is drawn off by ventilation with prewarmed and saturated air the substrate becomes wetter instead of becoming dryer as aimed at by the composting process. Indeed, it may be impossible, to scale down composting to laboratory size and thereby copying the natural process. However, great advances have been made in this regard in FINSTEIN's laboratory (HOGAN et al., 1989) by designing a *physical model* of the composting ecosystem. The progress has been mainly achieved by combining a dynamic control of the *conductive heat flux* with *ventilative cooling* by aeration in reference to temperature ceiling. Of the heat lost from the reactor to the environment, only 2.4% were transferred by conduction, the rest by ventilative cooling.

Laboratory studies on the self-heating of damp organic material and on composting have a long history. Of the numerous investigations to be found in the literature 17 have been selected to prepare Tab. 8, the earliest one by JAMES (1927), the last one by PALMISANO et al. (1993). 9 of the 17 references had been also analyzed this way by MILLER (1984). As can be easily seen, the alternatives of the main parameters listed in Tab. 6 have been com-

Tab. 7. Characteristics of Three Laboratory Composting Strategies (according to HOGAN et al., 1989)

Criterium	So-Called *Adiabatic Apparatus*	Constant Temperature Apparatus	*Physical Model* of HOGAN
Object of study	idealized self-heating	temperature as experimental varable	simulation of the composting process
Conduction	mimized by dynamic heating of surrounding fluid: control by temperature differential between reactor interior and surrounding water bath or incubator	maintenance of constant reactor temperature by cooling or heating, respectively, of the environment using conventional thermostat	minimized by dynamic control, based on instantaneous conductive heat flux from reactor surface to environment (incubator)
Ventilation	supply of O_2, removal of CO_2, inlet air equilibrated to temperature of surrounding fluid and saturated	supply of O_2, removal of CO_2, inlet air equilibrated to temperature of surrounding fluid and saturated	aeration to achieve ventilative cooling in reference to temperature ceiling, supply of O_2, removal of CO_2, inlet air like ambient
Δ H: enthalpy of outlet gas minus enthalpy of inlet air	negligible	negligible	substantial
Temperature gradient in reactor	negligible	negligible	substantial
Water status trend	becomes wetter	becomes wetter	becomes dryer
Eventual cause of microbial self-limitation	high temperatures	substrate depletion	substrate and/or water depletion
Reference	see Tab. 8	see Tab. 8	HOGAN et al. (1989)

bined in various ways. Only few remarks will be made to complement the information given in Tab. 8.

(1) *Performance:* In half of the studies composting was carried out as a static batch culture, in some cases the substrate was mixed with paddles. More interesting is the use of rotating drums, since these are also employed in large-scale composting. Special reference deserves the approach by SCHULZE (1962) who performed composting as a continuous culture, entirely in the thermophilic phase, in this way demonstrating that a succession of various phases is not necessary for effective composting.

(2) *Material to be composted:* A wide range of organic matter has been tested in laboratory units which again shows that the nature of the substrate is not critical as long as the other demands are met. In addition, laboratory composting allows the use of a standardized biomass of known composition which simplifies the monitoring by chemical analysis of the course of degradation of individual components such as carbohydrates, proteins, fats etc. This approach has been taken by CLARK et al. (1977, 1978).

(3) *Temperature regime:* Many authors allowed the biomass to undergo a self-heating process, in most cases unlimited, which led – depending on nutrient

Tab. 8. Selected Studies on Laboratory Composting (References in Chronological Order)

No.	References	Reactor	Performance	Material Composted	Temperature Regime	Temperature Range; Optimum
(1)	JAMES, 1927	Dewar flask 1 L	static	corn meal, cracked corn	self-heating	temperature maximal 62,5 °C
(2)	NORMAN et al., 1941	glass jar 1,1 L	static	oat straw + NH_4NO_3	self-heating	./.
(3)	WILEY and PEARCE, 1957	steel cylinder 57 L	stirred	mixed garbage + city refuse	self-heating	temperature maximal 70 °C
(4)	NIESE, 1959	Dewar flask 1 L	static	garbage + sewage sludge	self-heating	temperature maximal 75 °C
(5)[a]	WALKER and HARRISON, 1960	aspirator bottle 10 L	static	wool	self-heating	maximal 80 °C, heat output maximal at 60 °C
(6)	BARDTKE, 1961	steel cylinder 6 L	rotating	garbage + sewage sludge	self-heating	temperature maximal 60–70 °C
(7)[a]	SCHULZE, 1962	steel cylinder (drum) 208 L	intermittent rotation	table scraps + dried sewage sludge + paper	continuously thermophilic	44–68 °C
(8)[a]	BÅGSTAM et al., 1974 BAGSTAM and SVENSONS, 1976	horizontal steel cylinder 20 L	reciprocal rotating (with baffels)	spruce bark + urea & phosphate	isothermal or programed	35–55 °C, stepwise increase, optimum 45 °C
(9)[a]	SULER and FINSTEIN, 1977	horizontal glass cylinder 1,2 L	static, semi-continuous	garbage + news paper	isothermal	48–72 °C; optimum 57–61 °C
(10)[a]	CLARK et al., 1977, 1978	glass cylinder 1 L	static	synthetic mixture of protein, starch, fat, cellulose	isothermal	42–57 °C, optimum: 45 °C
(11)[a]	MOTE and GRIFFIS, 1979	steel basket (4 L) in air chamber	stirred by 3 blades	dairy manure + rice hulls + corn starch + cabbage	self-heating	up to 62 °C
(12)[a]	ASHBOLT and LINE, 1982	PVC cylinder 3,8 L	stirred by bar paddle	eukalyptus bark + fish waste	programed heating, then isothermal	up to 62 °C for several weeks(!)
(13)[a]	SIKORA et al., 1983	steel basket 8 L in air chamber	static	sewage sludge + wood chips	self-heating (a) limited (b) unlimited	22–72 °C 22–82 °C
(14)	BACH et al., 1984	horizontal PVC cylinder 16 L + 1 kg substrate	mixing paddle	dewatered sewage sludge + rice husks	isothermal	37–75 °C optimum: 60 °C
(15)	BACH et al., 1985	PVC cylinder 28,3 L	static	sewage sludge + rice husks	self-heating, limited	50, 60, 70 °C optimum: 60 °C
(16)	HOGAN et al., 1989	plastic cylinder 14 L	static	rice hulls + rice flour + NH_4Cl	self-heating	temperature maximal 55 °C
(17)	PALMISANO, 1993	plexiglas cylinder 19 L	tumbling vanes at reactor sides, intermittent rotation	rabbit chaw plus newspaper	self-heating	temperature maximal 58 °C

Tab. 8 (continued)

Heat Loss Control and Heat Exchange	Inlet Air[b] (a)	(b)	Air Flow	Outlet Air O_2; CO_2	Moisture Change	Volatile Solids Loss
Dewar flask plus insulation	–	–	pure O_2	./.[c]	./.	./.
dynamically heated water bath plus cooling system	+	+	10 L d^{-1}	./.	./.	./.
insulation by cotton duct eventually + water bath	–	–	50 L kg^{-1} h^{-1} (dry)	./.	becomes dryer	./.
Dewar flask within dynamically heated incubator	–	–	./.	./. increasing	stable or	17–53%
dynamically heated water bath	+	+	4,1 L kg^{-1} h^{-1} (dry)	< 13% CO_2	stable	./.
drum within small chamber in incubator room			pure O_2	./.	becomes dryer	./.
insulation by styrofoam layer	–	–	based on O_2 residual	5–10% O_2	stable	37–45%
cork layer insulation + electric heating tape + internal water circulation	+	+	59 L kg^{-1} h^{-1} (dry)	16–0,2% O_2	stable	13–16%
water bath, externally supplied heat	+	+	based on O_2 residual	optimum 10–18% residual O_2	slight increase	0–25%
water bath, constant at 48 °C	+	–	periodical evacuation + replacement with fresh air	6–25% residual O_2	./.	./.
dynamically heated water bath	+	+	recycled; O_2 replenish-ment	21–25% O_2	stable	./.
water bath, constant at 62 °C	+	–	16 L kg^{-1} h^{-1} (dry)	./.	condens-ed water recycled	14–26%
dynamically heated water bath	+	+	4,4 L h^{-1}	./.	33% decrease	12,8%
heater within reactor above substrate	–	+	0,8 L min^{-1}	./.	./.	./.
styrofoam insulator and varying ventilation	–	–	according to temperature feedback	./.	becomes dryer	40–60%
insolation + incubator; controlled conduction + ventilation cooling	–	(+)	according to temperature feedback control	variable according to ventilation	becomes dryer	17%
reactor within 37 °C incubator	–	+	0.7–1.01 L min^{-1}	./.	./.	50%

[a] References analyzed by MILLER (1984).
[b] Inlet air: (a) prewarmed; (b) water saturated.
[c] ./. Not given.

supply and heat loss – to temperature maxima of 70 or even 80 °C, a temperature regarded by most authorities as much too high for proper composting. On the contrary, by isothermal composting the optimal temperature could be determined; this may depend, however, to a certain degree on the material to be treated.

(4) *Heat loss control and heat exchange:* This item required the most sophisticated efforts of the investigators in this field. It soon became apparent that simple insulations – e.g., the vacuum of a Dewar flask – could not keep the high temperatures attained by self-heating for a reasonable time. In addition, aeration is necessary to supply oxygen making it essential to reduce further the conductive heat loss to the environment. The solution of this problem was best accomplished by dynamically heating the surrounding fluid as exercised probably first by NORMAN et al. (1941) (Fig. 6). The further development in this area led finally to the *physical model* of HOGAN et al. (1989).

The discussion of *laboratory composting* should not be concluded without emphasizing its great role in basic research in composting. Individual parameters can be varied at will, and their influence on the process can be studied under reproducible conditions. New substrates or substrate mixtures can easily be tested for their suitability for composting, and their requirements for the various parameters to be fulfilled can be determined. Also, laboratory composting allows a more accurate balancing of mass and energy in composting as well as – because of the homogenous, well known substrate – chemical characterization of the kinetics of the process. Finally, laboratory composting can be regarded as an enrichment culture for bacteria (including actinomycetes) and fungi adapted to the relevant parameters – substrate, moisture, aeration, temperature course – choosen at will by the "microbe hunter".

6 The Microorganisms of Composting

In this section, bacteria belonging to the order Actinomycetales will be discussed separately from the remaining bacteria. Further, because of traditional reasons, the genus *Thermoactinomyces* which belongs to the family Bacillaceae, will be dealt with together with the actinomycetes.

It is self-evident that the composting process depends on the activity of a very diverse community of microorganisms, bacteria and fungi. The role of invertebrate animals in the very final stages of composting at the large scale as well as in backyard composting of household garbage and yard waste will not be considered here. Common experience shows that the composting process can in most instances rely on the microorganisms dwelling on the waste material itself, i.e., there seems to be no need for an additional inoculum, a situation which composting shares with the treatment of wastewater. Thus, it is not surprising that attempts to improve the process by supplying "adapted cultures" (or other supplements such as enzymes) were of little or no effect NAKASAKI et al. (1985c). If augmentation of the process appears to be necessary the addition of composted material, preferentially from the same kind of waste as to be handled, should suffice. On the other hand, one is wondering about the fact that apparently all the microbes performing the various phases of the composting process are introduced along with the waste material, i.e., even thermophilic bacteria, actinomycetes and fungi, which would not be expected to belong to the indigenous population of the material. This seems to be a good example for proving the statement attributed to BEIJERINCK: *Everything is everywhere, and the environment selects.*

There exists a tremendous literature on the occurrence of certain groups of microorganisms and their role in the composting process; most of these findings have been discussed several times in recent reviews, particularly by FINSTEIN and MORRIS (1975), LACEY (1980), and MILLER (1991, 1993). Thus, the *principles of compost microbiology* are well established leaving only some very special questions for future research. The reasons for the still existing gaps in our knowledge are

(1) the highly variable nature of the starting materials of composting,
(2) the continuously changing conditions during the process leading to successions of microbial populations, and
(3) the specialization of microbiologists who are either primarily bacteriologists or mycologists.

Nevertheless, a few studies of a particular compost pile dealing with all microbial groups as well as following the various phases of the process have been carried out by the following investigators: CHANG and HUDSON (1967): composting straw plus NH_4NO_3, HAYES (1968): composting wheat straw plus horse manure, BÅGSTAM (1978): composting spruce bark plus urea, STREICHSBIER et al. (1982): composting grape mark, and NAKASAKI et al. (1985a): composting dewatered sewage sludge. Fig. 7, taken from CHANG and HUDSON (1967), may serve as an example of the outcome of such a study: While the mesophilic bacteria show only a small decrease of their numbers, the mesophilic actinomycetes and fungi disappear during the high temperature phase almost completely. On the contrary, at the outset of the thermophilic phase, the thermophilic bacteria exhibit a fast development, the thermophilic actinomycetes and fungi a somewhat later appearance. After the decline of the temperature, mesophilic representatives of each group reappear. Of course, one has to be cautious when interpreting such results; some remarks referring to this are made in the following (and further) paragraph(s).

Research in compost microbiology has followed up to now traditional approaches: quantitative determination and isolation of the microbes present employing respective media, methods and temperatures, followed by physiological characterization of the isolates to understand their particular role they play in this ecosystem. It should be born in mind, however, that colony counts of a certain microorganism or group of microorganisms do not necessarily give a true picture of their activities at the respective composting stage: mesophilic spore forming bacteria (genus *Bacillus*) may be counted (at 25 °C) during the thermophilic phase, and thermophiles are found long after their climax. Further, spore counts of actinomycetes and fungi need not to be in accordance with the amount of active mycelium. Regarding the methods employed, the time honored plating of dilutions on agar media is in common use. Alternatively, spores of actinomycetes and fungi which are liberated into the air can be captured by various methods of aerosol analysis. These procedures have been extensively used for the microbiological analysis of self-heated, moldy hay (GREGORY and LACEY, 1963) and various working places (EDUARD et al., 1990), and they are now also employed to detect allergic and pathogenic actinomycetes and fungi in the environment of composting plants (LACEY et al., 1992).

In the future more sophisticated methods as already employed by some investigators will certainly be applied to the study of the microbiology of composting: Determination of ATP (LEHTOKARI et al., 1983), estimation of fungal biomass using as indicators chitin, ergosterol, laccase (MATCHAM et al., 1985), evaluation of microbial biomass by measuring ergosterol, diaminopimelic acid and glucosamine (GRANT and WEST, 1986), quantification of molds by near infrared reflectance spectroscopy (ROBERTS et al., 1987), estimation of biomass and biological activity by determination of ATP content, rate of mineralization of ^{14}C labeled glutamate, amount of total lipids and lipid phosphate (DERIKX et al., 1990a), correlating emission of trace gases with the concentration of various lipids and fatty acids (HELMANN et al., 1997). Even molecular methods are ready for use to investigate the microbiology of compost (NEEF et al., 1995).

Because of the complexity of the ecosystem compost its microbiology can be described from various points of view. In this review the discussion of the literature will be organized in the following way:

(1) treating the various groups of microorganisms separately according to their taxonomy as they have been dealt with in many publications;

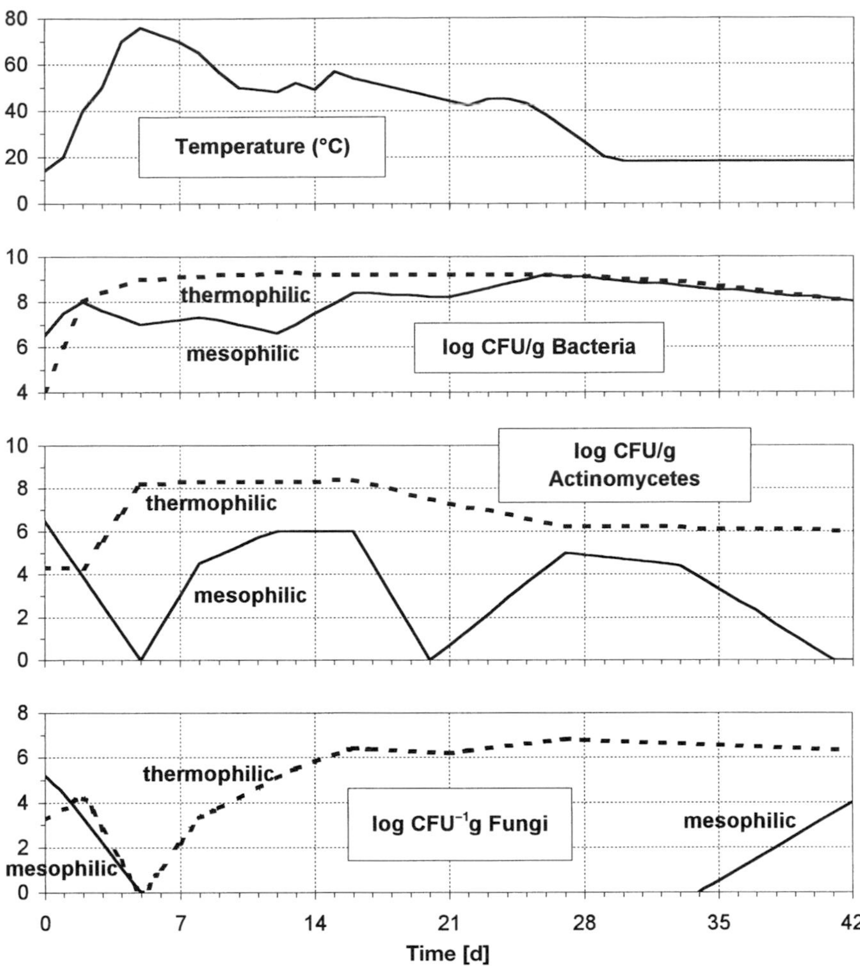

Fig. 7. Occurrence of mesophilic and thermophilic bacteria, actinomycetes and fungi during composting of straw plus ammonium nitrate as effected by the temperature course (according to CHANG and HUDSON, 1967).

(2) considering the process as a whole emphasizing the development of successive populations in accordance with the four phases mentioned above.

6.1 The Main Groups of Microorganisms Active in Composting

Compost microbiologists traditionally divide the microbial population of this habitat into the three groups

(1) bacteria,
(2) actinomycetes, and
(3) fungi;

often, however, investigators pay attention particularly to one or the other group, neglecting the others. Since high temperatures (40–70 °C) are the unique characteristic of the composting process, the thermophilic representatives of each group have generally found more attention than the mesophilic ones. Thereby the range of microorganisms having been of special interest to compost microbiologists becomes rather narrow. This, however, does not mean, that mesophilic microorganisms are less important in the composting process.

Inasmuch as temperature plays an important role in composting, both as a result of microbial activity and as selective agent determining the succession of microbial populations, it is important to remember the temperature relationships of the microorganisms involved. As shown in Fig. 8 microbes may be classified according to the temperature range of their growth. As it is readily observed, there is much overlapping between the groups; also, the groups have no taxonomic meaning, i.e., there are (besides few exceptions) no thermophilic families or genera, and even a species may comprise mesophilic and thermophilic strains. In spite of this, the classification – although appearing rather artificial – is still very useful in ecological studies and in applied microbiology (e.g., food microbiology). In the context of the microbiology of composting it is noteworthy that the same term has a different meaning in different groups of microorganisms, as strongly pointed out by COONEY and EMERSON (1964) (Tab. 9): A thermophilic fungus has a temperature range which is somewhat lower than that of a thermophilic actinomycete and much lower than that of a thermophilic bacterium. This fact is of utmost importance if the findings of various authors dealing with different groups of microorganisms are analyzed, because the results are significantly determined by the temperature used for the isolation of the respective microbial class. Unfortunately, since there is even variation within one temperature group, it is difficult to recommend just one temperature for the isolation of thermophilic actinomycetes or thermophilic fungi. Although five degrees of the Celsius scale appear rather negligible they may lead to quantitative and qualitative differences, i.e., in numbers of colony forming units and in species diversity. Still another relationship to temperature should be considered when analyzing the microbial succession during composting, namely the sensitivity toward heat under the prevailing conditions. This

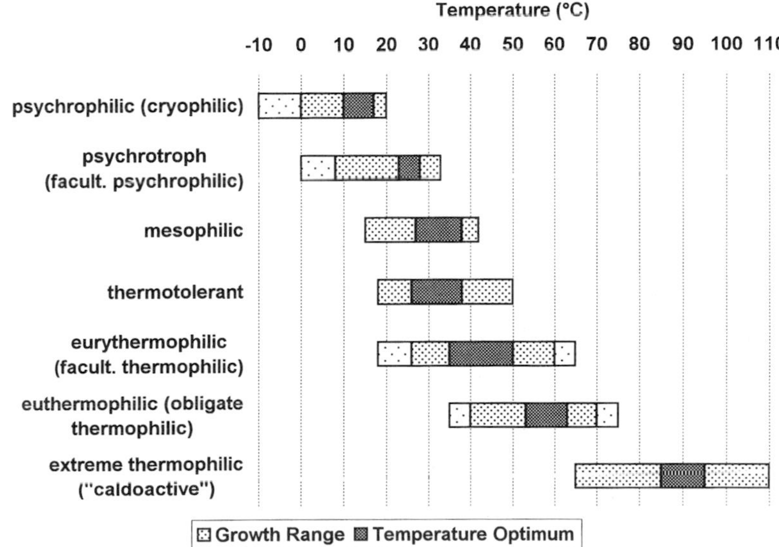

Fig. 8. Classification of microorganisms according to their temperature range (according to INGRAHAM, 1962 and CROSS, 1968)

Tab. 9. Temperature Relationships of Selected Microorganisms Pertinent to Composting

Micro-organism	Temperature Class	Cardinal Temperatures [°C]			Optimum Temperature for Isolation [°C]
		Minimum	Optimum	Maximum	
Bacteria					
Thermus	thermophilic	40	65–75	80	75
Bacillus	thermophilic	35	60–65	70–75	60
Thermo-actinomyces	thermophilic	35	45–55	70	55
Actinomycetes	thermotolerant	20	30	≤50	40/45
	thermophilic	30	40–45	≥50–60–(65)	45/50/55
Fungi	thermotolerant	<20	35–40	≤50	40
	thermophilic	≥20	40–45	≥50–55–(60)	40/45/55

property, after all, determines whether or not a particular microbe survives the high temperature phase and thus will continue to be active in the process as well as will show up when isolations will be carried out. Very little information is available regarding this aspect, except that fungi (including their conidia) are rather sensitive and may more or less completely disappear if the compost temperature exceeds 65–70 °C.

Actinomycetes and fungi are generally regarded as aerobic organisms, and thus they are isolated and cultivated under such conditions by microbiologists studying them – with just one exception: HENSSEN (1957a,b) isolated numerous actinomycetes and fungi (Tabs. 10, 12) from stable manure both under aerobic and anaerobic conditions, and pure cultures of several of them grew anaerobically as well as aerobically (described as "facultative aerobic"). Even more conspicuously, some actinomycetes [*Streptomyces thermoviolaceus* ssp. *pingens*, *Thermopolyspora polyspora* (present taxonomic position unsettled)] and *Pseudonocardia thermophila* grew at high temperatures (60–65 °C) only anaerobically whereas at 40 °C and 50 °C under either condition. The preference for anaerobic conditions varied among the actinomycetes and fungi under study and seemed to exist especially in freshly isolated cultures. The findings of HENSSEN (1957a,b) have been critically discussed by DEPLOEY and FERGUS (1975) who could not observe any growth of pure cultures of thermophilic actinomycetes (6 species) and fungi (12 species) in an atmosphere of pure nitrogen. For recognizable growth, fungi needed – depending on the medium – at least 0.2 or 0.7% of O_2, and actinomycetes 0.1% of O_2. It should be noted, however, that DEPLOEY and FERGUS (1975) – in contrast to HENSSEN (1957a,b) – excluded also CO_2 from the atmosphere by placing an NaOH solution into the anaerobic incubator. Under such conditions, i.e., alkalizing the pyrogallol solution with NaOH (the so-called Burri method) instead with Na_2CO_3, HENSSEN (1957a,b) also could not obtain growth with any of her cultures. Unfortunately, HENSSEN's method of isolating thermophilic actinomycetes and fungi from self-heated material has not been applied since then by other workers in this field; it certainly deserves to be scrutinized. This appears to be of special interest when imitating conditions which exist within the compost pile for isolation and *in vitro* cultivation of the microorganisms in question.

Finally, when reviewing the findings of earlier investigators, the problem arises as to the correct identity of the species name reported. In the following text and tables the author has used the names as given in the original paper, being aware that in one or the other case this name may be incorrect according to present day taxonomy, but difficult to emend. This note refers especially to certain thermophilic actinomycetes whose taxonomy has changed dramatically during the last four decades; e.g., the following two species have been known under various other names, some genera – marked by an asterisk – even having been completely dismissed:

(1) *Saccharomonospora viridis (Thermomonospora viridis, Thermoactinomyces viridis, Thermoactinomyces monosporus, Thermopolyspora* glauca, Micropolyspora* internatus)*;
(2) *Saccharopolyspora rectivirgula (Faenia* rectivirgula, Micropolyspora* faeni, Thermopolyspora* rectivirgula, Thermopolyspora* polyspora)*.

6.1.1 Bacteria

The procaryotic microorganisms are now divided into two kingdoms:

(1) *Bacteria* (formerly Eubacteria) and
(2) *Archaea* (formerly Archaebacteria) (WOESE, 1992)

The third kingdom, Eukarya, comprises the eukaryotic organisms, e.g., plants, fungi and animals. The tremendous large kingdom *Bacteria* is most profoundly described in the four volume treatise *"The Prokaryotes"* (BALOWS et al., 1992), of which especially Vol. 1 should be mentioned, including chapters on the ecophysiology and biochemistry of various bacterial groups. As an "introduction" to the bacteria there are many textbooks available, one of the best readable ones being *Brock's Biology of Microorganisms*, now in its 8th edition (MADIGAN et al., 1996). For the compost microbiologist only small parts of the legion of bacterial species are of interest. Bacteria of hygienic relevance (to be dealt with in Sect. 7.1), and some thermophilic bacteria, discussed in the following paragraphs have found special attention.

During the mesophilic phase of composting a rather heterogeneous population of bacteria develops leading very fast to elevated temperatures. There appears to exist no special compost microflora at this stage, rather most heterotrophic bacteria adhering to the incoming material will multiply. Therefore, a broad, but not specific species diversity which is determined by the contamination of the waste material will be found in the early phase of composting, e.g., members of non-sporulating bacteria such as gram-positive cocci and rods (*Micrococcus* sp., *Streptococcus* sp., *Lactobacillus* sp.) and gram-negative rods (Enterobacteriaceae, Pseudomonadaceae) as well as spore forming rods *(Bacillaceae)* (for references see FINSTEIN and MORRIS, 1975). Mesophilic bacteria isolated from sewage sludge compost even at the thermophilic stage of 60 °C have been characterized by NAKASAKI et al. (1985b).

Although conditions should favor the development of aerobic bacteria, there are certainly anaerobic niches in which facultative or even obligate anaerobes will grow as indicated by the formation of organic acids (decrease of pH in the early phase), gases of anaerobic metabolism (although only in traces) such as N_2O, H_2S and CH_4, and odorous substances (see Sect. 4.3).

The bacterial population of the thermophilic phase appears to consist primarily of species of *Bacillus* and *Thermus*. In the study of STROM (1985a,b) 87% of 652 randomly picked colonies were bacilli, *B. circulans* and *B. stearothermophilus* being the most frequent representatives of this genus. Few other isolates belonged to two genera of unidentified non-sporeforming bacteria, to the actinomycete genera *Streptomyces* and *Thermoactinomyces*, and to the mold species *Aspergillus fumigatus*. It should be noted, however, that STROM (1985a) isolated colonies after only 20 and 40 h of incubation of the plates, partly at 60 °C, which might have been a too short period of time and a too high temperature for the development of thermophilic actinomycetes and fungi. In a later paper STROM (1985b) cited two reports on a *Pseudomonas* ssp. growing at 50–55 °C in mushroom compost, and a coccobacillus in composting refuse–sludge mixtures at 50 °C. The situation becomes even more complex by the finding of DROFFNER and YAMAMOTO (1991) that prolonged environmental stress – although with nalidixic acid – is able to select mutants of *Escherichia* sp., *Salmonella* sp. and *Pseudomonas* sp. that grow at 54 °C.

In contrast to the belief that thermophilic sporeformers are the predominant bacterial population of composts at peak heating, BEFFA et al. (1996) recently found in composts of a temperature of 65–82 °C high numbers (10^7–10^{10} cells g^{-1} dry weight) of bacteria related to the non-sporeforming, gram-negative genus *Thermus*. These bacteria were isolated from enrichment cultures at 75 °C, whereas in parallel trials a similar number of oval sporeformers was isolated from compost (60–65 °C) after enrichment at 60 °C. It thus appears that members of the genus *Thermus* are even better adapted to high temperature compost than bacilli. Strains tentatively identified as *Thermus* were found earlier in sewage sludge compost by FUJIO and KUME (1991) along with strains of *B. stearothermophilus* after enrichment and isolation at 60 °C.

6.1.2 Actinomycetes

The actinomycetes (order Actinomycetales) form a morphological very diverse group of gram-positive bacteria, their DNA having an unusual high GC content in excess of 55 mol%. GOODFELLOW (1989) listed altogether 63 genera, their distinction based mainly on biochemical criteria – in contrast to earlier classifications of this group which were primarily established using morphological properties: morphology of substrate and aerial mycelium, formation of spores (conidia, arthrospores) either in more or less long chains (e.g., genus *Streptomyces*) or single/pairwise (e.g., genus *Thermomonospora, Microbispora*), morphology of spore surface as discernible by electron microscopy. However, even their taxonomy based primarily on cytochemistry is no longer tenable, and future concepts will rely heavily on RNA comparisons and DNA homology studies (ENSIGN, 1992). Only few genera of this large group are of special interest to the compost microbiologist, i.e., members of the family Streptomycetaceae (KORN-WENDISCH and KUTZNER, 1992) and some genera containing thermophilic species (CROSS, 1968; GREINER-MAI et al., 1987, 1988; KEMPF, 1996). Their ecology in relation to compost and other self-heated materials has been repeatedly reviewed, e.g., LACEY (1973, 1978, 1988) and CRAWFORD (1988).

In the context of this review it is important to consider the temperature relationships of actinomycetes (see also Tab. 9): The steady occurrence in temperate vs. warm to hot environments (i.e., soil vs. self-heated organic matter) leads to the distinction of mesophilic vs. thermophilic actinomycetes, the latter being defined as those which are capable to grow at ≥50 °C (CROSS, 1968). It should be stressed again, however, that there exists no clear borderline between these two groups: Whereas most streptomycetes which form an important part of the autochthonous microbial population of soils are definitely mesophilic (no growth above 45°), some species of this genus extend their temperature range from 28–55 or 60 °C (Tab. 4 in KORN-WENDISCH and KUTZNER, 1992). Being classified, therefore, as thermophilic, they often can be obtained from both, soil *and* composts using 28–37 °C or 50 °C for isolation. The 50 thermophilic streptomycetes studied by GOODFELLOW et al. (1987) (isolated at 55 °C, later characterized at 45 °C) were placed into 10 clusters, 8 of them containing strains from composts of various kind. Some clusters were given species names: *Streptomyces thermovulgaris, S. thermoviolaceus, S. macrosporus, S. megasporus, S. thermolineatus, S. albus*. On the other hand, thermophilic species of the genera *Saccharomonospora, Saccharopolyspora, Thermomonospora,* and *Thermoactinomyces* which can be considered as belonging to the autochthonous microbial community of self-heated material extend their temperature range from 30–55/65 °C. It appears, that many thermophilic actinomycetes are rather facultative thermophiles (eurythermophilic) and only few obligate thermophiles (euthermophilic).

Studies on thermophilic actinomycetes in self-heated organic material (hay, manure, compost) go back as far as to the work of WAKSMAN et al. (1939a,b), in fact, they were initiated by WAKSMAN's interest in the turnover of organic matter and formation of "humus" in nature (e.g., WAKSMAN, 1926, 1931). Of the numerous studies dealing with these organisms in the habitats just mentioned, 10 papers have been selected for preparing Tab. 10 (for further references, most of them, by the way, dealing with self-heated hay, straw and grain, see the reviews listed above). As it can be readily seen,

Tab. 10. Isolation of Thermophilic Actinomycetes from Composts and Self-Heated Hay by Various Investigators

Species	References[a]									
	1	2	3	4	5	6	7	8	9	10
Thermoactinomyces vulgaris	+	+		+		+	+	+	+	+
Thermoactinomyces sacchari		+								+
Thermoactinomyces thalpophilus					+					
Thermoactinomyces sp. amylase +			+							
Thermoactinomyces sp. amylase –			+							
Thermoactinomyces glaucus[b]					+	+				
Thermomonospora sp.		+		+			+		+	
Thermomonospora alba	+									
Thermomonospora chromogena				+				+		
Thermomonospora curvata	+		+		+	+		+		
Thermomonospora lineata (*curvata*)	+				+					
Thermomonospora fusca			+		+	+		+		
Thermopolyspora polyspora[c]						+	+			
Saccharomonospora viridis		+	+	+			+	+	+	+
Saccharopolyspora rectivirgula		+		+					+	+
Pseudonocardia thermophila					+	+				
Nocardia brasiliensis						+				
Microbispora bispora					+					
Streptomyces albus									+	
Streptomyces sp. (gray)		+	+	+			+	+	+	
Streptomyces sp. (green)									+	
Streptomyces thermovulgaris		+			+	+	+			
Streptomyces thermoviolaceus		+			+	+				
Streptomyces rectus					+	+				
Streptomyces violaceoruber						+				
No. of species isolated	4	8	7	5	1	1	5	6	7	5
Temperature of isolation	50	44 & 50	55	55	50 & 60	45	51	50		55

[a] References
(1) RÉSZ et al. (1977) municipal waste compost, (2) MILLNER (1982) sewage sludge plus wood chips compost, (3) VON DER EMDE (1987) municipal waste compost, (4) CROOK et al. (1988) domestic waste compost, (5) HENSSEN (1957a) rotting stable manure, (6) FERGUS (1964) compost for mushroom cultivation, (7) FERMOR et al. (1979) compost for mushroom cultivation, (8) AMNER et al. (1988) compost for mushroom cultivation, (9) LACEY (1978) hay, (10) GANGWAR et al. (1989) hay, wheat straw, compost.

[b] *Thermoactinomyces glaucus:* nomen dubium, possibly a streptomycete with green aerial mycelium or *Microtetraspora glauca* (?)

[c] *Thermopolyspora polyspora:* taxonomic position questionable; genus name not longer existent. Species of this former genus now placed into 3 different genera: *Saccharopolyspora, Saccharomonospora, Microbispora*.

some thermophilic actinomycetes appear to be widespread in self-heated material, including composts, the most common species being *Thermoactinomyces vulgaris, Thermomonospora curvata, Saccharomonospora viridis, Streptomyces thermovulgaris*, and other gray sporulating streptomycetes. A special citation deserves the PhD-thesis by VON DER EMDE (1987) on the role of actinomycetes in composting domestic refuse: At a composting temperature of 55 °C *Streptomyces* ssp., *Thermomonospora chromogena* and *Thermoactinomyces* sp. (amylase-positive) dominated, whereas at 75 °C *Thermomonospora fusca, Thermomonospora curvata, Thermoactinomyces* sp. (amylase-negative) and *Saccharomonospora* sp. were mainly found. Counts of *Saccharomonospora* sp. and *Thermomonospora chromogena* increased with extending composting time, and the occurrence of *Thermoactinomyces* sp. seemed to depend on the presence of attending microorganisms. However, the distribution of species as revealed in Tab. 10 does not only reflect the frequency of the species involved, but also conditions of the studies such as the medium (media) and methods employed as well as the experience and the particular interest of the investigator.

Thermophilic actinomycetes in various self-heated materials have been also determined in the laboratory of the author (KORN-WENDISCH and WEBER, 1992; KEMPF, 1996; KUTZNER and KEMPF, 1996): As depicted in Fig. 9, the habitats vary considerably in regard to the occurrence of particular genera. Tab. 11 shows another example from the author's laboratory demonstrating the variation of genus/species diversity in 2 kinds of compost. It appears that – as to be expected – the same parameters which determine the process of self-heating, e.g., the kind of organic matter, moisture content, aeration, temperature course also determine the prevalence of one or the other genus/species. However, in spite of numerous individual findings, at present the data are still too scarce to correlate definitely one of these conditions with the species composition to be attained. It should be mentioned that such an analysis is in fact rather laborious: KUTZNER and KEMPF (1996) recommended several media for the selective counting and isolation of one or the other genus of thermophilic actinomycetes. The need of several special media as well as certain selective agents (antibiotics, NaCl concentration) to assess separately the various genera of thermophilic actinomycetes in compost has been stressed before by ATHALYE et al. (1981) and AMNER et al. (1988). No doubt, a thorough analysis of the population of thermophilic actinomycetes in composts by the use of several media and methods (plating of dilutions, aerosol sampling) will still yield new representatives of this group of organisms as shown recently by the discovery of the ge-

Tab. 11. Thermophilic Actinomycetes in Compost of Stable Manure ("Mushroom Compost") and in Municipal Waste Compost (according to KORN-WENDISCH and WEBER, 1992)

Genus/Species	Mushroom Compost		Municipal Waste Compost	
	CFU[a]	%[b]	CFU[a]	%[b]
Thermoactinomyces	$8 \cdot 10^6$	0.8	$7 \cdot 10^4$	0.1
Thermomonospora fusca	$9 \cdot 10^5$	98.6	$3 \cdot 10^6$	3.3
Thermomonospora chromogena	$2 \cdot 10^6$	0.2	$3 \cdot 10^5$	0.2
Saccharomonospora viridis	$5 \cdot 10^6$	0.4	$6 \cdot 10^7$	46.8
Saccharopolyspora rectivirgula	$5 \cdot 10^3$	<0.001	$7 \cdot 10^4$	<0.001
Thermocrispum	n.d.[c]	–	$1 \cdot 10^7$	11.7
Streptomyces	$2 \cdot 10^4$	<0.001	$5 \cdot 10^7$	37.9

[a] CFU: colony forming units.
[b] %: in percent of total CFU thermophilic actinomycetes.
[c] n.d.: not detected.

Scale: Colony Forming Units in x 10^6/g
(Take notice of the different scales in the six graphs)

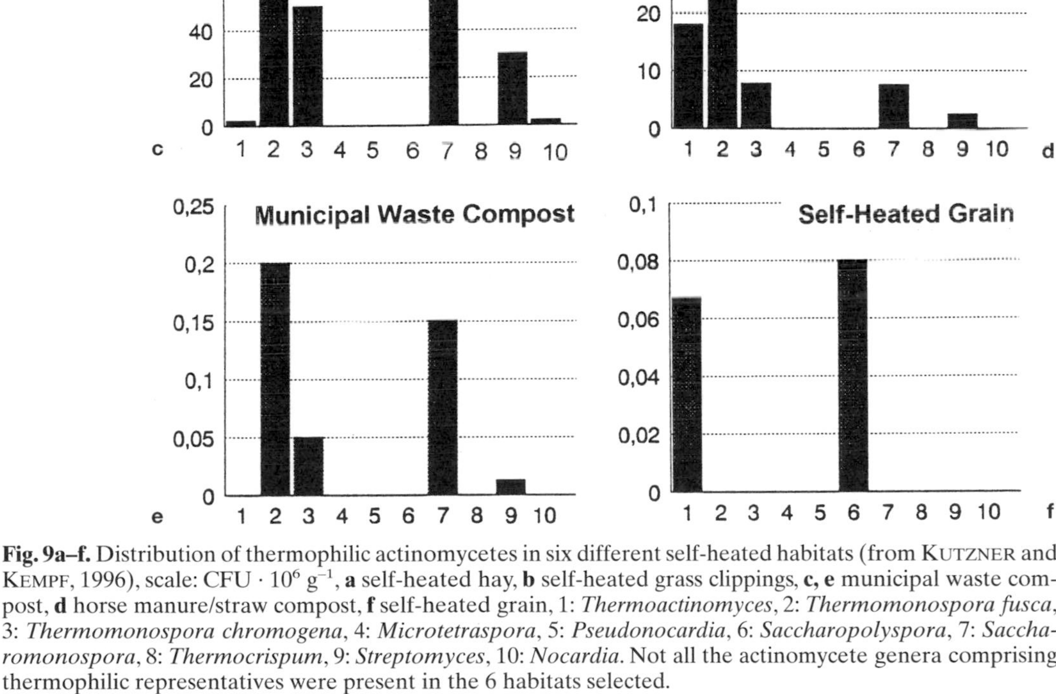

Fig. 9a–f. Distribution of thermophilic actinomycetes in six different self-heated habitats (from KUTZNER and KEMPF, 1996), scale: CFU · 10^6 g^{-1}, **a** self-heated hay, **b** self-heated grass clippings, **c, e** municipal waste compost, **d** horse manure/straw compost, **f** self-heated grain, 1: *Thermoactinomyces*, 2: *Thermomonospora fusca*, 3: *Thermomonospora chromogena*, 4: *Microtetraspora*, 5: *Pseudonocardia*, 6: *Saccharopolyspora*, 7: *Saccharomonospora*, 8: *Thermocrispum*, 9: *Streptomyces*, 10: *Nocardia*. Not all the actinomycete genera comprising thermophilic representatives were present in the 6 habitats selected.

nus *Thermocrispum* (KORN-WENDISCH et al., 1995) on media with novobiocin which is otherwise strongly selective for *Thermoactinomyces* sp., or the sporangia forming actinomycete resistant to kanamycin, which has not yet been studied further (AMNER et al., 1988).

Thermophilic actinomycetes do not only play a beneficial role in composting by decomposing various organic compounds and thereby contributing to the process of self-heating and probably to humification, some species of them are also of hygienic relevance. This aspect will be discussed in Sect. 7.2.

6.1.3 Fungi

The fungi – eukaryotic, heterotrophic, microorganisms (either saprophytic or parasitic) – comprise a tremendous number of species which are classified according to their life cycle and morphology into 5 classes:

(1) Myxomycetes,
(2) Phycomycetes,
(3) Ascomycetes,
(4) Basidiomycetes, and
(5) "Deuteromycetes" (fungi imperfecti).

A large number of species can be found in composting material, most of them belonging to the *Ascomycetes* and *Deuteromycetes* – due to their widespread occurrence in soil and on plant material from where they rapidly contaminate any dead deteriorating organic material. The general biology of fungi is treated in a wealth of textbooks and monographs which will not be cited here. However, monographs dealing with the ecology of fungi and their activities in nature might be of interest to compost microbiologists (FRANKLAND et al., 1982; CARROLL and WICKLOW, 1992; DIX and WEBSTER, 1995).

Because of their capacity to decompose dead organic matter of plant and animal origin, a pile of gathered organic waste serves as an excellent habitat for fungi – and *vice versa* – as an Eldorado for mycologists. Even a superficial inspection will provide the *microbe hunter* with a large collection of fungi, and if searching long enough with patience she or he will find most common fungi in composting material. For instance, VON KLOPOTEK (1962) identified altogether 106 species of 63 genera from composts; LEMBKE and KNISELEY (1985) isolated 60 species of 23 genera from the dust of waste recovery plants, and GÖTTLICH (1996), also from the dust of composting plants, representatives of 48 genera, 9 of them harboring thermophilic species.

Of course, only those species particularly adapted to the conditions of the compost pile – nutrients, temperature, water, aeration – will become predominant in this habitat. It is of interest that mycologists are accustomed to distinguish ecological groups of fungi just according to these parameters:

(1) *Substrate relationships* (GARRET, 1951): saprophytic sugar fungi, cellulose decomposers, lignin decomposers; further – but here of less relevance – root parasites, coprophilous fungi, predaceous fungi. (In fact, a similar grouping can be undertaken for bacteria and actinomycetes, e.g., copiotrophic, oligotrophic, fastidious, saprophytic, symbiotic, parasitic).
(2) *Temperature relationships* (CRISAN, 1973; DIX and WEBSTER, 1995, pp. 322–332; MOUCHACCA, 1995): psychrophilic, mesophilic, thermotolerant and thermophilic. Although there exists – as discussed above – overlapping between these groups the distinction is nevertheless quite useful in ecological studies. As in the case of bacteria and actinomycetes (see Sects. 6.1.1, 6.1.2), the thermophilic fungi found special interest of compost microbiologists. COONEY and EMERSON (1964) and EMERSON (1968) listed 17 "strictly" thermophilic fungi, several of them isolated from compost.
(3) *Water relationships* (AYRES and BODDY, 1986; DIX and WEBSTER, 1995, pp. 332–340): Only the groups adapted to low water potentials have been given special names: xerophilic (tolerant) fungi and osmophilic (tolerant) fungi. Representatives of the former group may become noticeable at the end of the maturation phase when the compost pile becomes rather dry.
(4) *Oxygen relationships:* Most fungi are rather aerobic and find excellent growth conditions in the aerated compost pile – anaerobic representatives (or rather facultative anaerobes) being confined to oxygen deprived niches.

The fungus population of composts comprises a legion of species. Most of them appear to be present there more or less by chance or accidentally because they were introduced by the original substrates, i.e., dead plant and animal remains; but some are well adapted to the composting conditions and certainly play an active role in the process. Among them, the thermotolerant/thermophilic species have been found the greatest interest of compost microbiologists, partly because of their peculiarity, partly because they are especially characteristic of the composting process. Of the numerous studies on thermophilic fungi in composts, 12 have been selected to prepare Tab. 12, in addition to two others dealing with the occurrence of these organisms in hay. A similar table has been prepared by STROM (1985b). Compost prepared for mushroom cultivation has found the particular interest of mycologists, one of the more recent studies being that of STRAATSMA et al. (1994). It can easily be observed that *Aspergillus fumigatus* and *Humicola lanuginosa* were isolated from all habitats investigated, *Chaetomium thermophile* was found in most composts, but not in hay, *Thermoascus aurantiacus* occurred in the five municipal waste composts, and *Humicola grisea* and *H. insolens* preferentially in "mushroom composts".

It makes little sense, to tabulate all the other (mesophilic) fungal species which have been observed by VON KLOPOTEK (1962), LEMBKE and KNISELEY (1985), GÖTTLICH (1996) or have been isolated once in a while from compost by other investigators, rather the succession of fungi and their role in the composting process deserve attention; these topics will be dealt with in the following paragraph.

6.2 Microbial Successions in Composting

There exists probably no other microbial habitat which would be more attractive for a study of microbial successions and interactions than a compost pile progressing through the various phases as described above. This is because in this ecosystem, as opposed to others, several environmental conditions – temperature, moisture, nutrient supply, substrate characteristics – change in time, and sometimes they exist in gradients within a compost pile. However, rather few investigators explored this aspect of composting – in contrast to the enormous amount of work done on microbial (in fact almost entirely fungal) successions in mesophilic terrestrial biotopes, especially in litter decomposition in forest soils or during dung decay (by coprophilous fungi). These findings have been reviewed numerous times, e.g., by HUDSON (1968), SWIFT and HEAL (1986), STRUWE and KJØLLER (1986), and FRANKLAND (1992). Also agricultural ecosystems have been studied in regard to fungal successions; e.g., hay (LACEY, 1980; BRETON and ZWAENEPOL, 1991), and grain (SINHA, 1992); here the two groups *field fungi* and *storage fungi* are distinguished. The population of 9 fungal species in self-heated sugar cane bagasse (up to 50 °C) has been analyzed by SANDHU and SIDHU (1980) over a period of 20 weeks.

On the contrary there exist only a limited number of studies related to microbial succession in composting. This is due to the fact that experience with all groups of microorganisms (bacteria, actinomycetes, fungi) is necessary – a rarely fulfilled demand because of the specialization of microbiologists. Further, studies of this kind are rather laborious and it is difficult to relate colony counts and identified isolates with their actual appearance during the time course of composting. Finally, the compost pile is a rather new man-made microbial ecosystem as compared with the habitats mentioned above (litter, herbivorous dung); thus inspite of its attraction as study subject it has until now found only limited consideration by microbiologists.

One of the best documented study on the microbial succession in composting has been carried out by CHANG and HUDSON (1967). Their results on the total counts of bacteria, actinomycetes and fungi are shown in Fig. 7 and have been briefly discussed. A very similar result, in fact, has been obtained by HAYES (1968) investigating composting of a wheat straw–horse manure mixture for two weeks (peak temperature 70 °C). A less comprehensive study on the microbial succession during grape mark composting has been undertaken by STREICHSBIER et al. (1982): Here the initial

Tab. 12. Isolation of Thermophilic Fungi from Compost and Self-Heated Hay by Various Investigators

Species[a] (*Basidiomycetes* see text)	References[b] 1	2	3	4	5	6	7	8	9	10	11	12	13	14
Phycomycetes														
*Absidia ramosa***			+		+					+				
*Absidia corymbifera***	+					+			+				+	+
*Mucor (Rhizomucor) miehei**		+							+					+
Mucor (Rhizomucor) pusillus*	+	+	+		+	+		+	+	+	+	+	+	+
Ascomycetes														
*Allescheria terrestris**					+									
*Chaetomium thermophile**	+	+		+	+	+		+	−	+	+	+		
Corynascus thermophilus									+					
Emericella nidulans									+					
*Myriumcoccum albomyces**			+											
*Talaromyces (Pen.) duponti**	+				+	+		+		+				
*Talaromyces emersonii**									+					
Talaromyces thermophilus			+	+	+				+			+	+	
*Thermoascus aurantiacus**	+	+	+	+	+	+			+					
*Thermoascus crustaceus**			+						+				+	
Thielavia (Sporotr.) thermophila								+						
Deuteromycetes														
*Aspergillus fumigatus***	+	+	+	+	+	+	+	+	+	+	+	+	+	+
Hormographiella aspergillata									+					
*Humicola grisea var. thermoidea**				+		+		+			+	+		
*Humicola insolens**					+	+		+		+	+	+		
*Humicola lanuginosa (Thermom.)**	+	+	+	+	+	+	+	+	+	+	+	+	+	+
*Humicola stellata**													+	
Humicola sp.*	+						+							
Malbranchea pulchella var. sulf. * *(Thermoideum sulfureum)*	+		+			+			+	+			+	
*Paecilomyces varioti*** (or sp.)			+				+		+				+	+
Scytalidium thermophilum					+				+					
*Sporotrichum thermophile**					+		+			+				
*Stilbella thermophila**								+	+			+		
Torula thermophila*	+	+		+								+		
Mycelia sterilia														
Papulaspora thermophila								+						
Myriococcum thermophilum									+					
Unidentified	+				+					+				
No. of species isolated	11	7	10	7	14	10	5	10	17	10	6	9	9	6
Temperature for isolation [°C]	48	45	40/ 50	44	50	48	50	45	45	45	51	48	40	50

[a] Species marked by * are listed in Tab. 1 of Emerson (1968); Species marked by ** are not regarded as "true thermophilic fungi" by Cooney and Emerson (1964), Straatsma et al. (1994) regarded as synonyms of *Scytalidium thermophilum*: *Humicola grisea var. thermoidea, Humicola insolens* and *Torula thermophila*

[b] References

(1) von Klopotek (1962) municipal waste compost, (2) Kane and Mullins (1973) municipal waste compost, (3) Knösel and Rész (1973) municipal waste compost, (4) Millner et al. (1977) sewage sludge compost, (5) De Bertoldi et al. (1983) municipal waste plus sewage sludge compost, (6) Gstraunthaler (1983) municipal waste compost, (7) Henssen (1957b) rotting stable manure, (8) Fergus (1964) compost for mushroom cultivation, Fergus and Sinden (1969) compost for mushroom cultivation, Fergus (1971) compost for mushroom cultivation, (9) Straatsma et al. (1994) compost for mushroom cultivation, (10) Chang and Hudson (1967) wheat straw compost, (11) Fermor et al. (1979) compost for mushroom cultivation, (12) Seal and Eggins (1976) "fermented" pig waste and straw, (13) Lacey (1980) self-heated hay, (14) Breton and Zwaenepoe (1991) self-heated hay

population consisted exclusively of yeasts, and, after the disappearance of the residual alcohol, of a mixture of bacterial species. Thermophilic fungi were represented by *Paecilomyces varioti, Aspergillus fumigatus, Mucor miehei, Humicola (Thermom.) lanuginosa, Rhizopus* sp. and *Myceliophthora thermophila*. Thermophilic and mesophilic species of *Streptomyces* effected the final humification.

The more detailed analysis of the fungal succession in wheat straw compost by CHANG and HUDSON (1967) is depicted in Fig. 10: Group 1 fungi are *primary saprophytes*, the first two species belonging to the category *saprophyic sugar fungi. Aspergillus fumigatus* deserves special mentioning as it reappears as soon as the temperature has dropped below 60 °C. As mentioned above, fungi in general are rather heat sensitive, thus they are not found – although probably not entirely eliminated – during the short period of maximum temperature (2nd to 7th day). Afterwards group 2 fungi appear, i.e., thermotolerant/thermophilic species which are very common in self-heated organic matter (see also Tab. 12). Somewhat later, 2 thermophilic species, classified as group 3a, come forth and persist until the end of the composting process, one of them, *Mycelia sterilia*, having its climax – at least as determined by the isolation procedure employed – after the temperature has fallen below 30 °C. Group 3b harbors four mesophilic fungi, among them three Basidiomycetes. This very remarkable result has been precisely commented on by HUDSON (1968), who pointed to the similarity between the coprophilous succession and the one to be observed in composting – with the exeption that in the former case only mesophilic conditions prevail. In both ecosystems, there is a short phycomycete phase, followed by Ascomycetes and fungi imperfecti (Deuteromycetes), and finally Basidiomycetes. – The particular role of Basidiomycetes in composting, especially in regard to the production of compost for mushroom cultivation, has been studied by HEDGER and BASUKI (1982), in fact with very similar results as obtained by CHANG and HUDSON (1967): The fungal succession in this wheat-straw compost consisted of a short colonization phase, characterized by *Rhizomucor pusillus* and other members of the Mucorales, followed by an extended period of de-

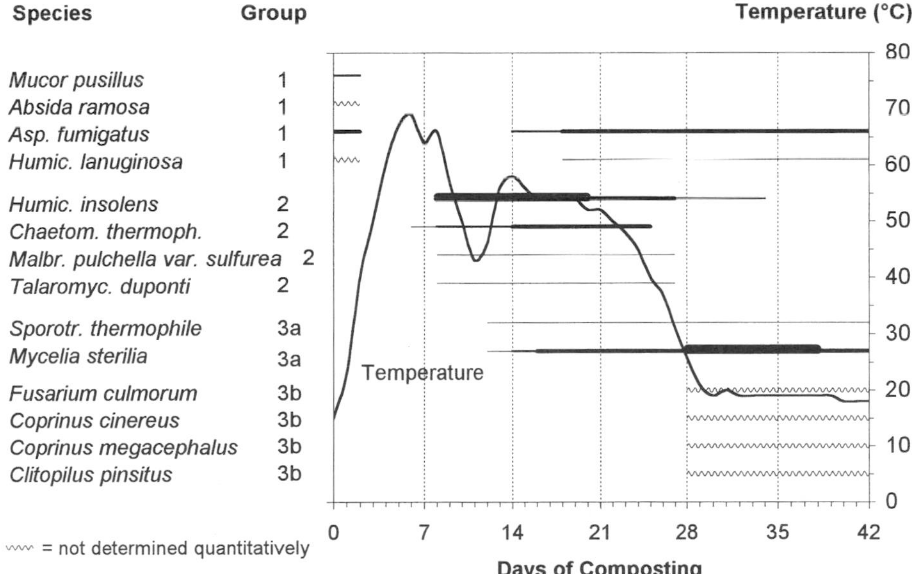

Fig. 10. Growth of fungi during composting of a straw–NH_4NO_3-mixture. Three methods were employed to determine the occurrence of the various species. To summarize the results of the three methods employed a somewhat arbitrary scale from "presence in low number" up to "high frequency of occurrence" was used for this drawing (redrawn from CHANG and HUDSON, 1967).

composition by thermophilic Deuteromycetes and *Ascomycetes*. Most activity was shown by the Deuteromycetes *Humicola insolens* and *Thermomyces lanuginosus* and by the ascomycete *Chaetomium thermophilum*. The final stage of the succession was dominated by a basidiomycete, *Coprinus cinereus*, associated apparently with *Mortierella wolfii*. Thermophilic basidiomycetes have also been found in "mushroom compost" by STRAATSMA et al. (1994): *Coprinus cinereus* and an undescribed taxon.

Much less than of fungal successions is known about actinomycete successions in composting. VON DER EMDE (1987) summarized her findings as follows: *Thermomonospora fusca*, *Thermomonospora curvata* and *Streptomyces* sp. seem to be primary decomposers, whereas *Thermomonospora chromogena* and *Saccharomonospora* sp. are rather characteristic of the advanced stage of composting, *Thermoactinomyces* sp. (amylase-negative) was typical of high temperature composting – as to be expected because it is the "actinomycete" with the highest temperature optimum. It appears that much more work needs to be done to obtain a complete picture of the particular environmental requirements (substrate, moisture, temperature) of all the actinomycetes which can be found in composts (see Sect. 6.1, Tab. 10, Sect. 6.1.2, Fig. 9) or in the aerosol of composting plants (Sect. 7.2, Fig. 12). The effect of the parameter moisture alone on the occurrence of actinomycetes in self-heated hay has been studied by FESTENSTEIN et al. (1965); however, the time course of the actinomycete development during the self-heating process has not been explored.

Our knowledge on the bacterial succession in composting is even more restricted – mainly due to the fact that the high temperature phase leads to a sharp decrease of the population of mesophilic non-spore forming bacteria indigenous of the starting material, and at the same time selects thermophilic species of the genus *Bacillus* and *Thermus* (STROM, 1985a,b; BEFFA et al., 1996). Particular fractions such as grass cuttings, leaves, food remains, manure and activated sludge certainly carry somewhat selected inocula, but these seem to have no specific effect on the commencement of the composting process: Any microbial population present at the outset will start to multiply if the conditions outlined (nutrients, water, oxygen) allow their growth. During this phase neither the nutrient supply nor the temperature exert a high selective pressure. Thus, within the first stage of composting a rather heterogenous microbial population develops, taking advantage of the more or less rich nutrient supply and a temperature of 30–45 °C which is the optimum for most mesophilic bacteria. – During the maturation phase, again mesophilic bacteria – together with actinomycetes and fungi – colonize the compost material. Noteworthy is the finding of myxobacteria at this stage which may play a role in cellulose degradation: *Myxococcus fulvus* (SINGH, 1947), *Sporophytophaga myxococcoides* and *Cytophaga hutchinsonii* (GODDEN and PENNINCKX, 1984). There is no doubt that many more "specialists" among the bacterial kingdom could be found if special methods for their isolation would be employed. – Although not investigated in detail, the mesophilic population of this phase should have a species composition which differs from the mesophilic one at the outset of the process. One would expect that the more the composting process approaches its end, i.e., its maturation, the more its population will become alike the microbial population of soil. – There are three other investigations on microbial successions in composting which deserve to be cited here: BÅGSTAM (1978, 1979), GODDON et al. (1983) and NAKASAKI et al. (1985a); in these papers, however, only total counts of the 3 microbial groups, i.e., bacteria, actinomycetes and fungi, are given.

Although not investigating the succession of microbial populations during composting, especially following the temperature course, the *isothermal incubation* of organic matter at various temperatures yields very similar results: As elaborated by WAKSMAN et al. (1939a) a wheat straw–horse manure mixture incubated at 28 °C gave rise to a population of mesophilic bacteria and fungi, at 50 °C mainly thermophilic fungi and actinomycetes developed, at 60 °C actinomycetes and bacteria, and at 75 °C only spore forming bacteria. A very similar result was obtained by KAILA (1952): at 20 and 35 °C a highly heterogenous microbial population showed up, at 50 °C thermophilic fungi predominated, and at 55/65 °C bacteria and actinomycetes. This result is actually also taken as proof that effective composting does not need a succession of various microbial populations but can, in fact, be carried out also isothermally, as, e.g., under mesophilic conditions (backyard composting), or under thermophilic conditions (continuously), as shown by SCHULZE (1962).

The succession of the various microbial populations is, of course, the result of the changing conditions during the composting process. In the discussion in the last paragraphs, mainly the temperature course has been made responsible for the observed sequence of mesophilic → thermophilic → mesophilic microorganisms. However, the changes of nutrient offer and perhaps to a certain degree also of moisture content participate in this process. Whereas the "early" population can take advantage of the readily available nutrients, its successors have to be satisfied with the more recalcitrant fraction of the organic matter, i.e., "hemicelluloses", cellulose and lignin: Thus, the populations of the late thermophilic and the following maturation phases consist of a high proportion of degraders of these macromolecules.

The degradation of lignocellulose by actinomycetes has been reviewed by MCCARTHY (1987). STUTZENBERGER et al. (1970) first investigated the cellulolytic activity in municipal waste composting, but since then up to now they have concentrated their work on the degradation of cellulose by *Thermomonospora curvata* (STUTZENBERGER, 1971; LIN and STUTZENBERGER, 1995). Cellulolytic activity has been found in *Thermoactinomyces* sp. (HÄGERDAL et al., 1978), and finally in various species of *Streptomyces*, e.g., *Streptomyces reticuli* (SCHLOCHTERMEIER et al., 1992). The distribution among thermophilic actinomycetes of the capacity to degrade a wide variety of biopolymers has also been shown by KEMPF and KUTZNER (1989) and KEMPF (1996). Studies on the decomposition of cellulose by fungi are legion, only 4 early reports will be cited: As shown by CHANG (1967) and FERGUS (1969) species of *Chaetomium* sp., *Humicola* sp., *Myriococcum* sp., *Sporotrichum* sp., *Torula* sp. and *Aspergillus fumigatus* proved to be especially active. Degradation of cellulose and lignocellulose by thermophilic/thermotolerant fungi has been studied by TANSEY et al. (1977) and ROSENBURG (1978).

The degradation of lignin appears to be a domain of a small group of fungi, particularly of *Basidiomycetes*. The role of these fungi in composting has been remarkably reviewed by HEDGER and BASUKI (1982). As also summarized by DIX and WEBSTER (1995, p. 324), weak ligninolytic activity is exhibited by *Talaromyces*, whereas other thermophilic *Ascomycetes* produce a soft rot of wood: *Allesheria* sp., *Thielavia* sp. and *Paecilomyces* sp. Of course, one of the most prominent lignin decomposers is the thermotolerant basidiomycete *Phanerochaete chrysosporium*, which has been found in woodchips piles. In contrast to the unequivocal role of fungi in lignin decomposition, there is uncertainty as to the ability of actinomycetes (and bacteria in general) to attack this macromolecule. Two points have to be emphasized (CRAWFORD, 1988):

(1) The mechanism of lignin degradation by actinomycetes, most extensively studied at the enzyme level in *Streptomyces viridosporus*, appears to be quite different from that in fungi. Various extracellular enzymes produce a water soluble, "acid precipitable, polymeric lignin" (APPL) which may be the end product of streptomycete lignin degradation.

(2) Various actinomycetes are able to utilize lignin monomers and related compounds such as coniferyl alcohol, cinnamic acid, vallinic acid and ferulic acid as carbon and energy sources; however, no relationship exists between this capability and the capacity to degrade lignin.

Since degradation products of lignin form the structural base of soil humus as proposed as early as more than a century ago by HOPPE-SEYLER (1889) *enzymatic combustion* of this biopolymer by microorganisms (KIRK and FARELL, 1987) leads inevitably to the question of their involvement in *humus* formation. Although great progress has been made since WAKSMAN (1938) summarized the knowledge on *humus* known at that time (see particularly WAKSMAN (1938) chapter VIII: humus formation in composts of stable manure and of plant residues) the specific participation of microorganisms (or their enzymes) is still a matter of uncertainty (HAIDER, 1991; SHEVCHENKO and BAILEY, 1996): Two hypotheses on huminification compete with each other, the *plant alteration hypothesis* and the *chemical polymerization hypothesis*, the former presently being more accepted: "The theory that considers lignin to be a preserved part of plants bonded to the soil matrix as only partly degraded and chemically modified "big fragments" is regarded now as the most probable hypothesis" (SHEVCHENKO and BAILEY, 1996).

In fact, both theories leave room for an active participation of microorganisms which degrade lignin to varying extents in this process. However, this aspect of compost microbiology – "organic matter changes"/formation of "humus" – will not be treated further in this review.

The change in substrate availability is not confined to the various macromolecules of plant origin, e.g., "hemicelluloses", cellulose, lignin, but includes even the microbial biomass itself. As discussed by HEDGER and BASUKI (1982), using as example composting for mushroom cultivation, basidiomycetes of the late composting stage, (in this case *Agaricus bisporus*) utilize dead as well as living cells of bacteria and of the fungus *Humicola grisea*. Also thermophilic actinomycetes have been shown to lyse living mycelium of *Humicola lanuginosa* (OKAZAKI and IIZUKA, 1970), and heat killed cells of 6 test bacteria (DESAI and DHALA, 1967).

The changing moisture conditions, i.e., the decrease of the water content, appears to be of little influence on the microbial succession. This is due to the fact that – except in the very late stage – the moisture content is above the threshold of the requirements for microbial growth, i.e., well above about 25% water content (relative to dry matter). Thus there are hardly conditions which could be selective for xerophilic fungi. The decrease of the moisture content may, however, accelerate the production of spores and conidia, and the resulting dry dust contaminating the environment may then carry a large load of germs. However, no experimental data are available for compost piles, and for some relevant information one has again to resort to studies on the storage of hay and grain: LACEY (1980) found that "*Aspergillus* ssp. are good indicators of the water content at which hay is baled, with *A. glaucus* group, *A. versicolor*, *A. nidulans* and *A. fumigatus* giving maximum numbers with water contents respectively of 25, 29, 31 and 40%." DIX and WEBSTER (1995, pp. 334–335) described the succession from *field fungi* on grain which require for growth a minimum water activity of at least 0.85 a_w, to *storage fungi* which are satisfied with a water activity between 0.85 and 0.65 a_w; again, species of *Aspergillus* are the ones which start with colonization of the rather dry grain (water content somewhat higher than 12%).

7 Hygienic Aspects of Composting

The hygienic aspects of composting, especially as they are concerned with bacteria and viruses, are treated in chapter 9, this volume, so they will be dealt with in this contribution only superficially. In Sect. 7.2., however, some results of a recent research project on this topic carried out in Germany will be communicated.

7.1 Inactivation of Pathogens

Depending on their origin, organic waste harbors germs of hygienic relevance, e.g., bacteria of the genera *Escherichia, Salmonella, Legionella*, viruses, and eggs of intestinal worms *(Ascaris)*. Although this seems to be especially true for sewage sludge and liquid animal manure, food residues and garden wastes are not free of these pathogens (GABY et al., 1972). Urban garbage, in addition, contains waste from the keeping of small animals (pets) (DE BERTOLDI et al., 1988). Persons employed in the collection and processing of waste are particularly exposed to these agents, but only recently have special measures been taken to protect these persons from direct contact with these materials to prevent infections and/or allergic reactions. In contrast, much more attention has been paid to the inactivation of pathogenic agents during the process of composting to allow safe application of the product. Thus, many studies have been carried out to determine the optimum temperature–time profile for destroying harmful germs. The control of pathogens during composting has recently been reviewed by FARRELL (1993) and STRAUCH (1996).

As shown by BURGE et al. (1981), "a temperature of 55 °C in the portion of the pile exhibiting the minimum pile temperature for a period of 2,5 days will provide adequate destruction of pathogenic microorganisms". These authors proposed the coliphage f2 as a standard organism for establishing time-by-temperature criteria for determining the level of inactivation achieved. According to DE BERTOLDI et al. (1983) a sufficiently safe endproduct is obtained if throughout the pile a temperature of 70 °C for 30 min or 65 °C for several hours is reached. Thus, self-heating to 55–65 °C for several days is considered to inactivate agents of hygienic relevance. A much longer survival of *Escherichia coli B* and *Salmonella typhimurium Q* during composting, however, has been demonstrated by DROFFNER and BRINTON (1995) employing DNA gene probes: In bench scale trials these bacteria survived for at least 9 d at 60–70 °C in a food waste compost or for 9 d, respectively 5 d, in wastewater sludge compost. In industrial compost the two bacteria survived even 59 d at about 60 °C. Alternatively, product control (instead of process control) should yield the absence of *Salmonella* or less than 10^3 coliforms per gram of dry compost, the number of coliforms serving as indicator for proper composting of sewage sludge (FARRELL, 1993). Of course, sporeforming bacteria, e.g., members of the genera *Bacillus* and *Clostridium*, are not affected. Although some of them are pathogenic, they are not epidemical and, therefore, of less relevance.

Regarding the kinetics of pathogen control, SAVAGE et al. (1973) found that in composting wastes from swine feeding in windrows, turned 20 times per week, a rapid rise of the temperatur was the most important prerequisite to reduce and finally eliminate *Salmonella* sp., other enteric bacteria and fecal streptococci. This was accomplished by adding straw to improve the structural properties of the waste and to widen the C–N ratio. In this study, the maximum temperature of 72 °C was attained at the 10th day and lasted till the end of the experiment (80th day). It took about 15 more days to eliminate the coliforms. Since *Salmonella* sp. is less heat resistant it had probably been killed earlier; (its complete destruction was tested only at day 40). The number of mesophilic bacteria decreased between 10^4- and 10^5-fold within 40 d, whereas the number of mesophilic actinomycetes and cellulolytic organisms increased 10^2- and 10^5-fold, respectively, within 35 d. A very similar result regarding the elimination of *Salmonella* sp. during composting has been reported more recently by SCHUMANN et al. (1993). It appears, however, that in the light of the findings of DROFFNER and BRINTON (1995) the data on the thermal inactivation of pathogens during composting have to be carefully reexamined.

Besides Enterobacteria, especially *Salmonella* ssp., members of the genus *Legionella* may be a potential risk to producers as well as users of compost: HUGHES and STEELE (1994) found up to $5 \cdot 10^5$ CFU g^{-1} compost: Of 33 "large scale" and 80 "home" composts tested 85% and 56% were positive. Also soil amended with compost contained up to 10^4 CFU g^{-1}. Among the isolates were *L. pneumophilia* (mostly of serogroups 2–14) and *L. longbeachae* serogroup 1. Although most species identified have not been implicated as causative agents of legionellosis in South Australia, the authors advise gardeners, especially those with impaired immunity, to avoid creating aerosols while watering plants and on the importance of handwashing to prevent ingestion of contaminated material. It appears that composting material represents an ecological niche for *Legionella* ssp. Unfortunately, the authors give no information on the composting conditions, i.e., the temperature–time course. Thus it is not known whether these bacteria survived the regular composting process or reinfected the once pasteurized compost.

The temperature–time requirements for the *pasteurization* of the material have been determined in numerous experiments. It appears, however, that pathogenic germs are not only inactivated by high temperatures but also by other mechanisms such as *competition* with and *antagonism* by the autochthonous microflora. By the same token, these means prevent the repopulation of mature compost once freed of pathogenic agents just by these bacteria. Suppression and even dying-off of *Salmonella* ssp. inoculated into unsterile compost comprising an actively growing population of gram-negative bacteria and actinomycetes have been demonstrated in three elaborate studies on this subject (Russ and Yanko, 1981; Hussong et al., 1985; Millner et al., 1987). In contrast, compost sterilized by irradiation (Hussong et al., 1985) or taken from 70 °C heated piles (Millner et al., 1987) did not exhibit this suppressive effect, apparently due to the absence of an effective microbial population.

7.2 Emission of Microorganisms from Composting Plants

Ever since the handling and composting of garbage, refuse, sewage sludge and biological wastes have been carried out *on the large scale*, it became apparent that persons employed in this endeavor may be adversely effected by pathogenic agents carried by these materials and – eventually even more important – developing during the composting process, e.g., thermophilic actinomycetes and thermophilic and/or thermotolerant fungi, notably *Aspergillus fumigatus*. Whereas the relevant viruses, bacteria and fungi inherent in the compost feedstocks are controlled by the composting process as described in Sect. 7.1, the dustmen and compost workers handling the contaminated materials are nevertheless exposed to these agents *before* as well as *during* the composting process. Thus the emission of germs via dust (aerosols) has become a major concern of environmental microbiologists. Since these aerosols are not confined to the plant area they are also of concern to public health officers. In fact, the neighborhood of compost plants is often alarmed by these health affecting agents, and the construction of new compost plants is often objected to on these grounds.

However, if care is taken to protect employees of composting facilities and the close neighborhood from contaminated aerosols, there appear to exist no specific risks originating from these plants. And since during the last decade great progress has been made in mechanizing the handling of compost feedstock as well as the composting process itself, and finally in biofiltration of the exhaust air, the direct contact of plant employes and neighborhood with contaminated aerosols is greatly reduced, and thereby also their health risk. To summarize the situation in advance: Clark et al. (1984) found in compost workers as compared with non-exposed persons: (1) excess of nasals, ear and skin infections, (2) a greater number of fungal colonies in cultures of anterior nares and throat swabs, (3) higher white blood cell counts and hemolytic complement, (4) higher antibody levels to compost endotoxin.

However, "serologically-detectable infections caused by fungi were uncommon among the workers studied, despite their exposure to fungal spores at the compost site". – On the contrary, Epstein and Epstein (1989) found no consistent difference between compost workers and workers not involved in compost activities as determined by antibody methods. These authors apparently agree with the statements by Dr. J. Rippon (after Epstein and Epstein, 1989): "There is in compost operations no significant level of bio-hazard risk associated with viable bacteria or fungi, dust or endotoxin"; however, "particular individuals hypersensitive to allergens or predisposed to opportunistic infections may be at risk": A summary of a $2^1/_2$ day workshop in 1993, organized by EPA, USDA and NIOSH, at which 25 experts discussed and analyzed questions related to bioaerosol associated health risks has been given by Millner et al. (1994); to make it short: "Composting facilities *do not* pose any unique endangerment to the health and welfare of the general public". This conclusion is based on the fact "that workers were regarded as the most exposed part of the community, and where worker health was studied, for periods of up to ten years on a composting site, no significant adverse health impacts were found". However, again, "immuno compromised individuals are at increased risk to infections by various opportunistic pathogens, such as *Asp. fumigatus*, and asthmatic and 'allergic' individuals to responses to bioaerosols from a variety of environmental and organic dust sources, including compost".

Because of the continuous public concern about potential health risks originating from composting installations, an extensive research project has been carried out in Germany in 1991–1993, and 6 workshops on this topic have been held alone in this country between 1994 and 1996 (Tab. 13). Some results of this undertaking which have been documented in four Ph.D. theses are depicted in Figs. 11–13, and can be summarized as follows: In general, the highest counts of relevant viruses, bacteria and fungi were found at such locations at which the

Tab. 13. Recent Workshops on the Emission of Germs from Compost Plants Held in Germany in 1994–1996

Göttingen	STALDER, K., VERKOYEN, C. (Eds.) (1994) *Gesundheitsrisiken bei der Entsorgung kommunaler Abfälle*, pp. 1–217. Göttingen, Die Werkstatt.
Hohenheim	BÖHM, R. (Ed.) (1994), *Nachweis und Bewertung von Keimemissionen bei der Entsorgung kommunaler Abfälle sowie spezielle Hygieneprobleme der Bioabfallkompostierung*, pp. 1–398. 5. Hohenheimer Seminar, Gießen: Deutsche Vet. Ges. e. V.
Berlin	Schriftenreihe WAR (TU-Darmstadt). Nr. 81 (1994), *Umweltbeeinflussung durch biologische Abfallbehandlungsverfahren*, pp. 1–321.
München	MÜCKE, W. (Ed.) (1995), *Keimbelastung in der Abfallwirtschaft*, pp. 1–171. TU München: Institut für Toxikologie und Umwelthygiene.
Delmenhorst	KRANERT, M. (Ed.) (1996), Hygieneaspekte bei der biologischen Abfallbehandlung. *Schriftenreihe des Arbeitskreises für die Nutzbarmachung von Siedlungsabfällen (ANS) e. V.*, 32, 1–370.
Darmstadt	Schriftenreihe WAR (TU-Darmstadt) Nr. 92, (1996), *Hygiene in der Abfallwirtschaft*, pp. 1–179.

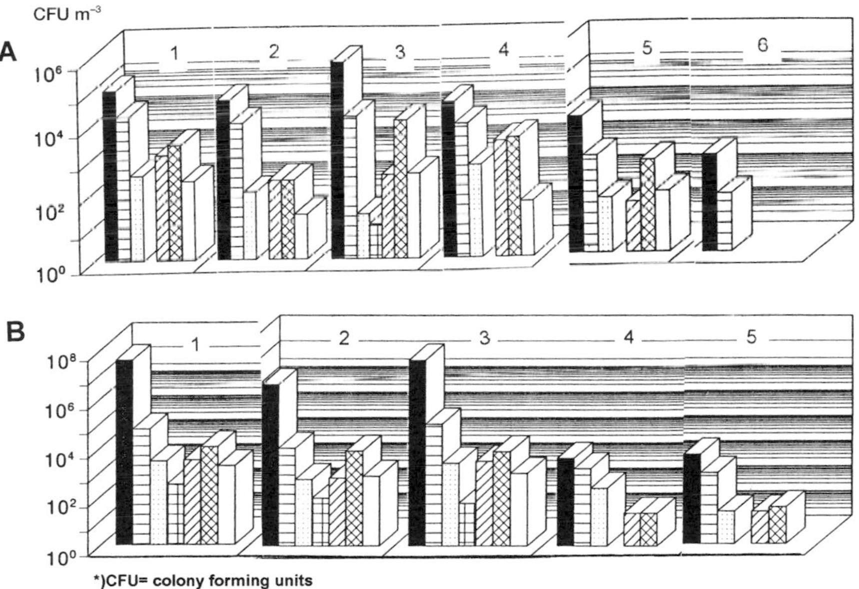

Fig. 11a, b. Bacterial population of aerosols at various localities of two compost plants (from SCHMIDT, 1994). **A** Compost plant A (Same as compost plant A in Fig. 12): 1: delivery hall, 2: machine station, 3: hand sorting facilities, 4: rotating drum exit, 5: windrow turning, 6: close neighborhood, **B** compost and recycling plant, 1: delivery hall, 2: porter house, 3: crane, upper level, 4: crane cabin, 5: close neighborhood, CFU: colony forming unit, ■ total count, ⊟ aerobic spore formers, □ *Staphylococcus aureus*, ⊞ group A streptococci, ▨ group D streptococci, ⊠ Enterobacteria, □ *Escherichia coli*.

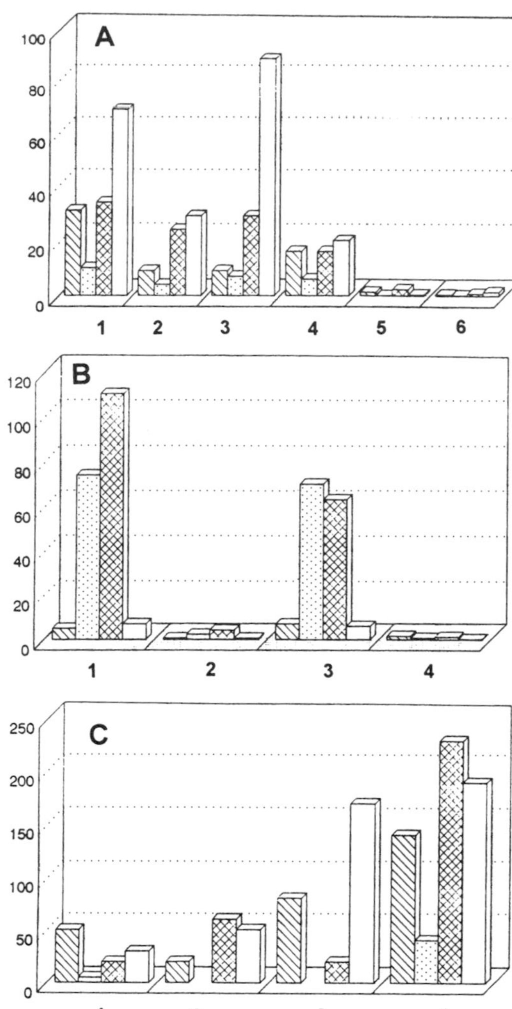

Fig. 12a–c. Occurrence of three actinomycete genera and *Thermoactinomyces* in the aerosol at selected localities in 2 compost plants (**A** and **B**) and one fand fill (**C**) (from KEMPF, 1996). **A** 1: delivery hall, 2: machine station, 3: hand sorting facilities, 4: rotating drum exit, 5: windrow turning, 6: close neighborhood, **B** 1: enclosed composting area during turning, 2: enclosed composting area while working pause, 3: driver cabin of front loader, 4: close neighborhood, **C** 1: entrance of landfill, 2: porter house (control station), 3: delivery of small loads, 4: damping place for large loads, **A** and **B**: CFU · 1,000 m^{-3} air; **C**: CFU m^{-3} air. Take notice of the different scales in the graphs, ▧ *Streptomyces*, ▨ *Saccharopolyspora*, ▩ *Saccharomonospora*, □ *Thermoactinomyces*.

highest density of dust was produced, i.e., at the delivery stations, the hand-sorting installations and the composting areas, especially during windrow turning. The aerosol contamination was much less in porter houses, the plant offices and the near outdoor neighborhood. It should be pointed out that – although data were collected over a period of 2 years and repeatedly at the particular locations – the results represent only momentary "snapshots" without statistical evaluation. Thus, the counts may not always be that high as shown in Figs. 11–13, but they may also not describe the "worst case".

Most viruses found by PFIRRMANN (1994) at the places indicated in the legends of Figs. 11–13 were identified as human Enteroviruses, whereas Herpes simplex isolates were rather rare. The bacterial population of the aerosols as determined by SCHMIDT (1994) is shown in Fig. 11. Most isolates belonged to the genus *Bacillus*; whereas *Salmonella* sp. could not be found in any sample (not shown), and *E. coli* did not occur in the crane cabin and the close neighborhood. Of the thermophilic actinomycetes, studied by KEMPF (1996), most attention was paid to the genera which are involved in allergic diseases (Fig. 12). There exists an extensive literature on the role of these organisms (as well as of fungi) as allergens and pathogens (Tab. 14), whose discussion, however, is out of the scope of this contribution. It may suffice to cite the comprehensive reviews by LACEY et al. (1972), LACEY and CROOK (1988), and LACEY et al. (1992). Fig. 12a contains a somewhat puzzling result: In this instance thermophilic actinomycetes were found already in the aerosol of the delivery hall, i.e., before the material had been subjected to composting. This was due to the growth of these organisms in the bin before collection of the bioorganic waste for composting.

The data presented in Fig. 13 are taken from the work of GÖTTLICH (1996): *Aspergillus fumigatus* is the species which received most attention by compost microbiologists, due to the fact that it is a potential pathogen as well as a potent agent of allergic reactions. This fungus is enormously widespread; it has been found in all studies used to prepare Tab. 12, and here it belongs to the most common species at all locations. Its occurrence in the air of compost

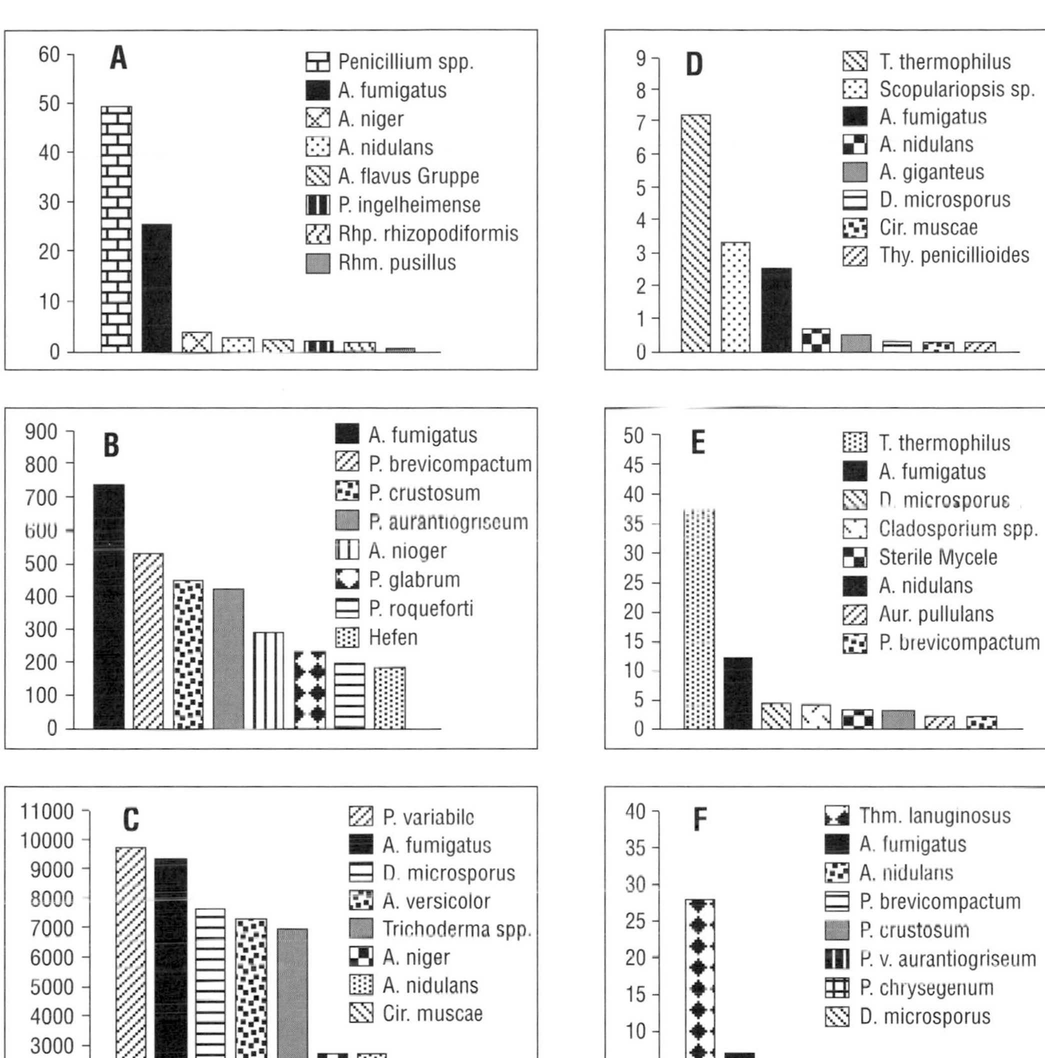

Fig. 13a–f. Predominating fungi in the aerosol at 6 localities of a compost plant (from GÖTTLICH, 1996), **A** delivery hall, **B** manual sorting, **C** prerotting hall during windrow turning, **D** late rotting phase (9th week), **E** sifting of the finished compost, **F** driver cabin of front loader during static maturation phase, **A–F**: CFU · 1,000 m^{-3} air. Take notice of the different scales in the graphs.

plants [and many other locations where bulks of organic matter are handled, e.g., wood chips (KOTIMAA, 1990)] has been reported by numerous authors, of which the following may be cited: MILLNER et al. (1980), KLEYN and WETZLER (1981), CLARK et al. (1983), KOTHARY and CHASE (1984), LEMBKE and KNISELY (1985), and BOUTIN et al. (1987). Finally, in bioaerosols from compost facilities FISCHER et al. (1998, 1999a) studied the occurrence of *Aspergillus fumigatus* and other molds as well as their mycotoxins which might also be of hygienic relevance.

Tab. 14. Actinomycetes and Fungal Spores in Air as Respiratory Allergens and Pathogens (condensed after LACEY et al., 1972 and LACEY and CROOK, 1988)

Species	Allergy		Infection
	Type I: Asthma Rhinitis	Typ III: Extrinsic Allergic Alveolitis	(Opportunistic)
Actinomycetes			
Thermoactinomyces vulgaris	+	+	–
Thermoactinomyces sacchari	–	+	–
Saccharopolyspora rectivirgula	+	+	–
Saccharomonospora viridis	–	+	–
Streptomyces albus	–	+	(?)
Nocardia asteroides	–	–	+
Fungi			
Absidia corymbifera	–	–	+
Absidia ramosa	–	–	+
Aspergillus fumigatus	+	+	+
Aspergillus flavus	+	–	+
Mucor pusillus	+	–	+
Alternaria tenuis	+	+	–
Penicillium sp.	+	+	–
Graphium sp.	–	+	–

In all the studies cited above the aerosol analysis has been carried out using various kinds of air samplers, notably the Andersen sampler, all-glass impingers or membrane filters. Selective media to determine particular species or genera can be employed with either method, circumventing to a certain degree the isolation and confirmative identification of the respective microorganisms. In the future, molecular methods may facilitate the detection and identification of microbial cells in aerosols using nucleic acid probes (NEEF et al., 1995) and PCR technology (ALVAREZ et al., 1995). For the evaluation of the health risk by high numbers of the various kinds of germs, especially for the controversial discussion about threshold values, the reader is referred to chapter 9, this volume.

8 Phytopathogenic Aspects of Composting

Because of the sources of certain starting materials for composting, i.e., dead plants, and the application of the produced compost as growth substrate for plants or as fertilizer, several relations exist between the composting process and the well-being of the plants cultivated in compost or in soil supplied with compost:

(1) compost should be free of pathogenic agents;
(2) fresh organic waste, i.e., immature compost, may exert toxic effects on plant growth;
(3) mature compost is able to control soil-borne plant pathogens;
(4) compost extracts have been shown to be effective against certain plant pathogens which infect leaves and stems of plants.

8.1 Inactivation of Plant Pathogens during Composting

In many instances the starting material for compost production contains diseased plants together with the respective pathogens, i.e., bacteria, fungi, viruses and nematodes. To prevent their spread when the compost is applied as fertilizer these agents have to be inactivated in the course of the compost process. This is especially important in regard to soilborne pathogens which may otherwise be introduced via compost into nurseries not yet subjected to these pathogens. Thus, the compost sold for this purpose has to be checked for the absence of these agents – or alternatively and easier to perform – the composting process itself has to be verified to be effective in inactivating selected test organisms, e.g., *Plasmodiophora brassicae, Fusarium oxysporum, Pythium ultimum, Sclerotinia sclerotiorum, Rhizoctonia solani*, and tobacco mosaic virus (TMV). This topic has been reviewed by Arbeitsgemeinschaft Universität/GH Kassel et al. (1997). Also in back yard composting the inactivation of plant pathogens becomes important to avoid a building up of a pathogen population in ones own private garden. Effective composting is also important to kill weed seeds; in this case tomato seeds serve as test subject.

The literature on the inactivation of phytopathogenic agents by composting has been exhaustively reviewed by BOLLEN (1993) and BOLLEN and VOLKER (1996): Most plant pathogenic fungi, bacteria and nematodes are inactivated by the heat generated in the compost pile. Soilborne viruses are not inactivated; however, since their vectors – either nematodes or fungi – are killed they do not need special attention of the compost practitioner. Only the TMV which does not need a vector is not affected; therefore, TMV infected residues should not be used in susceptible crops. – Microbial antagonism appears to play only a minor role in eradicating plant pathogens during composting, but toxic compounds (not yet identified) produced under anaerobic conditions prevailing in some composting systems (or niches) have been reported to inactivate sclerotia of *Sclerotium cepivorum*.

8.2 Adverse Effects of Fresh, Immature Compost on Plant Growth

Fresh, immature composts have often been observed to exert toxic effects on plant growth when applied as fertilizer to soil or used as container medium in nurseries. Therefore, a test on phytotoxicity using seeds of cress (*Lepidium sativum*) or radish (*Raphanus sativus*) belongs to the examinations routinely carried out to prove compost maturity (MATHUR et al., 1993). The adverse effects are mainly due to the microbial formation in fresh organic matter of volatile acids (e.g., acetic, propionic, and butyric acids) in the case of household garbage or of ammonium in the case of sewage sludge (KATAYAMA and KUBOTA, 1995). The production of organic acids and possibly other toxic compounds in the decomposition of plant residues (mainly straw) has been observed before by soil microbiologists (COCHRAN et al., 1977; LYNCH, 1977); it has been reviewed by BIDDLESTONE et al. (1987). However, the toxicity of immature compost observed in laboratory experiments need not necessarily show up in field trials where no difference in plant health and yield between mature and immature compost was observed (STÖPPLER-ZIMMER and PETERSEN, 1995). Fresh organic matter as fertilizer may exert also a negative effect by serving as nutrient source for plant pathogenic bacteria and fungi.

8.3 Control of Soilborne Plant Pathogens by Compost

Major contributions to this topic have been made by HOITINK and coworkers; the results of their studies have been summarized several times during the last decade, the 2 recent reviews being those of HOITINK and GEBUS (1994) and HOITINK et al. (1996): In the 1960's nurserymen observed more or less accidentally that compost – at that time those prepared from tree bark – could suppress soilborne plant pathogens such as *Pythium, Phytophthora, Fusarium*, and *Rhizoctonia*. The compost must not be too fresh, because this material can serve as growth substrate for these pathogens, and must not be too mature, in order to support the antagonistic microbial population, which of course must be present. It is assumed that the latter consists of mesophilic microbes which colonize the compost after peak heating. Various microorganisms are considered to serve as biocontrol agents: bacteria of the genera *Bacillus, Enterobacter, Flavobacterium, Pseudomonas, Streptomyces*, and fungi such as *Penicillium* ssp., species of *Trichoderma, Gliocladium virens* and others. Most of these organisms are killed during the high temperature phase of composting, thus the suppressive capacity of the compost relies on its recolonization by these microorganisms. Since this is a factor of uncertainty a specific inoculum of various bacteria and fungi can be introduced into the compost just after peak heating, i.e., before a significant natural recolonization has taken place (HOITINK, 1990). In Japan, PHAE et al. (1990) and NAKASAKI et al. (1996) produced composts with predictable biological control with a strain of *Bacillus* which has the advantage of surviving the high temperature phase. The biocontrol of phytopathogens by compost is due either to a "general suppression" exerted by competition, antibiosis and microbiostasis, or to a "specific suppression" as in the case of hyperparasitism of *Rhizotonia solani* by species of *Trichoderma* and of *Sclerotium rolfsii* by *Penicillium* ssp. Of the numerous authors reporting control by compost of root rots caused by *Pythium* sp. and *Rhizoctonia* sp. only SCHÜLER et al. (1989) and CRAFT and NELSON (1996) will be cited. As pointed out by HOITINK et al. (1996), "the future opportunities for both natural and controlled induced suppression of soilborne plant pathogens, using composts as the food stuff for biocontrol agents, appear bright."

8.4 Control of Foliar Diseases by Compost Water Extracts

As first observed by (WELTZIEN et al. (1987) and WELTZIEN (1992) at the University of Bonn compost extracts sprayed on leaves and stems can protect the treated plants from infection by respective pathogens. Host plant–pathogen systems successfully controlled by this means were grape wine–*Plasmopara viticola*, tomato–*Phytophthora infestans*, beans–*Botrytis fabae* and barley–*Erysiphe graminis*. The mechanism of this protective effect is still not completely understood: Induction of an unspecific resistance of the plants by turning on SAR genes (systemic acquired resistance) was first thought to be responsible for this phenomenon, but later work did not support this view. Hindrance of infection by the compost extracts could also be due to the stimulation of the autochthonous microbial population dwelling on the leaves and stems, inhibiting the pathogens by competition and antagonistic action. Further work is necessary to uncover the mechanism(s) of the pathogen control – last not least to make this method of biocontrol better reproducible as it appears to be at present.

9 Balancing the Composting Process

According to the definition of composting (see Sect. 1) one of the main aims of this process is the *stabilization* of organic matter, i.e., the degradation of that fraction of the input material which is easily decomposed by microorganisms. The microbial metabolism is strongly connected with the production of *new biomass* and *heat*, the latter resulting in high temperatures of the compost pile (or laboratory composter) because of their insulation. This heat actually removes part of the water by *ventilative cooling* thus leading to the *product compost* which is stable for storage. It is obvious that there exists a stoichiometric relationship between

(1) loss of organic matter,
(2) production of heat,
(3) aeration,
(4) heat transport to the surroundings, and
(5) removal of water.

In Sects. 9.1–9.4 the balance of mass as well as of energy of the composting process will be considered.

9.1 Mass Balance

To perform a mass balance, only few data are actually necessary: the loss of volatile solids (Δ v.s.), the loss of water (Δ H_2O), and the ash content which is assumed to remain constant. If the uptake of oxygen or the discharge of CO_2 are measured, the corresponding gas species (CO_2 or O_2, respectively) can be calculated, either from an empirical formula of Δ v.s. or an assumed RQ. Of course, the empirical formula of Δ v.s. remains an uncertainty as long as no chemical analysis of the organic matter has been carried out, at least of the main classes of chemical compounds such as carbohydrates, hydrocarbons, lipids, proteins, nucleic acids, and lignin. The empirical formulas of numerous organic materials have been collected from the literature by KROGMANN (1994, p. 139), a selection of these data is listed in Tab. 16.

The balance of a composting process can be best represented by columns comprising the various fractions of the composting material. Surprisingly, little use has been made of such graphs; one has been constructed by WILEY and PEARCE (1955) (Fig. 14). In this case, there was Δ v.s. = 1.724 kg (3.8 lbs), Δ H_2O = 1.8144 kg (4.0 lbs) and formation of CO_2 = 3.266 kg (7.2 lbs). From these data the authors calculated an empirical formula of that fraction of the starting material which had been decomposed, and also the theoretical amount of oxygen necessary to fulfill the stoichiometry. The authors, however, did not consider any nitrogen turnover and formation of new microbial biomass.

A mass balance of a rather hypothetical composting process contrived by the author is depicted in Fig. 15, Fig. 16, and the calculation is given in Tab. 15 (row 7a). The balance rests on the assumption of four succeeding microbial populations with a carbon turnover rate of 75% as shown in Tab. 5 for one population. Rows 1–6 in Tab. 15 relate to an input of 720 g glucose + 13.6 g NH_3; in row 7 these quantities are reduced to 350 g + 6.61 g as part of 1,000 g of fresh biomass to be composted as shown in Fig. 15. There is a decrease of volatile matter,

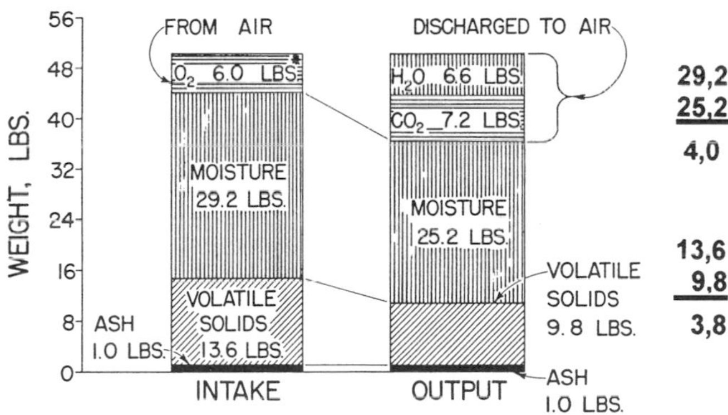

Fig. 14. Mass balance of a typical composting run (according to WILEY and PEARCE, 1955). For the loss of volatile solids (13.6 − 9.8 = 3.8 lbs) and the loss of water (29.2 − 25.2 = 4.0 lbs) the following stoichiometry was suggested:
(1) $C_{26}H_{40}O_{14} + 29\ O_2 \rightarrow 26\ CO_2 + 20\ H_2O$,
(2) in g: 576 + 928 → 1,144 + 360,
(3) reduced to: 100 + 161.1 → 198.6 + 62.5,
(4) in lbs: 1.27 + 2.04 → 2.52 + 0.79
(5) x2.94*): 3.73 + 6 → 7.41 + 2.32 [+4 = 6.32],
*) (multiplication to arrive at the calculated O_2 utilized)
(6) average 9 runs: 3.8 + 6.1 → 7.7 + 2.4.

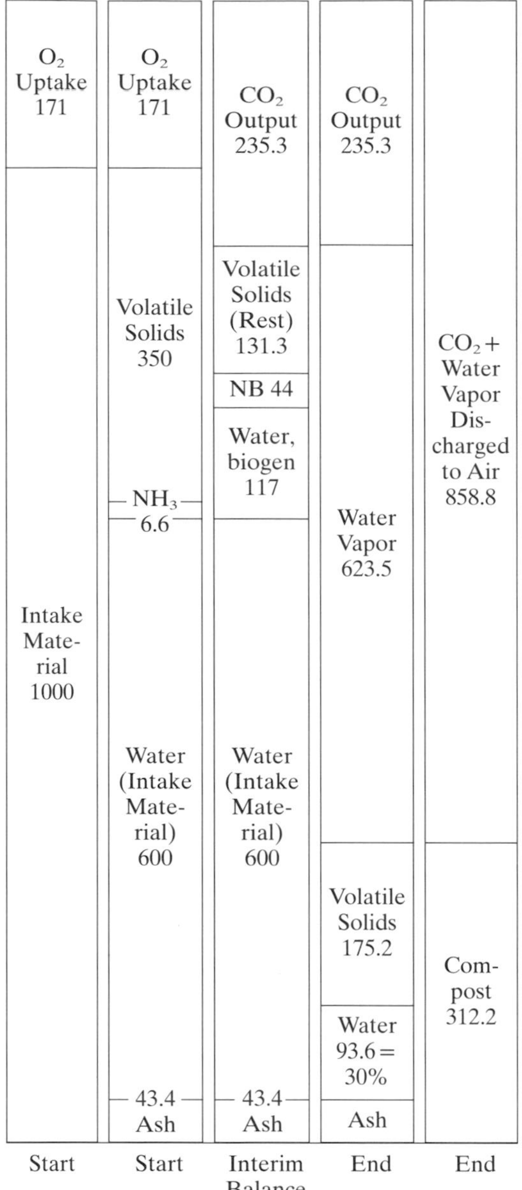

Fig. 15. Mass balance of a hypothetical compost run, reduced to 1 kg fresh organic waste (intake material). The stoichiometry of the various fractions, except water, is given in Tab. 15.

from 350 + 6.61 to 131.25 + 43.94 = 175.19, i.e., by about 49%. (It should be noted that actually 218.75 g of original organic matter have been "utilized", but 37.33 + 6.61 = 43.94 g ap-

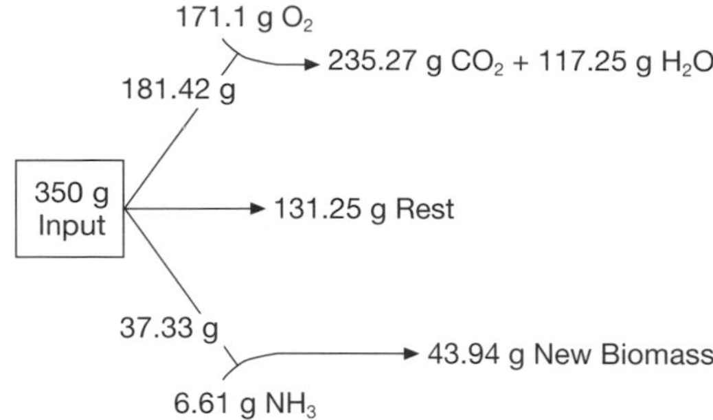

Fig. 16. Mass balance in composting, Eq. 7, in Tab. 15.

pear in new biomass). At the same time, the C–N ratio is reduced from 25.71 to 15.25; [the final value of 12.85 given in Tab. 4 is not reached because the new biomass has remained unlysed (Fig. 15)]. If the total balance is reduced to 1 M glucose (row 8 in Tab. 15), it comes very close to Eq. 6 in Tab. 3 (row 9 in Tab. 15), i.e., 4 *succeeding populations, each with a growth yield of* $Y_s = 0.502$ *lead to an overall yield coefficient of 0.2008 and thus correspond to one population with a growth yield of 0.188.* The same is actually true in regard to the reduction of the C–N ratio – here from 25.71 to 15.25 (Δ C–N = 11.79), whereas only one population with a growth yield of 0.188 leads to a Δ C–N of 13.93 (Fig. 5, Tab. 4).

Mass balances as those shown in Figs. 14, 15 can only be carried out if the *total mass* of the composting material is analyzed, e.g., the material of a laboratory composter or aliquots, *related to the original material*! If the entire mass is not known as in large scale composting, an equal amount, let say 1 kg, of the incoming material and of the finished product are analyzed; then the balance has to be performed *in reverse*, supposing the ash content remaining constant. Such a calculation is shown in Fig. 17, using the same data as those employed for Fig. 15.

Of the data used to construct the hypothetical balance in Fig. 15, those relating to the *new biomass* are, of course, speculative; they originate from the assumption of an overall growth yield of about 0.2. It may be questioned, however, whether or not such an amount of biomass is really present in mature compost. No efforts seem to have been undertaken, to relate in compost material microbial biomass

Tab. 15. Calculation of a Hypothetical Composting Process, See Row 7 for the Interpretation of the Data Implicated in Fig. 15), Input Volatile Solids: $C_{24}H_{48}O_{24} + 0.8\ NH_3$; NB: New Biomass [$C_5H_7O_2N$]

Calculated for	Carbon Substrate (Glucose)	Nitrogen Source NH_3	O_2	→	Rest of Substrate (Glucose)	NB	CO_2	H_2O
(1) Input [M]	4	0.8	11		1.5	0.8	11	13.4
(2) Input [g]	720	13.6	352		270	90.4	484	241.2
(3) Input [g]	100	2.0	49		38	12	67	34.0
(4) Consumed [g]	450	13,6	352			90.4	484	241.2
(5) Consumed [g]	100	3.02	78.22			20.08	107.56	53.6
(6) Degraded [g]	373.2		352				484	241.2
(7a) Input [g]	350	6.61	171.1		131.25	43.94	235.27	117.25
b) Consumed [g]	218.75	6.61	171.1			43.94	235.27	117.25
c) Degraded [g]	181.42		171.1				235.27	117.25
(8) Glucose consumed [(7b) ./. 218.75 · 100]								
a) [g]	100	3.02	78.22			20.08	107.55	53.6
b) [M]	1	0.32	4.40			0.32	4.40	5.36
$Y_s = 0.2008$								
(9) 1 M glucose	1	0.3	4.5			0.3	4.5	5.4
$Y_s = 0.188$ (Eq. 6 in Tab 3)								

A comparison of row 3 in this table with the legend to Fig. 14 shows that here much less oxygen is consumed and less CO_2 and H_2O are produced than in the calculation of WILEY and PEARCE (1955). This is due to the fact that these authors employed a more reduced substrate which, in addition, was completely decomposed to CO_2 and water, i.e., without formation of new biomass.

with viable and non-viable counts of microbial cells, spores and hyphae. Since the microbial biomass is part of the volatile solids, this uncertainty does not influence the overall mass balance; it has, however, an effect on the energy balance, although insignificant, because the more new biomass is retained in the compost, the less heat is produced from the decomposition of organic matter.

Mass balances in practice may greatly deviate from the hypothetical process depicted in Fig. 15 and Tab. 15, nevertheless, the stoichiometry, which actually also includes the fate of the water content (see Sect. 9.2), must be obeyed. Thus, the percentage of organic matter decomposed may vary widely, depending in the first place on the time allowed for degradation; in practice, however, where only *stabilization* is asked for, it is determined by the fraction of organic matter which can be easily attacked.

9.2 Energy Balance and Heat Transfer

The complete aerobic degradation of 1 M glucose to $CO_2 + H_2O$ leads to a heat output of 2872 kJ M^{-1}, or 15.95 kJ g^{-1} glucose (Tab. 2). Although the fraction of the composting substrate which is decomposed in the process of stabilization yields a certain amount of new biomass, it turns out that this does not influence the energy balance significantly: In the hypothetical composting process as depicted in Fig. 15 and Tab. 15, 218 g organic matter are decomposed, yielding 44 g new biomass, the rest appearing in CO_2 and H_2O. From an analytical point of view, however, the organic matter content has been reduced by Δ v.s. = 175 g. Assuming, for simplification, glucose (or a carbohydrate) as sole substrate, 218 g have a heat content of 3477 kJ, of which 880 kJ are incorporated into new biomass (44 g à 20 kJ g^{-1}) and 2597 kJ are dissipated as heat, i.e., 14.84 kJ g^{-1}

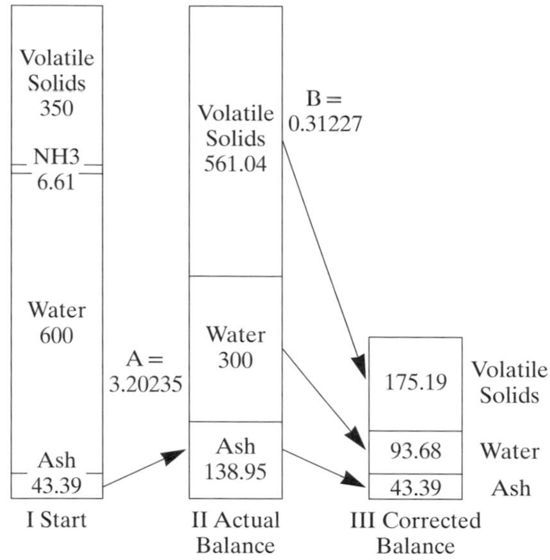

Fig. 17. Calculation of the mass balance, when equal amounts of starting material (I) as well as of mature compost (II) are analyzed, assuming the ash content remaining constant (same data as in Fig. 6):

(1) determine factor $A = \frac{\text{ash \% in II}}{\text{ash \% in I}} : \frac{138.95}{43.39} = 3.20235$,

(2) factor $B = \frac{1}{A} : \frac{1}{3.20235} = 0.31227$,

(3) vs. in II x factor B = vs. "corrected" in III: $561.04 \cdot 0.31227 = 175.19$,

(4) [vs. "corrected" in III/vs. in I] $\cdot$ 100 = % rest: $\frac{138.95}{43.39} \cdot 100 = 49.126$,

(5) Δ vs. = vs. in I – vs. in III: 356.61 – 175.19 = 181.42,

(6) degree of decomposition = 100 – % rest: 100 – 49.126 = 50.874% = Δ vs. as related to 100 g vs. in I; (proof: 50.874% of 356.61 = 181.42,

Δ v.s.: That is somewhat less than 15.95 kJ g^{-1} glucose. If less new biomass is formed (e.g., $Y_s = 0.1$), the deviation from the strict balance becomes even less. Thus, to allow for an inconsiderable simplification of the energy balance, the Δ v.s. observed in the process can well be employed to calculate the production of heat. Of course, formation of new biomass is concealed in energy balancing when the liberated heat is taken as difference between the enthalpy of combustion (H_{comb}) of the input and output material, as it has been done by STAHLSCHMIDT (1987).

As shown in Tab. 16 (lower part), heat formation in composting under practice conditions varies due to the variable nature of the input material, cellulosic materials yielding heat comparable to the balance carried out for glucose, organic matter with higher energy content such as fat containing wastes or sewage sludge leading to a higher heat output. In addition, exact calculation of the heat output is rather complicated, even under laboratory conditions, due to the multiple routes of heat transfer to the surroundings (see following text). Instead of relating the formation of heat to the mass of organic material being degraded, it can also be associated to the oxygen consumed (COONEY et al., 1968): As shown in Tab. 16 aerobic growth with glucose in synthetic

Tab. 16. Energy Content [ΔH_{comb}] and Formation of Heat During Degradation of Organic Matter (v.s.: Volatile Solids)

Substrate	Empirical Formula[a]	kJ g^{-1} ΔH_{comb}	g O_2 (g Substrate oxidized)$^{-1}$	kJ (g O_2 utilized)$^{-1}$ [b]	Peak Heat Output W kg^{-1} [c]
Glucose	$C_6H_{12}O_6$	15.64	1.07	14.63	
Cellulose	$[C_6H_{10}O_5]_n$	17.57			
Protein	$C_{16}H_{24}O_5N_4$	22.59	1.50	15.06	
Tripalmitin	$C_{51}H_{98}O_6$	38.95	2.87	13.57	
n-Decan	$C_{10}H_{22}$	47.39	3.49	13.58	
Bacterial biomass	$C_5H_7O_2N$	22.00	1.41	15.60	
Sewage sludge	$C_{10}H_{19}O_3N$	23.00	1.99	11.56	
Domestic garbage	$C_{54}H_{104}O_{37}N$	19.67	1.53	12.86	
Hedge trimmings	$C_{28}H_{40}O_{24}N$		0.83		
Grass clippings	$C_{23}H_{38}O_{17}N$	17.54	1.24	14.14	
Food waste	$C_{29}H_{40}O_{22}N$	16.92	1.15	14.71	
Author(s)	substrate	kJ g^{-1} Δ v.s.			
COONEY et al., 1968	glucose in synthetic broth			15.00	
CARLYLE and NORMAN, 1941	oat straw plus $(NH_4)NO_3$				12.5
WILEY and PEARCE, 1955	garbage with high fat content	22–28			
WILEY, 1957	domestic garbage				21.2
HELMER, 1973	domestic garbage				3.8–7.6
MOTE and GRIFFIS, 1982	cotton wool residues cow manure plus rice husks				20–28 27
HAUG, 1980	sewage sludge	23			
HANKE, 1985	oganic matter	11.66			
FINSTEIN et al., 1986	glucose, organic matter	15.60		14.00	
STAHLSCHMIDT, 1987	garbage plus sewage sludge	17.58			
BACH et al., 1987	sewage sludge plus rice husks				13.87
OI et al., 1988	sewage sludge	10.8–13.7			
KRANERT, 1989	sewage sludge plus tree bark	19–28 average 22.5)		12–18	7
HOGAN et al., 1989	rice manufactories waste	14,2–16.7			
MARUGG et al., 1993	grass clippings plus leaves	20–24.5			
MILLER, 1996	sewage sludge plus wood chips	15–22			

[a] Data collected by BIDLINGMAIER (1983) and KROGMANN (1994) from very diverse sources.
[b] Only the data from COONEY et al. (1968) and KRANERT (1989) were experimentally determined, all other data were calculated employing the value given by COONEY et al. (1968) and the amount of oxygen consumed for complete oxidation of the substrate to CO_2 and H_2O.
[c] Data collected by KRANERT (1989, pp. 52).

broth brought about a heat output of 15 kJ g^{-1} O_2. This value was taken to calculate the heat formed for the other substrates employing their empirical formula given in the second column assuming complete oxidation to CO_2 and H_2O. As to be expected, the values obtained this way vary much less than the values of kJ g^{-1} substrate decomposed. Finally, some authors characterized the composting process by recording the heat output in cal h^{-1} (Fig. 2) or W (kJ s^{-1}), and the last column of Tab. 16 presents data of peak heat output during composting, i.e., in the course of the steepest temperature increase. However, while the *temperature curve* of the composting process parallels the *heat output*, it should be stressed that there is no direct relationship between these two features: Rather the temperature curve is the result of the relation between heat formation and heat transfer to the surroundings, the latter being determined by the strength of insulation and the intensity of aeration. An elegant method for the determination of the *rate* of heat production in laboratory composting has been described by MOTE and GRIFFIS (1982).

A complete balance of the heat produced has to consider the various routes the heat takes:

(1) Part of it is used to elevate the temperature of the compost itself, including the composting unit (laboratory apparatus or installation). For the calculation, the mass and heat capacity of the individual components (dry matter, water, air, container wall) have to be taken into account.

(2) Depending on the insulation of the composting material a certain part of the heat is dissipated from the surface to the surroundings. There are three mechanisms of heat transfer involved in this kind of heat loss:
- conduction,
- convection, and
- radiation.

Although formula for the calculation of these three modes of heat transfer have been devised by FOURIER, NEWTON and STEFAN and BOLTZMANN, respectively they have found little application by compost workers (HOGAN, 1989). Instead, the overall heat dissipation rate from the surface (i.e., through the walls) is experimentally determined by following the lowering of the inner temperature with time of a biologically inert load in the absence of air circulation. This method has been employed by KRANERT (1989) and KROGMANN (1994), and the diagram of a curve showing this temperature decrease was given by VIEL et al. (1987). The heat lost by this route could also be considered as "difference" between the total heat produced and the sum of heat consumed for route (1) and (3) which may be estimated more easily.

(3) Finally, heat (together with water) is removed from the compost by increasing the enthalpy of the air used for aeration, often called *evaporative cooling*. This is especially true if the temperature of the pile is regulated by aeration as it is done employing the "Rutgers irategy" (FINSTEIN et al. 1986; FINSTEIN, 1989). The stoichiometry between the parameters describing this mechanism of heat removal is summarized in Tab. 17 (for the calculation of the enthalpy of humid air see any textbook on technical thermodynamics; it can also be derived from the Mollier-h,x-diagram for humid air). Attention may be directed to topics (7) and (8) in Tab. 17 which prove that much more air is needed for evaporative cooling to avoid overheating of the compost and "microbial suicide" than to satisfy the oxygen demand for aerobic degradation.

Pioneering work on the heat balance in composting has been carried out in the USA by FINSTEIN et al. (1986), and readers interested in this field are urged to consult the theses of MILLER (1984) and HOGAN (1989), as well as HOGAN et al. (1989). Much work in this field has been reviewed by MILLER (1996). Of the studies carried out in other countries the following deserve especial attention: BACH et al. (1987) in Japan, VIEL et al. (1987) in France and KRANERT (1989) and KROGMANN (1994) in Germany. The concern of these authors was the differentiation of the various routes of

Tab. 17. Stoichiometry betwenn Heat Production and Air Requirement for Removal of Heat and Water as well as for Meeting the Oxygen Demand o f Microorganisms (Conforming to the reasoning of HANKE, 1985, an STAHLSCHMIDT, 1987)

Example: Using thy hypothetical process of Tab. 15 and Fig. 15 (vs. = volatile solids)

(1) Degradation of organic matter produces 17,500 kJ kg^{-1} v.s.

(2) Evaporation of water at 60 °C needs 2.923 kJ kg^{-1} H_2O [= 1 kJ (0.342 g H_2O).

(3) 1 kg of dry air assuming inlet air 15 °C, 60% relative humidity, enthalpy = 31.31 kJ kg^{-1} dry air, outlet air 60 °C, 100% relative humidity, enthalpy = 464.56 kJ kg^{-1} dry air removes concomitantly heat (433.25 kJ) and water (148.28 g H_2O) (for 1,000 kJ: 2.308 kg dry air).

(4) To remove the heat produced by the degradation of 1 kg v.s. 17.5 · 72.308 = 40.392 kg of dry air are necessary.

(5) 40.392 kg dry air remove 40.392 · 148.28 = 5,989.32 g ~ 6 kg of H_2O, i.e., removal of the heat produced by the degradation of organic matter *alone* by *"evaporative cooling"* needs about 6 kg H_2O kg v.s. degraded.

(6) 6 kg H_2O kg v.s. degraded surmount in most cases the water content present in the input material; if this is the case, water has to be added during the composting process to avoid drying out before stabilization has been achieved.

(7) In the example discussed above (Tab. 15) 181.42 g v.s. were degraded, yielding 181.42 · 17.5 = 3,174.85 kJ. This heat can evaporate 1,085 kg H_2O. Since 0.623 kg H_2O were evaporated (Fig. 15) only 57% of the heat were removed by ventilative cooling, the rest by conduction, convection and radiation. About 5% of the heat were used to heat the compost matrix.

(8) For the removal of 0.623 kg H_2O by ventilative cooling an amount of about 4 kg dry air is required, whereas only 171 g O_2 = 0.763 kg dry air are needed to supply the oxygen for microbial metabolism, i.e., less than a 5th of the air requirement.

heat removal, i.e., the percentage of heat dissipated through the walls and the heat removed by evaporative cooling. In the study of VIEL et al. (1987), heat was drawn off in addition by a water jacket containing circulating water; the heat balance obtained is depicted in Fig. 18. Surprisingly, inspite of the thermal insulation, 75% of the heat generated by the microbial activity are actually dissipated through the walls of the reactor. This finding emphasizes the utmost importance of insulation of laboratory units used for studying the composting process.

The concern of some other studies was the relationship between production of heat and removal of water: VON HIRSCHEYDT and HÄMMERLI (1978) demonstrated the role of metabolic heat formation for the drying of the compost pile. SIKORA et al. (1991) performed a material balance of composting sewage sludge plus wood chips: twice as much heat was produced by volatile solids degradation than was used for the drying process. OI et al. (1988) showed that drying of sewage sludge via "aerobic solid state cultivation" is an economically feasible method to produce compost. RODRIGUEZ LEÓN et al. (1991) presented a rather complex formula for the calculation of air requirements for temperature control in solid state fermentations based on (1) heat transfer to air and surroundings, (2) heat formation as a function of specific growth rate, and (3) growth yield.

The usefulness of their reasoning in the field of compost science needs to be evaluated.

9.3 Heat Recovery from Composting Plants

As pointed out by DIAZ et al. (1986), "the possibility of using the heat generated in composting for space heating in greenhouses or in farm structures has intrigued composters for quite some time". In Tab. 18 five studies are listed, and a comprehensive list of references and personal communications concerning heat recovery from composting plants has been prepared by KRANERT (1988). BAINES et al. (1985) dealt with heat from aerobic treatment of liquid animal wastes.

Tab. 18. Recovery of Heat from Composting Plants

Author(s)	Substratea ΔH_{comb}	Method of Heat Recovery	Efficiency of Heat Recovery within Time
SCHUCHARDT (1983)	horse manure plus straw 19,700 kJ kg^{-1} DS 141,840 kJ (horse · d)$^{-1}$	plate shaped heat exchanger placed above and/or aside the pile	4,042.44 kJ (d · horse)$^{-1}$ =1,122 kWh =2.85% of input energy (8 d)
SCHUCHARDT (1984)	fresh hedge trimmings 18,000 kJ kg^{-1} DS	pipes placed within the pile	2,286 kJ kg^{-1} DS =12% of input energy (399.6 MJ m^{-3}) (8 months)
THOSTRUP (1987)	pig manure plus straw 20,964 kJ kg^{-1} Δ DS (continuous composting)	air–water heat exchanger (double cooling tower)	40% of input energy (1 kW m^{-3} manure) (residence time: 6 d)
VEROUGSTRAETE et al. (1989)	farm animal manure plus straw 18,500 kJ kg^{-1} DS	air–water heat exchanger (double cooling tower)	8,100 kJ kg^{-1} DS =43.4% of input energy (6 d)
KRANERT (1989)	sewage sludge plus tree bark 17,500 kJ kg^{-1} DS	PVC pipes placed within reactor	140–803 kJ kg^{-1} DS =0.8–4.6% of input energy
		from exhaust air	598–2,770 kJ kg^{-1} DS =3.4–15.8% of input energy (20 d)

[a] DS: dry solids input
Δ DS: dry solids degraded

Two methods of heat recovery from composting materials exist – besides the direct use of farm animal manure plus straw for preparing hotbeds for growing plants under glass:

(1) Heat transfer to heat exchangers (plates or pipes) being in direct contact with the composting material, either placed above or aside the pile or installed within the pile or reactor. A disadvantage of the method is the low heat conductivity of the composting material which is almost as low as insulating material, and, in fact, the basis of self-heating of the pile. Further, removing heat by this method leads to a temperature gradient in the close proximity of the exchanger resulting in condensation of water, both events disturbing to a certain degree the composting process. Installation of pipes is cumbersome and renders processing of the compost more difficult.

(2) Heat transfer from the exhaust air to heat exchangers installed in a so-called double cooling tower. This method removes the recovery of heat from the compost pile which remains undisturbed. To produce as much heat as possible the compost is intensely aerated; drying out is prevented by warming and humidifying the fresh air in the lower stage of the cooling tower before entering the reactor. As can be seen

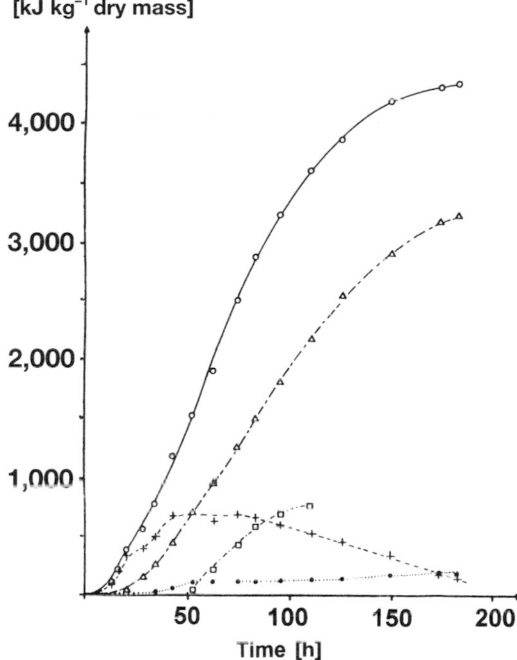

Fig. 18. Thermal balance of a laboratory composting experiment: cumulative energy vs. time plot (from VIEL et al., 1987), ○ total microbial heat generation, △ heat dissipated through walls, + heat used for warming up the compost matrix, □ heat removed through thermal exchanger, ● heat removed by effluent gas ("ventilative cooling").

> from the data in Tab. 18 this method recovers a much higher yield of heat which can be used for heating water for various purposes.

Although heat recovery from composting seems to be attractive, it has not been applied as wide as it may deserve. However, there is competition between heat from (aerobic) composting and (anaerobic) fermentation: Although both processes may yield the same amount of energy in terms of kJ kg^{-1} input material, the aerobic process is only able to supply warm water with a maximum temperature of about 60 °C, which, in addition is hard to store, whereas the anaerobic process yields methane which can produce rather hot water and steam or drive engines.

9.4 Mathematical Modeling of the Composting Process

The stoichiometry of the various factors describing the composting system as discussed above has invited several authors to model mathematically this biotechnological process: FINGER et al. (1976), NAKASAKI et al. (1987), MARUGG et al. (1993), STARK (1995), STROMBAUGH and NOKES (1996), VANDERGHEYNST et al. (1997), and KAISER (1996, 1998). Although the approaches of the various authors vary, they all appreciate more or less the stoichiometry stressed in Sects. 9.1, 9.2, and their results agree fairly well with experimental data. Of course, modeling goes beyond balancing as it includes the *kinetics* of the process which has not been dealt with in this review. As in other fields of biotechnology, modeling of composting will contribute to a more efficient control and regulation of the process. However, this aspect will not be discussed further.

10 Odor Formation and Control

The production of compost appears to lead inevitably to the emission of odors, i.e., odorous volatile substances that are either already present in the waste material (e.g., plant secondary metabolites such as terpenes), produced during composting by microbial activity (and that by aerobic as well as by anaerobic metabolism), or formed by chemical reactions, especially at high temperatures. In fact, odors seem to be the most conspicuous nuisance originating from compost plants and – as in the case of emission of germs – they are often the reason for the objection by the neighborhood of new compost plants to be installed. The formation of odors during composting has been reviewed by MILLER and MACAULEY (1988) and HAUG (1993: Chapt. 16 and 17), and it has been the subject of 4 theses which should be consulted for more detailed information, especially in regard to analytical methods: BONINSEGNI (1974), JAGER (1979), MAYER (1990) and

PÖHLE (1993). Some results of the two latter ones will be discussed in the following paragraphs.

For a long time, odors were semiquantitatively determined by olfactometry, i.e., the human nose served as sensor for these substances. Indeed, the human sense of smelling seems to be the most sensitive means to detect odors and, at the same time, to distinguish between pleasant and unpleasant ones. However, the determination of odors by probatory panels is cumbersome and somewhat subjective; even more important, it can only be carried out sporadically, whereas frequent, or even on-line, measurements would be necessary to obtain exact data. Total organic carbon (TOC) as determined by gas–liquid chromatography (GC), equipped with a flame ionization detector, has been employed to judge the quality of fresh air, but this method proved unsuitable to characterize the load of odorous substances in low, but still recognizable concentrations (MAYER, 1990). An even more sophisticated method makes use of a combination of GC and mass spectroscopy (MS), allowing the identification of individual substances, odorous ones being recognized by a combination of GC and the sniffling-port technique.

MAYER (1990) identified by GC–MS analysis the odorous substances formed in windrows. Altogether, 81 compounds could be identified [Tab. 5.3 in MAYER (1990)]; of which, however, only about 25% exhibited significant odorous activity, among them: diacetyl, acetoine, benzaldehyde, dimethyltrisulfide, myrcene, limonene, cineol, 2-methyl-isoborneol. Some compounds have to be considered as products of anaerobic degradation processes: short chain carboxylic acids, sulfides and mercaptans. In the plant investigated by MAYER (1990) the temperature in the center of the composting mass often increased up to 80–90 °C, leading to a characteristic pungent odor. 16 compounds of the "high temperature odor" could be identified [Tab. 7.5 in MAYER (1990)], among them: 2,4,6-trichloroanisol, 2-isobutyl-3-methoxy-pyrazine, dimethyltrisulfide, isovaleric acid, acetophenone, 2-isopropyl-3-methoxypyracin. The same odor could be reproduced by incubating the compost feed in small scale experiments at 80 °C over 4 weeks: In this case 67 odorous compounds were detected by the sniffling port technique [Tab. 9.1 in MAYER (1990)], 33 of them were identified. The substance primarily responsible for the "high temperature odor" proved to be 3-hydroxy-4,5-dimethyl-2(5H)-furanone (HDMF); its assay, however, required a special analytical method (therefore, it had not been detected in the foregoing analyses). In addition, alkylated pyridines and pyrazines contributed to the obnoxious smell. These compounds (and also

Tab. 19. Appearance of Odors During Composting (according to PÖHLE, 1993 and PÖHLE and KLICHE, 1996)

Phase	Main Components of Odor	Smell
Mesophilic phase	ethanol, propanol, butanol acetaldehyde acetic acid, butyric acid esters of acetic, propionic, butyric acid dimethyl ketone, ethyl methyl kentone limonene, pinene dimethyldisulfide	alcohol-like fruit-like ester-like
Thermophilic phase	dimethyl ketone, ethyl methyl ketone limonene, myrcene, pinene methanethiol, dimethylsulfide, dimethyldisulfide ammonium	*first:* like acetone and fresh hay *later:* moldy to unpleasant musty
Maturation phase	dimethylsulfide, dimethyldisulfide limonene ammonium	pungent musty

some others) are regarded as products of chemical processes, e.g., the Maillard reaction. The formation of this pungent odor took place, however, only if the compost material reached quickly the high temperature phase (≥80 °C) and if an acid reaction persisted, i.e., pH ≤6.5. In contrast, compost feed incubated at 60 °C produced a camphor-like, moldy smell. These differences correspond very well to odors from different parts of the compost pile, i.e., the hot center vs. the cooler surface layer. In the latter, 2-methyl-isoborneol, a metabolite of actinomycetes, was the main component of the mixture of odorous substances.

GC–MS analysis has also been employed by PÖHLE (1993). As reported by PÖHLE and KLICHE (1996) altogether 158 organic compounds were identified in the exhaust air of a pilot plant (organic waste), actually the same classes of compounds were found as by earlier authors (number of species in square brackets): aldehydes [12], alcohols [22], aliphatic hydrocarbons (including chlorinated ones) [29], aromatic hydrocarbons [15], carboxylic acids [4], carboxylic acid esters [31], ketones [21], oxygen containing heterocycles [5], sulfur containing compounds [10], terpenes [9]. A definite correlation existed between the *strength* of the odor as determined by olfactometry and the *quantity* of dimethyldisulfide and limonene, both compounds being present throughout the 3 weeks of composting, albeit not always defining the character of the smell. The occurrence of some other compounds correlated with the phases of the composting process (Tab. 19): Whereas in the early stage volatile alcohols, fatty acids, aldehydes and ketones predominated, the sulfur containing compounds originating from protein decay turned up during later phases. Ammonium was observed only in the cooling down stage. Sulfur containing compounds were also identified by GC–MS by DERIKX et al. (1990b) as the most conspicuous substances causing the nuisance emitted from compost production for mushroom cultivation: H_2S, COS, CH_3SH, $CS_2(CH_3)_2S$, $(CH_3)_2S_2$, $(CH_3)_2S_3$.

The formation of odorous compounds during composting is certainly a very complex process depending on various parameters such as

(1) input material, e.g., domestic garbage, sewage sludge, animal wastes, leaves and grass clippings; especially their chemical composition is of relevance, i.e., the content of carbohydrates, fats and proteins;
(2) process management, e.g., aeration, turning, control of moisture and temperature;
(3) composting stage, see Tab. 19.

There have been also attempts to relate odor emission to certain representatives of the microbial population (PÖHLE et al., 1993): The odorous compounds of the first phase are regarded as products of the fermentative metabolism of facultative anaerobes dwelling in oxygen deprived niches of the system; ketones may be formed by oxidation of secondary alcohols of *Pseudomonas* sp. or anaerobically by *Clostridium* sp. (not determined). Organic sulfur compounds originate from the degradation of proteins, more specifically of the sulfur containing amino acids methionine, cysteine and cystine. Protein brakedown by proteolytic bacilli appears to take place mainly in the thermophilic and maturation phase, as also exemplified by the appearance of ammonium. This study has been extended by PÖHLE et al. (1996) to odor formation by pure cultures of 6 bacterial and five fungal species growing on a synthetic organic waste: Rather complex spectra of volatile substances were found, *Aspergillus flavus* specifically produced furane, *Aspergillus fumigatus* led to a significant release of terpenes and yielded branched ketones. Formation of volatile compounds by fungi growing on wheat of 18.2% moisture content has been investigated by SINHA et al. (1988): Species of *Alternaria, Aspergillus* and *Penicillium* produced 3-methyl-1-butanol, 1-octen-3-ol and 3-octanone. The first 2 of the 3 substances were also found by FISCHER et al. (1999b) in a wide range of fungi, whereas some of the 109 identified volatile organic compounds (GC–MS) were rather species-specific and may thus serve as marker compounds for the selective detection of fungal species in indoor domestic and working environments.

The contribution to odor development of mesophilic streptomycetes which have their climax during the maturation phase consists in the formation of some volatile compounds identified by GERBER (1977): geosmin, 2-methyl-isoborneol and 2-isopropyl-3-methoxypyrazine. These substances possess respectively, an earthy, camphor-like, or musty smell, and are probably less responsible for the obnoxious odor emitted by compost plants; they may give, however, water and fish undesirable odors and tastes.

Since odor formation from composting installations (windrows, boxes, drums) appears to be unavoidable, efforts have been rather directed toward the control of its emission to protect the environment from this nuisance (KUCHTA and JAGER, 1993; WAR 81, 1994; DAMMANN et al., 1996). Thus the exhaust air of enclosed compost plants is passed through biofilters or – less frequently – biowashers. Open compost plants, however, keep this problem and can, therefore, be operated only in a safe distance (depending on the major wind direction and the topography) from residential areas.

11 Compost Maturity

According to the definition of composting given in Sect. 1, the process should lead to a product which meets the following requirement: *sufficiently stable for storage and application to land without adverse environmental effects*. This property necessitates that those fractions of organic matter which are easily degraded, i.e., some carbohydrates, proteins and lipids, have been mineralized more or less completely, and only materials with a low turnover, i.e., lignocellulose, humic substances, microbial cells, have been left. Thus, even after rewetting the compost (which should be rather dry at the end of the process) there will be no extensive microbial activity leading to self-heating of the material (under aerobic conditions) or – worse – to fermentation with the formation of putrescible odors (under anaerobic conditions).

The determination of whether or not the compost has reached this state of maturity is of importance in regard to

(1) the duration and termination of the process which effects the requirement of plant size, and
(2) the safe application as soil fertilizer.

Therefore, various tests have been worked out to evaluate compost ripeness by biological and chemical properties: Tab. 20. The methods employed have been described in numerous reviews, e.g., MOREL et al. (1985), JIMÉNEZ and GARCIA (1989), MATHUR et al. (1993) and BARBERIS and NAPPI (1996). The most comprehensive information can be gained from the theses by JOURDAN (1988) and BECKER (1997).

Although numerous and very diverse methods have been advocated for the purpose under consideration, it appears that after all the time-honored self-heating test employing Dewar flasks is the most suitable and reliable method to determine the maturity of compost (BRINTON et al., 1995). Alternatively, the uptake of oxygen has been widely used (JANOTTI et al., 1993; PALETSKI and YOUNG, 1995). The anaerobic counterpart, i.e., a test on the fermentation capacity of compost, has been suggested by JANN et al. (1959), but has found little application. Other workers in this field report on the usefulness of enzyme activities: GODDEN and PENNINCKX (1986) found a significant increase of alkaline phosphatase towards the end of composting, whereas invertase activity proved to be less sensitive. HERRMANN and SHANN (1993) found cellulase activity to be a good indicator of stability, and lipase activity to be a good indicator of compost maturity. The nearly complete degradation of cellulose can be tested for by the *Chaetomium* test (OBRIST, 1965): This fungus does not form aerial mycelium and fruiting bodies on agar medium containing mature compost as the only nutrient source. Since the formation of humic acids is a characteristic of the progressive ripening process, this chemical class of compounds has been favored as a measure of compost maturity by many workers in this field, e.g., GRUNDMANN (1991), CHEN and INBAR (1993), SENESI and BRUNETTI (1996) and CHEN et al. (1996). Finally, the earthy smelling substances produced by actinomycetes during the last phase (see Sect. 10) may serve as indicators of the completion of the composting process.

Tab. 20. Methods for the Determination of Compost Maturity

Desired Property	Test
Weak self-heating capacity	self-heating in Dewar-flasks
Weak respiratory capacity	O_2 uptake and/or CO_2 formation
Weak fermentative capacity	production of organic acids under anaerobic conditions (pH <7)
Cellulose nearly degraded	limited growth of *Chaetomium*, no formation of fruiting bodies
C–N ratio $\leq 10:1$	chemical analysis
NO_3-N >> NH_4^+-/NH_3-N	chemical analysis
Newly formed humic substances	chemical analysis
Plant compatibility (no phytotoxicity)	growth of cress or lettuce seedlings

12 References

ALVAREZ, A. J., BÜTTNER, M. P., STETZENBECK, L. D. (1995), PCR for bioaerosol monitoring: sensitivity and environmental interference, *Appl. Environ. Microbiol.* **61**, 3639–3644.

AMNER, W., MCCARTHY, A. J., EDWARDS, C. A. (1988), Quantitative assessment of factors affecting the recovery of indigenous and released thermophilic bacteria from compost, *Appl. Environ. Microbiol.* **54**, 3107–3112.

Arbeitsgemeinschaft Universität GH Kassel/Plan Co Tec/Universität Göttingen (1997), Phytohygiene der Bioabfallkompostierung. F&E-Vorhaben, *Abschlußbericht*, pp. 1–319, Osnabrück: Deutsche Bundesstiftung Umwelt.

ASHBOLT, N. J., LINE, M. A. (1982), A bench-scale system to study the composting of organic wastes, *J. Environ. Qual.* **11**, 405–408.

ATHALYE, M., LACEY, J., GOODFELLOW, M. (1981), Selective isolation and enumeration of actinomycetes using rifampicin, *J. Appl. Bacteriol.* **51**, 289–297.

ATKINSON, C. F., JONES, D. D., GAUTHIER, J. J. (1996), Putative anaerobic activity in aerated composts, *J. Ind. Microbiol.* **16**, 182–188.

AYRES, P. G., BODDY, L. (Eds.) (1986), *Water, Fungi, and Plants.* London: Cambridge University Press.

BACH, P. D., SHODA, M., KUBOTA, H. (1984), Rate of composting of dewatered sewage sludge in continuously mixed isothermal reactor, *J. Ferment. Technol.* **62**, 285–292.

BACH, P. D., SHODA, M., KUBOTA, H. (1985), Composting reaction rate of sewage sludge in an autothermal packed bed reactor, *J. Ferment. Technol.* **63**, 271–278.

BACH, P. D., NAKASAKI, K., SHODA, M., KUBOTA, H. (1987), Thermal balance in composting operations, *J. Ferment. Technol.* **65**, 199–209.

BÅGSTAM, G. (1978), Population changes in microorganisms during composting of spruce bark. I. Influence of temperature control, *Eur. J. Appl. Microbiol. Biotechnol.* **5**, 315–330.

BÅGSTAM, G. (1979), Population changes in microorganisms during composting of spruce bark. II. Mesophilic and thermophilic microorganisms during controlled composting, *Eur. J. Appl. Microbiol. Biotechnol.* **6**, 279–288.

BÅGSTAM, G., SVENSONS, H. (1976), Experiments made in bench scale composters. II: Composting of spruce bark, *Vatten* **1**, 45–53.

BÅGSTAM, G., EVEBO, L., LINDELL, T., SWENSSON, H. (1974), Experiments made in bench scale composters. I. Apparatus, *Vatten* **4**, 358–363.

BAILEY, J. E., OLLIS, D. F. (1977), *Biochemical Engineering Fundamentals.* Tokyo: McGraw-Hill, Kogakusha, Ltd.

BAINES, S., SVOBODA, I. F., EVANS, M. R. (1985), Heat from aerobic treatment of liquid animal wastes, in: *Composting of Agricultural and other Wastes* (GASSER, J. K. R., Ed.), pp. 147–166. London, New York: Elsevier Applied Science Publishers.

BALOWS, A., TRÜPER, H. G., DWORKIN, M., HARDER, W., SCHLEIFER, K.-H. (Eds.) (1992), *The Prokaryotes: A Handbook on the Biology of Bacteria.* Vols. I–IV., 2nd Edn. New York: Springer-Verlag.

BARBERIS, R., NAPPI, P. (1996), Evaluation of compost stability, in: *Science of Composting* (DE BERTOLDI, M., SEQUI, P., LEMMES, B., PAPI, T., Eds.), pp. 175–184. London: Blackie Academic & Professional.

BARDTKE, D. (1961), Die Vorgänge bei der Kompostierung, *Forum Städte-Hygiene* **7**, 135–137.

BATTLEY, E. H. (1987), *Energetics of Microbial Growth.* New York: John Wiley & Sons.

BECKER, G. (1997), Der Rottegrad als Gewährleistungskriterium für Kompostierungsanlagen. *Thesis*, Universität – GHS Essen, Germany.

BEFFA, T., BLANC, M., LYON, P.-F., VOGT, G., MARCHIANI, M. et al. (1996), Isolation of *Thermus* strains from hot composts (60 to 80 °C), *Appl. Environ. Microbiol.* **62**, 1723–1727.

BIDDLESTONE, A. J., GRAY, K. R., DAY, C. A. (1987), Composting and straw decomposition, in: *Envi-*

ronmental Biotechnology (FORSTER, C. F., WASE, D. A. J., Eds.), pp. 135–175. Chichester: Ellis Horwood.

BIDLINGMAIER, W. (1983), Das Wesen der Kompostierung von Siedlungsabfällen, in: *Müllhandbuch*, Kennz. 5305, pp. 1–23. Berlin: Erich Schmidt Verlag.

BOLLEN, G. J. (1993), Factors involved in inactivation of plant pathogens during composting of crop residues, in: *Science and Engineering of Composting* (HOITINK, H. A. J., KEENER, H. M., Eds.), pp. 301–318. Worthington, OH: Renaissance Publications.

BOLLEN, G. J., VOLKER, D. (1996), Phytohygienic aspects of composting, in: *The Sciene of Composting* (DE BERTOLDI, M., SEQUI, P., LEMMES, B., PAPI, T. Eds.), pp. 233–246. London: Blackie Academic & Professional.

BONINSEGNI, C. J.-C. (1974), Analyse von Gerüchen bei der Kehrichtkompostierung, *Thesis*, ETH Zürich, Switzerland.

BOUTIN, P., TORRE, M., MOLINE, J. (1987), Bacterial and fungal atmospheric contamination at refuse composting plants: a preliminary study, in: *Compost: Production, Quality and Use* (DE BERTOLDI, M., FERRANTI, M.-P., HERMITE, P. L., Eds.), pp. 265– 275. Amsterdam: Elsevier Applied Science.

BOWES, P. C. (1984), *Self-Heating: Evaluating and Controlling the Hazards.* London: Her Majesty's Stationary Office (HMSO).

BRETON, A., ZWAENEPOEL, P. (1991), Succession of moist hay mycoflora during storage, *Can. J. Microbiol.* **37**, 248–251.

BRINTON, W. F., EVANS, E., DROFFNER, M. L., BRINTON, R. (1995), Standardized test for evaluation of compost self-heating, *BioCycle* **36** (11), 64–69.

BURGE, W. D., COLACICCO, D., CRAMER, W. N. (1981), Criteria for achieving pathogen destruction during composting, *J. Water Poll. Control Fed.* **53**, 1683–1690.

CARLYLE, R. E., NORMAN, A. G. (1941), Microbial thermogenesis in the decomposition of plant materials. II. Factors involved, *J. Bacteriol.* **41**, 699–724.

CARROLL, G. C., WICKLOW, D. T. (Eds.) (1992), *The Fungal Community: Its Organization and the Role in the Ecosystem*, 2nd Edn. New York: Marcel Dekker.

CHANDLER, J. A., JEWELL, W. J., GOSSETT, J. M., VAN SOEST, P. J., ROBERTSON, J. B. (1980), Predicting methane fermentation biogradability, *Biotechnol. Bioeng. Symp.* **10**, 93–107.

CHANG, Y. (1967), The fungi of wheat straw compost. II. Biochemical and physiological studies, *Trans. Br. Mycol. Soc.* **50**, 667–677.

CHANG, J. C., HUDSON, H. J. (1967), The fungi of wheat straw compost. I. Ecological studies, *Trans. Br. Mycol. Soc.* **50**, 649–666.

CHARACKLIS, W. G. (1990), Energetics and stoichiometry, in: *Biofilms* (CHARACKLIS, W. G., MARSHALL, K. C., Eds.), pp. 161–192. New York: John Wiley & Sons.

CHEN, Y., INBAR, Y. (1993), Chemical and spectroscopical analyses of organic matter transformations during composting in relation to compost maturity, in: *Science and Engineering of Composting* (HOITINK, H. A. J., KEENER, H. M., Eds.), pp. 551–600. Worthington, OH: Renaissance Publications.

CHEN, Y., CHEFETZ, B., HADER, Y. (1996), Formation and properties of humic substance originating from composts, in: *The Science of Composting* (DE BERTOLDI, M., SEQUI, P., LEMMES, B., PAPI, T., Eds.), pp. 382–393. London: Blackie Academic & Professional.

CLARK, C. S., BUCKINGHAM, C. O., BONE, D. H., LARK, R. H. (1977), Laboratory scale composting: Techniques, *J. Environ. Engin. Div., ASCE* **103**, (No. EE5) 893–906.

CLARK, C. S., BUCKINGHAM, C. O., CHARBONNEAU, R., CLARK, R. H. (1978), Laboratory scale composting: Studies, *J. Environ. Engin. Div., ASCE* **104**, (No. EE1) 47–59.

CLARK, C. S., RYLANDER, R., LARSSON, L. (1983), Levels of gram-negative bacteria, *Aspergillus fumigatus*, dust and endotoxin at compost plants, *Appl. Environ. Microbiol.* **45**, 1501–1505.

CLARK, C. S., BJORNSON, H. S., SCHWARTZ-FULTON, H. J. W., HOLLAND, J. W., GARTSIDE, P. S. (1984), Biological health risks associated with the composting of wastewater treatment plant sludge, *J. Water Poll. Contr. Fed.* **56**, 1269–1276.

COCHRAN, V. L., ELLIOTT, L. F., PAPENDICK, R. I. (1977), The production of phytotoxins from surface crop residues, *Soil Sci. Soc. Am. J.* **41**, 903–908.

COONEY, D. G., EMERSON, R. (1964), *Thermophilic Fungi: An Account of their Biology, Activity and Classification.* San Francisco, CA: W. H. Freeman & Co.

COONEY, C. L., WANG, D. I. C., MATELES, R. I. (1968), Measurement of heat evolution and correlation with oxygen consumption during microbial growth, *Biotechnol. Bioeng.* **9**, 269–281.

CRAFT, C. M., NELSON, E. B. (1996), Microbial properties of composts that suppress damping-off and root rot of creeping bentgrass caused by *Phythium graminicola*, *Appl. Environ. Microbiol.* **62**, 1550–1557.

CRAWFORD, D. L. (1988), Biodegradation of agricultural and urban wastes, in: *Actinomycetes in Biotechnology* (GOODFELLOW, M., WILLIAMS, S. T., MORDARSKI, M., Eds.), pp. 433–499. San Diego, CA: Academic Press.

CRISAN, E. V. (1973), Current concepts of thermophilism and the thermophilic fungi, *Mycologia* **65**, 1171–1198.

CROOK, B., BARDOS, R. P., LACEY, J. (1988), Domestic waste composting plants as sources of airborne microorganisms, in: *Aerosols: Their Generation, Behavior and Application* (GRIFFITH, W. D., Ed.), pp. 63–68. London: Aerosol Society.

CROSS, T. (1968), Thermophilic actinomycetes, *J. Appl. Bacteriol.* **31**, 36–53.

CRUEGER, W., CRUEGER, A. (1984), *Biotechnologie –*

Lehrbuch der angewandten Mikrobiologie, 2nd Edn. München: Oldenbourg.

CURRIE, J. A., FESTENSTEIN, G. N. (1971), Factors defining spontaneous heating and ignition of hay, *J. Sci. Food Agric.* **22**, 223–230.

DAMMANN, B., WIESE, B., HEINIG, K., STEGMANN, R. (1996), Weitergehende Elimination von Gerüchen aus Kompostwerken, in: *Neue Techniken der Kompostierung* (STEGMANN, R., Ed.) *Hamburger Berichte 11*, pp. 459–476. Bonn: Economica Verlag.

DE BERTOLDI, M., VALLINI, G., PERA, A. (1983), The biology of composting: A review, *Waste Managem. Res.* **1**, 157–176.

DE BERTOLDI, M., ZUCCONI, F., CIVILINI, M. (1988), Temperature, pathogen control and product quality, *BioCycle* **29**, 43–50.

DE BERTOLDI, M., SEQUI, P., LEMMES, B., PAPI, T. (Eds.) (1996), *The Science of Composting*, Vol. **1**, pp. 1–672; Vol. **2**, pp. 673–1405. London: Blackie Academic & Professional.

DEPLOEY, J. J., FERGUS, C. L. (1975), Growth and sporulation of thermophilic fungi and actinomycetes in O_2–N_2 atmospheres, *Mycologia* **67**, 780–797.

DERIKX, P. J. L., DE JONG, G. A. H., OP DEN CAMP, H. J. M., DRIFT VAN DER, C., VAN GRIENSVEN, L. J. L. D., VOGELS, G. D. (1989), Isolation and characterization of thermophilic methanogenic bacteria from mushroom compost, *FEMS Microbiol. Ecol.* **62**, 251–258.

DERIKX, P. J. L., OP DEN CAMP, H. J. M., DRIFT VAN DER, C., GRIENSVEN, L. J. L. D., VOGELS, G. D. (1990a), Biomass and biological activity during the production of compost used as a substrate in mushroom cultivation, *Environ. Microbiol.* **56**, 3029–2034.

DERIKX, P. J. L., OP DEN CAMP, H. J. M., DRIFT VAN DER, C., GRIENSVEN, L. J. L. D., VOGELS, G. D. (1990b), Odorous sulfur compounds emitted during production of compost used as a substrate in mushroom cultivation, *Appl. Environ. Microbiol.* **56**, 176–180.

DESAI, A. J., DHALA, S. A. (1967), Bacteriolysis by thermophilic actinomycetes, *Antonie van Leeuwenhoek* **33**, 56–62.

DIAZ, L. F., GOLUEKE, C. G., SAVAGE, G. M. (1986), Energetics of compost production and utilization, *BioCycle* **27**, 49–54.

DIEKERT, G. (1997), Grundmechanismen des Stoffwechsels und der Energiegewinnung, in: *Umweltbiotechnologie* (OTTOW, J. C. G., BRIDLINGMAIER, W., Eds.), pp. 1–38. Stuttgart: Gustav Fischer.

DIX, N. J., WEBSTER, J. (1995), *Fungal Ecology*, pp. 1–499. London: Chapman & Hall.

DROFFNER, M. L., BRINTON, W. F. (1995), Survival of *E. coli* and *Salmonella* populations in aerobic thermophilic composts as measured with DNA gene probes, *Zbl. Hyg.* **197**, 387–397.

DROFFNER, M. L., YAMAMOTO, N. (1991), Prolonged environmental stress via a two step process selects mutants of *Escherichia, Salmonella* and *Pseudomonas* that grow at 54 °C. *Arch. Microbiol.* **156**, 307–311.

EDUARD, W., LACEY, J., KARLSSON, K., PALMGREN, U., STRÖM, G., BLOMQUIST, G. (1990), Evaluation of methods for enumerating microorganisms in filter samples from highly contaminated occupational environments, *Am. Ind. Hyg. Ass. J.* **51**, 427–436.

EMERSON, R. (1968), Thermophiles, in: *The Fungi, An Advanced Treatise*, Vol. III. *The Fungal Population* (AINSWORTH, G. C., SÜSSMAN, A. S., Eds.), pp. 105–128. New York, London: Academic Press.

ENSIGN, J. C. (1992), Introduction to the actinomycetes, in: *The Prokaryotes, A Handbook on the Biology of Bacteria.* 2nd Edn. (BALOWS, A., TRÜPER, H. G., DWORKIN, M., HARDER, W., SCHLEIFER, K. H., Eds.), pp. 811–815. New York: Springer-Verlag

EPSTEIN, E., EPSTEIN, J. I. (1989), Public health issues and composting, *BioCycle* **30**, 50–53.

FARRELL, J. B. (1993), Fecal pathogen control during composting, in: *Science and Engineering of Composting* (HOITINK, H. A. J., KEENER, H. M., Eds.), pp. 282–300. Worthington, OH: Renaissance Publications.

FERGUS, C. L. (1964), Thermophilic and thermotolerant molds and actinomycetes of mushroom compost during peak heating, *Mycologia* **56**, 267–284.

FERGUS, C. L. (1969), The cellulolytic activity of thermophilic fungi and actinomycetes, *Mycologia* **61**, 120–129.

FERGUS, C. L. (1971), The temperature relationships and thermal resistance of a new thermophilic *Papulaspora* from mushroom, *Mycologia* **63**, 426–431.

FERGUS, C. L., SINDEN, J. W. (1969), A new thermophilic fungus from mushroom compost: *Thielavia thermophila* spec. nov.: *Can. J. Biol.* **47**, 1635–1637.

FERMOR, T. R., SMITH, J. F., SPENCER, D. M. (1979), The microflora of experimental mushroom composts, *J. Hortic. Sci.* **54**, 137–147.

FERTIG, J. (1981), Untersuchungen von Wechselwirkungen zwischen Belüftung, Wärmebildung, Sauerstoffverbrauch, Kohlendioxidbildung und Abbau der organischen Substanz bei der Kompostierung von Siedlungsabfällen, *Thesis*, Universität Gießen, Germany.

FESTENSTEIN, G. N., LACEY, J., SKINNER, F. A., JENKINS, P. A., PEPYS, J. (1965), Self-heating of hay and grain in Dewar flasks and the development of farmer's lung antigens, *J. Gen. Microbiol.* **41**, 389–407.

FINGER, S. H., HATCH, R. T., REGAN, T. M. (1976), Aerobic microbial growth in semisolid matrices: Heat and mass transfer limitations, *Biotechnol. Bioeng.* **18**, 1193–1218.

FINSTEIN, M. S. (1989), Composting solid waste: Costly mismanagement of a microbial ecosystem, *ASM News* **55**, 599–602.

FINSTEIN, M. S. (1992), Composting in the context of municipal solid waste management, in: *Environmental Microbiology* (MICHELL, R., Ed.), pp.

355–374. New York, Brisbane, Toronto, Singapore: Wiley Liss.

FINSTEIN, M. S., MORRIS, M. L. (1975), Microbiology of municipal solid waste composting, *Adv. Appl. Microbiol.* **19**, 113–151.

FINSTEIN, M. S., MILLER, F. C., STROM, P. F. (1986), Waste treatment composting as a controlled system, in: *Biotechnology*, Vol. **8**, 1st Edn. (REHM, H. J., REED, G., Eds.), pp. 363–398. Weinheim: VCH.

FISCHER, G., SCHWALBE, R., OSTROWSKI, R., DOTT, W. (1998), Airborne fungi and their secondary metabolites in working places in a compost facility, *Mycoses* **41**, 383–388.

FISCHER, G., MÜLLER, T. OSTROWSKI, R., DOTT, W. (1999a), Mycotoxins of *Aspergillus fumigatus* in pure culture and in bioaerosols from compost facilities, *Chemosphere* **38**, 1745–1755.

FISCHER, G., SCHWALBE, R., MÖLLER, M., OSTROWSKI, R., DOTT, W. (1999b), Species-specific production of microbial volatile organic compounds (MVOC) by airborne fungi from a compost facility, *Chemosphere* **39**, 795–810.

FRANKLAND, J. C. (1992), Mechanisms in fungal succession, in: *The Fungal Community: Its Organization and Role in the Ecosystem*, 2nd Edn. (CAROLL, G. C., WICKLOW, D. T., Eds.), pp. 383–401. New York, Basel, Hongkong: Marcel Dekker.

FRANKLAND, J. C., HEDGER, J. N., SWIFT, M. J. (Eds.) (1982), *Decomposer Basidiomycetes: Their Biology and Ecology* (*Brit. Mycol. Soc. Symp.*, March 1979), Cambridge, London, New York: Cambridge University Press.

FUIJO, Y., KUME, S. J. (1991), Isolation and identification of thermophilic bacteria from sewage sludge compost, *J. Ferment. Bioeng.* **72**, 334–337.

GABY, N. S., CREEK, L. L., GABY, W. L. (1972), A study of the bacterial ecology of composting and the use of *Proteus* as an indicator organism of solid waste, *Dev. Ind. Microbiol.* **13**, 24–29.

GANGWAR, M., KHAN, Z. U., RANDHAWA, H. S., LACEY, J. (1989), Distribution of clinically important thermophilic actinomycetes in vegetable substrates and soil in north-western India, *Antonie van Leeuwenhoek* **56**, 201–209.

GARRET, S. D. (1951), Ecological groups of soil fungi: A survey of substrate relationships, *New Phytol.* **50**, 149–166.

GASSER, J. K. R. (Ed.) (1985), Composting of Agricultural and other Wastes, *Proc. Seminar*, Brasenose College, Oxford, UK, 19.–20. March, 1984, pp. 1–320. London, New York: Elsevier Applied Science Publisher.

GERBER, N. (1977) Three highly odorous metabolites from an actinomycete: 2-isopropyl-3-methoxy-pyrazine, methylisoborneol, and geosmin, *J. Chem. Ecol.* **3**, 475–482.

GLATHE, H. (1959), Die Selbsterhitzungsvorgänge in der Natur, *Zbl. Bakteriol. II* **113**, 18–31.

GLATHE, H. (1960), Die mikrobiologische Analyse im Dienste der Brandursachenermittlung, *Kriminalistik* **14**, 121–123.

GODDEN, B., PENNINCKX, M. (1984), Identification and evolution of the cellulolytic microflora present during composting of cattle manure: on the role of Actinomycetes sp., *Ann. Microbiol.* (Inst. Pasteur) **135 B**, 69–78.

GODDEN, B., PENNINCKX, M. (1986), On the use of biological and chemical indexes for determing agricultural compost maturity: extension to the field scale, *Agric. Wastes* **15**, 169–178.

GODDEN, B., PENNINCKX, M. PIERARD, A., LANNOYE, R. (1983), Evolution of enzyme activities and microbial populations during composting of cattle manure, *Eur. J. Appl. Microbiol. Biotechnol.* **17**, 306–310.

GÖTTLICH, E. (1996), Untersuchungen zur Pilzbelastung der Luft an Arbeitsplätzen in Betrieben zur Abfallbehandlung, *Thesis*, TU Stuttgart, Germany. (Stuttgarter Ber. Abfallwirtsch., **63**. Bielefeld: Erich Schmidt Verlag).

GOODFELLOW, M. (1989), Suprageneric classification of actinomycetes, in: *Bergey's Manual of Systematic Bacteriology*, Vol. **4**. (WILLIAMS, S. T., SHARPE, M. E., HOLT, J. G., Eds.), pp. 2333–2339. Baltimore, MA: Williams and Wilkins.

GOODFELLOW, M., LACEY, J., TODD, C. (1987), Numerical classification of thermophilic streptomycetes, *J. Gen. Microbiol.* **133**, 3135–3149.

GRANT, W. D., WEST, A. W. (1986), Measurement of ergosterol, diaminopimelic acid and glucosamine in soil: evaluation as indicators of microbial biomass, *J. Microbiol. Meth.* **6**, 47–53.

GREGORY, P. H., LACEY, M. E. (1963), Mycological examination of dust from mouldy hay associated with farmer's lung disease, *J. Gen. Microbiol.* **30**, 75–88.

GREINER-MAI, E., KROPPENSTEDT, R. M., KORN-WENDISCH, F., KUTZNER, H. J. (1987), Morphological and biochemical characterization and emended description of thermophilic actinomycetes species, *Syst. Appl. Microbiol.* **9**, 97–109.

GREINER-MAI, E., KORN-WENDISCH, F., KUTZNER, H. J. (1988), Taxonomic revision of the genus *Saccharomonospora* and description of *Saccharomonospora glauca* sp. nov., *Int. J. Syst. Bacteriol.* **38**, 398–405.

GRIFFIN, D. M. (1977), Water potential and wood-decay, *Ann. Rev. Phytopathol.* **15**, 319–329.

GRUNDMANN, J. (1991), Reifegradbestimmung von Komposten durch Huminstoffanalytik: Eignung und Methode, *Müll Abfall* **22**, 268–273.

GSTRAUNTHALER, G. (1983), Pilzsukzessionen während des Rotteverlaufs von Siedlungsabfällen, *Thesis*, Universität Innsbruck, Austria.

HÄGERDAL, B. G. R., FERCHAK, J. D., PYE, E. K. (1978), Cellulolytic enzyme system of *Thermoactinomyces* sp. grown on microcrystalline cellulose, *Appl. Environ. Microbiol.* **36**, 606–612.

HAIDER, K. (1991), Problems related to the humification processes in soils of temperate climates, *Soil Biochem.* **7**, 55–94.

HANKE, R. (1985), Grundlagen der Kompostierung – Forschungsarbeiten der VOEST-ALPINE AG., in: *Kompostierung von Abfällen 2* (THOMÉ-KOZ-

MIENSKY, Ed.), pp. 24–45. Berlin: EF-Verlag.

HARPER, E., MILLER, F. C., MACAULEY, B. J. (1992), Physical management and interpretation of an environmentally controlled composting ecosystem, *Austr. J. Exp. Agric.* **32**, 657–667.

HAUG, R. T. (1979), Engineering principles of sludge composting, *J. Water Poll. Control Fed.* **51**, 2189–2206.

HAUG, R. T. (1980), *Compost Engineering – Principles and Practice*, pp. 1–655. Ann Arbor, MI: Ann Arbor Science.

HAUG, R. T. (1993), *The Practical Handbook of Compost Engineering*, pp. 1–717. Boca Raton, FL: Lewis Publishers, CRC Press.

HAYES, W. A. (1968), Microbiological changes in composting straw/horse manure mixtures, *Mushroom Sci.* **7**, 173–186.

HEDGER, J. N., BASUKI, T. (1982), The role of basidiomycetes in composts: a model system for decomposition studies, in: *Decomposer Basidiomycetes: Their Biology and Ecology.* (FRANKLAND, J. C., HEDGER, J. N., SWIFT, M. J., Eds.), pp. 263–305 (*Brit. Mycol. Soc. Symp.*, March 1979). Cambridge: Cambridge University Press.

HELLMANN, B., ZELLES, L., PALOJÄRVI, A., BAI Q. (1997) Emission of climate-relevant trace gases and succession of microbial communities during open-windrow composting, *Appl. Environ. Microbiol.* **63**, 1011–1018.

HELMER, R. (1973), Thermische Vorgänge und Gasumsetzungen bei der Müllkompostierung, Versuche mit Modell-Rottezellen, *Müll Abfall Abwasser* **23**, 2–7.

HENSSEN, A. (1957a), Beiträge zur Morphologie und Systematik der thermophilen Actinomyceten, *Arch. Mikrobiol.* **26**, 373–414.

HENSSEN, A. (1957b), Über die Bedeutung der thermophilen Mikroorganismen für die Zersetzung des Stallmistes, *Arch. Mikrobiol.* **27**, 63–81.

HERRMANN, R. F., SHANN, J. R. (1993), Enzyme activities as indicators of municipal solid waste compost maturity, *Compost Sci. Util.* **1**, 54–63.

HIRSCHEYDT, A., VON, HÄMMERLI, H. (1978), Über den Zusammenhang zwischen Temperaturentwicklung und Wasserverlust während der ersten Rottephase in der Müllkompostierung, *Wasser Boden* **30**, 31–32.

HOGAN, J. A. (1989), Physical modeling of the composting microbial ecosystem, *Thesis*, Rutgers University, New Brunswick, N. J., USA.

HOGAN, J. A., MILLER, F. C., FINSTEIN, M. S. (1989), Physical modeling of the composting ecosystem, *Appl. Environ. Microbiol.* **55**, 1082–1092.

HOITINK, H. A. J. (1990), Production of disease suppressive compost and container media, and microorganism culture for use therein, *U.S. Patent* 4960348 (Feb. 13, 1990).

HOITINK, H. A. J., GREBUS, M. E. (1994), Status of biological control of plant diseases with composts, *Compost Sci. Util.* **2**, 6–12.

HOITINK, H. A. J., KEENER, H. M. (Eds.) (1993), Science and Engineering of Composting; *Proc. Int. Composting Res. Symp.*, Columbus, OH: 27.–29. March 1992, pp. 1–728. Worthington, OH: Renaissance Publications.

HOITINK, H. A. J., STONE, A. G., GREBUS, M. E. (1996), Suppression of plant diseases by composts, in: *The Science of Composting* (DE BERTOLDI, M., SEQUI, P., LEMMES, B., PAPI, T., Eds.), pp. 373–381. London: Blackie Academic & Professional.

HOPPE-SEYLER, F. (1889), Über Huminsubstanzen, ihre Entstehung und ihre Eigenschaften, *Hoppe-Seyler Z. Physiol. Chem.* **13**, 66–121.

HOWARD, A. (1948), *Mein landwirtschaftliches Testament, Das Indore-Verfahren zur Kompostherstellung*, pp. 48–58. Berlin: Siebeneicher Verlag.

HUDSON, H. J. (1968), The ecology of fungi on plant remains above the soil, *New Phytol.* **67**, 837–874.

HUGHES, M. S., STEELE, T. W. (1994), Occurrence and distribution of *Legionella* species in composted plant materials, *Appl. Environ. Microbiol.* **60**, 2003–2005.

HUSSAIN, H. M. (1972), Biologische Selbsterhitzung von Heu und ihre Weiterentwicklung bis zur Selbstentzündung, *Grundl. Kriminalistik* **8**, 359–375.

HUSSONG, D., BURGE, W. D., ENKIRI, N. K. (1985), Occurrence, growth and suppression of *Salmonellae* in composted sewage sludge, *Appl. Environ. Microbiol.* **50**, 887–893.

INGRAHAM, J. L. (1962), Temperature relationships, in: *The Bacteria, Vol. IV: The Physiology of Growth* (GUNSALUS, I. C., STANIER, R. Y., Eds.), pp. 265–296. New York, London: Academic Press.

JACKSON, D. V., MERILLOT, J. M., HERMITE, P. L. (Eds.) (1992), *Composting and Compost Quality Assurance Criteria*, pp. 1–433. Luxemburg, Brüssels, Office for Official Publications of the European Communities.

JAGER, J. (1979), Zur chemischen Ökologie der biologischen Abfallbeseitigung, *Thesis*, Universität Heidelberg, Germany.

JAMES, L. H. (1927), Studies in microbial thermogenesis. I. Apparatus, *Science* **65**, 504–506.

JAMES, L. H., BIDWELL, G. L., MCKINNEY, R. S. (1928), An observed case of spontaneous ignition in stable manure, *J. Agric. Res.* **36**, 481–485.

JANN, G., HOWARD, D. H., SALLE, A. J. (1959), Method for determination of completion of composting, *Appl. Microbiol.* **7**, 271–275.

JANOTTI, D. A., PANG, T., TOTH, B. L., ELWELL, D. L., KEENER, H. M., HOITINK, H. A. J. (1993), A quantitative respirometric method for monitoring compost stability, *Compost Sci. Util.* **1**, 52–65

JIMÉNEZ, E. I., GARCIA, V. P. (1989), Evaluation of city refuse compost maturity: A review, *Biol. Wastes* **27**, 115–142.

JOURDAN, B. (1988), Zur Kennzeichnung des Rottegrades von Müll- und Müllklärschlammkomposten, *Thesis*, Universität Stuttgart (Stuttgarter Berichte zur Abfallwirtschaft **30**, Bielefeld: Erich Schmidt Verlag.)

KAILA, A. (1952), Humification of straw at various temperatures, *Acta Agralia Fennica* **78**, 1–32.

KAISER, J. (1996), Modelling composting as a microbial ecosystem: A simulation approach, *Ecol. Model.* **91**, 25–37.

KAISER, J. (1998), Ein Simulationsmodell des Kompostierungsprozesses und seine Anwendung auf Grundfragen der Verfahrensgestaltung und Verfahrensführung, *Thesis*, TU Dresden, Germany.

KANE, B. E., MULLINS, J. T. (1973), Thermophilic fungi in a municipal waste compost system, *Mycologia* **65**, 175–182.

KATAYAMA, A., KUBOTA, H. (1995), Compost testing method for the compatibility of plants, *Lecture* S43 held at the 1st. Int. Symp. Biological Waste Management – A Wasted Chance? Bochum, Germany.

KEMPF, A. (1996), Untersuchungen über thermophile Actinomyceten: Taxonomie, Ökologie, Abbau von Biopolymeren, *Thesis*, TU Darmstadt, Germany.

KEMPF, A., KUTZNER, H. J. (1989), Screening of thermophilic actinomycetes for biopolymer degrading enzymes, in: DECHEMA *Biotechnology Conferences*, Vol. **3a**, pp. 159–162. Frankfurt/M.: DECHEMA.

KIRK, T. K., FARELL, R. L. (1987), Enzymatic combustion: The microbial degradation of lignin, *Ann. Rev. Microbiol.* **41**, 65–505.

KLEYN, J. G., WETZLER, T. F. (1981), The microbiology of spent mushroom compost and its dust, *Can. J. Microbiol.* **27**, 748–753.

KLOPOTEK, A. VON (1962), Über das Vorkommen und Verhalten von Schimmelpilzen bei der Kompostierung städtischer Abfallstoffe, *Antonie van Leeuwenhoek* **28**, 141–160.

KNÖSEL, D., RESZ, A. (1973), Pilze aus Müllkompost: Enzymatischer Abbau von Pektin und Zellulose durch wärmeliebende Species, *Städtehygiene* **24**, 143–148.

KORN-WENDISCH, F., KUTZNER, H. J. (1992), The family Streptomycetaceae, in: *The Prokaryotes – A Handbook on Habitats, Isolation and Identification of Bacteria*, Vol. **1**, 2nd Edn. (BALOWS, A., TRÜPER, H. G., DWORKIN, M., HARDER, W., SCHLEIFER, K.-H., Eds.), pp. 921–995. New York: Springer-Verlag.

KORN-WENDISCH, F., WEBER, C. (1992), Selective isolation of thermophilic actinomycetes, DECHEMA *Biotechnology Conferences*, Vol. **5**, pp. 1137–1141. Frankfurt/M.: DECHEMA

KORN-WENDISCH, F., RAINEY, F., KROPPENSTEDT, R. M., KEMPF, A., MAJAZZA, A., KUTZNER, H. J., STACKEBRANDT, E. (1995), *Thermocrispum* gen. nov., a new genus of the order Actinomycetales, and description of *Thermocrispum municipale* sp. nov. and *Thermocrispum agreste* sp. nov., *Int. J. Syst. Bacteriol.* **45**, 67–77.

KOTHARY, M. H., CHASE, T. (1984), Levels of *Aspergillus fumigatus* in air and in compost at a sewage sludge composting site, *Environ. Pollut.* **34** (Series A), 1–14.

KOTIMAA, M. H. (1990), Occupational exposure to fungal and actinomycete spores during handling of wood chips, *Grana* **29**, 153–156.

KRANERT, M. (1989), Freisetzung und Nutzung von thermischer Energie bei der Schlammkompostierung, *Thesis*, Universität Stuttgart, Germany. (Stuttgarter Ber. Abfallwirtschaft **33**, Bielefeld: Erich Schmidt Verlag).

KROGMANN, U. (1994), Kompostierung – Grundlagen zur Einsammlung und Behandlung von Bioabfällen unterschiedlicher Zusammensetzung, *Hamburger Berichte* **7**, pp. 1–437. Bonn: Economica Verlag.

KUCHTA, K., JAGER, J. (1993), Geruchsproblematik: Analytik, Behandlung und Erfolgskontrolle, in: *BMFT-Statusseminar: „Neue Techniken zur Kompostierung"* (KURTH, J., STEGMANN, R., Eds.), pp. 87–100. Berlin: Umweltbundesamt.

KUTZNER, H. J., KEMPF, A. (1996), Vorkommen von Actinomyceten in der Luft von Abfallbehandlungsanlagen, in: 50. Darmstädter Seminar: Hygiene in der Abfallwirtschaft, *Schriftenreihe WAR* **92**, 13–54 (Vertrieb: Institut WAR, Techn. Universität Darmstadt).

LACEY, J. (1973), Actinomycetes in soils, compost and fodders, in: *Actinomycetales: Characteristics and Practical Importance* (SY RES, G., SKINNER, F. A. Eds.), *Soc. Appl. Bacteriol. Symp. Ser. No. 2*, pp. 231–251. London, New York: Academic Press.

LACEY, J. (1978), Ecology of actinomycetes in fodders and related substances, *Zbl. Bakteriol. I. Supplement* **6**, 161–170.

LACEY, J. (1980), Colonization of damp organic substrates and spontaneous heating, in: *Microbial Growth and Survival in Extremes of Environment* (GOULD, G. W., CORRY, J. E. L., Eds.), *Soc. Appl. Bacteriol. Techn. Ser. No. 15*. pp. 53–70. London: Academic Press.

LACEY, J. (1988), Actinomycetes as biodeteriogens and pollutants of the environment, in: *Actinomycetes in Biotechnology* (GOODFELLOW, M., WILLIAMS, S. T., MORDARSKI, M., Eds.), pp. 359–432. San Diego, CA: Academic Press.

LACEY, J., CROOK, B. (1988), Fungal and actinomycete spores as pollutants of the work place and occupational allergens, *Ann. Occup. Hyg.* **32**, 515–533.

LACEY, J., PEPYS, J., CROSS, T. (1972), Actinomycete and fungus spores in air as respiratory allergens, in: *Safety in Microbiology* (SHAPTON, D. A., BOARD, R. G., Eds.), *Soc. Appl. Bacteriol. Techn. Series No. 6*, pp. 151–184. London: Academic Press.

LACEY, J., WILLIAMSON, P. A. M., CROOK, B. (1992), Microbial emmission from compost made for mushroom production and from domestic waste, in: *Composting and Compost Quality Assurance Criteria* (JACKSON, D. V., MERILLOT, J. M., HERMITE, P. L., Eds.), pp. 117–130. Luxemburg, Brussels: Office for Official Publications of the European Communities.

LAMANNA, C., MALLETTE, M. F. (1959), *Basic Bacteriology*, 2nd Edn. Baltimore, MA: The Williams & Wilkins Company.

LEHTOKARI, M., NIKKOLA, P., PAATERO, J. (1983), Determination of ATP from compost using the firefly luminescense technique, *Eur. J. Appl. Microbiol. Biotechnol.* **17**, 187–190.

LEINEMANN, B. (1998), Die Bildung klimarelevanter Spurengase während der Kompostierung in Abhängigkeit von der Bioprozeßführung und dem Inputmaterial, *Thesis*, TU Braunschweig, Germany.

LEMBKE, L. L., KNISELEY, R. N. (1985), Airborne microorganisms in a municipal solid waste recovery system, *Can. J. Microbiol.* **31**, 198–205.

LIN, C.-B., STUTZENBERGER, F. J. (1995), Purification and characterization of the major β-1,4-endoglucanase from *Thermomonospora curvata*, *J. Appl. Bacteriol.* **79**, 447–453.

LUONG, J. H. T., VOLESKY, B. (1983), Heat evolution during the microbial process – estimation, measurement, and application, *Adv. Biochem. Eng. Biotechnol.* **28**, 1–40.

LYNCH, J. M. (1977), Phytotoxicity of acetic acid produced in the anaerobic decomposition of wheat straw, *J. Appl. Bacteriol.* **42**, 81–87.

MADIGAN, M. T., MARTINKO, J. M., PARKER, J. (Eds.) (1996), *Brock Biology of Microorganisms*, 8th Edn. Englewood Cliffs, NJ: Prentice Hall.

MATCHAM, S. E., JORDAN, B. R., WOOD, D. A. (1985), Estimation of fungal biomass in a soil substrate by three independent methods, *Appl. Microbiol. Biotechn.* **21**, 108–112.

MARUGG, C., GREBUS, M., HANSEN, R. C., KEENER, H. M., HOITINK, H. A. J. (1993), A kinetic model of the yard waste composting process, *Compost Sci. Util.* **1**, 38–51.

MATHUR, S. P. (1991), Composting processes, in: *Bioconversion of Waste Materials to Industrial Products* (MARTIN, A. M., Ed.), pp. 147–186. London, New York: Elsevier Applied Science.

MATHUR, S. P., OWEN, G., DINEL, H., SCHNITZER, M. (1993), Determination of compost biomaturity. I. Literature review, *Biol. Agric. Hortic.* **10**, 65–85.

MAYER, J. (1990) Geruchstoffe bei der Heißrotte von Hausmüll, *Thesis*, Universität Tübingen, Germany.

MCCARTHY, A. J. (1987), Lignocellulose-degrading actinomycetes, *FEMS Microbiol. Rev.* **46**, 145–163.

MCKINLEY, V. L., VESTAL, J. R. (1984), Biokinetic analyses of adaptation and succession: Microbial activity in composting municipal sewage sludge, *Appl. Environ. Microbiol.* **47**, 933–941.

MCKINLEY, V. L., VESTAL, J. R. (1985a), Physical and chemical correlates of microbial activity and biomass in composting municipal sewage sludge, *Appl. Environ. Microbiol.* **50**, 1395–1403.

MCKINLEY, V. L., VESTAL, J. R. (1985b) Effects of different temperature regimes on microbial activity and biomass in composting municipal sewage sludge, *Can. J. Microbiol.* **31**, 919–925.

MILLER, F. C. (1984), Thermodynamic and matric water potential analysis in field and laboratory scale composting ecosystem, *Thesis*, Rutgers University, New Brunswick, NJ, USA.

MILLER, F. C. (1989), Matric water potential as an ecological determinant in compost, a substrate dense system, *Microb. Ecol.* **18**, 59–71.

MILLER, F. C. (1991), Biodegradation of solid wastes by composting, in: *Biological Degradation of Wastes* (MARTIN, A. M., Ed.), pp. 1–30. Essex, U.K.: Elsevier Science Publishers.

MILLER, F. C. (1993) Composting as a process based on the control of ecological selective factors, in: *Soil Microbial Ecology* (BLAINE METTING Jr., F., Ed.), pp. 515–544. New York, Hong Kong: Marcel Dekker.

MILLER, F. C. (1996), Heat evolution during composting of sewage sludge, in: *The Science of Composting* (DE BERTOLDI, M., SEQUI, P., LEMMES, B., PAPI, T., Eds.), pp. 106–115. London: Blackie Academic & Professional.

MILLER, F. C., MACAULEY, B. (1988), Odors arising from mushroom composting: A review, *Austr. J. Exp. Agric.* **28**, 553–560.

MILLNER, P. D. (1982), Thermophilic and thermotolerant actinomycetes in sewage sludge compost, *Dev. Ind. Microbiol.* **23**, 61–78.

MILLNER, P. D., MARSH, P. B., SNOWDEN, R. B., PARR, J. F. (1977), Occurrence of *Aspergillus fumigatus* during composting of sewage sludge, *Appl. Environ. Microbiol.* **34**, 765–772.

MILLNER, P. D., BASSETT, D. A., MARSH, P. B. (1980), Dispersal of *Aspergillus fumigatus* from sewage sludge compost piles subjected to mechanical agitation in open air, *Appl. Environ. Microbiol.* **3**, 1000–1009.

MILLNER, P. D., POWERS, K. E., ENKIRI, N. K., BURGE, W. D. (1987), Microbially mediated growth suppression and death of *Salmonella* in composted sewage sludge, *Microb. Ecol.* **14**, 255–265.

MILLNER, P. D., OLENCHOCK, S. A., EPSTEIN, E., RYLANDER, M. D., HAINES, J. et al. (1994), Bioaerosols associated with composting facilities, *Compost Sci. Util.* **2**, 6–57.

MOREL, J. L., COLIN, F., GERMON, J. C., GODIN, P., JUSTE, C. (1985), Methods for the evaluation of the maturity of municipal refuse compost, in: *Composting of Agricultural and Other Wastes* (GASSER, J. K. R., Ed.), pp. 56–72. London, New York: Elsevier Applied Science Publishers.

MOTE, C. R., GRIFFIS, C. L. (1979), A system for studying the natural composting process, *Agric. Wastes* **1**, 191–203.

MOTE, C. R. GRIFFIS C. L. (1982), Heat production by composting organic matter, *Agric. Wastes* **4**, 65–73.

MOUCHACCA, J. (1995), Thermophilic fungi in desert soils: A neglected extreme environment, in: *Microbial Diversity and Ecosystem Function* (ALLSOPP, D., COLWELL, R. R., HAWKSWORTH, D. L., Eds.), pp. 265–288. Wallingford, UK: CAB International.

NAKASAKI, K., KATO, J., AKIYAMA, T., KUBOTA, H. (1987), A new composting model and assessment of optimum operation for effective drying of composting material, *J. Ferment. Technol.* **65**, 441–447.

NAKASAKI, K., SASAKI, M., SHODA, M., KUBOTA, H. (1985a), Change in microbial numbers during thermophilic composting of sewage sludge with reference to CO_2 evolution rate, *Appl. Environ. Microbiol.* **49**, 37–41.

NAKASAKI, K., SASAKI, M., SHODA, M., KUBOTA, H. (1985b), Characteristics of mesophilic bacteria isolated during thermophilic composting of sewage sludge, *Appl. Environ. Microbiol.* **49**, 42–45.

NAKASAKI, K., SASAKI, M., SHODA, M., KUBOTA, H. (1985c), Effect of seeding during thermophilic composting of sewage sludge, *Appl. Environ. Microbiol.* **49**, 724–726.

NAKASAKI, K., SHODA, M., KUBOTA, H. (1985d), Effect of temperature on composting of sewage sludge, *Appl. Environ. Microbiol.* **50**, 1526–1530.

NAKASAKI, K., KUBO, M., KUBOTA, H. (1996), Production of functional compost which can suppress phytopathogenic fungi of lawn grass by inoculating *Bacillus subtilis* into grass clippings, in: *The Science of Composting* (DE BERTOLDI, M., SEQUI, P., LEMMES, B., PAPI, T., Eds.), pp. 87–95. London: Blackie Academic & Professional.

NEEF, A., AMANN, R., SCHLEIFER, K.-H. (1995), Detection of microbial cells in aerosols using nucleic acid probes, *Syst. Appl. Microbiol.* **18**, 113–122.

NIESE, G. (1959), Mikrobiologische Untersuchungen zur Frage der Selbsterhitzung organischer Stoffe, *Arch. Mikrobiol.* **34**, 285–318.

NIESE, G. (1969), Die Bestimmung der mikrobiellen Aktivität in Müll- und Müllkomposten durch die Messung der Sauerstoffaufnahme und der Wärmebildung, *Thesis*, Universität Gießen, Germany.

NORMAN, A. G., RICHARDS, L. A., CARLYLE, R. E. (1941), Microbial thermogenesis in the decomposition of plant materials. I. An adiabatic fermentation apparatus, *J. Bacteriol.* **41**, 689–697.

OBRIST, W. (1965), Enzym-Aktivität und Stoffabbau im rottenden Müll; Vorschlag für eine neue mikrobiologische Methode zur Bestimmung des Rottegrades, Int. Arbeitsgem. Müllforsch. (IAM), *Informationsblatt* **24**, 29–37.

OI, S., YAMADA, H., OHTA, H., TANIGAWA, H. (1988), Drying of sewage sludge by aerobic solid state cultivation, in: *Alternative Waste Treatment Systems* (BHAMIDIMARRI, R., Ed.), pp. 142–152. Essex, UK: Elsevier Appl. Sci. Publishers.

OKAZAKI, H., IIZUKA, H. (1970), Lysis of living mycelia of a thermophilic fungus, *Humicola lanuginosa*, by the cell wall-lytic enzymes produced from a thermophilic actinomycete, *J. Gen. Appl. Microbial.* (Tokyo) **16**, 537–541.

PALETSKI, W. T., YOUNG, J. C. (1995), Stability measurement of biosolids compost by aerobe respirometry, *Compost Sci. Util.* **3**, 16–24.

PALMISANO, A. C., MARUSCIK, D. A., RITCHIE, C. J., SCHWA, B. S., HARPER, S. R., RAPAPORT, R. A. (1993), A novel bioreactor simulating composting of municipal solid waste, *J. Microbiol. Methods* **18**, 99–117.

PAPENDICK, R. I., MULLA, D. J. (1986), Basic principles of cell and tissue water relations, in: *Water, Fungi and Plants* (AYRES, P. G., BODDY L., Eds.), *Brit. Mycol. Soc. Symp. No. 11*, pp. 1–25. Cambridge: Cambridge University Press.

PFIRRMANN, A. (1994), Untersuchungen zum Vorkommen von luftgetragenen Viren an Arbeitsplätzen in der Müllentsorgung und -verwertung, *Thesis*, Universität Hohenheim, Germany.

PHAE, C. G., SASKI, M., SHODA, M., KUBOTA, H. (1990), Characteristics of *Bacillus subtilis* isolated from composts suppressing phytopathogenic microorganisms, *Soil Sci. Plant Nutr.* **36**, 575–586.

PÖHLE, H. (1993), Geruchstoffemissionen bei der Kompostierung von Bioabfall, *Thesis*, Universität Leipzig, Germany

PÖHLE, H., KLICHE, R. (1996), Geruchstoffemissionen bei der Kompostierung von Bioabfall, *Zbl. Hyg.* **199**, 38–50.

PÖHLE, H., MIETKE, H., KLICHE, R. (1993), Zusammenhang zwischen mikrobieller Besiedlung und Geruchsemissionen bei der Bioabfallkompostierung, *BMFT-Statusseminar: „Neue Techniken zur Kompostierung"* (KURTH, J., STEGMANN, R., Eds.), pp. 153–168. Berlin: Umweltbundesamt.

PÖHLE, H., ROTTMAYER, K., BERGMANN, K., ALTER, T. (1996), Neue Erkenntnisse zur Entstehung, Freisetzung und Reduzierung von Gerüchen beim mikrobiologischen Abbau von Bioabfall, in: *Neue Techniken der Kompostierung* (STEGMANN, H., Ed.), *Hamburger Berichte 11*, pp. 477–492. Bonn: Economica Verlag.

PÖPEL, F. (1971), Energieerzeugung beim biologischen Abbau organischer Stoffe, *GWF-Wasser-Abwasser* **112**, 407–411.

POSTEN, C. H., COONEY, C. L. (1993), Growth of Microorganisms, in: *Biotechnology*, Vol. **1**, 2nd Edn., *Biological Fundamentals* (REHM, H.-J., REED G., PUHLER, A., STADLER, P., Eds.), pp. 111–162. Weinheim: VCH.

RESZ, A., SCHWANBECK, J., KNÖSEL, D. (1977), Thermophile Actinomyceten aus Müllkompost: Temperaturansprüche und proteolytische Aktivität, *Forum Städte-Hygiene* **28**, 71–73.

ROBERTS, C. A., MOORE, K. J., GRAFFIS, D. W., KIRBY, H. W., WALGENBACH, R. P. (1987), Quantification of mold in hay by near infrared reflectance spectroscopy, *J. Dairy Sci.* **70**, 2560–2564.

RODRIGUEZ LEÓN, J. A., TORRES, A., ECHEVARIA, J., SAURA, G. (1991), Energy balance in solid state fermentation processes, *Acta Biotechnol.* **11**, 9–14.

ROSENBURG, S. L. (1978), Cellulose and lignocellulose degradation by thermophilic and thermotolerant fungi, *Mycologia* **70**, 1–13.

RUSS, C. F., YANKO, W. A. (1981), Factors affecting *Salmonellae* repopulation in composted sludges, *Appl. Environ. Microbiol.* **41**, 597–602.

SANDHU, D. K., SIDHU, M. S. (1980), The fungal succession on decomposing sugar cane bagasse, *Trans. Brit. Mycol. Soc.* **75**, 281–286.

SAVAGE, J., CHASE, T., MACMILLAN, J. D. (1973), Population changes in enteric bacteria and other microorganisms during aerobic thermophilic windrow composting, *Appl. Microbiol.* **26**, 969–974.

SCHLEGEL, H. G., JANNASCH, H. W. (1992), Prokaryotes and their habitats, in: *The Prokaryotes, A Handbook on the Biology of Bacteria*, 2nd Edn. (BALOWS, A., TRÜPER, H. G., DWORKIN, M., HARDER, W., SCHLEIFER, K.-H., Eds.), pp. 82–85. New York: Springer-Verlag.

SCHLOCHTERMEIER, A., NIEMEYER, F., SCHREMPF, H. (1992), Biochemical and electron microscopic studies on the *Streptomyces reticuli* cellulase (Avicelase) in its mycelium-associated and extracellular forms, *Appl. Environ. Microbiol.* **58**, 3240–3248.

SCHMIDT, B. (1994), Bakteriologische Untersuchungen zur Keimemission an Arbeitsplätzen in der Müllentsorgung und -verwertung, *Thesis*, Universität Hohenheim, Germany.

SCHUCHARDT, F. (1977), Einfluß des ökologischen Faktors Struktur auf die Kompostierung von Flüssigmist–Feststoffgemischen, dargestellt am Beispiel des Feststoffverfahrens, *Thesis*, TU Berlin, Germany.

SCHUCHARDT, F. (1983), Versuche zum Wärmeentzug aus Festmist, *Landbauforsch. Völkerrode* **33**, 169–178.

SCHUCHARDT, F. (1984), Wärmeentzug bei der Kompostierung von Schnittholz, *Landbauforsch. Völkerrode* **34**, 189–195.

SCHÜLER, C., BIALA, J., BRUNS, C., GOTTSCHALL, R., AHLERS, S., VOGTMANN, H. (1989), Suppression of root rot on peas, beans and beetroots caused by *Pythium ultimum* and *Rhizoctonia solani* through the amendment of growing media with composted organic houshold waste, *J. Phytopathol.* **127**, 227–238.

SCHULZE, K. L. (1962), Continuous thermophilic composting, *Appl. Microbiol.* **10**, 108–122.

SCHUMANN, M., MIETKE, H., PÖHLE, H., KLICHE, R. (1993), Untersuchungen zur Überlebensfähigkeit von *Salmonella enteritides* bei der Bioabfallkompostierung, *BMFT-Statusseminar: „Neue Techniken zur Kompostierung"* (KURTH, J., STEGMANN, R., Eds.), pp. 207–217. Berlin, Umweltbundesamt.

SEAL, K. J., EGGINS, H. O. W. (1976), The uses of thermophilic fungi in the biodegradation of pig wastes, in: *Proc. 3rd Int. Biodegr. Symp.* Sessions XI, XII, XVI, Rhode Island, Kingston, USA, pp. 687–700. London: Applied Science Publishers.

SENESI, N., BRUNETTI, G. (1996), Chemical and physico-chemical parameters for quality evaluation of humic substances produced during composting, in: *The Science of Composting* (DE BERTOLDI, M., SEQUI, P., LEMMES, B., PAPI, T., Eds.), pp. 195–212. London: Blackie Academic & Professional.

SHEVCHENKO, S. M., BAILEY, G. W. (1996), Life after death: Lignin-humic relationships reexamined, *Crit. Rev. Environ. Sci. Technol.* **26**, 95–153.

SIKORA, L. J., RAMIREZ, M. A., TROESCHEL, T. A. (1983), Laboratory composter for simulation studies, *J. Environ. Qual.* **12**, 219–224.

SIKORA, L. J., WILLSON, G. B., COLACICCO, J. F. (1991), Materials balance in aerated static pile composting, *J. Water Poll. Control Fed.* **53**, 1702–1707.

SINGH, B. N. (1947), Myxobacteria in soils and composts; their distribution, number and lytic action on bacteria, *J. Gen. Microbiol.* **1**, 1–10.

SINHA, R. N., TUMA, D., ABRAMSON, D., MUIR, W. E. (1988), Fungal volatiles associated with moldy grain in ventilated and non-ventilated bin-stored wheat, *Mycopathologia* **101**, 53–60.

SMITH, J. F. (1993), The mushroom industry, in: *Exploitation of Microorganisms* (JONES, D. G., Ed.), pp. 249–271. London: Chapman & Hall.

STAHLSCHMIDT, V. (1987), Thermodynamics of refuse/sludge composting, in: *Compost: Production, Quality and Use* (DE BERTOLDI, M., FERRANTI, M.-P., L'HERMITE, P., Eds.), pp. 822–826. London: Elsevier Applied Sciences.

STARK, W. (1995) Grundlagen zur Modellierung und Regelung zwangsbelüfteter geschlossener Rottesysteme, *Oesterr. Wasser- Abfallwirtsch.* **48**, 274–280.

STEGMANN, R. (Ed.) (1996), *Neue Techniken der Kompostierung* (2. BMBF Statusseminar, Hamburg, Nov. 1996), *Hamburger Berichte 11*, pp. 1–499. Bonn: Economica Verlag.

STÖPPLER-ZIMMER, H., PETERSEN, U. (1995), New results concerning the application of compost in plant cultivation, *Lecture* S59 held at 1st Int. Symp. Biological Waste Management – A Wasted Chance? Bochum, Germany.

STRAATSMA, G., SAMSON, R. A., OLIJNSMA, T. W., OP DEN CAMP, H. J. M., GERRITS, J. P. G., VAN GRIENSVEN, L. J. L. D. (1994), Ecology of thermophilic fungi in mushroom compost, with emphasis on *Scytalidium thermophilum* and growth stimulation of *Agaricus bisporus mycelium*, *Appl. Environ. Microbiol.* **60**, 454–458.

STRAUCH, D. (1996), Occurrence of microorganisms pathogenic for man and animals in source-separated biowaste and compost: importance, control, limits, epidemiology, in: *The Science of Composting* (DE BERTOLDI, M., SEQUI, P., LEMMES, B., PAPI, T., Eds.), pp. 224–232. London: Blackie Academic & Professional.

STREICHSBIER, F., MESSNER, K., WESSELY, M., RÖHR, M. (1982), The microbiological aspects of grape marc humification, *Eur. J. Appl. Microbiol. Biotechnol.* **14**, 182–186.

STROM, P. F. (1985a), Effect of temperature on bacterial species diversity in thermophilic solid waste composting, *Appl. Environ. Microbiol.* **50**, 899–905.

STROM, P. F. (1985b), Identification of thermophilic bacteria in solid waste composting, *Appl. Environ. Microbiol.* **50**, 906–913.

STROM, P. F., MORRIS, M. L., FINSTEIN, M. S. (1980), Leaf composting through appropriate low-level technology, *Compost Sci.* **21**, 44–48.

STROMBAUGH, D. P., NOKES, S. E. (1996), Development of a biologically based aerobic composting simulation model, *Trans. ASAE* **39**, 239–250.

STRUWE, S., KJØLLER, A. (1986), Changes in population structure during decomposition, in: *Microbi-*

al Communities in Soil, Proc. FEMS Symp. 4.–8. Aug. 1985, Copenhagen (JENSEN, V., KJØLLER, A., SØRENSEN, L. H., Eds.), pp. 149–161. London, New York: Elsevier Applied Science Publishers.

STUTZENBERGER, F. J. (1971), Cellulase production by *Thermomonospora curvata* isolated from municipal solid waste compost, *Appl. Microbiol.* **22**, 147–152.

STUTZENBERGER, F. J., KAUFMANN, A. J., LOSSIN, R. D. (1970), Cellulolytic activity in municipal solid waste composting, *Can. J. Microbiol.* **16**, 553–560.

SULER, D. J., FINSTEIN, M. S. (1977), Effect of temperature, aeration and moisture on CO_2 formation in bench-scale, continuous thermophilic composting of solid waste, *Appl. Environ. Microbiol.* **33**, 345–350.

SWIFT, M. J., HEAL, O. W. (1986), Theoretical consideration of microbial succession and growth strategies: intellectual exercise or practical necessity?, in: *Microbial Communities in Soil, Proc. FEMS Symp.* 4.–8. August 1985, Copenhagen (JENSEN, V., KJØLLER, A., SØRENSEN, L. H., Eds.), pp. 115–131. London, New York: Elsevier Applied Science Publishers.

TANSEY, M. R., MURRMAN, D. N., BEHNKE, B. K., BEHNKE, E. R. (1977), Enrichment, isolation and assay of growth of thermophilic and thermotolerant fungi in lignin-containing media, *Mycologia* **69**, 463–476.

TEMPLE, K. L. (1981), Water availability terms, *ASM News* **47**, p. 43.

TERROINE, E. F., WURMSER, R. (1922), L'énergie de croissance. I. Le développement de l'*Aspergillus niger*, *Bull. Soc. Chim. Biol.* **4**, 519–567.

TOSTRUP, P. (1985), Heat recovery from composting solid manure, in: *Composting of Agricultural and other Wastes* (GASSER, J. K. R., Ed.), pp. 167–179. London, New York: Elsevier Applied Science Publishers.

UPDEGRAFF, D. M. (1972), Microbiological aspects of solid waste composting, *Dev. Ind. Microbiol.* **113**, 16–23.

VANDERGHEYNST, J. S., WALKER, L. P., PARLANGE, J.-Y. (1997), Energy transport in a high-solids aerobic degradation process: Mathematical modeling and analysis, *Biotechn. Progr.* **13**, 238–248.

VEROUGSTRAETE, E., NYNS, J., NAVEAU, H. P. (1985), Heat recovery from composting and comparison with energy from anaerobic digestion, in: *Composting of Agricultural and other Wastes* (GASSER, J. K. R., Ed.), pp. 135–146. London, New York: Elsevier Applied Sciences Publishers.

VIEL, M., SAYAG, D., PEYRE, A., LOUIS, A. (1987), Optimization of in-vessel co-composting through heat recovery, *Biol. Wastes* **20**, 167–185.

VON DER EMDE, K. (1987), Untersuchungen über das Vorkommen thermophiler Actinomyceten bei der Kompostierung von Hausmüll, *Thesis*, Universität Wien, Austria.

WAKSMAN, S. A. (1926), The origin and the nature of the soil organic matter or soil "humus"; I–V, *Soil Sci.* **22**, 123–162, 221–232, 323–333, 395–406, 421–436.

WAKSMAN, S. A. (1931), Decomposition of the various chemical constituents of complex plant materials by pure cultures of fungi and bacteria, *Arch. Microbiol.* **2**, 136–154.

WAKSMAN, S. A. (1938), *Humus: Origin, Chemical Composition, and Importance in Nature*, 2nd Edn., pp. 1–511. London: Ballière, Tindall and Cox.

WAKSMAN, S. A., CORDON, T. C., HULPOL, N. (1939a), Influence of temperature upon the microbiological population and decomposition processes in compost of stable manure, *Soil Sci.* **47**, 83–113.

WAKSMAN, S. A., UMBREIT, W. W., CORDON, T. C. (1939b), Thermophilic actinomycetes and fungi in soils and in composts, *Soil Sci.* **47**, 37–61.

WALKER, I. K., WILLIAMSON, H. M. (1957), The spontaneous ignition of wool. I. The causes of spontaneous fires in New Zealand wool, *J. Appl. Chem.* **7**, 468–480.

WALKER, I. K., HARRISON, W. J. (1960), The self-heating of wet wool, *N. Z. J. Agric. Res.* **3**, 861–895.

WAR 81 (1994), Umweltbeeinflussung durch biologische Abfallbehandlungsverfahren, *42. Darmstädter Seminar – Abfalltechnik*, pp. 1–321 (Vertrieb: Institut WAR, Technische Universität, Darmstadt).

WELTZIEN, H. C. (1992), Biocontrol of foliar fungal diseases with compost extracts, in: *Microbial Ecology of Leaves* (ANREWS, J. H., HIRANO, S., Eds.), *Brock Springer Series in Contemporary Bioscience*, pp. 430–450. Heidelberg, New York: Springer-Verlag.

WELTZIEN, H. C., KETTERER, N., SAMERSKI, C., BUDDE, K., MEDHIN, G. (1987), Untersuchungen zur Wirkung von Kompostextrakten auf die Pflanzengesundheit, *Nachr.bl. Dtsch. Pfl.schutzd.* **39**, 25–28.

WILEY, J. S. (1957), II. Progress report on high-rate composting studies. *Proc. 12th Ind. Waste Conf.*, Purdue University, Extension series No. **94**, 596–603.

WILEY, J. S., PEARCE, G. W. (1955), A preliminary study of high rate composting, *Proc. ASCE, J. San. Eng. Div.* **81**, Paper **846**, 1–28.

WILEY, J. S., PEARCE, G. W. (1957), A preliminary study of high-rate composting, *Trans. Am. Soc. Civil Eng.* **122**, 1009–1034.

WOESE, C. R. (1992), Prokaryote systematics: The evolution of a science, in: *The Procaryotes, A Handbook on the Biology of Bacteria*, 2nd Edn. (BALOWS, A. et al., Eds.), pp. 3–18. New York: Springer-Verlag.

3 Composting of Plant Residues and Waste Plant Materials

FRANK SCHUCHARDT

Braunschweig, Germany

1 Introduction

Composting is the biological decomposition of the organic compounds of wastes under controlled aerobic conditions. In contrast to uncontrolled natural decomposition of organic compounds the temperature in waste heaps can grow by self-heating to the range of mesophilic (25–40°C) and thermophilic microorganisms (50–70°C). The end product of composting is a biologically stable humus-like product for uses as soil conditioners, fertilizers, biofilter materials or fuel.

In addition to biogasification composting is part of integrated waste management systems with biological treatment (WHITE et al., 1995). Almost any organic waste can be treated by this method. The objectives of composting can be volume and mass reduction, stabilization, drying, elimination of phytotoxic substances and undesired seeds and plant parts, and/or sanitation. Composting is also a method for decontamination of polluted soils. The use of compost can play an important role in sustainable agriculture (SEQUI, 1996).

In any case composting of wastes is conducted with the objective of high economical effectiveness and has the goal of compost production with the lowest input of work and expenditure. The consequence of this approach is the effort to optimize the biological, technical, and organizational factors and elements which influence the composting process. Despite the long tradition of composting the scientific interest in the composting process is not older than 40 years.

The factors which influence the composting process are well known and have been published in numerous reviews and monographs (see Chapters 2 and 4, this volume). The time since 1970 is characterized by the development of new strategies, new composting processes and technologies and the optimization of existing processes against the background of an expanding market for composting plants and equipment. Among others, reasons for these developments are rising costs for sanitary landfill, improved environmental protection as well as new laws, ordinances, and regulations. The knowledge of limited resources and the idea of recycling of refuse back to soil also gave important impulse for the development in this field. In Germany, e.g., the deposit of organic refuse is forbidden beginning in the year 2005 and utilization of refuse is prescribed (TA Siedlungsabfall, 1993; KrW-/AbfG, 1994). Numerous publications, reviews, textbooks, and conferences in the last years, and some hundred internet addresses show the growing interest and activity in the field of composting (DEBERTOLDI, 1987; FINSTEIN, 1992; HAUG, 1993; HOITINK and KEENER, 1993; MILLER, 1993; SMITH, 1993; GASSER, 1995; THOMÉ-KOZMIENSKY, 1995; WHITE et al., 1995; DEBERTOLDI et al., 1996; EPSTEIN, 1997; WIEMER and KERN, 1993, 1994,1995, 1996, 1997).

2 Plant Materials for Composting

2.1 Origin and Amount of the Wastes

The origins of plant wastes and residues are agriculture, horticulture, landscape management, and forestry (Fig. 1). Usually plant production in agriculture and horticulture results in a raw or end product for private households or gardens, market or industry and in waste (e.g., straw from corn, potato haulms, roots and leaves from vegetable, and ornamental flowers). In most cases the wastes from this production remain at the place of growing in or on the soil and become part of the organic matter pool and the degradation and rotting processes by macro- and microorganisms (field composting). If the structure of the organic wastes hinders the following tillage of the soil an additional preparation like chopping is necessary. In some cases the wastes are collected because they are valuable as bedding or fuel material (straw) or feed (e.g., residues of vegetables). A large supply of carbon compounds to the soil can cause nitrogen immobilization and requires additional nitrogen fertilizing.

The plants and plant products in households are used as food or as ornamental plants with the consequence that a part or all of these materials become waste. This waste is the raw

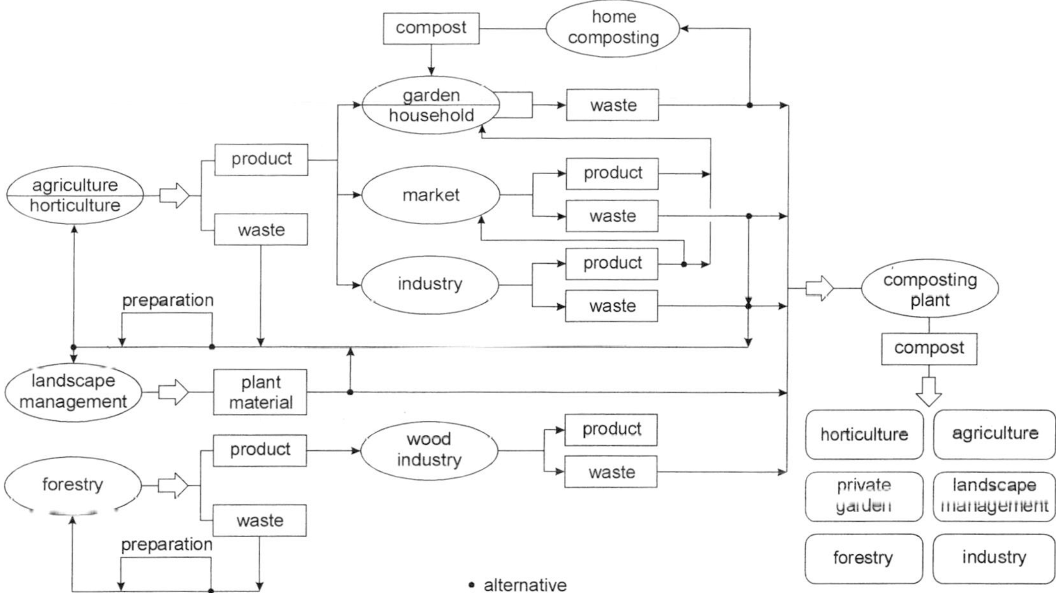

Fig. 1. Origin of plant wastes for composting.

product for composting, either for home composting in small piles or boxes or for composting plants.

In most cases the way of plant products is not direct from agriculture/horticulture to the private households, but indirect over markets or for further preparation over industry. On both ways a part of the plant material becomes waste for composting, organic fertilizer for agriculture/horticulture or landscape management or, if the wastes contain nutrients and have no hazardous components and impurities, foodstuff (e.g., brewers' grains, plant residues from oleaginous fruit processing).

The plant material from landscape management (including public areas) can directly be utilized at the place of origin or after transport to agricultural/horticultural areas without any preparation (e.g., cut grass, leaves) or after chopping (e.g., cut trees, branches and brushwood). The alternatives are collection and transport of the wastes to centralized or decentralized composting plants. Especially cut trees, branches and brushwood are valuable as bulking material in composting plants with a high input of wet organic wastes from households (biowaste, kitchen waste).

One problem of the direct use of plant wastes on agricultural/horticultural areas (field composting) can be the spread of plant diseases.

The utilization of plant wastes from forestry is similar to the one from landscape management. Plant wastes from forestry can remain at the place of origin (e.g., bark), with or without chopping, or can go to composting plants.

The circle of plant waste management is closed in principle, when the compost is used again in the area of the plant production. In reality the compost is used in a smaller area than the production area and the material circle is open and has an input of impurities, hazardous components, and pathogenic microorganisms. The mechanical technique can reduce and separate a part of the impurities. The composting process can inactivate or kill the pathogenic microorganisms (see Chapters 2 and 9, this volume) and can reduce organic hazardous components, but concentrates the heavy metal content.

2.2 Amount, Composition, and Availability of Plant Wastes

The amount, composition, and physical characteristics of plant wastes are influenced by numerous factors like origin, production process, preparation, season, collecting system, home composting, social structure, alternatives for use of production residues (Tabs. 1–6). The wide range of the waste amount and composition requires analyses for planning a composting plant and for estimation of the compost quality in each individual case. The organic fraction for composting in garbage from private households in Germany is approximately 42% (Umweltbundesamt, 1984). More than 72% of the kitchen waste from food consumption in Central Europe (Germany) are of vegetable origin, the remaining 28% are animal products (Tab. 7).

The content of heavy metals and organic compounds in the waste is of high importance particularly with regard to the use of compost as soil conditioner and fertilizer. Ways to reduce the heavy metal content in compost from biowaste are to collect the waste separately and to gain intensive information of the producers of the waste (BIDLINGMAIER, 1990).

Analyses of biowaste show a content of impurities (e.g., plastics, glass, metals, stones) between 0.5 and 5.0% depending on the social structure, buildings, and public relations work (FRICKE and TURK, 1991). In densely built-up areas the contamination is higher than in others. More than 90% of the impurities have a size >60 mm, and 90% of the biowaste have a size <60 mm (FRICKE, 1990).

Depending on climatic and cultural conditions (e.g., growing and harvest times, holidays, traditions) some plant wastes are not available during the whole year (Tab. 8). An important

Tab. 1. Specific Waste Amounts in Germany (FRICKE and TURK, 1991; Auswertungs- und Informationsdienst für Ernährung, Landwirtschaft und Forsten, 1991; THOMÉ-KOZMIENSKY, 1995; DOEDENS, 1996)

Waste		
Biowaste	[kg person^{-1} a^{-1}]	90–400[a]
	[kg person^{-1} a^{-1}]	40–200[b]
Kitchen waste	[kg person^{-1} a^{-1}]	60–90
Garden and green waste from private and public area		
Brushwood	[kg m^{-2} a^{-1}]	0.15–0.7
Grass	[kg m^{-2} a^{-1}]	1.5–3.0
Leaves	[kg m^{-2} a^{-1}]	0.4
City	[kg person^{-1} a^{-1}]	25
Country	[kg person^{-1} a^{-1}]	up to 330
Cut grass from landscape management/ fallow land	[kg m^{-2} a^{-1}]	4–40

[a] potential.
[b] in practice.

Tab. 2. Relation between Main Product and Residues of Agricultural Plants/Crops (Hydro Agri Duelmen GmbH, 1993)

Crop	Product (Grain, Bulb, Fruit) [Mg ha^{-1}]	Waste (Straw, Foliage) [Mg ha^{-1}]
Cereals		
Barley, wheat, rye	3.5– 9.0	4.0– 9.0
Grain maize	5.0– 9.5	5.0–10.0
Pulse crops		
Field pea	3.0– 5.0	2.0– 6.0
Vetch	1.0– 2.5	2.0– 4.0
Lupin	1.5– 4.0	2.0– 6.0
Oil plants, textile plants		
Rape	1.5– 4.5	5.0– 8.0
Rape-seed	1.0– 2.5	4.0– 7.0
White mustard	1.4– 2.2	6.0– 7.0
Sunflower	2.5– 4.0	7.0–10.0
Oil flax	1.7– 3.5	2.0– 4.0
Tuber and root fruits		
Potatoes	25.0– 60.0	5.0–12.0
Sugar beet	40.0– 80.0	30.0–60.0
Field beet	40.0–130.0	10.0–40.0
Swede turnip	50.0–100.0	10.0–17.0
Turnip	17.0– 40.0	3.0– 9.0

Tab. 3. Nutrient Content in Dry Substance of Some Wastes for Composting (BIDLINGMAIER, 1995; own analyses)

Waste	VS[a] [%]	C/N [–]	N [%]	P_2O_5 [%]	K_2O [%]	CaO [%]	MgO [%]
Kitchen waste	20–80	12–20	0.6–2.2	0.3–1.5	0.4–1.8	0.5–4.8	0.5–2.1
Bio waste	30–70	10–25	0.6–2.7	0.4–1.4	0.5–1.6	0.5–5.5	0.5–2.0
Garden and green waste	15–75	20–60	0.3–2.0	0.1–2.3	0.4–3.4	0.4–12	0.2–1.5
Garbage	25–50	30–40	0.8-1.1	0.6–0.8	0.5–0.6	4.4-5.6	0.8
Feces (human)	15–25	6–10	2	1.8	0.4	5.4	2.1
Wastewater sludge (raw)	20–70	15	4.5	2.3	0.5	2.7	0.6
Wastewater sludge (anaerobic stabilized)	15–30	15	2.3	1.5	0.5	5.7	1.0
Dung							
Cattle	20.3	20	0.6	0.4	0.7	0.6	0.2
Horse	25.4	25	0.7	0.3	0.8	0.4	0.2
Sheep	31.8	15–18	0.9	0.3	0.8	0.4	0.2
Pig	18.0	15–20	0.8	0.9	0.5	0.8	0.2
Liquid manure							
Cattle	10–16	8–13	3.2	1.7	3.9	1.8	0.6
Pig	10–20	5–7	5.7	3.9	3.3	3.7	1.2
Hens	10–15	5–7	9.8	8.3	4.8	17.3	1.7
Beet leaves	70	15	2.3	0.6	4.2	1.6	1.2
Straw	90	100	0.4	2.3	2.1	0.4	0.2
Bark fresh	90–93	85–180	0.5–1.0	0.02–0.06	0.03–0.06	0.5–1	0.04–0.1
Bark mulch	60–85	100–130	0.2–0.6	0.1–0.2	0.3–1.5	0.4–1.3	0.1–0.2
Wood chips	65–85	400–500	0.1–0.4	1.0	0.3–0.5	0.5–1.0	0.1–0.15
Leaves	80	20–60	0.2–0.5	—	—	—	—
Reed	75	20–50	0.4	—	—	—	—
Peat	95–99	30–100	0.6	0.1	0.03	0.25	0.1
Paunch manure	8.5–17	15–18	1.4	0.6	0.9	2.0	0.6
Grapes marc	81	50	1.5–2.5	1.0–1.7	3.4–5.3	1.4–2.4	0.2
Fruit marc	90–95	35	1.1	0.6	1.6	1.1	0.2
Tobacco	85–88	50	2.0–2.4	0.5–6.6	5.1–6.0	5.0	0.1–0.4
Paper	75	170–180	0.2–1.5	0.2–0.6	0.02–0.1	0.5–1.5	0.1–0.4

[a] VS Volatile solids.

Tab. 4. Heavy Metal Content in Dry Substance of Some Wastes for Composting (THOMÉ-KOZMIENSKY, 1995)

Waste	Zn [mg kg^{-1}]	Cu [mg kg^{-1}]	Cd [mg kg^{-1}]	Cr [mg kg^{-1}]	Pb [mg kg^{-1}]	Ni [mg kg^{-1}]	Hg [mg kg^{-1}]
Bio waste	50–470	8–81	0.1–1	5–130	10–183	6–59	0.01–0.8
Green waste	30–138	5–31	0.2–0.9	28–86	24–138	9–27	0.1–3.5
Paper	93	60	0.2	4	20	1	0.08
Paper printed	112	66	0.2	31	78	3	0.04
Paper collected	40	21	0.2	3	12	1	0.06
Paper sludge	150–1500	15–100	0.1–1.5	30–300	70–90	5–15	0.2–0.5
Bark	150–300	40–60	0.6–2.1	30–63	20–57	12–20	0.1–0.5
Bark mulch	40–500	10–30	0.1–2	500–1000	50–100	30–60	0.1–1
Wood chips	58–137	8–11	0.1–0.2	6–8	13–53	4	0.1
Grapes marc	60–80	100–200	0.5	2.5–7.4	10	1–4	0.02–0.04
Fruit marc	20–30	9.5	0.2	0.02–1	0.3–1	2–4	0.03
Brewers' grains	13	6	0.3	16	10	16	0.04
Oil seeds residues	4	1	0.03–0.05	0.1	0.1–0.4	1–3	0.01
Cacao hulls	89	7–12	0.25	0.5	0.4	0.3	0.02

Tab. 5. Bulk Density of Some Wastes for Composting (THOMÉ-KOZMIENSKY, 1994; BIDLINGMAIER, 1995)

Waste		Bulk Density [Mg m^{-3}]
Biowaste	kitchen and garden wastes (collected separately)	0.4–0.7
	vegetable wastes	0.45–0.6
Green waste	cut trees and brushwood	
	– not chopped	0.03–0.2
	– chopped	0.3–0.5
	grass	0.4
	leaves	0.05–0.2
Garbage	mixed collected	0.1–0.3
Wood wastes	shopped wood	0.3–0.4
	bark mulch	0.4–0.5
	pine bark	0.1 5–0. 3
	fir bark	0.1 5–0.3
	beech bark	0.3–0.4
Other organic wastes	straw	0.8–0.11
	paper/board	0.04–0.3
	paper sludge	0.8
	grapes marc	0.45
	fruit marc	0.3–0.5
	Brewers' grains	0.4–0.6
	cacao hulls	0.4–0.5
	wastes from canteen kitchen	0.8–1.0

Tab. 6. Properties of Plant Wastes for Composting (BIDLINGMAIER, 1992; own results)

Origin, Type	Stability of Structure; air pores	Moisture Content	Rate in Mixtures	Rotting Capability	Preparation
Biowaste	good – bad	too wet – good	50–100%	medium – good	grind, homogenize
Kitchen waste	bad	too wet	– 50%	good	none
Garden and green waste	good	good – medium	–100%	medium	shred, homogenize
Cut grass	bad	too wet, good when withered	– 50%	good	none
Leaves	medium – good	medium – too dry	– 80%	good	none
Cut wood	good	too dry	– 90%	bad	shred
Reed straw	good	too dry	– 70%	good	shred
Straw	good	too dry	– 50%	medium	shred roughly
Beet leaves	bad	middle	– 50%	good	none
Fruit marc	bad – medium	too wet – medium	– 30%	good	none
Paunch manure	medium	good when pressed	50–100%	good	dewatering
Mushroom substrate	good	good	–100%	good – medium	none
Bark	good	medium – too dry	–100%	medium	grind
Sawdust	bad	too dry	– 60%	bad	none
Paper	good	too dry	– 60%	good	grind

Tab. 7. Annual Consumption of Food and Kitchen Waste Composition (KAUTZ, 1995)

	Consumption	Residues of Preparation and Consumption	
	[kg person^{-1} a^{-1}]	[%]	[kg person^{-1} a^{-1}]
Pomaceous fruits	38.2	8	3.1
Stone fruits	13.7	12	1.6
Strawberries, other berries	6.6	3	0.2
Grapes	8.7	21	1.8
Bananas	14.1	35	4.9
Fresh fruits total	*81.3*	*14*	*11.4*
Oranges, tangerines	26.9	29	7.8
Grapefruit	2.9	37	1.1
Lemons	3.5	46	1.6
Citrus fruits total	*33.3*	*31*	*10.3*
Tomatoes	11.8	0.3	0.1
Cucumbers	8.9	26	2.3
Other fruit vegetable	3.8	27	1.1
Leaf and stem vegetable	12.1	21	2.5
Cauliflower	6.4	32	2.1
Other cabbage	9.6	22	2.1
Onion vegetables	10 3	10	1.0
Carrots	6.5	17	1.1
Other fresh vegetable	7.4	20	1.5
Vegetables total	*76. 8*	*18*	*13.8*
Potatoes	71.5	15	10.7
Bread, flour, baked goods	74.9	5	3.8
Coffee	4.8	370	17.8
Tea	0.3	500	1.4
Spoiled foods total	6.5	100	6.5
Potsoil	4.0	100	4.0
Cut flowers	1.3	100	13
Others total	*163. 3*	*28*	*45.5*
Plant products total	***354.7***	***23***	***81.0***
Beef and veal	14.5	15	2.2
Pork	20.6	15	3.1
Liver cattle/calf	1.2	15	0.2
Liver pig	1.2	15	0.2
Meat total	37.5	15	5.6
Fish	6.5	15	1.0
Eggs	16.0	11	1.8
Milk products	113.5	15	17.0
Animal products total	***173.5***	***18***	***31.1***
Total sum	**528.2**	**21**	**112.1**

Tab. 8. Availability of Plant Wastes during the Year in the Temperate Climatic Zone (FRICKE and TURK, 1991)

Type of Waste	Month											
	Jan	Feb	Mar	Apr	May	Jun	Jul	Aug	Sep	Oct	Nov	Dec
Green waste from public area												
Cut grass from lawns, meadows and waysides				—	—	—	—	—	—	—	—	
Cut plants from roadside						—	—	—	—	—		
Cut trees and brushwood	—	—	—							—	—	—
Cut trees from roadside		—	—									
Leaves	—	—							—	—	—	—
Plant waste					—	—	—	—	—	—	—	
Covering and removed material			—	—	—				—	—	—	
Cemetery plant waste	- -	- -	- -	- - / —	- - / —	- -	- -	- -	- -	- - / —	- - / —	- -
Christmas trees	—											
Cut reed from riverbanks and lakeside							—	—	—	—	—	—
Waterweed and submerged plants							—	—	—	—	—	
Wastes from household												
Kitchen waste	—	—	—	—	—	—	—	—	—	—	—	—
Small garden waste			—	—	—	—	—	—	—	—	—	—
Forestry waste												
Felling waste				—								
Windfall	- -	- -	- -	- -	- -	- -	- -	- -	- -	- -	- -	- -
Agricultural waste												
Straw						—	—	—	—	—	—	
Excess straw	- -	- -	- -	- -	- -	- -	- -	- -	- -	- -	- -	- -
Rotten hay						—	—	—	—	—	—	
Beet leaves and remainders										—	—	—
Potato haulms						—	—	—	—	—		
Vegetable leaves and stems						—	—	—	—	—	- -	- -
Plant production and market waste												
Fruit and vegetable waste from market	—	—	—	—	—	—	—	—	—	—	—	—
Plant residues from pharmaceutical production	- -	- -	- -	- -	- -	- -	- -	- -	- -	- -	- -	- -
Plant residues from oleaginous fruit processing	- -	- -	- -	- -	- -	- -	- -	- -	- -	- -	- -	- -
Plant residues from food and luxury food production	- -	- -	- -	- -	- -	- -	- -	- -	- -	- -	- -	- -
Marc									—	—	—	—
Brewers' grains	—	—	—	—	—	—	—	—	—	—	—	—
Bark waste	—	—	—	—	—	—	—	—	—	—	—	—

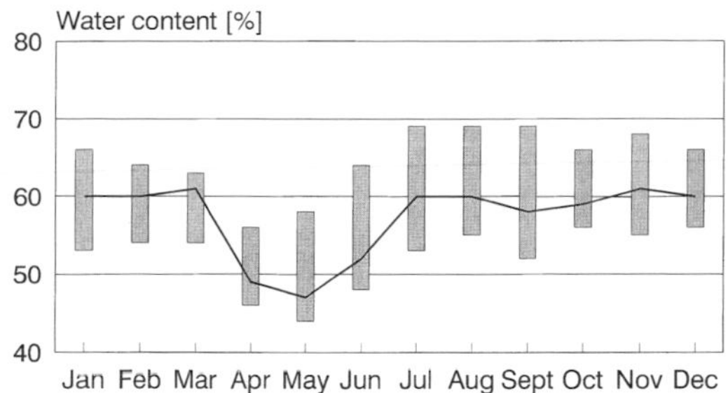

Fig. 2. Water content of biowaste during the year (FRICKE and TURK, 1991).

factor for the operation of a composting plant can be the fluctuation of the composition of the wastes, in particular the water content (Fig. 2).

3 Composting Process

3.1 Fundamentals, Principal Factors

A pile of rotting material, also during well-aerated composting, is characterized by aerobic and anaerobic microbial processes at the same time (Fig. 3) (HÜTHER et al., 1997a, b; GRONAUER et al., 1997). The relation between aerobic and anaerobic metabolism depends on the structure of the heap, its porosity, its water content, and the availability of nutrients.

The aerobic microorganisms in the rotting material need free water and oxygen for their activity. End products of their metabolism are water, carbon dioxide, NH_4 (with higher temperature and pH >7: NH_3), nitrate, nitrite (nitrous oxide as a product from nitrification), heat, and humus or humus-like products. The waste air from the aerobic metabolism in compost heaps contains evaporated water, carbon dioxide, ammonia, and nitrous oxide. The end products of the anaerobic microorganisms are methane, carbon dioxide, hydrogen, hydrogen sulfide, ammonia, nitrous oxide, nitrogen gas (both from denitrification) and water as liquid.

The mature compost consists of heavily or undegradable components (lignin, lignocellulosics, minerals), humus, microorganisms, water and mineral nitrogen compounds.

The organisms which take part in the composting process are microorganisms (bacteria, actinomyces, mildews) in the first place. They have their optimal growing conditions at different temperatures: psychrophilics between 15 and 20°C, mesophilics between 25 and 35°C, and thermophilics between 55 and 65°C. In the mature compost at temperatures below 30–35°C other organisms such as protozoa, collemboles, mites, earthworms join in the biodegradation.

A pile of organic wastes consists of the 3 phases "solid", "liquid", and "gaseous", whereby the microorganisms depend on free water for their metabolism (Fig. 4). Dissolved oxygen, from the gas phase in the heap, must be available for the activity of aerobic microorganisms. To make sure the oxygen transfer from the gas phase to the liquid phase and the carbon dioxide transfer from the liquid phase to the gas phase takes place a permanent, partial pressure gradient must be upheld which is only possible by a permanent exchange by the gas phase.

Specifications about an optimal water content for composting are only meaningful in combination with the respective waste, its structure and air pore volume (Tab. 9). In general, the water content can be higher the better and stabile the structure, air pores volume and water capacity of the waste material (also during the rotting process). Theoretically the water content for composting can be 100%, provided the oxygen supply is sufficient for microbial activity.

Tab. 9. Optimal Water Content and Structure of Wastes for Composting

Waste	Water Content [%]	Structure	Air Pores Volume [%]
Woodchips, cut trees and brushwood	75–90	good	>70
Straw, hay, cut grass	75–85	good	>60
Paper	55–65	middle	<30
Kitchen waste	50–55	middle/bad	25–45
Sewage sludge	45–55	bad	20–40

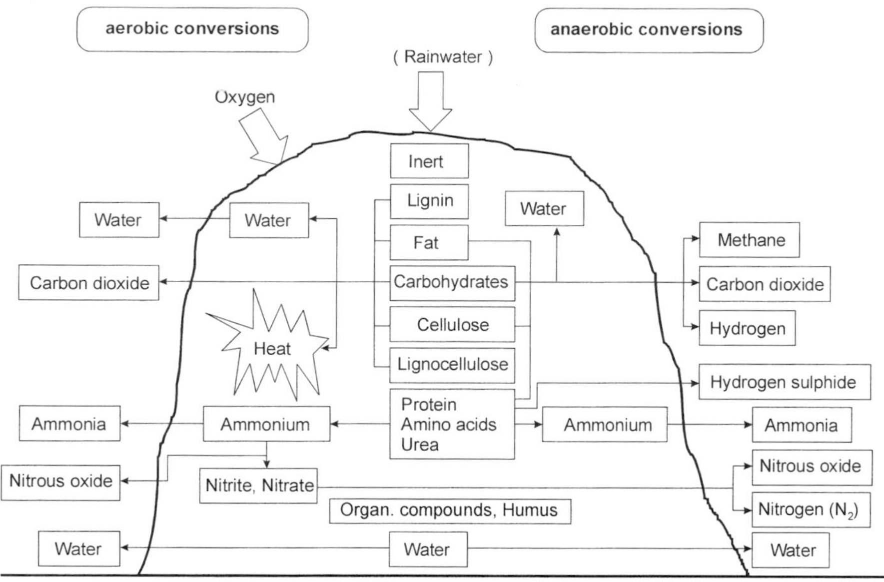

Fig. 3. Products of the microbial activity in a compost heap.

Beneath a sufficient content of free water the microorganisms need a C/N ratio in the substrate of 25–30, whereby the carbon should be widely available. With lower C/N ratios the danger of nitrogen losses by ammonia gas increases (especially when the temperature rises and the pH is higher than 7). If the C/N ratio is higher the composting process needs a longer time to stabilize the waste material.

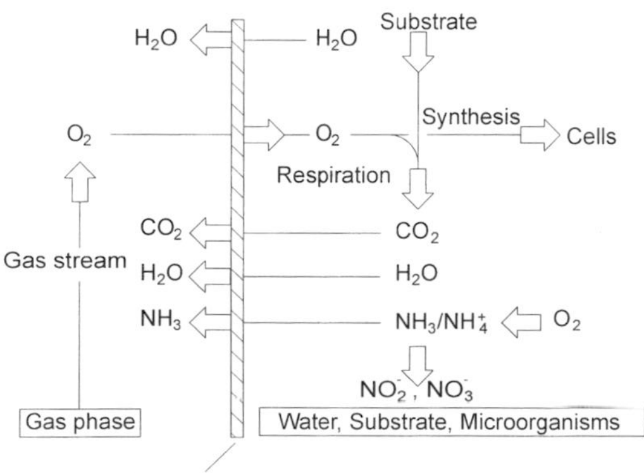

Fig. 4. Metabolism of aerobic microorganisms at the border gas/water.

Fig. 5 shows the correlation between the factors influencing the rotting process. The structure, i.e. consistence, configuration and geometry of the solids of the waste determines the pore volume (filled with water or air) and the streaming resistance of a compost heap. These in turn influence the gas exchange and the oxygen and carbon dioxide concentration in the air pores and liquid phase and therewith the exothermic microbial activity with the result of growing temperature by build up of heat. Microbial activity is influenced by the water content, nutrients (C/N ratio, availability) and pH. The mass and the volume of the heap influence the temperature by its heat capacity and the heat losses by irradiation. By the heat convection in the heap, conditioned by the temperature difference between material and ambient the gas exchange is influenced. The gas exchange and the temperature level influence the water evaporation and therewith the proportion of water filled pores.

One effect of the activity of the different microbial groups is a characteristic temperature curve (Fig. 6). After a short lag phase temperature increases exponentially to 70–75°C. At 40°C there is often a lag during the change-over from mesophilic to thermophilic micro-

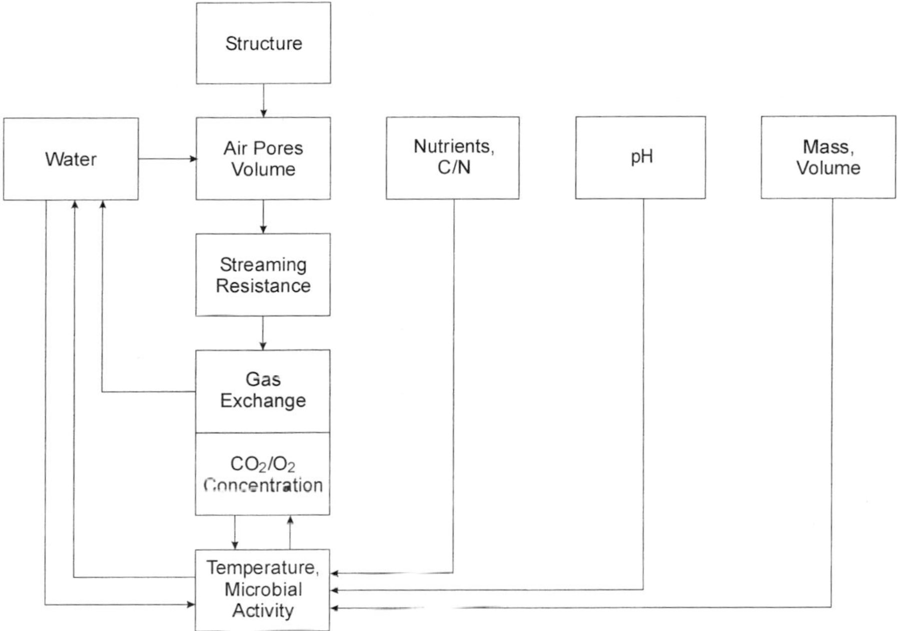

Fig. 5. Factors influencing the composting process.

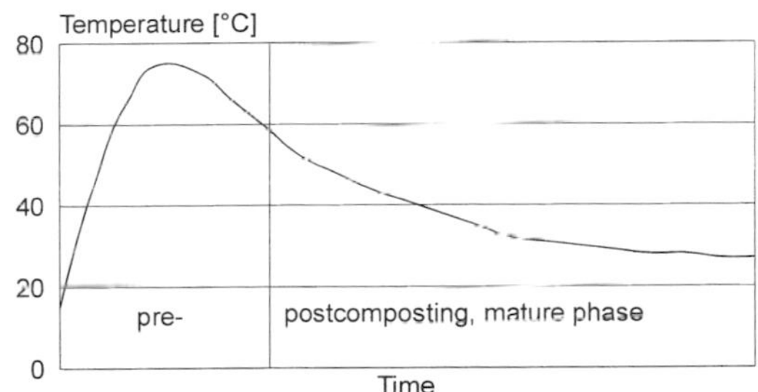

Fig. 6. Characteristic temperature curve during composting process.

organisms. After reaching the maximum the temperature declines slowly to the level of the atmosphere. The progression of the temperature curve depends on numerous factors such as kind and preparation of the waste, relation of surface to volume of the heap, air temperature, wind velocity, aeration rate, C/N ratio, process technique, mixing frequency.

The first phase of the composting process, until a temperature up to about 60°C is called the pre- and main composting, the second phase is called the post-composting or mature phase. Both phases are characterized by different processes (Tab. 10).

Tab. 10. Phases and Characteristics of the Composting Process

Pre- and Main Composting	Postcomposting
Degradation of easily degradable compounds	Degradation of heavily degradable compounds
Inactivation of pathogenic microorganisms and weed seeds	Composition of high molecular compounds (humus)
High oxygen demand	Low oxygen demand
Emissions of odor and leakage water	Low emissions
Time: 1–6 weeks	Time: 3 weeks to 1 year

Frequently the composting considers both phases by division into different process techniques, especially when the wastes present the risk of strong odor emissions:

- pre- and main composting in closed reactors or roofed and frequently mixed or forced aerated windrows;
- post-composting/mature phase in windrows.

3.2 Consequences for Composting of Plant Materials

The aim of composting of wastes under economical conditions is to optimize the factors which influence the rotting process. The most important factor is, under a given composition of waste, to ensure the gas exchange in the heap. This can be done by the following measures:

- adaptation of the height of the heap to the structure, the water content and the oxygen demand (high during pre- and main composting, low during the mature phase),
- turning (mixing, loosening) of the windrows,
- construction of windrows in thin, ventilatable layers,
- mixing and loosening of the rotting material in reactors (in rotating drums, with tools),
- forced aeration,
- reduction of the streaming resistance by addition of bulking material with rough structure or forming of pellets.

4 Composting Technologies

4.1 Basic Flow Sheet

The production of compost consists of the steps "preparation/conditioning" of the raw material and the actual "composting" (Fig. 7). To produce a marketable product it will be necessary to prepare the compost to an end product.

The aim of raw material preparation and conditioning is the modulation of optimal conditions for the following composting process and the separation of impurities to protect the technical equipment, to reduce the input of heavy metals and hazardous organic components (if the impurities contain these components), and to fulfill quality requirements of compost. The basic steps of raw material preparation and conditioning are:

- disintegration of rough wastes (e.g., wood, trees and brushwood, long grass) by chopping, crushing or grinding to enlarge the surface for the activity of the microorganisms,
- dehydration or (partial) drying of water-rich, structureless wastes (e.g., sludge, fruit rests) if these wastes are too wet for the composting process,
- addition of water (fresh water or wastewater, sludge) if the wastes are too dry for the composting process,
- mixing of components (e.g., wet and dry wastes, N rich and C rich wastes, wastes with rough and fine structure),
- separation of impurities by hand or automatic (glass, metals, plastics).

The products of preparation/conditioning of the wastes are waste air (depending on the composition and the conditions of storage with bad smell and dust) and possibly leakage water beneath the raw material for the following composting step.

The basic steps of the composting process may be:

- aeration to exchange the respiration gases oxygen and carbon dioxide and to withdraw water (the only essential step during composting),
- mixing to compensate irregularities in the compost heap (e.g., dry zones at the surface, wet zones at the bottom, cool zones, hot zones) and to renew the structure for better aeration,
- moisturing of dry material to save microbial activity,

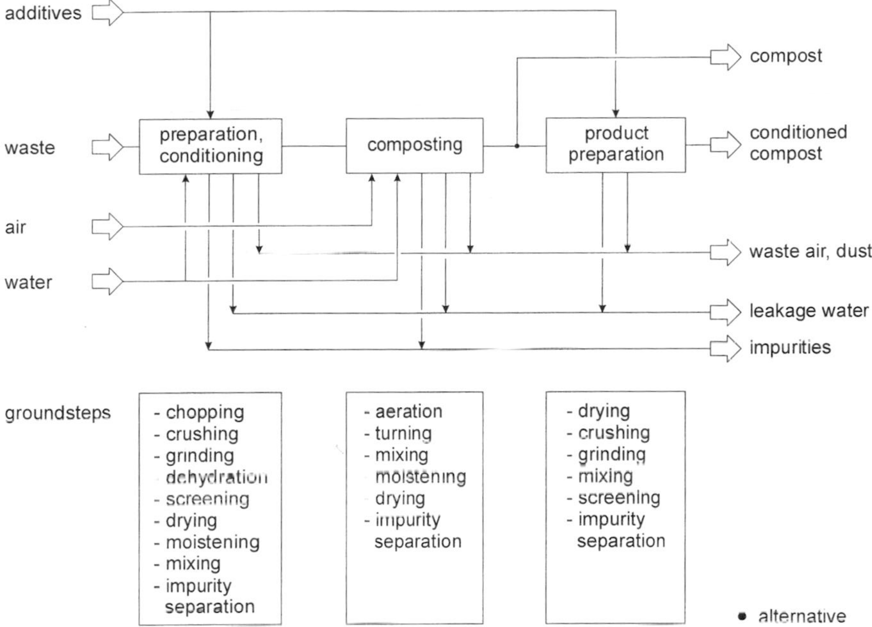

Fig. 7. Basic flow sheet of compost production.

- drying of wet material by aeration or/and mixing to increase the free air pore space for microbial activity or to improve the structure of the compost for packaging,
- separation of impurities by hand.

The products of the composting process are a biologically stabilized compost, waste air and leakage water (when the material is very wet). It may be necessary to prepare the compost for transport, storage, selling, and its placement. In this case the basic steps can be:

- drying of wet compost to prevent a clumpy, smeary product and leakage water during storage,
- disintegration of clots in the compost by crushing/grinding to exclude problems which may occur when the fertilizer is packed,
- mixing of the compost with additives to bind moisture and to increase the air pore volume (e.g., perlite, peat) or to influence the composition (e.g., mineral fertilizer),
- sieving of the compost to classify different fractions for marketing or to separate impurities,
- separation of impurities by hand or automatically.

4.2 Machines and Equipment

4.2.1 Disintegration

Disintegration (crushing, chopping, grinding), especially of wooden, bulky wastes, is necessary to enlarge the surface for the microorganisms and to ensure the function of the following machines and equipment in the process (e.g., turning machine or tools, screens, belt conveyor). The intensity of disintegration depends on the velocity of the biodegradation of the waste, the composting process, the dimensions of the heap, the composting time, and the intended application of the final product (Tab. 11).

Tab. 11. Disintegration of Waste Material for Composting

	Disintegration and Structure			
	None	Rough	Medium	Fine
Velocity of biodegradation				
High	●	—	●	
Low			●	●
Composting process				
Windrow	●	—	●	
In vessel			●	●
Dimension of heap				
Triangular				
Low, <1.5 m			●	●
High, >1.5 m		●	●	
Trapezoid				
Low, <1.5 m		●	●	
High, >1.5 m		●	●	
Composting time				
Short			●	●
Long		●	●	
Compost application				
Garden	●	—	●	
Horticulture			●	●
Agriculture			●	●
Recultivation			●	●

If the structure of the waste material for composting is too fine there is the danger of insufficient exchange of respiratory gases which creates anaerobic conditions and emissions of bad odors. If the structure of the wastes is too rough there is the danger of low biological activity and/or a high rate of gas exchange with the consequence of low temperature and high water evaporation. In most cases a mixture of rough, medium and fine structures is favorable, because then air pores are evenly distributed in the heap. The bulking material with a rough structure can be used several times after sieving of the compost.

For disintegration of organic wastes the following machines are mainly used in practice:

- cutting mills (Fig. 8),
- cracking mills (Fig. 9),
- hammer mills (Fig. 10),
- screw mills (Figs. 11, 12).

Cutting, cracking and hammer mills are high-speed machines with the disadvantage of high energy demand and high wear in contrast to slow speed screw mills. Because of the high capacity of some of these machines ($>100\ m^3\ h^{-1}$ disintegrated material) and/or

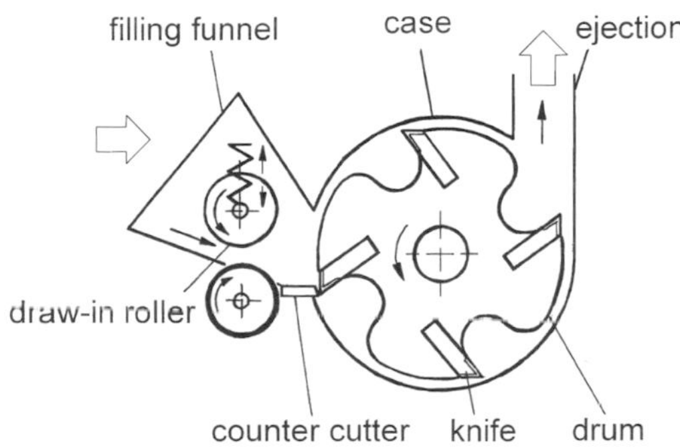

Fig. 8. Cutting mill (RELOE and SCHUCHARDT, 1993).

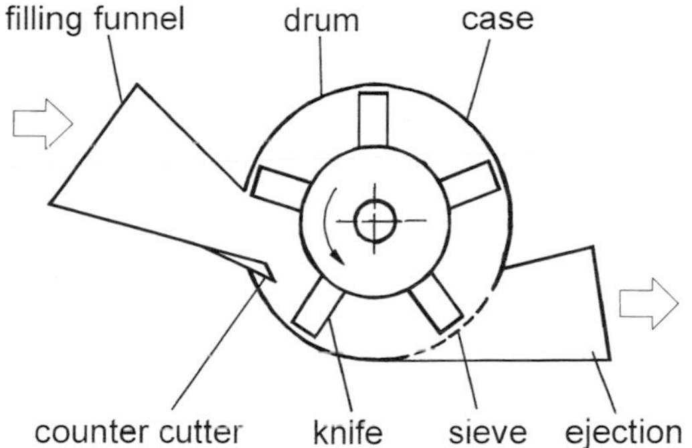

Fig. 9. Cracking mill (RELOE and SCHUCHARDT, 1993).

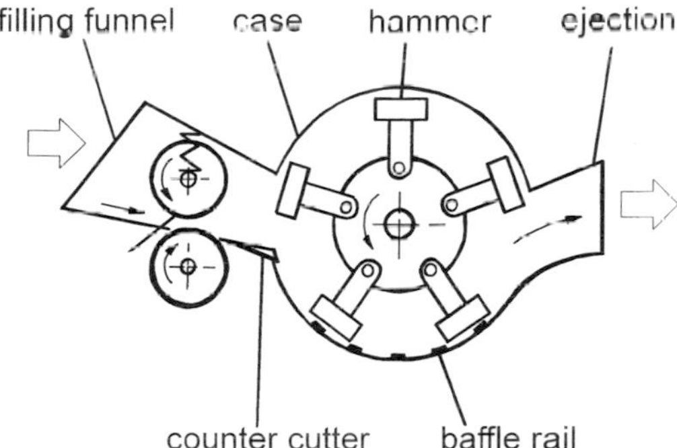

Fig. 10. Hammer mill (RELOE and SCHUCHARDT, 1993).

their usage at the place of origin of the wastes (e.g., cut trees and brushwood), they are constructed as mobile machines (Figs. 13–15) but stationary ones are also in use.

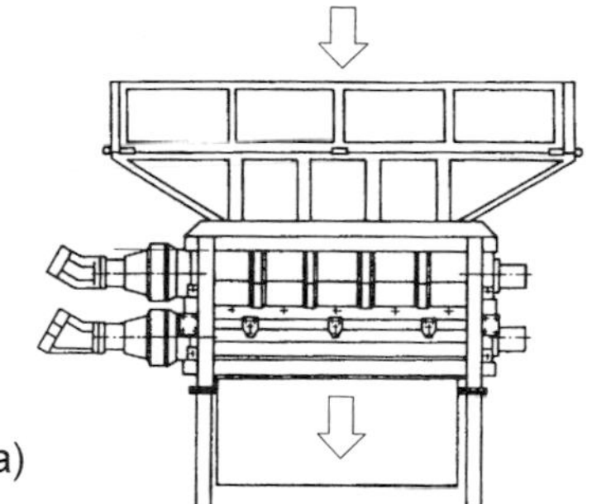

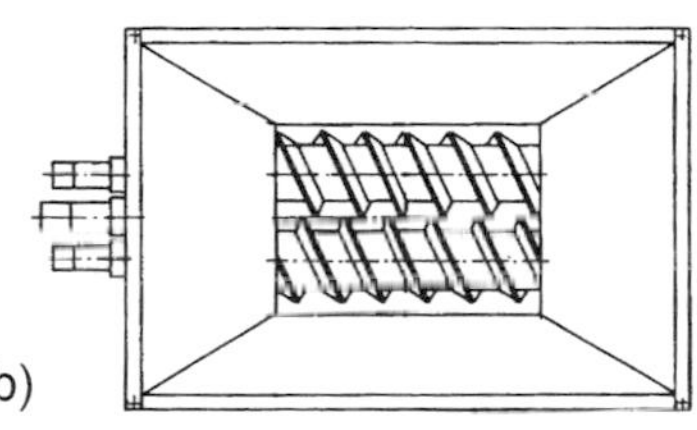

Fig. 11a, b. Screw mill; **a** view, **b** topview (drawing: Buehler AG, Uzwil).

4.2.2 Screening

The effect of screening of the raw wastes or of compost is the separation of particles with a required size of granulation. These particles can be the organic raw material for composting, the compost itself, or impurities. Because of the fibrous structure, adhesive characteristics, and sometimes high moisture contents screening can be very difficult. In practice, drum and plain screens (with hole plates, wire grates, stars

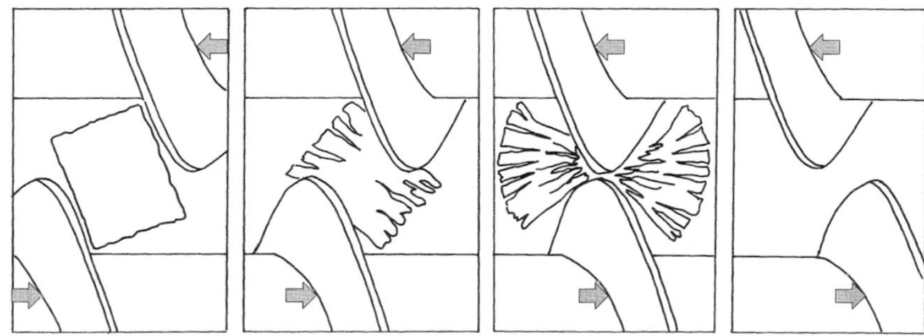

Fig. 12. Function of a screw mill.

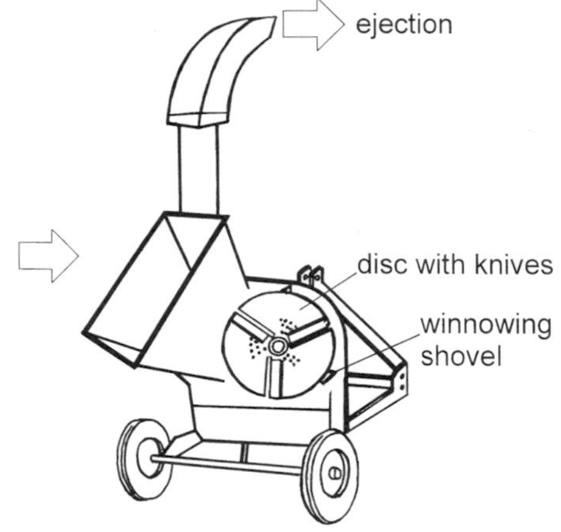

Fig. 13. Mobile cutting mill (MATTHIAS, 1992).

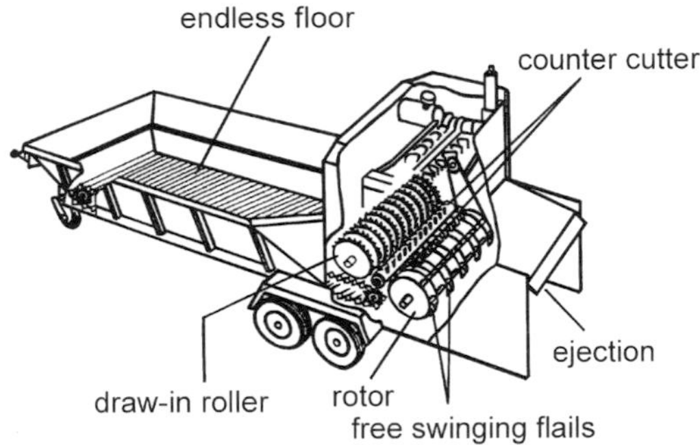

Fig. 14. Mobile shredder machine (MATTHIAS, 1992).

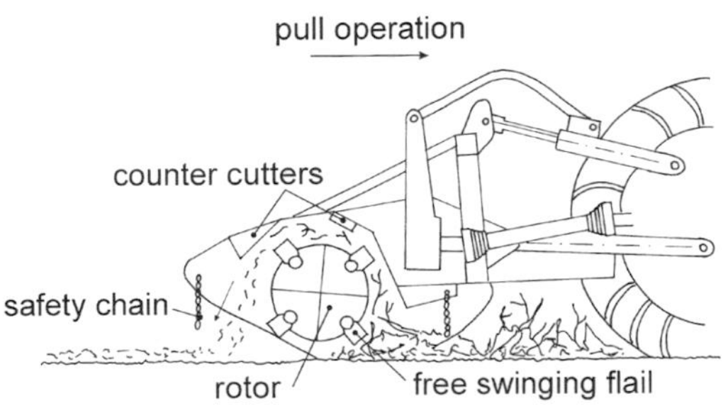

Fig. 15. Mulching machine (drawing: Willibald GmbH, Frickingen).

or profile iron) are usually in use. As described for disintegration aggregates screens can be stationary or mobile (Figs. 16, 17).

The screen residue, if necessary after separation of impurities and disintegration, can be used as an inoculum or as mulching material. The homogenous size of granulation is a precondition for compost sales. The size of the sieve holes depends on the further application of the compost or the separation of impurities. The screen residues are used for the following purposes:

>80 mm:	impurities (e.g., plastic bags, bottles, cans)
>40 mm:	mulching material for fruit and viticulture and landscape gardening; impurities; inoculum; bulking material
<24 mm:	compost for landscape gardening and agriculture
<10 mm:	compost for gardening

4.2.3 Selection of Impurities, Sorting

The aim of sorting is the selection of impurities such as plastic, paper, glass, metals (Fe, non-Fe), stones, textiles, and other unwanted materials. The techniques for sorting depend on the kind of waste, the capacity of the composting plant, the quality requirements of the compost, and the costs for manpower. In composting plants sorting will be done by

- pick-out (by hand or automatically),
- density sorting (e.g., ballistic separator, slope plate conveyor separator, throw separator),
- a magnetic separator.

4.2.4 Mixing

To ensure a homogenous rotting material of different kinds of wastes or additives, mixing of these materials is necessary. In practice the following solutions are known:

- mixing of wastes with a manure spreader (in small composting plants),
- composition of heaps layer by layer (different wastes) by a wheel loader (e.g., green waste and biowaste),

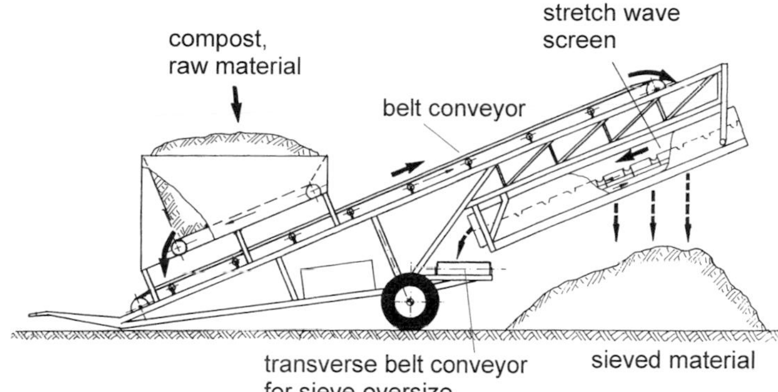

Fig. 16. Plain screen (RELOE and SCHUCHARDT, 1993).

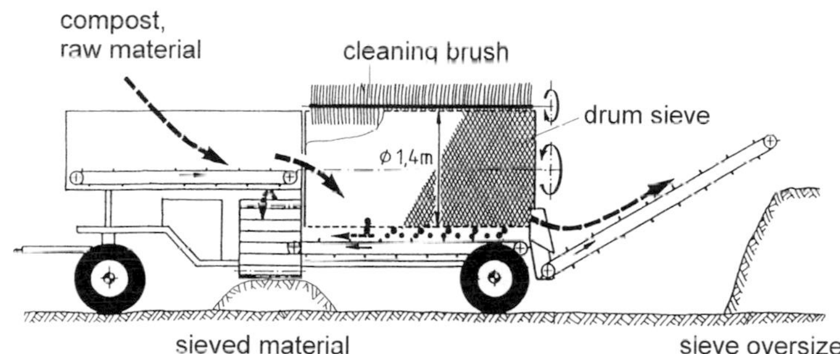

Fig. 17. Mobile drum screen (RELOE and SCHUCHARDT, 1993).

- mixing of wastes in stationary or mobile charge or continuous-flow mixers (e.g., mixing trailer, drum mixer; dry wastes and sludge).

The retention time in the mixer can be between a few seconds and a few days. In some composting processes the drum mixer has also the function of a precomposting reactor with forced aeration and self-heating and a retention time of more than 1 d. Screen drums can also be used as mixing equipment.

5 Composting Systems

Composting systems can be classified as

- non-reactor composting systems (Fig. 18) with
  - field composting and
  - windrow composting, and
- reactor composting (Fig. 19).

5.1 Non-Reactor Composting

5.1.1 Field Composting

During field composting, the simplest way of composting organic wastes, all microbial activity takes place in a thin layer at the surface or within a few cm of the soil (acre or grassland area). This system is applicable for sludge and green wastes (grass, straw, brushwood). To ensure a quick and uniform decomposition the green wastes have to be minced. Mulching machines (Fig. 15) can be used if the wastes are growing at the same area, otherwise collected wastes will be spread out by a manure spreader after chopping.

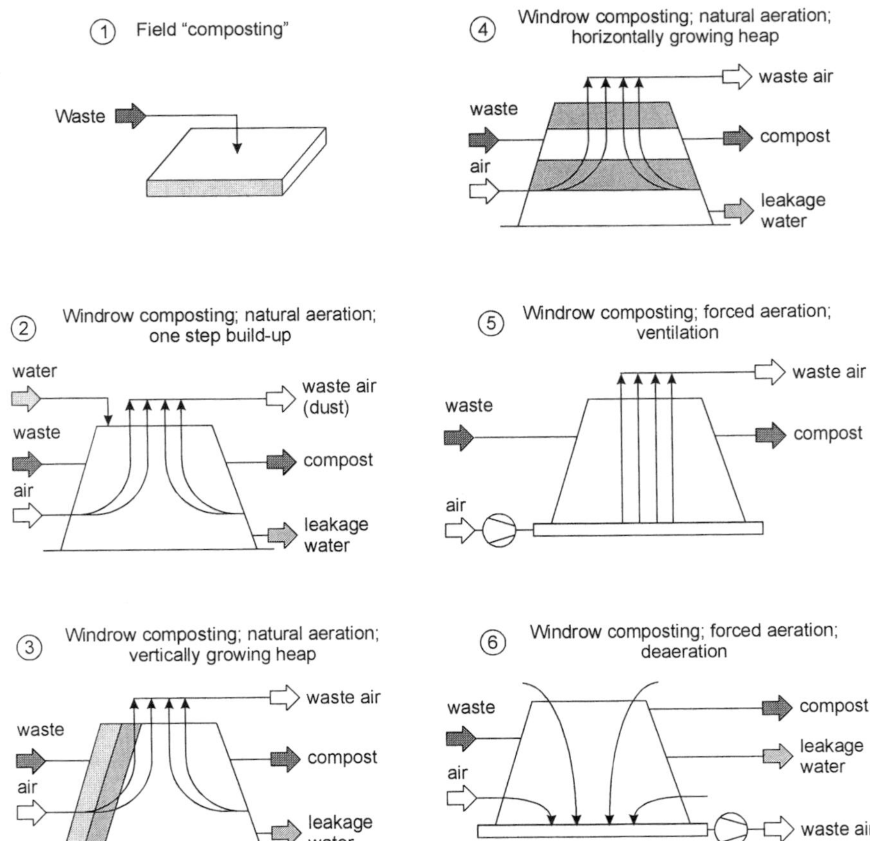

Fig. 18. Classification of non-reactor composting systems.

Because of the large surface between waste material and air there is no self-heating and, therefore, no thermal disinfecting and killing of weed seeds. Therefore, only wastes without hygienic problems and weed seeds should be utilized in this kind of composting.

In the narrower sense of the definition of composting field composting is not composting, because there is no self-heating and no real process control.

5.1.2 Windrow Composting

The characteristic of non-reactor windrow composting is the direct contact between the waste material and the atmosphere and, therefore, an interdependence of both. The composting process influences the atmosphere by odor, greenhouse gases, spores, germs, and dust. The atmosphere, carrier of the respiration gas oxygen, can influence the composting process by

- rain water,
  - advantage: water addition if the material for composting is or became too dry with the effect of higher intensity of the biodegradation process;
  - disadvantages: blocking of free air-space; intensified anaerobic conditions with odor emissions; debasement of compost quality; higher amount of leakage water;

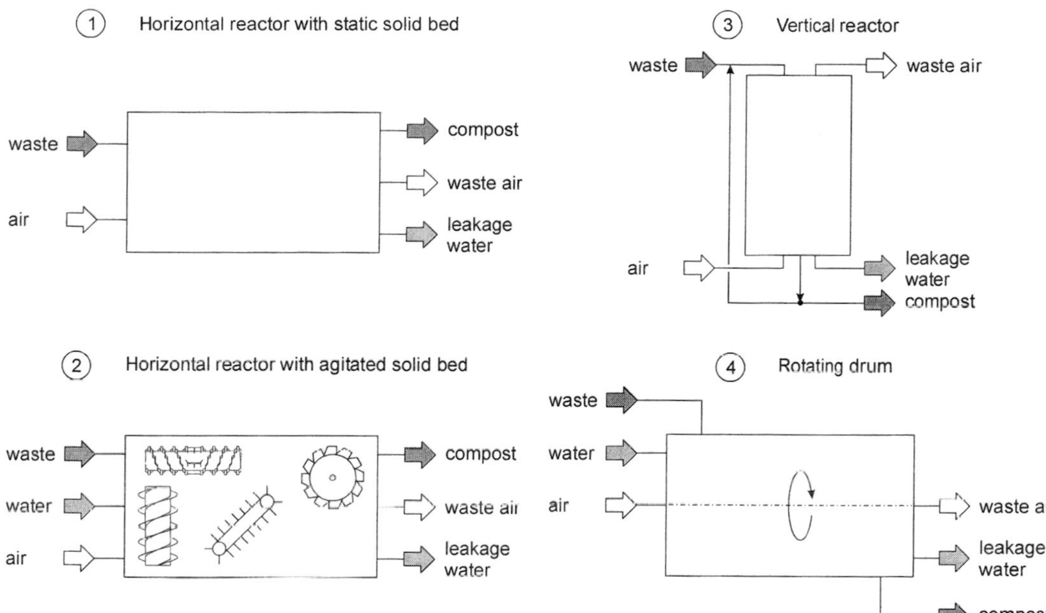

Fig. 19. Reactor composting systems.

- air temperature,
  - advantages: high air temperatures can increase the evaporation rate of very wet wastes with the effect of higher free air space and can shorten the lag phase at the start;
  - disadvantages: high air temperatures can increase the evaporation rate with the effect of water shortage; low air temperatures can delay or inhibit self-heating;
- air humidity,
  - advantages: low air humidity can increase the evaporation rate of very wet wastes; high air humidity reduces the evaporation rate;
  - disadvantages: low air humidity can increase the evaporation rate with the effect of water shortage; high air humidity can increase the evaporation rate with the effect of surplus;
- wind velocity can intensify the effects of air temperature and of air humidity.

The contact between waste material and atmosphere can be influenced by covering the piles with mature compost material, straw or special textile or fleece materials which allow a gas exchange but reduce an infiltration of rain water.

The geometric shape of a compost pile can be a

- triangular windrow,
- trapezoid windrow or
- tabular windrow (Fig. 20).

The height, width, and shape of a windrow depend on the waste material, climatic conditions, and the turning equipment.

Windrow composting (Fig. 18) can be classified as

- windrow composting with natural aeration and
  - 1-step buildup or
  - vertically growing heap or
  - horizontally growing heap and
- windrow composting with forced aeration by
  - ventilation or
  - deaeration.

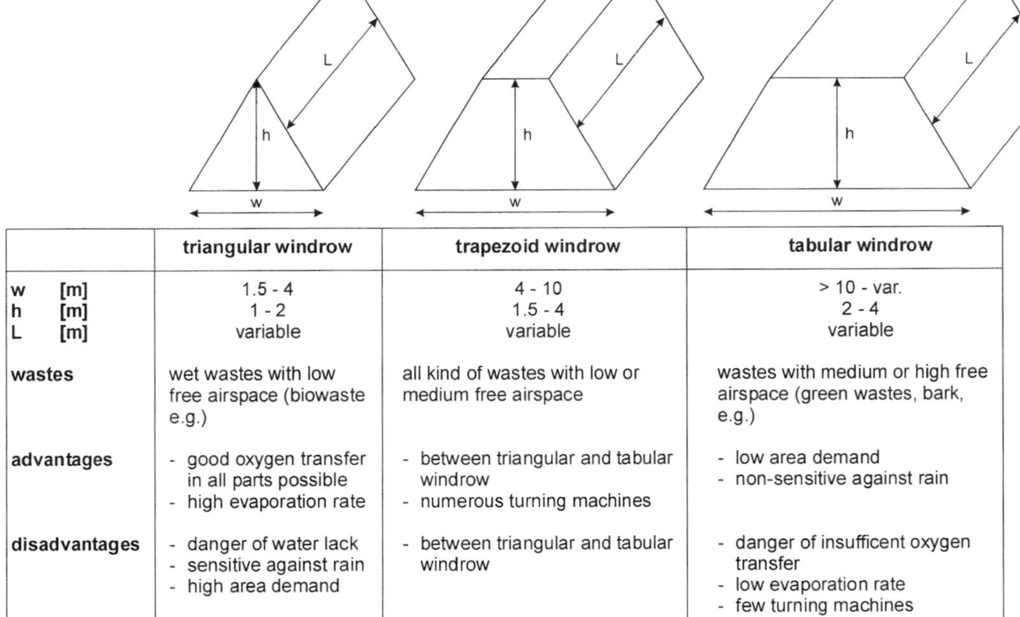

	triangular windrow	trapezoid windrow	tabular windrow
w [m] h [m] L [m]	1.5 - 4 1 - 2 variable	4 - 10 1.5 - 4 variable	> 10 - var. 2 - 4 variable
wastes	wet wastes with low free airspace (biowaste e.g.)	all kind of wastes with low or medium free airspace	wastes with medium or high free airspace (green wastes, bark, e.g.)
advantages	- good oxygen transfer in all parts possible - high evaporation rate	- between triangular and tabular windrow - numerous turning machines	- low area demand - non-sensitive against rain
disadvantages	- danger of water lack - sensitive against rain - high area demand	- between triangular and tabular windrow	- danger of insuffficent oxygen transfer - low evaporation rate - few turning machines

Fig. 20. Geometric shapes of windrows.

Furthermore, windrow composting can be classified as composting with and without mechanical turning of the heaps.

5.1.2.1 Windrow Composting with Natural Aeration and One-Step Buildup

This is a most widely used way of composting with the lowest costs. After preparation of the wastes the windrow will be built up in one step. As a result of microbial activity the temperature rises to 75°C with typical isothermal zones. The aeration of the windrow is natural and caused by heat convection (chimney effect). The exchange of the respiration gases is influenced by the temperature in the heap, the structure of the waste material, the free airspace, and the climatic conditions. The waste air contains all gases from the biodegradation processes, odor components, and, if the surface of the windrow is dry and wind is blowing or the windrow is turned, dust with spores and germs. Because of the high oxygen demand at the beginning of the decomposition process the danger of anaerobic conditions especially in the center of the heap is high. Rain water infiltrates into the windrow if there is no roof or rain protection and can increase the amount of leakage water.

5.1.2.2 Windrow Composting with Natural Aeration and Vertically Growing Heap

A growing trapezoid windrow can be build up vertically layer by layer. This offers the possibility, if the layer is thin enough (<0.5 m) and the time between two layers is long enough (>1 week), to ensure a sufficient oxygen supply. The self-heating in the upper layer will be provided by the heat from the lower layers of the heap. To reduce odor emissions from the upper layer with fresh material this layer can be covered by a thin layer (10–20 cm) of mature compost, which has the function of a biofilter. The amount of leakage water can be reduced because of higher evaporation rates at the surface of the upper layer and the higher water capacity of the lower layers.

5.1.2.3 Windrow Composting with Natural Aeration and Horizontally Growing Heap

This system is the same as described above, with the difference of horizontal layers in a triangular windrow. A turning can start after finishing the build up of the windrow. The area demand is higher than for composting in windrows which are built-up in one step (Sect. 5.1.2.1).

5.1.2.4 Windrow Composting with Forced Aeration by Ventilation

In this system a defined air stream is pressed through perforated pipes or plates into the waste material from the bottom, the zone with the highest oxygen demand. In dependence of the oxygen demand of the microorganisms the air flow can be adapted. Precondition for a uniform oxygen supply is a uniform structure of the waste and a relatively low streaming resistance. A disadvantage of this system can be a water lack at the bottom of the heap caused by the aeration air which transports water to the surface. If the aeration pipe is not integrated in the floor, mixing of the heaps with turning machines is not possible.

5.1.2.5 Windrow Composting with Forced Aeration by Deaeration

In this system the windrow is deaerated at the bottom by a blower, in contrast to the system with forced aeration by ventilation. This is the only system of non-reactor composting processes which allows collecting of the waste air and its treatment. The disadvantage is the transport of the air from the surface to the bottom and its growing content of carbon dioxide which can cause a lack of oxygen in this zone of the heap.

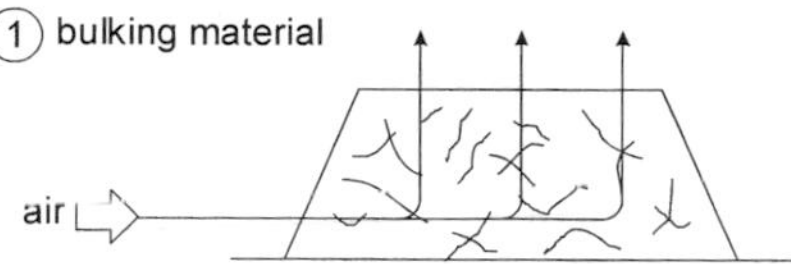

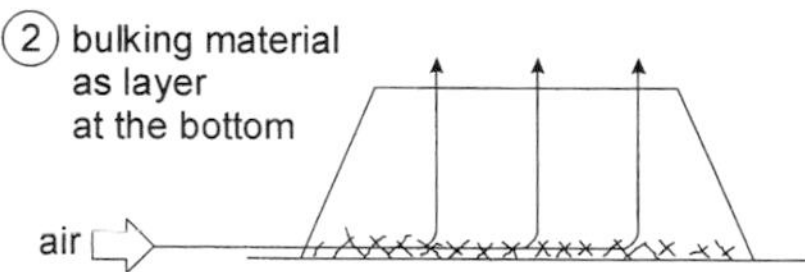

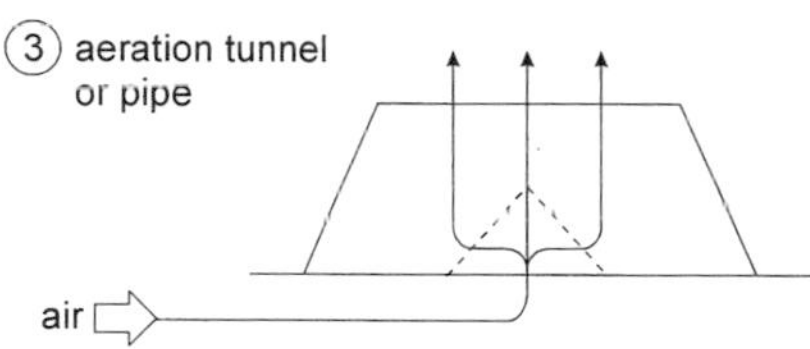

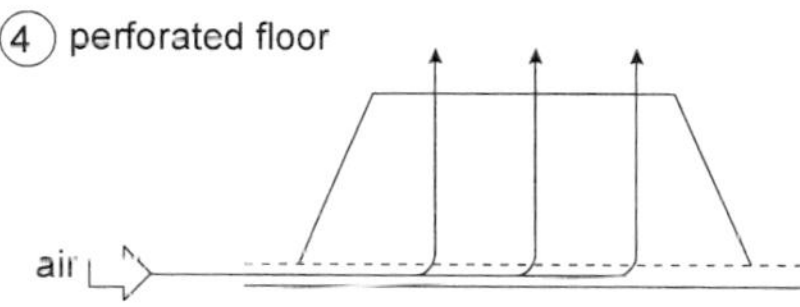

Fig. 21. Possibilities to support natural aeration in compost windrows.

Natural aeration in windrows can be supported by (Fig. 21)

- addition of bulking material to the waste,
- bulking material as aeration layer at the bottom of the windrow (20–30 cm),
- aeration tunnel at the bottom of the windrow, and
- perforated floor.

5.1.3 Windrow Turning Machines

To ensure a high quality of the compost windrows are agitated from time to time by turning. The effects of turning are

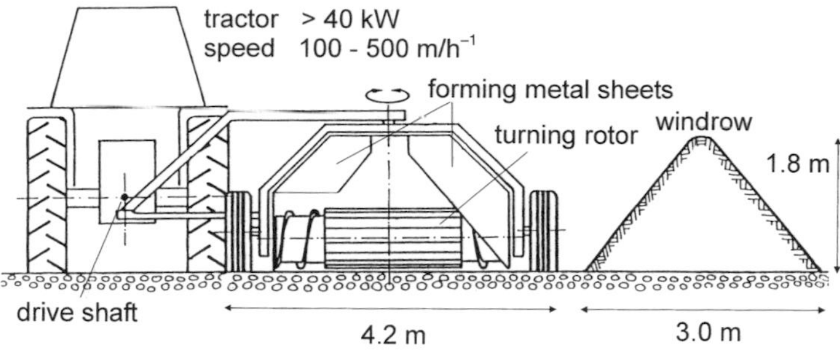

Fig. 22. Tractor driven windrow turning machine (RELOE and SCHUCHARDT, 1993).

- mixing of the material
  - for homogenization (dry or wet zones at the surface, wet zones at the bottom) and
  - killing of pathogenic microorganisms and weed seeds,
- renewing the structure and free airspace,
- increase of evaporation to dry the waste material or the mature compost.

The turning frequency depends on the kind and structure of the waste and the quality requirements of the compost. It can be between several times a day (at the start of the process when the oxygen demand is high or for drying mature compost) and some weeks.

Machines and equipment for turning are

- front–end loader of a tractor,
- wheel loader shovel,
- wheel loader shovel with rotor,
- manure spreader,
- tractor-driven windrow turning machines (Fig. 22), and
- self-driving windrow turning machines (Fig. 23).

The mixing quality of front–end loaders and wheel loaders is relatively poor and requires a well experienced driver. The compression of the (wet) wastes can be a disadvantage.

5.2 Reactor Composting

Every composting in a closed space (e.g., foil, textile, container, box, bin, tunnel, shed) with forced exchange of respiration gases is a reactor composting. The composting reactor (Fig. 19) can be classified according to the manner of the material flow as

- horizontal flow reactors with
  - static solids bed or
  - agitated solids bed,

Fig. 23. Self-driving windrow turning machine (drawing: Backhus GmbH, Edewecht).

- vertical flow reactors, and
- rotating drums.

With a few exceptions all reactors have a controlled forced aeration and the possibility to collect and to treat the waste air and the leakage water. An addition of water or other additives is possible only when the waste material is mixed in the reactor. If possible, composting processes are used only for precomposting because of the high costs of reactor composting compared to windrow composting reactors.

5.2.1 Horizontal Flow Reactors with Static Solids Bed

This is a batch system where the waste material will be filled by a wheel loader or transport devices into a horizontal reactor or will be covered by foil or textile material. The forced aeration, positive or negative mode or alternating, issues from pipes at the bottom of the material or from holes in the floor. The waste air with odor components and water can be treated in a biofilter or biowasher. In some systems a part of the waste is going in a cycle. The air flow rate can be controlled by the temperature in the material or the oxygen/carbon dioxide concentration in the air. The retention time in the reactor is between several days (precomposting) and some weeks (mature compost). The end product can be inhomogenous (partially too dry) and not biologically stabilized because there is no turning/mixing and no water addition to the waste material and a forced aeration only from one direction.

5.2.2 Horizontal Flow Reactors with Agitated Solids Bed

In this reactor type the waste material can be turned mechanically and water can be added in contrast to horizontal flow reactors with static solids bed. The devices for mixing of the wastes in tabular windrows or channels can be horizontally or vertically working rotors or screws, scraper conveyors or shovel wheels. A fully automatic function of the whole process is possible.

5.2.3 Vertical Flow Reactors

In this process the waste material will be treated in a vertical reactor, with or without stages, with a mass flow from the top to the bottom as a plug flow system or, if the material from the outlet at the bottom is filled back at the top, as a mixed system. The forced aeration is from the bottom or from vertical pipes in the material. This process can also run fully automatic.

5.2.4 Rotating Drum Reactor

The waste material is filled into a horizontal slowly rotating drum with forced aeration. The filling capacity is approximately 50%. On the helical way from one end of the drum to the other (plug flow) the material is mixed intensively and self-heating starts after a short time. Water addition is possible. Rotating drum reactors can also have the function of a mixing equipment.

6 Compost Quality

With regard to the use as fertilizer and soil conditioner compost must fulfill some quality requirements such as

- low contents of heavy metals and organic contamination,
- favorable contents of nutrients and organic matter,
- favorable C/N ratio,
- pH value neutral or alkaline,
- optimally mature,
- without interference of plant growing,
- mostly free from impurities (e.g., plastics, glass, metals, stones),
- mostly free from germinable seeds and plant parts,
- low content of stones,
- typical smell of forest soil,
- dark brown.

Because compost quality is a key factor in compost marketing and in waste management it is determined in some countries by regulations, norms and rules (HAUG, 1993; BIDLINGMAIER and BARTH, 1993; AMLINGER and LUDWIG-BOLTZMANN, 1996; FOLLIET-HOYTE, 1996; EPSTEIN, 1997).

7 References

AMLINGER, F., LUDWIG-BOLTZMANN, L. (1996), Biowaste compost and heavy metals: a danger for soil and environment?, in: *The Science of Composting* (DEBERTOLDI, M., SEQUI, P., LEMMES, B., PAPI, T., Eds.), pp. 314–328. London: Blackie.

Auswertungs- und Informationsdienst für Ernährung, Landwirtschaft und Forsten (1991), *Technik der Landschaftspflege, AID paper 1092*, Bonn: Auswertungs- und Informationsdienst für Ernährung, Landwirtschaft und Forsten.

DEBERTOLDI, M. (Ed.) (1987), *Compost: Production, Quality and Use.* London: Elsevier.

DEBERTOLDI, M., SEQUI, P., LEMMES, B., PAPI, T. (Eds.) (1996), *The Science of Composting*. London: Blackie.

BIDLINGMAIER, W. (1990), Möglichkeiten der Schwermetallreduzierung im Kompostrohstoff durch unterschiedliche Formen der getrennten Sammlung, in: *Müll-Handbuch* (HOESEL, G., BILITEWSKI, B., SCHENKEL, W., SCHNURER, H., Eds.), Code No. 5326. Berlin: Erich Schmidt Verlag.

BIDLINGMAIER, W. (1992), Charakteristik fester Abfälle im Hinblick auf ihre biologische Zersetzung, in: *Müll-Handbuch* (HOESEL, G., BILITEWSKI, B., SCHENKEL, W., SCHNURER, H., Eds.), Code No. 5303. Berlin: Erich Schmidt Verlag.

BIDLINGMAIER, W. (1995), Anlageninput und erzeugte Kompostqualität, in: *Biologische Abfallbehandlung II* (WIEMER, K., KERN, M., Eds.), pp. 109–120. Witzenhausen: M.I.C. Baeza Verlag.

BIDLINGMAIER, W., BARTH, J. (1993), Anforderungsprofile für Komposte im europäischen Vergleich, in: *Biologische Abfallbehandlung* (WIEMER, K., KERN, M., Eds.), pp. 105–130. Witzenhausen: M.I.C. Baeza Verlag.

DOEDENS, H. (1996), Einfluß der Sammellogistik und des Gebührensystems auf die Bioabfallmengen, in: *Biologische Abfallbehandlung* (WIEMER, K., KERN, M., Eds.), pp. 127–138. Witzenhausen: M.I.C. Baeza Verlag.

EPSTEIN, E. (1997), *The Science of Composting.* Lancaster, Basel: Technomic Publication.

FINSTEIN, M. S. (1992), Composting in the context of municipal solid waste management, in: *Environmental Microbiology* (MITCHELL, R., Ed.), pp. 355–374. New York: Wiley-Liss.

FOLLIET-HOYTE, N. (1996), Canadian compost standards, in: *The Science of Composting* (DEBERTOLDI, M., SEQUI, P., LEMMES, B., PAPI, T., Eds.), Part 1, pp. 247–254. London: Blackie.

FRICKE, K. (1990), Stellenwert der biologischen Abfallbehandlung in integrierten Entsorgungskonzepten, in: *Biologische Verfahren der Abfallbehandlung* (DOTT, W., FRICKE, K., OETJEN, R., Eds.), pp. 1–58. Berlin: Erich Freitag Verlag.

FRICKE, K., TURK, T. (1991), Stand und Stellenwert der Kompostierung in der Abfallwirtschaft, in: *Bioabfallkompostierung – Flächendeckende Einführung* (WIEMER, K., KERN, M., Eds.), pp. 13–98. Witzenhausen: M.I.C. Baeza.

GASSER, J. K. R. (Ed.) (1995), *Composting of Agricultural and Other Wastes*. London, New York: Elsevier.

GRONAUER, A., HELM, M., SCHOEN, H. (1997), Verfahren und Konzepte der Bioabfallkompostierung, in: *Bioabfallkompostierung* (GRONAUER, A., HELM, M., SCHÖN, H., Eds.), No. 139, pp. 3–132. München: Bayerisches Landesamt für Umweltschutz.

HAUG, R. T. (1993), *The Practical Handbook of Composting*. Boca Raton, FL: Lewis.

HOITINK, H. J., KEENER, H. M. (Eds.) (1993), Science and Engineering of Composting, *Proc. Int. Composting Res. Symp.*, Columbus, OH, March 27–29, 1992.

HÜTHER, L., SCHUCHARDT, F., WILLKE, T. (1997a), Emissions of ammonia and greenhouse gases during storage and composting of animal manure, *Proc. Int. Symp. Ammonia and Odour Emissions from Animal Production* (VOERMANS, A. M., MONTENY, G., Eds.), pp. 327–334. Vinkeloord, NL, 6–10 October 1997.

HÜTHER, L., SCHUCHARDT, F., AHLGRIMM, H.-J. (1997b), Factors influencing nitrous oxide emissions during storage and composting of animal excreta, *Proc. 7th Int. Workshop on Nitrous Oxide Emissions*, pp. 163–170, Köln, Germany, 21–23 April 1997.

Hydro Agri Duelmen GmbH (Ed.) (1993), *Faustzahlen für die Landwirtschaft*, 12th Edn. Münster-Hiltrup: Landwirtschaftsverlag.

KAUTZ, O. (1995), Schwermetalle in Bioabfällen, in: *Biologische Abfallbehandlung* (THOMÉ-KOZMIENSKY, K. J., Ed.), pp. 80–92. Berlin: Erich Freitag Verlag.

KrW-/AbfG (1994), *Gesetz zur Förderung der Kreislaufwirtschaft und Sicherung der umweltverträglichen Beseitigung von Abfällen* (Kreislaufwirtschafts- und Abfallgesetz – KrW-/AbfG) vom 27. 9. 1994, BGBl. I, p. 2705.

MATTHIAS, J. (1992), Mechanische Aufbereitung als Mittel zur Verbesserung der Kompostierbarkeit pflanzlicher Reststoffe, *Thesis*, Göttingen, Germany: Georg-August-Universität.

MILLER, F. C. (1993), Composting as a process based on the control of ecological selective factors, in: *Soil Microbial Ecology* (BLAINE, F., METTING, J., Eds.), pp. 515–544. Basel, NewYork: Marcel Dekker.

RELOE, H., SCHUCHARDT, F. (1993), Kompostierungstechnik und -verfahren, in: *Kompostierung und landwirtschaftliche Kompostverwertung* (KTBL, Ed.), pp. 171–200. Münster-Hiltrup: Landwirtschaftsverlag.

SEQUI, P. (1996), The role of composting in sustainable agriculture, in: *The Science of Composting* (DEBERTOLDI, M., SEQUI, P., LEMMES, B., PAPI, T., Eds.). London: Blackie.

SMITH, J. F. (1993), The mushroom industry, in: *Exploitation of Microorganisms* (JONES, D. G., Ed.), pp. 249–271. London: Chapman & Hall.

TA Siedlungsabfall (1993), *Dritte Allgemeine Verwaltungsvorschrift zum Abfallgesetz* (TA Siedlungsabfall), Technische Anleitung zur Verwertung, Behandlung und sonstigen Entsorgung von Siedlungsabfällen vom 14. 5. 1993, BAnz., p. 4967.

THOMÉ-KOZMIENSKY, K. J. (1994), Qualitätskriterien und Anwendungsempfehlungen für Kompost, *Abfallwirtschaftsjournal* **6**, 477–494.

THOMÉ-KOZMIENSKY, K. J. (Ed.) (1995), *Biologische Abfallbehandlung*. Berlin: Erich Freitag Verlag.

Umweltbundesamt (1984), *Daten zur Umwelt*, Berlin: Umweltbundesamt.

WHITE, P., FRANKE, M., HINDLE, P. (1995), *Integrated Solid Waste Management: a Lifecycle Inventory*. London: Blackie.

WIEMER, K., KERN, M. (Eds.) (1993), *Biologische Abfallbehandlung*. Witzenhausen: M.I.C. Baeza.

WIEMER, K., KERN, M. (Eds.) (1994), *Verwertung biologischer Abfälle*. Witzenhausen: M.I.C. Baeza.

WIEMER, K., KERN, M. (Eds.) (1995), *Biologische Abfallbehandlung II*. Witzenhausen: M.I.C. Baeza.

WIEMER, K., KERN, M. (Eds.) (1996), *Biologische Abfallbehandlung* Witzenhausen: M.I.C. Baeza.

WIEMER, K., KERN, M. (Eds.) (1997), *Bio- und Restabfallbehandlung*. Witzenhausen: M.I.C. Baeza.

4 Technology and Strategies of Composting

UTA KROGMANN

New Brunswick, NJ, USA

INA KÖRNER

Hamburg, Germany

1 Objectives

"Composting is a method of solid waste management whereby the organic component of the solid waste stream is biologically decomposed under controlled conditions to a state in which it can be handled, stored, and/or applied to land without adversely affecting the environment" (GOLUEKE, 1977). Composting technologies that range from simple windrow systems to more complicated in-vessel systems and composting strategies like oxygen control (USEPA, 1981), temperature control (FINSTEIN et al., 1986), or biodrying (RICHARD, 1998) all have the common goal to optimize certain components of the composting process. Composting technologies are selected based on different needs and objectives such as targeted feedstock, required environmental controls, existing socioeconomic factors, and targeted compost end uses.

1.1 Feedstock

Generally, all feedstocks that are biodegradable can be composted. A list of various feedstocks is given below:

- **MSW** (mixed municipal solid waste)
- **Residual waste** (residuals after source separation of biowaste and dry recyclables)
- **Yard waste and other green wastes**
  - Grass clippings
  - Plant residuals
  - Brush and tree trimmings
  - Leaves
  - Cemetery wastes
  - Christmas trees
  - Seaweed and other aquatic plants
- **Agricultural wastes**
  - Excess straw
  - Spoiled hay and silage
  - Beet leaf residuals
  - Dead animals (not allowed in some countries)
  - Solid and liquid manure
- **Biowaste** (source separated food and yard waste)
- **Sewage sludge (biosolids)**
- **Paper products**
- **Market wastes**
- **Processing residuals**
  - Residuals from pharmaceutical industry
  - Residuals from the food processing and beverage industry

- Residuals from vegetable oil production
- Fish processing wastes
- Paunch contents from slaughterhouses
- Bark
- Sawdust and shavings

- **Forestry wastes**
  - Residuals from windbreaks
  - Logging residuals

Feedstocks can be of plant (i.e., vegetable scraps, yard waste) or animal origin (i.e., fish waste, dead animals); feedstocks may be processed by natural (i.e., animal manure) or by industrial processes (i.e., sewage sludge, paper, cardboard, biodegradable plastics). Feedstocks can be characterized by chemical, physical, and biological parameters including moisture content, particle size, organic matter content, nutrient content, heavy metal content, and synthetic organics content. Comprehensive data about the characteristics of various feedstocks can be found in SEEKINS (1986), ATV (1989), SEEKINS and MATTEI (1990), RYNK (1992), RASP (1993), SIHLER et al. (1993, 1996) and MÜLLER et al. (1997). Food waste and raw sewage sludge are examples of feedstocks requiring careful selection of the most appropriate composting methods. These wastes are very moist, easily degradable, and high in proteins; as a result, they are more difficult to handle than other wastes and require more controlled composting systems in most cases.

1.2 Environmental Controls

During composting, environmental controls are necessary to control condensates from active aeration, leachates, and air emissions, including odors and bioaerosols. Leachates, and to a lesser extent condensates, contain high concentrations of dissolved organic matter and nutrients which should not be released into surface and groundwater. Odorous compounds (low molecular weight alcohols, aldehydes and ketones, low molecular weight carbonic acids, nitrogen containing compounds, sulfur containing compounds, and etheric oils) are released during composting (SCHILDKNECHT and JAGER, 1979). Such odors can be a nuisance to neighbors and have resulted in the shutdown of several composting facilities. Bioaerosols (colloidal particles with bacteria, fungi, actinomycetes, or viruses attached) can be a potential health hazard to compost facility workers and neighbors (BÖHM, 1995). Environmental controls in composting facilities vary with technology and strategy and are a key factor in the selection of a composting technology. For example, comparisons between odor emissions of windrows and more controlled, enclosed compost facilities showed a reduction in odor emissions (measured as odor units) by 65–75% (BIDLINGMAIER and GRAUENHORST, 1996).

1.3 Socioeconomic Factors

Different countries, regions, and municipalities vary widely with regard to laws and regulations, culture, economy, waste quantities, and the availability of open space. Each of these factors influences the selection of optimum compost technologies. In urban areas with high property prices and neighbors nearby, more controlled technologies and strategies are needed in comparison to more rural areas. However, it should be noted that also in rural areas, especially adjacent to residential areas, increasing process control is desired. High population densities in many Central European countries have led to the acceptance of higher costs for solid waste management technologies than in other parts of the world. This generally allows the design and construction of more enclosed facilities in this region. Developing countries make use of simpler technologies for economic reasons. As a result, less developed countries often rely on more labor intensive technologies than industrialized countries.

1.4 Compost End Uses

The feedstock is the main factor which determines the chemical, physical and biological characteristics of the end product compost including its nutrient and contaminant content. For example, if a nutrient rich compost is needed, biowaste, sewage sludge, and manure are a more suitable feedstock than municipal solid waste (MSW) or yard waste. The most efficient way to reduce contaminants (heavy metals and synthetic organics) in compost is the implementation of source separation (see Sect. 4). The choice of composting technology and processing strategy also influences final characteristics of the compost. For example, the moisture content can be controlled by aeration, temperature, and pile size; pathogen reduction is affected by temperature; and the salt content in compost is reduced in open windrows due to leaching by rainwater (KROGMANN, 1994).

2 Influencing Factors

To enhance the composting process and to minimize environmental impacts the microbial decomposition must be controlled during the composting process. In comparison to a homogeneous system such as an activated sludge process, composting is a heterogeneous and solid substrate system with a limited moisture level (HAUG, 1993). Homogeneous systems are normally modeled using Monod kinetics, assuming that the mass transport of the substrate is not rate limiting and the kinetics are controlled by the limiting substrate. In a heterogeneous system like composting, many additional mass transport processes such as transport of both oxygen and solubilized substrate are limiting. Since these processes are very complex and are not understood in detail, the following main influencing factors are typically controlled: biodegradability of feedstock (Sect. 2.1), moisture content (Sect. 2.2), oxygen content, material structure and aeration (Sect. 2.3), temperature (Sect. 2.4), nitrogen content and transformation (Sect. 2.5), and pH (Sect. 2.6).

2.1 Biodegradability of Feedstock

Chemically, feedstocks consist of an inorganic fraction and several, more or less biodegradable organic compounds including

(1) lignin, hemicelluloses and cellulose,
(2) sugars and starches,
(3) fats and waxes, and
(4) proteins.

Feedstocks are mostly of plant origin. Generally, the resistance of these organic compounds to microbial decomposition increases in the following order: sugar, starch, proteins, fats, hemicelluloses, cellulose, lignin and other high-molecular phenolic compounds (POINCELOT, 1975). Waxes are difficult to degrade. KROGMANN (1994), who reviewed the organic composition of plants, foodstuff, and biodegradable organic wastes, confirmed that cellulose (20–50% of dry solids), hemicelluloses (10–25% of dry solids), and lignin (15–30% of dry solids) are the main components of lignocellulytic wastes like branches, tree stumps, etc. Foodstuff can contain significantly higher portions of sugars and starches (up to 84% dry solids), proteins (up to 81% dry solids), and fats (up to 63%). For sewage sludge, the following composition was reported: 37% proteins, 4.7% lipids, 2.6% cellulose, and 6.9% lignin (EPSTEIN, 1997). It should be emphasized that there is a considerable range in the organic composition of the same feedstock, which can change, for example, for tree branches even within the same genus like pine.

Degradation rates of biodegradable organic wastes are measures of stoichiometric oxygen consumption and also are key factors influencing the energy balance during composting. In a typical composting facility, food wastes achieve organic matter degradation rates of more than 60% (KROGMANN, 1994), biowaste about 50% (KROGMANN, 1994), lignocellulytic plant materials about 35–45% (KROGMANN, 1994), and raw sewage sludge about 70–80% (HAUG, 1993). A feedstock with a low degradation rate can cause insufficient self-heating during the winter, as HAUG and ELLSWORTH (1991) showed for a raw sludge (degradation rate of only 48% within 60 d) amended with pine sawdust (degradation rate of only 11.2% within 90 d). To ensure sufficient self-heating, a more biodegradable waste had to be added (HAUG, 1993).

2.2 Moisture Content

Moisture is essential to the decomposition process, since most microbial decomposition occurs in thin liquid films on the surface of particles. Microorganisms absorb dissolved nutrients and water also serves as a medium for distribution within the heterogeneous compost substrate. During the composting process the minimal moisture content depends on the requirements of the microorganisms for water, whereas the maximum moisture content for composting to occur is determined by the competition between air and water in pores (O_2 supply).

The minimum required moisture content for microbial degradation ranges between 12 and 25% (BIDLINGMAIER, 1983; SCHUCHARDT, 1988; GOLUEKE, 1989). The optimum moisture content is feedstock specific and varies between 40% and 70% with higher optimum moisture contents for coarser feedstocks (BIDLINGMAIER, 1983). During composting, the maximum tolerable moisture content of coarser feedstocks (i.e., wood and bark 74–90%) exceeds the tolerable moisture content of less structured feedstocks (i.e., paper 55–65%, food waste and grass 50–55%) (BIDLINGMAIER, 1983; KUTZNER and KEMPF, 1994). At the end of the composting process, the finished compost should not have a moisture content of more than 35–45% to avoid storage, transport and handling problems (LAGA, 1995).

For some dry wastes like MSW (initial moisture content: 30–45%), the addition of water or a moister feedstock may be necessary (BIDLINGMAIER, 1983). Generally, feedstocks tend to be more often too wet than too dry (typical initial moisture contents: food waste 40–75%, sewage sludge 87–98%, yard waste 40–85%). The addition of a dry bulking agent like wood chips, shredded bark, sawdust, or recycled compost is a common practice to lower the initial moisture content. The bulking agent is usually drier than the initial feedstock and can be used to reduce the moisture content of the combined feedstock. Additionally, the optimum moisture content for the feedstock and bulking agent mixture is usually raised since, in most cases, the bulking agent increases the coarseness of the feedstock.

The addition of a bulking agent increases the compost mass significantly. The ratio between the wet mass of the bulking agent (G_B) and the wet mass of the feedstock (G_R) is called the recycling ratio R ($R = G_B/G_R$). For example, the moisture content of a biowaste (75%) is adjusted to 50% and 60%, respectively, through the addition of a bulking agent (30%). The resulting recycling ratio is 1.25 $[(R \cdot G_R \cdot 0.3 + G_R \cdot 0.75)/(R \cdot G_R + G_R) = 0.5 \Rightarrow R = 1.25]$ and 0.5, respectively. This means that the addition of the bulking agent ($G_B = R \cdot G_R$) increases the initial composting mass (G_R) by 125% and 50%, respectively.

Biodrying has recently been suggested in order to reduce the amount of bulking agent needed (RICHARD, 1998). The initial composting phase is operated like a sequential batch reactor with sequential addition of the wet feedstock which is dried by the already composting feedstock.

During composting the initial moisture content changes. Small amounts of water (i.e., 0.6 g g^{-1} $C_6H_{12}O_6$ degraded) are generated as a metabolic endproduct and organic matter is degraded resulting in an increase in the moisture content. More importantly, the increased temperature during composting reduces the moisture content via evaporation. Evaporation is the major energy release mechanism during composting (MILLER et al., 1982). How the moisture content changes during composting is especially dependent on the temperature throughout the compost and the aeration rate (see Sect. 2.3).

If too much moisture is lost, water needs to be added. To compensate for the moisture loss during composting, water is added directly – most efficiently during turning – or sometimes the aeration air is saturated with water to increase overall moisture levels in the compost. However, the effect of aeration with water-saturated air is limited since the temperature of the aeration air will increase while moving through the compost and the air will not remain water-saturated at elevated temperatures and potentially dry the compost. In addition, since evaporation is the major energy release mechanism during composting the aeration air's ability to control the temperature is reduced.

If the water holding capacity of the compost is exceeded, leachate is released. Analyses of leachates from a biowaste windrow composting facility under a roof showed a COD of 18,000–68,000 mg O_2 L^{-1}, a BOD_5 of 10,000–46,000 mg O_2 L^{-1}, and NH_4-N of 310–855 mg L^{-1} (FRICKE, 1988). These concentrations indicate that treatment of the leachate is necessary.

Forced-pressure or vacuum-induced aeration can generate a vertical moisture gradient in the compost material that increases from the air inlet to the exhaust air outlet because the aeration air is already water-saturated when it moves towards the outlet. This can slow down the microbiological degradation near the air inlet if the moisture decreases below optimum. The moisture gradient can be changed by reversing the direction of the active aeration, frequent turning (see Sect. 3.3.1.1), or recirculation of large aeration air amounts (see Sect. 3.3.3.1).

To control the moisture content during composting, the moisture content can be estimated based on the experience of the operator, measured in a compost sample, or calculated based on the initial moisture content of the feedstock and the humidity and temperature of the inlet air and exhaust air.

2.3 Oxygen Content, Material Structure, and Aeration

During aerobic metabolism a sufficient supply of the electron acceptor oxygen is essential (example for stoichiometric oxygen demand in Tab. 1). The recommended oxygen content in the exhaust gas leaving the compost varies between 5% (STROM et al., 1980) and 18% (DE BERTOLDI et al., 1983). According to GOLUEKE (1977), in particles of 1–2 mm, anaerobic niches are usually present.

Regardless of the feedstock or the selected technology, a minimum free pore space of 20–30% is recommended for a sufficient supply of oxygen (BIDLINGMAIER, 1983; HAUG, 1993). Grinding and shredding reduce the structure and porosity of the feedstock and increase feedstock surfaces which enhances the microbiological degradation. If the feedstock has a stable structure, the feedstock can be finer without adversely affecting the oxygen supply (woody material <1 cm, food waste >2.5–5 cm) (GOLUEKE, 1977).

Besides supplying oxygen, aeration has the function of drying the compost (see Sect. 2.2) and controlling temperatures in the compost that could be detrimental for the microorganisms (see Sect. 2.4). A sample calculation (Tab. 1) shows that drying can require 10-fold more aeration air than would be stoichiometrically necessary. Temperature control requires about the same aeration rates as drying (HAUG, 1993).

At the beginning of the composting process, the high degradation rate (see Sect. 3.1) results in a high oxygen demand compared to the average oxygen demand. Oxygen can be provided by forced-pressure or vacuum-induced aeration, by natural ventilation (diffusion and convection), and to a lesser extent by turning.

If active aeration is used, the air flow rate (i.e., restricting air flow with a baffle, ventilator with variable rotation frequency), frequency

Tab. 1. Aeration Demand Based on Different Objectives (Example Calculation)

Objective	Total Air Demand [L air g^{-1} dry feedstock]
Supply according to the stoichiometric oxygen demand	2.54
Supply according to the stoichiometric oxygen demand, 50% utilization rate	5.08
Drying (initial moisture content 65%)	11.40
Drying (initial moisture content 80%)	27.60

Assumptions: organic matter (initial) – 65%; degradation rate – 54.3%; stoichiometric oxygen demand – 2 g O_2 per g degraded organic matter; moisture content (end) – 35%; air – 1.2 g L^{-1} at 25 °C, 10^5 Pa, 23.4 wt.-% O_2; air temperature (inlet) – 20 °C; air temperature (outlet) – 60 °C.

and length of aeration periods, the direction (forced-pressure, vacuum-induced), type (fresh, exhaust air), and condition (temperature, moisture content) of the aeration air can be varied.

2.4 Temperature

During composting, energy is released as heat due to microbial degradation of the feedstock. The rate at which the temperature of the feedstock increases depends on the energy balance of the whole system, and this varies with type and amount of feedstock, aeration, and/or insulation. For example, a backyard composting pile is relatively cool because only small amounts of feedstock are added and all heat energy is released due to radiation, convection, and conduction to the surrounding air before the temperature in the composting pile can increase.

The temperature affects the microorganism population, pathogen reduction, oxygen diffusion, oxygen solubility in water, degradation of the proteins, release of ammonia via the exhaust gas, nitrification rate, etc.

Most composting experiments have concluded that the optimum temperature during the high-rate decomposition period is about 55 °C. At temperatures over 60 °C, the diversity of the microorganisms is significantly reduced. At 70 °C the total biological activity is 10–15% less than at 60 °C, whereas at 75–80 °C no significant biological activity was detected (STROM, 1985). Laboratory experiments with biowaste, food waste, and food waste mixed with wood chips showed no significant advantages of temperature control if only short temperature increases up to 70 °C were measured (KROGMANN, 1994). During curing, the temperature optimum is around 40 °C (JERIS and REGAN, 1973), which is an optimum temperature for nitrification (DE BERTOLDI et al., 1983).

In large-scale composting facilities with active aeration, the temperature is controlled via aeration during high-rate degradation (see Sect. 2.3) (FINSTEIN et al., 1986). A thermistor in the compost enables feedback control of the blowers. However, with temperature control measures must be taken to prevent excessive substrate drying (see Sect. 2.2).

2.5 Nutrients

The amount of nutrients necessary for the composting process depends on the chemical composition of the decomposing microorganisms and additional elements that are involved in the metabolism. With the exception of nitrogen, biodegradable wastes generally contain enough macronutrients including carbon, sulfur, phosphorus, potassium, magnesium, and calcium, and micronutrients to sustain the composting process (GLATHE and FARKASDI, 1966). Very uniform feedstocks can create exceptions. For example, MSW with a high paper content was too low in phosphorus (BROWN et al., 1998).

Suitable carbon/nitrogen ratios (C/N) at the beginning of the composting process are between 20:1 to 30:1 for most wastes. Of greater importance is the actual availability of the carbon and nitrogen which is often neglected as a factor. Lignin, some aromatics and cellulose bedded in lignin are difficult to degrade. With the exception of keratin (structure protein, e.g., in hair) and a few similarly resistant components, nitrogen is considered very easily degradable (GOLUEKE, 1977). For woody feedstocks, C/N ratios of 35:1 to 40:1 are considered optimum (GOLUEKE, 1977). Too high C/N ratios slow down the microbial degradation and too low C/N ratios result in the release of nitrogen as ammonia (GOLUEKE, 1977). The most important method of controlling this ratio is by varying the composition of the feedstock.

2.6 pH

The pH value in most ecosystems is in the range between 5 and 9. For composting, the optimum pH range is between 7 and 8. Low-molecular organic acids produced as anaerobic intermediate products in collection containers can reduce the pH in biowaste to 5 (KROGMANN, 1994). During composting, the pH increases due to the volatilization of the organic acids and the release of bases like ammonium, pyridine, and pyracine (FINSTEIN and MORRIS, 1975). FERNANDES et al. (1988) enhanced the initial degradation of fat containing feedstocks at a pH less than 6 by the addition of CaO. In most cases, the addition of alkaline materials is not considered to be necessary (FINSTEIN and MORRIS, 1975; HAUG, 1993).

2.7 Process Control

In a composting facility, the initial feedstock properties are adjusted and during the composting process the previously discussed factors are automatically or manually controlled according to predetermined set points. Process control is facility specific, however, one example for box composting (see Sect. 3.3.3.2) will be presented (Fig. 1).

The objective of phase I is to reach the operating temperature of 40 °C as quickly as possible. The main degradation of the feedstock takes place in phase II as indicated by the highest aeration rates combined with the highest CO_2 concentrations. For maximum degradation during this phase, aeration should control the temperatures between 40 and 50 °C. Hygienization (pasteurization effect for the reduction of pathogens) takes places in phase III with temperatures over 60 °C at reduced aeration rates over a period of 3 d. In phase IV, the aeration rate is increased again to cool the compost to minimize odors when the composting box is opened. After 10 d of high-rate degradation in the composting box, stabilization and curing will subsequently be conducted in windrows.

3 Composting Systems

3.1 Overview

Processing in a composting facility (Fig. 2) includes steps for pre-processing of the initial feedstock and post-processing of the final compost (see Sect. 3.2). The composting process itself is divided into three phases: high-rate degradation, stabilization, and curing. According to the German grading system, self-heating and respiration rates are measures of

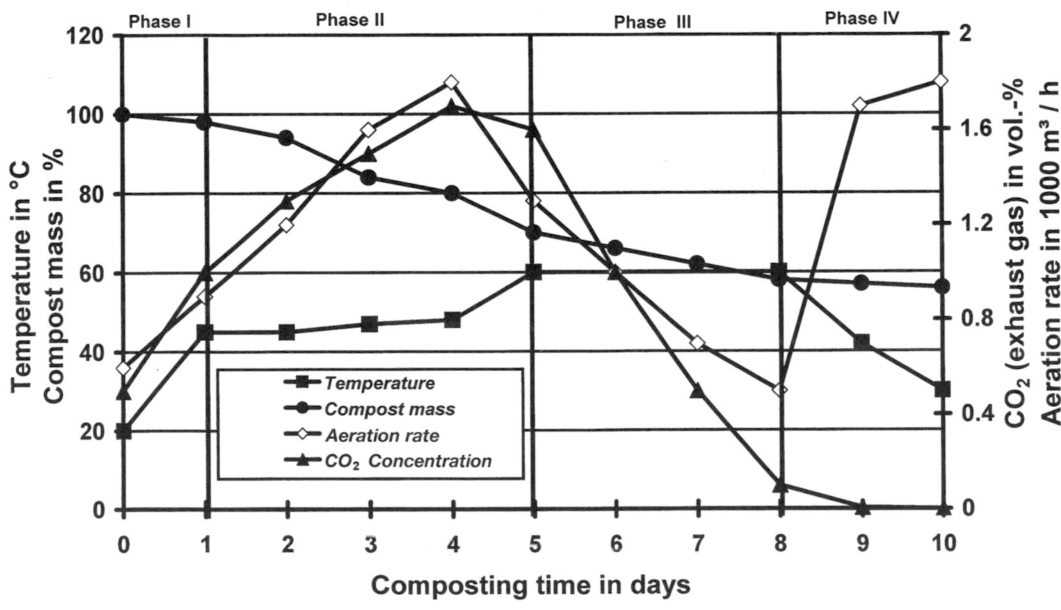

Fig. 1. Example for process controll with different set point temperatures in each phase of composting in Herhof boxes (Herhof, 1995).

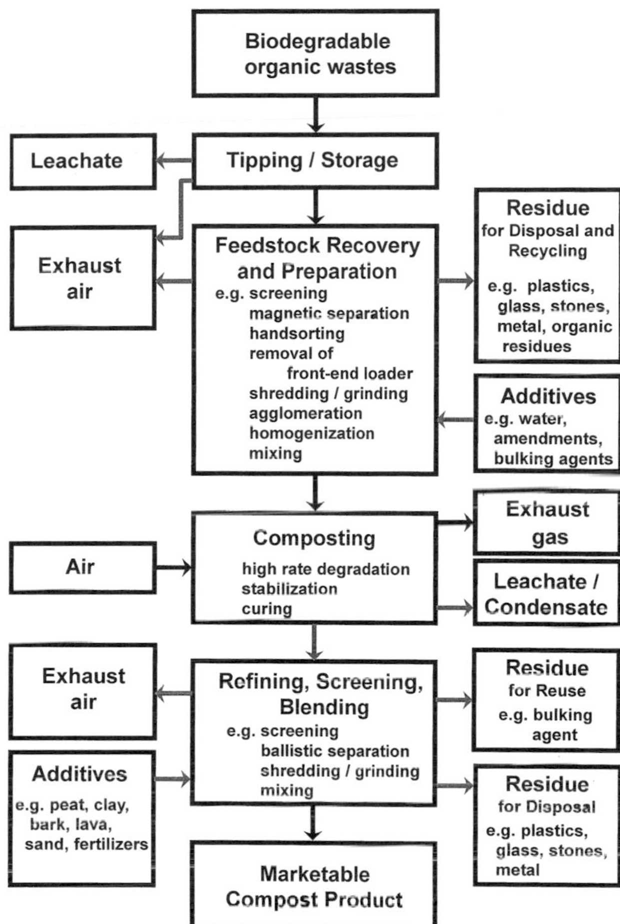

Fig. 2. Processing steps in a composting facility.

compost stability; a class of stability of I indicates a fresh feedstock and a class of stability of V a finished compost. High-rate degradation is the thermophilic part of the composting process where the feedstock is degraded to a "fresh compost" (class of stability: II). During the high-rate degradation phase, volume and mass are reduced by degradation of the easily degradable material, that is usually responsible for both vector attraction (disease spreading organisms like flies, rats, etc.) and the most intense odor emissions released from composting facilities. Additionally, pathogens are destroyed due to the thermophilic temperatures. During stabilization the "fresh compost" is degraded to "active compost" (class of stability: III–IV). The temperature decreases, the degradation continues and the organic matter is further stabilized. During curing, the compost matures to "finished compost" (class of stability: V). Ambient temperatures are reached and the previously started humification continues. However, it should be noted that even a "finished compost" might not be completely stable.

Malodorous compost metabolic products such as sulfur and nitrogen containing compounds may be deodorized by treating the exhaust gases in biofilters or other treatment processes. Liquid emissions (leachate, condensate) are returned to the compost for watering the feedstock or treated by standard wastewater treatment methods before release into surface waters.

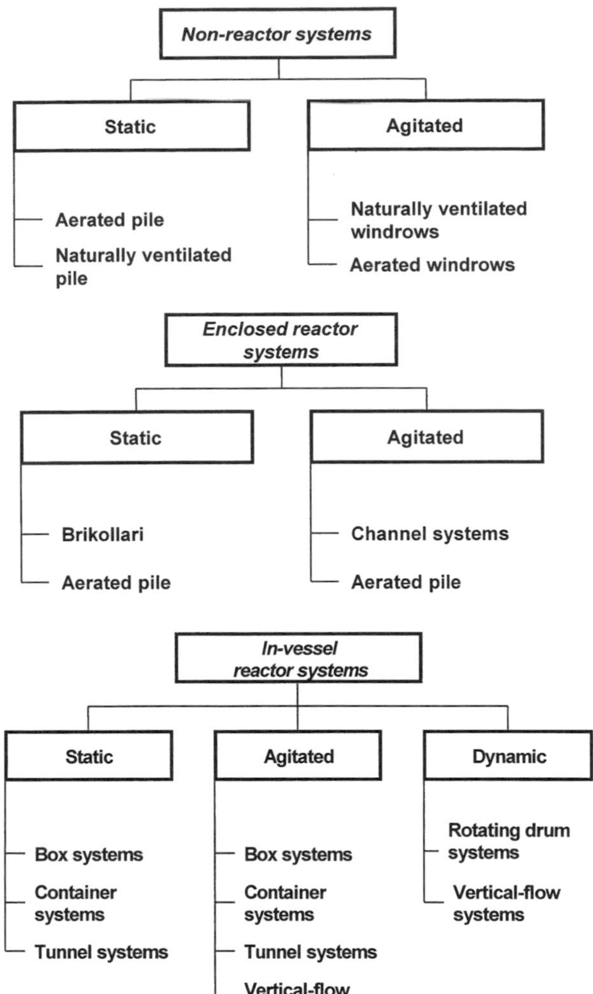

Fig. 3. Overview of common composting systems on market.

Currently, there are many different composting systems on the market (Fig. 3). The main difference is whether the high-rate degradation takes place outside (open or also called non-reactor system), in an enclosed building (enclosed reactor system), or in an in-vessel reactor system. In non-reactor systems, the exhaust gas of the composting process escapes in most cases into the surrounding environment without deodorization (exception: i.e., aerated static pile with vacuum-induced aeration). Reactor systems enable the treatment of the exhaust gas. In an in-vessel reactor system, the free air space above the compost is minimal which lowers the exhaust gas quantity compared to an enclosed reactor system. Especially in Central Europe, where countries are densely populated and where residents are sensitive regarding odors, reactor composting systems have become the system of choice during the last 10 years.

Another distinction is how and in which direction the waste moves through the composting system. In a static system, the substrate is not moved after placement. In dynamic systems the waste is moving continuously for longer intervals. Dynamic systems include only a few in-vessel systems including towers (vertical flow) and rotating drums (horizontal flow). In agitated systems, the waste is at rest for most of the time but is moved or turned at certain time intervals for short periods for homogenization, fluffing, and to a lesser extent for aeration. In some systems, the compost is agitated in place, in other systems the agitation is combined with movement of the compost.

Due to higher biological activity and higher odor emissions, high-rate degradation and stabilization require more control compared to curing. Therefore, the discussed enclosed and in-vessel composting systems are in many cases only used for high-rate degradation and stabilization, while curing takes place in non-reactor systems.

3.2 Pre- and Post-Processing

As the first step in a composting facility, biodegradable organic wastes are unloaded on the tipping floor. Very putrescible organic wastes like biowaste must be processed as soon as possible while other wastes like tree stumps might be stored for several months before being further processed. The purpose of the subsequent feedstock recovery and preparation (Fig. 2) is to recover recyclable materials (i.e., metals from MSW), to remove non-compostable materials, to improve the final compost quality, and to prepare the waste to adjust its biological, chemical and physical properties. For example, commonly sludges are too wet and yard wastes are too bulky. Since mixed MSW contains up to 45% (MANSER and KEELING, 1996) of non-compostable materials in comparison to source-separated biowaste with 0.5–7% (KROGMANN, 1994), MSW composting facilities need more extensive unit separation processing to remove non-compostable materials.

Basic unit separation processes are divided into size fractionation and sorting. For size fractionation, screens separate the waste into large materials (film, plastics, large paper, cardboard, miscellaneous), mid-sized materials (recyclables, most organics, miscellaneous), and fines (organics, metal fragments, miscellaneous). Sorting refers to separation by material properties like density or electromagnetic characteristics (i.e., plastics and paper by air classification, metals, glass, gravel by ballistic separation, ferrous metals by magnetic separation, non-ferrous metals by eddy current separation, recyclables, inerts by handsorting, oversized items by front–end loader) (RICHARD, 1992; VAUCK and MÜLLER, 1992). Unit processes used for creating sufficient feedstock surfaces are shredding and grinding (i.e., by shear shredder, hammer mill). Some composting processes require the generation of briquettes (agglomeration) which are made in special presses (see Sect. 3.3.2.3). The last stage of pre-processing is usually the adjustment of moisture and C/N ratio and the addition of a bulking agent. Drum mixers can be used for homogenizing and mixing of two and more feedstock streams. Not all of the pre-processing steps listed in Fig. 2 need to be carried out, and they also may differ in how and when they are carried out.

To ensure the workers' health and safety due to the waste's release of odors (i.e., etheric oils and anaerobic metabolic products from waste storage) and bioaerosols, the pre-processing areas in many composting facilities are vented. In many cases, the vented air is used as aeration air for the composting process.

Which unit process is chosen for post-processing depends on the proposed use of the compost. Controlling the particle size can be accomplished by size fractionation and by shredding and grinding. In most facilities, the compost is screened to remove larger, not degradable particles and to produce a compost with specified particle sizes according to the recommended use (i.e., potting soil 0–10 mm; mulch 10–30 mm) (STÖPPLER-ZIMMER et al., 1994). If the overs contain only small amounts of inert materials, the overs can be recycled as bulking agent. In addition, other inert material can be removed by ballistic separation or air classification. For example, air classification of plastics is more efficient in this process stage, because the compost is drier than the feedstock. For some uses (i.e., growing media), the compost is blended with additives (i.e., organic components like peat or bark and inorganic components like clay, lava, or sand) (HAUKE et al., 1996). The selection and order of different unit processes depend on the physical and chemical characteristics of the feedstock and the proposed compost end use.

3.3 Composting Technologies

3.3.1 Non-Reactor Systems

3.3.1.1 Windrow Composting

Windrow composting is the oldest and simplest composting technology. Windrows, which are elongated piles, can be used for the entire composting process or for curing only. Windrow composting is a non-reactor process requiring frequent turning by specialized equipment. Windrows are naturally ventilated as a result of diffusion and convection (see Sect. 2.3). Less often, windrows are aerated by forced-pressure or vacuum-induced aeration similar to aerated static piles (see Sect. 3.3.1.2). The aeration pipes are placed in a bed under the windrow if turning is also performed.

The turning machine is used to increase the porosity, to break up clumps, and to homogenize the compost and thereby equalize moisture and temperature gradients in the windrow. The turning machine lifts, turns, reforms, and sometimes moistens the windrow. The effect of turning on the oxygen supply of the windrow is minimal. A simple, however, not very effective turner is the front–end loader. Straddle turners are commonly used, which straddle the windrow and simply turn it over. Some achieve this by means of a rotating drum with welded scrolls or teeth; the rotating drum spans the side frame of the machine at ground level. Others use a wide, back inclined steel plate conveyor that also spans the frame of the machine. Turning machines drive self-powered over the windrow or are powered by a vehicle that drives next to the windrow.

Only a few facilities work without turning. Turning frequencies decrease from high-rate degradation to curing. During curing turning is often omitted. In most cases, higher turning frequencies lead to a decrease in retention time but also to an increase in operating costs. For example, the retention time to produce an "active compost" (class of stability: III–IV) made from leaf, grass, and brush wastes was reduced from 4–5 months to 2–3 months when the turning frequency was increased from once per month to 7 times per month (MICHEL et al., 1996).

The turning equipment and the aeration type (natural versus active) determine the windrow dimensions like shape (i.e., triangle or trapezoid), height, and width. For example, the width of naturally ventilated, triangle windrows for biowaste varies between 3.0 and 4.0 m and the height between 1.0 and 2.5 m (KERN, 1991). The height of naturally ventilated windrows of leaves were recommended to be limited to 1.5 m when not turned and to 2.0 m when turned once (FINSTEIN et al., 1986). With active aeration, heights of between 2. 5 m and 3.0 m are feasible. The windrow length depends on waste quantities and the available space.

For frequently turned, naturally ventilated windrows of biowaste, retention times of 12–20 weeks are reported (KERN, 1991), while for windrows of yard waste, 4–36 months are recommended (TCHOBANOGLOUS et al., 1993).

3.3.1.2 Static Pile Composting

The main difference between the windrow and the static pile system is that static piles are not agitated or turned. The lack of agitation in static pile composting requires the maintenance of adequate porosity over an extended period of time even more than so in windrows. Either the feedstock itself must have sufficient porosity (i.e., certain yard waste feedstocks) or a bulking agent must be added to feedstocks without any structure (i.e., sewage sludge) (see Sect. 2.3).

In most cases, the static pile has the shape of a truncated pyramid. Typical dimensions are between 12 and 15 m at the base with a height of 3 m (TCHOBANOGLOUS et al., 1993).

In the US, the static aerated pile is the most common sewage sludge composting system (GOLDSTEIN and STEUTEVILLE, 1996). The technology was developed in the 1970s in Beltsville, Maryland (USA). Piles are often covered with a layer of "matured compost" to prevent heat loss from the upper layer and provide a minimum odor treatment (USEPA, 1981). The timer controlled blowers maintained an oxygen level of 5–15%. The realization of unfavorable temperatures in the static pile resulted in the development of the Rutgers process, which adopted temperature-con-

Fig. 4. Covered curing and storage area (Rethman Kreislaufwirtschaft GmbH & Co. KG/Germany).

trolled blowers (in most cases below 60 °C in the pile) (FINSTEIN et al., 1986). In the initial and the final composting phase (see phase I in Fig. 1) the temperature feedback must be overridden to ensure minimum aeration if the temperatures in the pile are below the set point (LENTON and STENTIFORD, 1990). The aeration of the initial process in Beltsville was vacuum-induced while the Rutgers process used forced-pressure aeration. Typical retention times in aerated static piles are 21 d followed by 6–8 weeks of curing in windrows (Fig. 4).

In piles without active aeration, a few facilities try to enhance the natural ventilation by placing aeration pipes (loops or open ended pipes) in the piles (LYNCH and CHERRY, 1996; FERNANDES and SARTAJ, 1997). The pipes enhance the natural ventilation inside the pile.

In another modification, the feedstock can be stacked in open composting cells. To compensate for vertical moisture and temperature gradients in the piles, the compost is moved from one cell to another.

3.3.1.3 Other Systems

Other less often used, special applications of the previously discussed composting systems include mat composting and vermicomposting.

Mat composting is almost more a storage and pre-processing measure than a composting technology. It is a simple process to mix different yard waste feedstocks like grass, branches, leaves, hedge cuttings which are delivered to the composting site over a period of 3–12 months. The first 0.5 m high mat layer consists of mostly bulky yard wastes. Additional yard waste layers are spread on this initial layer up to a final height of about 1.5 to 3.0 m. A front–end loader regularly, fluffs the yard waste with a fork-like device. Due to the lack of effective turning of the entire mat, the bulky particle size of the feedstock, and the permanent addition of a thin layer of fresh yard waste, temperatures in the mat reach only 40–50 °C. To continue high-rate degradation, stabilization, and curing the yard waste in the mat is reformed to windrows (KERN, 1991).

Vermicomposting is a simple, very unique technology. Biodegradable organic wastes are inoculated with earthworms (i.e., *Eisenia foetida*) which break down and fragment the waste. Low, medium and high technology vermicomposting systems – all without agitation – are available. Vermicomposting is relatively labor intensive and, for large-scale production, requires large land areas to cool the waste to avoid temperatures which are detrimental to earthworms.

The temperature suitable for growing earthworms ranges between 22 and 27 °C (WESEMANN, 1992). As a result, in traditional open vermicomposting systems, the waste is placed in beds or windrows only up to a height of about 0.5–1.2 m. At elevated temperatures, worms are only found in a fairly localized zone towards the outside of the compost pile (LODGSON, 1994). Sometimes vermicomposting is only implemented for curing where temperatures are low. Another reason for combining high-rate degradation (3–15 d) without worms with vermicomposting for stabilization and curing is the need for pathogen reduction (RIGGLE and HOLMES, 1994). At the end of the process, the worms are separated from the castings by screening. The end product is a fine, peat-like material (LODGSON, 1994).

Retention times of 0.5 to 1.0 year are reported for techniques relying solely on vermicomposting (DOMINGUEZ et al., 1997; WESEMANN, 1992).

3.3.2 Enclosed Reactor Systems

The major difference between the enclosed composting reactor systems and the non-reactor systems is that the composting takes place in an enclosed building. The main advantage of an enclosed composting reactor system is that the off-gases of the composting process can be collected and treated thereby reducing odor emissions from the composting facility. However, the warm, humid off-gases of the composting process condense on the cooler building roof, walls, and pipes and can cause corrosion.

3.3.2.1 Channel, Cell and Windrow Composting

In an enclosed building, the feedstock is placed in triangle or trapezoid windrows, however, in most cases the feedstock is divided by walls which are used as tracks for a turning machine. In these uncovered channels, the feedstock is stacked up to a height of about 2.0–2.5 m (HAUG, 1993; LENTZEN, 1995; SCHMITZ and MEIER-STOLLE, 1995). The length (i.e., 50 m, 65 m, 220 m) and the number of channels depend on the capacity of the facility and the proposed retention time. A facility can start with only two channels and additional channels can be added as the capacity of the facility increases. Short channels (about twice the width) are called cells.

Both aeration (forced-pressure or vacuum-induced) and turning are used to control the composting process. Air is supplied from manifolds below each channel and aeration as well as water addition are controlled separately for each channel. In many cases, each channel is divided into several aeration areas each aerated by its own controlled fan.

Due to frequent turning, the compost moves from the beginning to the end of the channel. Turning frequencies of every day (HAUG, 1993) or every other day (LENTZEN, 1995) are reported. Some systems compensate for the volume loss during turning to keep the height of the compost constant (SCHMITZ and MEIER-STOLLE, 1995) while other systems compensate for the volume loss when restacking the compost in another channel (LENTZEN, 1995). Typical retention times in these channels are 6–8 weeks for all composting phases (SCHMITZ and MEIER-STOLLE, 1995). However, some facilities compost for 21 d in channels followed by 6 months curing outside in static piles (HAUG, 1993).

3.3.2.2 Aerated Pile Composting with Automatic Turning Machines

The difference between channel composting and aerated pile composting systems with automatic turning machines is that the feedstock is not placed in channels which can be individually controlled but in one large pile with heights up to 1.8 and 3.3 m (KUGLER et al., 1995). The frame for the turning machine spans the whole width of the composting hall (i.e., about 35 m in the Bühler AG facility in Bassum, Germany).

A forced and/or vacuum-induced aeration system is placed under the compost bed similar to the aerated static pile bed. The composting hall is subdivided into several aeration areas that can be separately aerated and moistened depending on the progress of the composting process. In most cases, the feedstock is conveyed from the pre-processing area to one end of the composting hall, moves through the composting hall via a turning machine, and exits the composting hall at the opposite side (flow systems). The volume loss is compensated in most systems by the turning machine. The aeration rate can be reduced from, i.e., 18,000 m^3 Mg^{-1} source-separated biowaste for forced-pressure aeration to 7,000 m^3 Mg^{-1} source-separated biowaste for a combination of vacuum-induced and forced-pressure aeration with a heat exchanger (KUGLER et al., 1995).

In non-flow systems, the height of the pile decreases during composting. In these systems the feedstock is stacked in one area in the composting hall, where it remains for the whole retention time (BORLINGHAUS, 1995; BRÖHL, 1995).

In most cases, water is added during turning. Turning machines vary with the manufacturer, but they are typically floor-independent and movable over the entire composting hall [i.e., impeller (Fig. 5), diagonally working screws, double spindle agitators, vertically working screws (BORLINGHAUS, 1995; BRÖHL, 1995; HODZIC, 1995; KUGLER et al., 1995)]. The turning frequencies range from once a day to once a week which, of course, affects the moisture gradient in the pile. A few systems do not turn for the first four weeks (ZACHÄUS, 1995). Retention times range between 4 and 12 weeks, followed by windrow composting for the shorter retention times.

3.3.2.3 Brikollari Composting

The brikollari process is a unique type of static composting with facilities in Germany (LINDER, 1995). Biowaste, well mixed and amended with a bulking agent, is compressed into blocks (30–60 kg) that are stacked crosswise on pallets (Fig. 6). Channels pressed into the surface of the blocks provide aeration via natural diffusion. The coarse structure of the blocks ensures aeration even within the block.

The compaction is very energy intensive in comparison to other pre-processing steps, which in turn is compensated for by the low space requirements. An automatic transport system moves the stacks with a mass of 1.2 to 1.4 Mg each to a multi-floor high-rack warehouse for high rate and stabilization degradation. Dividers separate the warehouse into several areas which can all be ventilated separately depending on the progress of the composting process.

Retention times in the warehouse range from 5–6 weeks. As a result of a moisture content of about 20% at the end of the composting process, the blocks can be marketed as "active compost" (class of stability: III–IV) after grinding, can be stored before processed further, or can be immediately cured after the addition of water. Retention time for curing in windrows is about 8–10 weeks (LINDNER, 1995).

Fig. 5. Turning machine "Wendelin" in aerated pile composting facility in Bützberg/Germany (Bühler AG/Switzerland).

Fig. 6. Brikollari system, biowaste blocks crosswise stacked on pallets (Rethman Kreislaufwirtschaft GmbH & Co. KG/Germany).

3.3.3 In-Vessel Reactor Systems

In comparison to enclosed reactor systems, in-vessel reactor systems have minimal free air space above the compost and this reduces the volume of exhaust gases that must be treated. Additionally, the aeration system can be better controlled (i.e., exhaust air recirculation, conditioning of the aeration air).

3.3.3.1 Tunnel Composting

Tunnel reactors include static and agitated composting systems with different levels of process control and have been used as composting systems for MSW, sewage sludge, and manure for many years. Recently, there has been special interest in tunnel reactors for composting manure in the mushroom industry because these reactors are better controlled than previous MSW tunnel composting systems and their design and operation is based on many years of experience. Several manufacturers adapted this concept for composting MSW and biowaste (Grundmann, 1995; Nieveen, 1995). A relatively homogeneous temperature and moisture profile over the height of the compost can be maintained due to large amounts of recirculating exhaust gases. As a result, less turning is needed to homogenize the compost.

The number of tunnels is chosen according to the proposed capacity of the facility. Typical tunnel reactor lengths are 30–50 m and widths and heights 4–6 m. Each tunnel reactor is separately controlled, and depending on the process control fresh air, recirculated exhaust gases, or a mixture of both are supplied from below the bed. Tunnels are often equipped with nozzles in the ceiling that are used to moisten the compost with process waters (condensate, leachate). The tunnels are provided with hatches or doors to feed or remove the compost. Adapted from the mushroom industry, special plastic slide nets are used to empty the tunnel with a movable net winder with integrated skimming roller. In most cases, feeding is carried out automatically by band conveyor systems. If the retention times in the tunnel system are short (i.e., 1 week), agitation may not be carried out (Zachäus, 1995). One way of turning is to empty the tunnel and to backfill another tunnel, which also compensates for the volume and mass loss (Grundmann, 1995). Turning can also be accomplished within one tunnel by means of turning devices movable into each tunnel – e.g., on rails.

The retention time of the compost in the different tunnel systems varies between 1 and 7 weeks (Grundmann, 1995; Zachäus, 1995). If composting in the tunnel system is of short duration, additional curing in windrows is required.

3.3.3.2 Box and Container Composting

In box and container composting systems, the reactor units are very similar to tunnels, however shorter. Composting boxes with volumes of about 50–60 m^3 (7 m · 3 m · 3 m) (HERHOF, 1995) up to volumes of 250 m^3 (i.e., System Schmutz & Hartmann, Switzerland) are found. The whole front area of a box reactor consists of a door.

In comparison to composting boxes, the smaller composting containers with volumes of about 20–25 m^3 (approximately 5–6 m · 2 m · 2 m) are movable. Next to the pre-processing area, the containers that can be opened at the top are filled. A truck or crane moves the container back to the composting area (Fig. 7) where each container is connected to the aeration, the air conditioning, and the leachate collection system with quick-couplings.

The boxes and containers are filled by a front–end loader or automatically by specialized band conveyors (Fig. 8). In the composting area, the aeration and air conditioning system is individually controlled for each reactor. Forced-pressure aeration, vacuum-induced aeration, or a combination of both are possible. A homogeneous moisture and tem-

Fig. 7. Kneer system, composting containers (von Ludowig GmbH/ Germany).

Fig. 8. Loading of Herhof boxes with a front-end loader in the composting facility in Beselich (Herhof Umwelttechnik GmbH/ Germany).

perature profile is accomplished due to air recirculation and the addition of fresh air. Some reactors have segmented aeration floors which permits control of the different aeration areas individually. Some composting boxes pull air from a box filled with fresh feedstock and force it through a composting box with more matured feedstock.

The retention times in container or box reactors are usually between 7 to 14 d followed by curing for about 12 weeks in windrows resulting in total retention times of 13–14 weeks (GÖTTE, 1995; ZACHÄUS, 1995). The retention time is shorter if the compost after the first run is backfilled twice to a box for a second and a third run with a total retention time of 6–7 weeks. In some systems, agitation is carried out between 1 and 4 times a week with retention times of 14–50 d in the box. Curing in windrows may follow (ZACHÄUS, 1995).

3.3.3.3 Rotating Drum Composting

Composting in a rotating drum is a dynamic process; this system has been used for composting MSW all over the world. The rotating drum acts like a ball mill and the constant movement of the rotating drum ensures homogenization and fluffing. As a result of the thorough aeration, odor emissions from rotating drums are low. The removal of inerts is more effective after leaving the rotating drum, because the compost is drier and well homogenized.

Rotating drums are angled to move the material and volumes vary between 30 and 500 m^3 (HAUG, 1993; KASBERGER, 1995). The reactors are not completely filled to ensure sufficient mixing. Feedstock inlets and compost outlets are located on opposite ends of the rotating drum and the compost tumbles slowly through the reactor. To prevent possible short-circuiting, the rotating drum can be divided into different cells.

The rotational speed [i.e., 0.1 to 1 rpm by Dano/Denmark for the composting of MSW (HAUG, 1993)] and whether the rotation is continuous or intermittent [i.e., 45 min in 24 h on the average for biowastes (KASBERGER, 1995)] can be controlled. Especially for wet feedstocks like biowaste, high rotational speeds can result in compaction. In most cases, aeration is provided by a fan, less often the rotating drum is perforated allowing aeration via natural diffusion. In the last case, the rotating drum is vented to minimize odor emissions.

The retention time in the rotating drum system ranges from 1–10 d (KERN, 1991; KASBERGER, 1995). At lower retention times only a portion of the high-rate degradation is performed while at higher retention times high-rate degradation and stabilization are conducted in the rotating drum. In any case, additional windrow composting is required which may be between 2 and 3 months (HAUG, 1993, KASBERGER, 1995).

3.3.3.4 Vertical Flow Composting

Vertical flow reactors, mostly cylindrical towers and less often rectangular reactors, are agitated or dynamic composting systems. They are often used to compost sewage sludge and less often MSW and biowaste. Some towers are divided into separate vertical compartments by interior floors. Tower volumes of 400 up to 1,800 m^3 with total material depths of 6–9 m can be found (HAUG, 1993).

Vertical reactors are commonly fed at the top on a continuous or intermittent basis as the feedstock in the tower moves slowly from the top to the bottom during composting. If interior floors are used, the material is transported vertically to the next floor by flaps or movable grates. The floors enhance agitation during the movement down through the reactor. Without the floors, the movement through the reactor is more like a plug flow.

In most cases, the reactors are aerated by forced-pressure aeration countercurrent to the compost flow. Large compost depths result in a high pressure drop in the compost which affects the efficiency of the aeration system. In addition, due to the lack of effective agitation, the compost can become compacted. This can potentially result in anaerobic niches. In particular, odor problems have led to the closure of a significant number of tower facilities (LOLL, 1996).

After movement through the vertical composting reactor, the material is either cured in open windrows or filled into another composting tower for a second run. Typical retention times in vertical reactors range between 14 and 20 d (HAUG, 1993). Stabilization and curing in windrows follows.

4 Compost Quality and End Uses

Compost can have various end uses that range from serving as a fuel source to a soil conditioner. More common uses are amendments for growing media and most importantly soil conditioners in agriculture, horticulture, silviculture, landscaping, and reclamation. Depending on the feedstock, compost provides various levels of macronutrients including N, P, K, Ca, and Mg and micronutrients required for plant growth. However, the main effect of compost as a soil conditioner is not the addition of nutrients to land, but the role in the soil humus balance and the soil structure (DE BERTOLDI et al., 1983). In this role, compost improves the soil's water drainage and water holding capacity, acts as pH buffering agent, helps regulate the temperature, aids in erosion control, and improves air circulation by increasing the void space. Currently, compost's ability to suppress plant diseases finds growing interest. Microorganisms in compost have been shown to act as antagonists of plant pathogens (HOITINK et al., 1993).

Compost also contains heavy metals. Heavy metals are natural components of soil forming rocks with a density of more than 5 kg L^{-1}. In low concentrations, some of the heavy metals such as zinc and copper are essential nutrients. However, higher concentrations (and for some heavy metals even low concentrations) are especially problematic for the environment, because they can concentrate in soil, plants, humans, and animals to toxic concentrations. Source separation of compost feedstocks reduces the heavy metal content in composts (Tab. 2). The heavy metal content in source-separated composts can be affected by season and population density (KROGMANN, 1999).

Another category of contaminants in composts are synthetic organics. Since some synthetic organics are ubiquitous, they can be detected even in biowaste composts. Which synthetic organics are found depends mostly on the feedstock. Their origin in biowaste derives mainly from dust and particulates (KRAUß et al., 1991). Synthetic organics in relevant concentrations in compost include chlorinated hydrocarbons like polychlorinated biphenyls

Tab. 2. Heavy Metals in Composts

	Biowaste		Yard Waste	MSW	
Number of Facilities (Samples)	Germany[a] 52 (450–490)	USA[b] 6 (?)	Germany[c] 3 (12)	Germany[c] 1 (3)	USA[d] 7 (7)
As, mg kg^{-1} [e]	–	–	7.1	7.2	7.7
Pb, mg kg^{-1}	77.6	74	87.7	396	234
Cd, mg kg^{-1}	0.78	1.1	0.53	2.9	3.3
Cr, mg kg^{-1}	33.7	15	57.0	98	76.0
Cu, mg kg^{-1}	43.2	64	48.7	346	281
Ni, mg kg^{-1}	19.1	8	28.3	62	34
Hg, mg Mg^{-1}	330	1,000	170	2,970	–
Zn, mg kg^{-1}	233	292	185	998	665

[a] FRICKE et al. (1992)
[b] RICHARD and WOODBURY (1992)
[c] KEHRES (1990)
[d] HE et al. (1995)
[e] not a heavy metal

Tab. 3. Compost Quality Criteria in Selected Countries

	Germany (LAGA, 1995)[f]	The Netherlands (AMLINGER, 1998)	Spain (I FONTANALS, 1998)	USA (USEPA, 1993)
Regulation/ Guideline	LAGA M 10	KIWA guideline BRI.K265/02	Ordinance, 28 May 1998	40 CFR Part 503
Feedstocks	biowaste, yard waste, other green wastes	food and yard waste with less than 20% bulking agent	all, must be indicated	sewage sludge and sewage sludge products[a]
Compost types	fresh compost, finished compost, specialty compost	biowaste compost	as fertilizer	
Hygiene	test with indicator organisms	4 d >55 °C, <2 plants L^{-1} compost, test of eelworms, rhizomanie virus, *Plasmodiophora brassicae*	*Salmonella:* absence in 25 g compost, fecal streptococci: 1×10^3 MPN g^{-1}, total enterobacteria: 1×10^3 CFU g^{-1}	Class A[b]: 3 d >55 °C (in-vessel) or 15 d >55 °C (windrow, 5 times turning) Class B[b]: 5 d >40 °C, 4 h >55 °C within these 5 d
Foreign matter	<0.5% dry matter (cat. I[c]), <1.0% dry matter (cat. II[c])	<0.2% dry matter (>2 mm)	plastic particles and other inerts <10 mm	
Stones		<2% dry matter (<5 mm)		
Phytotoxicity	not phytotoxic for proposed use, no nitrogen fixation	100% of control (15% compost)		
Stability	fresh compost: II–III, finished compost: IV–V			
pH	must be reported	<6.5		
Salt content/ Conductivity	<5 g KCl L^{-1}	<5.5 mS cm^{-1}		
Particle size and distribution	fine: < 5 mm (I), <12 mm (II) medium: <25 mm coarse: <40 mm		90% (<25 mm)	
Moisture content	Bulk: <45% of fresh matter, bagged: <35% of fresh matter	% organic matter +10% (<40% organic matter), % organic matter +6% (>40% organic matter)	<40%	

Tab. 3. Continued

	Germany (LAGA, 1995)[f]			The Netherlands (AMLINGER, 1998)			Spain (I FONTANALS, 1998)		USA (USEPA, 1993)		
Organic matter	≥35% of dry matter			≥20%			>25%				
Heavy metals [mg kg^{-1}]		cat. I[c]	cat. II[c]		a[d]	b[d]				a	b[e]
	Pb	150	250	Pb	100	65	Pb	300	Pb	300	840
	Cd	1.5	2.5	Cd	1.0	0.7	Cd	10	Cd	39	85
	Cr	100	200	Cr	50	50	Cr	400	Cr	–	–
	Cu	100	200	Cu	60	25	Cu	450	Cu	1,500	4,300
	Ni	50	100	Ni	20	10	Ni	120	Ni	420	420
	Hg	1	2	Hg	0.3	0.2	Hg	7	Hg	17	57
	Zn	400	750	Zn	200	75	Zn	1,100	Zn	2,800	7,500
									As	41	75
									Mo	–	75
									Se	36	100
Nutrients	macronutrients N, P, K, Ca, Mg must be reported			macronutrients N, P, K, Ca, Mg must be reported			organic N (if >1.1%) must be reported				

[a] No federal regulations for other feed stocks, however, individual states can have stricter regulations and regulations for other feedstocks.
[b] Class A: additional *Salmonella* or fecal coliform requirements; production regulated, but use unregulated; Class B: additional general requirements, management, and site restrictions.
[c] Related to 30% organic matter. Category I: agriculture (field crops, pasture); horticulture (vegetables, asparagus, fruits, shrubs, tree nursery), viticulture; others uses including floriculture, container plants, plants for decks. Category II: landscaping, silviculture, other uses including noise protection walls, lawn grid stones, roof gardens.
[d] a: compost, b: very clean compost.
[e] Additional loading limits.
[f] Beyond the LAGA Guideline, biowaste ordinance with slight modifications exists since the 1998.

(PCB) and polychlorinated dibenzodioxins and -furans, and polycyclic aromatic hydrocarbons (PAH). The concentrations of synthetic organics in MSW compost are higher than in biowaste compost (FRICKE et al., 1992).

To ensure the appropriate use of compost, many countries have developed compost quality standards with regard to beneficial properties, like organic matter content and nutrients, and unwanted contaminants, like heavy metals. Northern and Central European countries like. The Netherlands and Germany set more defined and stricter standards with regard to contaminants than the USA and southern European countries like Spain (Tab. 3).

5 References

AMLINGER, F. (1998), European Survey on the Legal Basis for Separate Collection and Composting of Organic Waste, in: *Compost – Quality Approach in the European Union, EU Symposium* (Federal Ministry for Environment, Youth and Family Affairs, Ed.), Proc. EU Symp., 29–30 Oct., Vienna. Vienna: Federal Ministry for Environment, Youth and Family Affairs.

ATV (1989), Bau und Betrieb von Pflanzenabfall-Kompostierungsanlagen, Bericht der ATV/VKS/ANS-Arbeitsgruppe 3.2.1 "Kompostierung", *Korrespondenz Abwasser* **36**, 190–195.

DE BERTOLDI, M., VALLINI, G., PERA, A., ZUCCONI, F. (1983), Comparison of three windrow, systems, *BioCycle* **24**, 45–50.

BIDLINGMAIER, W. (1983), Das Wesen der Kompostierung von Siedlungsabfällen, in: *Müll-Hand-*

buch Vol. 4, KZ 5305 (Hösel, G., Bilitewski, B., Schenkel, W., Schnurer, H., Eds.). Berlin: Erich Schmidt Verlag.

Bidlingmaier, W., Grauenhorst, V. (1996), Geruchsemissionen von Kompostierungsanlagen, in: *Neue Techniken der Kompostierung* (Stegmann, R., Ed.), pp. 441–458. Bonn: Economica Verlag.

Böhm, E. (1995), Keimemissionen bei der Kompostierung, in: *Biologische Abfallbehandlung* (Thomé-Kozmiensky, K. J., Ed.), pp. 506–516. Berlin: EF-Verlag für Energie- und Umwelttechnik.

Borlinghaus, W. (1995), Das DYNACOMP-Verfahren, in: *Abfallwirtschaft: Neues aus Forschung und Praxis – Herstellerforum Bioabfall – Verfahren der Kompostierung und anaeroben Abfallbehandlung im Vergleich* (Wiemer, K., Kern, M., Eds.), pp. 193–203. Witzenhausen: M. I. C. Baeza-Verlag.

Bröhl, U. (1995), Die MABEG Bioabfallkompostierung "System Ernst", das DYNACOMP-Verfahren, in: *Abfallwirtschaft, Neues aus Forschung und Praxis – Herstellerforum Bioabfall – Verfahren der Kompostierung und anaeroben Abfallbehandlung im Vergleich* (Wiemer, K., Kern, M., Eds.), pp. 132–142. Witzenhausen: M. I. C. Baeza-Verlag.

Brown, K. H., Bouwkamp, J. C., Gouin, F. R. (1998), Carbon : Phosphorus Ratio in MSW Composting, *Comp. Sci. Util.* **6**, 53–58.

Dominguez, J., Edwards, C. A., Subler, S. (1997), A comparison of vermicomposting and composting, *BioCycle* **38**, 57–59.

Epstein, E. (1997), *The Science of Composting*. Lancaster, PA: Technomic Publishing Co.

Fernandes, L., Sartaj, M. (1997) Comparative static pile composting using natural, forced and passive aeration methods, *Comp. Sci. Util.* **5**, 65–77.

Fernandes, F., Viel, M., Sayag, D., André, L. (1988), Microbial breakdown of fats through in-vessel co-composting of agricultural and urban wastes, *Biol. Wastes* **26**, 33–48.

Finstein, M. S., Morris, M. L. (1975), Microbiology of municipal solid waste composting, *Adv. Appl. Microbiol.* **19**, 113–151.

Finstein, M. S., Miller, F. C., Strom, P. F. (1986), Waste treatment composting as a controlled system, in: *Biotechnology* 1st Edn., Vol. 8 (Rehm, H.-J., Reed, G., Eds.), pp. 363–398. Weinheim: VCH.

i Fontanals, F. G. (1998), Organic waste management in Catalonia (Spain): Source separate collection and treatment, in: *Compost – Quality Approach in the European Union, EU Symposium* (Federal Ministry for Environment, Youth and Family Affairs, Ed.), *Proc. EU Symp.*, 29–30 Oct., Vienna. Vienna: Federal Ministry for Environment, Youth and Family Affairs.

Fricke, K. (1988), Grundlagen zur Biomüllkompostierung unter besonderer Berücksichtigung von Rottesteuerung und Qualitätssicherung, *Thesis*, Gesamthochschule Kassel, Witzenhausen, Germany. Göttingen: Verlag Die Werkstatt.

Fricke, K., Nießen, H., Turk, T., Vogtmann, H., Hangen, H. O. (1992), Situationsanalyse Bioabfall 1991 – Teil 2, *Müll und Abfall* **24**, 649–660.

Glathe, H., Farkasdi, G. (1966), Bedeutung verschiedener Faktoren für die Kompostierung, in: *Müll-Handbuch* Vol. 4, KZ 5040 (Hösel, G., Bilitewski, B., Schenkel, W., Schnurer, H., Eds.). Berlin: Erich Schmidt Verlag.

Goldstein, N., Steuteville, R. (1996), Steady climb for biosolids composting, *BioCycle* **37**, 68–78.

Golueke, C. G. (1977), *Biological Reclamation of Solid Waste*. Emmaus: Rodale Press.

Golueke, C. G. (1989), Putting principles into successful practice, in: *BioCycle Guide to Yard Waste Composting*. Emmaus: JG Press.

Götte, M. (1995), ML-Bio-Containerkompostierung, in: *Abfallwirtschaft: Neues aus Forschung und Praxis – Herstellerforum Bioabfall – Verfahren der Kompostierung und anaeroben Abfallbehandlung im Vergleich* (Wiemer, K., Kern, M., Eds.), pp. 144–159. Witzenhausen: M. I. C. Baeza-Verlag.

Grundmann, J. (1995), Das Rottetunnelverfahren der Deutsche Babcock Anlagen, in: *Abfallwirtschaft: Neues aus Forschung und Praxis – Herstellerforum Bioabfall – Verfahren der Kompostierung und anaeroben Abfallbehandlung im Vergleich* (Wiemer, K., Kern, M., Eds.), pp. 44–52. Witzenhausen: M. I. C. Baeza-Verlag.

Haug, R. T. (1993), *The Practical Handbook of Compost Engineering*. Boca Raton, FL: CRC Press.

Haug, R. T., Ellsworth, W. F. (1991), Measuring compost substrate degradability, *BioCycle* **37**, 56–62.

Hauke, H., Stöppler-Zimmer, H., Gottschall, R. (1996), Development of compost products, in: *The Science of Composting* Part I (De Bertoldi, M., Sequi, P., Lemmes, B., Papi, T., Eds.), pp. 477–494. London: Blackie Academic & Professional.

He, X. T., Logan, T. J., Traina, S. J. (1995), Physical and chemical characteristics of selected U.S. municipal solid waste composts, *J. Environ. Qual.* **21**, 543–552.

Herhof (1995), *Grunddaten der Planung von Kompostierungsanlagen mit HERHOF-Rotteboxen* (Herhof Umwelttechnik GmbH, Ed.). Solms-Niederbiel, Germany: Eigenverlag.

Hodzic, A. (1995), Biorapid-Kompostierungssystem, in: *Abfallwirtschaft: Neues aus Forschung und Praxis – Herstellerforum Bioabfall – Verfah-*

ren der Kompostierung und anaeroben Abfallbehandlung im Vergleich (WIEMER, K., KERN, M., Eds.), pp. 99–111. Witzenhausen: M. I. C. Baeza-Verlag.

HOITINK, H. A. J., BOEHM, M. J., HADAR, Y. (1993), Mechanisms of suppression of soilborne plant pathogens in compost-amended substrates, in: *Science and Engineering of Composting: Design, Environmental, Microbiological and Utilization Aspects* (HOITINK, H. A. J., KEENER, H. M., Eds.), pp. 601–621. Ohio State University; Worthington, OH: Renaissance Pubs.

JERIS, J. S., REGAN, R.W. (1973), Controlling environmental parameters for optimum composting, *Comp. Sci.* **1/2**, 10–15.

KASBERGER, P. (1995), Das dynamisch gesteuerte LESCHA-Verfahren, in: *Herstellerforum Bioabfall – Verfahren der Kompostierung und anaeroben Abfallbehandlung im Vergleich* (WIEMER, K., KERN, M., Eds), pp. 120–131. Witzenhausen: M. I. C. Baeza-Verlag.

KEHRES, B. (1990), Zur Qualität von Kompost aus unterschiedlichen Ausgangsstoffen, *Thesis*, Gesamthochschule Kassel, Witzenhausen, Germany.

KERN, M. (1991), Untersuchungen zur vergleichenden Beurteilung von Kompostierungsverfahren: Technik-Umweltrelevanz-Kosten, in: *Abfallwirtschaft: Bioabfallkompostierung – flächendeckende Einführung* (WIEMER, K., KERN, M., Eds.), pp. 235–278. Witzenhausen: M. I. C. Baeza-Verlag.

KRAUß, P., HAGENMEIER, H., BENZ, T., HOHL, J., HUMMLER, M. et al. (1991), Organische Schadstoffe im Kompost, in: *Stuttgarter Berichte zur Abfallwirtschaft* Vol. 4 (Institut für Siedlungswasserbau, Wassergüte- und Abfallwirtschaft der Universität Stuttgart, Ed.), p. 109. Bielefeld: Erich Schmidt Verlag.

KROGMANN, U. (1994), Kompostierung – Grundlagen zur Einsammlung und Behandlung von Bioabfällen unterschiedlicher Zusammensetzung, *Thesis*, Technical University of Hamburg-Harburg, Germany. Bonn: Economica Verlag.

KROGMANN, U. (1999), Effect of season and population density on source-separated waste composts, *Waste Manag. Res.* **17**, 109–123.

KUGLER, R., HOFER, H., LEISNER, R. (1995), Das Wendelin-Tafelmieten-Kompostierungsverfahren, in: *Abfallwirtschaft: Neues aus Forschung und Praxis – Herstellerforum Bioabfall – Verfahren der Kompostierung und anaeroben Abfallbehandlung im Vergleich* (WIEMER, K., KERN, M., Eds.), pp. 13–23. Witzenhausen: M. I. C. Baeza-Verlag.

KUTZNER, H. J., KEMPF, T. A. (1994), Kompostierung aus mikrobiologischer Sicht – Ein Essay, in: *Proc. 5. Hohenheimer Seminar*, Nachweis und Bewertung von Keimemissionen bei der Entsorgung von kommunalen Abfällen sowie spezielle Hygieneprobleme der Bioabfallkompostierung (DVG – Deutsche veterinärmedizinische Gesellschaft e.V., Ed.), pp. 281–303, 5.–6. Oktober. Gießen: Gahmig Druck.

LAGA (1995), LAGA- M 10: Qualitätskriterien und Anwendungsempfehlungen für Kompost, in: *Müll-Handbuch* Vol. 4, KZ 6856 (HÖSEL, G., BILITEWSKI, B., SCHENKEL, W., SCHNURER, H., Eds.). Berlin: Erich Schmidt Verlag.

LENTON, T. G., STENTIFORD, E. I. (1990), Control of aeration in static pile composting, *Waste Manag. Res.* **8**, 299–306.

LENTZEN, M. (1995), Dynamische Zeilenkompostierung der PASSAVANT-Werke AG, in: *Abfallwirtschaft: Neues aus Forschung und Praxis – Herstellerforum Bioabfall – Verfahren der Kompostierung und anaeroben Abfallbehandlung im Vergleich* (WIEMER, K., KERN, M., Eds.), pp. 160–169. Witzenhausen: M. I. C. Baeza-Verlag.

LINDER, H. (1995), Das Brikollare-Verfahren, in: *Abfallwirtschaft: Neues aus Forschung und Praxis – Herstellerforum Bioabfall – Verfahren der Kompostierung und anaeroben Abfallbehandlung im Vergleich* (WIEMER, K., KERN, M., Eds.), pp. 110–182. Witzenhausen: M. I. C. Baeza-Verlag.

LOGSDON, G. (1994), Worldwide progress in vermicomposting, *BioCycle* **35**, 63–65.

LOLL, U. (1996), Stand und Perspektiven bei der biologischen Klärschlammbehandlung, in: *Biologische Abfallbehandlung III – Kompostierung, Anaerobtechnik, mechanisch-biologische Abfallbehandlung, Klärschlammverwertung* (WIEMER, K., KERN, M., Eds.), pp. 475–506. Witzenhausen: M. I. C. Baeza-Verlag.

LYNCH, N. J., CHERRY, R. S. (1996), Design of passively aerated compost piles, in: *The Science of Composting* Vol. 2 (DE BERTOLDI, M., SEQUI, P., LEMMES, B., PAPI, T., Eds.), pp. 947–973. London: Blackie Academic & Professional.

MANSER, A. G. R., KEELING, A. A. (1996), *Practical Handbook of Processing and Recycling Municipal Waste*. Boca Raton, FL: Lewis Publishers.

MICHEL, F. C., FORNEY, L. J., HUANG, A. J.-F., DREW, S., CZUPRENSKI, M. et al. (1996), Effects of turning frequency, leaves to grass mix ratio and windrow vs. pile configuration on the composting of yard trimmings, *Comp. Sci. Util.* **4**, 26–43.

MILLER, F. C., MACGREGOR, S. T., PSARIANOS, K. M., CIRELLO, J., FINSTEIN, M. S. (1982), Direction of ventilation in composting wastewater sludge, *J. WPCF* **57**, 111–113.

MÜLLER, H. M., SCHULTKIES, V., BILITEWSKI, B. (1997), Organische und anorganische Schadstoffe im Papier, in: *Müll-Handbuch* Vol. 6, KZ 8614.8 (HÖSEL, G., BILITEWSKI, B., SCHENKEL, W., SCHNURER, H., Eds.). Berlin: Erich Schmidt Verlag.

NIEVEEN, H. O. (1995), Das GICOM Verfahren, in:

Abfallwirtschaft: Neues aus Forschung und Praxis – Herstellerforum Bioabfall – Verfahren der Kompostierung und anaeroben Abfallbehandlung im Vergleich (WIEMER, K., KERN, M., Eds.), pp. 75–82. Witzenhausen: M. I. C. Baeza-Verlag.

POINCELOT, R. P. (1975), The Biochemistry and Methodology of Composting, *Bulletin 754*, New Haven: The Connecticut Agricultural Experiment Station.

RASP, H. (1993), Abfall- und Reststoffe aus der industriellen Produktion als Dünge- und Bodenverbesserungsmittel im Landbau – Eine Übersicht, in: *Müll-Handbuch* Vol. 4, KZ 6555 (HÖSEL, G., BILITEWSKI, B., SCHENKEL, W., SCHNURER, H., Eds.). Berlin: Erich Schmidt Verlag.

RICHARD, T. L. (1992), Municipal solid waste composting: Physical and biological processing, *Biomass and Bioenergy* **3**, 163–180.

RICHARD, T. L. (1998), Composting strategies for high moisture manures, in: *Proc. Manure Management in Harmony with the Environment and Society Conference*, Feb. 10–12, pp. 135–138. Ames, IA: The Soil and Water Conservation Society, West North Central Region.

RICHARD, T. L., WOODBURY, P. B. (1992), The impact of separation on heavy metal contaminants in municipal solid waste composts, *Biomass and Bioenergy* **3**, 195–211.

RIGGLE, D., HOLMES, H. (1994), Expanding horizons for commercial vermiculture, *BioCycle* **35**, 58–62.

RYNK, R. (Ed.) (1992), *On-Farm Composting Handbook*. Ithaca, NY: Northeast Regional Agricultural Engineering Service.

SCHILDKNECHT, H., JAGER, J. (1979), Zur chemischen Ökologie der biologischen Abfallbeseitigung, *Research Report* 10302407. Bonn: Bundesministerium des Innern.

SCHMITZ, T., MEIER-STOLLE, G. (1995), Das Biofix- und Kompoflex-Verfahren, in: *Abfallwirtschaft: Neues aus Forschung und Praxis – Herstellerforum Bioabfall – Verfahren der Kompostierung und anaeroben Abfallbehandlung im Vergleich* (WIEMER, K., KERN, M., Eds.), pp. 183–192. Witzenhausen: M. I. C. Baeza-Verlag.

SCHUCHARDT, F. (1988), Verlauf von Kompostierungsprozessen in Abhängigkeit von technisch-physikalischen und chemischen Rahmenbedingungen, *Proc. Seminar Herstellung und Vermarktung von Komposten nach Gütekriterien*, Haus der Technik, Essen.

SEEKINS, B. (1986), *Usable Waste Products for the Farm – An Inventory for Maine*. Augusta, ME: Maine Department of Agriculture, Food and Rural Resources.

SEEKINS, B., MATTEI, L. (1990) *Update: Usable Waste Products for the Farm*. Augusta, ME: Maine Department of Agriculture, Food and Rural Resources.

SIHLER, A., BIDLINGMAIER, W., TABASARAN, O. (1993), Analysenübersicht von Komposten und deren Ausgangsstoffen, *Research Report*, University of Stuttgart, Germany.

SIHLER, A., CLAUß, D., GROSSI, G., FISCHER, K. (1996), Untersuchung organischer Abfälle auf Schadstoffe und Charakterisierung anhand eines Handbuches, in: *Neue Techniken zur Kompostierung* (STEGMANN, R., Ed.). Bonn: Economica Verlag.

STÖPPLER-ZIMMER, H., BERGMANN, D., HAUKE, H., MARSCHALL, A. (1994), *Kompost mit Gütezeichen für den Garten- und Landschaftsbau mit öffentlichem Grün und Rekultivierung*. Köln: Bundesgütegemeinschaft Kompost e.V.

STROM, P. (1985), Effect of temperature on bacterial species diversity in thermophilic solid-waste composting, *Appl. Environ. Microbiol.* **50**, 899–905.

STROM, P. F., MORRIS, M. L., FINSTEIN, M. S. (1980), Leaf composting through appropriate, low-level technology, *Comp. Sci./Land Util.* **11/12**, 44–48.

TCHOBANOGLOUS, G., THEISEN, H., VIGIL, S. (1993), *Integrated Solid Waste Management*. New York: McGraw-Hill.

USEPA (1981), Composting Processes to Stabilize and Disinfect Municipal Sewage Sludge, *EPA 430/9-81-0111*. Washington, DC: Office of Water Program Operations.

USEPA (1993), 40 CFR Part 503 – Standards for the Use or Disposal of Sewage Sludge, *Fed. Register* **58**, 9387–9401.

VAUCK, W. R. A., MÜLLER, H. A. (1992), Grundoperationen chemischer Verfahrenstechnik, 9th Edn. Leipzig: Deutscher Verlag für Grundstoffindustrie.

WESEMANN, W. (1992), Wurmkompostierung, in: *Handbuch der Kompostierung – Ein Leitfaden für Praxis, Verwaltung, Forschung* (AMLINGER, F., Ed.), pp. 106–110. Vienna: Bundesministerium für Land- und Forstwirtschaft, Bundesministerium für Wissenschaft und Forschung.

ZACHÄUS, D. (1995), Kompostierung, in: *Biologische Abfallbehandlung* (THOMÉ-KOZMIENSKY, K. J., Ed.), pp. 215–353. Berlin: EF-Verlag für Energie- und Umwelttechnik GmbH.

5 Anaerobic Fermentation of Wet or Semi-Dry Garbage Waste Fractions

NORBERT RILLING

Hamburg, Germany

1 Introduction

During the last 30 years the amount of solid wastes has rapidly increased. Now one of the primary aims in waste management is to reduce the amount of waste to be disposed by avoidance, reduction, and utilization (KrW-/AbfG, 1994).

By means of separate collection and biological treatment of biowaste the amount of municipal solid waste (MSW) to be incinerated and/or landfilled will significantly be reduced.

Biological treatment of garbage waste fractions can be carried out aerobically (composting) or anaerobically (anaerobic digestion). Each technique is appropriate for a certain spectrum of wastes. Today most biological waste is composted as this technology is already well-developed with quite a lot of experience at hand, but anaerobic processes advance in their importance for the utilization of solid organic waste.

Anaerobic digestion which is typically conducted inside a closed vessel where temperature and moisture are controlled, is particularly suited to wastes with a high moisture content and a high amount of readily biodegradable components. In Sects. 2, 3 the characteristics of the anaerobic process will be discussed under ecological and economical aspects.

2 Basic Aspects of Biological Waste Treatment

The leading aim of separate collection and treatment of biological wastes is stabilization of the waste by microbial degradation. The product is a compost that, in case of a corresponding low pollution, can be used as a fertilizer or soil conditioner and in that way led back into the natural cycle.

In contrast to the commonly established composting processes the technique of anaerobic fermentation of waste is relatively young and dynamic. With great scientific expenditure process developments and optimizations are pursued so that it may be assumed that the technological potential of biowaste fermentation has not yet been fully exhausted.

2.1 Biochemical Fundamentals of Anaerobic Fermentations

Biogas is produced everywhere, where organic matter is microbially degraded in the absence of oxygen. In nature this process can be observed in marshlands, in marine sediments, in flooded rice fields, in the paunch of ruminants, and in landfill sites (MAURER and WINKLER, 1982).

Anaerobic degradation is effected by various specialized groups of bacteria in several successive steps, each step depending on the preceding one. For industrial scale applications of anaerobic fermentation processes it is necessary to have a thorough knowledge of these interactions to avoid substrate limitation and product inhibition.

The whole anaerobic fermentation process can be divided into three steps:

(1) hydrolysis,
(2) acidification,
(3) methane formation.

At least three groups of bacteria are involved. The degradation is represented in Fig. 1.

First of all, during *hydrolysis*, the mostly water-insoluble biopolymers such as carbohydrates, proteins, and fats are decomposed by extracellular enzymes to water-soluble monomers (e.g., amino acids, glycerin, fatty acids, monosaccharides) and thus made accessible to further degradation.

In the second step (*acidification*) the intermediates of hydrolysis are converted into acetic acid (CH_3COOH), hydrogen (H_2), carbon dioxide (CO_2), organic acids, amino acids, and alcohols by different groups of bacteria. Some of these intermediate products (acetic acid, hydrogen, and carbon dioxide) can directly be used by the methanogenic bacteria while most of the organic acids and alcohol are decomposed into acetic acid, hydrogen and carbon dioxide during the acidogenesis. Only these products, as well as methanol, methylamine, and formate, can be transformed into carbon dioxide (CO_2) and methane (CH_4) by methan-

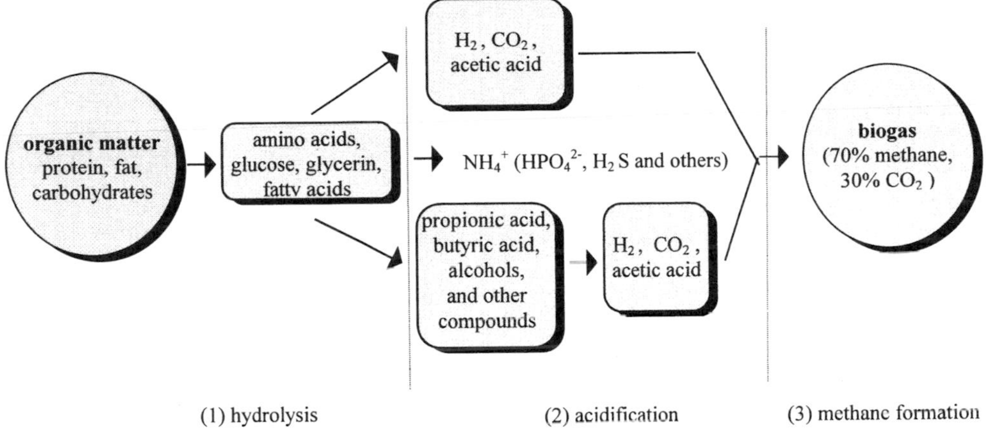

Fig. 1. Scheme of the three-stage anaerobic process of organic matter (SAHM, 1981).

ogenic bacteria during the third and last step, the *methane formation*.

2.1.1 Hydrolytical and Acid Forming (Fermentative) Bacteria

The first group of bacteria is very heterogenous: besides obligatory anaerobic bacteria also facultative anaerobic bacterial strains occur. The high molecular weight compounds of the biomass (proteins, polysaccharides, fats) are decomposed into low molecular weight components by enzymes, which are excreted by fermentative bacteria. This first step is inhibited by lignocellulose containing materials, which are degraded only very slowly or incompletely. Subsequently, acid forming bacteria transform the hydrolysis products into hydrogen, carbon dioxide, alcohols, and organic acids such as acetic, propionic, butyric, lactic, and valeric acid. The formation of the acids decreases the pH value.

2.1.2 Acetic Acid and Hydrogen Forming (Acetogenic) Bacteria

The group of acetogenic microorganisms represents the link between fermentative and methanogenic bacteria. They decompose alcohols and long chain fatty acids into acetic acid, hydrogen, and carbon dioxide. It is a characteristic of acetogenic bacteria that they can only grow at a very low hydrogen partial pressure. For this reason they live in close symbiosis with methanogenic and sulfidogenic bacteria, which use hydrogen as energy source (SAHM, 1981).

2.1.3 Methane Forming (Methanogenic) Bacteria

The group of methanogenic bacteria is formed by extreme obligatory anaerobic microorganisms which are very sensitive to environmental changes. They transform the final products of acidic and acetogenic fermentation into methane and carbon dioxide. Methane is produced to about 70% by the degradation of acetic acid and to about 30% by a redox reaction from hydrogen and carbon dioxide (ROEDIGER et al., 1990).

The slowest step is rate-determining for the whole process of anaerobic fermentation. While methanogenesis of acetic acids is the rate-limiting step concerning the anaerobic fermentation of easily degradable substances, hydrolysis can be rate-limiting in case of sparingly degradable substances. Because of the complexity of the anaerobic degradation mechanisms and the stringent requirements of the microorganisms, process operation is very important for fermentation processes. In order to reach an optimized, undisturbed anaerobic degradation the speed of decomposition of the consecutive steps should be equal.

2.2 Influence of Milieu Conditions on Fermentation

The activity of microorganisms in anaerobic fermentation processes is mainly dependent on water content, temperature, pH level, redox potential and inhibitory factors.

2.2.1 Water Content

Bacteria take up the available substrates in dissolved form. As a consequence, there is an interdependency between biogas production and water content of the initial material. When the water content is below 20% by weight hardly any biogas is produced. With increasing water content the biogas production is enhanced reaching its optimum at 91–98% by weight (KALTWASSER, 1980).

2.2.2 Temperature

The process of biomethanation is very sensitive to changes in temperature, and the degree of sensitivity is dependent on the temperature range. For the predominant portion of methane bacteria the optimum temperature range is between 30 and 37 °C. (MAURER and WINKLER, 1982). Here temperature variations of ±3 °C are of minor effect on the fermentation (WINTER, 1985). In the thermophilic range, however, i.e., at temperatures between 55 and 65 °C, a fairly constant temperature has to be kept, since deviations by only a few degrees cause a drastic reduction of the degradation rates and thus of biogas production.

2.2.3 pH Level

The pH optimum for methane fermentation is between pH 6.7 and 7.4. If the pH level of the media drops <6, because the balance is disturbed and the acid producing dominate the acid consuming bacteria, the medium will have an inhibitory or toxic effect on the methanogenic bacteria. In addition a strong ammonia production during the degradation of proteins may inhibit methane formation, if the pH level exceeds 8. Normally, acid or ammonia production vary only slightly due to the buffer effect of carbon dioxide bicarbonate (CO_2 HCO^{3-}) and ammonia ammonium (NH_3 NH_{4+}), which are formed during fermentation, and the pH level is kept constant normally between 7 and 8.

2.2.4 Redox Potential and Oxygen

Methane bacteria are very sensitive to oxygen and reduce their activity in the presence of oxygen. The anaerobic process, however, shows a certain tolerance of smaller quantities of oxygen. Even permanent limited oxygen introduction normally is tolerated (MUDRAK and KUNST, 1991). The redox potential can be used as an indicator of the process of methane fermentation. Methanogenic bacterial growth requires a relatively low redox potential. HUNGATE (1966) (cited by BRAUN, 1982) found −300 mV to be the minimum value. The change of redox potential during the fermentation process is caused by a decrease of oxygen as well as by the formation of metabolites like formate or acetate.

2.2.5 Inhibitory Factors

The presence of heavy metals, antibiotics, and detergents can have an inhibitory effect on the process of biomethanation. Referring to investigations of KONZELI-KATSIRI and KARTSONAS (1986) Tab. 1 lists the limit concen-

Tab. 1. Inhibition of Anaerobic Digestion by Heavy Metals (KONZELL-KATSIRI and KARTSONAS, 1986)

Heavy Metal	Inhibition [mg L^{-1}]	Toxicity [mg L^{-1}]
Copper (Cu)	40–250	170–300
Cadmium (Cd)	–	20–600
Zinc (Zn)	150–400	250–600
Nickel (Ni)	10–300	30–1,000
Lead (Pb)	300–340	340
Chromium III (Cr)	120–300	200–500
Chromium VI (Cr)	100–110	200–420

trations (mg L^{-1}) for inhibition and toxicity of heavy metals in anaerobic digestion.

2.3 Gas Quantity and Gas Composition

Biogas is a mixture of various gases. Independent on the fermentation temperature a biogas is produced which consists of 60–70% methane and 30–40% carbon dioxide. Trace components of ammonia (NH_3) and hydrogen sulfide (H_2S) can be detected. The caloric value of the biogas is about 5.5–6.0 kWh Nm^{-3}. This corresponds to about 0.5 L of diesel oil.

If the chemical composition of the substrate is known, the yield and the composition of the biogas can be estimated from Eq. (1) referring to SYMONS and BUSWELL (1933):

$$C_nH_aO_b + \left(n - \frac{a}{4} - \frac{b}{2}\right) H_2O \rightarrow \left(\frac{n}{2} + \frac{a}{8} - \frac{b}{4}\right) CH_4 + \left(\frac{n}{2} - \frac{a}{8} + \frac{b}{4}\right) CO_2 \quad (1)$$

Tab. 2 shows the mean composition and specific quantity of biogas dependent on the kind of the degraded substances.

For the anaerobic digestion of the organic fraction of municipal solid waste an average biogas yield of 100 Nm^3 t^{-1} moist biowaste with a methane content of about 60% by volume may be assumed.

2.4 Comparison of Aerobic and Anaerobic Waste Treatment

Professional biological treatment must include the separation of interfering matter, sanitation, and the microbial degradation of the readily and medium degradable substances so that the final product is biologically stable, compatible with roots, and as far as possible free from pollutants and can be applied as soil improver in horticulture and agriculture. Aerobic composting and anaerobic fermentation are in principle available as processes for the biological treatment of organic residues.

Composting is suitable for the stabilization of rather dry and solid waste, while anaerobic processes are applied to very humid waste (e.g., kitchen garbage) which is easy to degrade. Both systems have advantages and disadvantages. In the following they will be listed and compared with regard to their general aspects (Tab. 3).

A big advantage of anaerobic fermentation is the production of biogas which can be used as a source of energy. Either local users can be found for the gas recovered from the process, or it can be cleaned and upgraded for inclusion in a gas supply network. Compared to this, during composting the whole energy is released as heat and cannot be used. In addition to that intensive composting demands a lot of energy for artificial aeration of the waste material.

The technical expenditure of the anaerobic fermentation is higher than that of composting, but if standards, especially concerning the reduction of emissions (odor, germs, noise,

Tab. 2. Gas Yields

Substance	Gas Yield [Nm^3 kg^{-1} TS]	CH_4 Methane Content [Vol. %]	CO_2 Carbon Dioxide Content [Vol. %]
Carbohydrates	0.79	50	50
Fat	1.27	68	32
Protein	0.70	71	29
Municipal solid waste (MSW)	0.1–0.2	55–65	35–45
Biowaste	0.2–0.3	55–65	35–45
Sewage sludge	0.2–0.4	60–70	30–40
Manure	0.1–0.3	60–65	35–40

Tab. 3. Comparison of Aerobic and Anaerobic Waste Treatment (according to RILLING, 1994a)

Characteristics	Anaerobic Digestion	Aerobic Composting
Phases	solids, liquid	solids, liquid, gas
Degradation rate	up to 80% volatile solids	up to 50% volatile solids
Energy consumption	excess	demand
Technical expenditure	in the same range	
Duration of the process	1–4 weeks (only anerobic stage)	4–16 weeks (depending on the process)
Posttreatment	generally necessary (post-composting about 2–8 weeks)	generally none
Floor space required	comparatively low	comparatively low or high (depending on the process)
Odor emission	comparatively low	comparatively high
Stage of development	little experience but increasing	much experience
Costs	in the same range	
Sanitation	external process step	integrated
Suitability of wastes	wide (wet and dry wastes)	small (dry wastes)

dust), are raised, the technical expenditure of composting can be expected to grow as well.

The duration of anaerobic and aerobic treatment very much depends on the substrate and the process applied so that it cannot be compared in general. The same is valid for the floor space required.

A major advantage of anaerobic digestion in comparison with aerobic composting is the ability to exert total control of gaseous and liquid emissions, as well as the potential to recover and use methane gas generated as the wastes degrade. For composting the problem of odor control has not yet been sufficiently solved. Large volumes of olfactory irritating waste air have to be treated costly in order to avoid irritation in the surroundings. In contrast, there is hardly any odorous emission in anaerobic fermentation, as this biological treatment takes place in closed reactors. Malodorous waste air is produced only during loading and unloading the reactor.

No general statement can be made, if anaerobic or aerobic fermentation is the more favorable process, as in every individual case many factors must be considered. This means that in the field of waste treatment, as well as in the field of wastewater treatment, several different processes with similar intentions can have their qualification – depending on the operational area.

3 Processes of Anaerobic Waste Treatment

Anaerobic fermentation has succesfully been applied for many years as a treatment for wastewater, sewage sludge, and manure. However, anaerobic digestion of municipal solid waste is a relative young technique which has been developed in the past 10–15 years.

Only biodegradable household wastes – i.e., those of organic or vegetable origin – can be processed in anaerobic digestion plants. Garden waste or green wastes, as they are often called, can also be included, but woodier material (like branches) is less suitable because of its relatively longer decomposition time under anaerobic conditions.

As a result of anaerobic fermentation combined with an additional postcomposting step a material is produced in most cases that is similar to the compost produced with aerobic processes. It can be used as fertilizer, soil conditioner, or peat substitute.

While composting is widely used for high dry matter-containing wastes, anaerobic digestion turned out to be a good alternative for wet organic wastes (Fig. 2). At present the anaerobic fermentation technique is mainly used in Western Europe where more than 30 com-

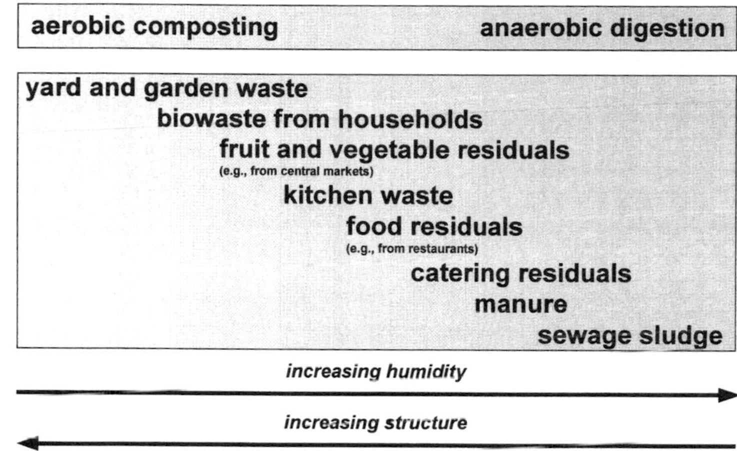

Fig. 2. Suitability of wastes for aerobic composting and anerobic digestion (according to KERN et al., 1996).

panies offer anaerobic treatment plants commercially for the digestion of putrescible solid waste.

3.1 Procedures of the Anaerobic Waste Fermentation

Generally, the following steps are required for the anaerobic treatment of organic waste (RILLING, 1994a):

(1) delivery and storage of the biological waste,
(2) pre-processing of the incoming biological waste,
(3) anaerobic fermentation,
(4) storage and treatment of the digester gas,
(5) treatment of the process water,
(6) post-processing of the digested material.

Fig. 3 shows the possible treatment steps used in biowaste fermentation. In principle, all fermentation processes can be described as a combination of a selection of these treatment steps. The process technology demanded for the realization of the different steps of the treatment differs very much depending on the anaerobic process chosen. In general, the gas production grows and the detention time decreases with an increasing energy input for the preparation of the material and the fermentation itself (mesophilic/thermophilic).

3.1.1 Delivery and Storage

Both in composting and fermentation the wastes are pretreated, before the actual biological stabilization takes place. The supplied biowastes are quantitatively and qualitatively recorded by weighing and visually inspected at an acceptance station and unloaded into a flat or deep bunker or a collecting tank which serves as an intermediate storage for a short time and permits the continuous feeding to the subsequent pretreatment plant.

3.1.2 Preprocessing

The purpose of pretreatment is to remove pollutants and interfering matter as well as the homogenization and conditioning of the biowaste. The kind of pretreatment depends upon the specific system of the anaerobic fermentation process.

Dry fermentation processes use dry pre-processing, where sieves, shredders, grinders, homogenization drums, metal separators, ballistic

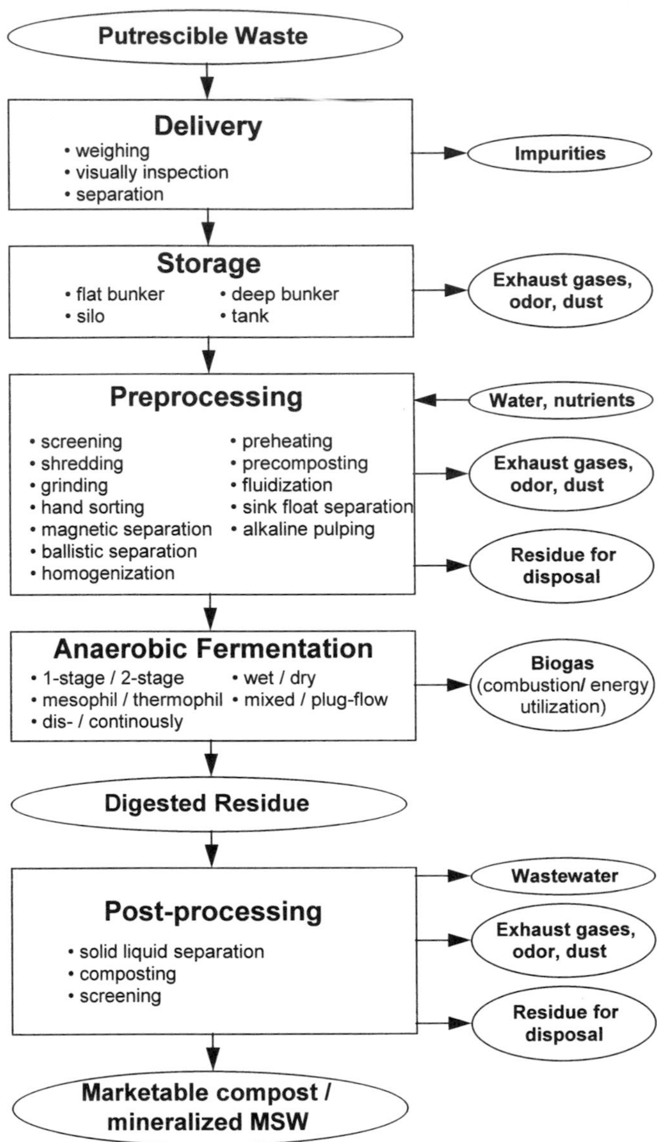

Fig. 3. Possible treatment steps used in biowaste fermentation.

separators, and hand sorting sections can be combined. In wet fermentation processes the biowastes are additionally mixed with water, homogenized and shredded, and by means of sink–float separation other foreign substances can be removed.

3.1.3 Anaerobic Fermentation

Having separated any recyclable or unwanted materials from the incoming wastes, the organic material is shredded and fed into the digestor. Shredding results in a material that can be handled more easily. In addition, a larger surface area will more easily be broken down by the bacteria. Shredding in a drum can be combined with a pre-composting step. On the other hand due to shredding the material will lose structure. The desired particle size after shredding may be between 5 and 40 mm, in special cases up to 80 mm. Yard and garden waste (especially branches, etc.) have to be shredded separately before composting.

If very wet wastes, like sewage sludge, are included, the addition of further water may not be necessary, but in the case of household organic wastes, water is added in most cases to dilute the solids.

Low-structured wastes with high moisture content are predestined for the fermentation, but also wastes rich in structure can preferably be degraded anaerobically by means of dry fermentation processes. Heat is needed to adjust the required process temperatures of about 35 °C (mesophilic process) or 55 °C (thermophilic process), and in some cases water must be added. During fermentation organic degradation takes place anaerobically, i.e., under exclusion of oxygen, in closed, temperature-regulated containers. Depending on the process operation the material consistency may vary between well-structured matter and past thick sludge and fluid suspension. The optimum pH value is in the neutral range.

The output of the fermentation reactor is a wet, organically stabilized fermentation residue and the biogas produced. After dewatering a soil-improving product comparable to compost can be obtained from the fermentation residue by aerobic post-composting. The wastewater generated during draining can in part be recirculated into the pretreatment unit to adjust the water content. Surplus wastewater has to be treated and discharged. With only minor energy loss the biogas can be used in decentralized heating power stations to produce electrical power and heat so that, in general, the fermentation process can be operated energy-autarkical, and the surplus can be marketed by feeding it into the public power and heat supply mains.

When the fermentation is confined only to the easily degradable organic waste components, energy can be produced with minimized technical expenditure and the odor and energy intensive prefermentation can be omitted. In a subsequent composting step the medium and difficult to degrade organic substances, which can only be degraded anaerobically to a limited extent, are aerobically decomposed at low cost. Thus when investigating the question "fermentation or composting" in many cases the answer may suggest the demand for fermentation and composting.

3.1.4 Post-Processing

To complete the stabilization and disinfection of the digested residue some kind of refining process is needed before it can be used for agriculture or horticulture. After possible dewatering and/or drying, the anaerobically fermented waste is generally transferred to aerobic biological post treatment and matured for about 2–4 weeks to become a good, marketable compost.

After drying and, in a few cases gas purification, the biogas can be used as an energy source.

Depending on local regulations the excess process water is transferred to a wastewater treatment plant or may be applied directly to farmland as a liquid fertilizer.

3.2 Process Engineering of Anaerobic Fermentation of Biowastes

Anaerobic fermentation processes are generally suitable for the biological treatment of readily degradable substances with a low structure and a high water content as, e.g., kitchen waste. At present, several processes for the anaerobic fermentation of organic solid waste are under development. The processes differ in the number of biodegradation stages (one- or two-stage processes), separation of liquid and solids (one- or two-phase system), water content (dry or wet fermentation), feed method (continuous or discontinuous), and in the means of agitation. The most important characteristics of anaerobic fermentations are compiled in Tab. 4.

Tab. 4. Characteristics of Anaerobic Waste Treatments (according to RILLING, 1994a)

Characteristics		
Stages of biodegradation	one-stage	two-stage
Separation of liquid and solids	one-phase	two-phase
	dry fermentation	wet fermentation
Total solids content	25–45%	<15%
Water content	55–75%	>85%
Feed method	discontinuously	continuously
Agitation	none	stirring, mixing, percolation
Temperature	mesophilic (30–37 °C)	thermophilic (55–65 °C)

3.2.1 One or Two-Stage Process Operation

Anaerobic fermentation of biowaste can be achieved by either one-stage or two-stage fermentation. In the *one-stage process* (Tab. 5) hydrolysis, acidification, and methane formation take place in one reactor, so that it is not possible to achieve optimum reaction conditions for the overall process due to slightly different milieu requirements during the different stages of the fermentation phases. Therefore, the degradation rate is reduced and the retention time increases. The basic advantage of one-stage process operation is the relatively simple technical installation of the anaerobic digestion plant and lower costs.

In *two-stage processes* (Tab. 5) hydrolysis and acidification on the one hand and methane formation on the other hand take place in different reactors so that, e.g., mixing and adjustment of the pH value can be optimized separately, permitting higher degradation degrees and loading rates. As a result the detention time of the material can be reduced significantly. This, of course, involves a more sophisticated technical design and operation resulting in higher costs.

During the first stage the organic fraction is hydrolyzed. As a result dissolved organics and mainly organic acids as well as CO_2 and low concentrations of hydrogen are produced. At the second stage the highly concentrated water is supplied to an anaerobic fixed film reactor or sludge blanket reactor or other appropriate systems where methane and CO_2 are produced as final products. Reciprocal inhibiting effects are excluded so that high process stability with a better methane yield is obtained.

3.2.2 Dry and Wet Fermentation

Different anaerobic digestion systems can handle waste with different moisture contents. According to this the processes are classified as dry fermentation processes with a water

Tab. 5. Comparison of One- and Two-Stage Processes

Process Operation	One-Stage	Two-Stage
Operational Reliability	in the same range	
Technical equipment	relatively simple	very extensive
Process control	compromize solution	optimal
Risk of process instability	high	minimized
Retention time	long	short
Degradation rate	reduced	increased

content between 55 and 75% and wet fermentation processes with a water content of >85%. Tab. 6 shows advantages and disadvantages of dry and wet fermentations.

With the *dry fermentation* process no or only small quantities of water are added to the biowaste. As a consequence, the material streams to be treated are minimized. The resulting advantages are smaller reactor volumes and easier dewatering of the digested residue. On the other hand, operating with high dry matter contents involves higher requirements on mechanical pretreatment and conveying, gas tightness of charge and discharge equipment and, if planned, on mixing in the reactors. Bridging of the material and the possibility of clogging have to be encountered. Because of the low mobility when applying dry fermentation, a defined residence time can be reached by approximation to the plug flow, which is particularly important under the aspect of product hygiene in the thermophilic operation mode. Compared to wet fermentation the degradation rates of dry fermentation processes are lower due to the larger particle size and reduced substrate availability.

When *wet fermentation* processes are applied, the organic wastes are ground to a small particle size and mixed with large quantities of water so that sludges or suspensions are obtained. This allows the use of simple and established mechanical conveying techniques (pumping) and the removal of interfering substances by sink–float separation. At the same time the reactor contents can easily be mixed, which permits controlled degassing and defined concentration equalization in the fermenter. As a consequence, the degradation performance of the microorganisms is optimized. The mean substrate concentrations and thus also the related degradation rates are lower than in plug flow systems, since for completely mixed systems the concentrations in the system are equal to the outlet concentrations. Mixing is limited by the shear sensitivity of methane bacteria. On the other hand a too low degree of mixing may result in swim and sink layers. Homogeneity and a fluid consistency permit easier process control. By fluidizing the biowaste the mass to be treated increases, dependent on the total solids content of the substrate, until the 5-fold with the consequence that the aggregates and reactors have to be dimensioned much larger. Especially the fluidization and dewatering of the fermentation suspension require considerable technical and energetic expenditure. On the condition that the degradation degrees are the same, by recycling the liquid phase from the dewatering step to the fluidization of the input material, it is possible to reduce the wastewater quantity to an amount comparable to that of the dry fermentation and to keep a considerable part of the required thermal energy within the system.

Tab. 6. Comparison of Wet and Dry Fermentations

Process Mode	Dry	Wet
Total solids content	high 25–45%	low 2–15%
Reactor volume	minimized	increased
Conveying technique	expensive	simple
Agitation	difficult	easy
Scumming	little risk	high risk
Short circuit flow	little risk	high risk
Solid–liquid separation	simple	expensive
Variety of waste components	small	wide

3.2.3 Continuous and Discontinuous Operation

When an anaerobic process is run in *continuous operation mode* the reactor is fed and discharged regularly. Completely mixed and plug flow systems are available. Substrate is fed into the reactor to the same extent as putrefied material is discharged. Therefore, the substrate must be flowable and uniform. By steady provision of nutrients in the form of raw biodegradable waste stable process operation with a constant biogas yield can be obtained. Depending on the reactor design and the means of mixing short circuits may occur and, therefore, the retention time cannot be guaranteed for each part of the substrate in completely mixed systems.

In the *discontinuous operation mode* (batch process) the fermentation vessel is completely filled with raw garbage mixed with inoculum (e.g., digestate from another reactor) and then completely discharged after a fixed detention time. Batch digesters are easy to design, comparatively low in cost, and suitable for all wet and dry organic wastes (Tab. 7).

3.2.4 Thermophilic and Mesophilic Operation

The optimal process temperatures for methane fermentation are in the mesophilic temperature range at about 35 °C and in the thermophilic temperature range at about 55 °C.

Reactors with mesophilic operation are heated to 30 to 40 °C. According to experience in this way of operation the process stability is high. Minor temperature variations have only a small effect on mesophilic bacteria.

The advantages of mesophilic process operation result from the lower amount of heat to be supplied and the related higher net energy yield. In addition, higher process stability is reached since there exists a wide spectrum of mesophilic methane bacteria which show a lower sensitivity against temperature changes.

The *thermophilic* range requires temperatures between 50 and 60 °C. Under certain circumstances the thermophilic process operation allows a faster substrate turnover so that the residence times may be shorter. The higher expenditure of energy to maintain the process temperature is a disadvantage.

When the process is run under thermophilic conditions for a defined residence time, sanitation in the reactor is possible, which otherwise has to be achieved in a separate treatment step or by aerobic after-composting. On the other hand, the net energy yield is lower due to the higher heat requirement, and the temperature sensitivity of the microorganisms reduces the process stability. In Tab. 8 the advantages and disadvantages of mesophilic and thermophilic process operations are listed.

Tab. 7. Comparison of Continuous and Discontinuous Feed

Process Operation	Continuous	Discontinuous
Retention time	shorter	longer
Technical equipment	extensive	simple

3.2.5 Agitation

For a high degradation activity of the bacteria it is necessary to provide the active biomass with sufficient degradable substrate. Simultaneously, the metabolic products of the organisms have to be removed (DAUBER, 1993). These requirements can be met by mechanical mixing or other agitation of the reactor content. Another possibility is to install a water recirculation system, where the process water, which ensures the nutrient provision and the removal of metabolic products, trickles through the biowaste in the reactor (RILLING and STEGMANN, 1992). Other processes use compressed biogas for a total or partly mixing of the material.

3.3 Survey of Anaerobic Fermentation Processes

During the past years several anaerobic processes for the utilization of solid waste have been developed. Each of them has its own ben-

Tab. 8. Comparison of Mesophilic and Thermophilic Process Operation

Process Operation	Mesophilic (35 °C)	Thermophilic (55 °C)
Process stability	higher	lower
Temperature sensitivity	low	high
Energy demand	low	high
Degradation rate	reduced	increased
Detention time	longer or the same	shorter or the same
Sanitation	no	possible

efits. Currently the European market offers at least 30 different processes or process variants. Fig. 4 shows (without claiming to be complete) the division of the processes according to water content and number of fermentation stages.

It illustrates that most of the offered processes are one- or two-stage wet fermentation systems. The number of dry fermentation processes is comparatively small. One primary reason for this situation is the higher technical effort required for dry fermentation. While wet fermentation can be based on the well-known and successful technology of sewage sludge treatment and the digestion of manure, dry fermentation requires the employment of new, innovative technologies, especially in the field of gas tight filling and emptying as well as in conveyer systems.

As an example of a one-step wet fermentation system Fig. 5 shows the flow chart of the DBA-WABIO process, as an example of a two-step wet fermentation system Fig. 6 shows the flow chart of the BTA-process and as examples of one-step dry fermentation systems Fig. 7 shows the flow chart of the ATF process and Fig. 8 that of the DRANCO process.

3.4 Feedstock for Anaerobic Digestion

The feedstock for an anaerobic digestion plant can either be organic wastes which have separately been collected and delivered to the plant ready for processing or alternatively municipal solid waste (MSW) or a fraction of MSW (e.g., < 100 mm) from a mechanical sorting plant where the other fraction is a kind of refuse-derived fuel. A further source of organic waste are "green wastes" collected at centralized collection points.

At least the purity of the raw material fed into the anaerobic digestion process dictates the quality of the horticultural product at the end of the process.

The range of application of the anaerobic digestion process is very wide. Principally, any organic material can be digested, i.e. (RILLING, 1994b):

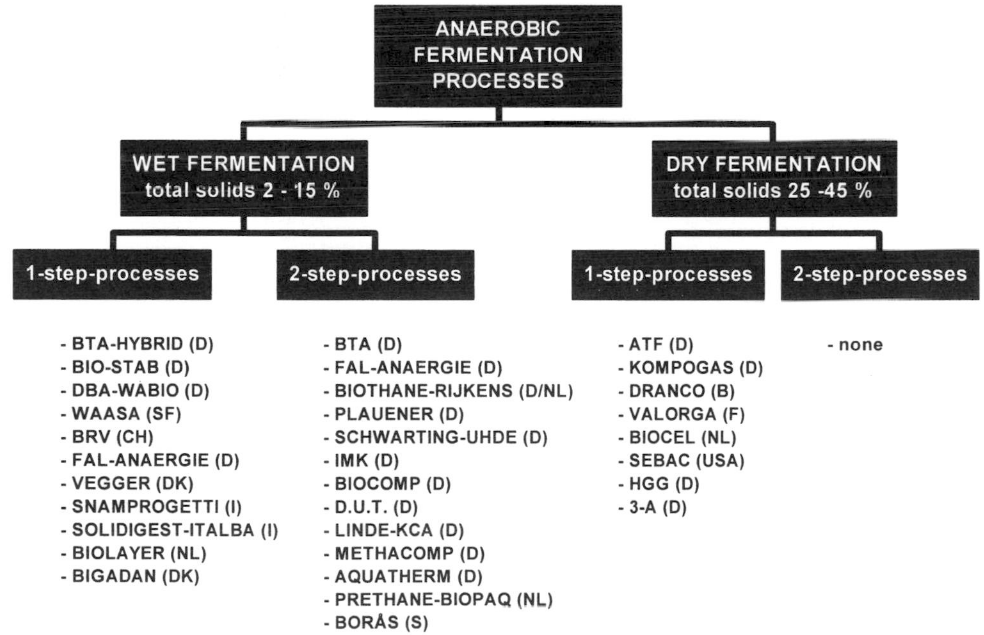

Fig. 4. Processes of anaerobic waste fermentation (according to RILLING, 1994a, updated).

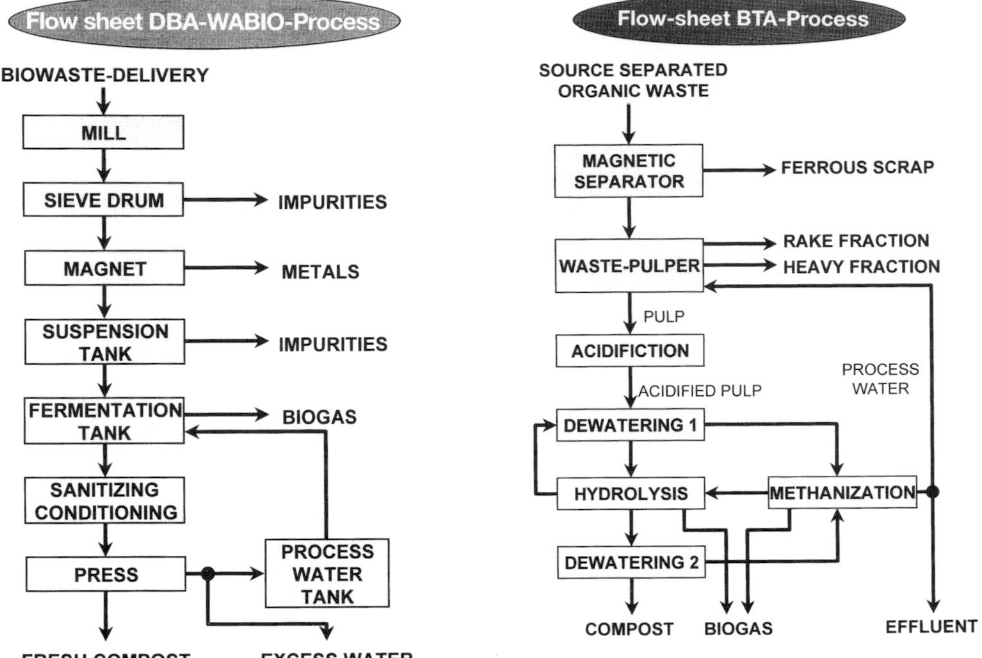

Fig. 5. Example of a one-step wet fermentation plant (DBA-WABIO)

Fig. 6. Example of a two-step wet fermentation plant (Biotechnische Abfallbehandlung, BTA).

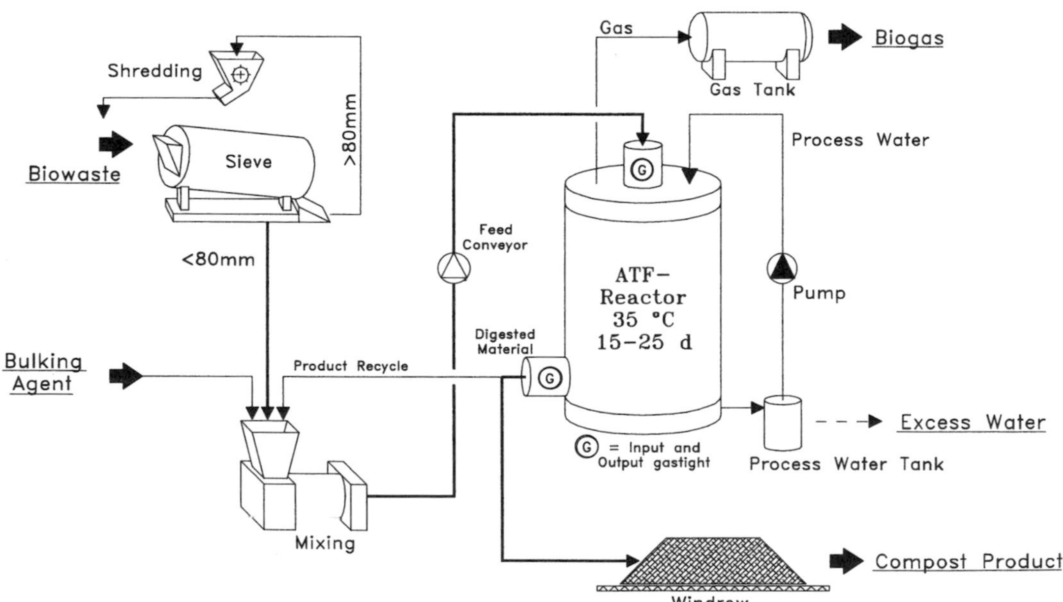

Fig. 7. Example of a one-step dry fermentation plant (ATF process) (RILLING et al., 1996).

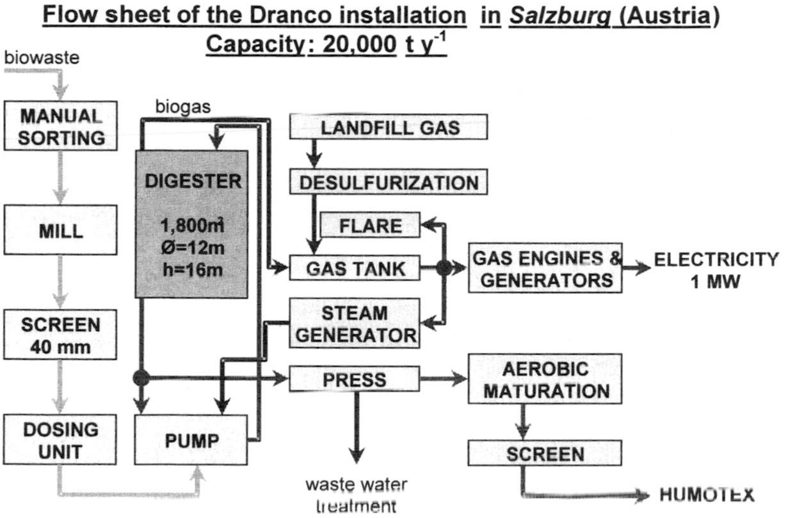

Fig. 8. Example of a one-step dry fermentation plant (DRANCO process) (SIX et al., 1995).

- organic municipal solid waste,
- waste out of central markets (e.g., fruit, vegetable, and flower residuals),
- slaughterhouse waste (paunch manure),
- residues from the fish processing industry,
- food waste from hotels, restaurants, and canteens,
- bleaching soil,
- drift material such as seaweed or algae,
- agricultural waste,
- manure,
- beer draff,
- fruit or wine marc,
- sewage sludge.

4 Conclusions

The separate collection of biological wastes and their biological treatment are an important part of waste utilization. There is no universal process for the biological treatment of biological waste, not even in the future. It can be concluded that besides composting anaerobic digestion turns out to be a useful technology both in its economical and ecological aspects.

The basic aims of the anaerobic biological waste treatment are:

- reduction of the waste volume,
- reduction of organic substance for stabilization of the waste,
- conservation of nutrients and recycling of the organic substance (fertilizer or soil improvement),
- energy recovery by the generation of biogas,
- odor reduction,
- sanitation.

Anaerobic digestion is a sustainable treatment technique for many kinds of organic wastes. Based on the expectations of future rising market shares, the increasing number of the processes offered shows the dynamic and innovative force in anaerobic fermentation technology.

5 References

BRAUN, R. (1982), *Biogas – Methangärung organischer Abfallstoffe*, Berlin, Heidelberg. New York: Springer-Verlag.

DAUBER, S. (1993), Einflußfaktoren auf die anaeroben biologischen Abbauvorgänge, in: *Anaerobtechnik – Handbuch der anaeroben Behandlung von Abwasser und Schlamm* (BÖHNKE, B., BISCHOFSBERGER, W., SEYFRIED, C. F., Eds.), pp. 62–93. Berlin, Heidelberg, New York: Springer-Verlag.

HUNGATE, R. E. (1966), *The Rumen and Its Microbes*. New York, London: Academic Press.

KALTWASSER, B. J. (1980), *Biogas – Regenerative Energieerzeugung durch anaerobe Fermentation organischer Abfälle in Biogasanlagen*. Wiesbaden, Berlin: Bauverlag.

KERN, M., MAYER, M., WIEMER, K. (1996), Systematik und Vergleich von Anlagen zur anaeroben Abfallbehandlung, in: *Biologische Abfallbehandlung III* (Wiemer, K., Kern, M., Eds.), pp. 409–437. Witzenhausen: M.I.C. Baeza.

KONZELI-KATSIRI, A., KARTSONAS, N. (1986), Inhibition of anaerobic digestion by heavy metals, in: *Anaerobic Digestion of Sewage Sludge and Organic Agriculturals Wastes* (BRUCE, A. M., KONZELI-KATSIRI, A., NEWMAN, P. J., Eds.), pp. 104–119. London, NewYork: ElsevierApplied Science Publishers.

KrW-/AbfG (1994), *Gesetz zur Förderung der Kreislaufwirtschaft und Sicherung der umweltverträglichen Beseitigung von Abfällen.*

MAURER, M., WINKLER, J.-P. (1982), *Biogas – Theoretische Grundlagen, Bau und Betrieb von Anlagen.* Karlsruhe: C. V. Müller.

MUDRACK, K., KUNST, S. (1991), *Biologie der Abwasserreinigung*. Stuttgart: Gustav Fischer.

RILLING, N. (1994a), Anaerobe Trockenfermentation für Bioabfall, *Ber. Abwassertechn. Ver.* **44**, 985–1002.

RILLING, N. (1994b), Untersuchungen zur Vergärung organischer Sonderabfälle, in: *Anaerobe Behandlung von festen und flüssigen Rückständen*, DECHEMA-Monographien, Bd. 130 (MAERKL, H., STEGMANN, R., Eds.), pp. 185–205. Weinheim: VCH.

RILLING, N., STEGMANN, R. (1992), High solid content anaerobic digestion of biowaste, in: *Proc. 6th Int. Solid Wastes Congr. ISWA 92*, Madrid, Spain.

RILLING, N., ARNDT, M., STEGMANN, R. (1996), Anaerobic fermentation of biowaste at high total solids content – experiences with ATF-system, in: *Management of Urban Biodegradable Wastes* (HANSEN, J. A., Ed.), pp. 172–180. London: James & James.

ROEDIGER, H., ROEDIGER, M., KAPP, H. (1990), *Anaerobe alkalische Schlammfaulung*. München: Oldenbourg.

SAHM, H. (1981), Biologie der Methanbildung, *Chem. Ing. Technol.* **53**, 854–863.

SYMONS, G. E., BUSWELL, A. M. (1933), The methane fermentation of carbohydrates, *J. Am. Chem. Soc.* **55**, 2028.

SIX, W., KAENDLER, C., DE BAERE, L. (1995), The Salzburg plant: a case study for the biomethanization of biowaste, in: *Proc. 1st Int. Symp. Biol. Waste Manag. – A Wasted Chance?* Bochum, Germany.

WINTER, J. (1985), Mikrobiologische Grundlagen der anaeroben Schlammfaulung, *gwf Wasser Abwasser* **126**, 51–56.

6 Landfill Systems, Sanitary Landfilling of Solid Wastes – Long-Term Problems with Leachates

KAI-UWE HEYER
RAINER STEGMANN
Hamburg, Germany

List of Abbreviations

AOX	halogenated organic hydrocarbons
BOD_5	biological oxygen demand (in 5 d)
C_0	concentration at test beginning
C_E	limiting value
COD	chemical oxygen demand
E_h	redox potential
k	permeability
LSR	landfill simulation reactor
MSW	municipal solid waste
TASI	TA Siedlungsabfall
$T_{1/2}$	half-life
T_E	time period for reaching the limiting value
TKN	total Kjeldahl nitrogen
TOC	total organic carbon
TS	total (dry) substance/Trockensubstanz

1 Introduction

The sanitary landfill plays a most important role in the framework of solid waste disposal and will remain an integral part of the new strategies based on integrated solid waste management. The quality of landfill design, according to technical, social and economic developments, has improved dramatically in recent years. Design concepts are mainly devoted towards ensuring minimal environmental impact in accordance with observations made concerning the operation of old landfills. The major environmental concern associated with landfills is related to discharge of leachate into the environment, and the current landfill technology is primarily determined by the need to prevent and control leachate problems. In order to reduce the emissions of gases that cause global warming the control of landfill gas gains more and more importance.

2 Biochemical Processes in Sanitary Landfills

To understand the requirements for landfill systems and sanitary landfilling of solid wastes the main chemical, physical and biological factors which influence the processes in landfills and the leachate quality will be described.

The mechanisms which regulate mass transfer from wastes to leaching water, from which leachate originates, can be divided into three categories (ANDREOTTOLA, 1992):

- hydrolysis of solid waste and biological degradation,
- solubilization of soluble salts contained in the waste, and
- wash out of fines.

The first two categories of mechanisms, which have greater influence on the quality of leachate produced, are included in the more general concept of waste stabilization in landfills. In this chapter the various phases of waste stabilization processes are described and the mechanisms that lead to mass transfer from wastes to leachate are discussed.

2.1 Aerobic Degradation Phases

The first phase of aerobic degradation of organic substances is generally of limited duration due to the high oxygen demand of waste relative to the limited quantity of oxygen present inside a landfill (Phase I, Fig. 1). The only layer of a landfill involved in aerobic metabolism is the upper layer where oxygen is trapped in fresh waste and is supplied by diffusion and rainwater. In this phase it was observed that proteins are degraded into amino acids, then into carbon dioxide, water, nitrates and sulfates, typical catabolites of all aerobic processes (BARBER, 1979). Carbohydrates are converted to carbon dioxide and water and fats are hydrolyzed to fatty acids and glycerol and are then further degraded into simple catabolites through intermediate formation of volatile acids and alkalis. Cellulose, which constitutes the majority of the organic fraction of wastes, is degraded by means of extracellular enzymes into glucose which is used subsequently by bacteria and converted to carbon dioxide and water. This stage, due to the exothermal reactions of biological oxidation, may reach elevated temperatures if the waste is not compacted. Usually the aerobic phase is quite short and no substantial leachate generation will take place.

In very old landfills, when only the more refractory organic carbon remains in the landfilled wastes, a second aerobic phase may appear in the upper layer of the landfill. In this phase the methane production rate is very low and air will start diffusing from the atmosphere, giving rise to aerobic zones and zones with redox potentials too high for methane formation (CHRISTENSEN and KJELDSEN, 1989).

2.2 Anaerobic Degradation Phases

Three different phases can be identified in the anaerobic decomposition of waste. The first phase of anaerobic degradation is acid fermentation, which causes a decrease in leach-

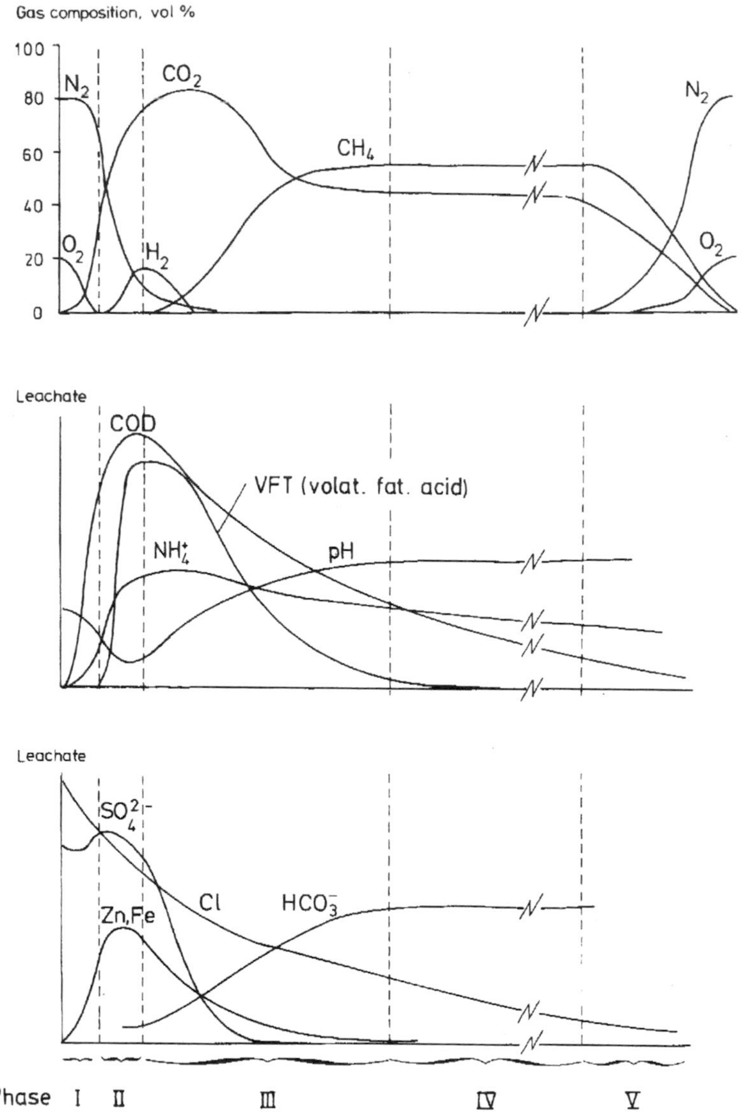

Fig. 1. Illustration of developments in leachate and gas in a landfill cell (CHRISTENSEN and KJELDSEN, 1989).

ate pH, high concentrations of volatile acids and considerable concentrations of inorganic ions (e.g., Cl^-, SO_4^{2-}, Ca^{2+}, Mg^{2+}, Na^+). The initial high content of sulfates may slowly be reduced as the redox potential drops and metal sulfides are gradually generated that are of low solubility and precipitate iron, manganese and other heavy metals that were dissolved by the acid fermentation (CHRISTENSEN and KJELDSEN, 1989). The decrease in pH is caused by the high production of volatile fatty acids and the high partial pressure of CO_2. The increased concentration of anions and cations is due to lixiviation of easily soluble material consisting of original waste components and degradation products of organic substances. Initial anaerobic processes are elicited by a population of mixed anaerobic microbes, composed of strictly anaerobic bacteria and facultative anaerobic bacteria. The facultative anaerobic bacteria reduce the redox potential so that methanogenic bacteria can grow. In fact, the latter are sensitive to the presence of oxygen and require a redox potential below -330 mV. Leachate from this phase is characterized by high BOD_5 values (commonly $>10{,}000$ mg O_2 L^{-1}), high BOD_5/COD ratios (commonly >0.7) and acidic pH values (typically 5–6) and ammonia (often 500–1,000 mg L^{-1}) (ROBINSON, 1989), the latter due to hydrolysis and fermentation of proteinaceous compounds in particular.

The second intermedial anaerobic phase (Phase III, Fig. 1) starts with slow growth of methanogenic bacteria. This growth may be inhibited by an excess of organic volatile acids which are toxic to methanogenic bacteria at concentrations of 6,000–16,000 mg L^{-1} (STEGMANN and SPENDLIN, 1989). The methane concentration in the gas increases, while hydrogen, carbon dioxide and volatile fatty acids decrease. Moreover, the concentration of sulfate decreases owing to biological reduction. Conversion of fatty acids causes an increase in pH values and alkalinity with a consequent decrease in solubility of calcium, iron, manganese and heavy metals. The latter are probably precipitated as sulfides. Ammonia is released and is not converted in the anaerobic environment.

The third phase of anaerobic degradation (Phase IV, Fig. 1) is characterized by methanogenic fermentation elicited by methanogenic bacteria. The pH range tolerated by methanogenic bacteria is extremely limited and ranges from 6–8. At this stage, the composition of leachate is characterized by almost neutral pH values, low concentrations of volatile acids and total dissolved solids while biogas presents a methane content which is generally higher than 50%. This confirms that solubilization of the majority of organic components has decreased at this stage of landfill operation, although the process of waste stabilization will continue for several years and decades.

Leachates produced during this phase are characterized by relatively low BOD_5 values and low ratios of BOD_5/COD. Ammonia continues to be released by the first stage acetogenic process.

In Tab. 1 the ranges of leachate concentrations depending on the degradation phase for some relevant parameters are presented. EHRIG (1990) compiled leachate concentrations from German landfills from the 1970s and 1980s. According to his evaluation the organics (COD, BOD_5, TOC) as well as AOX, SO_4, Ca, Mg, Fe, Mn, Zn and Cr are determined by the biochemical processes in the landfill, there are striking differences between the acid phase and the methanogenic phase.

KRUSE (1994) investigated 33 landfills in Northern Germany, the leachate concentrations mainly derive from the late 1980s and early 1990s. He defined three characteristic periods according to the BOD_5/COD ratio:

- acid phase: $BOD_5/COD \geq 0.4$
- intermediate phase: $0.4 > BOD_5/COD > 0.2$
- methanogenic phase: $BOD_5/COD \leq 0.2$

Between the two investigations there are significant differences concerning the organic parameters. In the younger landfills (KRUSE, 1994) leachate concentrations of COD, BOD_5 and TOC are lower than those determined by EHRIG (1990) some 10 years before. This can be explained by developments in the technology of waste landfilling. In many younger landfills waste deposition and compaction in thin layers in combination with an aerobic pretreated bottom layer were carried out. This led to a reduction of the period for the acid phase and to an accelerated conversion of organic leachate components to the gaseous phase, the degradation of organics to methane and carbon dioxide.

2.3 Factors Affecting Leachate Composition

The chemical composition of leachate depends on several parameters, including those concerning waste mass and site localization and those deriving from design and management of the landfill. Of the former the main factors influencing leachate quality are:

Tab. 1. Constituents in Leachates from MSW Landfills (EHRIG, 1990; KRUSE, 1994)

Parameter	Unit	Leachate from MSW Landfills (EHRIG, 1990)				Leachate from MSW, Landfills (KRUSE, 1994)					
		Acid Phase		Methanogenic Phase		Acid Phase		Intermediate Phase		Methanogenic Phase	
		Range	Medium	Range	Medium	Range	Medium	Range	Medium	Range	Medium
pH value	–	4.5–7	6	7.5–9	8	6.2–7.8	7.4	6.7–8.3	7.5	7.0–8.3	7.6
COD	mg O_2 L^{-1}	6,000–60,000	22,000	500–4,500	3,000	950–40,000	9,500	700–28,000	3,400	460–8,300	2,500
BOD_5	mg O_2 L^{-1}	4,000–40,000	13,000	20–550	180	600–27,000	6,300	200–10,000	1,200	20–700	230
TOC	mg L^{-1}	1,500–25,000	7,000	200–5,000	1,300	350–12,000[b]	2,600[b]	300–1,500[b]	880[b]	150–1,600[b]	660[b]
AOX	µg L^{-1}	540–3,450	1,674	524–2,010	1,040	260–6,200	2,400	260–3,900	1,545	195–3,500	1,725
Organic N[a]	mg L^{-1}	10–4,250	600	10–4,250	600						
NH_4—N[a]	mg L^{-1}	30–3,000	750	30–3,000	750	17–1,650	740	17–1,650	740	17–1,650	740
TKN[a]	mg L^{-1}	40–3,425	1,350	40–3,425	1,350	250–2,000	920	250–2,000	920	250–2,000	920
NO_2—N[a]	mg L^{-1}	0–25	0.5	0–25	0.5						
NO_3—N[a]	mg L^{-1}	0.1–50	3	0.1–50	3						
SO_4	mg L^{-1}	70–1,750	500	10–420	80	35–925	200	20–230	90	25–2,500	240
Cl	mg L^{-1}	100–5,000	2,100	100–5,000	2,100	315–12,400	2,150	315–12,400	2,150	315–12,400	2,150
Na[a]	mg L^{-1}	50–4,000	1,350	50–4,000	1,350	1–6,800	1,150	1–6,800	1,150	1–6,800	1,150
K[a]	mg L^{-1}	10–2,500	1,100	10–2,500	1,100	170–1,750	880	170–1,750	880	170–1,750	880
Mg	mg L^{-1}	50–1,150	470	40–350	180	30–600	285	90–350	200	25–300	150
Ca	mg L^{-1}	10–2,500	1,200	20–600	60	80–2,300	650	40–310	150	50–1,100	200
Total P[a]	mg L^{-1}	0.1–30	6	0.1–30	6	0.3–54	6.8	0.3–54	6.8	0.3–54	6.8
Cr[a]	mg L^{-1}	0.03–1.6	0.3	0.3–1.6	0.3	0.002–0.52	0.155	0.002–0.52	0.155	0.002–0.52	0.155
Fe	mg L^{-1}	20–2,100	780	3–280	15	3–500	135	2–120	36	4–125	25
Ni[a]	mg L^{-1}	0.02–2.05	0.2	0.02–2.05	0.2	0.01–1	0.19	0.01–1	0.19	0.01–1	0.19
Cu[a]	mg L^{-1}	0.004–1.4	0.08	0.004–1.4	0.08	0.005–0.56	0.09	0.005–0.56	0.09	0.005–0.56	0.09
Zn	mg L^{-1}	0.1–120	5	0.03–4	0.6	0.05–16	2.2	0.06–1.7	0.6	0.09–3.5	0.6
As[a]	mg L^{-1}	0.005–1.6	0.16	0.005–1.6	0.16	0.0053–0.11	0.0255	0.0053–0.11	0.0255	0.0053–0.11	0.0255
Cd[a]	mg L^{-1}	0.0005–0.14	0.006	0.0005–0.14	0.006	0.0007–0.525	0.0375	0.0007–0.525	0.0375	0.0007–0.525	0.0375
Hg[a]	mg L^{-1}	0.0002–0.01	0.01	0.0002–0.01	0.01	0.000002–0.025	0.0015	0.000002–0.025	0.0015	0.000002–0.025	0.0015
Pb[a]	mg L^{-1}	0.008–1.02	0.09	0.008–1.02	0.09	0.008–0.4	0.16	0.008–0.4	0.16	0.008–0.4	0.16

[a] Parameter more or less independent of the biochemical degradation phase.
[b] DOC.

2.3.1 Waste Composition

The nature of the waste organic fraction influences the degradation of waste in the landfill considerably and thus also the quality of the leachate produced. In particular, the presence of substances which are toxic to the bacterial flora may slow down or inhibit biological degradation processes with consequences for the leachate quality. The inorganic content of the leachate depends on the contact between waste and leaching water as well as on pH and the chemical balance at the solid–liquid interface. In particular, the majority of metals are released from the waste mass under acid conditions.

2.3.2 pH

pH influences chemical processes which are the basis of mass transfer in the waste leachate system, such as precipitation, dissolution, redox and sorption reactions. It will also affect the speciation of most of the constituents in the system. Generally, acid conditions, which are characteristic of the initial phase of anaerobic degradation of waste, increase solubilization of chemical constituents (oxides, hydroxides and carbonated species), and decrease the sorptive capacity of waste.

2.3.3 Redox Potential

Reducing conditions, corresponding to the second and third phases of anaerobic degradation, will influence solubility of nutrients and metals in leachate.

2.3.4 Landfill Age

Variations in leachate composition and in quantity of pollutants removed from waste are often attributed to landfill age, defined as time measured from the deposition of waste or time measured from the first appearance of leachate. Landfill age obviously plays an important role in the determination of leachate characteristics governed by the type of waste stabilization processes. It should be underlined that variations in composition of leachate do not depend exclusively on landfill age but on the degree of waste stabilization and volume of water which infiltrates the landfill. The pollutant load in leachate generally reaches maximum values during the first years of operation of a landfill (2–3 years) and then gradually decreases over the following years. This trend is generally applicable to organic components, main indicators of organic pollution (COD, BOD_5, TOC), microbiologic population and to main inorganic ions (heavy metals, Cl, SO_4, etc.).

3 Leachate Problems in Landfills

The most typical detrimental effect of leachate discharge into the environment is that of groundwater pollution. To prevent this, the 1st step in landfill design development was to site the landfill far from the groundwater table and/or far from groundwater abstraction wells. Thus more attention was focused on studying the hydrogeology of the area in order to identify the best siting of the landfill.

A further step in landfill technology was to site the landfill in low permeability soil and/or to engineer impermeable liners to contain wastes and leachate. Containment, however, poses the problem of leachate treatment. Nowadays the leachate control strategies involve not only landfill engineering but the concept of waste management itself.

Leachate pollution is the result of a mass transfer process. Waste entering the landfill reactor undergoes biological, chemical and physical transformations which are controlled, among other influencing factors, by water input fluxes. In the reactor three physical phases are present:

- the solid phase (waste),
- the liquid phase (leachate), and
- the gas phase.

The liquid phase is enriched by solubilized or suspended organic matter and inorganic ions from the solid phase. In the gas phase mainly carbon (prevalently in the form of CO_2 and CH_4) is present.

Discharge of leachate into the environment is nowadays considered under more restrictive views. The reasons for this are (STEGMANN et al., 1992):

- many severe cases of groundwater pollution at landfills,
- the greater hazard posed by the size of landfill which is larger than in the past,
- the need to comply with more and more restrictive legislation regarding quality standards for wastewater discharges,
- with an integrated waste management strategy the volume of refuse will be reduced but more hazardous waste may need to be landfilled, e.g., combustion residues, hazardous components consequent to separate collection, etc.,
- more and more often landfills are located on the ground or on a slope and in both cases accumulation of leachate may be a negative factor with respect to geotechnical stability.

The leachate problem accompanies landfill from its beginning to many decades after closure. Therefore, leachate management facilities should also last and their effectiveness be ensured over a long period of time – so far, this still remains to be proven.

4 Sanitary Landfilling and Leachate Control Strategies

In view of all aspects mentioned above leachate control strategies involve the input (waste and water), the reactor (landfill) and the output (leachate and gas). That is one of the reasons why the German Technical Instructions on Waste from Human Settlements (TASI, 1993) laid down standards for the disposal, including the collection, treatment, storage, and landfilling of wastes from human settlements. State-of-the-art technology is required and with the so-called multi-barrier concept, it is the waste to be dumped itself which forms the most important barrier. The other barriers are the geological barrier of the landfill site, the base sealing with an effective drainage system and a surface sealing, after a landfill section has been completely filled. With these major aspects the Instructions on Waste from Human Settlements define two classes of landfills:

(1) landfill class I:
  - particularly high standards for mineralization levels of the waste to be dumped,
  - relatively low standards for landfill deposit sites and landfill sealing (base and surface sealing);

(2) landfill class II:
  - lower standards for mineralization levels of the waste to be dumped,
  - considerably higher standards for landfill deposit sites and landfill sealing than apply to landfill class I.

4.1 Control of Waste Input/Pretreatment before Deposition

The first step in the waste input control strategy should be that of reducing to a minimum the amount of waste to be landfilled. This could be obtained by waste avoidance, separate collection activities, recycling centers for recyclables, incineration and mechanical-biological pretreatment for residual municipal solid waste (MSW) or composting for biowaste. Separation of the hazardous fraction of municipal waste such as batteries, expired medicines, paint, mercury lamps, pesticides, etc., would reduce leachate concentrations of heavy metals, halogenated hydrocarbons and other toxic compounds.

Another step is that of reducing waste to a non-leachability level. This could be achieved for MSW by incineration followed by fixation of the solid residues.

Pretreatment could also aim to reduce the biodegradability of waste to be landfilled. This would reduce or even eliminate the need for process water in the biostabilization. One way of reaching this aim is to pretreat waste by mechanically sorting organic matter and paper. This material could then be either composted or anaerobically digested.

With regard to its properties the waste can be separated into different fractions: a light fraction of high calorific value, in some cases also a mineral fraction, and a fraction rich in organics can be gained. There is also a reutilization potential of some of these fractions. The mechanical-biological pretreatment of MSW can be applied within a waste management concept as a sole process or in combination with thermal pretreatment (Fig. 2).

To describe the effects of future biological and thermal pretreatment on the leachate emissions, landfill simulation experiments have been carried out under anaerobic conditions. The test system ensures that the typical landfill phases such as the acid phase and stable methane phase subsequently take place in the reactor (STEGMANN, 1981). By choosing

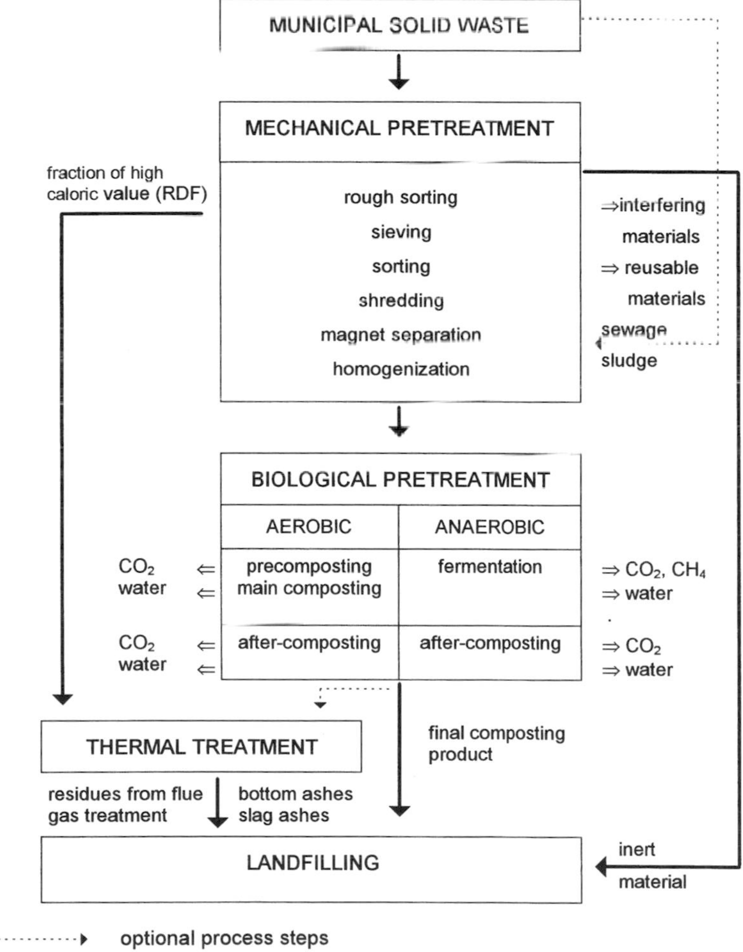

Fig. 2. Scheme of waste pretreatment before landfilling (LEIKAM et al., 1997).

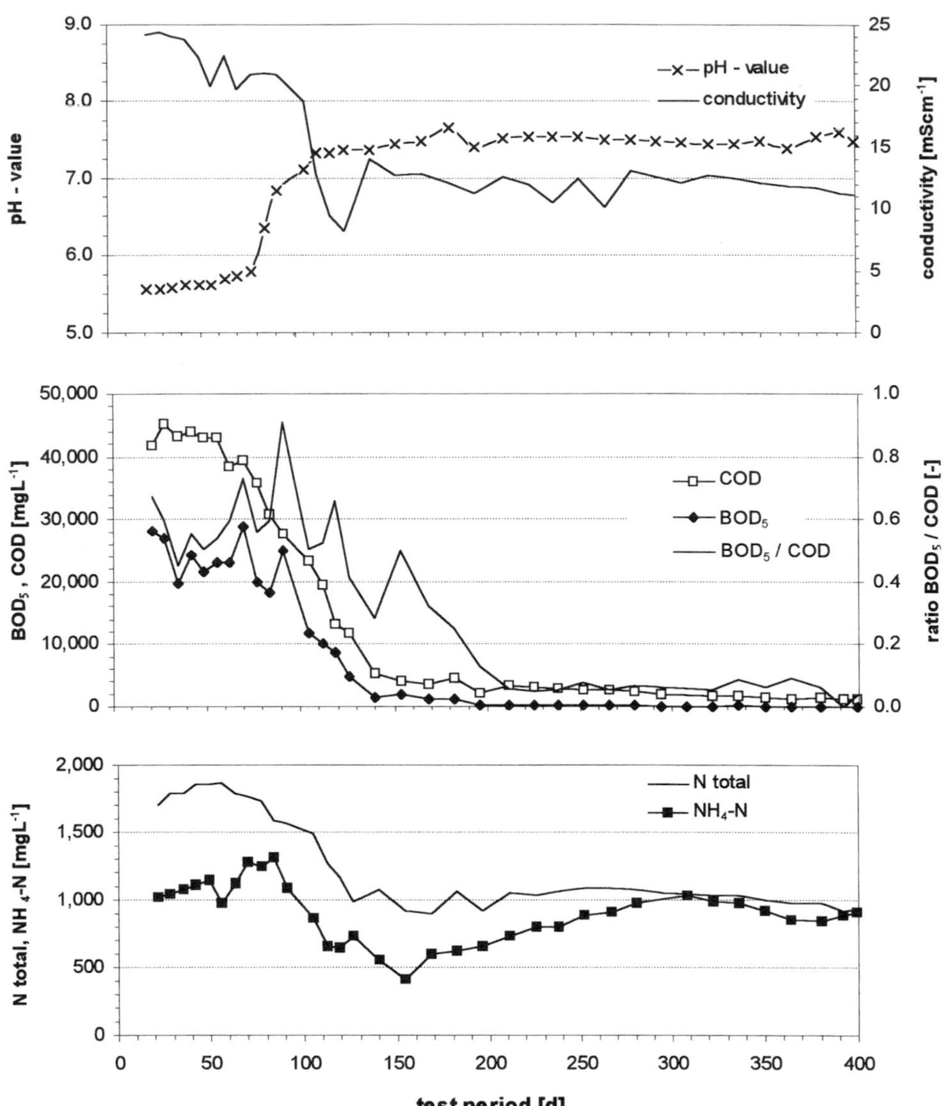

Fig. 3. pH, nitrogen and organic components in leachate during landfill simulation tests in LSR with untreated MSW (water flow–dry matter ratio 1.6:1 after 400 d of testing) (LEIKAM et al., 1997).

appropriate milieu conditions, an enhanced biological degradation process is achieved. By this means, the maximum emission potential represented by gas production and leachate load can be determined within reasonable periods of time.

Untreated municipal solid waste and biological as well as thermal pretreated MSW samples were investigated in these landfill simulation reactors (LSR). Fig. 3 shows curves of selected leachate parameters for untreated MSW, Fig. 4 for mechanical-biologically pre-

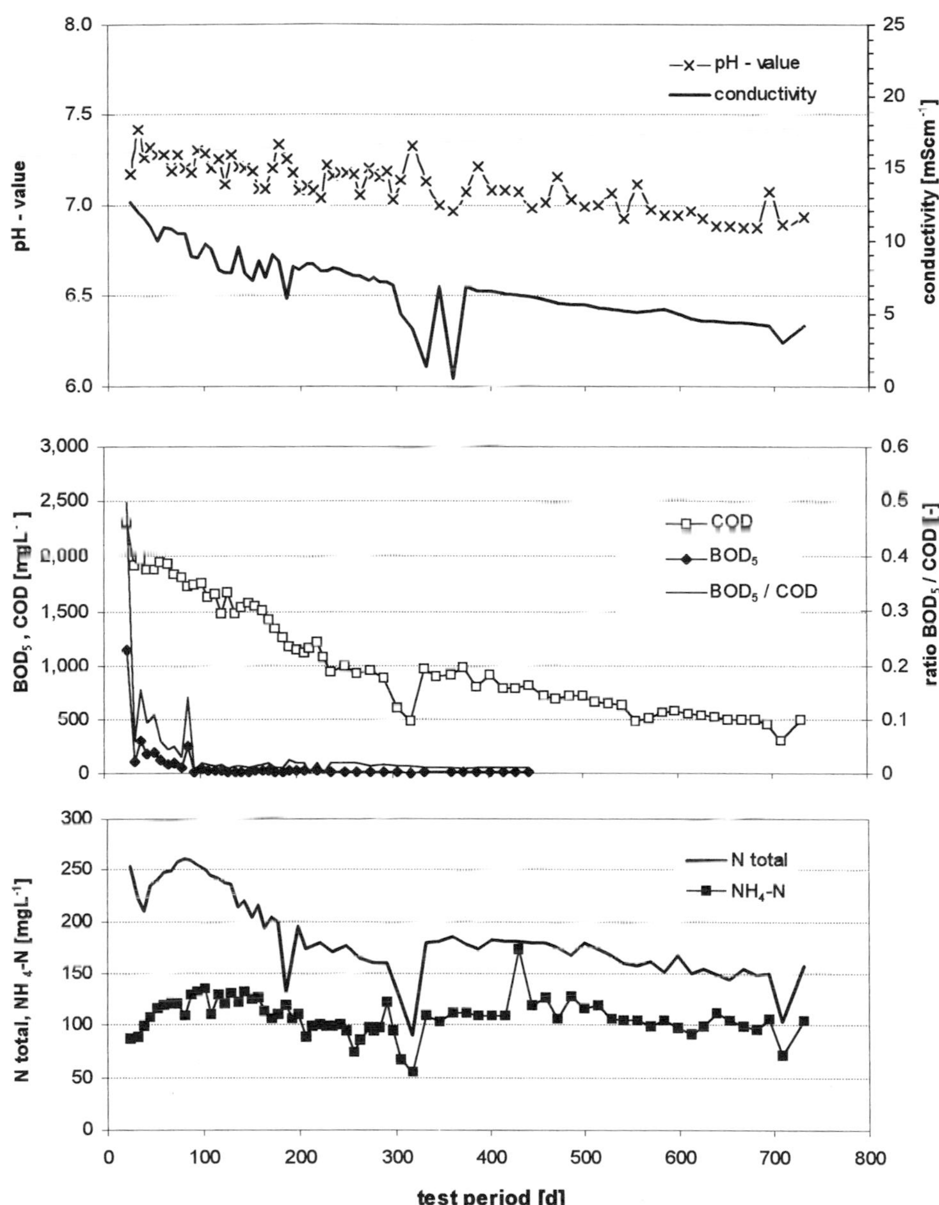

Fig. 4. pH, nitrogen and organic components in leachate during landfill simulation tests in LSR with 4-months biologic pretreated MSW (water flow–dry matter ratio 2.6:1 after 700 d of testing) (LEIKAM et al., 1997).

treated MSW. The untreated waste was only homogenized. The pretreated waste was derived from the same waste type but sieved (grain size <80 mm) and the sieve undersize was composted in a container for 4 months.

In Fig. 3 the typical high concentrations occur within the acidic phase of the untreated waste. With pH values of 5.5 there are COD concentrations of 45,000 mg O_2 L^{-1} and BOD_5 concentrations up to 30,000 mg $O_2\,L^{-1}$.

For pretreated waste the acidic phase with this highly loaded leachate is omitted. After 250 d of testing in the LSR, the COD content in the leachate was below 100 mg O_2 L^{-1}, the corresponding BOD_5 concentration was only about 20 mg O_2 L^{-1}. It has to be considered that the water flow rate after 250 d of testing corresponds to a water flow–dry matter ratio of 1:1.

The positive effect of biological pretreatment becomes also evident when the concentrations of the parameter nitrogen are compared. While the total nitrogen content in the leachate of untreated waste adjusts to about 1,000 mg L^{-1}, these concentrations are below 200 mg L^{-1} for the pretreated waste. Besides a lower expenditure for leachate purification a shorter period of landfill aftercare may be expected (see also Sect. 6).

When landfill conditions are simulated in laboratory tests, leaching processes are very important, particularly during the methanogenic period which occurs in landfills over years and decades. For this reason, the comparison of the leachate emissions of untreated and pretreated waste can be done assessing the emitted quantities. The leaching process and the release of substances in the water phase can be determined using the specific leachate generation rate, the ratio of water flow related to dry matter. A water flow–dry matter ratio of 1.0 means that the same amount of water has penetrated the same amount of waste.

Out of the leachate constituents that were determined in the LSR tests the released loading rates of the parameters COD, total nitrogen and chloride are presented (Fig. 5). To assess the quantity of freights the emissions of bottom ash from a MSW incineration being tested in the same test device are added.

Comparing the emitted loads the efficiency of mechanical-biological pretreatment becomes obvious. For a water flow–dry matter ratio of 1.0 the mass transfer of the pretreated MSW was 90% lower regarding the parameters COD and nitrogen. Because of the high degree of inertization of the bottom ash from MSW incineration the leachate load is reduced by 98%. However, the difference in the emitted loads for biologically pretreated MSW and the bottom ash is relatively small compared to the untreated waste.

4.2 Control of Water Input

The strategy for water input control is strictly related to the quality of the waste to be landfilled. In the case of non-biodegradable waste and according to its hazardous potential for the environment, prevention of water infiltration can be adopted as the main option (normally by means of top sealing). On the contrary, in the case of biodegradable waste, a water input must be assured until a high degree of biostabilization is achieved. In this case the water input should be limited to the strictly necessary amount and minimization techniques should be applied. The most important parameters in this regard are (STEGMANN et al., 1992):

- siting of landfills in low precipitation areas, if possible,
- usage of cover and topsoil systems suitable for vital vegetation and biomass production,
- vegetation of the topsoil with species which optimize the evapotranspiration effect,
- surface lining in critical hydrological conditions,
- limitation on sludge disposal,
- surface water drainage and diversion,
- high compaction of the refuse in place,
- measures to prevent risks of cracking owing to differential settlement.

Furthermore, utilization of mobile roofs or intermediate covers in the landfill operation area could represent a useful minimization technique.

The two final surface sealing systems that must be applied according to the regulations of the TASI (TASI, 1993) are shown in Fig. 6.

4.3 Landfill Reactor Control

The main option in controlling leachate quality through controlling the landfill reactor is the enhancement of the biochemical processes (when biodegradable wastes are deposited).

One of the main objectives is to convert and transport as much carbon as possible from the

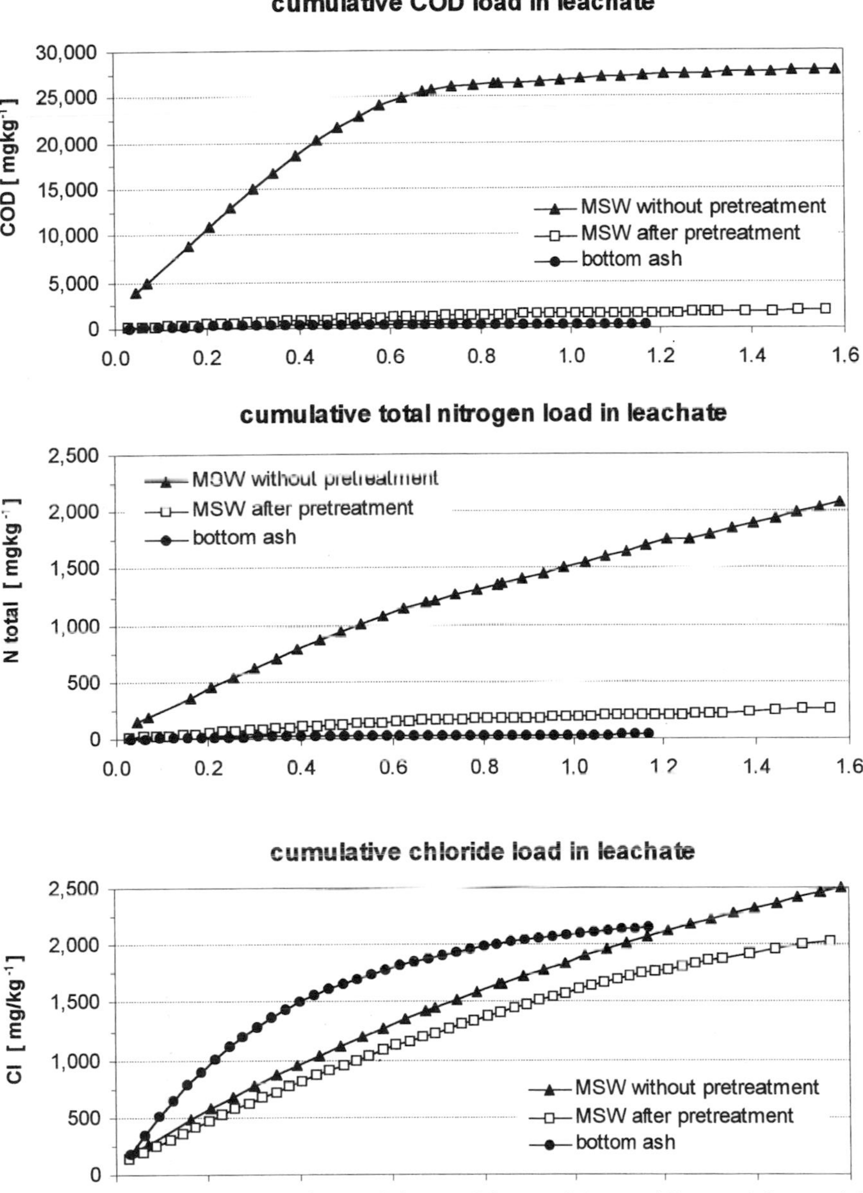

Fig. 5. COD, nitrogen and chloride load for untreated, biologically pretreated and thermally pretreated MSW in LSR tests (LEIKAM et al., 1997).

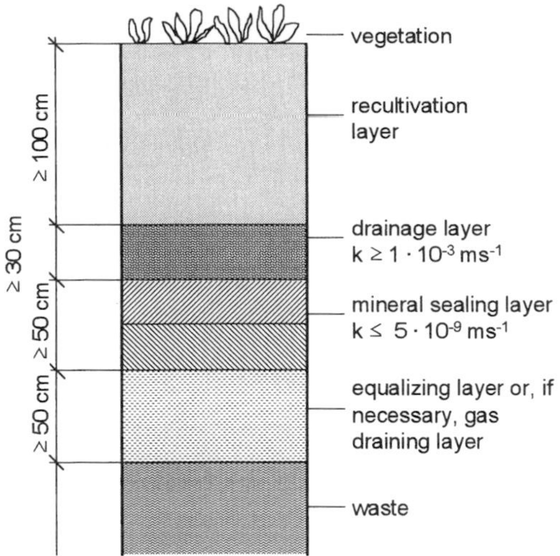

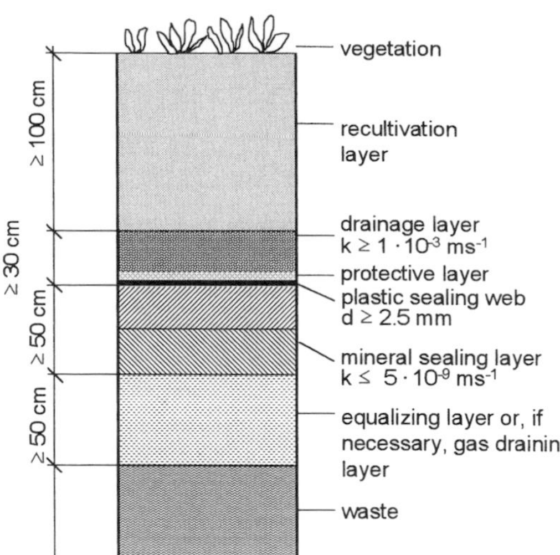

Fig. 6. Landfill surface sealing systems (TASI, 1993).

solid phase into the gas phase rather than into the liquid phase. This is achieved by accelerating the methane generation step.

4.4 Control of Leachate Discharge into the Environment

As mentioned above, this is the parameter traditionally controlled and nowadays the regulations are more restrictive. The following tools are adopted.

4.4.1 Lining

The lining system should be based on the multi-barrier effect (double or triple liners). Quality control of material and construction should be improved to ensure higher safety and durability.

4.4.2 Drainage and Collection Systems

A rational drainage and collection system is important to avoid accumulation of leachate inside the landfill. The main problems of drainage systems are proper choice of material, clogging, durability and maintenance.

In Fig. 7 the base sealing systems of landfill class I and II according to the requirements of the TA Siedlungsabfall (TASI, 1993) are presented. They must be arranged on the landfill base and on the sloping areas. For landfill class II, the geological barrier should comprise naturally arranged, slightly permeable bedrock of several meters in thickness. If this requirement is not met a homogenous equalizing layer of at least 3 m and a permeability of $k \leq 1 \cdot 10^{-7}$ m s^{-1} has to be installed.

The landfill base must be at least 1 m above the highest expected groundwater level, once settlement under the load of the landfill has come to an end.

The surface of the base sealing system is to be formed in the manner of a roof profile. Once the sealing bearing surface has finished settling, the surface of the sealing layer must exhibit a transverse gradient of $\geq 3\%$ and a longitudinal gradient of $\geq 1\%$. The perforated pipes (collectors), additionally capable of being rinsed and monitored, must be provided for the collection and discharge of leachate. The leachate must be chanelled by means of free flow into drainage shafts that are to be installed outside the dumping area.

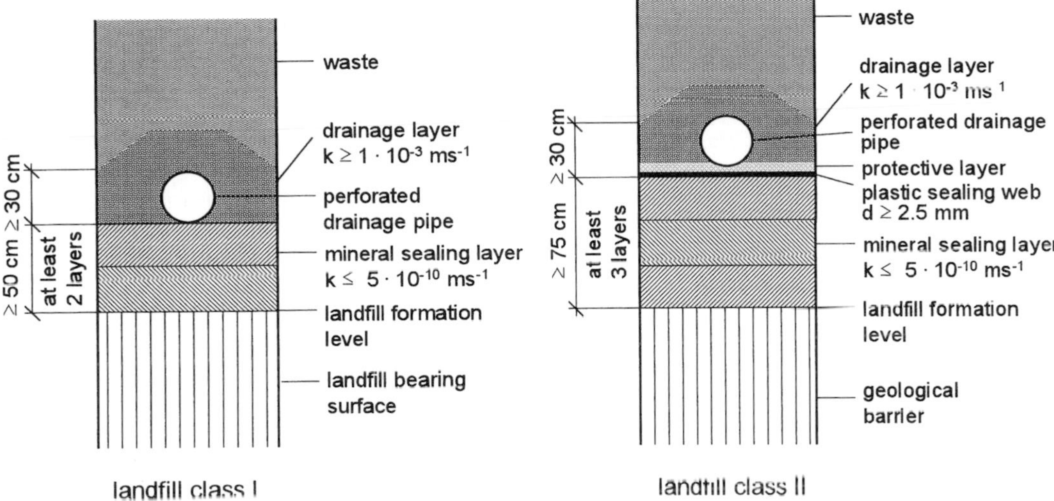

Fig. 7. Landfill base sealing systems (TASI, 1993).

4.4.3 Treatment

Leachate has always been considered a problematic wastewater from the point of view of treatment as it is highly polluted and the quality and quantity are modified with time in the same landfill. Nowadays according to the increasingly restrictive limits for wastewater discharge, complicated and costly treatment facilities are imposed. Normally a combination of different processes is required.

4.4.4 Environmental Monitoring

This aspect is of extreme importance for the evaluation of landfill operational efficiency and for the observation of the environmental effects on a long-term basis. The following monitoring facilities must in general be provided and checked at regular intervals for proper operation (TASI, 1993):

- groundwater monitoring system with at least one measuring station in the inflowing current of groundwater and a sufficient number of measuring stations in the current of groundwater flowing out of the landfill area,
- measuring facilities for monitoring settlements and deformations in the landfill body and the landfill sealing systems,
- measuring facilities for recording meteorological data like precipitation, temperature, wind, evaporation,
- measuring facilities for collecting the leachate and water quantities that are necessary for analyzing the water balance,
- measuring facilities for recording the quality of leachate and other waters,
- measuring facilities for monitoring the temperature at the landfill base,
- if the generation of landfill gas is to be expected, it shall be necessary to provide facilities fôr measuring landfill gas and install gas level indicators for the purpose of emission control.

5 Leachate Production

In most climates rain and snowfall will cause infiltration of water into the landfilled waste and, after saturation of the waste, generation of leachate. Determination of the amount of water infiltrating a covered landfill cell can be made from the hydrological balance of the top cover, paying attention to precipitation, surface runoff, evapotranspiration and changes in moisture content of the soil cover. However, the balancing must be timewise discretized in order to identify the periods of excessive water leading to infiltration into the waste below (see also RAMKE, 1991).

The water content of the waste being landfilled is usually below saturation (actually field capacity) and will result in absorption of infiltrating water before drainage in terms of leachate is generated. The water absorption capacity of the landfilled waste and its water retention characteristics are very difficult to specify due to the heterogeneity of the waste. Furthermore these characteristics may change over time as the waste density is increasing and the organic fraction which dominates the water retention is degraded in the landfill. It has often been observed that new landfills produce leachate soon after the waste has been filled in, although the total water absorption capacity should not be depleted that quickly (SPILLMANN and COLLINS, 1986). Based on observations of open landfill sections it can be concluded that a share of leachate migrates through the landfill on preferential routes.

Accounting for these aspects is still very rudimentary and definitely the less developed component of the hydrological cycle of the landfill. It should be emphasized also that landfills in areas with an annual deficit of rainfall may produce leachate in the wet season.

6 Long-Term Problems with Leachate

The characteristics of landfill leachate are relatively well known, at least for the first 20–30 years of life of the landfill, the period from which actual data are available. On the other hand the leachate composition of later phases of the landfill is hardly known and the basis for making good estimates is rather weak.

For this reason, several landfills in a German joint research project "Landfill Body" (EHRIG, 1997) were investigated. It was the aim to describe the present stage of stability of landfills of different ages, their corresponding emissions and the future development of emissions. All available data of the sites were evaluated, leachate and gas samples were taken and drillings were carried out to get solid waste samples from different depths and different periods, e.g., from 6 up to 33 years of deposition.

As a main focus of the research program long-term experiments in test lysimeters were carried out to assess the long-term emissions, that the solid waste samples from the old landfills will release in future. Leachate and landfill gas samples from landfill simulation reactors were taken regularly to determine their change of element concentrations with time.

6.1 Lysimeter Tests in Landfill Simulation Reactors

As already mentioned, the tests in LSR are supposed to show the long-term behavior of emissions, which will develop in the landfill body in the future. For this reason, the test devices are designed and carried out to describe the events in real landfills under aerobic and anaerobic conditions on a laboratory scale. Furthermore, an acceleration of chemical, biological and physical processes can be caused by modified water balances and optimal conditions for microorganisms, like constant temperatures or homogenous moisture distribution (HEYER et al., 1997; STEGMANN, 1981).

The LSR lysimeters were filled with 35–90 kg of waste (dry substance) at a water content of 50% and a density of 0.84–1.28 kg L^{-1} (wet waste). 2 L of leachate were analyzed and replaced weekly by fresh water to accelerate the decomposition and release of elements.

On the basis of four LSR lysimeters from two landfills (landfill A: age of deposition: 8–18 years and landfill G: age of deposition: 28–31 years), the range of emissions in the water phase that can occur in the landfill in future will be presented.

Fig. 8 shows concentrations of COD and TKN on a logarithmic scale over a test period of more than 1,000 d:

- Like the other parameters, the COD concentrations show a very similar qualita-

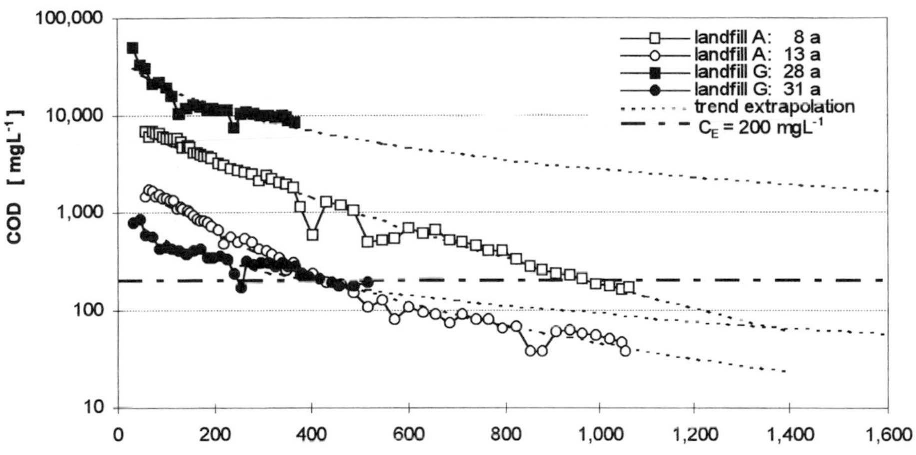

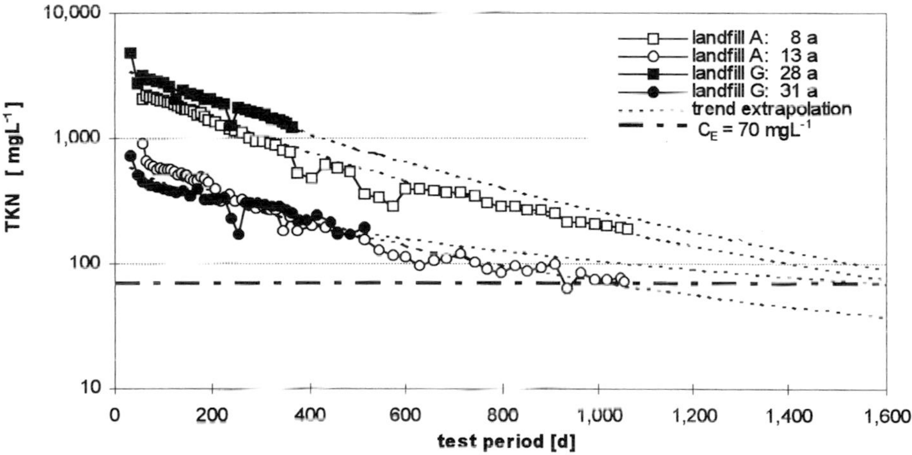

Fig. 8. COD and TKN: LSR leachate concentrations, waste samples landfill A and G (HEYER et al., 1998).

tive decline. It is a gradual asymptotic course, which can be described as a function of dilution and mobilization. The BOD_5 concentrations are very low as expected, because almost all waste samples were in the stabilized methane phase when the tests started. The BOD_5/COD ratio was lower than 0.1.

- There are striking differences in the magnitude of concentrations according to the age of deposition and the conditions within the landfill body before the sampling and because of varying waste compositions. The waste samples of landfill A, with COD concentrations from 2,000–8,000 mg O_2 L^{-1} at the beginning of the tests, show the probable range of leachate pollution to be expected in the future. The older waste samples of landfill G (28–31 years of deposition) produce lower leachate pollution in general, although one waste sample has beginning concentrations up to 50,000 mg O_2 L^{-1}. But this is not representative for a bigger part of the landfill, only for a small layer like a "hot spot".
- The course of nitrogen emissions is comparable to the organic parameters.

However the decline of nitrogen in the leachate occurs more slowly as a higher portion of organic compounds becomes hydrolyzed over the LSR test period. More than 90% of the TKN nitrogen is emitted as ammonia.

- The faster and regular decrease of chloride is mainly due to dilution (Fig. 9). Only a small share is released long-term when medium and hard degradable fractions become hydrolyzed. The halogenated organic hydrocarbons AOX have a similar decline and heavy metal concentrations are generally very low because of the anaerobic conditions with negative redox potentials E_h and neutral pH values.

6.2 Application of LSR Results to Landfill Conditions

The quantitative course of emissions will be demonstrated with balances of freights in the leachate phase. Within this research program the release of emissions was divided into three parts:

- **phase A:** emission freight from the moment of deposition until the moment of sampling,

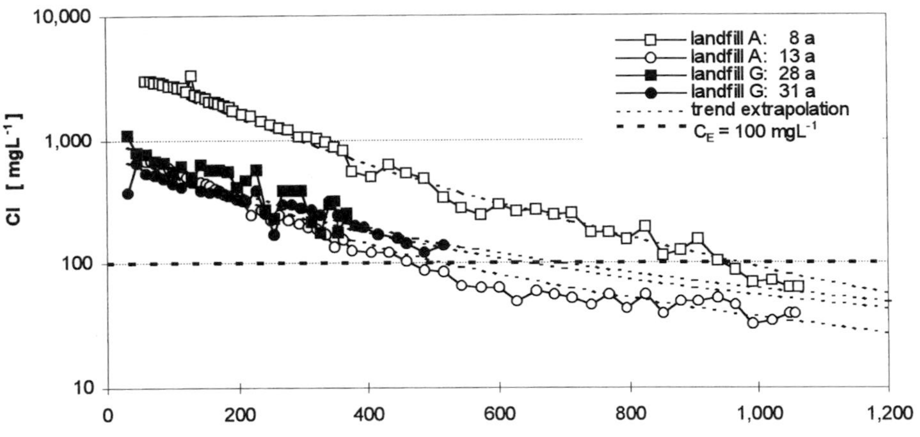

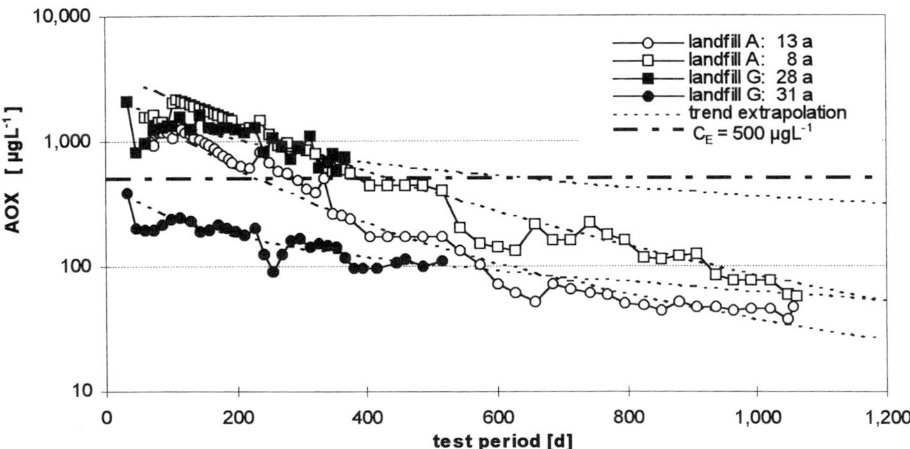

Fig. 9. Cl and AOX: LSR leachate concentrations, waste samples landfill A and G (HEYER et al., 1998).

- **phase B:** emission freight (in the LSR after sample taking) in the aftercare phase until limiting values of concentrations are reached, and
- **phase C:** remaining emission freight that could be released long-term without control after the end of the aftercare phase.

6.2.1 Phase A + B + C

The total amount of freights can hardly be measured directly on landfills. According to a great number of laboratory tests, the emission potentials of German MSW in the leachate phase are in the following range (LEIKAM and STEGMANN, 1996; KRUSE, 1994):

- COD: 25,000–40,000 mg O_2 kg^{-1}
- TKN: 2,000– 4,000 mg kg^{-1}
- AOX: 6,000–12,000 µg kg^{-1}
- Cl: 2,500– 4,000 mg kg^{-1}

The emission potentials are related to the total solids TS at the moment of deposition.

6.2.2 Phase B + C

The long-term emission behavior of phase B and C can be described very well because of the long test periods in the landfill simulation reactors. To define the end of phase B (end of aftercare) and the beginning of phase C, limiting concentration values were selected according to Swiss and German regulations:

- COD: $C_{E\text{-}COD} = 200$ mg O_2 L^{-1} [51. Anhang, RAbwVwV, (GMBl., 1996)]
- TKN: $C_{E\text{-}TKN} = 70$ mg L^{-1} [51. Anhang, RAbwVwV, (GMBl., 1996)]
- AOX: $C_{E\text{-}AOX} = 500$ µg L^{-1} [51. Anhang, RAbwVwV, (GMBl., 1996)]
- Cl: $C_{E\text{-}Cl} = 100$ mg L^{-1} [Lim. value, Switzerland, (BELEVI and BACCINI, 1989)]

The total freights (A + B + C), the freights after sampling on landfills (B + C) and the freights after the end of aftercare (C) for the two parts of landfill A and G are shown in Fig. 10. For the four standard parameters it shows the average freights of the investigated landfill bodies representing the different ages of deposition and the degree of stabilization.

For instance the average remaining COD potential at the moment of sampling (phase B + C) is about 4,000–6,000 mg O_2 kg^{-1} TS, that are 10–20% of the possible potential at deposition (phase A + B + C: 25,000–40,000 mg O_2 kg^{-1} TS, mean value 32,500 mg O_2 kg^{-1} TS). When the limiting concentration of 200 mg O_2 L^{-1} is reached and the aftercare will hypothetically end, a COD freight of 600–1,000 mg O_2 kg^{-1} TS could still be released long-term (phase C). The LSR tests of waste samples from further landfill parts all show a similar behavior.

6.3 Prognosis of Periods of the Long-Term Course of Emissions

The course of leachate emissions in time mainly depends on:

- the potential of substances that can be mobilized,
- the water balance in the landfill, mainly the water flux, and
- the mobilization behavior.

In Figs. 8 and 9 the leachate concentrations of some relevant compounds have already been shown. These LSR tests do not allow final and generalizing predictions for the emission development with time. One reason are the specific conditions of each landfill site, e.g., the climate, the surface cover, the waste composition or inhibition effects. Another reason is the high water exchange rate in the LSR tests to simulate an accelerated conversion, mobilization and dilution in the landfill. However, a possible development of emissions for the two landfills A and G shall be expounded.

The course of emissions with time can be described with an exponential function. With the idealized conditions in the LSR test devices and the setting of a water balance, which is approximately 100 times higher than at the landfill, periods T_E can be estimated, until a limiting value C_E is reached. The estimations are based on the following assumptions:

- constant climatic leachate generation of 250 mm per year (this means no impermeable surface sealing, only a permeable soil cover),
- a standard height of 20 m,
- the dry densities in the LSR tests are similar to the landfill with approximately 0.75 Mg TS m^{-3},
- uniform percolation through the landfill body.

The periods T_E are compiled in Tab. 2 together with the concentrations C_0 at the beginning of the LSR tests and the values of half-life $T_{1/2}$:

According to German standards for COD in the leachate, the estimation results in a period of 80–360 years with a mean of 160 years, until the limiting concentrations of 200 mg O_2 L^{-1} will be reached. Chloride shows similar periods. All investigations and tests point to nitrogen to be the parameter with the longest release of relevant concentrations into the leach-

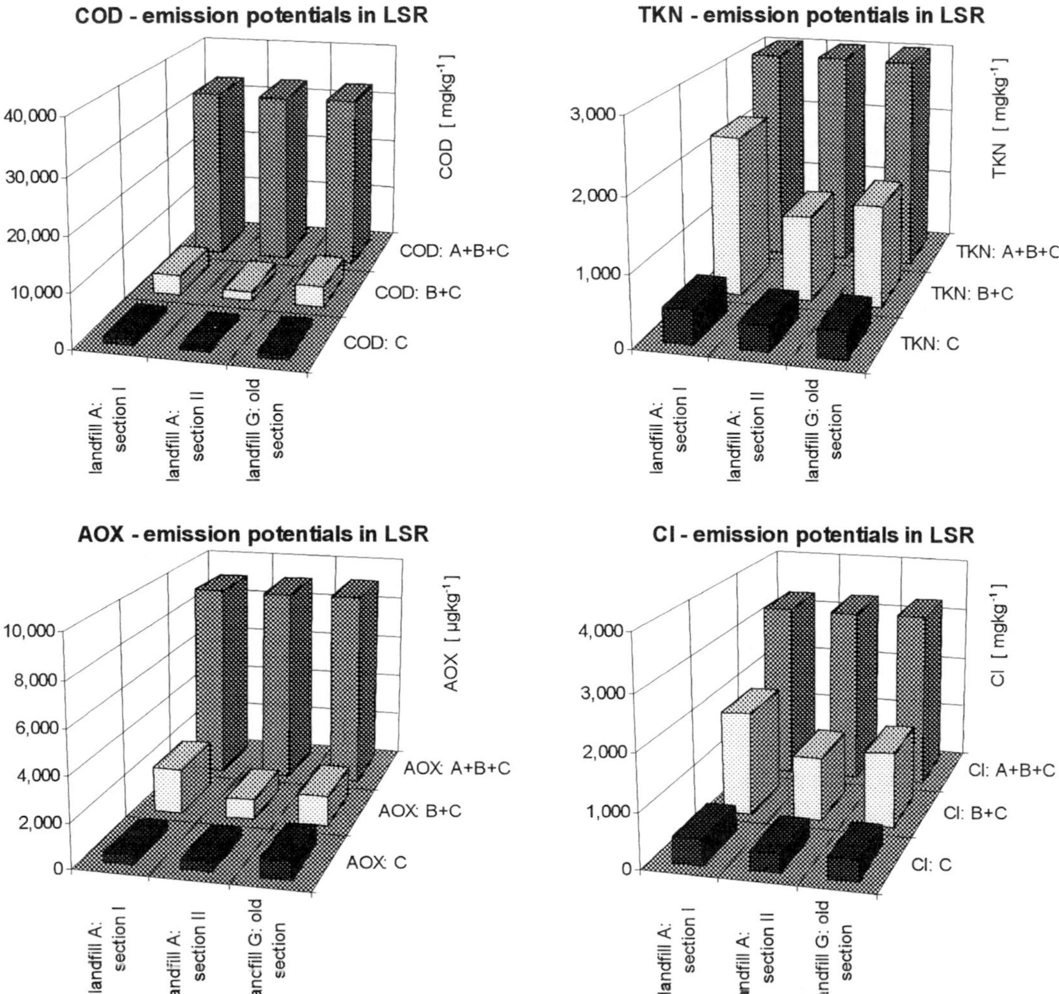

Fig. 10. Emissions into the leachate path divided into 3 phases, comparison between landfill A, section I (age of deposition: 8–13 years), landfill A, section II (age of deposition: 16–18 years) and an old section of landfill G (age of deposition: 28–31 years) (HEYER et al., 1998).

Tab. 2. Estimations of Periods T_E for Reaching the Limiting Values C_E (HEYER et al., 1977)

Parameter	C_E Limiting Values	C_0 Concentrations at Test Beginning [$mg\ L^{-1}$]	$T_{1/2}$ Half Life [a]	T_E Periods [a]
COD	$C_E = 200\ mg\ O_2\ L^{-1}$ mean	500–49,000 5,100	25–96	80–360 160
TKN	$C_E = 70\ mg\ L^{-1}$ [a] mean	200–4,700 1,200	40–150	120–450 250
Cl	$C_E = 100\ mg\ L^{-1}$ mean	340–2,950 1,200	40–90	90–250 160
AOX	$C_E = 500\ \mu g\ L^{-1}$ mean	390–2,380 $\mu g\ L^{-1}$ 1,600 $\mu g\ L^{-1}$	10–80	30–210 80

[a] Total amount of nitrogen, sum of ammonia, nitrite and nitrate.

ate phase. 250 years on average are possibly necessary until the concentration of 70 mg L^{-1} will be reached.

The plausibility of these estimations is quite difficult to jugde. Each landfill has a different water balance which can vary with the seasons or change because of surface covers, lining systems or damages of these technical barriers.

The wide range of values for the half-life and the resulting periods of relevant emissions seem to be unsatisfactory at first glance. The tendency shows that the values for the half-life increase with time. In the first years after deposition the easily and medium degradable organics are decomposed biologically and their products rule the emission behavior. In this period a distinctive decrease from high to low emission quantities can be stated directly in landfills and in the laboratory tests. The decrease of organic pollution is not proportional to the water balance but faster as long as there is sufficient humidity in the waste for gas generation.

The medium- and long-term emissions are ruled by hard degradable and inorganic compounds with increasing half-life. Instead of neglecting biological conversion, the main processes are now chemical-physical with leaching and diluting effects. These processes are directly dependent on the water balance as the chosen model predicts.

All in all the estimations of periods with relevant emissions that determine the necessary aftercare are on the safe side. Though a distinctive decrease of pollution in the emission paths of landfills can be stated in the first years after deposition, long-term emissions because of increasing half-life will probably occur (HEYER and STEGMANN, 1997).

7 Leachate Treatment

Since more and more landfills all over the world will be sealed at their bases by means of mineral and/or artificial liners and leachate will be collected, there is a great need for appropriate leachate treatment facilities. The high concentrations of organic and inorganic constituents in leachate have to be treated due to the requirements in different countries. The tendency is to reach relatively low effluent concentrations in organics, ammonia, halogenated hydrocarbons, heavy metals, and fish toxicity. It is still discussed, if it makes sense to treat leachate down to low COD values where at concentrations of 1,000 mg $O_2\ L^{-1}$ probably mainly humic- and fulvic-like acids are removed. The potential effects of these components in rivers, lakes, etc., are probably limited but cannot be predicted in detail.

Removal of nitrogen especially from leachate from the methanogenic phase (low BOD_5 and high COD concentrations) is still a problem. Nitrification can be achieved when the treatment plants are designed for this purpose. Denitrification can be practiced with "young" leachates; "old" leachates do not have enough degradable organics to supply denitrifying bacteria with the carbon needed for the reduction of nitrate.

The trace organics have to be looked at in more detail. Also in MSW landfills halogenated hydrocarbons and other organics like aromatic compounds are landfilled with the daily life products. These components can also be found in leachate and have to be removed by means of special treatment.

Due to the kind of treatment there will still remain a variety of components in the leachate; these will be mainly salts, organic trace components, low amounts of heavy metals and the above mentioned humic- and fulvic-like acids.

To meet the requirements for low concentrations of the effluent, many leachate treatment plants consist of more than one step. It may start in the landfill body itself by practicing the enhanced biological degradation which results from the early stage of landfilling in low BOD_5 concentrations. The removal of organics and nitrification will mainly take place by means of biological processes while further treatment requires chemical-physical methods.

Leachate treatment is already and will be very costly especially if the required effluent standards of Germany have to be met. This is especially true when the total time of leachate treatment after the landfill has been closed (50–100 years or more?) is respected.

Concerning the different approaches in the different countries, it has to be kept in mind that legislation, landfill management, operation, etc., may be different. In addition it should be pointed out that there is not only one solution for the treatment of leachate to obtain the required final concentrations.

7.1 Biological Leachate Treatment

In general, biological leachate treatment is a highly favorable procedure that should also be used in those cases when chemical-physical treatment is required. Biologic leachate treatment is a relatively low cost process and organics are degraded mainly to CO_2, water, and biomass. All the substances that have been eliminated using biologic degradation do not have to be treated by means of high cost chemical-physical procedures.

As already mentioned, with increasing landfill age or by means of enhancement techniques, leachate treatment will mainly focus on the nitrification of ammonia. Biological denitrification can be achieved when an external organic substrate is added to the leachate.

For the design and operation of biologic leachate treatment plants specific points have to be considered:

- foam production during certain periods,
- precipitation of constituents that results in clogging of pipes, etc.,
- low leachate temperatures during biological treatment due to long retention times in the reactors,
- low phosphorus concentrations in the leachate,
- low BOD_5 and high ammonia concentrations in old leachate,
- halogenated hydrocarbons.

Biological systems can be distinguished in anaerobic and aerobic treatment processes that are realized by means of different techniques. Some of them are (EHRIG and STEGMANN, 1992):

- anaerobic biological treatment:
  - parts of the landfill body used as a reactor,
  - anaerobic sludge bed reactor;
- aerobic biological treatment:
  - aerated lagoons,
  - activated sludge process,
  - rotating biological contactor,
  - trickling filter.

7.2 Chemical-Physical Leachate Treatment

Chemical-physical treatment mainly results in the separation and concentration of pollutants from the leachate. The concentrate has to be incinerated, landfilled or further treated. So it is not a "real" treatment process compared to biologic methods. Other chemical processes such as wet oxidation as well as ozone, UV, and H_2O_2 treatment may also result in a conversion of organics to mainly CO_2 and water.

The oxidation processes are expected to improve the biological treatment of the organics and/or to totally oxidize them. By means of reverse osmosis, organic and inorganic components of the leachate are accumulated in the concentrate due to their chemical-physical characteristics. Evaporation or incineration of leachate remove also organics and inorganics from the liquid phase. Precipitation using organic and/or inorganic flocculants results mainly in a removal of organics and an increase of the salt content when inorganic flocculants are used. Activated carbon removes mainly organics from the water and gas phase.

When considering chemical-physical treatment the whole process including energy requirement, gas treatment, residue removal, quality of the treated leachate, stability and efficiency of the process as well as costs have to be considered.

7.3 Leachate Recirculation

Leachate recirculation has often been practiced in the past, with the aim of totally solving the leachate problem. Experience shows that under Central European climatic conditions this is not possible and that on the contrary a buildup of leachate in the landfill has been observed. In some cases water migrated over the edge of the landfill pit.

Leachate recirculation may be practiced only with biologically treated leachate in a controlled way. The main aim of leachate recirculation is the maximization of evaporation. So leachate recirculation should be practiced dependent upon the actual evaporation rate over the year; since this procedure is somewhat theoretical the rate of leachate recirculated with time may be related to the average evaporation rates and the water retention potential in the upper 10 cm of waste and/or cover soil. A certain amount of leachate may penetrate into the landfill, which in general will not cause adverse effects (see also STEGMANN et al., 1992).

8 References

ANDREOTTOLA, G. (1992), Chemical and biological characteristics of landfill leachate, in: *Landfilling of Waste: Leachate* (CHRISTENSEN, T. H., COSSU, R., STEGMANN, R., Eds.), pp. 65–88. London, New York: Elsevier.

BARBER, C. (1979), *Behavior of Wastes in Landfills, Review of Processes of Decomposition of Solid Wastes with Particular Reference to Microbiological Changes and Gas Production*. Stevenage, UK: Water Research Centre, Stevenage Laboratory Report LR 1059.

BELEVI, H., BACCINI, P. (1989), Long-term behavior of municipal solid waste landfills, *Waste Manag. Res.* **7**, 483–499.

CHRISTENSEN, T. H., KJELDSEN, P. (1989), Basic biochemical processes in landfills, in: *Sanitary Landfilling: Process, Technology and Environmental Impact* (CHRISTENSEN, T. H., COSSU, R., STEGMANN, R., Eds.), pp. 29–49. London: Academic Press.

EHRIG, H.-J. (1990), Qualität und Quantität von Deponiesickerwasser, *Entsorgungspraxis spezial* **1**, 100–105.

EHRIG, H.-J. (1997), Einführung in das Verbundvorhaben Deponiekörper, in: *Verbundvorhaben Deponiekörper, Proceedings 2. Statusseminar*, Wuppertal, pp. 1–5, Umweltbundesamt, Projektträgerschaft Abfallwirtschaft und Altlastensanierung des BMBF.

EHRIG, H.-J., STEGMANN, R. (1992), Biological processes, in: *Landfilling of Waste: Leachate* (CHRISTENSEN, T. H., COSSU, R., STEGMANN, R., Eds.), pp. 185–202. London, New York: Elsevier.

GMBl. (1996). Anhang 51: Oberirdische Ablagerung von Abfällen. Allgemeine Rahmen-Verwaltungsvorschrift über Mindestanforderungen an das Einleiten von Abwasser in Gewässer, Neufassung am 31. 7. 1996 im Gemeinsamen Ministerialblatt Nr. 37, 47. Jahrgang, 729–785, vom November 1996.

HEYER, K.-U., STEGMANN, R. (1997), Untersuchungen zum langfristigen Stabilisierungsverlauf von

Siedlungsabfalldeponien, in: *Verbundvorhaben Deponiekörper, Proceedings 2. Statusseminar*, Wuppertal, pp. 46–78, Umweltbundesamt, Projektträgerschaft Abfallwirtschaft und Altlastensanierung des BMBF.

HEYER, K.-U., ANDREAS, L., BRINKMANN, U. (1997), SAV 3: Beprobung von Abfallstoffen in Deponiesimulationsreaktoren (DSR), in: *Verbundvorhaben Deponiekörper, Proceedings 2. Statusseminar*, Wuppertal, pp. 345–358, Umweltbundesamt, Projektträgerschaft Abfallwirtschaft und Altlastensanierung des BMBF.

HEYER, K.-U., STEGMANN, R., KABBE, G., DOHMANN, M. (1998), Emissionsverhalten von Deponien und Altablagerungen in den alten Bundesländern, in: *Entwicklungstendenzen in der Deponietechnik*, 1. Hamburger Abfallwirtschaftstage 28.–29. Januar 1998, Hamburger Berichte, Band 12 (STEGMANN, R., RETTENBERGER, G., Eds.), pp. 25–50. Bonn: Economica Verlag.

KRUSE, K. (1994), Langfristiges Emissionsgeschehen von Siedlungsabfalldeponien, Heft 54 der Veröffentlichungen des Instituts für Siedlungswasserwirtschaft. Braunschweig: Technische Universität.

LEIKAM, K., STEGMANN, R. (1996), Stellenwert der mechanisch-biologischen Restabfallbehandlung. *Abfallwirtschaft J.* **9**, 39–44.

LEIKAM, K., STEGMANN, R. (1997), *In-situ* Stabilisierung von Altdeponien, in: *Verbundvorhaben Deponiekörper, Proceedings 2. Statusseminar*, Wuppertal, pp. 153–174, Umweltbundesamt, Projektträgerschaft Abfallwirtschaft und Altlastensanierung des BMBF.

RAMKE, H.-G. (1991), Hydraulische Beurteilung und Dimensionierung der Basisentwässerung von Deponien fester Siedlungsabfälle, Wasserhaushalt, hydraulische Kennwerte, Berechnungsverfahren, Thesis. Braunschweig: Technische Universität, Leichtweiß-Institut für Wasserbau.

ROBINSON, H. D. (1989), Development of methanogenic conditions within landfill, *Proc. 2nd Int. Landfill Symp. Sardinia '89*, Porto Conte, October 9–13.

SPILLMANN, P., COLLINS, H.-J. (1986), Physikalische Untersuchungen zum Wasser- und Feststoffhaushalt, in: *Wasser- und Stoffhaushalt von Abfalldeponien und deren Wirkung auf Gewässer* (SPILLMANN, P., Ed.). Weinheim: VCH.

STEGMANN, R. (1981), Beschreibung eines Verfahrens zur Untersuchung anaerober Umsetzungsprozesse von festen Abfallstoffen im Deponiekörper, *Müll Abfall* **2**.

STEGMANN, R., SPENDLIN, H. H. (1989), Enhancement of degradation: German experiences, in: *Sanitary Landfilling: Process, Technology and Environmental Impact* (CHRISTENSEN, T. H., COSSU, R., STEGMANN, R., Eds.), pp. 61–82. London, New York: Academic Press.

STEGMANN, R., CHRISTENSEN, T. H., COSSU, R. (1992), Landfill leachate: an introduction, in: *Landfilling of Waste: Leachate* (CHRISTENSEN, T. H., COSSU, R., STEGMANN, R., Eds.), pp. 3–14. London, New York: Elsevier.

TASI (1993), Technical Instructions on Waste from Human Settlements (TA Siedlungsabfall), Dritte Allgemeine Verwaltungsvorschrift zum Abfallgesetz vom 14. Mai 1993, Technische Anleitung zur Verwertung, Behandlung und sonstigen Entsorgung von Siedlungsabfällen, *Bundesanzeiger Nr. 99a*.

7 Sanitary Landfills – Long-Term Stability and Environmental Implications

MICHAEL S. SWITZENBAUM

Amherst, MA, USA

List of Abbreviations

BOD	biochemical oxygen demand
COD	chemical oxygen demand
EPA	Environmental Protection Agency
MSW	municipal solid waste
RCRA	Resource Conservation and Recovery Act
TOC	total organic carbon
VFA	volatile fatty acids
VOC	volatile organic compounds

1 Introduction

Our society generates significant quantities of municipal solid waste. According to TCHOBANOGLOUS et al. (1991) solid wastes comprise all the wastes arising from human and animal activities that are normally solid and that are discarded or unwanted. Municipal solid waste is normally assumed to include all community wastes (residential, commercial, institutional, construction and demolition, and municipal services) and does not include industrial and agricultural sources. While municipal solid waste is only a relatively small fraction of the total amount of solid waste generated, proper management is essential for the control of disease vectors and for protection of the environment.

In the United States, about $1.9 \cdot 10^{11}$ kg ($210 \cdot 10^6$ t) of municipal solid waste are generated per year. The per capita generation rate is about 2 kg (4.4 lb) per person per day (USEPA, 1997b). These rates vastly increased over the past 30–40 years, but are now starting to level off or even slightly decrease (Fig. 1). Generation rates in the United States are considerably higher than generation rates in European countries.

Due to recycling efforts, the net generation rate has been decreasing (Fig. 2). Net generation is the amount of solid waste remaining from the total amount generated after materials (such as newspapers, glass, and aluminum cans) have been recovered by recycling activities.

Municipal solid waste is a heterogeneous and diverse mixture of materials which society uses. General categories include: food wastes, paper, cardboard, plastics, textiles, rubber, leather, yard wastes, wood, glass, tin cans, and ferrous metals.

The standard method of reporting waste generation is in terms of mass. However, mass

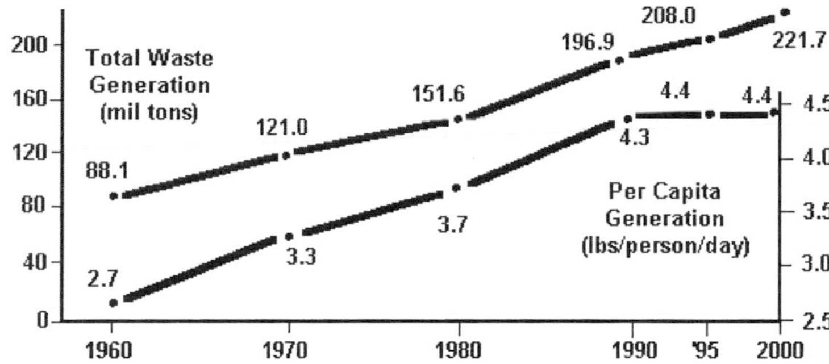

Fig. 1. Municipal solid waste generation rates 1960–2000 (USEPA, 1997a).

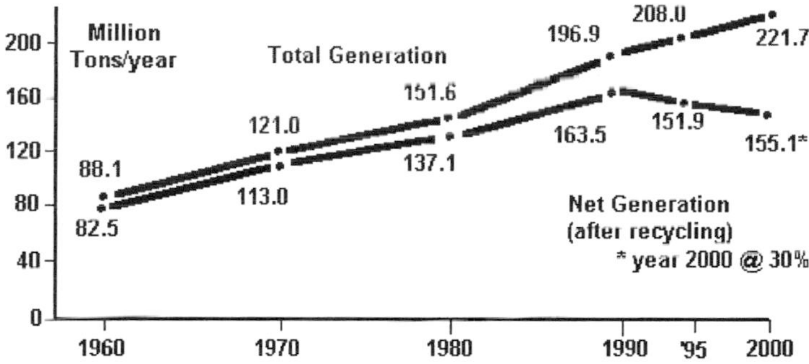

Fig. 2. Total and net waste generation 1960–2000 (USEPA, 1997a).

data are of limited value for certain applications such as landfill design since average landfill requirements depend on volumetric measurements. In addition to the original density of any materials making up the solid waste mixture, volume also depends on how much the waste has been compacted (Tab. 1).

The comparative percentages of various components of the waste stream are variable. Composition varies geographically, and is also influenced by efforts in source reduction and recycling (Tab. 2).

In summary, large amounts of municipal solid waste are generated and the waste is diverse in nature. These wastes must be managed for the control of disease vectors and for the protection of the environment.

Tab. 1. Density of Municipal Solid Waste as Influenced by Compaction (according to TCHOBANOGLOUS et al., 1991)

Component	Density [kg m^{-3}]
Residential	130
In compactor truck	300
In landfill (normal)	450
In landfill (well compacted)	590
Baled	700

2 Integrated Waste Management

TCHOBANOGLOUS et al. (1991) defined solid waste management as the discipline associated with the control of generation, storage, collection, transfer and transport, processing and disposal of solid wastes in a manner that is in accord with the best principles of public health, economics, engineering, conservation,

Tab. 2. Comparison Data – Solid Waste Stream Composition (% by mass) (according to O'LEARY and WALSH, 1992)

Solid Waste Component	Medford, WI 1988 Study	Franklin 1990 Estimate	IDNR Guidelines	Cal Recovery 1988 Study	GBB-Metro Des Moines
Food waste	22.80	8.39	8.00	8.80	10.00
Yard waste	11.80	19.80	20.00	28.20	11.00
Other organic	16.20	7.92	11.00	8.90	7.80
Subtotal	50.80	36.11	39.00	45.90	28.80
Newsprint	3.10	5.15		7.30	5.20
Corrugated	6.20	7.31		6.70	10.80
Mixed paper	5.60	24.39		19.10	32.70
Subtotal	14.90	36.85	37.00	33.10	48.70
Ferrous Metal	5.20	7.45		3.20	4.90
Aluminum	1.60	1.34		1.00	0.30
Other metal	0.20	0.20		0.70	0.20
Subtotal	7.00	8.99	6.00	4.90	5.40
Plastics	7.10	7.92	7.00	7.30	9.40
Glass	8.00	8.25	5.00	4.90	2.40
Inorganics	12.20	1.88	6.00	3.90	5.30
Subtotal	27.30	18.05	18.00	16.10	17.10
Total	100.00	100.00	100.00	100.00	100.00

aesthetics, and other environmental considerations, and that is also responsive to public attitudes. Integrated solid waste management, according to TCHOBANOGLOUS et al. (1991) involves the selection and application of suitable techniques, technologies and management programs to achieve specific waste management objectives and goals. Integrated solid waste management refers to the complimentary use of a variety of waste management practices to safely and effectively handle the municipal solid waste stream with the least adverse impact on human health and the environment.

The United States Environmental Protection Agency has adapted a hierarchy in waste management, which can be used to rank management actions (USEPA, 1988). The hierarchy, as is shown in Fig. 3, is in order of preference. Source reduction is at the highest level, while landfilling is at the lowest level.

An integrated approach will contain some or all of the following components:

(1) source reduction: reduction in amount of waste generated,
(2) recycling: separation and collection of waste materials,
(3) transformation (incineration): alterations to recover energy or other products (such as compost),
(4) landfilling: land disposal of wastes, those not able to be recovered combusted, or otherwise transformed.

Note that while land disposal of wastes is ranked lowest in the hierarchy, it is still a widely used waste management strategy. There are certain materials, which cannot be reduced at the source, recycled or transformed. Therefore, they are landfilled.

3 Land Disposal

While efforts have been made in source reduction, recycling, and incineration, most of the solid waste generated in the United States still ends up in landfills (56.9% in 1995). However, the percentage is decreasing as shown in Tab. 3. The data reflect progress made in achieving the goals of the EPA Hierarchy.

Land disposal was promoted as the "modern means" of solid waste disposal in the mid 1960s as a response to the air pollution problems associated with uncontrolled combustion of municipal solid waste. Unfortunately, in many cases, land disposal was not properly conducted. Rather than disposal into well-engineered and operated landfills, solid waste was merely buried in uncontrolled dumps. In addition, there was no distinction between municipal solid waste and hazardous waste – both were placed in these "dumps". This often resulted in an adverse impact on the environment. In fact, many of the sites listed on the National Priority List of the United States Environmental Protection Agency are abandoned dumps (or orphaned landfills). A schematic illustrating the potential impact of improperly constructed dumps is shown in Fig. 4. As can be seen, there are several detrimental impacts including groundwater contamination from leachate, surface water contamination from runoff, as well as impacts from gas migration (which include damage to plants, and potential explosions in confined areas, and atmospheric contamination).

As of result of the problems from improperly constructed dumps, new federal regulations were promulgated. Subsequent landfill disposal criteria are listed in the Resource Conservation and Recovery Act (RCRA) under Subtitle D (Federal Register, 1991). A summary of USEPA regulations for MSW landfills is shown in Tab. 4.

The cost of landfilling has greatly increased due to new federal regulations. This can be seen in Fig. 5. Average tipping fees greatly increased following promulgation of RCRA.

At the same time, many old landfills have been closed, often because they were causing environmental damage. Fig. 6 shows the trends in terms of the number of landfills in the United States. While the number of landfills has decreased, the trend is to build larger landfills (often regional) which are often owned by private businesses (Denison et al., 1994).

In landfills, solid waste is disposed of in thin layers which are compacted to minimize volume, and then covered (usually daily) with a thin layer of material (usually soil) to minimize environmental problems.

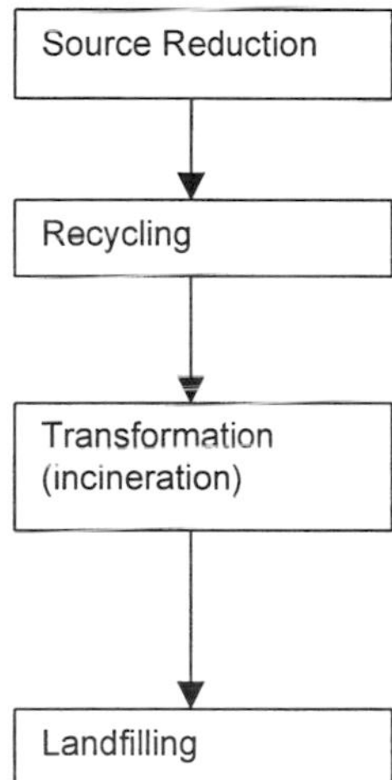

Fig. 3. Waste management hierarchy (USEPA, 1988).

Tab. 3. Waste Management Practices, 1960–2000 (as a percent of generation) (after USEPA, 1997b)

	1960	1970	1980	1990	1995	2000
Generation	100	100	100	100	100	100
Recovery for recycling/composting	6.4	6.6	9.6	17.2	27.0	30
Discards after recovery	93.6	93.4	90.4	82.8	73.0	70
Combustion	30.3	20.7	9	16.2	16.1	16.2
Discards to landfill	60.3	72.6	81.4	66.7	56.9	53.7

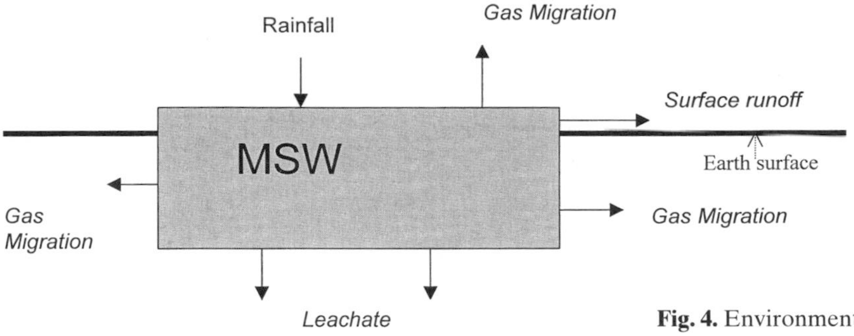

Fig. 4. Environmental impacts from dumps.

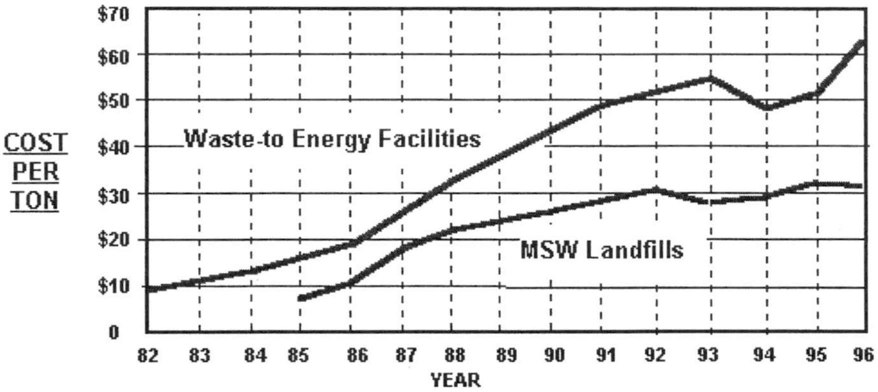

Fig. 5. Landfill tipping fees, national averages in the USA (USEPA, 1997a).

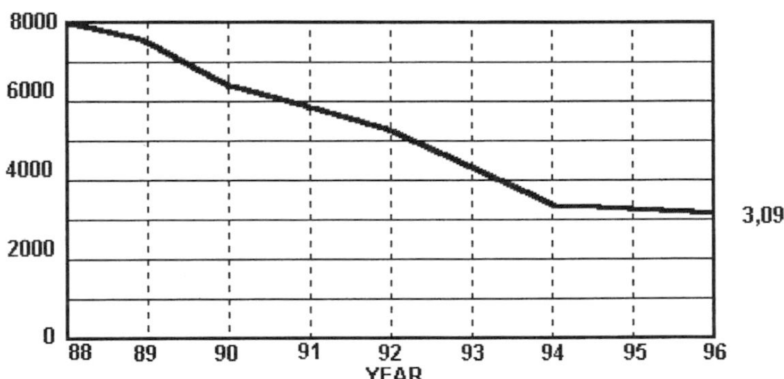

Fig. 6. Number of landfills in the United States (USEPA, 1997a).

Intermediate cover is usually placed on the top and exposed sides of the compacted solid waste. The covered compacted material is called a cell. A series of adjacent cells is called a lift. A lift is usually 3–3.5 m in height. The bottom lift will be placed above the bottom liner (usually a composite of clay and a flexible membrane) and leachate collection and removal system. A landfill will consist of several vertical lifts and will be 15–30 m in height. The landfill cap is placed above the top lift and is a combination of soil and synthetic materials. A schematic diagram showing selected engineering features of a modern sanitary landfill is shown in Fig. 7.

Tab. 4. Summary of US Environmental Protection Agency Regulations for Municipal Solid Waste Landfills (from TCHOBANOGLOUS and O'LEARY, 1994)

Item	Requirement
Applicability	– all active landfills that receive MSW after October 9, 1993 – certain requirements also apply to landfills which received MSW after October 9, 1991, but closed within 2 years – certain exemptions for very small landfills – some requirements are waived for existing landfills – new landfills and landfill cells must comply with all requirements
Location requirement	– airport separation distances of 1.524 m, 3.048 m, and in some instances greater than 3.048 m are required – landfills located on floodplains can operate only if flood flow is not restricted – construction and filling on wetlands is restricted – landfills over faults require special analysis and possibly construction practices – landfills in seismic impact zones require special analysis and possibly construction practices – landfills on unstable soils zones require special analysis and possibly construction practices
Operating criteria	– landfill operators must conduct a random load checking program to ensure exclusion of hazardous waste – daily cover with 0.1524 m of soil or other suitable materials is required – disease vector control is required – permanent monitoring probes are required – probes must be tested every 3 months – methane concentrations in occupied structures cannot exceed 1.25% – methane migration offsite must not exceed 5% at the property line – Clean Air Act criteria must be satisfied – access must be limited by fences or other structures – surface water drainage run-on to the landfill and runoff from the working face must be controlled for 24 year rainfall events – appropriate permits must be obtained for surface water discharges – liquid wastes or wastes containing free liquids cannot be landfilled – extensive landfill operating records must be maintained
Liner design criteria	– geomembrane and soil liners or equivalents are required under most new landfill cells – groundwater standards may be allowed as the basis for liner design in some areas
Groundwater monitoring	– groundwater monitoring wells must be installed at many landfills – groundwater monitoring wells must be sampled at least twice per year – a corrective action program must be initiated where groundwater contamination is detected
Closure and postclosure care	– landfill final cover must be placed within 6 months of closure – the type of cover is soil or geomembrane and must be less permeable than the landfill liner – postclosure care and monitoring of the landfill must continue for 30 years
Financial assurance	– sufficient financial reserves must be established during the site operating period to pay for closure and postclosure care amounts

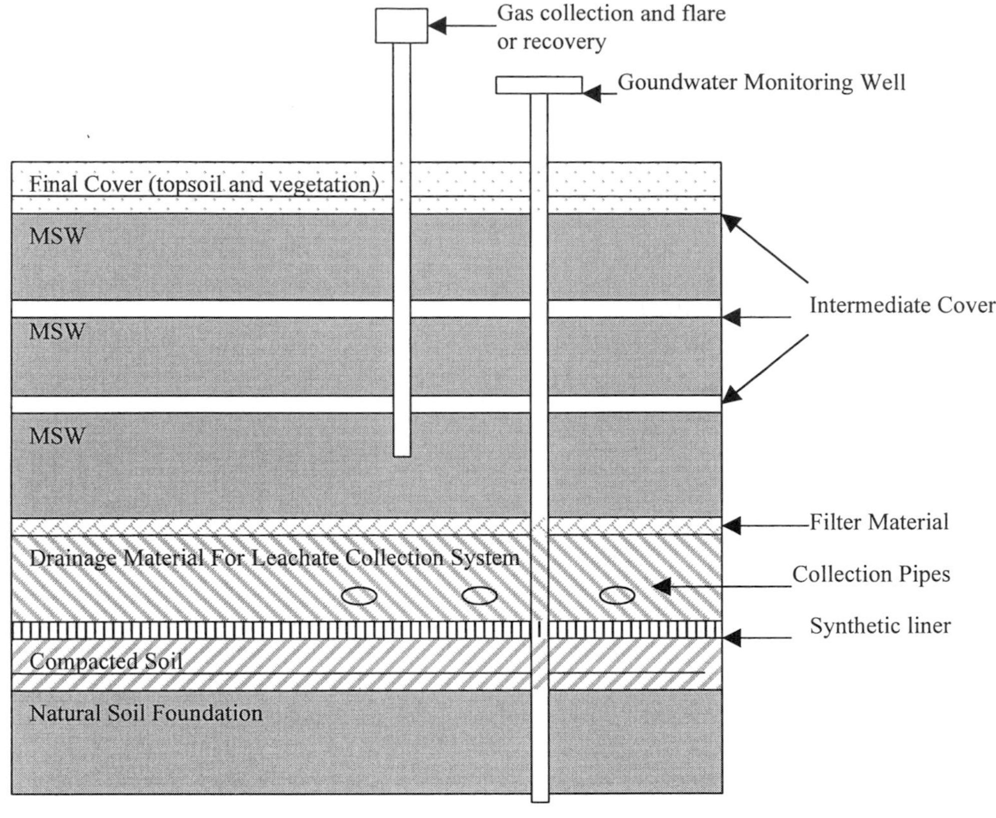

Fig. 7. Profile of landfill engineered features (not to scale).

The design of a modern sanitary landfill focuses on the prevention of leachate and gas migration. Vents and collection systems ensure that any gas produced will be captured and either recovered, flared, or dissipated in the atmosphere. The landfill cap is designed to prevent precipitation from entering the landfill. Liners prevent contamination of the subsurface. Leachate collection and removal systems prevent a buildup of hydraulic head on the liner.

In should be noted that the strategy of complete entombment of municipal solid waste is somewhat controversial. Many believe that even though landfills may be carefully designed and operated, all landfills will eventually leak. Subtitle D requirements of RCRA can only cause a delay in, but not prevent the generation and migration of leachate and methane (DENISON et al., 1994). In this regard, several investigators have promoted the adaptation of "wet-cell design and operation" (i.e., leachate recycle) to accelerate the decay of the organic fraction of landfilled waste and thus avoid the problem of releases after the post-closure period (POHLAND, 1980). Wet cell design and operation also increased the methane yield in earlier years which made gas recovery more feasible, and presented the potential for reuse of the physical property of the landfill (DENSION et al., 1994).

There are three common configurations for landfills – area, ramp, and trench methods. The area method involves constructing a landfill above ground, while the trench method involves excavation. The ramp method is a variation of the area method as it is built above ground, but on a slope.

4 Leachate and Gas Management

The RCRA landfill design criteria included provisions for leachate and gas management to prevent groundwater contamination, and to avoid any potential problems from gas migration including explosions and fires, vegetation damage, and atmospheric contamination. The rate and extent of gas generation are influenced by numerous factors, but are primarily controlled by the products of microbial reactors in the landfills. Typical constituents found in MSW landfill gas are shown in Tab. 5.

Methane and carbon dioxide are the principal gaseous products. Smaller amounts of nitrogen, oxygen, hydrogen, and trace compounds [many of which are volatile organic compounds (VOC)] can also be found in landfill gas. Some of the trace gases, although present in small amounts can be toxic and could present risks to public health (TCHOBANOGLOUS et al., 1991). The occurrence of significant quantities of VOC is associated with older landfills, which accepted industrial and commercial wastes containing VOC. In newer landfills, the disposal of hazardous wastes is not allowed, and as such, the concentrations of VOC in the landfill gas are low. Gas migration is controlled by landfill caps, vents, and recovery systems, or flaring systems.

Leachate is formed when water passes through the landfilled waste materials. Leachate is a mixture of organic and inorganic, soluble and colloidal solids. It includes the products from decomposition of materials as well as soluble constituents leached from the landfilled materials. Leachate generation rates are determined from the amount of water contained in the original material and the amount of precipitation, which enters the landfill. Other factors include climate, site topography, final cover material, vegetative cover, site operating procedures, and type of waste in the landfill.

Tab. 5. Typical Constituents Found in MSW Landfill Gas (according to TCHOBANOGLOUS et al., 1991)

Component	Percent (Dry Volume Basis)
Methane	45–60
Carbon dioxide	40–60
Nitrogen	2–5
Oxygen	0.1–1.0
Sulfides, and S compounds	0–1.0
Ammonia	0.1–1.0
Hydrogen	0–0.2
Carbon monoxide	0–0.2
Trace constituents	0.01–0.6

Landfill leachate characteristics vary widely form season to season and from year to year as well as from one landfill to another. Typical data on the composition of leachate are shown in Tab. 6. Particle size, degree of compaction, waste composition, landfilling technique, cover configuration, site hydrology, climate, and age of the tip are among the factors which affect leachate quantity and quality (POHLAND and HARPER, 1986). Leachate characteristics are directly related to the natural processes occurring inside the landfill.

The effects of climate on leachate production are fairly well understood. Rainwater provides both the mode of transport of contaminants out of the landfill, as well as the moisture required for microbial activity. Depending on the porosity of the cover material (as well as the integrity of the cap, if one exists) high levels of rainfall often result in the production of large quantities of leachate. Large amounts of rainwater effectively reduce the contaminant concentration through dilution of the leachate. In more arid regions, evapotranspiration may reduce the expected leachate volume.

POHLAND and HARPER (1986) have described five relatively well-defined phases, which occur as landfill wastes are degraded. These phases are summarized in Tab. 7.

Under favorable conditions, generally dictated by the presence of sufficient moisture to support microbial activity, landfills function as large scale anaerobic reactors. Complex organic material is converted to methane gas through the biological methane fermentation process. As a result, leachates produced in younger landfills are generally characterized by the presence of substantial amounts of volatile fatty acids (VFA), which are the precursors of methane. These VFA make up the majority of the COD of young leachates.

Tab. 6. Typical Data on the Composition of Leachate from New and Mature Landfills (according to TCHOBANOGLOUS et al., 1991)

Constituent	Value [mg L^{-1}] New Landfill (<2 years) Range	Typical	Mature Landfill (>10 years)
BOD_5	2,000–30,000	10,000	100–200
TOC	1,500–20,000	6,000	80–160
COD	3,000–60,000	18,000	100–500
Total suspended solids	200–2,000	500	100–400
Organic nitrogen	10–800	200	80–120
Ammonia N	10–800	200	20–40
Nitrate	5–40	25	5–10
Total P	5–100	30	5–10
Ortho P	4–80	20	4–8
Alkalinity as $CaCO_3$	1,000–10,000	3,000	200–1,000
pH	4.5–7.5	6	6.6–7.5
Total hardness as $CaCO_3$	300–10,000	3,500	200-500
Calcium	200–3,000	1,000	100–400
Magnesium	50–1,500	250	50–200
Potassium	200–1,000	300	50–400
Sodium	200–2,500	500	100–200
Chloride	200–3,000	500	100–400
Sulfate	50–1,000	300	20–50
Total iron	50–1,200	60	20–200

Along with a high organic content, leachates may also contain considerable amounts of heavy metals (CHIAN and DEWALLE, 1977). The oxidation–reduction potential and pH inside a biologically active landfill are such that metals tend to be solubilized. High concentrations of iron, calcium, manganese, and magnesium have been reported by several researchers (POHLAND and HARPER, 1986; LEMA et al., 1988; KENNEDY et al., 1988). Other potentially more toxic metals have also been measured in significant concentrations in leachates (POHLAND and HARPER, 1986).

Leachate migration is controlled by a combination of landfill cap, bottom liner, and leachate collection and removal.

The nature of leachate contaminants generally mandates that some type of treatment process be employed prior to ultimate discharge of the wastewater. Numerous studies have been performed to evaluate the performance of biological and/or physical chemical treatment processes for removing organic and inorganic leachate contaminants. In general, leachates produced in younger landfills are most effectively treated by biological processes since the wastewater is mostly composed of volatile fatty acids, which are readily amenable to biological treatment. Both aerobic and anaerobic processes have been used for the treatment of landfill leachate (POHLAND and HARPER, 1986; IZA et al., 1992).

On the other hand, contaminants in older leachates tend to be more refractory and inorganic in nature, making physical-chemical treatment more applicable. In some cases, both biological and physical-chemical treatment are combined.

As mentioned previously, a particularly attractive alternative for leachate treatment is leachate recycle. Leachate recycle has been shown to be an effective *in situ* treatment alternative (POHLAND, 1980). More recently, POHLAND (1998) has noted that bioreactor landfills with *in situ* leachate recycle and treatment prior to discharge provide accelerated waste stabilization in a more predictable and shorter time period than that achieved in a conventional landfill.

Tab. 7. Summary of Stage of Landfill Stabilization (according to POHLAND and HARPER, 1986)

Phase	Description
I. Initial adjustment	– initial waste placement and moisture accumulation – closure of each fill section and initial subsidence – environmental changes are first detected, including onset of stabilization process
II. Transition	– field capacity is exceeded and leachate is formed – anaerobic microbial activity replaces aerobic – nitrate and sulfate replace oxygen as the primary electron acceptor – a trend toward reducing conditions is established – measurable intermediates (VFA) appear in leachate
III. Acid formation	– intermediary VFA become predominant with the continued hydrolysis and fermentation of waste organics – pH decrease, causing mobilization and possible complexation of heavy metals – nutrients such as N and P are released and assimilated – hydrogen produced from anaerobic oxidations may control intermediary fermentation products
IV. Methane fermentation	– acetic acid, carbon dioxide, and hydrogen produced during acid formation are converted to methane – pH increases from buffer level controlled by VFA to one more characteristic of the bicarbonate system – redox potential is at a minimum – leachate organic strength decreases dramatically due to conversion of organic matter to methane gas
V. Final maturation	– relative dormancy following active biostabilization – measurable gas production all but ceases – more microbial resistant organics may be converted with the possible production of humic-like substance capable of complexing and remobilizing heavy metals

5 Summary and Conclusions

Landfilling of municipal solid waste emerged approximately 35 years ago as the principal means of solid waste management. It emerged as an alternative to incineration, which became less popular due to air quality concerns. Unfortunately, landfills were not designed and/or operated in a proper manner and, therefore, caused significant impact on the environment.

In response to this situation, new laws (RCRA) were passed and strict regulations were established for the solid waste landfill design and operation. This includes provisions to control gas and leachate migration. In addition, the United States Environmental Protection Agency developed a hierarchy that relegated landfilling to a low priority. Higher emphasis is now placed on source reduction, recycling and transformation processes.

As a result of these new regulations and policies, the amount of municipal solid waste being landfilled has decreased. The number of landfills has also decreased (but newer landfills are generally larger in size). However, it is recognized that landfilling will remain a significant solid waste management tool, as there are certain components in the solid waste stream that can be economically recovered or transformed.

While the new regulations are focussed on controlling gas and leachate migrations and subsequent environmental impact, there are concerns about the long-term stability of landfill and integrity of the liner systems. As a result, several investigators have promoted leachate recycle of the "wet cell concept" as a means of accelerating the rate of solid waste decomposition.

6 References

CHIAN, E. S. K., DEWALLE, F. B. (1977), Evaluation of leachate treatment, Vol. 1. Characterization of leachate, *USEPA Report no. EAP-600/2-77-186a*. Cincinnati, OH.

DENISON, R. A., RUSTON, J., TYRENS, J., DIEDRICH, R. (1994), Environmental prospectives, in: *Handbook of Solid Waste Management* (KREITH, F., Ed.). New York: McGraw-Hill.

Federal Register (1991), Solid Waste Disposal Facility Criteria: Final Rule, *40CFR Parts 257 and 259* (October 9, 1991).

IZA, J. M., KEENAN, P. J., SWITZENBAUM, M. S. (1992), Anaerobic treatment of municipal solid waste landfill leachate: Operation of a pilot scale hybrid UASB/AF reactor, *Water Sci. Technol.* **25**, 225.

KENNEDY, K. J., HAMODA, M. F., GUIOT, S. G. (1988), Anaerobic treatment of leachate using fixed film and sludge bed systems, *J. Water Pollut. Control Fed.* **60**, 1675–1683.

LEMA, J., MENDEZ, R., BLAZQUEZ, R. (1988), Characteristics of landfill leachates and alternatives for their treatment: a review, *Water Air Soil Pollut.* **40**, 223–250.

O'LEARY, P. R., WALSH, P. W. (1992), *Solid Waste Landfills*. Madison, WI: Dept of Engineering Professional Development, University of Wisconsin.

POHLAND, F. G. (1980), Leachate recycle as landfill management option, *J. Environ. Eng.* (ASCE) **106**, 1057–1069.

POHLAND, F. G. (1998), *In situ* anaerobic treatment of leachate in landfill bioreactors, *Proc. 5th Latin American Seminar on Anaerobic Wastewater Treatment*, October 1998, Viña del Mar, Chile.

POHLAND, F. G., HARPER, S. R. (1986), Critical review and summary of leachate and gas production from landfills, *USEPA Report no. EPA 600/2-86/073*. Cincinnati, OH.

TCHOBANOGLOUS, G., THEISSEN, H., VIGIL, S. (1991), *Integrated Solid Waste Management: Engineering Principles and Management Issues*. New York: McGraw-Hill.

TCHOBANOGLOUS, G., O'LEARY, P. (1994), Landfilling, in: *Handbook of Solid Waste Management* (KREITH, F., Ed.). New York: McGraw-Hill.

USEPA (1997a), *The EPA Municipal Solid Waste Factbook*, Version 4.0 software. Washington, DC.

USEPA (1997b), *Characterization of Municipal Solid Waste in the United States*, 1996 Update. Washington, DC: OSW Publications Distribution Center.

USEPA (1988), The Solid Waste Dilemma: An Agenda for Action, *EPA/530-SW-88.052*. Washington, DC.

8 Combined Mechanical and Biological Treatment of Municipal Solid Waste

PETER SCHALK

Emmendingen, Germany

1 Introduction

Agenda 21 is a declaration of intent to the World family of nations regarding the global problem of preserving natural resources and promoting environmental protection. The object must be to produce long lasting, and thus future-oriented concepts of waste disposal and transfer these concepts into practical application. The re-orientation of waste management is concerned with the following central ideas (ALEXIOU et al., 1999):

- Conservation and management of resources,
- waste avoidance (quantity and toxicity),
- waste recovery (materials, energy),
- safe disposal (landfill, incineration).

The treatment and utilization of municipal solid waste is gaining increasing importance worldwide from economic and ecological aspects. Domestic waste consists of difficult to define mixtures of organic fractions containing recyclable material, non recyclable material, and mineral components. Mechanical processing in combination with biological conversion produces utilizable substance groups from the waste mixture and reduces the amount of waste considerably. The residing organic fraction can bepartially converted using the biological processes of composting or digestion. Mechanical-biological waste treatment leads to enduring waste management and can also be linked to existing disposal concepts (landfill, incineration).

2 Concept of Mechanical-Biological Waste Treatment

2.1 Waste Composition

The composition of municipal solid waste largely depends on the following factors:

- District structure and social standard of living,
- basic legal conditions,
- waste organization and logistics (separate collection of specific waste flows, etc.),
- available recovery and disposal equipment (composting, sanitary landfill disposal, etc.).

Waste composition in Europe is exemplarily illustrated in Fig. 1.

Extensive pure grading recovery (recycling) requires separate collection of the waste flows (paper, plastics, glass, biowaste, etc.). The basis for this is a high-level of organization and logistics, together with responsible environmental awareness of the waste producers.

The proportion of the organic fractions within municipal solid waste varies considerably in European countries. Of the approximately 60 mio. t of organic waste in the European Community, 9–10 mio. t are collected separately and are treated biologically by composting or digestion. The compost produced is mainly used as a soil conditioner and organic fertilizer in landscaping, horticulture, and agriculture.

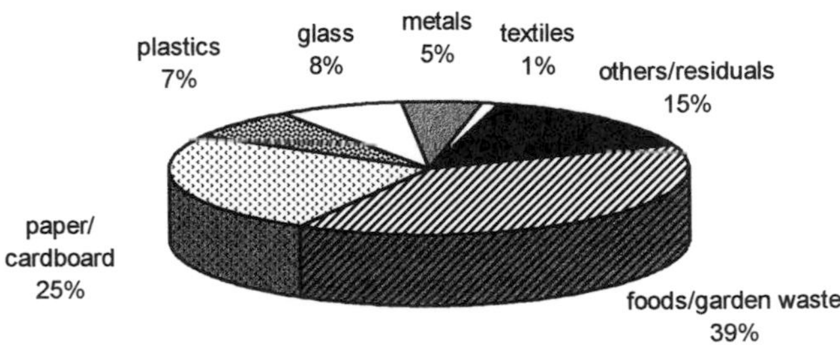

Fig. 1. Waste composition in Europe (AVNIMELCH, 1997).

2.2 Waste Disposal Concepts with Mechanical-Biological Treatment

In most countries, industrial and household waste are disposed of in landfill sites or incineration plants. Landfill disposal, in particular of waste containing organic fractions, can produce significant emissions (outgassing of odors and methane, release of leachate). For this reason, there are specific requirements concerning the location and the operation management of landfill sites. During management of the landfill site and for a long time after its closure site after-care is necessary (degassing, leachate purification), which means ecological and financial risk.

With incineration, energy from the waste is utilized, and electricity and heat are generated. Because it is scarcely possible to determine the amount of toxic substances in the waste, relatively complicated and cost-intensive waste gas purification systems are needed for low emission incineration. Because of the higher investment and operating cost of incineration, landfill disposal of the waste is still predominant in most countries.

In some countries, concepts and systems for source separation of waste have been introduced, particularly for paper, glass, plastics, and biowaste. For this purpose, appropriate preparation and recovery plants have been established with the aim of setting up a recycling business instead of simply waste disposal. The amount of residual waste for disposal has thereby been reduced by up to 40% depending on the proportion of collection and recovery.

Despite the introduction of a separate waste collection service (recyclable substances, biowastes), the composition of the residual waste reveals similar disposal characteristics as the original domestic waste. This situation has led to the development of systems of mechanical-biological waste treatment, which can be linked appropriately to existing disposal concepts (landfill, incineration). This combination of methods has been used on a commercial scale for about 5 years.

The mechanical-biological pretreatment before landfill disposal allows use of the existing landfill capacities over the longer term (Fig. 2). This treatment optimizes and – compared with landfill disposal – accelerates the biological processes. A largely stabilized output with a

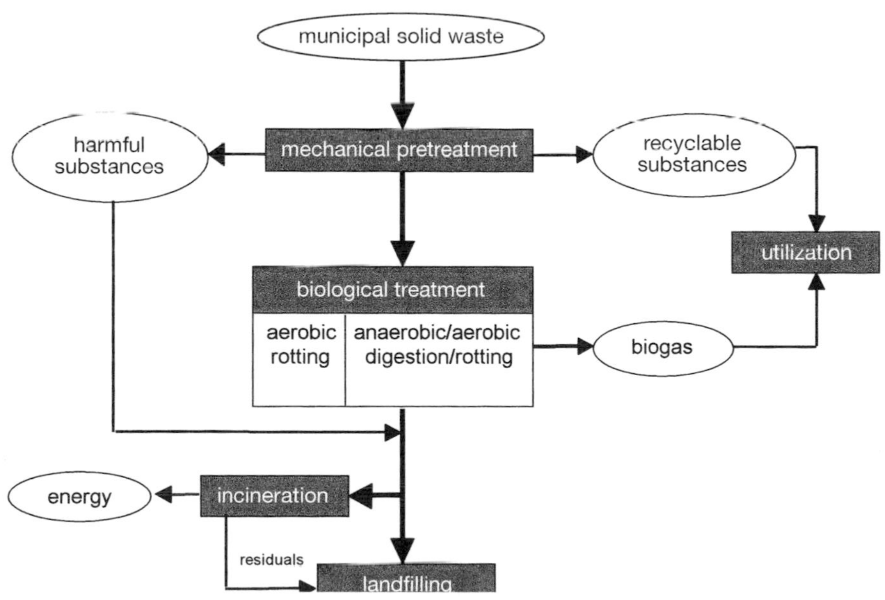

Fig. 2. Basic concept of mechanical-biological waste treatment.

low emission potential finally is deposited on the landfill site. Recycling of substances and biological degradation reduces the amount of waste for disposal, so that the landfill operating time is extended correspondingly.

Mechanical-biological waste treatment can also be used for waste pretreatment before incineration. In addition to reducing the total waste mass, the waste mixture is homogenized and prepared for more efficient incineration. Biological stabilization produces a fraction with a high caloric value (substitute fuel) able to be transported and stored (waste-to-energy). Mechanical-biological waste treatment combined with incineration has definite cost advantages compared with a system using incineration only.

Extensive investigations into mechanical-biological waste treatment and their possible combination have been presented at scientific conferences regularly (e.g., WIEMER and KERN, 1999; GALLENKEMPER et al., 1999). An updated overview is available concerning plants for composting, digestion, and mechanical-biological waste treatment (WIEMER and KERN, 1998).

3 Mechanical-Biological Waste Treatment

3.1 Mechanical Pretreatment

The waste must be decompacted after transport to the treatment plant and homogenized for further mechanical conditioning. The system technology for mechanical treatment depends on the composition of the waste and the requirements for the subsequent biological stage. Mechanical pretreatment mainly consists of the following steps:

- Shredding of waste,
- sieving of over-sized particles,
- magnetic separation of metals,
- homogenization of the sieved fraction.

Depending on the composition of the waste, different equipment is used for shredding and separating the waste mixture, e.g., cutting and hammer mills, shredders, etc. Sieving normally takes place in screening drums with different sieve sizes (40–120 mm). More than 90% of the organic fraction from the residual waste passes through the sieve up to a sieve size of 80 mm (Fig. 3). The screen overflow con-

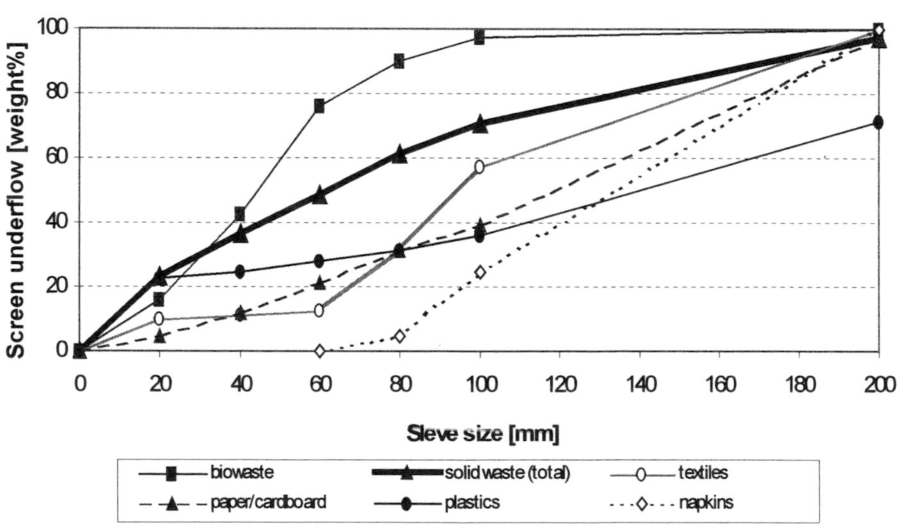

Fig. 3. Screen underflow for different sieve sizes (LEIKAM and STEGMANN, 1995).

sists mainly of paper and plastics which can be reused to produce materials or energy.

Materials (metals, light plastics) are extracted by electromagnetic or mechanical processing, and harmful substances (bulky refuse, stones, etc.) are ejected. This fraction of waste cannot be converted in the subsequent biological treatment. With some systems, a fraction with a high calorific value (mainly plastics) is separated and used as a substitute fuel for energy utilization (waste-to-energy). Mechanical pretreatment reduces the amount of waste to 70–85% which is then processed further by biological treatment.

3.2 Biological Treatment

Biological treatment is either aerobic (rotting) or anaerobic (digestion with short subsequent rotting). During rotting with active aeration the organic substances are broken down into carbon dioxide and water leading to self-heating of the material. In the first phase (pre-rotting and intensive rotting) easily degradable substances are converted with high oxygen consumption. After 2–4 weeks less intensive after-rotting occurs which can last 4–8 months. The following rotting systems are generally used (KERN, 1999):

- Windrow or stack rotting,
- container rotting,
- tunnel rotting,
- drum rotting.

At some sites, rotting is performed in open air at the landfill site (low-level-treatment). Normally there are only reduced possibilities for process control (aeration, moistening, off-gas treatment, etc.). An alternative are encapsulated plants (technical plant treatment) mainly for pre-rotting with designed process optimization and off-gas purification. For later landfill disposal the subsequent rotting (final rotting) takes place either in an encapsulated plant or outdoors on the landfill site.

The rotting process requires a certain structural content and a corresponding water content (approximately 50–60% by weight). To optimize biological degradation structural material must be added.

Conversion of wet, compost waste (e.g., kitchen waste) can be carried out by the anaerobic digestion process. In the absence of oxygen, biogas (55–65% by volume of CH_4) is produced, from which electricity and heat are generated in a combined heat and power generating plant.

Currently, about 1 mio. t of waste are subjected to digestion in Europe and there is an increasing trend (DE BAERE, 1999). Two thirds of this is biowaste and one third mixed household waste. The systems and combined methods can be classified as follows:

- Process stages (hydrolysis, methanation): single stage – multi stage,
- operating mode: continuous – discontinuous,
- process temperature: mesophilic (35 °C) – thermophilic (55 °C),
- process moisture: dry (DS >25%) – wet (DS <15%).

Because of the lower investment, cost, 90% of the waste treatment plants are built as single-stage processes and – with respect to the process temperature – predominantly in the mesophilic temperature range. Thermophilic systems may have a higher conversion rate, but constant temperatures must be provided. About 60% of the waste treatment plants make use of dry anaerobic digestion, whereas about 40% of the waste treatment plants apply wet digestion. The selection of a dry or wet digestion process is governed by waste composition and final rotting requirements (structural stability).

Biogas production primarily depends on the degradability of the organic waste components and on the retention time in the digester (5–25 d). Some waste treatment configurations require pretreatment by shredding of the waste (macerating, pulping) in order to accelerate the degradation process. Depending on the proportion of the organic fraction, the biogas yield constitutes about 60–140 $Nm^3\ t^{-1}$ of the waste (55% by volume of CH_4). This covers the energy requirement of the plant. Excess energy can go to third-party users.

Encapsulation of the digestion reactor prevents volatilization of odiferous substances (mainly volatile fatty acids). The final rotting of the dewatered digestion residue (non-digestable material plus surplus sludge) is considerably simpler with respect to off-gas purification compared to exclusively aerobic waste composting. The stabilized discharge from the biological treatment is used either for incineration or sent to a landfill site.

Currently, a change-over takes place in waste management in Germany from pure waste disposal to the recycling of commodities. In addition to waste avoidance the aim is to achieve the greatest possible recovery of waste materials. This includes firstly a recovery of material and secondly the utilization of the energy content of the non-recyclable proportion of the wastes (waste-for-recovery). In this respect, mechanical-biological waste treatment concentrates on the recovery of energy from waste products (WIEMER et al., 1995). Mechanical screening stages eliminate inert and harmful substances. The waste then undergoes a multi-stage biological stabilization process, including biological drying. This produces a product of high caloric value which can be marketed as substitute fuel with certain product qualities.

In addition to co-generation district refuse heating plants, energy-intensive industrial processes are also possible customers (e.g., cement manufacture, steel production). The disposal of waste by incineration or landfill plays a subordinate role when quantities are concerned.

4 The BIOPERCOLAT® Procedure

Wehrle-Werk AG, Emmendingen, a medium-sized company in Germany, is active in the field of energy and environmental technologies. The BIOPERCOLAT® procedure (Fig. 4) is an innovative mechanical-biological waste treatment procedure for residual waste and biowaste, combining mechanical separation

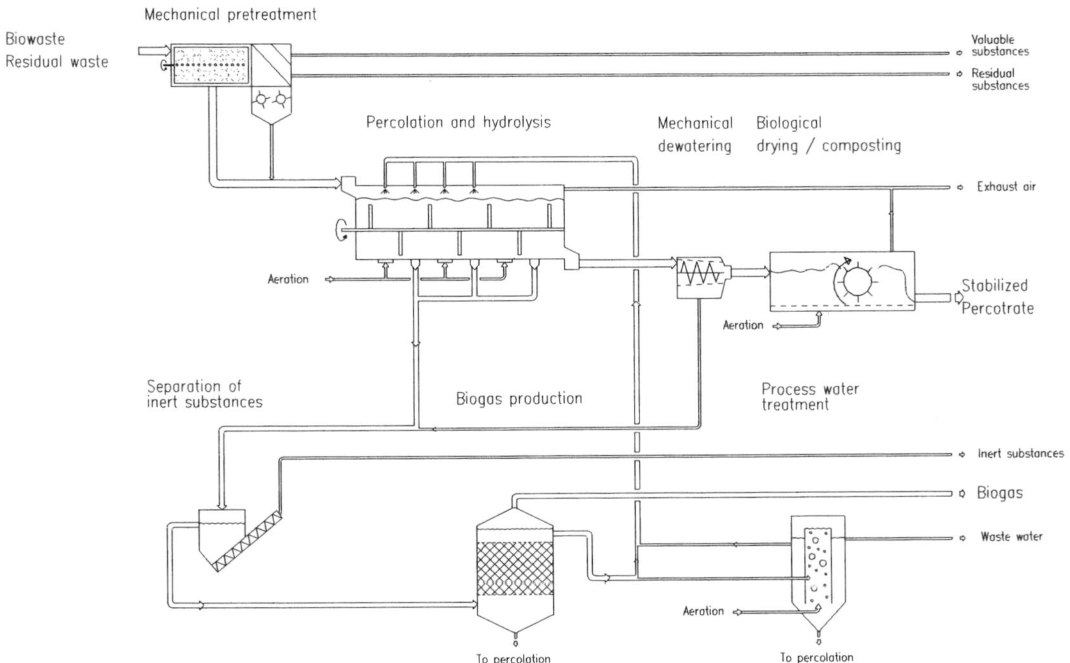

Fig. 4. Scheme of the BIOPERCOLAT® procedure.

and preparation of waste with aerobic processing and anaerobic digestion. It is a two-stage procedure with the following main units:

- Mechanical pre-treatment,
- percolation and hydrolysis,
- composting/biological drying,
- anaerobic digestion with biogas utilization,
- process water treatment.

During mechanical pretreatment the solid waste is shredded and sieved, and recyclable substances are removed. The screen underflow, which contains the biogenic part enters the percolator. Easily soluble and odoriferous substances are washed out (percolation). Solid organic waste substances, which can be hydrolyzed within the retention time of the waste in the percolator are also washed out. Waste circulation and waste hydrolysis are supported by a horizontal stirring device. The percolation leachate is converted anaerobically into biogas (retention time 3–4 d) after separation of inert substances (sand, etc.).

The anaerobic stage consists of a biofilm digester which is operated in the mesophilic temperature range. The biogas is used for energy production in a combined heat and power generator. The process provides enough energy for self-sufficiency through biogas utilization. The surplus power and heat can be delivered to third parties.

The effluent of the anaerobic digester is reused as process water for the percolation. Part of the circuit water and the excess wastewater are treated in an aerobic wastewater plant. The organic fractions and nitrogen compounds are removed by denitrification and nitrification.

The residing solid waste (Percotrate) is dewatered by an intensity screw press after 2–3 d of retention within the percolation. The odorless Percotrate is either composted (biowaste compost) or stabilized by further biological drying (energetic waste utilization).

5 Results of Pilot Plant Studies

Since May 1997, an entire BIOPERCOLAT® plant according to Fig. 4 has been operated continuously as a mobile demonstration plant at the Kahlenberg landfill site, Germany. Mechanical-biological treatment is carried out with the non-recyclable domestic waste fraction, including the biowaste. The intention is mainly focused on weight reduction and stabilization of the mentioned waste fraction before final disposal. The organic proportion of the waste is converted to biogas. The dewatered and dried residues allow for material and energy utilization (waste-for-recovery) (WELLINGER et al., 1999).

Initially, water-soluble organic and inorganic substances from primary residual waste are washed out with the percolation water to reach COD concentrations of $>100{,}000$ mg L^{-1}. After 1 h of percolation the formation of organic acids begins in the aerated percolator (hydrolysis). The leachate from percolation is converted to biogas in an anaerobic filter. The methane content of the biogas is constantly about 65–70% by volume.

The specific biogas production is about 70–80 Nm^3 t^{-1} (65% by volume CH_4) of treated residual waste. The plant operation is self-sufficient in energy, and even excess electricity, and heat for sale are generated.

The same fraction of non-recyclable waste was also treated in a single-stage procedure. At thermophilic temperatures, the digester produced 100 Nm^3 t^{-1} (approximately 60% by volume CH_4) of biogas. The retention time of the solid waste in the digester was about 25 d. With regard to a commercial plant (50,000 t a^{-1}), a digester volume of about 6,700 m^3 would be required for a single-stage treatment, compared to about 3,500 m^3 for the BIOPERCOLAT® process (Tab. 1).

The BIOPERCOLAT® process provides 75% of the biogas potential within 4 d of retention time in the anaerobic reactor. The percolator itself extracts and produces easily convertable substances and accelerates the subsequent digestion. The stirring device of the percolator macerates the organic waste substances and supports hydrolyzation and extraction

Tab. 1. Comparison of the Digestion of a Single-Stage Procedure with BIOPERCOLAT® (exemplary)

Input: Municipal solid waste 50,000 t a^{-1}
Specific weight of solid waste 0.6 t m^{-3}

	Single-Stage Procedure	BIOPERCOLAT®
Temperature range	thermophilic	mesophilic
Biogas production	100 Nm^3 t^{-1}	75 Nm^3 t^{-1}
Retention time	25 d	4 d
Digester volume	6,700 m^3	3,500 m^3

of solubles. This signifies a special feature essential to the design and investment costs of the plant.

The solid residues of percolation are dewatered by a screw press to reach 60% dry matter. The so-called Percotrate can be stabilized by biological drying to obtain a higher caloric value (substitute fuel). Percolation prior to drying supports an effective drying within 8–10 d of retention time. The dried percotrate with a residual moisture content of 10–15% has a calorific value of 12,000–18,000 kJ kg^{-1}. The municipal solid waste is reduced to about 30% dried percotrate which can be used for energy utilization (Fig. 5).

From the original non-recyclable municipal waste fraction about 12% of inert substances are separated or mechanically and another 13% are removed by biological degradation. This process and the dewatering leads to excess water which is treated by aerobic wastewater treatment. A mass balance of the whole treatment procedure is given in Fig. 5.

The BIOPERCOLAT® procedure provides biogas by biological conversion and essentially reduces the amount of residual solid waste mass. The percolation process allows for very high flexibility in reaction to the composition of solid waste and its changes. The separation of soluble organic compounds from the solid waste for anaerobic digestion results in an overall short treatment and retention times. The BIOPERCOLAT® procedure achieves a compact and cost-effective plant, which also reduces operating costs.

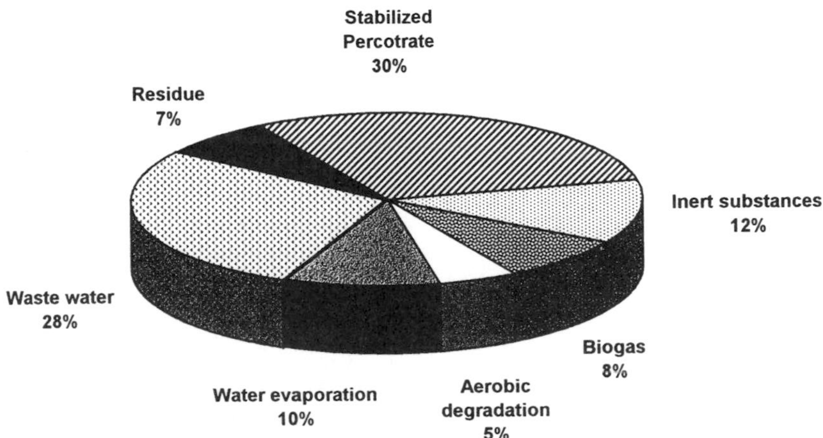

Fig. 5. Mass balance of the BIOPERCOLAT® procedure (municipal solid waste).

6 References

ALEXIOU, I. E., ZATARI, T. M., DALEY, C. M., PAPADIMITRIOU, E. K., BOSSANYI, M. (1999), Implementation of Agenda 21 for waste management: The impact on biotechnology of organic solid waste, in: *Int. Symp. Anaerobic Digestion of Solid Waste* (MATA-ALVAREZ, J., Ed.), pp. 382–385. Barcelona: Avda. Can Sucarrats Rubi.

AVNIMELCH, Y. (1997), Environmental aspects of msw composting, in: *Int. Conf. Organic Recovery and Biological Treatment into the Next Millenium*, Harrogate, UK.

DE BAERE, L. (1999), Anaerobic digestion of solid waste: state-of-the-art, in: *Int. Symp. Anaerobic Digestion of Solid Waste* (MATA-ALVAREZ, J., Ed.), pp. 290–299. Barcelona: Avda. Can Sucarrats Rubi.

GALLENKEMPER, B., BIDLINGMAIER, W., DOEDENS, H., STEGMANN, R. (1999), 6. Münsteraner Abfallwirtschaftstage, Tagungsband, *Münsteraner Schriften zur Abfallwirtschaft*, Fachhochschule Münster, Germany.

KERN, M. (1999), Stand und Perspektiven der biologischen Abfallbehandlung in Deutschland (Vortrag beim Kasseler Abfallforum 1999), in: *Bio- und Restabfallbehandlung III* (WIEMER, K., KERN, M., Eds.), pp. 293–321. Witzenhausen: M.I.C. Baeza-Verlag.

LEIKAM, K., STEGMANN, R. (1995), Mechanical and biological treatment of residual waste before landfilling, in: *Sardinia '95* (CHRISTENSEN, T. H., COSSU, R., STEGMANN, R., Eds.), pp. 947–955. Cagliari: CISA.

WELLINGER, A., WIDMER, C., SCHALK, P. (1999), Percolation – a new process to treat msw, in: *Int. Symp. Anaerobic Digestion of Solid Waste* (MATA-ALVAREZ, J., Ed.), pp. 315–322. Barcelona: Avda. Can Sucarrats Rubi.

WIEMER, K, TÄUBER, U., FROHE, R., MAYER, M., KERN, M. (1995), Mechanisch-biologische Restabfallbehandlung nach den Trockenstabilatverfahren, 2nd Edn. 1996. Witzenhausen: M.I.C. Baeza-Verlag.

WIEMER, K., KERN, M. (1998), *Kompost-Atlas 1998/1999*, Kompostierung – Anaerobtechnik – MBA-Aggregate. Witzenhausen: M.I.C. Baeza-Verlag.

WIEMER, K., KERN, M. (1999), Bio- und Restabfallbehandlung – biologisch – mechanisch – thermisch. Witzenhausen: M.I.C. Baeza-Verlag.

9 Hygienic Considerations on Aerobic/Anaerobic Treatment of Wastewater, Sludge, Biowaste, and Compost

DIETER STRAUCH
WERNER PHILIPP
REINHARD BÖHM
Stuttgart, Germany

List of Abbreviations

ATS	aerobic thermophilic stabilization of sludge
ATV	German Wastewater Engineering Organization
CEC	Commission of the European Communities
CFR	Code of Federal Regulations (USA)
DM	Deutsche Mark
EAA	exogenic allergic alveolitis
EEC	European Economic Community
EPA	Environmental Protection Agency (USA)
EU	European Union
FS	fecal streptococci
HRT	hydraulic retention time
IEAE	International Atomic Energy Agency
LAGA	German (Laender) Working Group for Waste
LASI	German (Laender) Guidelines for Health and Safety Standards at Work
MGRT	minimum guaranteed retention time
ODTS	organic dust toxic syndrome
OSS	ordinance on sewage sludge
PFRP	processes to further reduce pathogens
PSRP	processes to significantly reduce pathogens
RCS	Reuter centrifugal sampler
TBC	total bacterial count
TFC	total fecal count
TGE	transmissible gastroenteritis
VKS	German Association of Public Cleansing Enterprises
WHO	World Health Organization
WP	working party

1 Wastewater and Sludge

1.1 Introduction

It is an undeniable fact, which meanwhile is also acknowledged by the most persistent sceptics, that sewage and sludge from municipal treatment plants do contain pathogens. These microbial organisms derive from acutely sick persons, from known and often unknown carriers of pathogens in the population. The quantity of pathogenic microorganisms is also increased and their spectrum enlarged by other to the sewerage system connected sources like stockyards, slaughterhouses and related enterprises like butchers, meat industry and other sources from trade, industry, and agriculture.

It is well known that a certain percentage of the human population and the pet and farm animals are afflicted with infections. As it is nowadays possible to isolate pathogens from raw sewage without difficulty, it seems almost idle to speculate about the percentage of infected individuals. A wealth of factors plays a role which is not exactly ascertainable. Therefore, the estimates vary considerably within a factor of 10 (0.5–5% of the population).

Depending on the type of infection these individuals excrete the pathogens via feces and/or urine, secretions from nose, pharynx, vagina, mucous membranes or parts of skin and mucosa. Thus they reach the sewers and the sewage treatment plants. Therefore, it is understandable that at their collection of epidemiologic data some authors consider the results of microbiological investigations of sewage samples as reflection of the epidemiological situation whithin the population of certain drainage areas (STRAUCH, 1983). These pathogens present in sewage may be divided into 6 groups: bacteria, viruses, protozoa, helminths, yeasts, and other fungi, which are listed in Tabs. 1–4. A very comprehensive survey on the problems of sanitation and disease, including all health aspects of excreta and wastewater management can be found in the literature (FEACHEM et al., 1983; MARA and CAIRNCROSS, 1989).

It cannot be accepted for reasons of public health that these pathogens are washed into surface waters without any treatment of the sewage. Therefore, in many countries large parts of the municipal sewage are very intensively cleaned by mechanical, biological, chemical, and physical methods in sewage treatment plants (see Chapters 9–11, Volume 11a of *Biotechnology*).

For a long time it was thought that by the standard sewage treatment methods the pathogens would be eliminated from the sewage. But it has been shown that this is not correct. The influence of these purification technologies on pathogens is discussed in detail by STRAUCH (1983). Summarizing the present knowledge of the effects of the various steps in a sewage treatment plant on the pathogens in the wastewater, it can be said that most pathogenic agents survive the treatment processes of wastewater, but numerically reduced. Some of them are adsorbed by or enclosed in fecal particles and reach the formed sludge during the various sedimentation or filtration processes. Therefore, sewage sludge is rightly described as a concentrate of pathogens (STRAUCH, 1998).

The conclusion of this knowledge is that wastewater and sludge must be sanitized before the effluents can be discharged into the receiving waters or the sludge is to be utilized in agriculture, horticulture or forestry.

Tab. 1. Bacterial Pathogens to be Expected in Sewage and Sewage Sludge (STRAUCH, 1991)

Primary Pathogenic Bacteria	Secondary Pathogenic Bacteria
Salmonella spp.	*Escherichia*
Shigella spp.	*Klebsiella*
Escherichia coli	*Enterobacter*
Pseudomonas aeruginosa	*Serratia*
Yersinia enterocolitica	*Citrobacter*
Clostridium perfringens	*Proteus*
Clostridium botulinum	*Providencia*
Bacillus anthracis	
Listeria monocytogenes	
Vibrio cholerae	
Mycobacterium spp.	
Leptospira spp.	
Campylobacter spp.	
Staphylococcus	
Streptococcus	

Tab. 2. Viruses Excreted by Humans which can be Expected in Sewage and Sewage Sludge (STRAUCH, 1991; HURST, 1989)

Virus Group	Number of Types	Diseases or Symptoms Caused
Enterovirus		
Polio virus	3	poliomyelitis, meningitis, fever
Coxsackie virus A	24	herpangina, respiratory disease, meningitis, fever
Coxsackie virus B	6	myocarditis, congenital heart anomalies, meningitis, respiratory disease, pleurodynia, rash, fever
Echovirus	34	meningitis, respiratory disease, rash, diarrhea, fever
New "numbered" enteroviruses	4	meningitis, encephalitis, respiratory disease, acute hemorrhagic conjunctivitis, fever
Adenovirus	41	respiratory disease, eye infections
Reovirus	3	not clearly established
Hepatitis A virus	1	infectious hepatitis
Rotavirus	4	vomiting and diarrhea
Astrovirus	5	gastroenteritis
Calicivirus	2	vomiting and diarrhea
Coronavirus	1	common cold
Norwalk agents	1	vomiting and diarrhea
Small round viruses	2	vomiting and diarrhea
Adeno-associated virus	4	not clearly established, but associated with respiratory disease in children
Parvovirus	2	one type possibly associated with enteric infection

Tab. 3. Parasites to be Expected in Sewage and Sewage Sludge (STRAUCH, 1991)

Protozoa	Cestodes	Nematodes
Entamoeba histolytica	*Taenia sagginata*	*Ascaris lumbricoides*
Giardia lamblia	*Taenia solium*	*Ancylostoma duodenale*
Toxoplasma gondii	*Diphyllobothrium latum*	*Toxocara canis*
Sarcocystis spp.	*Echinococcus granulosus*	*Toxocara cati*
		Trichuris trichiura

Tab. 4. Pathogenic Yeasts and Fungi to be Expected in Sewage and Sewage Sludge (STRAUCH, 1991)

Yeasts	Fungi
Candida albicans	*Aspergillus* spp.
Candida krusei	*Aspergillus fumigatus*
Candida tropicalis	*Phialophora richardsii*
Candida guillermondii	*Geotrichum candidum*
Crytococcus neoformans	*Trichophyton* spp.
Trichosporon	*Epidermophyton* spp.

Wastewater is often reused in developing countries. In the last years a discussion came up, that hygienic standards applied to waste reuse in the past, based solely on potential pathogen survival, have been stricter than necessary. A meeting of sanitary engineers, epidemiologists and social scientists, convened by the World Health Organization (WHO), the World Bank and the International Reference Center for Waste Disposal, held in Switzerland in 1985, "proposed a more realistic approach to the use of treated wastewater and excreta, based on the best and most recent evidence" (MARA and CAIRNCROSS, 1989). The guidelines proposed by this group for the safe use of wastewater and excreta in agriculture and aquaculture, which were obviously adapted to the environmental conditions of the moment in developing countries of the tropics and subtropics, may be right for a certain period in these areas. But they should not be transferred to industrialized countries.

This is shown by a "memorandum on the threat posed by infectious diseases", which was published by 10 German scientific societies dealing with microbiology, hygiene, infectiology, hospital hygiene, environmental medicine, and public health (Rudolf Schülke Foundation, 1996). They pointed out that in the first half of this century, due to improved hygienic conditions and to progress made in diagnosis, therapy and immunoprophylaxis, successes which hitherto had been unknown were scored in the prevention, diagnosis and control of infectious diseases. Therefore, one succumbed to the illusion that the infectiologic situation could be controlled at any time, by virtue of the superior innovative abilities of modern society. Deceptive complacency and ignorance increasingly governed attitudes towards the problem of communicable diseases.

In the meantime new pathogens have emerged worldwide, also in the developed countries; old pathogens which were believed to be under control appear on the scene again; resistance even to newly developed antibiotics is rapidly increasing. In October 1992 a document was published with the title *Emerging Infections – Microbial Threats to Health in the United States* (LEDERBERG and SHOPE, 1992). It created a great stir. The authors, who are members of the prestigous Institute of Medicine of the US National Academy of Science, declared unequivocally that the presumed situation was tantamount to a fatal misconception of global dimensions. This report produced a similar effect in expert circles as the publication of the Club of Rome "*The Limits of Growth*" in 1972. The tenor of the report is that infectious diseases will present in the future one of the greatest hazards to mankind and that the extent of the impending menace can only be kept under control by a continuous global and concerted action by both the developing and the developed countries (Club of Rome, 1972).

The 10 German scientific societies stressed, that the report of LEDERBERG and SHOPE in 1992 conjured up for the public images of a scientifically corroborated scenario. From the perspective of global interactions, continent-related situation, national and local salient features not only in USA, but in all industrialized countries, this heralded the beginning of a – albeit tentative – reappraisal of the situation (Rudolf Schülke Foundation, 1996).

Faced with this scenario it is necessary to do everything possible to prevent that pathogens are freely distributed by discharging wastewater into receiving waters and that sewage sludges in agriculture and horticulture are used without a preceding sanitation by disinfection or other effective measures.

The more intensified treatment of wastewaters is, the more rapidly. The amount of the concentrated end product, sewage sludge, is growing. In the USA an alternate term for sludge was coined some years ago by the Water Environment Federation: *biosolids*. This was thought to have a more pleasant connotation than sludge and to emphasize the potential of the sludge as safe, recyclable, natural fertilizer and soil conditioner.

In Europe the European Water Pollution Control Association (EWPCA) used a questionnaire, which was sent to their national member organizations, to find out what they consider as their future tasks, scope, and activities in the field of sludge management. As a result of the answers the Editorial Committee of EWPCA decided to publish a special issue of its official publication "European Water Pollution Control". In that issue the opinions of the following countries are discussed: Belgium (OCKIER et al., 1997), France (BÉBIN, 1997), Germany (BERGS and LINDNER, 1997), Italy (RAGAZZI, 1997), Netherlands (DIRKZWAGER et al., 1997), UK (DAVIS and HALL, 1997), and USA (BASTIAN, 1997). There are papers with basic information on production, treatment, and disposal of sludge in the European Union (EU), they also describe national attidudes, preferred disposal or recycling options, environmental standards and future trends for both sludge disposal and treatment and the papers show the large differences in these fields between the countries (VAN DE KRAATS, 1997).

1.2 Disinfection of Wastewater

Disinfection technologies have the function to destroy or inactivate pathogens in wastewater. The nature of wastewater should not be changed unfavorably by disinfection processes. In some countries such as various States of the USA, as a rule, municipal wastewater is disinfected after biological treatment. In Germany, e.g., only wastewaters from special isolation wards in hospitals, tuberculosis sanatoriums, etc. must be disinfected at present. For treatment of wastewater from enterprises for genetical engineering and animal rendering plants (knacker's yards) special demands have to be fulfilled which are laid down in specific laws and ordinances.

The disinfection of biologically treated municipal wastewater should always be taken into account, if an increased risk exists for the transmission of infectious diseases or their causative microorganisms directly via drinking water or bathing waters and indirectly via contaminated water which is used for the production of food or irrigation.

In a special publication the German Wastewater Engineering Organization (ATV) has issued a paper on disinfection of wastewater as an instruction leaflet (ATV, 1997). In that instruction the disinfection of biologically treated wastewater by UV irradiation, ozonization, membrane filtration, and chlorination is described in detail.

Before wastewater is discharged into the receiving water as a permanent solution chemical disinfection is rejected because the utilization of additional chemicals in the wastewater could result in uncontrollable reactions and impairment of the receiving water. Further, the side effects of wastewater disinfection are discussed. Details are available in the above mentioned instruction (ATV, 1997).

Discussions are continuing on how far treatment of wastewater should go, how much it may cost and what is/are the good treatment method(s) for removing parasites and pathogens. For that reason the Editorial Committee of the European Water Pollution Control Association (EWPCA) decided to devote a special issue of its official publication "European Water Management" to parasites and pathogens (VAN DE KRAATS, 1998). In that issue the following topics are discussed: occurrence of pathogenic microorganisms in sludge and effects of various treatment processes for their removal (STRAUCH, 1998; SCHWARTZBROD et al., 1998), disinfection of secondary effluents from sewage treatment plants (POPP, 1998), effects of wastewater on the presence of salmonellae in coastal waters (HEINEMEYER and HELMUTH, 1998), removal of pathogens from wastewater effluents by microfiltration technology (DORAU, 1998), economic and political consequences of pathogens in inland and coastal waters (LOPEZ-PILA, 1998).

1.2.1 Sanitation of Sewage Sludge

As mentioned in Sect. 1.1 pathogens occur in sewage and are enriched in sludge, which raises public health concerns not only for humans but also for domestic animals and agricultural livestock. This has been proved in Switzerland, where fertilization of pastures with undisinfected sludge resulted in a spread of salmonellosis in dairy cows. In a very detailed study in 1983 the economic losses in one year caused by salmonellosis in the states of the Federal Republic of Germany were calculated for the sectors of human medicine to 108 million DM and for livestock in agriculture to another 132 million DM. The calculated expenses for salmonellosis control in the areas of man, food, animals, feed, and water were about 480 million DM (KRUG and REHM, 1983). For damage caused by other groups of bacteria, viruses (HURST, 1989), and parasites in sewage sludge no figures are available. Therefore, the agricultural use of hygienically dubious sewage sludge poses a risk of losses for the whole national economy which can only be prevented by sanitation measures such as disinfection of sludge before its use in agriculture, horticulture, and forestry.

1.2.2 Council Directive on Sewage Sludge

The Commission of the European Communities has issued in 1986 the "Council Directive 86/278/EEC *on the Protection of the Environment, and in Particular of the Soil, when Sew-*

age Sludge is Used in Agriculture". The used term "treated sludge" means sludge which has undergone biological, chemical or heat treatment, long-term storage or any other appropriate process to reduce significantly its fermentability and the health hazards resulting from its use (Article 2, Council Directive 86/278/EEC). Sludge shall be treated before being used in agriculture. Member states may nevertheless authorize, under conditions laid down by them, the use of untreated sludge if it is injected or worked into the soil (Article 6, Council Directive 86/278/EEC). The use of sludge is prohibited on grassland or forage crops if the grassland is to be grazed or the forage crop must be harvested before at least three weeks have elapsed. The use of sludge is prohibited on soil in which fruit and vegetable crops are growing, except fruit trees and on ground intended for the cultivation of fruit and vegetable crops which are normally in direct contact with the soil and are normally eaten raw, for a period of 10 months preceding the harvest of the crops and during the harvest itself (Article 7, Council Directive 86/278/EEC). These articles, which must be seen in the context of the occurrence of pathogens in the sludge, are formulated without much binding regulations.. However, Article 12 permits member states to take more stringent measures than those provided in the Council Directive 86/278/EEC. The problem of the occurrence of pathogens in sludge is not directly mentioned (EEC, 1986).

1.2.3 German Ordinances on Sewage Sludge

This first German ordinance came into force on April 1, 1983. The definition of sanitized sludge is given in § 2 (2): Sludge is sanitized, when it was treated by chemical or thermal conditioning, thermal drying, heating, composting, chemical stabilization, or another process in such a way that pathogenic microorganisms have been destroyed, or, based on its ascertainable origin does not need such a treatment (OSS, 1983).

On July 1, 1992 the second German ordinance came into force. In that ordinance the term "sanitized sludge" is not longer used. The application of sewage sludge on soils for agricultural or horticultural use is allowed only in such a way that public welfare is not impaired [§ 3 (1)] (OSS, 1992).

The German legislator made a serious mistake with his opinion that it is not longer necessary to sanitize the sewage sludge used in agriculture. As indicated above, there are many possibilities that pathogens are spread into the environment, even if the sludge is ploughed into the soil. Therefore, it must be demanded that all kinds of sewage sludge used for agricultural or horticultural purposes must be disinfected on the premises of the sewage treatment plants before they are delivered for any agricultural or horticultural use.

The first ordinance stated indirectly that the use of sewage sludge on pasture and forage land is not longer allowed if it is not "hygienically safe" (sanitized, disinfected). Since the term "hygienically safe" was not defined, a working group was established to elaborate such a definition and also to define the appropriate technologies to achieve "hygienically safe" sludge.

1.3 Recommendations of the Working Group "Disinfection of Sewage Sludge"

1.3.1 Definition

Sewage sludge is considered as sanitized when it was treated by a sanitation process for which by an appropriate investigation it has been proved, that

- the number of indigenous or seeded salmonellas has been reduced by at least 4 powers of 10 (logs) and
- indigenous or seeded eggs of ascaris are rendered noninfectious (system control).

Furthermore, the sanitation process must result in a sewage sludge in which directly after the treatment 1 g contains

- no salmonellae
- not more than 1,000 enterobacteriaceae

can be detected (process control).

The identification of salmonellae, enterobacteriaceae, and *Ascaris* eggs has to be done according to implementation instructions, which are published in ATV-VKS (1988b).

Furthermore, a constant "operational control" must be carried out. In addition to the system and process control the operational control shall supervise that the process conditions are continually met which are necessary for the sanitation of sludge by the relevant process. The "operational control" is performed by continuous and controllable registration of the process conditions which are representative for each system with regard to the process aim "sanitation". These conditions have to be laid down system-specific.

1.3.2 Definition of Sludge Disinfection Processes

The whole authentic text of the recommendations would need too much space. Therefore, only the definitions of the "process description" are given. The details of "process control" and "operational control" must be looked up in the original publications (ATV-VKS, 1988a, b in German; STRAUCH, 1991, 1998 in English).

1.3.3 Sludge Pasteurization (Prepasteurization)

During pasteurization raw sewage sludge is heated to temperatures below 100 °C, but at least to 65 °C, for at least 30 min by administration of heat. This is done prior to a stabilization process as so-called prepasteurization. A comminution of larger particles is necessary prior to the pasteurization process. To ensure that all sludge particles are exposed to the reaction temperature and time their size may not exceed 5 mm.

Other temperature–time combinations can also be used as, e.g.,

- 70 °C, 25 min,
- 75 °C, 20 min,
- 80 °C, 10 min.

Even at still higher temperatures a reaction time lower than 10 min is not allowed.

1.3.4 Aerobic Thermophilic Stabilization of Sludge (ATS)

In the course of the ATS process caused by air (oxygen) supply, exothermal microbial degradation and metabolic processes result in a rise of the temperature and an increase of the pH up to values about 8. Provided that the reaction vessel has a good insulation, the air supply is correctly calculated and the sludge has a sufficient concentration of organic dry matter, temperatures can be reached in the ATS process, which besides stabilization also ensure sanitation of the sludge.

ATS processes should be operated at least as two-stage reactors (2 vessels connected in series) to avoid the microbiological disadvantages of hydraulic short circuits. Detention times in the complete system of at least 5 d are to be calculated when the reactors have the same volume.

Considering the batch type operation (e.g., 1 h feeding per day) and 23 h stabilization (reaction time) and the temporary decrease of temperature inevitably connected with this type of operation the following reaction times and temperatures are necessary:

- 23 h at 50 °C or
- 10 h at 55 °C or
- 4 h at 60 °C.

1.3.5 Aerobic Thermophilic Stabilization of Sludge (ATS) with Subsequent Anaerobic Digestion

In the aerobic thermophilic first stage the sludge is sanitized. The sanitation is ensured by sufficiently high temperatures in the sludge which can be produced by additional heating with extraneous energy and exothermic microbial processes during the partial stabilization. The subsequent anaerobic mesophilic or thermophilic stabilization ensures the necessary security of the sanitation process.

After treatment in this two-stage process the sludge is considered sanitized if in the first stage either the conditions of prepasteurization have been fulfilled or if the reaction temperature of at least 60 °C has continuously been kept for at least 4 h. During these 4 h no raw sludge may be added. In the second (anaerobic) stage a process temperature of at least 30 °C must be kept.

1.3.6 Treatment of Sludge with Lime as $Ca(OH)_2$ (Lime Hydrate, Slaked Lime)

$Ca(OH)_2$ (calcium hydroxide, lime hydrate, slaked lime) is used for sanitation of liquid sludge before its use or for conditioning of the sludge before dewatering. In both cases the addition of lime results in an increase of the pH as a function of the amount of lime added and the properties of the sludge. The wet addition of lime as lime milk (Tab. 5) should be preferred compared to lime powder because of the better mixing and sanitizing effect.

The initial pH of the lime–sludge mixture must be 12.6 at least and the mixture must be stored for at least 3 months (reaction time) before use.

In particular cases the supervising authority must decide whether it is necessary to store the sludge for 3 months after liming to destroy *Ascaris* eggs. This may depend on their content in the sludge. In many cases it is possible to refrain from a 3 months storage because the numbers of worm eggs are negligible.

1.3.7 Treatment of Sludge with Lime as CaO (Quicklime, Unslaked Lime)

On addition of CaO to dewatered sludge the lime–sludge mixture heats up to temperatures to between 55 and 70 °C by exothermic reactions of the calcium oxide with still available water and when the insulation is sufficient. The initial pH of the lime–sludge mixture must reach 12.6 at least and the temperature of the whole mixture must be 55 °C at least during 2 h.

According to present knowledge liming with CaO also initiates high temperatures by exothermic reactions besides the high pH values mentioned above. The combination of high pH (>12) and high temperatures (between 60 and 70 °C) also destroys *Ascaris* eggs as well as reo-, polio, parvo- and ECBO-viruses (STRAUCH, 1984).

1.3.8 Composting of Sludge in Windrows

The sanitation of sludge by composting in windrows with bulking material (e.g., municipal refuse, straw, saw dust, wood shavings) is caused by the heat generated during composting by microbial processes. This temperature and the reaction time as well as microbial metabolic substances with antibiotic effects are important.

Tab. 5. Treatment of Sludge with Lime as $Ca(OH)_2$

Dry Matter [%]	Lime Milk [kg $Ca(OH)_2$ m^{-3}]	Remarks
1–3	6	homogeneous mixture of lime and sludge; pH 12.5
3–5	10	after liming; 24 h reaction
6	12	time at pH ≥12 before
7	14	delivery; for destruction
8	16	of *Ascaris* eggs a reaction
9	18	time of 3 months is necessary
10	20	

Sufficient aeration of the mixture of sludge and bulking material by technical means as, e.g., turning of the windrows or forced aeration of static windrows or piles (which are not turned) are required for the sanitizing effect. It must be ensured that the effective temperatures are present in each part of the composting material during the necessary reaction time.

The initial water content of the composting material must be 40–60% and the reaction temperature in the windrow must be 55 °C at least during 3 weeks.

1.3.9 Composting of Sludge in Reactors

The sanitation of sludge by composting in reactors (in-vessel composting) with bulking material (e.g., saw dust, wood shavings, bark, reflux material) is caused by the same factors as mentioned in Sect. 1.3.8.

Requirements for the sanitizing effect are the same as described in Sect. 1.3.8. The steadiness of desired temperature profiles in the reactors can be influenced and controlled by the techniques of aeration, filling, and emptying. As described in Sect. 1.3.8 it must be ensured that the effective temperatures are reached in each part of the composting material for the necessary reaction time.

The initial water content of the composting material should not exceed 70%. The complete mixture should be exposed to a temperature of at least 55 °C during a passage time through the reactor of at least 10 d. Besides, the composting material should not pass the "hot zone" with a temperature of at least 65 °C faster than 48 h.

The reactor passage shall be followed by a phase of curing the material in windrows or piles of at least 2 weeks with at least one turning or in a second reactor in which the necessary security of the sanitation process is ensured by measurements.

The achievement of sanitation depends to a large extent on an undisturbed operation of the composting process. When disturbances of operations are connected with a decrease of temperatures the batch of compost either must be used as reflux material and thus pass the reactor for a second time or it must again be composted in a windrow under the conditions described for composting of sludge in windrows.

The following technologies for sanitation of sewage sludge are not contained in the report of the Working Group "Disinfection of Sewage Sludge" (ATV-VKS, 1988a, b). They are not discussed here in detail. Further information can be gained from original publications, mentioned in Sects. 1.3.10–1.3.14.

1.3.10 Thermal Drying

Thermal drying removes moisture from sludge by means of evaporation. Drying is a means of disinfection due to the high temperatures used. Three types of drying are used:

- convection dryers,
- contact dryers, and
- vacuum dryers.

Thermal drying eradicates all pathogens and is an exceptionally safe method of disinfection in this respect. However, the investment costs are rather high compared with other disinfection processes (HUBER et al., 1987).

1.3.11 Irradiation

Irradiation with γ rays from radioactive sources (Co 60, Cs 137) or with β rays, i.e., accelerated electrons, has been performed in a few treatment plants. However, due to the high costs and the disapproval of γ radiation by the public this method of disinfection could not be carried out on a larger scale, though quite a lot of research work was invested in these technologies (HUBER et al., 1987; IAEA, 1975; LESSEL, 1985).

In Sects. 1.3.12–1.3.14 some further processes are briefly mentioned. For them partial results of hygienic investigations are available. Perhaps they may find broader application in the future.

1.3.12 Long-Term Storage of Sewage Sludge

Wastewater treatment plants with sufficient storage capacity could sanitize anaerobically or aerobically stabilized sludge by long-term storage. This is confirmed by investigations of various authors (PIKE, 1983). Based on very intensive studies under practical conditions in our own laboratories the following recommendations are possible:

- the minimum long-term storage time of these conventionally treated sewage sludges must be at least 1 year,
- it is recommended that after 1 year the hygienic safety is proved by microbiological analyses (not more than 1,000 enterobacteriaceae and no salmonellae in 1 g of sludge (STRAUCH et al., 1987; HAIBLE, 1989).

1.3.13 Processing of Sewage Sludge in Reed Beds

Dewatering of sludge in reed beds has been used in some sewage treatment plants in Germany for several years. In that case problems of hygiene play an important role.

Taking into account the research results of various authors, the following recommendations are given:

- In reed beds dewatered sewage sludge must be kept in the beds for at least 15 months,
- when the beds are cleared of the sludge, it has to be stored for further 6 months before it can be considered sanitized,
- it is recommended that at the end of the storage period the hygienic safety is proved by microbiological investigations as described for long-term storage of sludge. More detailed results are available in the final report on these investigations (PHILIPP et al., 1987; PHILIPP, 1988).

In a recent publication on 10 years of sewage sludge refinement in reed beds (PAULY et al., 1997) the authors explain that the processing of large amounts of sludge in reed beds does not pose any technical problems and that it has become obvious that a high degree of economic efficiency can also be achieved in large sewage treatment plants. The method is considered to be environmentally friendly and the product ensures that the sludges can be utilized further. In a personal communication one of the authors has written that a safe sanitation of the sludges dewatered in reed beds is "process inherent" because the drying phase in the beds is 1 year, followed by a further storage period of another year. This is in agreement with our own results and recommendations to achieve sanitized sludge by the reed bed drying process (PAULY et al., 1997; VON BORKE, personal communication).

1.3.14 Joint Fermentation of Sewage Sludge and Biowaste in Wastewater Treatment Plants

This technology shall only be mentioned without great detail because the legal background of such procedures is not yet entirely clear, at least in Germany. But experts voice the opinion that this technology in existing sewage treatment plants will gain importance in connection with the provivions of the German Recycling and Wastes Act (BERGMANN, 1997; see Chapters 22–27 of Volume 11a of *Biotechnology* and the chapters on solid waste treatment of this volume).

The hygienic problems of that procedure are yet unknown and it must be considered that according to German legislation the final products will not be subject to the Ordinance on Biowaste (present draft) but to the requirements of the Ordinance on Sewage Sludge in which the utilization of such mixtures in agriculture, horticulture, and forestry is regulated more restrictively.

1.4 Conclusions

In many countries sludge from municipal sewage treatment plants is utilized in agriculture, horticulture, and sometimes also in forestry. It must be considered in such cases that these sludges are reflecting the hygienic situation of the population which is served by the treatment plants. Most of the pathogenic microorganisms are excreted by man and animals via feces, urine and/or sputum, and they reach the treatment plants via the sewers. The sewage treatment processes reduce the numbers of pathogens only insufficiently. Therefore, considerable amounts of pathogens still occur in the final product of sludge treatment utilized as fertilizer or soil improver.

According to the Council Directive on Sewage Sludge, the member states of the European Community have "to regulate the use of sewage sludge in agriculture in such a way as to prevent harmful effects on soil, vegetation, animals and man" (EEC, 1986). In the First German Ordinance on Sewage Sludge it was said that "sewage sludge is hygienically safe if it was treated by chemical or thermic conditioning, thermic drying, heating, composting, chemical stabilization or another process in such a way that pathogenic microorganisms are destroyed" (OSS, 1983). But in the Second German Ordinance on Sewage Sludge it is only said, that "sewage sludge is only permitted to be applied on soils for agricultural or horticultural use in such a way that the public welfare is not impaired" (OSS, 1992).

The EU definition as well as the two German definitions imply from the point of view of hygiene that in any case the aims declared by these legislators can only be achieved by an effective sanitation process (EEC, 1986; OSS, 1983, 1992). Therefore, a German Working Party has defined seven treatment processes which ensure a safe sanitation of sewage sludges prior to their utilization in agriculture or horticulture (ATV-VKS, 1988a, b). These treatments contain a "system control", "process control" and an "operational control". When these rules are observed the treated sludge is sanitized in accordance with the definitions. Besides these seven defined sanitation processes two further technologies were established which also can achieve sanitation, but did not find wide use due to reasons of environmental protection or economics.

In recent years some additional processes were introduced which also may achieve a sanitation in accordance with the described regulations. Three of those are shortly discussed, but it must be pointed out that the third one has not yet been investigated from the viewpoint of hygiene.

To protect the environment more effectively from microbial contamination all legislators should make efforts to specify the up to now very vague regulations for sewage sludge utilization in agriculture and horticulture in such a way that an effective sanitation is stipulated in any case. For that purpose the Danish regulations are recommended to the legislators as worth emulating (BENDIXEN and AMMENDRUP, 1992; BENDIXEN, 1997; STRAUCH, 1998).

A very valuable source of information is the final report of the activities of the Community Concerted Action Programme on the Treatment and Use of Sewage Sludge and Liquid Agricultural Wastes. This program was financed by the Commission of the European Communities (CEC) and was effective from 1972–1990. Five Working Parties (WP) were established

- WP 1: Processing of Organic Sludges and Liquid Agricultural Wastes;
- WP 2: Chemical Contamination of Sludge and Soils;
- WP 3: Hygienic Aspects Related to Treatment and Use of Organic Sludge and Sanitary Aspects of Spreading of Slurries and Manures;
- WP 4: Agricultural Value of Organic Sludge and Liquid Agricultural Wastes;
- WP 5: Environmental Effects of Organic Sludge and Liquid Agricultural Wastes (HALL et al., 1992).

The contents of that report is a real treasurehouse for everybody interested in the problem of sewage sludge and liquid agricultural wastes. Besides the research results of the WPs the report contains a discussion of the remaining problems and the overall state of the art as well as lists of all their publications during these 18 years.

In the USA the use and disposal of sewage sludge and domestic sewage are regulated in Chapter I of Title 40 of the Code of Federal Regulations (CFR), Part 503. Since these regulations are very extensive it was necessary that the US Environmental Protection Agency

(EPA) issue three voluminous publications to explain the rather complicated rules for all institutions which are concerned (EPA 1992, 1994, 1995). Details are not discussed here because this would go too far. But it shall be mentioned that these publications contain very detailed directions for reducing pathogens in biosolids as well as options for reducing the potential for biosolids to attract vectors. In a subpart D of Part 503 alternatives concern the designation of biosolids (sludge from sewage and domestic sewage) as Class A or Class B in regard to pathogens. Class A contains processes to further reduce pathogens (PFRP: composting, heat drying, heat treatment, thermophilic aerobic digestion, β-ray irradiation, γ-ray irradiation, pasteurization). Class B are processes to significantly reduce pathogens (PSRP: aerobic digestion, air drying, anaerobic digestion, composting, lime stabilization).

For further reading the following publications are recommended:

Treatment and use of sewage sludge (ALEXANDRE and OTT, 1980); water pollution control technologies for the 1980s (SCHMIDTKE and EBERLE, 1979); characterization, treatment and use of sewage sludge (L'HERMITE and OTT, 1981); disinfection of sewage sludge: technical, economical and microbiological aspects (BRUCE et al., 1983), processing and use of sewage sludge (L'HERMITE and OTT, 1984); inactivation of microorganisms in sewage sludge by stabilization processes (STRAUCH et al., 1985); processing and use of organic sludge and liquid agricultural wastes (L'HERMITE, 1986); epidemiological studies of risks associated with agricultural use of sewage sludge: knowledge and needs (BLOCK et al., 1986); sewage sludge treatment and use: new developments, technological aspects and environmental effects (DIRKZWAGER and L'HERMITE, 1989); treatment and use of sewage sludge and liquid agricultural wastes (L'HERMITE, 1991); waste managemant (BILITEWSKI et al., 1997); pollution science (PEPPER et al., 1996).

2 Disposal of Wastewater in Rural Areas

2.1 Introduction

The aim of wastewater disposal is to return the used water into the natural water circulation, combined with greatest possible conservation of resources. Therefore, in rural areas the natural water cycle and the circulation of nutrients on agricultural land can be closed as far as sufficient installations for sewage treatment are available (SCHULZE, 1994).

In Germany about 95% of the population are connected to biological sewage treatment plants. But deficiencies still exist in sparsely populated rural areas. The municipalities are obliged to dispose of their wastewater in such a way that the public well-being is not impaired. To carry out their obligations the local authorities have to collect the wastewater to supply it to a treatment plant. The treated wastewater must be disposed of according to the legal requirements.

To ensure that the state of the technological development of today is considered, cost saving solutions are looked for which allow a disposal of wastewater in rural areas, which gives no cause for concern under public health considerations.

It is a proven fact that pathogenic microorganisms are occurring in sewage and sewage sludge (see Sect. 1.1, Tabs. 1–4). As is shown in Sect. 1.1, parts of the pathogens survive the sewage treatment processes and are discharged with the effluent into the receiving waters, where by self-purification a certain further reduction of the pathogens is achieved. But in rural areas this is very often different. Sometimes only a small receiving water is available or the purified wastewater is percolated in particular cases on agricultural fields with permission of the relevant authority. This causes an additional hygienic risk since in these small receiving waters the diluting effect and the self-purification are smaller than in larger receiving waters.

Since the technology of wastewater treatment has been clearly improved good purification efficiency is also possible in small treatment plants. The result is that the decentralized wastewater treatment, which until now was only considered as a temporary solution in the German Federal State of Baden-Württemberg is now considered as a permanent solution on condition that the effluent can be discharged into a suitable receiving water or – in particular cases – with special permission be used for seepage on agricultural fields. It is not allowed to discharge rain water into small sewage treatment plants (DIETRICH, 1996).

2.2 Demands on Farms for Decentralized Wastewater Disposal

Only solitary farms are allowed to dispose of the wastewater from their leak-proof cesspool or small sewage treatment plant on their self-managed agricultural fields. Prerequisite is that a minimum of 0.5 ha arable land is available in addition to the area which is necessary for the utilization of the animal manure. Besides, the soil layer must be so thick that no direct seepage into the underground can occur. Furthermore, the wastewater must immediately be incorporated into the soil after it has been spread on the field. In other cases the agricultural utilization of wastewater is not acceptable due to concerns about the hygienic threat to groundwater and the soil in general.

Application of wastewater on grassland is only permissible when it had been treated before in a three-compartment septic tank or another equal pretreatment such as, e.g., passage through a biogas plant. After disposal on grassland it is forbidden to use the grassland as pasture or the grass for green forage for at least 8 weeks. Haymaking and production of silage are not included in that restriction. These more stringent measures for the utilization of wastewater on grassland compared to plowland are objectively based on a higher hygienic and epidemiological risk.

2.3 Discharge of Fecal Wastewater into Urine or Slurry Pits

According to the previous regulations fecal wastewater had to be treated first in a small wastewater treatment plant and only the overflow could be stored together with urine or slurry. In the new administrative instruction of the German Federal State Baden-Württemberg (VwV, 1997) it is prescribed that in case of disposal of wastewater into urine or slurry pits or storage tanks an additional capacity of 5 m^3 for fecal wastewater and 10 m^3 for other domestic wastewaters (kitchen, bathing, laundry) must be available per capita of occupant.

For storage and utilization the following situations must be distinguished in principle

- domestic wastewater can directly be discharged into the urine or slurry pit without addition of fecal wastewater,
- when domestic together with fecal wastewater is discharged into the urine or slurry pit, the contents can only be utilized on plowland,
- in case that no or too little plowland is available (minimum of 0.5 ha, see also Sect. 2.2) the utilization of domestic together with fecal wastewater on grassland is only allowed when the mixture has previously been treated in a three-compartment septic tank or another equal pretreatment like passage through a biogas plant.

2.4 Integration of Agricultural Biogas Plants as a Pretreatment Equal to a Three-Compartment Septic Tank

The utilization of domestic wastewater together with animal slurry in agricultural biogas plants is possible as long as the substrate is not too much diluted. This method is also environmentally friendly (climate protection by reduced CO_2 emission and utilization of methane, preservation of resources, better fertilization properties of biogas slurry) and can be economical: the wastewater is co-utilized for heat, electric current and fertilizer. There is no accumulation of sludge to be transported and disposed of in sewage treatment plants, no energy requirement for the biogas production, no necessity to build long sewers to central wastewater treatment facilities, a low requirement for space and discharge of effluent into receiving waters is unnecessary (UM-Arbeitskreis, 1997).

The utilization of discharge from cess pits and small sewage treatment plants into receiving waters causes serious concerns whereas the digestion process in biogas plants with its "biocidal potency" minimizes the sanitary risk. A demonstrable reduction of germs occurs during mesophilic digestion (35 °C). It could be shown that a temperature stable serovar of *Salmonella senftenberg* (strain W_{775}) could not be isolated after 7 d of anaerobic treatment in biogas plants.

If foreign wastewater e.g., from restaurants and/or conference venues will be utilized in mesophilic biogas plants, it is stipulated that these wastewaters are prepasteurized for 1 h at 70 °C before they are introduced into the biogas plant. This is not necessary, if it is guaranteed that for each single particle in the biogas reactor a reduction time of 20 d is kept.

If the biogas plants are operated in the temperature range of >55 °C prepasteurization is not required provided the reduction time for each particle of 24 h in the biogas plant is guaranteed (UM-Arbeitskreis, 1997).

From a hygienic point of view the cofermentation of wastewater in an agricultural biogas plant is to be preferred to any other treatment. Thus the cofermentation of domestic wastewater from farms in a biogas plant represents an economically and ecologically interesting procedure that is safe from the hygienic point of view.

2.5 Aspects of Hygiene and Environment

Besides fecal germs biogenic wastes can also contain pathogens such as salmonellas and other bacteria as well as viruses. In connection with cofermentation different serovars of salmonellae, the viruses of hog cholera, and Aujeszky's disease are highly relevant from a hygienic and epidemiological point of view. Their causative organisms can be found in wastes, especially food wastes when pigs are slaughtered during the incubation period (before they develop clinical signs of the disease) and parts of raw meat and bones are present in wastes (BECKER, 1987; BLAHA, 1988; see also Sect. 3).

Since cofermentation is mainly carried out in farms with cloven-footed animals (cattle, pigs) an absolutely effective inactivation of the pathogens must be ensured in biogas plants to avoid hygienic risks like spreading of infectious diseases in the surroundings of a farm. Therefore, research work was done in bench-scale anaerobic reactors to investigate the influence of mesophilic (35 °C) and thermophilic (55 °C) temperatures of the slurry on the survival of fecal streptococci, salmonellas and *Escherichia coli*. These preliminary investigations were followed by bacteriologic controls in the biogas plant of a farm (GRUNWALD, 1995). Further investigations in large scale biogas plants with and without cofermentation followed (BÖHM et al., 1998).

Tab. 6 shows a survey on the decimal reduction rate of indicator bacteria and various pathogens in a slurry during anaerobic treatment and storage in slurry tanks. Tab. 7 gives an overview of the inactivation times for animal viruses in a slurry at various storage and digestion temperatures (BENDIXEN and AMMENDRUP, 1992). With rising temperatures during storage of slurry as well as in biogas plants the survival times of the investigated bacteria and viruses become shorter. In the mesophilic range (35 °C) the survival time varied from days to several weeks. At 55 °C the survival times, depending on the pathogens tested, varied from a few minutes or hours and in an exeptional case (porcine parvovirus) up to several days. Since parvovirus is extremely thermoresistant a Danish group of researchers is recommending that in stipulating sanitation demands for biogas reactors it seems more reasonable to use a less resistant virus, such as a reo- or picornavirus, which better represent the pathogenic animal viruses (LUND et al., 1996).

New results on the behavior of fecal streptococci and salmonellas in biogas plants with different modes of operation, size and construction and anaerobic treatment of varying organic substrates were published recently (BENDIXEN, 1997). For a period of 2–3 years 10 large biogas plants in Denmark were systematically investigated with regard to the inactivation of fecal streptococci (FS) as indicator or-

Tab. 6. Decimation Times (T-90 Average) for some Pathogenic Bacteria and Indicator Bacteria in Slurry with Biogas Digestion and Conventional Storage (BENDIXEN and AMMENDRUP, 1992)

Bacteria	Biogas System		Slurry System	
	53 °C T-90 [h]	35 °C T-90 [d]	18–21 °C T-90 [w]	6–15 °C T-90 [w]
Salmonella typhimurium	0.7	2.4	2.0	5.9
Salmonella dublin	0.6	2.1		
Escherichia coli	0.4	1.8	2.0	8.8
Clostridium perfringens type C	ND	ND	ND	ND
Bacillus cereus	ND	ND		
Erysipelothrix rhusiopathiae	1.2	1.8		
Staphylococcus aureus	0.5	0.9	0.9	7.1
Mycobacterium paratuberculosis	0.7	6.0		
Coliform bacteria		3.1	2.1	9.3
Group of D streptococci		7.1	5.7	21.4
Streptococcus faecalis	1.0	2.0		

ND: not determined

Tab. 7. Inactivation Times for Animal Viruses in Slurry at Various Storage and Digestion Temperatures (BENDIXEN and AMMENDRUP, 1992)

Virus Manure Type	5 °C	20 °C	35 °C	40 °C	45 °C	50 °C	55 °C
Swine influenza							
Pig slurry	9 w	2 w	>24 h	>24 h	n.p.	>2.5 h	1 h
Porcine parvovirus							
Pig slurry	>40 w[a]	>40 w[a]	21 w	9 w	>19 d[a]	5 d	8 d
Bovine virus diarrhea							
Cattle slurry	3 w	3 d	3 h	50 min	20 min	5 min	5 min
Infectious bovine rhinotracheitis							
Cattle slurry	>4 w[a]	2 d	24 h	3 h	1.5 h[b]	40 min	10 min
Aujeszky's disease							
Pig slurry	15 w	2 w	5 h	2 h	45 min	20 min	10 min
Foot-and-mouth disease, pigs							
Pig slurry	>14 w	2 w	24 h	10 h	5 h	1 h	1 h
Foot-and-mouth disease, cattle							
Cattle slurry	n.p.	5 w	>24 h[a]	n.p.	n.p.	n.p.	>60 min[a]
Classical swine fever							
Pig slurry	>6 w[a]	2 w	4 h	>3 h[a]	>3 h[a]	immediately	immediately
TGE in pigs							
Pig slurry	>8 w[a]	2 w	24 h	>5 h[a]	2.5 h	1 h	30 min

[a] time of complete inactivation not reached.
[b] infectious virus verified through inoculation tests on calves. After 2.5 h of digestion at 45 °C similar inoculation tests on calves came out negatively.
n.p. not performed.
TGE transmissible gastroenteritis.

ganisms. An adequate pathogen reducing effect could be achieved in the digestion and sanitation tanks of the biogas plants, provided they were operated correctly and respected the criteria of the official requirements. The FS method may be used to check the sanitation effect achieved by the treatment in a tank. The effect is expressed numerically by the $\log_{10}$ reduction of the numbers of FS measured in the biomass before and after treatment. It was demonstrated that properly directed and well-functioning thermophilic digestion tanks ensure the removal of most pathogenic microorganisms from organic waste and slurry. The removal of pathogens by the treatment in mesophilic digestion tanks is incomplete. Systematic studies of the inactivation of bacteria and viruses in slurry and in animal tissues gave evidence that the pathogen reducing effect is enhanced in the microbial environment of thermophilic digestion tanks. The sanitation criteria are specified in Tab. 8, e.g., as combinations of temperature and time for the processing of biomass in digestion tanks.

It remains to be recorded that not only in thermophilic but also in mesophilic biogas plants a reduction of germs can be observed. Besides the measurable parameters "temperature" and "time" of persistence of the germs further "biocidal effects" like pH, ammonia content, free fatty acids, organic dry substance, competing microflora which exerts its influence on the survivability of the inserted pathogens limit the numbers of bacterial and viral pathogens especially in the mesophilic temperature range.

Despite the measured partially good reduction rates of the investigated microorganisms it is advisable that each authority which has to issue the notice of approval for biogas plants demand that the biogenic material which is cofermented with slurry in mesophilic plants must be sanitized by heat treatment at 70 °C for at least 1 h before it is added to the slurry.

In Germany an ordinance is in preparation which contains the regulation that in biogas plants with cofermentation of biogenic material other than agricultural slurry an official pro-

Tab. 8. Controlled Sanitation Equivalent to 70 °C in 1 h (BENDIXEN, 1997)

Temperature [°C]	Minimum Guaranteed Retention Time (MGRT) in a Thermophilic Digestion Tank[a]	Retention Time (MGRT) by Treatment in a Seperate Sanitation Tank[b]	
	[h]	Before or after Digestion in a Thermophilic Reactor Tank[c] [h]	Before or after Digestion in a Mesophilic Reactor Tank[d] [h]
52.0	10		
53.5	8		
55.0	6	5.5	7.5
60.0		2.5	3.5
65.0		1.0	1.5

The treatment should be carried out in a digestion tank at a thermophilic temperature or in a sanitation tank combined with digestion in a thermophilic or a mesophilic tank. The specific temperature – MGRT combinations should be respected.

[a] Thermophilic digestion is defined here as a treatment at 52 °C or more. The hydraulic retention time (HRT) in the digestion tank must be at least 7 d.

[b] Digestion may take place either before or after sanitation.

[c] The thermophilic digestion temperature must be at least 52 °C. The hydraulic retention time (HRT) must be at least 7 d.

[d] In this connection the mesophilic digestion temperature must be in the range of 20–52 °C. The hydraulic retenton time (HRT) must be at least 14 d.

cess and product control must also be carried out. With these controls the sanitary effects shall be supervised, not only those of anaerobic digestion on microorganisms pathogenic for man and animals, but also on phytopathogenic microbes (see Sect. 3.2).

According to German veterinary legislation in farms keeping cloven-footed animals the installations for the sanitation of food wastes for feeding animals must be operated in a special building which is located at a sufficient distance from the farm. The same procedure should be used for all biogenic wastes from households and other sources which are to be digested together with animal slurries.

If slurry is digested in biogas plants alone or together with biogenic wastes only positive effects are to be expected, apart from ammonia emissions (Tab. 9). With regard to the additional energy production the cofermentation is to be considered the method of choice for an environmentally friendly treatment of slurry (KTBL, 1998).

3 Hygienic Considerations on Biowaste, Composting and Occupational Health Risks

3.1 Introduction

The origin and nature of organic wastes always cause a hygienic risk in storage, collection, handling, processing, and utilization. Those risks are existing either if the organic wastes are collected and processed source separated (biowastes) or if they are collected together with other wastes from households or processing industries. Most organic materials collected from households or processing industries may contain pathogenic microorganisms which can affect the health of man, animals or plants as well as seeds or parts of plants which will cause undesired side-effects in agricultural or horticultural use. Hygienic principles must be followed in collection, transport, processing, storage and distribution of raw materials as well as of the final product. Tab. 10 gives a survey on the hygienic relevance of some organic wastes originating from households.

Tab. 9. Problems of Slurry Management and Effects of Anaerobic Digestion

Problems of Slurry	Effects of Anaerobic Treatment
Odor	degradation of odorous substances
Demulsification	more proportionate structure and smaller size of particles
	narrowing of the C/N ratio, increase in mineral nitrogen, faster availability for plants
Pathogens	inactivation by thermophilic digestion ($\geq 55\,^{\circ}C$) or preheating for 1 h at $70\,^{\circ}C$
Quality of pasture	avoidance of quality losses
Degassing of ammonia	indifferent effect; increase by improper storage
Release of methane and nitrous oxide	reduced release of methane and nitrous oxide

Tab. 10. Hygienic Relevance of Different Biological Wastes Originating from Households

Type of Waste	A	B
Meat leftovers (raw or insufficiently heated)		
– Meat cuttings, tendons, rinds, etc.	+	–
Food of animal origin		
– Egg shells	+	–
– Several meat and dairy products	+	–
– Raw milk products		
– Leftovers from fish and shellfish	+	–
Other wastes (animal and man)		
– Dirty packing material for meat and products of animal origin	+	–
– Used litter and wastes from pets	+	+
– Used paper handkerchiefs and sanitary pads	+	–
– Diapers	+	–
Household wastes from		
– Potatoes	–	+
– Carrots	–	+
– Onions	–	+
– Tomatoes	–	+
– Cucumbers	–	+
– Salad	–	+
– Cabbage	–	+
– Beans	–	+
– Cut flowers	–	+
– Balcony and indoor plants	–	+
Garden wastes		
– Boughs and plant material	–	+
– Fruits	–	+
– Dead leaves and lawn trimmings (fecal contamination)	(+)	+
Other wastes (plant origin)		
– Paper	–	–
– Paperboard	–	–
– Organic packing material (e.g., wood wool)	–	–

A May contain pathogens of man and animal.
B May contain plant pathogens and/or weed seeds.

3.2 Hygienic Risks

Three main types of risks mainly related to pathogens for man and animals have to be considered in collection and processing of organic wastes (BÖHM et al., 1996; BÖHM, 1995b):

- occupational health risks,
- risks concerning product safety, and
- environmental risks.

Occupational health considerations in collecting and processing organic wastes are not the subject of this contribution and have been treated by BÖHM et al. (1996) and GRÜNER (1996). Hygienic risks due to the product compost itself will mainly be regarded here. This includes the direct transmission of pathogens to man or plants and animals of agricultural importance as well as introducing them into the biocenosis and environment by the application of compost.

A compilation of bacterial, fungal, parasitic, and viral pathogens for man, animals and plants which may be present in organic wastes is shown in Tabs. 11–13. Since it is impossible to supervise the product compost for each of the pathogenic agents which may occur other strategies have to be used in order to assure the hygienic safety of the processed material. The first step in such a strategy is to find out a representative indicator organism which may be used for checking the product for hygienic safety as well as for evaluating the composting process for its capability to inactivate pathogens which are of epidemiological relevance. The second step which is necessary in this connection is to define hygienic requirements for the composting process itself, since due to the high volume of the product to be controlled as well as to the inhomogeneity of the distribution of pathogens in the material only compost processed in a validated process should be distributed to the consumer. Therefore, the following two steps are necessary to assure hygienic safety of the product compost:

- hygienic validation of the composting process,
- investigation of the final product for the presence of representative indicator organisms.

3.3 Process Validation

The validation of the composting process with respect to hygienic safety for animals, man, and plants may be done in several ways. The German LAGA M10 (Länderarbeitsgemeinschaft Abfall, 1995) offers a relatively broad approach in solving this problem.

Process safety concerning the inactivation of relevant transmissible agents for man and animals is validated in two steps. The first step is the validation of the process as designed by the producer of the technical equipment in a basic procedure, the second step is a bringing into service validation of a composting plant with the input material under practical conditions. In both validation procedures *Salmonella senftenberg* W 775 (H_2S negative) is used as test organism exposed in specially designed test carriers (RAPP, 1995; BÖHM et al., 1998). The test organisms used with respect to phytohygienic safety are tobacco mosaic virus, *Plasmodiophora brassicae*, and seeds of *Lycopersicon lycopersicum* (L) breed St. Pierre (BRUNS et al. 1994; POLLMANN and STEINER, 1994). Testing is done twice, in summer and in winter. Concerning the phytohygienic validation the bringing into service procedure is repeated at least every 2 years as a consecutive validation.

This is a very complete and safe system. Due to economical considerations the system should be simplified and only a one-step procedure should be the aim, which must be the brought into service validation. A scheme how this validation could be organized taking into account the annual throughput of material in the plants is shown in Tab. 14 from the draft of the German Biological Wastes Ordinance. The validation with pathogens and seeds may be regarded as "direct process validation" and must be accompanied by continuous recording of measurable process data like temperature, pH value, humidity, etc., in order to detect deviations and disturbances of the process over

Tab. 11. A Survey on Obligatory and Facultative Pathogens for Man and Animals which had Been Isolated from Biological and Household Wastes

Bacteria	*Citrobacter* sp., *Clostridia* sp., *Enterobacter* sp., *Escherichia coli, Klebsiella* sp., *Proteus* sp., *Pseudomonas* sp., *Salmonella* sp., *Serratia* sp., *Staphylococcus* sp., *Streptococcus* sp., *Yersinia* sp.
Fungi	*Aspergillus* sp., e.g., *Aspergillus fumigatus*
Viruses	adenovirus, coxsackie virus, ECHO virus, enterovirus, hepatitis A virus, herpesvirus suis, paramyxovirus, parvovirus, pestivirus, poliomyelitis virus, reovirus

Tab. 12. A Selection of Plant Pathogenic Bacteria and Viruses (Menke, 1992) modified[a])

Pathogen	Susceptible Plant	Type
Xanthomonas campestris	white cabbage, turnip cabbage, swede cauliflower	B
Pseudomonas marginalis	salad, endive	B
Pseudomonas phaseolicola	beans	B
Pseudomonas lacrimans	cucumber	B
Pseudomonas tabaci	tobacco	B
Corynebacterium michiganense	tomato	B
Corynebacterium sepedonicum	potato	B
Erwinia phytophthora	potato; carrot	B
Erwinia amylovora	pomaceus fruits, flowers	B
Agrobacterium tumefaciens	various hosts	B
Potato virus Y	potato; tobacco; tomato	V
Potato virus X	potato, tomato, tobacco, paprica, eggplants	V
Aucuba virus	potato, tobacco, tomato	V
Tobacco ring spot virus	potato, tobacco, beans, cucumber	V
Rattle virus	potato, tobacco	V
Tobacco mosaic virus	tobacco, tomato, paprica	V
Tobacco necrosis virus	tobacco, beans	V
Horse bean mosaic virus	horse-beans, peas	V
Pea mosaic virus	peas, horse-beans	V
Bean mosaic virus	beans, runner beans	V
Yellow bean mosaic virus	beans, peas	V
Cauliflower mosaic virus	several cabbage species	V
Cucumber mosaic virus	cucumber, melon, pumpkin, spinach, peas, beans, salad, tomato, celery	V
Aucuba mosaic virus	cucumber, melon	V
Cabbage ring spot virus	cauliflower, white cabbage, horse raddish, spinach, tobacco, rhubarb, flowers	V
Lettuce mosaic virus	salad, endive	V
Beet mosaic virus	spinach, root beet, leaf beet, peas	V
Onion mosaic virus	onion, leek	V

B Bacteria.
V Viruses.
[a] More details about transmission and resistance may be obtained from the original paper, which also contains informations concerning parasitic nematodes and weed seeds.

the whole year, which may result in an insufficient microbicidal effect. The system of process validation has to be completed by a continuous supervision of the final product, at least twice a year.

3.4 Hygienic Safety of the Product

As mentioned above, the investigation of the final product compost in order to detect every pathogen which may be present in the material is impossible. Therefore, representative indicator organisms have to be determined from the point of view of human and animal health as well as for the purpose of safe plant breeding and production. Those indicator organisms must fulfill several requirements:

- they have to be present in the raw materials with a high probability,
- the transmission via compost must be a factor in epidemiology,
- the indicator should not be involved in the biotechnological process of composting itself,

Tab. 13. A Selection of Plant Pathogenic Fungi (MENKE, 1992, modified[a])

Pathogen	Susceptible Plant
Plasmodiophora brassicae	cabbage
Phoma apiicola	cabbage
Peronospora brassicae	celery
Peronospora spinaciae	spinach
Peronospora destructor	onion
Marssonina panattoniana	salad
Sclerotinia minor	salad
Botrytis cinerea	salad
Bremia lactucae	salad, endive
Cercospora beticola	turnip
Aphanomyces raphani	radish
Alternaria porri	carrot
Septoria apii	celery
Albugo tragoponis	scorconera
Albugo candida	horse raddish
Turburcinia cepulae	onion
Sclerotium cepivorum	onion
Botrytis allii	onion
Uromyces appendiculatus	bean
Mycosphaerella pinodes	peas
Ascochyta pinodella	peas
Erysiphe polygoni	peas
Cladosporium cucumernum	cucumber
Sclerotinia sclerotiorum	cucumber
Septoria lycopersici	tomato
Alternaria solani	tomato
Didymella lycopersici	tomato
Rhizoctonia solani	potato
Phytophthora infestans	potato, tomato
Synchytrium endobioticum	potato
Verticillium albo-atrum	potato

[a] More details concerning transmission and resistance may be taken from the original paper, which also contains informations concerning parasitic nematodes and weed seeds.

- the indicator should not be an organism which is generally present in soil and soil related materials, and
- the method for isolation and identification must be simple, definite and reliable if applied to a substrate with a complex microbiological matrix as compost.

With respect to public health and veterinary requirements several indicators and parameters are under discussion:

- *Salmonella* spp.,
- enterococci (streptococci of group E),
- *Staphylococcus aureus*,
- enterobacteriaceae,
- *Escherichia coli*,
- *Clostridium perfringens*,
- sulfite reducing clostridia,
- eggs of nematodes, and
- larvae of nematodes.

Compost is a product of a microbial degradation process but the knowledge about the microbiological ecology of compost and compost related materials is still limited. Therefore, the isolation and identification techniques common in clinical microbiology should not be used without careful validation in combination with the involved sample materials. The variety of species present in environmental and compost samples by far exceeds the limited number of species to be taken into account in secreta and excreta as well as in body fluids and the variability in species is high and not yet fully understood. Moreover, microbial parameters which are used in the field of water hygiene and food inspection are not applicable to substrates like compost because most of those indicators belong to the indigenous flora of agricultural soils (BÖHM, 1995a). If the limited reliability and applicability of methods adopted from clinical microbiology and water inspection for the intended field of use is taken into account as well as the fact that the exclusion of organisms which generally may be found in normal soils makes no sense for a substrate and fertilizer such as compost, the following microbial parameters are inappropriate: enterococci, *Staphylococcus aureus*, enterobacteriaceae, *Escherichia coli*, *Clostridium perfringens*, and sulfite reducing clostridia.

The only parameter which seems to be useful and reliable in this connection is the absence or presence of salmonellas. Salmonellas are found at a rate of up to 90% in biowaste bins. Due to mixing the content of many sources during transport the waste delivered at the compost plant contains salmonellas with a high probability in various concentrations. Since it is known that the probability to identify a positive sample is basically related to the amount of investigated material a compromise between feasibility and reliability has to be

Tab. 14. Example of a Validation and Supervision Strategy for Composting Plants and Composts

Investigated Parameter		Direct Validation of the Composting Process	Indirect Process Supervision	Supervision of the Final Product
Hygienic safety concerning risks for man, animals and plants		– new constructed compost plants (within 12 months after opening of the plant) – already validated plants if new technologies have been invented or if the process has been significantly modified (within 12 months after invention or modification) – existing plants without validation within the last 5 years before this validation strategy was invented (within 18 months)	– continuous registration of temperature at 3 representative locations in the process, responsible for the inactivation of the microorganisms and seeds – recording of process data (e.g., turning of windrows, moisture of material, starting and finishing data)	regular investigation of the final product for hygienic safety[b, c]
Number of test trials		2 test trials at open air composting plants, at least 1 in winter	continuous data recording to be filed for at least 5 years	continuously all over the year at least – semiannual (plants with $\leq$3,000 t a^{-1} throughput) – quarterly (plants $>$3,000 t a^{-1} throughput)
Number of test organisms	human and veterinary hygiene	1 test organism (*Salmonella senftenberg* W 775, H_2S-negative)		no salmonella in 50 g compost detectable
	phytohygiene	3 test organisms (*Plasmodiophora brassicae*, tobacco mosaic virus, tomato seeds		less than 2 seeds capable of germinating and/or reproducible parts of plants in 1 L of compost
Number of samples Sample per test trial: Human and veterinary hygiene Phytohygiene		 24[a] 36[a]		throughput of the plants in t a^{-1} 1. $\leq$3,000 (6 samples per year) 2. $>$3,000–6,500 (6 samples per year plus one more sample for every 1,000 t throughput) 3. $>$6,500 (12 samples per year plus one more sample for every 3,000 t)
Total		60		

[a] At small plants half the number of samples ($\leq$3,000 t a^{-1}).
[b] Every statement concerning the hygienic safety of the product is always based on the result of the supervision of the final product together with the result of the validation of the process.
[c] Every sample is a "mixed sample" (about 3 kg) based on 5 single samples of the final product.

found. It is proposed to check 50 g of compost for the presence or absence of salmonellas with the method described in principle in the LAGA M10 (Länderarbeitsgemeinschaft Abfall, 1995) using a pre-enrichment in buffered peptone water and an enrichment step (Edel and Kampelmacher, 1969; Rappaport et al., 1956; Vassiliadis, 1983).

The question whether nematodes or nematode eggs are a useful indicator in this connection is not easy to answer. With respect to nematodes pathogenic for man and/or animals the experience shows, that even eggs of *Ascaris suum* are less thermoresistant than salmonellas. Therefore, if salmonellas won't survive the composting process *Ascaris* eggs and all other nematode eggs won't either. But nematodes may be an indicator for insufficient storage conditions for the final product which plant pathogenic nematodes may have invaded. In order to identify this situation eggs or larvae of such species have to be identified properly. This requires special expertise which is generally not available in the involved laboratories. Therefore, a general parameter "free of nematode eggs and/or larvae" seems not to be useful.

This leads to the problem of indicator organisms from the point of view of phytohygiene. No virus, fungus or bacterium pathogenic for plants has been found until now which is of comparable importance as salmonellas are for the above mentioned purpose. The only indicator which is widely distributed in biological wastes from households are tomato seeds. Even knowing, that this indicator will not cover totally all requirements, it seems to be reasonable and feasible to define the term "phytohygienic safety" of the product as follows: The final product (compost) should not contain more than two seeds capable to germinate and/or reproducible parts of plants in 1 L. A suitable test method is described by Bundesgütegemeinschaft Kompost (1994).

3.5 Conclusions

In order to assure hygienic safety of composts used in agriculture and horticulture a three-step control system is recommended which is based on approved methods and which is designed to minimize the costs and labor on one side and to come to an optimal product safety by using additive effects on the other side. Fig. 1 summarizes the strategies to be applied in order to reach this aim.

3.6 Occupational Health Risks

3.6.1 Introduction

The collecting, handling, and processing of source separated biological wastes generated a public discussion concerning occupational health risks due to bioaerosols. It is hard to understand, that in spite of those public concerns only a few data exist resulting from medical examinations of populations exposed to such aerosols in composting plants. Moreover,

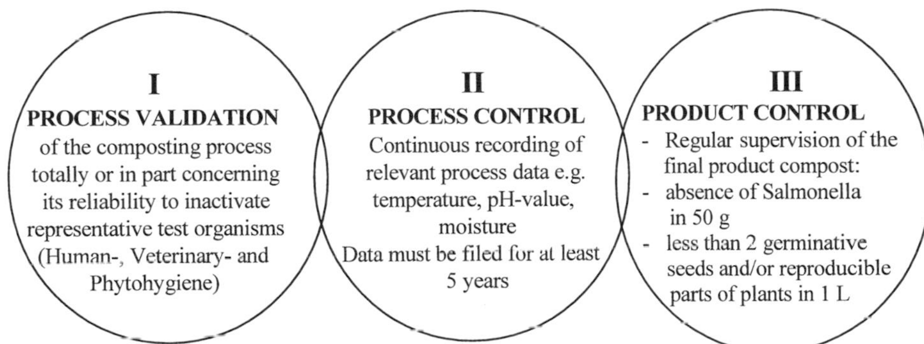

Fig. 1. Hygienic requirements for the production of compost.

results of recent investigations showed that there are, in principle, no differences between aerosols generated from biowastes, residual wastes and the traditional unseparated household wastes, which is especially the fact in collection, delivery and sorting of such material. Therefore, if health risks for workers in such facilities exist, they could not be new; they must have existed for decades.

3.6.2 Microbial Risks Connected with Biological Waste Treatment

First one must distinguish between primary and secondary sources concerning the emission of microorganisms, their components and/or their metabolites. Collection and handling of household wastes, especially of source separated biowastes, represents a sequence of different steps of which each one has a special importance with respect to propagation and the emission of microorganisms. Simplified it can be said that every step in which the collected material comes into contact with the environment and sets free microorganisms it is followed by a phase in which biological and physical influences cause a microbiological process in the material itself, which will influence the hygienic state of the material in a positive or negative way. This has been reported in more detail by BÖHM et al. (1996). It must be kept in mind, that not only the waste but the persons handling the material may also be a primary source of microorganisms, because they are biotopes themselves. Moreover, the involved personnel represents a secondary source of germs because of dirty skin, clothes and tools. This leads to an extremely high microbial load of the air especially in small rooms, like driver cabins. Another secondary source may be rotating parts of machines, fans and similar equipment. If the waste air from a composting plant is treated in biowashers or biofilters the microbial flora of such equipment may also be set free as a secondary aerosol.

The involved personnel is either exposed to the microorganisms and related materials by direct contact with the wastes or via aerosol.

The following risk factors must be considered:

(1) bacteria:
- endotoxins,
- exotoxins,
- enzymes,
- metabolites;

(2) fungi:
- glucans,
- mycotoxins,
- metabolites (volatile);

(3) viruses.

From the large variety of germs and biological products that may be present in the collected and treated wastes only a very limited number of parameters may be detected due to principal and technical reasons. The most frequently measured parameters are the total germ count (bacteria) and the total fungal count. Both parameters are relatively unspecific, and it must be questioned whether any causal relationship may exist to occupational health effects. Even if more specific parameters are determined (like the amount of airborne *Aspergillus fumigatus* spores or thermophilic *Actinomyces* spp.) no dose–effect relationship can be found, which could explain why negative health effects can be found in workers in one case and not in another. The microbial load in waste treatment plants mainly depends on the kind of material and the way of handling. In thermal utilization of waste the microbial load is mainly caused by the germs present in the waste itself. In biological waste treatment additional sources such as the involved biotechnological process and the biological waste air treatment have to be considered.

Besides a broad variety of saprophytic microorganisms wastes also contain pathogenic agents which affect human health. Those germs are present in unseparated household wastes and probably also in residual wastes resulting from separate collection of biowastes. With regard to pathogenic microorganisms the untreated wastes are the main source, therefore, their handling is mainly related to specific risks caused by pathogens transmitted via contact or stab wounds (e.g., *Clostridium tetani*).

The aerogenic distribution of microorganisms, their components or metabolites may happen at every stage of treatment up to the final thermal or biological processing and in the latter case also the handling of the resulting product. Obligatory pathogenic germs are transmitted via aerosol. Generally, widely distributed only in rare cases microorganisms propagating in the collected material or involved in the biological degradation of the organic matter are released into the air. Salmonellae could not be isolated from the air of a biocompost plant to date even if they may be present in the collected material up to 10^6 cfu g^{-1} moist weight according to SCHERER (1992). Therefore, the risk for a grown up healthy person to catch an aerogenic infection is negligible compared to the risks of developing an allergic or a more complex multifactorial clinical picture like the organic dust toxic syndrome (ODTS) as stated by MALMROS (1994).

The following pathophysiological reactions and symptoms may result from exposure to organic aerosols from waste treatment:

- exogenic allergic alveolitis (EAA),
- organic dust toxic syndrome (ODTS),
- allergies caused by inhaled fungal spores,
- allergic symptoms caused by ingestion of organic allergens,
- allergies caused by contact to organic allergens, and
- intoxication by volatile components.

3.6.3 Microbial Emissions in Composting Plants

The first step in which occupational risks due to aerosols may occur is collecting of the wastes on the street. Until now no representative data have been elaborated. Nevertheless, some aspects may be covered by existing investigations from related fields. Especially the situation of opening and closing the bin had been studied by HAUMACHER (personal communication) (Tab. 15). Specific requirements concerning the collection frequency (1 or 2 weeks) with regard to occupational risks do not seem to exist. This confirms the results of STREIB et al. (1989) who measured comparable data in collecting household wastes, source separated biowastes and wet wastes at several collection frequencies and different temperatures.

The delivery and sorting areas are generally locations with high bioaerosol emissions. A detailed documentation of data measured in biological waste treatment plants has been published by BÖHM (1995b). Therefore, only a few recent data are presented in Tabs. 16 and 17. Aerobic spore formers dominate the airborne bacterial flora isolated by the techniques applied. There may be different results if other substrates are processed but there are only limited data available from comparative inves-

Tab. 15. Total Fungal Count in the Air (cfu m^{-3}) during Opening and Closing of Two Types of Biowaste Bins at Different Storage Times Measured with the RCS-plus Sampler (HAUMACHER, personal communication)

Type of Biowaste	Bins	Winter Storage Time [d] 7	14	Spring Storage Time [d] 7	14
Mixed organic household and garden waste	aerated	$5.65 \cdot 10^3$	$3.90 \cdot 10^3$	$1.35 \cdot 10^4$	$2.07 \cdot 10^4$
	not aerated	$5.40 \cdot 10^3$	$1.06 \cdot 10^3$	$1.21 \cdot 10^4$	$4.07 \cdot 10^3$
Biowaste from household	aerated	$1.10 \cdot 10^4$	$2.71 \cdot 10^3$	$1.18 \cdot 10^4$	$8.84 \cdot 10^3$
	not aerated	$9.55 \cdot 10^3$	$2.60 \cdot 10^3$	$1.09 \cdot 10^4$	$1.55 \cdot 10^4$
Background				$2.97 \cdot 10^3$	$3.08 \cdot 10^3$

RCS Reuter-centrifugal-sampler (Biotest AG, Dreieich).
cfu Colony forming unit.

Tab. 16. Airborne Microorganisms in the Delivery and Sorting Areas of Six Different Composting Plants Collected with the Sartorius MD 8 Sampler

Sampled Area	Microbiological Parameter	n	Minimum	Median	Maximum
Delivery	total bacterial count	40	$9.2 \cdot 10^3$	$1.7 \cdot 10^5$	$3.7 \cdot 10^6$
Sorting[a]	total bacterial count	40	$3.7 \cdot 10^3$	$6.9 \cdot 10^4$	$1.3 \cdot 10^6$
Background	total bacterial count	9	$1.3 \cdot 10^3$	$1.3 \cdot 10^4$	$2.8 \cdot 10^4$
Delivery	enterobacteria	40	$1.4 \cdot 10^1$	$4.9 \cdot 10^2$	$3.7 \cdot 10^3$
Sorting[a]	enterobacteria	40	n.n.	$1.1 \cdot 10^2$	$6.9 \cdot 10^3$
Background	enterobacteria	9	n.n.	n.n.	$1.1 \cdot 10^1$
Delivery	*E. coli*	40	n.n.	n.n.	$8.4 \cdot 10^1$
Sorting[a]	*E. coli*	40	n.n.	n.n.	$3.2 \cdot 10^1$
Background	*E. coli*	9	n.n.	n.n.	n.n.
Delivery	aerobic spore formers	40	$6.9 \cdot 10^3$	$1.3 \cdot 10^5$	$9.2 \cdot 10^5$
Sorting[a]	aerobic spore formers	40	$6.9 \cdot 10^2$	$6.0 \cdot 10^4$	$2.8 \cdot 10^5$
Background	aerobic spore formers	9	$1.3 \cdot 10^3$	$4.5 \cdot 10^3$	$1.3 \cdot 10^4$
Delivery	total bacterial count	40	$3.7 \cdot 10^3$	$8.1 \cdot 10^5$	$3.7 \cdot 10^6$
Sorting[a]	total bacterial count	40	$1.7 \cdot 10^3$	$6.9 \cdot 10^5$	$3.7 \cdot 10^6$
Background	total bacterial count	9	$1.3 \cdot 10^3$	$4.5 \cdot 10^3$	$2.2 \cdot 10^4$
Delivery	*Aspergillus fumigatus*	40	$1.4 \cdot 10^3$	$3.7 \cdot 10^5$	$3.7 \cdot 10^6$
Sorting[a]	*Aspergillus fumigatus*	40	$9.2 \cdot 10^2$	$4.1 \cdot 10^5$	$4.2 \cdot 10^6$
Background	*Aspergillus fumigatus*	9	$2.7 \cdot 10^2$	$2.8 \cdot 10^3$	$1.5 \cdot 10^4$

n Number of measurements.
n.n. Below detection level.
[a] Sorting cabins only in 4 plants.
cfu Colony forming unit.

Tab. 17. Distribution of Particle Sizes of Bioaerosols Collected with the Andersen Sampler in the Delivery and Sorting Areas of 6 (4[a]) Composting Plants

Area	Microbiological Parameter	Stage 1 >8 µm [%]	Stage 2 4.7–8 µm [%]	Stage 3 3.3–4.7 µm [%]	Stage 4 2.1–3.3 µm [%]	Stage 5 1.1–2.1 µm [%]	Stage 6 0.6–1.1 µm [%]
Delivery	TBC	34.4	16.6	11.9	15.2	11.3	10.6
	TFC	19.6	14.2	15.7	14.8	15.7	19.9
Sorting[a]	TBC	17.2	15.0	16.6	15.7	22.0	13.4
	TFC	11.8	12.2	11.8	16.1	19.7	28.3

[a] Sorting cabins only in 4 plants.
TBC Total bacterial count,
TFC Total fungal count.
deposition in the respiratory tract.
Stage 1 Nose.
Stage 2 Pharynx.
Stage 3 Trachea and primary bronchi.
Stage 4 Secondary bronchi.
Stage 5 Terminal bronchi.
Stage 6 Alveoli.

tigations. This group of bacteria has no importance as pathogens or allergens but within limitations they can be regarded as indicators for the emission of dust. The most important airborne fungus in this regard is *Aspergillus fumigatus*.

Thus, from the point of view of occupational health these locations seem to be risk areas even if no epidemiological proofs have been found until now. Technical measures have to be undertaken for the areas of delivery and sorting in order to keep the emission of dust and microorganisms into the air as low as possible or to keep these areas free of working personnel by applying a high degree of automation. Concerning the delivery the investigations of SCHMIDT (1994) show that there exists – under comparable conditions – a tendency to lesser emission of microbial aerosols in deep shelters than in flat shelters. In the sorting area the number of airborne microorganisms can be reduced significantly by using effective ventilation techniques in combination with strict hygienic measures such as frequent cleaning of the room, the equipment and the protective clothes of the personnel, in order to avoid the formation of secondary aerosols.

If possible, hand sorting should generally be avoided. It seems to be better to use electronic detectors which will indicate the presence of undesired materials during the collection of wastes. These have been shown to be very effective in order to reduce the heavy metal contents of compost from biowastes as could be demonstrated by KRAUSS et al. (1996).

Concerning the emission of bioaerosols resulting from the biotechnological treatment and handling of the final product the following considerations may be taken into account.

A basic difference exists concerning the emission of microorganisms and microbial products between aerobic and anaerobic biotechnological treatment. In composting the setting free of aerosols is unavoidable while biogas production takes place in a closed vessel. Therefore, from this point of view anaerobic treatment is more advantageous than composting. Microbial aerosols may be avoided or also reduced by composting in closed systems and by modern waste air treatment, but gener-

Tab. 18. Airborne Microorganisms Collected in Six Composting Plants Collected during Turning of Windrows and Screening with a Sartorius MD 8 Sampler

Sampled Area	Microbiological Parameter	n	Minimum	Median	Maximum
Turning	total bacterial count	48	$9.2 \cdot 10^3$	$2.4 \cdot 10^5$	$5.6 \cdot 10^6$
Screening	total bacterial count	36	$3.7 \cdot 10^3$	$1.0 \cdot 10^6$	$1.7 \cdot 10^7$
Background	total bacterial count	9	$1.3 \cdot 10^3$	$1.3 \cdot 10^4$	$2.8 \cdot 10^4$
Turning	*Escherichia coli*	48	n.n.	n.n.	$1.6 \cdot 10^2$
Screening	*Escherichia coli*	36	n.n.	n.n	$7.4 \cdot 10^1$
Background	*Escherichia coli*	9	n.n.	n.n.	n.n.
Turning	enterobacteria	48	n.n.	$6.4 \cdot 10^1$	$1.4 \cdot 10^3$
Screening	enterobacteria	36	n.n.	$1.6 \cdot 10^2$	$1.1 \cdot 10^4$
Background	enterobacteria	9	n.n.	n.n.	$1.1 \cdot 10^1$
Turning	aerobic spore formers	48	$9.2 \cdot 10^2$	$1.9 \cdot 10^5$	$2.8 \cdot 10^6$
Screening	aerobic spore formers	36	$1.3 \cdot 10^3$	$5.4 \cdot 10^5$	$1.1 \cdot 10^7$
Background	aerobic spore formers	9	$1.3 \cdot 10^3$	$4.5 \cdot 10^3$	$1.3 \cdot 10^4$
Turning	total fungal count	48	$1.4 \cdot 10^3$	$3.7 \cdot 10^5$	$1.8 \cdot 10^7$
Screening	total fungal count	36	$9.2 \cdot 10^2$	$2.2 \cdot 10^3$	$5.2 \cdot 10^5$
Background	total fungal count	9	$1.3 \cdot 10^3$	$4.5 \cdot 10^3$	$2.2 \cdot 10^4$
Turning	*Aspergillus fumigatus*	48	$1.4 \cdot 10^3$	$3.7 \cdot 10^5$	$1.8 \cdot 10^7$
Screening	*Aspergillus fumigatus*	36	$9.2 \cdot 10^2$	$2.2 \cdot 10^3$	$5.2 \cdot 10^5$
Background	*Aspergillus fumigatus*	9	$2.7 \cdot 10^2$	$2.8 \cdot 10^3$	$1.5 \cdot 10^4$

n Number of measurements.
n.n. Below detection level.

ally composting in windrows leads to significantly high emissions of bioaerosols as shown in Tab. 18. If the anaerobically treated material is used in a liquid state no further processing which may generate aerosols will happen. On the other hand, if the solids are separated and composted the same situation as in genuine composting of biowastes could be expected, but until now no experimental data are available.

In composting plants it was found that during turning of windrows the aerosol fraction with particles which may be inhaled into the lung is larger than at other places of the plants (Tab. 19). It must be considered that in this place the number of airborne *Aspergillus fumigatus* spores is also relatively high. Therefore, no steady working places should be in such areas and turning of windrows should be done by automatic devices. Drivers' cabins should be equipped with air conditioners in combination with an appropriate dust filter system.

Tab. 20 demonstrates that this measure is effective in reducing the microbial aerosol in the cabin for at least one log step. This technical measure has to be accompanied by keeping the cabin strictly clean and by always wearing clean protective clothes.

With regard to occupational health risks the screening area is less critical with regard to aerosols. Even if the emissions are generally high at this place the emitted particles are relatively large and the measured values for *Aspergillus fumigatus* are in the background range. The aerobic spore formers are the most numerous bacteria collected from the air in the screening area.

Tab. 19. Distribution of Particle Size of Bioaerosols Collected with the Anderson Sampler in Six Composting Plants during Turning and Screening of Compost

Area	Microbiological Parameter	Stage 1 >8 μm [%]	Stage 2 4.7–8 μm [%]	Stage 3 3.3–4.7 μm [%]	Stage 4 2.1–3.3 μm [%]	Stage 5 1.1–2.1 μm [%]	Stage 6 0.6–1.1 μm [%]
Turning	TBC	24.3	13.1	14.6	13.1	20.0	14.4
Screening	TBC	47.5	11.2	11.4	9.8	9.5	10.2
Turning	TFC	18.9	10.7	10.4	10.7	17.9	31.5
Screening	TFC	50.4	16.8	7.6	8.4	8.4	8.4

TBC Total bacterial count,
TFC Total fungal count.
Deposition in the respiratory tract:
Stage 1 Nose.
Stage 2 Pharynx.
Stage 3 Trachea and primary bronchi.
Stage 4 Secondary bronchi.
Stage 5 Terminal bronchi.
Stage 6 Alveoli.

Tab. 20. Total Bacterial and Total Fungal Count in Driver Cabins with and without Fine Dust Filter in the Ventilation System

	Total Bacterial Count			Total Fungal Count		
	n	Without Filter	With Filter	n	Without Filter	With Filter
Composting plant 1	9	$1.8 \cdot 10^5$	$1.1 \cdot 10^4$	9	$3.3 \cdot 10^3$	$6.0 \cdot 10^3$
Composting plant 2	8	$6.2 \cdot 10^5$	$4.8 \cdot 10^4$	8	$3.3 \cdot 10^5$	$4.7 \cdot 10^4$

All results in cfu m^{-3} air.
n Number of measurements.

3.6.4 Health Effects Found in Personnel of Waste Treatment Plants

Negative health effects may occur in workers at waste treatment plants due to several reasons which are not always connected with bioaerosols. BITTIGHOFER (1998) listed the following factors which will cause health effects and/or health risks in workers (Tab. 21).

Even if factors not connected with the working place could be excluded, which is nearly impossible in most cases, one could never differentiate between microorganisms transmitted to any person via aerosol or via contact, whether they had been isolated from the skin or from the mucosa. Nevertheless, the effects observed concerning the respiratory symptoms resulting from the inquiry of workers in waste treatment plants by several authors are summarized in Tab. 22 (BITTIGHOFER, 1998).

Tab. 23 shows the results of investigations concerning the change of immunological and hematological parameters from workers handling wastes. The results differ and no relationship to any microbiological parameter could be found. The only tendency which could be confirmed in most cases was that the more microorganisms can be found in the air the higher the risk of health effects. Until a lack of causal relationships could be confirmed by scientific investigations all technical measures should be taken to keep the concentration of aerosols as low as possible in order to minimize the unspecific risk. For prevention of infections by certain pathogens it is obligatory according to LASI-LV 1 (1995) and LASI-LV 13 (1997) to vaccinate the personnel in waste treatment plants against tetanus and it is recommended to vaccinate against polyomyelitis and diphtheria. Vaccination against hepatitis B is strictly recommended for all workers handling untreated wastes.

Addendum: On September 23–25, 1998, the 3rd International Conference on Bioaerosols, Fungi and Mycotoxins was held in Saratoga Springs, N.Y., USA. The conference dealt with health effects, assessment, prevention and control. The procedings will be available some time in 1999. Contact address: ECKARDT JOHANNING, M.D., M. Sc., Medical Director, Eastern N.Y. Occupational and Environmental Health Center, Mount Sinai School of Medicine, 155 Washington Ave., Albany, NY 12210, USA.

Tab. 21. Factors Generally Influencing Health and/or Well-Being of Workers (BITTIGHOFER, 1998)

Factor	Variables
1. Germs	– spectrum, concentration, cultivability – seasonal influences
2. Toxins	– spectrum, measurability – concentration
3. Allergens	– spectrum, measurability – concentration
4. Dust	– composition – concentration
5. Gases	– composition, measurability – concentration
6. Climate	– effective temperature – related measurable factors
7. Working place	– individual ergonometric situation – organisation and duration of work – social climate
8. Private life	– living conditions – social situation, recreational activity etc.

Tab. 22. Summary of Results from the Evaluation of Questionnaires (Respiratory Disorders) Filled in by Workers of Waste Treatment Plants (BITTIGHOFER, 1998, shortened)

Disorder Irritation of mucosae	Author Type	SIGSGAARD et al., 1994 (C)	MARTH et al., 1997 (C)	GLADDING et al., 1992 (S)	BITTIGHOFER et al., 1998 (S)
Conjunctivitis	I	–	+	+	+
Burning eyes syndrome	Q	0		–	+
Itching nose	Q	+	–	+	0
Running nose	Q	+	–	–	+
Hoarseness	Q	–	+	–	–
Sore throat	Q	0	–	+	+
Cough	Q	+	+	+	+
Pains in the chest	Q	0	–	+	+
Dyspnoea	Q	0	–	–	+
Chronic bronchitis	Q	+	–	–	0

I Clinical investigation.
Q Questionnaire.
C Composting plant.
S Waste sorting plant.
0 No difference to control group.
+ Difference to control group.
– Not investigated.

Tab. 23. Summary of Immunological and Hematological Parameters Determined in Workers of Waste Treatment Plants (BITTIGHOFER, 1998, shortened)

Author Parameter	CLARK et al., 1984 (C)	SIGSGAARD et al., 1994 (C)	MARTH et al., 1997 (C)	COENEN et al., 1997 (W)	BÜNGER et al., 1997 (C)	BITTIGHOFER et al., 1998 (S)
Immunoglobulins						
IgG (total)	0	+	–	+	–	+
IgG (specific)	0	–	–	–	+	1
IgE (total)	–	+	0	–	–	+
IgE[a] (specific)	–	–	0	–	+	+
Blood sedimentation	–	–	–	–	–	0
CRP	0	–	–	–	–	–
Leukocytes	+	–	–	–	–	0
Eosinophils	+	+	–	–	–	–

C Composting plant.
W Waste collecting.
S Waste sorting.
0 No difference to control group.
+ Difference to control group.
– Not investigated.
1 Results see BÜNGER et al. (1997)
[a] CLARK et al. (1984) *Aspergillus fumigatus*; BÜNGER et al. (1997) *Saccharopolyspora rectivirgula, S. hirsuta, S. viridis, Streptomyces thermovulgaris*; MARTH et al. (1997) fungi only IgE.

4 References

ALEXANDRE, D., OTT, H. (Eds.) (1980), Treatment and use of sewage sludge, *Proc. 1st Eur. Symp.*, Cadarache/France, 13–15 February 1979, Brussels; Commission of the European Communities, DG XII, SL/54/80/255/81.

ATV (1997), *Abwassertechnische Vereinigung, Desinfektion von biologisch gereinigtem Abwasser, Merkblatt M 205*, Hennef: GFA.

ATV-VKS (1988a), Entseuchung von Klärschlamm, 2. Arbeitsbericht, AG 3.2.2, *Korresp. Abwasser* **35**, 71–74.

ATV-VKS (1988b), 3. Arbeitsbericht, AG 3.2.2, *Korrespondenz Abwasser* **35**, 1325–1333.

BASTIAN, R. K. (1997), The biosolids (sludge) treatment, beneficial use, and disposal situation in the USA, *Eur. Water Pollut. Control* **7**, 62–79.

BÉBIN, J. (1997), The sludge problem in France: technical advances, changes in regulations, and French involvement in CEN/TC 308, *Eur. Water Pollut. Control* **7**, 18–28.

BECKER, Y. (Ed.) (1987), *African Swine Fever*, pp. 145–150. Dordrecht: Martinus Nijhoff.

BENDIXEN, H. J. (1997), Hygiene and sanitation requirements in Danish biogas plants, in: *ALTENER- Workshop – The Future of Biogas in Europe* (HOLM-NIELSEN, J. B., Ed.), pp. 50–59, Esbjerg: Inst. of Biomass Utilization and Biorefinery.

BENDIXEN, H. J., AMMENDRUP, S. (1992), *Safeguards Against Pathogens in Danish Biogas Plants. Practical Measures to Prevent Dissemination of Pathogens and Requirements for Sanitation, The Danish Veterinary Service*. Copenhagen: Ministry of Agriculture.

BERGMANN, D. (1997), Gemeinsame Behandlung von Bioabfall und Klärschlamm, *Korrespondenz Abwasser* **44**, 1778–1783.

BERGS, C.-G., LINDNER, K.-H. (1997), Sewage sludge use in the Federal Republic of Germany, *Eur. Walter Pollut. Control* **7**, 47–52.

BILITEWSKI, B., HÄRDTLE, G., MAREK, K., WEISSBACH, A., BOEDDICKER, H. (1997), *Waste Management*, New York, Heidelberg: Springer-Verlag.

BITTIGHOFER, P. M. (1998), Arbeitsmedizinische Relevanz der ermittelten Daten, in: *Aktuelle Bewertung der Luftkeimbelastung in Abfallbehandlungsanlagen* (BÖHM, R., MARTENS, W., BITTIGHOFER, P. M., Eds.), Herford: Wissenschaftliche Expertise (angefertigt für die SOLU-Stiftung).

BITTIGHOFER, P. M., GRÜNER, C., PFAFF, G. (1998), *Belastung und Gesundheitszustand von Beschäftigten in Wertstoffsortieranlagen und Deponien. Abschlußbericht (in preparation)*. Dortmund, Berlin: Bundesanstalt für Arbeitsschutz und Arbeitsmedizin.

BLAHA, T. (Ed.) (1988), *Angewandte Epizootiologie und Tierseuchenbekämpfung*. Jena: Gustav Fischer.

BLOCK, J. C., HAVELAAR, A. H., L'HERMITE, P. (Eds.) (1986), *Epidemiological Studies of Risks Associated with the Agricultural Use of Sewage Sludge: Knowledge and Needs*. London, New York: Elsevier.

BÖHM, R. (1995a), Die Problematik der Festsetzung mikrobiologischer Grenz- und Richtwerte in der Umwelthygiene, in: *Grenzwerte und Grenzwertproblematik im Umweltbereich* (ARNDT, V., BÖKER, R., KOHLER, A., Eds.), pp. 75–85. Ostfildern: Verlag Günter Heimbach.

BÖHM, R. (1995b), Keimemissionen bei der Kompostierung, in: *Biologische Abfallbehandlung* (THOMÉ-KOZMIENSKY, K. J., Ed.), pp. 508–526. Berlin: EF-Verlag für Energie und Umwelttechnik.

BÖHM, R., FACK, TH., PHILIPP, W. (1996), Anforderungen an die biologische Abfallbehandlung aus der Sicht des Arbeitsschutzes, in: *Biologische Abfallbehandlung III* (WIEMER, K., KERN, M., Eds.), pp. 335–365. Witzenhausen: M. C. Baeza-Verlag.

BÖHM, R., FRANK-FINK, A., MARTENS, W., PHILIPP, W., WEBER, A., WINTER, D. (1998), *Abschlußbericht zum Forschungsvorhaben 02-WA 9257/5 "Veterinär- und seuchenhygienische Untersuchungen zur Überprüfung von Gülleaufbereitungsverfahren und der erzeugten Gülleaufbereitungsprodukte"*. Stuttgart: Institut für Umwelt- und Tierhygiene, Universität Hohenheim.

BRUCE, A. M., HAVELAAR, A. H., L'HERMITE, P. (Eds.) (1983), Disinfection of Sewage Sludge: Technical, Economic and Microbiological Aspects. Dordrecht, Boston, MA: Reidel.

BRUNS, C., GOTTSCHALL, R., MARCINISZYN, E., SCHÜLER, C., ZELLER, W. et al. (1994), *Phytohygiene der Kompostierung – Sachstand, Prüfmethoden, F- und E-Vorhaben, Tagungsband "BMFT-Statusseminar, Neue Techniken der Kompostierung"*, pp. 191–206. Berlin: Umweltbundesamt, FB III.

Bundesgütegemeinschaft Kompost e.V. (1994), *Methodenhandbuch zur Analyse von Kompost Nr. 222*, Methode 8. Köln.

BÜNGER, J., BITTIGHOFER, P. M., GRÜNER, C., STALDER, K., HALLIER, E. (1997), Spezifische Antikörper gegen Aktinomyzeten als Biomarker der Exposition in Abfallverwertungsanlagen, *Verh. Dtsch. Ges. Arbeitsmed. Umweltmed.* **37**, 583–586.

CLARK, S. C., BJORNSON, H. S., SCHWARTZ-FULTON, J. et al. (1984), Biological health risks associated with the composting of wastewater treatment plant sludge, *J. Water Pollut. Control Fed.* **56**, 1269–1276.

Club of Rome (1972), *The Limits of Growth*. Paris. (German Edn.: *Die Grenzen des Wachstums*. Stuttgart: Deutsche Verlagsanstalt).

COENEN, G. J., DAHL, S., EBBENHOE, J. N., IVENS, U. T., STEINBAEK, E. I., WÜRTZ, H. (1997), Immunoglobulins and peak expiratory flow measurements in waste collectors in relation to bioaerosol exposure, *Ann. Agric. Environ. Med.* **4**, 75–80.

DAVIS, R. D., HALL, J. E. (1997), Production, treatment and disposal of wastewater sludge in Europe from a UK perspective, *Eur. Water Pollut. Control* **7**, 9–17.

DIETRICH, H. (1996), *Lecture, ALB-Fachtagung, Abwasserbehandlung im ländlichen Raum*, 7/8 March. Stuttgart: Regierungspräsidium Stuttgart.

DIRKZWAGER, A. H., DUVOORT VAN ENGERS, L. E., VAN DEN BERG, J. J. (1997), Production, treatment and disposal of sewage sludge in the Netherlands, *Eur. Water Pollut. Control* **7**, 29–41.

DIRKZWAGER, A. H., L'HERMITE, P. (Eds.) (1989), *Sewage Sludge Treatment and Use: New Developments, Technological Aspects and Environmental Effects*. London, New York: Elsevier.

DORAU, W. (1998), Removal of pathogens from wastewater effluents by microfiltration technology, *Eur. Water Management* **1**, 42–52.

EDEL, W., KAMPELMACHER, E. H. (1969), *Salmonella* isolation in nine European laboratories using a standardized technique, *Bull. World Health Organ.* **41**, 297–306.

EEC (1986), Council Directive of 12 June 1986 on the protection of the environment, and in particular of the soil, when sewage sludge is used in agriculture (86/278 EEC), *Offic. J. Eur. Commun.* **L 181/6-12**, 4. 7. 86.

EPA (1992), *U.S. EPA, Environmental Regulations and Technology – Control of Pathogens and Vector Attraction in Sewage Sludge, EPA/625/R-92/013*. Washington, DC: 20460.

EPA (1994), *A Plain English Guide to the EPA Part 503 Biosolids Rule, EPA 832/R-93/003*. Washington, DC: 20460.

EPA (1995), *U.S. EPA, A Guide to the Biosolids Risk Assessments for the EPA Part 503 Rule, EPA 832-B-93-005*. Washington, DC: 20460.

FEACHEM, R. G., BRADLEY, D. J., GARELICK, H., MARA, D. D. (1983), *Sanitation and Disease, Health Aspects of Excreta and Wastewater Management*. Chichester: John Wiley & Sons.

GLADDING, T. L., COGGINS, P. C. (1997), Exposure to microorganisms and health effects of workers in UK materials recovery facilities – a preliminary report, *Ann. Agric. Environ. Med.* **4**, 137–141.

GRÜNER, C. (1996), Gesundheitszustand und Belastung von Beschäftigten im Abfallbereich. Erste Ergebnisse und Schlußfolgerungen für die Praxis, in: *Biologische Abfallbehandlung III* (WIEMER, K., KERN, M., Eds.), pp. 315–334. Witzenhausen: M. C. Baeza-Verlag.

GRUNWALD, R. (1995), Hygienisch-mikrobiologische Untersuchungen zur gemeinsamen anaeroben Fermentation von Gülle und Speiseresten in Biogasanlagen. *Thesis*. Stuttgart: Institut für Umwelt- und Tierhygiene, Universität Hohenheim.

HAIBLE, C. (1989), Hygienisch-mikrobiologische Untersuchungen über die Langzeitlagerung von Klärschlamm, *Thesis*. Universität Giessen, Germany.

HALL, J. E., NEWMAN, P. J., L'HERMITE, P. (Eds.) (1992), *Treatment and Use of Sewage Sludge and Liquid Agricultural Wastes*. Luxembourg: Office for Official Publications of the European Communities.

HEINEMEYER, E.-A., HELMUTH, R. (1998), Effects of Wastewater on the presence of salmonellae in coastal waters, *Eur. Water Management* **1**, 32–41.

HUBER, J., SIGEL, O., BRUNNER, P. H. (1987), *Survey of Sewage Sludge Disinfection Processes*. Brussels: CEC SL/121/87/XII/ENV/15/87.

HURST, C. J. (1989), Fate of viruses during wastewater sludge treatment processes, *CRC Crit. Rev. Environ. Control* **18**, 317–343.

IAEA (1975), *International Atomic Energy Agency, Radiation for a Clean Environment*. Vienna: STI/PUB/402.

KRAUSS, P., KRAUSS, M., WILKE, M., WALLENHORST, T. (1996), Herkunft und Verbleib organischer und anorganischer Schadstoffe im Bioabfall, in: *Biologische Abfallbehandlung III* (WIEMER, K., KERN, M., Eds.), pp. 335–365. Witzenhausen: M. C. Baeza-Verlag.

KRUG, W., REHM, N. (1983), Nutzen-Kosten-Analyse der Salmonellenbekämpfung, Vol. 131, *Schriftenr. d. BM f. Jugend, Familie u. Gesundheit*. Stuttgart: Kohlhammer.

KTBL (1998), *Positionspapier der KTBL-Arbeitsgruppe Kofermentation*, Arbeitspapier 249. Darmstadt: KTBL.

Länderarbeitsgemeinschaft Abfall (LAGA) (1995), LAGA-Merkblatt M10: Qualitätskriterien und Anwendungsempfehlungen für Kompost, Stand 15. 2. 1995, in: *Müll-Handbuch*, Kennzahl 6856,1-59 (HÖSEL, G., BILITEWSKI, B., SCHENKEL, W., SCHNURER, H. E., Eds.), Berlin: Erich Schmidt Verlag.

LASI-LV 1 (1995), *Leitlinien für den Arbeitsschutz in Wertstoffsortieranlagen*. Wiesbaden: Länderausschuß für Arbeitssicherheit.

LASI-LV 13 (1997), *Leitlinien für den Arbeitsschutz in biologischen Abfallbehandlungsanlagen*. Wiesbaden: Länderausschuß für Arbeitssicherheit.

LEDERBERG, J., SHOPE, R. E. (1992), *Emerging Infection – Microbial Threats to Health in the United States*. Washington, DC: National Academy of Science, Institute of Medicine, Oaks St. C.

LESSEL, T. (1985), Ein Beitrag zur Optimierung des Verfahrens zur Gamma-Bestrahlung von Klärschlamm, *Berichte zur Wassergütewirtschaft und Gesundheitsingenieurwesen Nr. 54*. München:

Technische Universität.

L'HERMITE, P. (Ed.) (1986), *Processing and Use of Organic Sludge and Liquid Agricultural Wastes*. Dordrecht, Boston, MA: Reidel.

L'HERMITE, P. (Ed.) (1991), *Treatment and Use of Sewage Sludge and Liquid Agricultural Wastes*. London, New York: Elsevier.

L'HERMITE, P., OTT, H. (Eds.) (1981), *Characterization, Treatment and Use of Sewage Sludge*. Dordrecht, Boston, MA: Reidel.

L'HERMITE, P., OTT, H. (Eds.) (1984), *Processing and Use of Sewage Sludge*. Dordrecht, Boston, MA: Reidel.

LOPEZ-PILA, J. M. (1998), Some economic and political consequences of pathogens in inland and coastal waters, *Europ. Water Management* **1**, 70–78.

LUND, B., JENSEN, V. F., HAVE, P., AHRING, B. (1996), Inactivation of virus during anaerobic digestion of manure in laboratory biogas reactors, *Antonie van Leeuwenhoek* **69**, 25–31.

MALMROS, P. (1994), Occupational health problems owing to recycling of waste, in: *Nachweis und Bewertung von Keimemissionen bei der Entsorgung von kommunalen Abfällen sowie spezielle Hygieneprobleme der Bioabfallkompostierung* (BÖHM, R., Ed.), pp. 36–50. Giessen: Dtsch. Vet. Med. Ges.

MARA, D., CAIRNCROSS, S. (1989), *Guidelines for the Safe Use of Wastewater and Excreta in Agriculture and Aquaculture*. Geneva: World Health Organization.

MARTH, E., REINTHALER, F. F., SCHAFFLER, K., JELOVCAN, S., HASELBACHER, S. et al. (1997), Occupational health risks to employees of waste treatment facilities, *Ann. Agric. Environ. Med.* **4**, 143–147.

MENKE, G. (1992), Hygienische Aspekte der Bioabfallkompostierung. Teil 1: Phytohygiene, in: *Proc. Symp. Umweltministeriums Baden-Württemberg und der LG-Stiftung "Natur und Umwelt", 26. March 1991, Stuttgart, "Bioabfallkompostierung – Chance der Abfallverwertung oder Risiko der Bodenbelastung?"*, pp. 68–73 and Tab. 1–5. Stuttgart: LG-Stiftung "Natur und Umwelt".

OCKIER, P., BONCQUET, W., VERCAUTEREN, F., BARTHOLOMEEUSEN, W. (1997), Production, treatment and disposal of sewage sludge in Belgium, *Eur. Water Pollut. Control* **7**, 53–61.

OSS (1983), *Klärschlammverordnung – AbfKlärV v. 25 June 1982*, BGBl I, S. 734.

OSS (1992), *Klärschlammverordnung – AbfKlärV v. 15 April 1992*, BGBl I, S. 912.

PAULY, U., BLAU, S., VON BORKE, P., VON SYDOW, R. (1997), Zehn Jahre Klärschlammvererdung in Schilfbeeten – Neue Wege der Klärschlammverarbeitung und -verwertung, *Korrespondenz Abwasser* **44**, 1812–1822.

PEPPER, J. L., GERBA, C. P., BRUSSEAU, M. L. (1996), *Pollution Science*. New York, London: Academic Press.

PHILIPP, W. (1988), Untersuchungen über den Einsatz von Pflanzen zur Klärschlammentwässerung, Teilvorhaben 2: Seuchenhygienische Untersuchungen, *UFOPLAN, Forschungsbericht 10301227*. Berlin: Umweltbundesamt.

PHILIPP, W., LANG, A., STRAUCH, D. (1987), Zur Frage der Wirkung von Schilf auf die Entseuchung von Klärschlamm in Entwässerungsbeeten, *Forum Staedte-Hyg.* **38**, 325–328.

PIKE, E. B. (1983), Long-term storage of sewage sludge, in: *Disinfection of Sewage Sludge* (BRUCE, A., HAVELAAR, A., L'HERMITE, P., Eds.), pp. 212–225. Dordrecht: Reidel.

POLLMANN, B., STEINER, A. M. (1994), A standardized method for testing the decay of plant diaspores in biowaste composts by using tomato seed, *Agribiol. Res.* **47**, 27–31.

POPP, W. (1998), Disinfection of secondary effluents from sewage treatment plants requirements and applications, *Eur. Water Management* **1**, 27–31.

RAGAZZI, M. (1997), Production, treatment and disposal of sludge in Italy, *Europ. Water Pollut. Control* **1**, 42–46.

RAPP, A. (1995), Hygienisch-mikrobiologische Untersuchungen zum Verhalten von ausgewählten Bakterien und Viren während der längerfristigen Speicherung von Flüssigmist in Güllegemeinschaftsanlagen, *Thesis*. Stuttgart: Universität Hohenheim.

RAPPAPORT, F., KONFORTI, N., NAVON, B. (1956), A new enrichment medium for certain salmonellae, *J. Clin. Pathol.* **9**, 261–266.

Rudolf Schülke Foundation (Ed.) (1996), Memorandum on the Threat Posed by Infectious Diseases. Wiesbaden: mhp-Verlag.

SCHERER, P. A. (1992), Hygienische Aspekte bei der getrennten Abfallsammlung, in: *Getrennte Wertstofferfassung und Biokompostierung* (THOMÉ-KOZMIENSKY, K. J., SCHERER, P. A., Eds.), pp. 135–161. Berlin: EF-Verlag für Energie- und Umwelttechnik.

SCHMIDT, B. (1994), Bakteriologische Untersuchungen zur Keimemission an Arbeitsplätzen in der Müllentsorgung und -verwertung, *Thesis*. Stuttgart: Universität Hohenheim.

SCHMIDTKE, N. W., EBERLE, S. H. (Eds.) (1979), *Water Pollution Control Technologies for the 80's*. Burlington, Ontario: Canada Centre for Inland Waters.

SCHULZE, D. (1994), *Ratgeber für die kommunale Abwasserentsorgung. Mercedes oder Vernunftwesen?* Berlin: Verlag für Bauwesen.

SCHWARTZBROD, J., GASPARD, P., THIRIAT, L. (1998), Pathogenic microorganisms in sludge and effects of various treatment processes for their removal, *Eur. Water Management* **1**, 64–69.

SIGSGAARD, T., MALMROS, P., NERSTING, L., PETERSEN, C. (1994), Respiratory disorders and atopy in Danish refuse workers, *Am. J. Respir. Crit. Care. Med.* **149**, 1407–1412.

STRAUCH, D. (1983), Ursachen und mögliche Auswirkungen des Vorkommens pathogener Agentien in kommunalem Klärschlamm, *Schweiz. Arch. Tierheilkd.* **125**, 621–659.

STRAUCH, D. (1984), Use of lime treatment as disinfection process, in: *Processing and Use of Sewage Sludge* (L'HERMITE, P., OTT, H., Eds.), pp. 220–223. Dordrecht: Reidel.

STRAUCH, D. (1991), Survival of pathogenic microorganisms and parasites in excreta, manure and sewage sludge, *Rev. Sci. Tech. Off. Int. Epiz.* **10**, 813–846.

STRAUCH, D. (1998), Pathogenic microorganisms in sludge. Anaerobic digestion and disinfection methods to make sludge usable as fertilizer, *Eur. Water Management* **1**, 12–26.

STRAUCH, D., HAVELAAR, A. H., L'HERMITE, P. (Eds.) (1985), *Inactivation of Microorganisms in Sewage Sludge by Stabilization Processes*. London, New York: Elsevier.

STRAUCH, D., PHILIPP, W., HAIBLE, CH. (1987), Vorläufige Ergebnisse von Untersuchungen über die mögliche entseuchende Wirkung der Langzeitlagerung von organischen Düngern und von Klärschlamm. *Forum Staedte-Hygiene* **38**, 329–332.

STREIB, R., HEROLD, K., BOTZENHART, K. (1989), Keimzahlen ausgewählter Mikroorganismen in ungetrenntem Hausmüll, Biomüll und Naßmüll bei unterschiedlichen Standzeiten und Außentemperaturen. *Forum Staedte-Hygiene* **40**, 290–292.

UM-Arbeitskreis (1997), *Bericht des UM-Arbeitskreises zur "Mitbehandlung von Abwasser und Fäkalschlamm in landwirtschaftlichen Biogasanlagen"*. Stuttgart: Ministerium für Umwelt und Verkehr Baden-Württemberg.

VAN DE KRAATS, J. (1997), Editorial, *Eur. Water Pollut. Control* **7**, 3–4.

VAN DE KRAATS, J. (1998), Editorial, *Eur. Water Management* **1**, 3.

VASSILIADIS, P. (1983), The Rappaport–Vassiliadis (RV) enrichment medium for the isolation of salmonellas: an overview, *J. Appl. Bacteriol.* **54**, 69–74.

VwV (1997), *Verwaltungsvorschrift des Ministeriums für Umwelt und Verkehr Baden-Württemberg über die Abwasserbeseitigung im ländlichen Raum vom 21 November 1997*, Az.: 51-8950.11. Stuttgart.

10 Future Settlement Structures with Minimized Waste and Wastewater Generation

RALF OTTERPOHL
Hamburg, Germany

1 Differentiating Sanitation Systems – The Basic Step to Resources Management

Future sanitation concepts should produce a low-polluted organic fertilizer for use in agriculture instead of waste for deposition or incineration. Sanitation and biowaste treatment can go hand-in-hand with the production of a special quality of fertilizer, containing many trace substances which are needed for soil fertility. However, there are some possibly harmful compounds as well, e.g., residues of pharmaceuticals. Degradation of these compounds is critical observed. It may very well be necessary to keep these substances away from the water cycle. Appropriate treatment is required with first priority given to hygienic aspects – alternative concepts can and should also be better solutions in this respect.

In the long run, the type of wastewater and waste management for production of secondary organic fertilizers strongly affect the soil quality (ARRHENIUS, 1992, personal communication; PIMENTEL, 1997). Care for soil quality with source control and reuse of matter originating from the soil will also decrease accumulation of these substances in the final receiving waters: the oceans. Water saving technologies are necessary in many regions of the world. Once again – as a highly welcome side effect – source control will save water.

During construction of human settlements or single houses the installation of new sanitation systems can be taken into consideration. One of many technical solutions is the anaerobic treatment of separated blackwater in biogas plants. Combination with digestion of organic househould wastes results in a mixture which is suitable for anaerobic digestion. Anaerobic treatment is very advantageous, especially for wet biowaste, and it has become more economical during the last years (STEGMANN and HUPE, 1997). There are also options for anaerobic treatment of municipal wastewater with some major advantages (ZEEMANN and LETTINGA, 1988), but for low-diluted mixed wastewater in cold climates the backwater flow has to be collected separately to allow economic digestor sizes.

Separation of differently, polluted watewater streams and their appropriate treatment for reuse is common in industry and is fundamental for new concepts (Tab. 1).

Blackwater (toilet water) and solid kitchen waste (group 1) contains nearly all of the nutrients nitrogen, phosphorus, and potassium. In blackwater the majority of nutrients is concentrated in urine, thus making separate treatment feasible (LARSEN and GUJER, 1996). Blackwater should be protected from pollution at source by use of biodegradable toilet cleaning chemicals and especially by avoiding copper or zinc pipes for drinking water. Further precautions will have to be taken with respect to pharmaceuticals that have to be designed for degradability in biological treatment processes. If the blackwater is kept anaerobic and digested in an anaerobic reactor more residual pharmaceuticals in faeces and urine are bioconverted or degraded than by conventional aerobic treatment (DALHAMMER, personal communication).

The greywater (group 2) contains little nutrients, if phosphorous-free detergents are used. This fraction can easily be treated to re-

Tab. 1. Classification of Domestic Waste and Wastewater for Adequate Treatment Processes

Classification	Treatment	Type of Cycle
Kitchen waste and low-diluted feces with urine	anaerobic or composting	food cycle
Grey wastewater (greywater) from bathrooms, laundry, and kitchen (little nutrients)	aerobic with biofilm plants	water cycle
Stormwater runoff	local discharge or infiltration	water cycle
Non-biodegradable solid waste (small fraction with reuse of packages)	processing to raw materials	raw materials

usable quality as it did not have contact with toilet wastewater. However, a low content of fecal bacteria has to be taken into account (washing of diapers, showering). Contrary to common belief greywater often has a high COD concentration due to lesser dilution when water consumption is reduced. Mechanical pretreatment is necessary for most biological treatment technologies. Biofilm reactors should be preferred for treatment, since activated sludge may disintegrate, if the nutrient concentrations supplied with the wastewater are too low. Biofilm systems like trickling filters, rotating disk or sandfilters (technical or as constructed wetland) can reuse the nutrients released by lysis of biomass. An additional effort has to be made in the production of household chemicals. They have to be designed to be non-toxic and completely biodegradable. The technology is available and prices of products will have to reflect damage to the water cycle or problems caused in treatment.

Avoidance of central stormwater sewers is an important step towards economically feasible source control sanitation. Since a couple of years stormwater infiltration has become increasingly popular in many countries. The advantages are obvious in recharge of groundwater in a short circuit of the local water cycle. Unfortunately, stormwater runoff is often loaded with a wide variety of organic and inorganic chemicals (FÖRSTER, 1996). Direct infiltration into the groundwater should thus be avoided.

Infiltration through swales with biologically active soil can be a fairly good treatment, but even here a considerable amount of the initial concentrations can reach the groundwater – especially when the flow does not pass the whole area, but mostly the rim near sealed surfaces (MEISSNER, 1998). This indicates that precautions have to be taken to protect the groundwater. The flow of groundwater must be considered: There may be a difference, if stormwater infiltrates near a river where it discharges all year long, or to a place with groundwater reservoirs with little exchange other than evaporation. Surface runoff in ditches directed to receiving waters might be safer depending on the situation. Stormwater has to be kept clean by means of source control in any case. Roofs and gutters should not be made of zinc plated metal or copper. Air and road pollution also have to be minimized. Parking lots can be equipped with pans to collect dripping oil from the engines. Basic considerations about disadvantages of the conventional sanitation systems can be found in OTTERPOHL et al., 1997; more concepts are discussed in OTTERPOHL et al., 1999.

2 A Pilot Project for Vacuum Biogas Systems in Urban Areas

An integrated sanitation concept with vacuum toilets, vacuum sewers, and a biogas plant for blackwater is implemented for the new settlement "Flintenbreite" located within the city of Lübeck (Baltic Sea, Germany). The area with a total of 3.5 ha is not connected to the central sewerage system. The authors have designed this system for the local construction company developing this area in cooperation with the city of Lübeck. The settlement will be settled by about 350 people and is a pilot project to demonstrate the concept in practice. However, all components of the project are in use in different fields of application for many years and, therefore, they are well developed. Vacuum toilets are used in ships, airplanes, and trains. There are already some implementations in flat buildings for saving water. Unified vacuum sewerage is used by hundreds of communities. Anaerobic treatment is applied in industrial wastewater treatment, in biowaste treatment, on many farms, and for feces in ten thousands of cases in South East Asia and elsewhere.

The system in Lübeck mainly consists of:

- vacuum closets (VC) with collection pipes and anaerobic treatment,
- co-treatment of organic household waste in decentralized/semicentralized biogas plants,
- recycling of digested anaerobic sludge to agriculture with storage for further growth periods,
- use of biogas in a heat and power generator (heat for houses and digestor) in

addition to other fuel (in this case: natural gas),

- decentralized treatment of gray wastewater in constructed wetlands (energetically highly efficient),
- stormwater collection for reuse and collection of excess stormwater in a through-and-drain trench for retention and infiltration (GROTEHUSMANN, 1993).

Heat for the settlement will be produced by a combined heat and power generating engine which is switched to use biogas when the storage is filled. It will also be used to heat the biogas plant. In addition there will be passive solar systems to support heating of the houses and active solar systems for warm water production. A sketch of such a system is presented in Fig. 1. It does not show all details, but gives an idea of the concept with collection and treatment of feces.

At the digestor a vacuum pumping station is installed. The pumps have a reserve unit in case of failure. The pressure in the system is 0.3 bar operating both the vacuum toilets and the vacuum pipes. The pipes are dimensioned with 50 mm to allow good transport by the air. They have to be installed deep enough to be protected against freezing and must have down-bows about every 30 m to to create plugs of the transported matter. Noise is a concern with vacuum toilets but modern units are comparable to flushing toilets.

Feces mixed with shredded biowaste (only blackwater for mixing) will be hygienized by heating the feed to 55 °C for 30 min. The energy is reused by a heat exchanger which preheats the incoming flow. The digestor is operated at mesophilic conditions around 37 °C with a capacity of 50 m^3. This is half the size required for mesophilic operation (around 37 °C). Thermophilic operation may cause problems from high concentrations of NH_4/NH_3 around 2,000 mg/L^{-1}. Under mesophilic conditions the proportion of NH_3 is lower at the same pH value. Another concern is the amount of sulfur in the biogas. This can be minimized by controlled input of oxygen into the digestor or into the gas flow.

The biogas plant is designed as production unit for liquid fertilizer. It is important to consider pathways of pollutants from the beginning. One important source for heavy metals are copper-or zinc-plated pipes for drinking water. Instead of these materials polyethylene pipes will be used. The sludge will not be dewatered for obtaining a good composition of the

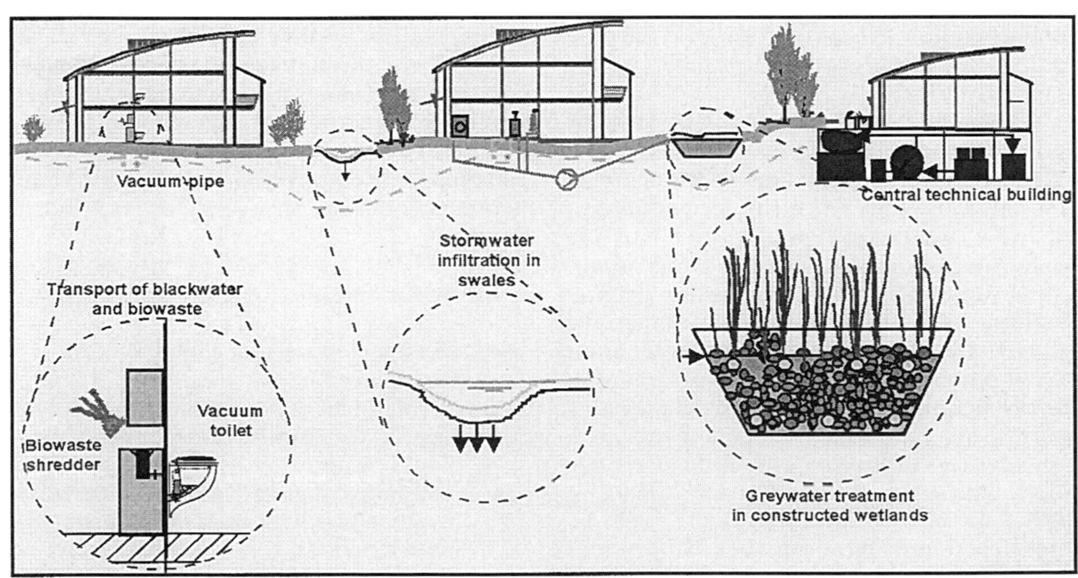

Fig. 1. Vacuum biogas system integrated into a new settlement.

fertilizer and for not having to treat the sludge-water. The relatively small amount of water added to the blackwater keeps the volumes small enough for transportation. There will be a 2 weeks storage tank for the collection of the digestor effluent. Biogas will be stored in the same tank within a balloon. This will permit more flexibility in operation. The fertilizer will be pumped off by a truck and transported to a farm with a storage tank for 8 months operation. Such tanks are often available anyway or can be built with little investment.

Decentralized treatment of gray wastewater should be done using biofilm processes. Appropriate technologies with very limited space are aerated sandfilters, rotating disk plants, and trickling filters with infiltration of treated greywater within the stormwater storage and infiltration system. Constructed wetlands are also a suitable solution for urban areas. They can be integrated with gardens and parks. Greywater is relatively easy to treat because of its low nutrient contents. There may even be a lack of nutrients during the start-up of the greywater treatment system. The microorganisms can reuse nutrients released by lysis as soon as there is a sufficient biofilm. Several projects on a technical scale have demonstrated the feasibility and the good to excellent performance of decentralized greywater treatment. These plants allow reuse of the water in toilet flushing. This is not economically feasible for the Lübeck project because of the low water consumption of the vacuum toilets. Greywater in "Flintenbreite" is treated in decentralized, vertically fed constructed wetlands with sizes of 2 m^2 per inhabitant. They are relatively cheap in construction and especially in operation. The pumping wells will serve as a grit chamber for grease control and has filters for larger particles above the waterline. The effluent will preferably be infiltrated into the drain trench system for stormwater.

The infrastructure of "Flintenbreite" including the integrated sanitation concept is prefinanced by the construction company and a private company. Participating companies, planners, and house and flat owners are financially integrated and will have the right to vote on decisions. Part of the investment is covered by a connection fee, just as in the traditional system. Money is saved by not having to construct a flushing sewerage system, by lower freshwater consumption, and by the coordinated construction of all pipes and lines (vacuum sewers, local heat and power distribution, water supply, phone and TV lines), which is essential for the economic feasibility of this concept. The fees for wastewater and biowaste charged cover operation, interest rates on additional investment, and rehabilitation of the system. Part of the operation cost has to be paid for a part-time operator, but this also offers local employment. The company cares for operation of the whole technical structure including heat and power generation and distribution, active solar systems, and an advanced communication system.

Material and energy intensity of the structure in comparison to a traditional system are presently studied with the MIPS method at the Wuppertal Institute in Germany (RECKER-ZÜGL, 1997). Material and energy intensity for the decentralized system are less than half as for a conventional central system serving a medium-densely populated area (see Tab. 2). For the central system most of the material intensity results from the construction of the sewerage system. The predicted effluent values are based on averages of measurements of greywater effluent qualities and are presented in comparison to average values of a modern treatment plant with an advanced nutrient removal and good performance.

Tab. 2 indicates some major advantages of the new system which justify further research. The cumulated savings of emissions to the sea and of energy and material usage for an average lifetime of 70 years for the 350 inhabitants would be:

- about 245,000 m^3 of freshwater,
- 70 t of COD,
- 1,470 kg of P,
- 13 t N,
- 32 t of K,
- 5,250 MWh of energy and
- about 56,000 t of material usage.

The saved emissions can replace fertilizer production from fossil resources and also synthesis of nitrogen (80 t N). This can be calculated as another 1,470 MWh of energy saved (BOISEN, personal communication). Other refer-

Tab. 2. Estimated Emissions, Energy Consumption and Material Intensity of the Proposed System Compared to a Traditional System

Advanced Traditional Sanitation (WC-S-WWTP) Concept		New Sanitation System	
Emissions		**Emissions**[a]	
COD	3,6 kg $(P\cdot a)^{-1}$	COD	0,8 kg $(P\cdot a)^{-1}$
BOD_5	0,4 kg $(P\cdot a)^{-1}$	BOD_5	0,1 kg $(P\cdot a)^{-1}$
Total N	0,73 kg $(P\cdot a)^{-1}$	Total N	0,2 kg $(P\cdot a)^{-1}$
Total P	0,07 kg $(P\cdot a)^{-1}$	Total P	0,01 kg $(P\cdot a)^{-1}$
Total K[b]	[>1,7 kg $(P\cdot a)^{-1}$]	Total K[b]	[<0,4 kg $(P\cdot a)^{-1}$]
Energy		**Energy**	
Water supply (wide variation)	–25 kWh $(P\cdot a)^{-1}$	water supply (20% water savings)	–20 kWh $(P\cdot a)^{-1}$
Wastewater treatment (typical demand)	–85 kWh $(P\cdot a)^{-1}$	vacuum system	–25 kWh $(P\cdot a)^{-1}$
Consumption		greywater treatment	–2 kWh $(P\cdot a)^{-1}$
		transport of sludge (2 per month, 50 return)	
Consumption	*–110 kWh $(P\cdot a)^{-1}$*	*consumption*	*–67 kWh $(P\cdot a)^{-1}$*
		biogas	110 kWh $(P\cdot a)^{-1}$
		substition of fertilizer	60 kWh $(P\cdot a)^{-1}$
		surplus	*170 kWh $(P\cdot a)^{-1}$*
Total	**–110 kWh $(P\cdot a)^{-1}$**	**Total**	**103 kWh $(P\cdot a)^{-1}$**
Material Intensity[c]	3.6 t $(P\cdot a)^{-1}$	material intensity[c]	1.3 t $(P\cdot a)^{-1}$

[a] Measurement HH-Allermöhe.
[b] Assumption, no data.
[c] MIPS study (RECKERZÜGL, 1997).

ences quote energy demands up to ten times of this value for production of mineral fertilizer. These numbers are important with respect to a large global population and decreasing fossil resources.

3 Further Options for Integrated Sanitation Concepts Based on Biogas Plants

The interest in the integrated concept described above has dramatically increased since its first publication (OTTERPOHL and NAUMANN, 1993) and the start of planning the project in Lübeck. There are other projects where this type of concept will be applied. The system in general can well be cheaper than the traditional system. This depends on the possibility to locally infiltrate stormwater, which at present is becoming the standard approach. It also depends on the size of the area that is served and on the number of inhabitants. An optimum size may be an urban area with about 500–2,000 inhabitants. Smaller units are feasible, if the blackwater–biowaste mixture is only collected and then transported to a larger biogas plant which would preferably be situated on a farm. Greywater can be treated in an existing wastewater treatment plant, if the sewerage system is nearby. In some cases this is the most economic way. If a percentage of the population is served by separate blackwater treatment, nutrient removal can be improved. Above a certain proportion nitrification would be obsolete.

The size of cities is of concern because of transport distances. However, even in metropolitan areas there would be possibilities to deal with this problem. The liquid fertilizer or the raw blackwater–biowaste mixture can either be pumped or be transported by rail (avoiding peak load times for passenger transportation). These are questions of long-term planning in close connection to city conceptions. From the point of view of sustainable sanitation cities of the future should be developed in the shape of stars with rural areas in between. This is advantageous for food production and transport and provides a closer contact of citizens to nature.

The proposed system is based on vacuum toilets, but there are other ways to collect blackwater. Urine separation toilets and a type of pressure flush toilet with a lid instead of a water siphon to prevent smells can be used as well (LANGE and OTTERPOHL, 1997). Both are developed in Sweden. The latter type needs a pipe gradient >5% to a collection pit. Further transport could be done by a vacuum or a pressure system. Systems based on biogas plants should have a heat and power generator, if heat is needed around the plant, typically in the settlement served by the system (colder climates). A charming concept could be the local production of cold-pressed biofuel with fertilizer from the digester. There are engines that can be run with different mixtures of biofuel and biogas.

4 References

ARRHENIUS, E. (1992), Population, development and environmental disruption anissue on efficient natural resource management, *Ambio* **21**.

GROTHEHUSMANN, D. (1993), Alternative urban drainage concept and design, *Proc. 6th Int. Conf. Urban Storm Drainage, Niagara Falls, Canada.*

GUJER, W., LARSEN, T. A. (1998), Technologische Anforderungen an eine nachhaltige Siedlungswasserwirtschaft, Wasserwirtschaft in urbanen Räumen, *Schriftenreihe Wasserforsch.* **3**, 65–83.

LANGE, J., OTTERPOHL, R. (1997), *Abwasser – Handbuch zu einer zukunftsfähigen Wasserwirtschaft.* Pfohren, Germany: Mallbeton Verlag.

LARSEN, T. A., GUJER, W. (1996), Separate management of anthropogenic nutrient solutions, *Water Sci. Technol.* **34**, 87–94.

MEISSNER, E. (1998), Ergebnisse von Feldversuchen zur Versickerung von Niederschlagswasser, 27. Abwassertechnisches Seminar „Dezentrale Abwasserbehandlung für ländliche und urbane Gebiete“, *Berichte aus Wassergüte- und Abfallwirtschaft,* TU München, Nr. 138.

OTTERPOHL, R., NAUMANN, J. (1993) *Kritische Betrachtung der Wassersituation in Deutschland, Symposium Umweltschutz, wie?* Köln: Kirsten Gutke-Verlag.

OTTERPOHL, R., GROTTKER, M., LANGE, J. (1997), Sustainable water and waste management in urban areas, *Water Sci. Technol.* **35**, 121–133 (Part 1).

OTTERPOHL, R., ALBOLD, A., OLDENBURG, M. (1991), Source control in urban sanitation and waste management: 10 options with resource management für different social and geographical conditions, *Water Sci. Technol.* **3/4** (Part 2).

PIMENTEL, D. (1997), Soil erosion and agricultural productivity: The global population food problem, *Gaia* **6**.

RECKERZÜGL, T. (1997), Vergleichende Materialintensitäts-Analyse zur Frage der zentralen oder dezentralen Abwasserbehandlung anhand unterschiedlicher Anlagenkonzepte, *Thesis,* Universität-Gesamthochschule Paderborn, Germany.

STEGMANN, R., HUPE, K. (1997), Biologische Bioabfallverwertung: Kompostierung kontra Vergärung, *Studie des Ingenieurbüros für Abfallwirtschaft,* Hamburg-Harburg.

ZEEMAN, G., LETTINGA, G. (1998), The role of anaerobic digestion of domestic sewage in closing the water and nutrient cycle at community level, *Int. Wimek Congr. "Options for Closed Water" Sustainable Water Management,* Wageningen 1998.

II Waste Gas Treatment

General Aspects

11 Process Engineering of Biological Waste Gas Purification

MUTHUMBI WAWERU
VEERLE HERRYGERS
HERMAN VAN LANGENHOVE
WILLY VERSTRAETE
Gent, Belgium

1 Introduction

Process engineering of biological waste gas purification aims at the selection and operation of biological waste gas purification technologies with the ultimate aim of assuring mass transfer and biodegradation of one or more pollutants in a waste gas stream. Biodegradation of the pollutants occurs when the microorganisms use the pollutants as a carbon source or an electron donor. In some special cases, the microorganisms using a particular substrate such as glucose, ethanol, etc., can also oxidize another pollutant. The latter is due to unspecific metabolism by the enzymes of organisms and is called co-metabolism (ALEXANDER, 1981). The extent to which biological waste gas purification can occur is determined mainly by the physico-chemical characteristics of the pollutant(s), the intrinsic capabilities of the microbial physiology and ecology, and the operating and environmental conditions.

When selecting the bioreactor technology, focus is placed on the operational and control requirements needed to ensure an optimal chemical and physical environment for mass transfer and biodegradation in order to achieve a high and constant removal efficiency of the pollutant.

2 Biological Waste Gas Purification Technology

2.1 General Characteristics

Biological waste gas purification technology currently includes bioreactors known as: biofilters, biotrickling filters, bioscrubbers, and membrane bioreactors. The mode of operation for all these reactors is very similar. Air containing volatile compounds is passed through the bioreactor where the volatile compounds are transferred from the gas phase into the liquid phase. Microorganisms, such as bacteria or fungi, grow in this liquid phase and are involved in the removal of the compounds acquired from the air. The microorganisms responsible for the biodegradation normally grow as a mixture of different organisms. Such a mixture of different bacteria, fungi, and protozoa depends on a number of interactions and is often referred to as a microbial community. Microorganisms are generally organized in thin layers called biofilms. In most cases the pollutants in the air (such as toluene, methane, dichloromethane, ethanol, carboxylic acids, esters, aldehydes, etc.; TOLVANEN et al., 1998) act as a source of carbon and energy for growth and maintenance of the microorganisms. It must be noted that some waste gases, such as those produced during composting, are composed of many (often up to several hundreds) different chemicals such as alcohols, carbonyl compounds, terpenes, esters, organosulfur compounds, ethers, ammonia, hydrogen sulfide, and many others (TOLVANEN et al., 1998; SMET et al., 1999). The remarkable aspect of the microbial community is that it generally develops to a composition so that all these different chemicals are removed and metabolized simultaneously. Microorganisms also require essential nutrients and growth factors in order to function and produce new cells. The latter include nitrogen, phosphorus, sulfur, vitamins, and trace elements. Most often these nutrients and growth factors are not present in the waste gas and have to be supplied externally.

There are fundamental differences between the four types of reactors mentioned above. They range from the way microorganisms are organized (i.e., immobilized or dispersed) to the state of the aqueous phase in the reactor (i.e., mobile or stationary). The latter significantly influences the mass transfer characteristics of the system. A short description of each of the four types of bioreactors for biological waste gas purification currently in use is given below (see also Fig. 1).

2.2 Technology Types

2.2.1 Biofilter

In a biofilter the air is passed through a bed, packed with organic carrier materials, e.g., compost, soil, or wood bark. The compounds in the air are transferred to a biofilm which grows on the filter materials. The nutrients necessary for growth of the microorganisms are supplied by the organic matter. On top of the biofilm, there is a thin liquid layer. An important control parameter is the moisture content of the overall carrier matrix, which must be between 40 and 60% (w/w). To avoid dehydration, the air is generally humidified before entering the biofilter. If the waste gas contains high levels of solid particles (i.e., the waste gas is an aerosol), an aerosol removal filter can be installed before the humidification chamber. This prevents clogging of the biofilter by the particles.

2.2.2 Biotrickling Filter

A biotrickling filter is very similar to a biofilter. In this case, pollutants are also transferred from the gas phase to a biofilm, which grows on a packing material. However, the packing materials are made of chemical inert materials, such as plastic rings. Because nutrients are not available in these materials, they have to be supplied to the microorganisms by recirculating a liquid phase through the reactor in co- or countercurrent flow.

2.2.3 Bioscrubber

A bioscrubber consists of two reactors. The first part is an absorption tower, where pollutants are absorbed in a liquid phase. This liquid phase goes to a second reactor, which is a kind of activated sludge unit. In the latter, microorganisms growing in suspended flocs in the water, degrade the pollutants. The effluent of this unit is recirculated over the absorption tower in a co- or countercurrent way to the flow of the waste gas.

2.2.4 Membrane Bioreactor

In a membrane bioreactor, the waste gas stream is separated from the biofilm by a membrane which is selectively permeable to the pollutants. One side of the membrane is in contact with a liquid phase supplemented with nutrients while the other side is in contact with the waste gas stream. The nutrient rich liquid phase is inoculated with microorganisms capable of degrading the pollutant. These microorganisms organize themselves and form a biofilm attached onto the membrane. As the pollutants migrate through the selectively permeable membrane, they enter the nutrient rich liquid phase and are consequently degraded. The liquid phase is maintained in a reservoir where the nutrients are refreshed, oxygen is supplied, and pH and temperature are controlled. Different types of membranes can be used, such as polar or hydrophobic ones. They can be installed in different configurations, i.e., tubular or flat sheets. In Fig. 1 a close-up of the site of biological activity in the four types of bioreactors is illustrated. Note that in a bioscrubber the microorganisms are in the second reactor fully suspended as flocs or granules in the liquid.

3 Performance Parameters

Different biological waste gas purification technologies can be compared based on the performance using a set of parameters. These parameters include

- empty bed contact time [s],
- surface loading rate [$m^3\ m^{-2}\ h^{-1}$],
- mass loading rate [$g\ m^{-3}\ h^{-1}$],
- volumetric loading rate [$m^3\ m^{-3}\ h^{-1}$],
- elimination capacity [$g\ m^{-3}\ h^{-1}$],
- removal efficiency [%].

The subsequent outline particularly relates to the biofilter type of reactor.

3.1 Empty Bed Contact Time or True Contact Time

The residence time of the gas in a bioreactor can be calculated in two different ways;

(1) Superficial residence time or empty bed residence time based on the total volume of the reactor and referred to as empty bed contact time (EBCT)

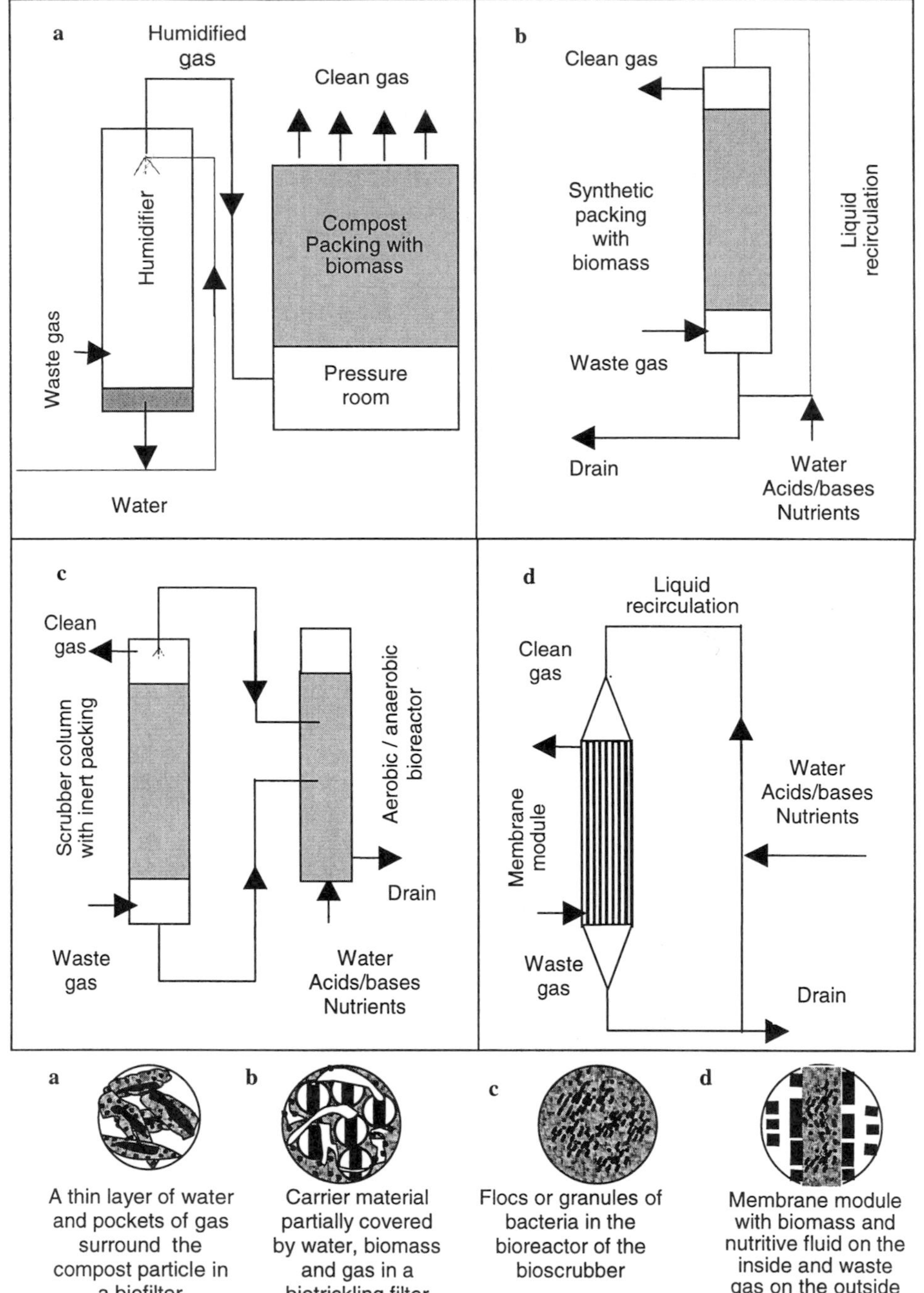

Fig. 1a–d. Schematic representation of four different types of bioreactors used in biological waste gas purification and close-up view of their respective microbial configurations; (**a**) biofilter, (**b**) biotrickling filter, (**c**) bioscrubber, (**d**) membrane bioreactor.

$$EBCT = \frac{V \cdot 3{,}600}{Q} \quad [\mathrm{s}] \tag{1}$$

where V = volume of the filter material in the reactor [m^3]
and Q = waste gas flow rate [$m^3\,h^{-1}$].

(2) True residence time τ, which is based on the free space in the reactor and defined as

$$\tau = \frac{\varepsilon \cdot V \cdot 3{,}600}{} \quad [\mathrm{s}] \tag{2}$$

where ε = porosity of the packing materials (without dimension).

In many cases the exact porosity needed to calculate the true residence time, is not known. Hence, most often the empty bed contact time is used. The EBCT is typically used for comparison of gas residence times in different reactor technologies or under different loading conditions. However, one has to keep in mind that this gives an overestimation of the true residence time. Due to preferential currents through the larger voids in the packing, there can be a considerable deviation of the actual residence time from the calculated residence time.

The residence time in the reactor is useful as an indicator of the time available for mass transfer of the pollutant from the gas phase to the liquid phase through the biofilm. The latter is often the factor limiting the microbial degradation.

3.2 Surface Loading Rate (B_A)

The surface loading rate indicates the amount of air that is passed through the bioreactor per unit surface area per unit time.

$$B_A = \frac{Q}{A} \quad [m^3\,m^{-2}\,h^{-1}] \tag{3}$$

where A = total surface of the packing or filter material in the bioreactor [m^2].

One can also express the velocity of the gas [$m\,h^{-1}$] through the empty reactor. However, the reactor is normally filled with packing materials, which results in a higher velocity gas compared to the surface loading rate.

3.3 Mass Loading Rate (B_V)

The mass loading rate gives the amount of pollutant which is introduced into the bioreactor per unit volume and per unit time.

$$B_V = \frac{Q \cdot C_{g\text{-}in}}{V} \quad [g\,m^{-3}\,h^{-1}] \tag{4}$$

where $C_{g\text{-}in}$ = concentration of the pollutant in the inlet waste gas stream [$g\,m^{-3}$].

3.4 Volumetric Loading Rate (v_S)

The volumetric loading rate is the amount of waste gas passed through the reactor per unit reactor volume.

$$v_S = \frac{Q}{V} \quad [m^3\,m^{-3}\,h^{-1}] \tag{5}$$

3.5 Elimination Capacity (EC)

The elimination capacity EC gives the amount of pollutant removed per volume bioreactor per unit time. An overall elimination capacity is defined by Eq. (6).

$$EC = \frac{Q \cdot (C_{g\text{-}in} - C_{g\text{-}out})}{V} \quad [g\,m^{-3}\,h^{-1}] \tag{6}$$

where $C_{g\text{-}out}$ = concentration of the pollutant in effluent waste gas [$g\,m^{-3}$].

3.6 Removal Efficiency (RE)

Removal efficiency is the fraction of the pollutant removed in the bioreactor expressed as a percentage. It is defined as

$$RE = \frac{(C_{g\text{-}in} - C_{g\text{-}out})}{C_{g\text{-}in}} \cdot 100\ [\%] \tag{7}$$

It should be noted that the different parameters are interdependent. There are only 4 independent design parameters: reactor height, volumetric loading rate, gas phase concentration at inlet ($C_{g\text{-in}}$) and outlet ($C_{g\text{-out}}$).

4 Characteristics of the Waste Gas Stream

There are several characteristics of the waste gas stream that have to be known when considering the implementation of biological waste gas purification technologies. Tab. 1 gives a list of characteristics of the waste gas stream which are essential in order to correctly design the biological purification system.

Physical parameters such as relative humidity and temperature are important, because they have considerable influence on the microbial degradation of the pollutant. Different microorganisms have different optimal ranges of temperature and relative humidity for growth. Temperature also has an effect on the partitioning of the pollutant between the gas and the liquid phase. The waste gas flow rate influences the volumetric loading rate of pollutant on the biologically active phase and, hence, the elimination capacity. Equally important are the identity and the concentration of the pollutant and/or odor units in the waste gas stream, because they influence the overall efficiency of the biological waste gas system.

Tab. 1. Important Characteristics of the Waste Gas Stream

Parameter	Unit
Relative humidity	%
Temperature	°C
Waste gas flow rate	$m^3 h^{-1}$
Pollutant identity (chemistry)	
Pollutant concentration	$g m^{-3}$
Odor concentration	$ou m^{-3}$

Odor unit (ou) is the amount of (a mixture of) odorous compounds present in 1 m^3 of odorless gas (under standard conditions) at the panel threshold (CEN, 1998).

It is also important to establish the chemical composition of the waste gas stream before starting with the design of the treatment system. The microbial degradability of the pollutant in the waste gas stream is largely dependent on its chemical identity. The pollutant may be of organic or of inorganic nature. Typical organic pollutants which are often encountered in the waste gas streams are ethers, ketones, fatty acids, alcohols, hydrocarbons, amines, and organosulfur compounds. Valuable information about the biodegradability of chemicals can be obtained, e.g., from VAN AGTEREN et al. (1998). Waste gases can also contain inorganic compounds such as NH_3, NO_2, NO, H_2S, and SO_2. Some of these compounds might be present at toxic levels or they reduce the degradation capacity by, e.g., acidifying the biofilter material. Therefore, these compounds either have to be eliminated before the waste gas stream enters the bioreactor or control of pH has to be installed. Different compounds can also affect each other's degradation, without being toxic to the microorganisms. SMET et al. (1997) reported that isobutyraldehyde (IB) had to be removed by a first layer of the biofilter before a *Hypohomicrobium*-based microbial community in a subsequent layer could develop and metabolize the dimethyl sulfide (DMS) present in the waste gas. In separate batch experiments it was shown that the same *Hypohomicrobium* sp. switched its metabolism from using IB to the consumption of DMS, when IB concentrations decreased.

When bioreactors are used for the abatement of odor problems, the odor concentration of the waste gas has to be determined as well. The odor concentration in odor units per cubic meter [$ou m^{-3}$] corresponds to the number of times a waste gas sample has to be diluted with reference air before the dilution can be distinguished from the reference air by 50% of a standard panel. In this respect, the European Committee for Standardisation (CEN) is currently involved in standardizing the determination of odor compounds by dynamic olfactometry. This will improve the reproducibility of olfactometric measurements basically by using panels standardized with respect to 1-butanol (detection threshold of 40 ppbv) (CEN, 1998). Although the evaluation of bioreactor performance aimed at odor reduction has to be based on olfactometric measurements, design and optimization always require chemical characterization of the overall process.

The concentration of the pollutants and/or odor units largely depends on the source of the waste gas stream. Waste gas streams from, e.g., hexane oil extraction processes have a pollutant concentration in the range of a few g m^{-3}. On the other hand, for waste gas streams polluted with offensive odors, concentrations of the odorous compounds can be in the range of mg m^{-3} or less (SMET et al., 1998).

The characteristics of the waste gas stream determine to a large extent the type of bioreactor system that can be used. Tab. 2 gives a first indication of the suitability of bioreactors for waste gas purification in relation to the characteristics of the waste gas stream. Note the preponderant importance of the Henry coefficient. Chemicals which dissolve easily in water (hydrophilic substances) can be retained efficiently by means of scrubbing with water. Chemicals which are poorly water soluble (high Henry coefficient) are better dealt with by means of a biofilter. In Tab. 2, the membrane reactor is not mentioned; depending on the nature of the membranes, it can be suited to handle a range of compounds (STERN, 1994).

5 Process Principles

Several processes take place in biological waste gas cleaning systems. They include partitioning of the pollutant from the gaseous to the liquid phase followed by its diffusion from the bulk liquid to the biofilm. The process of microbial degradation of the pollutant takes place in the biofilm, and the end products diffuse back into the bulk liquid. Mass transfer is the combined migration of compounds from the gaseous to the liquid phase and from the bulk liquid to the biofilm (Fig. 2).

Tab. 2. Range of Pollutant Concentration, Henry Coefficient, and Concomitant Operating Parameters of Biofilters, Biotrickling Filters, and Bioscrubbers (after VAN GROENESTIJN and HESSELINK, 1993)

	Biofilter	Biotrickling Filter	Bioscrubber
Pollutant concentration [g m^{-3}]	<1	<0.5	<5
Henry coefficient (dimensionless)	<10	<1	<0.01
Surface loading rate [m^3 m^{-3} h^{-1}]	50–200	100–1,000	100–1,000
Mass loading rate [g m^{-3} h^{-1}]	10–160	<500	<500
Empty bed contact time [s]	15–60	30–60	30–60
Volumetric loading rate [m^3 m^{-3} h^{-1}]	100–200		250–580
Elimination capacity [g m^{-3} h^{-1}]	10–160		
Removal efficiency [%]	95–99		85–95

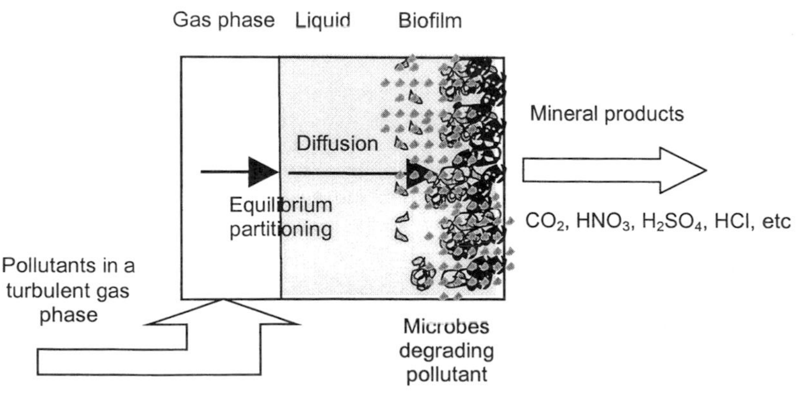

Fig. 2. Schematic view of the sequence of processes leading to microbial degradation of pollutants in a biofilter.

5.1 Equilibrium Partitioning of the Pollutant

The first step towards microbial degradation of the pollutant is the partitioning of the gaseous pollutants to the liquid phase. In a bioscrubber and biotrickling filter this is obvious, but also in a biofilter a small water layer will normally be present on top of the microbial biofilm.

When describing gas/liquid mass transfer, interfacial resistance between the liquid and the gas is often neglected. For practical reasons, it is normally assumed that straightforward partitioning of the pollutant between the two phases occurs and the resulting concentration in both the gas and the liquid phase is at equilibrium. Equilibrium partitioning largely depends on the Henry constant of the pollutant.

The concentrations of the pollutant in both the gas phase and the liquid phase are related by Eq. (8) (SANDER, 1999).

$$K_H = \frac{C_g}{C_l} \tag{8}$$

where
K_H = dimensionless Henry constant,
C_g = gas phase concentration [mol m^{-3} or g m^{-3}],
C_l = liquid phase concentration [mol m^{-3} or g m^{-3}].

For pollutants with high Henry constants, the partitioning of the pollutant to the liquid phase is very poor. In Tab. 3 some Henry constants are compared for different kinds of pollutants. The Henry constant varies with temperature and with salinity of the water. DEWULF et al. (1995) carried out several measurements of Henry constants. The authors found that, in general, the Henry constants increase with a decrease in temperature while they increase with an increase in salinity as expressed by Eq. (9):

$$\ln K_H = a\,\frac{1}{T} + bZ + c \tag{9}$$

where
K_H = dimensionless Henry constant,
a, b, c = constants,
T = absolute temperature [K],
Z = salt concentration [g L^{-1}].

In a biofilter, there is a low water content (40–60%) and, therefore, the gas/liquid mass transfer takes place with less interfacial resistance than in a biotrickling filter or a bioscrubber where the water content is higher. In a membrane bioreactor there is no gas/liquid interface. Therefore, this reactor may be very suitable to treat pollutants with high Henry coefficients, provided the membrane is quite apolar such as, e.g., in the case of silicone membranes (DE SMUL and VERSTRAETE, 1999). However, it should be noted that in a membrane bioreactor, two mass transfers have to be dealt with: gas/membrane and membrane/biofilm. For the gas/membrane mass transfer, an equation similar to the Henry equation can be used.

$$S = \frac{C_m}{C_g} \tag{10}$$

where
S = solubility ratio (dimensionless),
C_g = gas phase concentration [mol m^{-3} or g m^{-3}],
C_m = pollutant concentration in the membrane [mol m^{-3} or g m^{-3}].

Methods to enhance mass/liquid mass transfer have been explored. Addition of a surface active reagent, as, e.g., silicone oil to the liquid

Tab. 3. Dimensionless Henry Constants for Pollutants Treatable with a Biotechnological Waste Gas Treatment System (HOWARD and MEYLAN, 1997)

Compound	Ethanol	Butanone	Isobuteraldehyde	Dimethylsulfide	Trichloroethene	Limonene	Hexane
K_H (25 °C)	0.00021	0.0023	0.0074	0.0658	0.403	0.82	74

phase gives good results in lab-scale biofilters (BUDWILL and COLEMAN, 1997). In a biotrickling filter, an intermittent circulation can be used to enhance the transfer of the poorly water soluble pollutant into the biofilm. DE HEYDER et al. (1994) used this approach for the removal of ethene from air and obtained an increase in removal of ethene by a factor of 2.25.

5.2 Diffusion

Migration of the pollutant from the bulk liquid to the biologically active phase (biofilm) occurs by means of diffusion. Diffusion can be described using Fick's law:

$$J = -D \frac{dC_l}{dx} \tag{11}$$

where
J = mass flux [mol m^{-2} s^{-1} or g m^{-2} s^{-1}],
D = diffusion coefficient [m^2 s^{-1}],
C_l = liquid concentration [mol m^{-3} or g m^{-3}],
x = distance in the biofilm [m].

The value of the effective diffusion coefficient D varies over different orders depending on the medium. In Tab. 4 the diffusion coefficient of a few compounds in different media is given.

The diffusion is much slower in water than in air and furthermore slower in a membrane than in water. In porous membranes, diffusion occurs through the fluid in the pores, and in dense membranes it occurs through the membrane material itself. When choosing a membrane, a study should be made to determine an appropriate material with high diffusion characteristics for the given pollutant.

In the bioreactors without membranes, the pollutant has to pass through the water phase before it reaches the biofilm. The limit between the water phase and the biofilm is as yet vaguely defined. The diffusion coefficient will vary between that in water and that in a biofilm (DEVINNY et al., 1999). Roughly, the diffusion in a biofilm is about 0.5–0.7 of that in water. Most important in the whole concept is that the concentration gradient needed for diffusion flux will be maintained by a constant input from the gas phase and the removal of pollutants by the microbial degradation in the biologically active phase. Fig. 3 illustrates the concentration gradient that exists between the gas phase through the liquid phase to the biofilm. Although microorganisms have high affinity for most biodegradable substrates, they can have difficulties in exhausting substances below levels of 1 μg L^{-1} of water (VERSTRAETE and TOP, 1992). This so-called lower microbial threshold means that for substances with high Henry constants, the corresponding levels (C_{thr} H) remain in the gas phase. Various approaches have been described to derive appropriate kinetic parameters from such performance curves (DE HEYDER et al., 1997b).

5.3 Microbial Degradation of the Pollutant

Microbial metabolism of pollutants readily occurs when the pollutants are used as a source of energy. For instance, toluene is used as electron and carbon donor by several organotrophic bacteria; they use oxygen as elec-

Tab. 4. Diffusion Coefficients of some Compounds in Air, Water, and Membrane Materials (after REID et al., 1987)

Compound	D_{air} [m^2 s^{-1}]	D_{water} [m^2 s^{-1}]	Membrane Material	$D_{membrane}$ [m^2 s^{-1}]
Oxygen			natural rubber	$2.5 \cdot 10^{-10}$
Oxygen (25 °C)	$1.40 \cdot 10^{-5}$	$2.50 \cdot 10^{-9}$	polydimethyl siloxane (35 °C)	$4.0 \cdot 10^{-9}$
Ethanol	$1.24 \cdot 10^{-5}$	$1.13 \cdot 10^{-9}$	poly(vinyl acetate)	$1.5 \cdot 10^{-13}$
CO_2	$1.64 \cdot 10^{-5}$	$2.00 \cdot 10^{-9}$	PMDA-MDA (20 °C)	$9.0 \cdot 10^{-13}$
Benzene	$1.20 \cdot 10^{-5}$	$1.30 \cdot 10^{-9}$	poly(vinyl acetate)	$4.8 \cdot 10^{-17}$

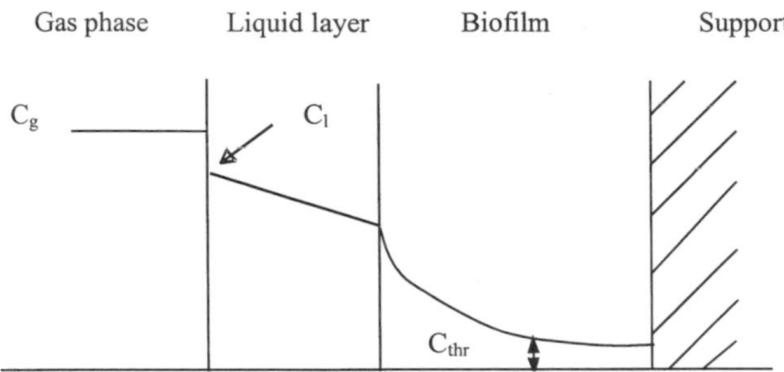

Fig. 3. Concentration gradient of the pollutant between the gas phase through the liquid phase to the biofilm. C_g: concentration in the gas phase, C_l: concentration in the liquid phase, C_{thr}: lower threshold level.

tron acceptor. Ammonium is used as electron donor by the lithotrophic nitrifying bacteria; they use oxygen as electron acceptor and carbon dioxide as carbon source to build up cell biomass (FOCHT and VERSTRAETE, 1977). In some cases, the pollutant can be a cosubstrate. For instance, trichloroethene can be metabolized together with toluene (MU and SCOW, 1994).

Sufficient availability of nutrients such as minerals, vitamins, and growth factors is essential for proper growth of the microbial community. Hence, the microbial biomass acts as a kind of (bio)catalyst that constantly maintains itself (e.g., by dying off and regrowing).

The energy released during degradation of the pollutants will be used for maintenance metabolism and for growth of the microorganisms according to the modified Monod equation given below.

$$\frac{dC_l}{dt} = \left(\frac{\mu}{Y_{XS}}\right) X \tag{12}$$

where

C_l = concentration of the substrate dissolved in the liquid [g m^{-3}],

μ = growth rate [g g^{-1} h^{-1}],

Y_{XS} = yield of dry cell weight per mass of substrate metabolized [g g^{-1}],

m = maintenance energy consumption [g substrate (g cell dry weight)$^{-1}$ h^{-1}],

X = dry weight of biomass in the biofilm or suspension [g m^{-3}].

It must be recognized that in practice it is difficult to get values for these parameters. Therefore, it is recommended that reactor performance should be designed based on pilot experiments.

In a waste gas treatment reactor, growth of the bacteria is often minimal. This means that all the released energy mainly serves for maintenance metabolism of the bacteria. The advantage is that there is little or no waste sludge in contrast to, e.g., wastewater treatment systems. Often one deliberately tries to minimize the growth of excess biomass, e.g., by limiting the supply of mineral nutrients such as phosphate (WUBKER and FRIEDRICH, 1996) or by seeding the reactor with protozoa which graze on the bacteria (COX and DESHUSSES, 1999). Microbial degradation of the pollutant depends on a multitude of factors. The most important external factors are temperature, nutrient availability, toxicity of the gaseous components. There is also a series of internal factors directly related to the way microorganisms develop and work. For instance, different species, each with different capabilities can cooperate and achieve very effective pollutant removal. DE HEYDER et al. (1997a) described the stimulation of ethene removal by *Mycobacterium* sp. in the presence of an active nitrifying population. VEIGA et al. (1999) identified two different bacterial species (*Bacillus* and *Pseudomonas*) and a fungus (*Trichosporon*) as cooperative functional agents in a biofilter treating alkylbenzene gases.

The microorganisms (biomass) can be introduced into the bioreactor in several ways. In some cases, natural sources such as night soil, aquatic sediments, or sludge from wastewater treatment plants are used as inoculum. Moreover, in the case of a biofilter, the carrier material (compost, wood bark, etc.) already has a naturally occurring microbial community. In other cases, specific bacteria or mixtures of isolated strains of naturally occurring bacteria that can metabolize the pollutant in the waste gas stream are introduced into the bioreactor (KENNES and THALASSO, 1998). Such seeding is referred to as bioaugmentation. To accelerate the removal of a particular recalcitrant pollutant, one could make use of genetically modified microorganisms with improved degradation capacities. For bioscrubbers in which the removal occurs in an activated sludge reactor system, bioaugmentation as described for wastewater systems by VAN LIMBERGEN et al. (1998) could be applied.

Generally, inoculation of the reactor with appropriate bacteria decreases the start-up period for these reactors significantly (SMET et al., 1996). However, the operating conditions and the prevailing environmental factors generally exert selective pressure on the microorganisms present in the bioreactor resulting in the development of a specific microbial community. This community may be quite different from the enrichment culture which was introduced into the reactor (BENDINGER, 1992). The development of the structure and function of the microbial community in the bioreactor affects the microbial degradation rate and, hence, the extent of pollutant removal. The microorganisms are surrounded by an extracellular organic layer, which has normally a negative charge and serves many functions including adhesion, protection, carbon storage, and ion exchange (BISHOP and KINNER, 1986).

Most often, biological waste gas treatment results in non re-usable end products. In the case of SO_2 scrubbing, however, a special approach has been developed (DE VEGT and BUISMAN, 1995; VERSTRAETE et al., 1997). As schematized in Fig. 4, by a sequence of biotechnological reactors the SO_2 is recovered as sulfur powder.

6 Reactor Performance

The overall performance of the bioreactor is determined to a large extent by mass transfer as governed by equilibrium partitioning at the gas/liquid phase and diffusion from the bulk liquid to the bioactive phase (biofilm) combined with microbial degradation of the pollutant. As mentioned before, this is expressed in removal efficiency [%] or elimination capacity [g m^{-3} h^{-1}]. Fig. 5 shows a typical elimination capacity as a function of the mass loading rate. Such a performance diagram can experimentally be determined by changing the concentration in the influent gas flow and measuring the resulting elimination capacity. In most studies these curves are determined during short-term experiments. This means that no significant growth of biomass is allowed to occur during the experiment. In Fig. 5, two main regions can be distinguished. At the lower mass loading rate the elimination capacity is equal to the mass loading, and the removal efficiency is 100%. In this range the reactor kinetics are first-order kinetics; the microbial metabolism normally represented by the Monod equation becomes a simple first order equation.

$$-\frac{dC_l}{dt}=\frac{K\cdot C_l}{K_S+C_l} \qquad (13)$$

with

C_l = concentration in the liquid [mol m^{-3} or g m^{-3}] substrate,

K = maximum conversion rate [g substrate (g cell dry weight)$^{-1}$ h^{-1}],

K_S = substrate level at which the biomass works at half-maximum velocity [mol m^{-3} or g m^{-3}].

When $K_S \geq C_l$, Eq. (13) becomes

$$-\frac{dC_l}{dt}=kC_l \qquad (14)$$

All pollutants which are fed to the biofilter are removed from the air. Low mass loading can be achieved by a low concentration of the pollutants in the gas phase and by a low gas flow rate. When the mass loading increases, com-

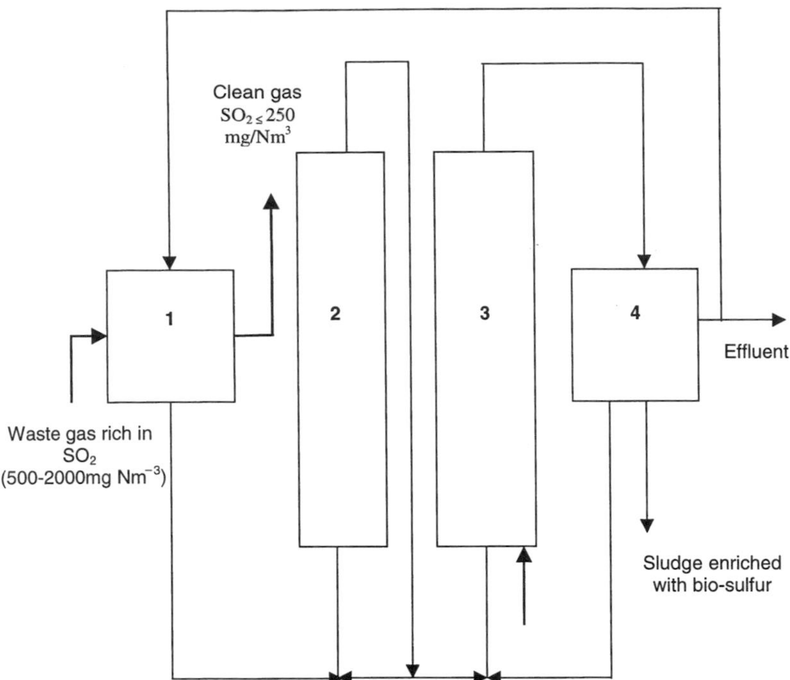

Fig. 4. Conversion of SO_2 by means of a sulfate reduction and subsequent sulfide oxidation reactor to biologically formed elemental sulfur (after GROOTAERD et al., 1977). 1: absorption of SO_2 gas, 2: sulfate reduction, 3: partial oxidation of hydrogen sulfide, 4: separation of sludge enriched with bio-sulfur.

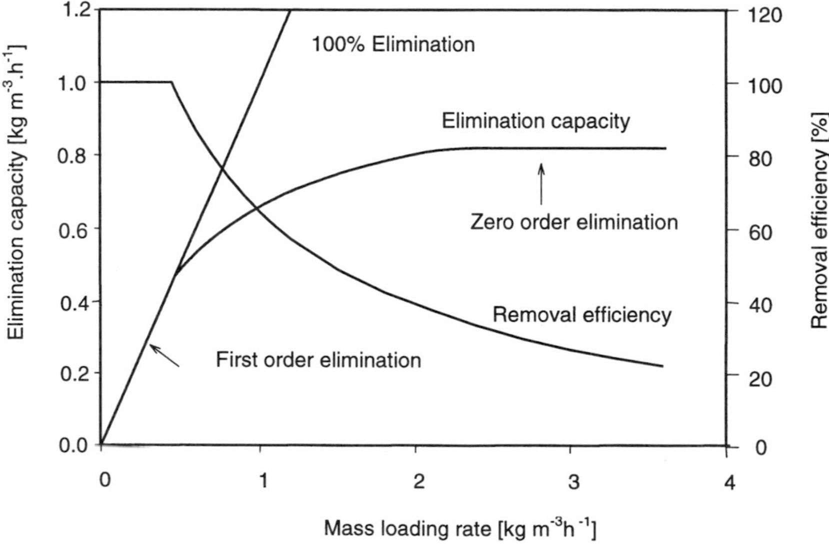

Fig. 5. Typical curve for the elimination capacity of a biofilter.

plete removal of the pollutants is not possible anymore. At even higher mass loading, the elimination capacity does not increase further and remains at a steady value. This is the region of zero-order kinctics: the elimination capacity is independent of the mass loading rate. At this stage the removal efficiency [%] will decrease by further increasing the mass loading rate. Two phenomena explain the incomplete removal of the pollutants: diffusion and reaction limitation. In the case of diffusion limitation, not all the pollutants diffuse into the biofilm and not all microorganisms can take part in the degradation of the pollutant. When the diffusion rate is slower than the degradation rate, e.g., for pollutants with a high Henry coefficient, the concentration in the liquid phase will be low compared to the gas phase. In the case of reaction limitation, the pollutants diffuse into the complete biofilm, but the pollutants are not removed rapidly and sufficiently enough by the biocatalyst. Indeed, microbial metabolism might be hampered by other limiting factors such as shortage of nutrients, presence of toxicants, etc.

7 Reactor Control

The environmental factors prevailing in the bioreactor such as pH, temperature, oxygen level, water content affect the ability of the microorganism to metabolize the pollutant. Prevailing environmental factors determine to a large extent the composition of the microbial community in the bioreactor. Most species of microorganisms exhibit optimal growth over a certain pH range (DEVINNY et al., 1999). Often, the pH range of about 6–8 is suitable for most microorganisms, but some species can tolerate lower or higher pH. Microbial activity is strongly influenced by temperature. Some microorganisms operate optimally in the mesophilic temperature range (15–40 °C) while others do so in the thermophilic temperature range (40–60 °C).

The microbial activity and the mass transfer of the pollutant from the gas phase to the biofilm are related to the water content in the bioreactor. This is especially the case for biofilters that operate optimally at 40–60% relative humidity (DEVINNY et al., 1999). Excess water in a biofilter may lead to loss of nutritive supplements (SMET et al., 1996). Moreover, wet pockets may be formed in which the diffusion of both pollutant and oxygen used as an electron acceptor for the microorganisms becomes limiting.

In practice, it is quite difficult to implement control and mitigation actions in biofilter systems. On the contrary, due to the circulation of a liquid in the other reactor types, it generally is possible to optimize the latter reactor systems for temperature, pH, and nutrient supply. Eventually, if required, one can also supplement a cosubstrate to the needs for the bacteria.

8 Perspectives

Biological waste gas purification processes have strong competitors such as activated carbon sorption and incineration. Also other physical-chemical techniques, with a small footprint, are more frequently marketed for odor treatment of air, e.g., ozonization and UV treatment. The main advantages of the biocatalytic removal of pollutants are the low investment and operation costs. The main disadvantages are the often slow start-up and the limited reliability, e.g., as a result of changing environmental conditions and autointoxication.

The major efforts in the near future for bioprocess engineers should, therefore, be directed to develop reactors with a controllable microbial biomass. Adequate natural or even possibly genetically modified organisms should be available as ready to use industrial biocatalysts. They should, as is the case for activated carbon, upon introduction in the reactor be immediately operational. Moreover, the overall environment and performance in the reactor should be monitored on-line and, if necessary, adjusted. These aspects, mass production of biocatalysts with a guaranteed quality and implementation of process control, are crucial for the future of biological waste gas treatment.

Acknowledgement
This paper was in part supported by a grant of the Fonds Wetenschappelijk Onderzoek, Belgium.

9 References

ALEXANDER, M. (1981), Biodegradation of chemicals of environmental concern, *Science* **11**, 132–138.

BISHOP, P. L., KINNER, E. N. (1986), Aerobic fixed film processes, in: *Biotechnology*, 1st Edn. Vol. 8 (REHM, H.-J., REED, G., Eds.), pp. 113–176. Weinheim: VCH.

BENDINGER, B. (1992), Microbiology of biofilters for the treatment of animal-rendering plant emissions: occurrence, identification, and properties of isolated coryneform bacteria, *Thesis*, University of Osnabrück, Germany.

BUDWILL, K., COLEMAN, R. N. (1997), Effect of silicone oil on biofiltration of *n*-hexane vapors. *Forum for Applied Biotechnology, Mededelingen Faculteit Landbouwkundige en Toegepaste Biologische Wetenschappen* **62**, 1521–1527.

CEN (Comité Européen de Normalisation) (1998), Odors (Doc. Nr. *Cen/Tc 292/WG 2*) Air Quality-Determination of Odor Concentration by Means of Dynamic Olfactometry.

COX, H., DESHUSSES, M. (1999), Biomass control in waste air biotrickling filters by protozoan predation, *Biotechnol. Bioeng.* **62**, 216–224.

DE HEYDER, B., OVERMEIRE, A., VAN LANGENHOVE, H., VERSTRAETE, W. (1994), Ethene removal from a synthetic waste gas using a dry biobed, *Biotechnol. Bioeng.* **44**, 642–648.

DE HEYDER, B., VAN ELST, T., VAN LANGENHOVE, H., VERSTRAETE, W. (1997a), Enhancement of ethene removal from waste gas by stimulating nitrification, *Biodegradation* **8**, 21–30.

DE HEYDER, B., VANROLLEGHEM, P., VAN LANGENHOVE, H., VERSTRAETE, W. (1997b), Kinetic characterization of mass transfer limited biodegradation of a low water soluble gas in batch experiments – necessity for multiresponse fitting, *Biotechnol. Bioeng.* **55**, 511–519.

DE SMUL, A., VERSTRAETE, W. (1999), The phenomenology and the mathematical modeling of the silicone-supported chemical oxidation of aqueous sulfide to elemental sulfur by ferric sulphate, *J. Chem. Technol. Biotechnol.* **74**, 456–466.

DE VEGT, A. L., BUISMAN, C. J. N. (1995), Thiopaq bioscrubber: An innovative technology to remove hydrogen sulfide from air and gaseous streams, *Proc. GRI Sulfur Recovery Conf.* Austin, TX, Sept. 24–27, 1995.

DEVINNY, J. S., DESHUSSES, M. A., WEBSTER, T. S. (1999), *Biofiltration for Air Pollution Control.* New York: Lewis Publishers.

DEWULF, J., DRIJVERS, D., VAN LANGENHOVE, H. (1995), Measurement of Henry's law constant as function of temperature and salinity for the low temperature range, *Atmospher. Environ.* **29**, 323–331.

FOCHT, D., VERSTRAETE, W. (1977), Biochemical ecology of nitrification and denitrification, *Adv. Microb. Ecol.* **1**, 135–214.

VAN GROENSTIJN, J. W., HESSELINK, P. G. M. (1993), Biotechniques for air pollution control, *Biodegradation* **4**, 283–301.

GROOTAERD, H., DE SMUL, A., DRIES, J., GOETHALS, L., VERSTRAETE, W. (1997), Epuration biologique des eaux de lavage des fumées riches en SO_2, *Procédés d'une journée d'étude organisée au Faculté Polytechnique de Mons*, 13–15 mai 1997.

HOWARD, P. H., MEYLAN, W. M. (1997), Handbook of physical properties of organic chemicals. London: Lewis Publishers.

KENNES, C., THALASSO, F. (1998), Waste gas biotreatment technology, *J. Chem. Technol. Biotechnol.* **72**, 303–319.

MU, Y. D., SCOW, K. M. (1994), Effect of trichloroethylene (TCE) and toluene concentrations on TCE and toluene biodegradation and the population density of TCE and toluene degraders in soil, *Appl. Environ. Microbiol.* **60**, 2662–2665.

REID, R. C., PRAUSNITZ, J. M., POLING, B. E. (1987), The properties of gases and liquids. Singapore: McGraw-Hill.

SANDER, R. (1999), Compilation of Henry's law constants for inorganic and organic species of potential importance in environmental chemistry. http://www.mpch-mainz-mpg.de. /+sander/res/henry.html.

SMET, E., CHASAYA, G., VAN LANGENHOVE, H., VERSTRAETE, W. (1996), The effect of inoculation and the type of carrier material used on the biofiltration of methyl sulfides, *Appl. Microbiol. Biotechnol.* **45**, 293–298.

SMET, E., VAN LANGENHOVE, H., VERSTRAETE, W. (1997), Isobutyraldehyde as a competitor of the dimethyl sulfide degrading activity in biofilters, *Biodegradation* **8**, 53–59.

SMET, E., LENS, P., VAN LANGENHOVE, H. (1998), Treatment of waste gases contaminated with odorous sulfur compounds, *Crit. Rev. Environ. Sci. Technology*, **28**, 89–117.

SMET, E., VAN LANGENHOVE, H., DE BO, I. (1999), The emission of volatile compounds during the aerobic and the combined anaerobic/aerobic composting of biowaste, *Atmospher. Environ.* **33**, 1295–1303.

STERN, S. A. (1994), Polymers for gas separations: the next decade, *J. Membrane Sci.* **94**, 1–65.

TOLVANEN, O. K., HANNINEN, K. I., VEIJANEN, A., VILLBERG, K. (1998), Occupational hygiene in biowaste composting, *Waste Managem. Res.* **16**, 525–540.

VAN AGTEREN, M. H., KEUNING, S., JANSSEN, D. B. (1998), *Handbook on Biodegradation and Biological Treatment of Hazardous Organic Compounds*. Dordrecht: Kluwer Academic Publishers.

VAN LIMBERGEN, H., TOP, E. M., VERSTRAETE, W. (1998), Bioaugmentation in activated sludge: Current features and future perspectives, *Appl. Microbiol. Biotechnol.* **50**, 16–23.

VEIGA, M. C., FRAGA, L. A., AMOR, L., KENNES, C. (1999), Biofilter performance and characterization of a biocatalyst degrading alkylbenzene gases, *Biodegradation* **10**, 169–176.

VERSTRAETE, W., TOP, E. (1992), Holistic environmental biotechnology, in: *Microbial Control of Pollution* (FRY, J. C., GADD, G. M., HERBERT, R. A., JONES, C. W., WATSON-CRAIK, I. A., Eds.). Society for General Microbiology Symposium Vol. 48. Cambridge: Cambridge University Press.

VERSTRAETE, W., TANGHE, T., DE SMUL, A., GROOTAERD, H. (1997), Anaerobic biotechnology for sustainable waste treatment, in: *Biotechnology in the Sustainable Environment* (SAYLER, G. S., SANSEVERION, J., DAVIS, K. L., Eds.), Proc. Conf. Biotechnology in the Sustainable Environment, April 14–17, 1996, Knoxville, TN. pp. 343–359. New York: Plenum Press.

WUBKER, S., FRIEDRICH, C. (1996), Reduction of biomass in a bioscrubber for waste gas treatment by limited supply of phosphate and potassium ions, *Appl. Microbiol. Biotechnol.* **46**, 475–480.

12 Microbiological Aspects of Biological Waste Gas Purification

KARL-HEINRICH ENGESSER

THORSTEN PLAGGEMEIER

Stuttgart, Germany

List of Abbreviations

BTEX	benzene, toluene, ethylbenzene, and xylenes
CNM	Corynebacteria/*Nocardia*/Mycobacteria
DCM	dichloromethane
DLVO	Derjaguin, Landau, Verwey, Overbeck
DOC	designed oligospecies culture
FAME	fatty acid methyl esters
GC	gas chromatograph
MTBE	methyl-*t*-butyl ether
MVOC	microbially produced volatile organic compounds
PEP	phosphoenol pyruvate
TCE	trichloroethylene

1 Introduction

Biofiltration in general has been the subject of numerous papers dealing with the different odoral, biological and process engineering aspects of this technique (Cox and Deshusses, 1998; Wu et al., 1999; Deshusses 1997; Wani et al., 1997; Plaggemeier et al., 1997; Engesser, 1992; Engesser et al., 1996; Li et al., 1996; Devinny et al., 1999). In this chapter, an overview about the biological fundamentals of biofiltration is given. It emphasizes the importance of analysis of the microbial flora in respective plants in order to gain knowledge for optimizational efforts in practical applications (Engesser et al., 1997). The shortage of sound data within this field definitely limits all efforts to improve all-day usability of biofiltration plants.

2 The Microbial Community

Like in natural habitats, in biological waste gas purification reactors not a pure culture, but a mixed culture is established. The overall performance of a biofilter may in most cases not only be attributed to one single species, but to a consortium consisting of primary degraders feeding on the pollutant, secondary degraders utilizing by-products of the pollutant degradation process, and predators (Fig. 1). The biomass may be aggregated in a biofilm (as in biological trickling filters or biofilters) or living planktonically in a liquid medium (as in bioscrubbers). Bacteria, fungi, and protozoa have been identified as members of microbial communities in biological waste gas reactors. In some cases they have also been used for inoculation (Tab. 1).

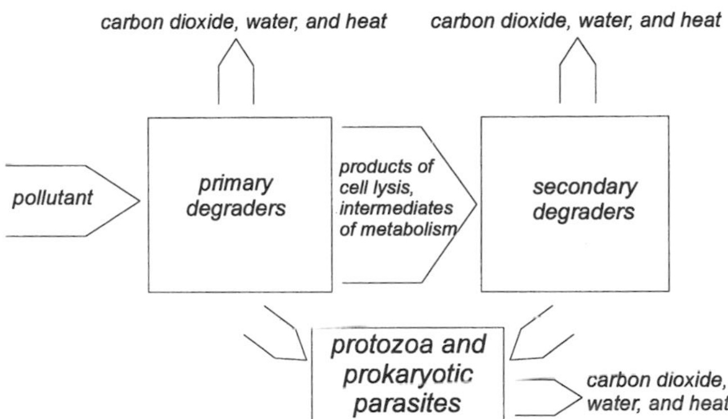

Fig. 1. Ecological relationships in complex mixed culture biocenoses. Primary degraders feeding on the pollutant oxidize and grow on it. Secondary degraders use cell lysis products and metabolic intermediates for growth, while predators feed on primary as well as on secondary degraders. Eventually, the pollutant is converted to carbon dioxide, water, heat, and biomass.

2.1 Biosuspensions

In bioscrubber systems, a cell suspension is used for the absorption of typically highly water soluble compounds such as ethanol from waste gas streams. The "loaded" cell suspension will be regenerated by microbial metabolic action in an aerated tank.

Conditions in this aerated tank are pretty close to conditions in continuously stirred industrial fermenters. Like in these, the cell suspension may be addressed as being homogeneous in terms of substrate concentration and composition of the microbial community. A bioscrubber system is close to a chemostat with biomass recycling: Cells and dissolved substrate enter an aerated reactor, in which the substrate is degraded, and a cell suspension – ideally substrate free – leaves the reactor.

Thereby, in contrast to fixed film systems (see Sect. 2.2), in suspended biomass systems it is possible that a single species – the one with the highest growth rate under the set conditions – will overgrow other species. Theoretically, mixed populations in bioscrubbers should at least be less complex than the populations in fixed film reactors. Population dynamics are still possible, but rather due to changes in the operation parameters and, therefore, changing selection pressure than as a result of the principle itself.

2.2 Biofilms

Being immobilized in biofilms offers a couple of advantages to microorganisms compared with living in "solubilized" state in a liquid medium. It allows them to stay in a desired microenvironment instead of being rinsed away. Furthermore, it protects them (at least those living in deeper biofilm layers) from toxic substances or predating protozoa. However, being surrounded by other organisms in a biofilm means that essential substances such as the carbon substrate or oxygen reach them rather by diffusional than by convectional transport, and the effective concentrations are lower than for planktonical microorganisms.

The settling of surfaces takes place in five major steps:

(1) adsorption of macromolecules to the surface,
(2) transport of microorganisms to the surface,

Tab. 1. Bacterial and Fungal Species Used for the Inoculation[a] of Biological Waste Gas Reactors or Found during the Process[b]

Strain	Reactor Type/Substrate	Reference
Hyphomicrobium MS3[a]	compost biofilter/dimethyl sulfide	SMET et al. (1996)
Methylobacterium sp., *Pseudomonas* sp., *Flavobacterium* sp., *Alcaligenes* sp.[b]	peat–perlite biofilter/methanol	SHAREEFDEEN et al. (1993)
Pseudomonas putida, P. mendocina[a]; *Burkholderia cepacia, B. vietnamiensis*[b]	Ralu rings trickling filter/ Solvesso 100	STOFFELS et al. (1998)
Methylomonas fodinarum[a]	glass tubes trickling filter/methane	SLY et al. (1993)
Thiobacillus thioparus TK-m[a]; Metazoa: Rotatoria, Nematoda; Protozoa: Mastigophora, Ciliata[b]	polypropylene pellets trickling filter/dimethyl sulfide, methyl mercaptan, hydrogen sulfide	TANJI et al. (1989)
Pseudomonas putida 54G[b]	flat plate bioreactor/toluene, ethanol	VILLAVERDE and FERNÁNDEZ-POLANCO (1999)
Cladosporium sphaerospermum[b]	biofilter/toluene	WEBER et al. (1995)
Pseudomonas fluorescens[a]	bioscrubber/ammonïa, butanal	WECKHUYSEN et al. (1994)
Pseudomonas putida, Rhodococcus sp., *Arthrobacter paraffineus*[a]	peat biofilters/toluene	WU et al. (1999)
Pseudomonas acidovorans[b]	peat biofilter/dimethyl sulfide	ZHANG et al. (1991)
Pseudomonas acidovorans DMR-11, *Hyphomicrobium* sp. I55[a]	peat biofilter/dimethyl sulfide	ZHANG et al. (1992)
Methylosinus trichosporium OB3b[a]	celite biofilter/methane, trichloroethene, dichloroethane	SPEITEL Jr. and MCLAY (1993)
Pseudomonas putida Tol1A[a]	compost–perlite biofilter/ dichloromethane,	ERGAS et al. (1994)
Thiobacillus ferrooxidans[a]	bubble column/hydrogen sulfide	HALFMEIER et al. (1993)
Methylobacterium spp.[a]	membrane reactor/nitrogen monoxide	HINZ et al. (1994)
Pseudomonas fluorescens[a]	Raschig rings trickling filter/ propionaldehyde	KIRCHNER et al. (1987)
Pseudomonas putida R1[b]	biofilm reactor/toluene	PEDERSEN et al. (1997)
Tsukamurella sp., *Pseudomonas* sp., *Sphingomonas* sp., *Xanthomonas* sp.[b]	peat biofilter/styrene	ARNOLD et al. (1997)
Thiobacillus thioparus DW44[a]	peat biofilter/sulfur containing gases	CHO et al. (1992)
Pseudomonas corrugata[a], Protozoa: *Tetrahymena pyriformis, Vorticella microstoma*[a]	Pall rings trickling filter/toluene	COX and DESHUSSES (1999)
Exophiala jeanselmei[a]	perlite biofilter/styrene	COX et al. (1997)
Mycobacterium E3[a]	activated carbon trickling filter/ ethene	DE HEYDER et al. (1994)
Pseudomonas putida PPO1, *P. putida* ATCC 33015[a]	dry peat/perlite, biofilter/benzene, toluene, xylene	OH and BARTHA (1997a)
Actinomyces[a]	lava stones biofilter/styrene	POL et al. (1998)
Xanthobacter Py2[a]	membrane bioreactor/propene	REIJ and HARTMANS (1996)

[a] Species was used for inoculation.
[b] Species was found during the process.

(3) their initial adhesion,
(4) their ongrowth by formation of special polysaccharide structures, and finally
(5) the biofilm formation (VAN LOOSDRECHT and HEIJNEN, 1996).

The initial reversible adhesion of microorganisms to the surface is of particular importance which is described by the DLVO theory. This theory accounts for attracting van der Waals forces and for repulsing electrostatic forces (FLETCHER, 1996). The initial adhesion can be influenced by hydrophobic interactions (ROSENBERG and DOYLE, 1990) as well as by surface active compounds (NEU, 1996). In the following step, a glycocalyx typically made of polysaccharide compounds is formed by bacterial excretion, which connects cells and ongrowth substratum in a mechanically stable manner.

The biofilm – especially under low shearing forces as they dominate in biotrickling filters and biofilters – is of a rugged, filamentous shape. Bacteria grow in matrix enclosed microcolonies (Fig. 2). In other regions the biofilm is less dense and may even be penetrated by water or gas channels (COSTERTON et al., 1994). Substrate, salts, and oxygen may thereby reach the biofilm organisms more easily (DE BEER et al., 1994; ZHANG et al., 1995) than they would in a flat shaped biofilm due to the larger specific surface.

Biofilms offer a plenty of different ecological niches, which are divergent in terms of substrate and oxygen availability, water activity, possibly pH value, and shear forces. Each of these niches may result in an own complex biocenosis. In the course of time conditions change, usually due to surplus biofilm ongrowth. Subsequently, other strains will prove to be the "fittest" and possibly overgrow the former. As long as no steady state is reached (which seems to be rather the exception than the rule in biological waste gas purification) the biocenosis will continue to change.

Besides bacteria, fungi may play a significant role in fixed film biological waste gas reactors (see, e.g., KRAAKMAN et al., 1997). The formation of an air mycelium will possibly allow them to take up the substrate directly from the gas phase (TALBOT, 1999). In the case of sparingly soluble compounds, this may significantly ease the uptake process, since a dissolution step in the water phase is omitted.

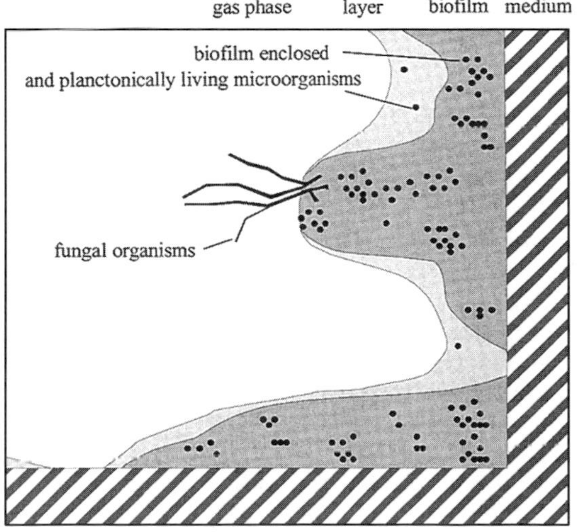

Fig. 2. Schematic biofilm structure in biological trickling filters, deduced from the findings of COSTERTON et al. (1994). In this type of biological waste gas purification reactors, a biofilm sticks on a chemically inert support medium being wetted and supplied with nutrients by an aqueous medium. The biofilm consists mainly of a polysaccharide matrix in which microorganisms thrive in microcolonies. Compared to the number of organisms living sessile, only a few will be found living planctonically in the medium phase. The biofilm itself has a rugged, uneven shape. It is not yet clearly understood, if direct contact of the biofilm and the gas phase is possible, nor if fungal organisms are able to extend an air mycelium into the gas phase and, therefore, obtain carbon substrate directly from the waste gas without a dissolution step (TALBOT, 1999).

3 Microorganisms as Biochemical Agents

3.1 Microbial Degradative Reactions

Organisms need energy for multiplying, but also for maintenance requirements (like repair mechanisms or to maintain concentration gradients between cytosol and cellular environment). Energy may be obtained from light (photolithoautotrophic metabolism) or from energy rich chemical compounds (chemotrophic metabolism). Of course, in biological waste gas purification plants, only the latter are of interest.

The chemotrophic metabolism can be divided into subgroups: chemolithoautotrophic and chemoorganoheterotrophic. An example for the first is the oxidation of hydrogen sulfide to elemental sulfur, sulfite, or finally to sulfate, a reaction sequence which is essential for cleaning waste gas streams from landfill gas treatment stations. The reaction energy of this process is conserved within the cells and the carbon needed for building up cell mass is derived from atmospheric carbon dioxide. Representative organisms for this biochemical process are, e.g., *Thiobacillus ferrooxidans* or *Thiobacillus thiooxidans*.

The other main metabolic scheme is the chemoorganoheterotrophic way. Here, organic substrates are consumed with concomitant production of CO_2 and water as well as "biohydrogen" stored in reducing equivalents. The carbon necessary for building up cellular structures is derived from the organic substrate as well. An example for this process is *Pseudomonas fluorescens*, consuming toluene as sole source of carbon and energy. This reaction type represents the most frequently used biochemical strategy and works with nearly all types of organic molecules.

A problem exists, if the substrate molecules do not consist of carbon chains containing at least five carbon atoms. In this case, gluconeogenetic reactions consuming additional energy are to be used, which diminishes the yield of the overall reaction. This is true, e.g., for the degradation of ethanol, acetic acid, isopropanol, and similar compounds. In the case of production of C_2 units such as acetyl-CoA, so-called anaplerotic sequences are to be used in order to permit synthesis of C_4 units, which can be used for gluconeogenesis after transformation to PEP or as building blocks in synthesis. An example for ethanol degrading organisms is *Escherichia coli*, but other organisms like pseudomonads and *Candida* yeasts are capable as well.

Metabolic reaction pathways of microorganisms are generally designed to proceed up to the total degradation of the carbon skeleton. Frequently, however, incomplete metabolism is observed, which means that metabolic intermediates can theoretically accumulate in the medium and cause unexpected problems. In the case of ethanol degradation, e.g., ethanol is converted by so-called "acetic acid bacteria" preferentially to acetic acid, which persists for some time outside the cellular environment, although being *per se* perfectly biodegradable by many other strains.

There can be different reasons for this strange behavior. First, degradation rates of different molecular species of a biochemical pathway can be misbalanced, which automatically causes formation of metabolic pools of different concentration. These pools can be leaky due to unspecific transport and, therefore, lead to excretion into the surroundings of the cell.

Second, structural reasons can cause a dramatic slow down of the transformation process of individual molecular species, which is true for many xenobiotic structures, e.g., trichloroethylene or dichlorinated benzene (see Sect. 4.1). In these cases, specialized cultures (DOC systems or "designed oligospecies culture" systems) might be of substantial value.

Such systems can explain the existence of microbial consortia or even of microbial syntrophic systems, depending on the mutual metabolic actions of each of the partners in the system. A mixed culture, e.g., was isolated degrading the compound 2-carboxybiphenylether. Careful analysis reveiled two active strains, one degrading the ether to salicylate and presumably catechol, the latter compound being the source of carbon and energy for the ether cleaving organism. Salicylate was the substrate for the second organism and at the same time toxic for the ether cleaving strain.

This makes sense when explaining the obligatory mutualistic relationship of both strains. Whether the strains will use the more waste air-related substrate, biphenylether, employing the same strategy still has to be clarified (results to be published). From a strategic point of view such systems tend to fuse and, therefore, assemble their mutual genetic traits in order to create a stable degradative unit, which is more effective than the original two species in culture. This is due to the fact that the original system still has care for the transport of interchangeable molecules, which has to be quickly enough not to intoxify the first strain and to feed the second strain with a rate as high as possible. Despite this theoretical consideration, active mixed cultures in waste air treatment plants frequently are very complex with nearly no tendency to evolve to a naturally generated oligospecies system. This is only partially due to the nature and number of substrates. It is easy to imagine that the number of different strains in a biofiltration system should depend on the kind and number of different substrates in the waste air supplied to the system. In contrast to this assumption, even if one substrate (toluene) is absolutely predominant, a complex biocenosis is still found consisting of a couple of active strains (results to be published). Usage of a more "convenient" substrate for the microbial community should elicit the selection even of more strains. In addition, these considerations only take into account culturable strains. It seems feasible that more species are in biofilms which cannot be detected using normal methods of cultivation. This is even more true, as merely cometabolizing strains per definition do not "grow" with the substrate of choice and will, therefore, not be detected by conventional methods. The degree of complexity may, therefore, have been underestimated up to now. To solve these problems, more sophisticated methods of strain detection have to be applied, which are being developed at the moment (see Sect. 6).

The question to be clarified is, therefore, whether in complex substrate mixtures and possibly also in oligosubstrate mixtures, strategies like "sequential metabolism" or "holometabolism" represent the preferred way of substrate utilization.

A different group of substrates is composed of compounds with intrinsic resistance to biodegradation. Here, incomplete types of metabolism frequently are the method of choice, enforced by the nature of the substrate.

An example of such an incomplete metabolism which results in complete or incomplete (co-)metabolism of complex substrates is the degradative action of a methane or methanol degrading consortium. It was described to act also on MTBE (methyl-*t*-butyl ether). Strains of this type frequently have been described to be extraordinarily well performing cometabolizers, i.e., strains using the enzyme apparatus elicited by growth on methane for the simultaneous metabolism of related or even quite unrelated compounds. It is, therefore, easily conceivable, that also molecules with xenobiotic structural elements like MTBE can at least be attacked in peripheral reaction steps in order to split "the xenobiotic bonds", i.e., the ether bond. The strains gain energy from the C_1 unit originating from the MTBE, and leave the *t*-butanol unit untouched (KIM and ENGESSER, unpublished results). In this case, evolution seems to be ongoing. Other authors, however, have described a full metabolism on MTBE (FORTIN and DESHUSSES, 1999).

The first case would be an example of so-called incomplete cometabolism. Full cometabolism can be demonstrated analyzing the degradation of phenyl substituted cyclohexane. In this case, a consortium consisting of two strains was isolated. One degraded the phenyl moiety down to cyclohexylcarboxylic acid and excreted it into the medium. The second strain took up the latter molecule to finally oxidize it to carbon dioxide and water. This phenomenon can be termed "relaxed consortial metabolism". It is highly probable that the first strain uses enzymes originally designed for the degradation of biphenyl. This is why the process of partial degradation of phenylcyclohexane to cyclohexylcarboxylic acid can be called cometabolic, although the presence of biphenyl before or during the reaction is not necessary, as it would be the case in a strictly cometabolic process.

4 Problems of Microbial Metabolism and Perspectives of a Corresponding Solution

4.1 Metabolism of Xenobiotic Structures

Many common chemicals with rather high volatility are more or less biodegradable with acceptable velocity. In general alkanols and polyols (as long as they show volatility), alkanals, ketones, esters, some ethers (Tab. 2), lower molecular weight carboxylic acids, and nitrogen containing compounds (amines, nitriles) belong to this category of substances.

However, not only in bioremediation procedures applied to water and soil, but also in waste air cleaning processes, frequently chemical structures have to be dealt with which are not easily biodegradable or biodegradable only by specialized cultures (see, e.g., Fig. 3 for the initial degradation reactions of benzene). As this criterion of biodegradability is an absolute prerequisite for biological waste gas purification methods, it is worthwhile to consider the reasons for more or less retarded metabolism and/or incomplete metabolism. Two cases have to be distinguished here: the overall degradation rate of a compound may be slowed down and/or the different individual metabolic steps may be hampered differentially.

In biological waste air cleaning dealing with compounds exhibiting substantial and sufficient vapor pressure in order to be volatile, a reduced number of classes of xenobiotic compounds can be envisaged. They all bear one or more structural elements, which are not known to nature as normal "natural" structural elements. (The Greek word "xenos" means "strange to something".) The most important structural elements with respect to this definition are the halogens, the ether bond, nitro substituents, thioethers, mixed mercaptans, and organosilicones.

In the chemical industry and in waste air (derived as stripped air from groundwater bio-

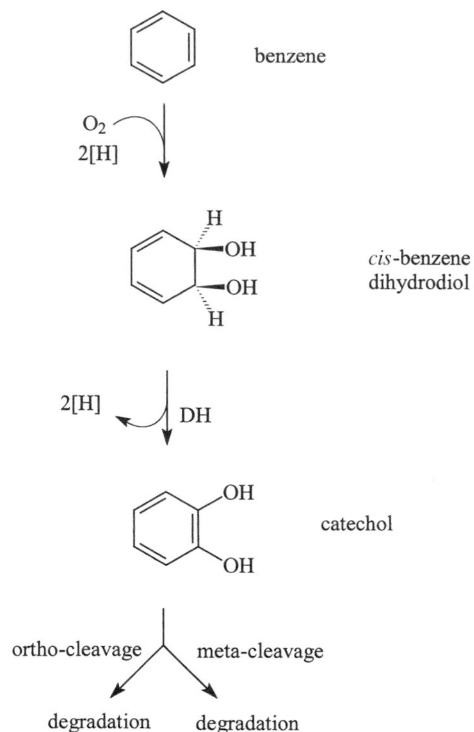

Fig. 3. Degradational scheme of benzene turnover by normal flora in biofiltration plants (alternative monohydroxylation pathways – as realized for toluene – are thinkable, but not proven).

remediation) frequently chlorobenzene is found accompanied by toluene. This combination of substrates at first glance seems to be ideal in terms of a chlorobenzene metabolism adjusted by the toluene metabolic pathway using a tailor-made cometabolic process. The opposite is true, however.

It has been demonstrated that the metabolism of toluene is managed in nearly all cases by enzymes of one of the five aerobic peripheral pathways described so far (Bertoni et al., 1996; Keener et al., 1998; Whited and Gibson, 1991; Olsen et al., 1994; Johnson and Olsen, 1995) which converge at the stage of methyl-substituted catechols (*o*-dihydroxybenzenes). These metabolites are, as a rule, metabolized via a meta-type cleavage reaction pathway (Murray et al., 1972) eventually yielding pyruvate and acetaldehyde, which in turn is metabolized to acetate. This enzymatic sequence

Tab. 2. Selection for Biochemically Oxidizable Waste Gas Hydrocarbons (Non-Xenobiotics)

Compound Class	Compound	Reference
Aliphatics	methane	SLY et al. (1993)
	iso-butane	BARTON et al. (1998)
	n-pentane	BARTON et al. (1998)
	n-hexane	MORGENROTH et al. (1996), CHERRY and THOMPSON (1997), PLAGGEMEIER et al. (1997)
	1,3-butadiene	CHOU and LU (1998)
Alkanols	methanol	SHAREEFDEEN et al. (1993), DE HOLLANDER et al. (1998)
	ethanol	CIOCI et al. (1997), AURIA et al. (1998)
	propanol	KIARED et al. (1997)
	butanol	WÜBKER and FRIEDRICH (1996), HEINZE and FRIEDRICH (1997)
	ethanol/butanol mixt.	BALTZIS et al. (1997)
Alkanals	butanal	WECKHUYSEN et al. (1993)
	isobutanal	SMET et al. (1997)
Ketones	methyl ethyl ketone	CHOU and HUANG (1997)
Esters	ethyl acetate	DESHUSSES et al. (1999), SCHÖNDUVE et al. (1996)
	isopropyl acetate	WU et al. (1998)
Ethers	diethyl ether	RIHN et al. (1997), ZHU et al. (1998), ALONSO et al. (1999), CHOU and HUANG (1999), KIM (1999)
BTEX	benzene	SORIAL et al. (1997), ZHOU et al. (1998), YEOM and YOO (1999)
	ethyl benzene	SORIAL et al. (1997), KLEINHEINZ and STJOHN (1998)
	toluene	SORIAL et al. (1997), MATTEAU and RAMSAY (1997)
	p-xylene	ROUHANA et al. (1997)
	BTEX mixture	KENNES et al. (1996), MALLAKIN and WARD (1996), SORIAL et al. (1997), ABUMAIZAR et al. (1997), NGUYEN et al. (1997), OH and BARTHA (1997a), TAHRAOUI and RHO (1997), QUINLAN et al. (1999), LU et al., (1999)

is initiated by a catechol-2,3-dioxygenase, the so-called metapyrocatechase. This enzyme can handle catechol as well as methyl substituted catechols, but is destroyed in a suicidal way by chlorinated catechols, the intermediates of haloarene metabolism. These compounds in turn are productively metabolized via a specialized ortho-cleavage pathway or ortho-II-pathway. At the same time, alkyl substituted catechols can only be metabolized via these metabolic pathways in rare cases employing highly specialized bacteria. The result of simultaneous metabolism of a mixture of chlorinated and alkylated benzenes is often disastrous: In the worst case, a lot of halo and methyl substituted catechols are generated. Due to their oxygen susceptibility and their general chemical instability they tend to form radicals which oligomerize and at the end stick to organic matrices of the waste air cleaning apparatus. This would, therefore, result in the production of hazardous waste and cause enormous financial problems during the waste disposal process at the end of the lifespan of the filter system. These costs can be prohibitive. This irreversible attachment of activated derivatives of the compounds to be destroyed is called "humification". It is currently under debate whether it represents an accepted form of bioremediation strategy to be applied in everyday use.

There are solutions offered by basic research on the nature of the underlying microbial consortia. Up to now, only partial solutions have been created for the metabolism of 4-substituted catechols, which can successfully be handled via ortho-cleavage pathways, irrespective whether they are methyl or chloro substituted (ENGESSER et al., 1989). Work for the 3-substituted series is still ongoing, but solutions are emerging. Whereas the load with rather high amounts of toluene accompanied with low amounts of chlorobenzene can be very stressful for the system, the opposite case is much less sensitive as the products of methylcatechol accumulation in the course of ortho-pathway usage are methyllactones, which fortunately are biodegradable and nontoxic. Total degradation of such mixtures, therefore, affords the enrichment of haloarene degradative strains and of methyllactone degraders.

Another example for the degradative metabolism of a really resistant compound has been found during the investigation of waste air containing fluorinated benzenes originating from a chemical factory. The biodegradative theory claims fluoroaromatics, and in particular fluorinated benzenes without substitution with carboxyl groups to be highly resistant (Fig. 4). Fluorinated metabolites are generated which are hardly biodegradable, yielding dead end metabolites. Alternatively, the phenomenon of "humification", i.e., the irreversible addition of activated forms of fluorobenzenes to the filter matrix has to be envisaged (the more toluenes are around!). Therefore, a careful analysis should show that the fluorine stoichiometry is in accordance with the expected values. Unfortunately, this has not been done. Although a biochemical strategy is feasible which could cope with the known obstacles of fluoroarene biodegradation (ENGESSER and SCHULTE, 1989; ENGESSER et al., 1990; LYNCH et al., 1997) this mechanism has still to be proven in order to exclude pseudodegradation or humification as the leading mechanism of this very special example of biological waste air filtration, which in this case would represent a true filtration and no degradative action (Fig. 5). Here, much more research is necessary.

Biological waste air purification, however, can also show a lot of successful examples of cleaning up different kinds of waste air containing compounds of more or less xenobiotic character. Whereas a lot of different classes of compounds have been described to be degradable under conditions of biological waste gas treatment, the more recalcitrant compounds afford special attention. In addition, frequent problems of low solubility the structural problem of inherent degradability can play a major role. Some solutions are described in the following, and some still problematic compounds are identified.

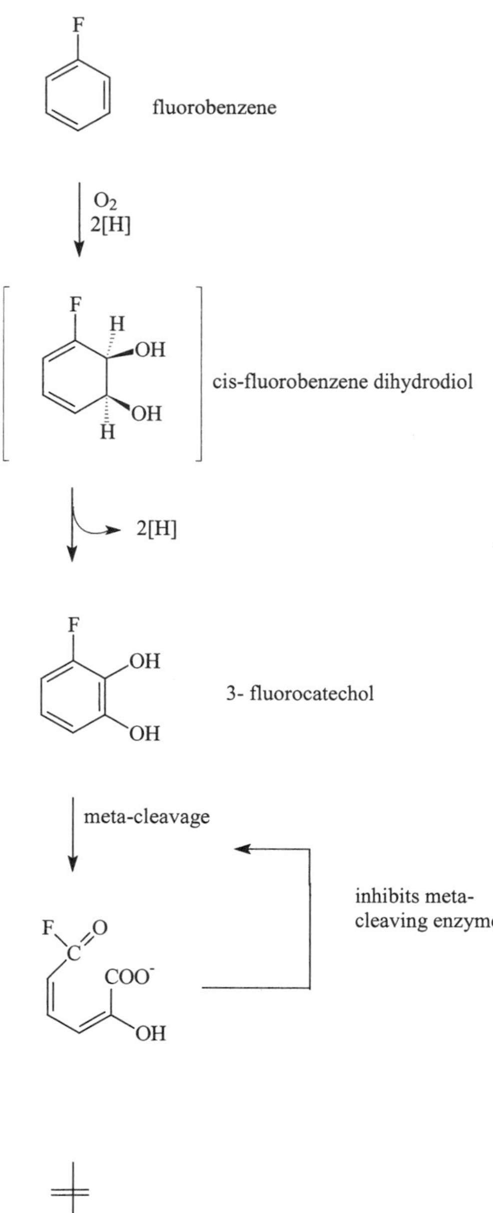

Fig. 4. Metabolism of fluorobenzene by normal benzene degrading organisms leads to formation of dead end products like fluorocatechol, fluoro substituted hydroxymuconic semialdehydes and presumably also humification products of fluorocatechol.

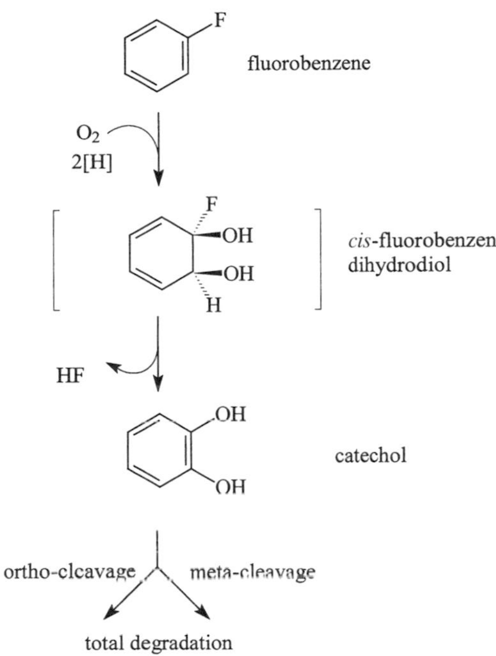

Fig. 5. Hypothetical productive metabolism of fluorobenzene by specialized fluorobenzene degraders in biofiltration of chemical waste air. Initial defluorination reaction liberates fluorine in the first step and yields well metabolizable compounds.

Dichloromethane, e.g., has been described to be an excellent substrate for specialized strains like *Hyphomicrobium* sp. (FERGAS et al., 1994). In strain *Hyphomicrobium* GJ21, e.g., it has been shown that stable conditions can be achieved, in which no net production of cell biomass can be detected (DIKS et al., 1994). These conditions are favorable, as they avoid excess biomass formation leading to the so-called "clogging" (see Chapter 15, this volume). ZUBER et al. (1997) found that in biotrickling filters loads of up to 2 g m^{-3} DCM can be degraded at high efficiency (99.5%), at a rather high flow rate up to 100 m^3 $(m^3\ h^1)^{-1}$. Compared to other principles, this method proved to be superior. There also have been theoretical considerations (OKKERSE et al., 1999) which adjust this observation of a non-clogging run of DCM biotrickling filters leading to the assumption that even higher amounts of DCM could be degraded without the danger of a clogged system.

For trichloroethylene, biological degradation theoretically can take place following two ways: anaerobically and aerobically. Anaerobic reductive dehalogenation has been demonstrated in the clean-up of groundwater using an externally added electron donor substrate. During biofiltration this may happen in deeper zones of the biofilm where oxygen is more or less depleted. The second strategy consists of a cometabolism-based mechanism of oxygenic attack on trichloroethylene yielding an unstable epoxide, which is hydrolyzed spontaneously or enzymatically at the very end to glyoxylic acid – a well utilizable substrate rendering the obligatory nature of a cometabolic way of metabolism somewhat obscure.

An example of anaerobic dechlorination is given by GERRITSE et al. (1995), who described in addition the aerobic dechlorination of lower chlorinated ethenes generated from higher chlorinated ethenes.

For aerobic dechlorination – the mechanism with probably the highest relevance for waste air filtration – there are numerous examples of cometabolism-based dechlorination reactions. Substrates used as replenisher for biotransformation energy demands have been reported to be propene (SAEKI et al., 1999; REIJ et al., 1995), propane (WILCOX et al., 1995), methane (UCHIYAMA et al., 1995), phenol (HOPKINS et al., 1993; SHURTLIFF et al., 1996), cumene (DABROCK et al., 1992; TAKAMI et al., 1999), methanol (FITCH et al., 1996), toluene and biphenyl (FURUKAWA et al., 1994; LANDA et al., 1994), dimethylsulfide (TAKAMI et al., 1999), and isoprene (EWERS et al., 1990). These processes are, therefore, highly unspecific. They have in common the attack of an oxygenase at the double bond presumably forming an epoxide. This has been shown with a methane degrading system, *Methylosinus trichosporium* OB3b (VLIEG et al., 1996). The usefulness of such systems depends on the tolerance the strains exhibit to the epoxides, which can also mechanistically depend on the degradation mechanism of the parent compound (EWERS et al., 1990).

Finally, some reports demonstrate the action of such systems in waste air cleaning (SUKESAN and WATWOOD, 1997; SUKESAN and WATWOOD, 1998; COX et al., 1998; PRESSMAN et al., 1999). According to the results mentioned above,

TCE degradation was tremendously stimulated (about 10-fold, SUKESAN and WATWOOD, 1998) by the addition of propane or methane. Before industrial-scale application, however, more research seems necessary.

Chlorobenzene and *o*-dichlorobenzene have also been demonstrated to be subject to biological waste air filtration (OH and BARTHA, 1994). An overview of the biological degradation of this class of compounds in general is given by ENGESSER and FISCHER (1991). However, generalizations are difficult to make, as pure culture conditions in bench-scale experiments are sometimes barely applicable to practical conditions.

From the group of alkylethers, diethylether was reported to be degradable in biotrickling filters (CHOU and HUANG, 1999). CHOU and HUANG (1999) reported maximum elimination capacities of about 100 g m^{-3} h^{-1}. They made, however, no comments on other ethers such as tetrahydrofuran, diisopropylether, or di-*n*-butyl ether, which are of practical importance.

m-Cresol has been found to be degradable in biotrickling filters (GAI, 1997). Due to the rather high solubility of this toxic molecule, degradation appeared mainly in the heel (draining tank) of the reactor. The trickling bed exhibited mainly the function of an absorption column.

Very recently methyl-*t*-butyl ether was described to be subject to biotrickling filter treatment being thereby removed from waste air (FORTIN and DESHUSSES, 1999). An aerobic microbial consortium was enriched after prolonged operation of the filter. A maximum elimination capacity of 50 g m^{-3} h^{-1} was achieved. This demonstrated the usefulness of the biotrickling filter technique to handle even a xenobiotic molecule as MTBE. The biological reaction was recognized to be the rate limiting step in the overall reaction, again emphasizing the importance analysis and monitoring (FORTIN and DESHUSSES, 1999).

Alkylated and polyalkylated benzenes have also been the subject of investigation for their suitability to be degraded in waste air cleaning operations (HEKMAT et al., 1997; HEKMAT et al., 1998; OH and BARTHA, 1997a; ORIGGI et al., 1997; NGUYEN et al., 1997; SORIAL et al., 1997). Besides more theoretical considerations elimination capacities were also given, which were in the order of 40–50 g m^{-3} h^{-1} in laboratory-scale fermenters. This again asks for more practice oriented reactors presenting more practically applicable values for dimensioning real-world plants. Own experiences with hexane as sole substrate in 150 L reactors eliminating about 10–20 g m^{-3} h^{-1}, cast some doubt on these elimination capacities. Considerating that the low solubility of these compounds often should be the rate limiting factor of the degradation process, the biological waste air purification method in general seems to be strongly limited.

JUTRAS et al. (1997) showed that gasoline vapors can be treated at a field scale biofiltration plant. They replaced a combustion plant with a biofilter apparatus, removing 90% of total petroleum hydrocarbons and more than 90% of the BTEX (benzene, toluene, ethylbenzene and xylenes) compounds. However, they did not comment on the specific elimination capacity.

Styrene has frequently been investigated as a substrate for biological waste air purification (COX et al., 1996a, b, 1997; POL et al., 1998; ARNOLD et al., 1997; SORIAL et al., 1998; CHOU and HSIAO, 1998). Reported styrene elimination capacities were quite low in the range of about 10–12 g m^{-3} h^{-1}. COX et al. (1997) published a much higher rate, 3- to 6-fold increased, using a yeast as the active microorganism. Biological degradability of styrene has been proven in many laboratory examinations (see, e.g., PANKE et al., 1998; VELASCO et al., 1998; ITOH et al., 1996; O'CONNOR and DOBSON, 1996; COX et al., 1996a; MARCONI et al., 1996; HARTMANS, 1995; CORKERY et al., 1994; WARHURST et al., 1994; COX et al., 1993; WARHURST and FEWSON, 1994; UTKIN et al., 1991). Therefore, despite many suspicions, styrene can be totally degraded and is not eliminated by some sort of activation-sorption process, which has been deduced from its chem-ical instability and tendency to polymerize.

Even nitrobenzene has been reported to be subject to biological waste air filtration (OH and BARTHA, 1997b). A microorganism was isolated which used nitrobenzene as a sole source of energy, carbon, and nitrogen. The nitro group obviously was reduced and liberated as ammonia. The filter showed ammonia emissions at pH 8,7 so that pH and salinity control are not necessary. A maximum elimination capacity of ca. 50 g m^{-3} h^{-1} was reported.

α-Pinene, a member of the class of terpene cyclic aliphatics, has been shown to be degraded in a biofiltration column (KLEINHEINZ et al., 1999). Elimination was established with reasonably high efficiency at waste gas concentrations of up to 1,8 g m^{-3}, which is a fairly high value for a compound showing only moderate biodegradability. No comment has been made on the other members of this class (e.g., the waste gas ingredients limonene, myrcene, α-terpineol).

For the elimination of inorganic compounds, mainly ammonia and hydrogensulfide including its alkyl derivatives have to be taken into account.

Ammonia as a waste air ingredient was degraded in an activated carbon fiber-based biofilter (YANI et al., 1998). The maximum elimination capacity was 62 g m^{-3} h^{-1}, which is a very high value.

HARTIKAINEN et al. (1996) gave an example how to inoculate a peat filter matrix which does not show nitrifying organisms nor exhibits any nitrification potential. They added nutrients, shifted the pH value to neutral values, and inoculated nitrifiers from activated sludge – a process that not necessarily is the best. They showed rather low, but reasonable elimination rates of nearly 2 g m^{-3} h^{-1} after adaptation. Overloading the filter resulted in accumulation of ammonia and – according to its inhibitory power – nitrite.

In a granular activated carbon filter for removal of ammonia in deodorizing systems a biomonitoring procedure for the detection of *Nitrobacter* sp. was applied (MALHAUTIER et al., 1997). The authors got first insight into the spatial distribution of the bacteria, revealing a higher concentration at the gas inlet, which is in accordance with the assumption of higher growh rates at higher substrate concentrations.

SAKANO and KERKHOF (1998) showed that in a laboratory biofilter with ammonia as the sole substrate, chemotrophic and heterotrophic florae could be analyzed using gene probes for the strain backgrounds as well as for the main gene regulating ammonia oxidation, i.e., *amoA*. With respect to the heterotrophic part of the flora, there was a shift during operation from strains mainly belonging to the β- and γ-subclass of Proteobacteria to strains only belonging to the latter, with an overall decline by 38%. The ammonia oxidizers did not show such a decrease in biodiversity. Instead, a new *amoA* gene was discovered, closely affiliated with *Nitrosococcus oceanus* and the particulate methane monooxygenase genes belonging to the γ-subdivision of Proteobacteria.

Oxidation of hydrogen sulfide can occur in a "normal" autotrophic way or in a mixotrophic way. Autotrophic organisms solely feed on the inorganic substrate and use it as source of energy, whereas the carbon is derived from inorganic carbon dioxide. This energy consuming-process can be circumvented by certain specialized mixotrophic microorganisms. They oxidize hydrogen sulfide, as a main source of energy, but instead of carbon dioxide they use an organic substrate to gain carbon. This uncommon phenomenon has been demonstrated to be active in a hydrogen sulfide degrading biofilter. The organism of choice was *Thiobacillus novellus* CH3 (CHUNG et al., 1998). Under mixotrophic conditions, a maximum elimination capacity of 25(–40) g m^{-3} h^{-1} (CHUNG et al., 1996b; calculated under the assumption of a filter material water content of about 50%) was noted, which is remarkable (CHUNG et al., 1996c). Besides *Thiobacillus* sp., also a *Pseudomonas* strain, *Pseudomonas putida* CH11, was isolated from the filter, exhibiting rather limited oxidation power when active with hydrogen sulfide. Its main product was elemental sulfur S^0 (CHUNG et al., 1996a).

Other authors (SHOJAOSADATI and ELYASI, 1999) reported eliminative capacities in a biofilter system under autotrophic regimes of about 22 g m^{-3} h^{-1}.

The plant was, however, only of pilot size, which makes it difficult to up-scale these observations.

4.2 Problems of Regulation

Enzymes as biocatalysts are responsible for all the biotransformation reactions in a biofiltration device. Due to regulatory constraints, not all enzymes of a cell are present at any time of its life cycle. Instead, the so-called "induction" causes the synthesis of certain enzymes only when they are needed. This synthesis on demand is especially valid for peripheral enzymes of degradative pathways, whereas so-called "constitutive enzymes", which are synthesized continuously, frequently are enzymes of the central pathways of sustaining and anabolic character.

Induction as a regulatory phenomenon will only take place at a certain minimum concentration of the compound to be degraded (threshold value). This means, that compounds like toluene will not be degraded at very low concentrations (ROCH and ALEXANDER, 1997). This is not because the corresponding enzymes are missing, but due to lacking induction.

Compounds can act as anti-inducers, i.e., they can block the induction of other molecules. As shown in Sect. 4.4, O'CONNOR et al. (1995) demonstrated that degradative pathways of aromatic catabolism can also be subject to the effect of so-called "repression". The styrene degradation pathway in *Pseudomonas putida* CA-3, e.g., is split into an upper part transforming styrene to phenylacetaldehyde, and a lower part catabolizing the latter compound. Both parts could be inhibited by glutamate and citrate, whereas glucose had no adverse effect. This inhibitory mechanism is clearly not based on a structural analogy, but due to complex signal elements binding low molecular weight compounds.

Furthermore, not all compounds can elicit an inductional event. It was shown, e.g., that 2,4-dichlorophenol is well metabolizable when supplied together with an inducer of the set of degradative enzymes, 2,4-dichlorophenoxyacetic acid (PIEPER et al., 1989). However, 2,4-dichlorophenol itself can become an inducer by a mutational event, thereby being easily biodegradable in such a modified strain. The enzymes in this case are not induced by this molecule, but instead the regulation is switched off and constitutive production takes place. This is, however, accompanied by a waste of energy, as the respective enzymes are produced irrespectively of their actual need.

Phenomena of this kind may explain why a biofiltration apparatus will not work properly, but more research and development should be carried out.

4.3 Membrane Reactors and Activated Carbon as New Applications for Treatment of Sparingly Soluble Compounds

When substrates are only sparingly soluble in water, the transfer of substrate from the gaseous phase into the water of the biofilm can be rate limiting. The basic idea behind is the assumption that the active biocenosis is attached to a membrane which should eliminate or draw back the water-related transfer (REIJ et al., 1998; ERGAS et al., 1999; REISER et al. 1994; see also Chapter 16, this volume). In a pure culture study of REIJ et al. (1997), a *Xanthobacter* Py2 was selected to study biofilm formation and reactor performance at propene concentrations of 200–350 ppm. Even at low concentrations of propene (9–30 ppm) the biofilm, once formed at higher substrate concentrations, remained in an active state.

Such a system could also be advantageous when strongly fluctuating concentrations of lipophilic substrates are offered, which is frequently the case in industrial practice of biological waste air filtration. In contrast, WEBER and HARTMANS (1995) used an activated carbon system as a buffering system in order to transiently adsorb a surplus of substrate (up to 1,000 mg m^{-3}) while keeping maximum concentrations below 300 mg m^{-3}.

These ideas still afford basic and applied research and development and are not treated here in detail.

4.4 Handling of Mixtures of Compounds

Mixing two or more substrates in biological waste air cleaning can be advantageous or disadvantageous, depending on the respective point of view. Spiking up a waste air stream of toluene with ammonia, e.g., can result in a strong increase of the corresponding elimination capacity (MORALES et al., 1998). Carbon dioxide balances showed that biomass increase was the explanation for this effect. A possible increase in performance due to the positive cometabolic contribution of ammonia monooxygenase generating a broader metabolic base for initial oxidation of toluene was not considered to be relevant. This effect could be the more interesting as the primary oxidation of substrates of this type is often a rate limiting step of the biological oxidation machinery.

DE HEYDER et al. (1997) observed a similar phenomenon adding ammonia to an ethene metabolizing culture of *Mycobacterium* E3. They could not differentiate between the two possible effects causing enhanced removal rate of ethene.

A disadvantageous effect of addition of a second substrate has been demonstrated by SMET et al. (1997). They observed that a biofilter inoculated with *Hyphomicrobium* degraded dimethylsulfide at reasonable rates of 40 g m^{-3} h^{-1}. However, addition of isobutyraldehyde as a second substrate caused a sequential degradation profile with the aldehyde being degraded first. As toluene did not show such an effect, it seems to be rather specific, although it does not seem reasonable that both substrates compete for the same active site of the primary enzyme.

As already mentioned in Sect. 4.1, addition of a second substrate can also exert deleterious influences such as suicide destruction of key enzymes, misrouting of substrates, thereby generating dead end metabolites of unknown biodegradability or – even worse – the irreversible attachment of reactive metabolites formed to the filter matrix. Alternatively, simple inhibition phenomena may occur according to different kinetics (SCHINDLER and FRIEDL, 1995; KAR et al., 1997; BIELEFELDT and STENSEL, 1999).

O'CONNOR et al. (1995) revealed that degradative pathways of aromatic catabolism can be subject to the effect of so-called "repression". The styrene degradation pathway in *Pseudomonas putida* CA-3, e.g., was shown to split into an upper part transforming styrene to phenylacetaldehyde, and a lower part catabolizing the latter compound. Both parts could be inhibited by glutamate and citrate, whereas glucose had no adverse effect.

There are numerous indications of substrate mixture inhibitory effects, which are unexplained in biochemical and physical terms. Therefore, these effects are not discussed *in extenso*, but should certainly be investigated in the future more intensely.

4.5 Influence of Toxicity on the Purifying Biology

Toxic or mutagenic substrates in biological waste air filtration can be disadvantageous for the active biocenosis of the biofiltration apparatus and for the surrounding environment. That biofiltration can reduce the hazard for the environment has been shown by KLEINHEINZ and BAGLEY (1998). Using "microtox" and "Ames" assays, they demonstrated a substantial reduction of toxicity and mutagenicity of the clean gas due to the degradative action of the bioflora.

Toxic substrates can be and are as a rule also toxic for the active flora of the biofiltration apparatus. Employing three test systems including a bench scale flat biofilm reactor, JONES et al. (1997) showed that toluene led to decreased culturability on toluene media. This correlated with a decreased specific toluene degradation rate, particularily at higher toluene concentrations. The authors observed similar effects using alkylethers as test substrates, sometimes observing drastic reductions in culturable cells after prolonged treatment with these substrates (unpublished data). These problems are not discussed in detail here.

5 Microbiology of Biological Waste Gas Treatment Facilities under Practical Aspects

5.1 Overall Process of Compound Transfer, Transport, and Degradation

The overall process of biological waste gas purification consists of three important steps:

(1) interphase mass transfer of the pollutant,
(2) diffusion of the pollutant, and
(3) subsequent degradation.

Each of these steps can be rate determining, therefore, the overall performance of a biological waste gas purification system can be interphase mass transfer, diffusion, or reaction limited. The biology of these systems has a main impact on the degradation processes, i.e., the reaction, but it does also influence the processes of mass transport.

5.2 Biological Influence on the Gas-to-Liquid Transfer of Gaseous Compounds

Usually, a certain compound is bioavailable only, if it is dissolved in water. For many well-soluble compounds of concern in biological waste gas purification, e.g., ethanol or certain acidic substances, interphase mass transfer is fast, and they may accumulate in the aqueous phase. For non-polar compounds like toluene or *n*-hexane, solubility is much lower and interphase mass transfer difficult. However, the mass transfer from the gas to the liquid phase and subsequently to the biofilm may be eased, if the solubility of the pollutant is enhanced by, e.g., biosurfactants secreted by microorganisms to improve the bioavailability of these sparingly soluble compounds (GOMA and RIBOT, 1978). Such mechanisms have been described for a lot of different microorganisms: *Pseudomonas* spp. and *Acinetobacter* spp. (PHALE et al., 1995; ZHANG and MILLER, 1995; ARINO et al., 1996), bacilli (COOPER et al., 1981; MULLIGAN et al., 1989), microorganisms of the CNM (i.e., Corynebacteria/*Nocardia*/Mycobacteria) and *Rhodococcus* taxonomic group (GUTNICK and ROSENBERG, 1977; MARGARITIS et al., 1979; RAMSAY et al., 1988; VOGT SINGER et al., 1990), and yeasts (CAMEOTRA et al., 1983; CAMEOTRA et al., 1984). Chemically, these biosurfactants can be quite divergent: glycolipids, lipopeptides, fatty acids, phospholipids, and neutral lipids were all described to have tenside-like properties (HOMMEL, 1990). CESÁRIO et al. (1996) found an enhancement of the ethene transfer coefficient in a biphasic bioreactor system in the presence of *Mycobacterium parafortuitum* E3 cells. They concluded that the cells themselves or an excretion product affect the interfacial area.

Many microorganisms belonging to *Mycobacteria* and *Rhodococci* form a lipophilic waxy-like cell wall with unpolar surface fimbriae. It is yet unclear, if these fimbriae enable these cells to take up unpolar substances directly from the gas phase. By bypassing a real solution step, the mass transfer of sparingly soluble compounds could be significantly alleviated.

5.2.1 Biological Influence on the Diffusion of Dissolved Compounds

To be degraded, a dissolved compound has to be transported from the gas–liquid interfacial area to the degradational active biology. This transport is – at least in biofilms – rather due to diffusional than to convective processes. The effective diffusivity of a substance will often be lowered by biofilm structures compared to a layer of pure water. However, the mass transport conditions inside a biofilm are far from uniform (DE BEER et al., 1994; RASMUSSEN and LEWANDOWSKI, 1998), since the biofilm usually is not a flat layer, but a complex structure of biomass clusters separated by water or biopolymer filled instertitial voids (COSTERTON et al., 1994) with different physical properties like density, porosity, and effective diffusivity (ZHANG et al., 1995).

5.2.2 Microbial Uptake of Dissolved Compounds

In general, compounds have to be taken up by microbial cells to be metabolized (although some fungi secrete exoenzymes for the conversion of polymeric substances into monomers). Small compounds may diffuse through the cell membrane following a concentration gradient into the cytosol where they are attacked by microbial enzymes, but usually substrates are taken up actively. Such transport mechanisms often require energy, but the cells are thereby able to select for certain substances, and the concentration inside the cell may be higher than outside. These mechanisms are substantial for the bioconversion of poorly soluble compounds.

5.2.3 Microbial Degradative Reaction

As already outlined, organisms need energy for multiplying, but also for maintenance requirements (like repair mechanisms or to maintain concentration gradients between cytosol and the cellular environment). Energy may be obtained from light or from energy rich chemical compounds. Of course, in biological waste gas purification plants, only the latter play a role of interest. Most degradative processes in biofilter-like reactors are aerobic: A more or less reduced organic compound is metabolized to carbon dioxide and water, plus possibly inorganic salts. Nitrate and sulfate may be formed from compounds containing nitrogen and sulfur. For aerobic metabolism, oxygen is needed, however, deeper layers may show oxygen depletion. In these anoxic zones anaerobic respiration may take place, denitrification or sulfate respiration, giving rise to odorous compounds such as ammonia, hydrogen sulfide, or other sulfur containing compounds. If the redox potential drops below zero, fermentation processes (where in a disproportion reaction a part of the substrate is oxidized while the other is reduced) will occur.

Biological reactions usually are substrate concentration dependent, with the reaction rate following Michaelis–Menten enzyme kinetics. For very low substrate concentrations the reaction rate follows first-order kinetics, i.e., the reaction rate rises linearly with an increase of substrate concentration. For very high substrate concentrations, the conversion reaction reaches a maximum rate. Further increase of the substrate concentration does not influence the reaction rate (zero-order kinetics). Some relevant pollutants in biological waste gas purification exhibit concentration-dependent toxic effects on the microflora, most often because of their solvent character. In this case, an increase of concentration will lead to a decrease of the reaction rate. Intermediates of an incomplete degradation may also inhibit microbial metabolism. An important factor are pH drops to microbially unfavorable values, because of the formation of acidic substances during microbial metabolism.

However, finally the relevant substrate concentration for the reaction rate as well as for possible toxic effects is the one the cells are confronted with. This means, an unpolar toxic substance may reach the cells only at non-toxic concentrations because of its high Henry coefficient, while more polar substances accumulate in the aqueous phase, possibly up to toxic levels.

5.3 Biological Waste Gas Reactors as Open Systems

While much is known about physical circumstances in waste gas purification reactors, we lack a deeper understanding of the biological processes and their impact on the overall performance of these reactors. We work with a very complex system, which biology is inhomogeneous on a macroscopic and a microscopic biofilm level. The biology may differ in species composition, and the metabolic status of each single cell may also be different. The biocenosis in one filter may differ from another, even if the waste gas is similar. Therefore, one filter may fail while the other is successful, although both have the same reactor design.

These problems are due to the fact that biofilters are open systems. Microorganisms enter the reactor mainly via the waste gas to be cleaned. One can easily calculate from the number of germs per cubic unit of waste gas and the gas stream forced through the reactor, how many microorganisms from the environment enter the reactor per process hour. Many operators rely on these microorganisms as inoculum and possibly the autochthonous microflora (the microflora already settling on the biofilter material) to make the filter function. More often, complex and rather undefined microbial inocula like sewage sludge are used, expecting that the best degrader for a certain task will succeed and overgrow the other organisms. However, positive effects on the elimination capacity after inoculation with sewage sludge are possibly due mainly to its fertilizing effect.

It would be very advantageous to design an overall system composed of an engineering solution of a well-working reactor compartment and a defined biocenosis, that allows for a fast start-up and a reliable high performance. A certain purification efficiency could be predicted more reliably. More than that, inoculation with well-known strains would minimize the risk of accidentally selecting pathogenic germs like those possibly originating from sewage sludge inocula.

Some biological research has focused on inoculation with pure cultures. However, the inoculum strains were often found to be unable to establish, although under laboratory culture conditions they competed with other species. Laboratory enrichment conditions should be chosen as similar as possible to the latter application. This includes enrichment at lower substrate concentrations and screening for more sessile (for the use in biofilters or biotrickling filters) or more planktonic strains (for the use in bioscrubbers). But even if such strains were chosen as inoculum source, it seemed like they were overgrown by other, not actively inoculated and, therefore, accidental strains. Especially fixed bed reactors contain plenty of different ecological niches offering different conditions in terms of carbon substrate availability, oxygen availability, water activity, and many other parameters. In addition, substrate loadings and compounds often vary during months of opera-tion. Though DARWIN's statement that only the fittest will survive is true, in biological waste gas reactors the biocenosis usually does not reach a stable state. It tends to fluctuate during the whole process time, since the selection criterion is not stable. These fluctuations may, but do not necessarily, lead to unstable purification efficiency of the whole process.

Biological waste gas reactors are also open systems under the aspect that substrate is constantly brought into the reactor. This substrate is converted to carbon dioxide, water, possibly inorganic salts, heat, and biomass (Fig. 1). Since most of this biomass is retained in the system, it accumulates and – especially in the case of fixed bed reactors – leads to clogging of the packing within a certain process duration. This phenomenon is well-known for biotrickling filters, while in biofilters it is observed less often, possibly due to the usually lower specific substrate loadings applied to biofilters. Nevertheless, if not all of the substrate entering the reactor will somehow leave it, be it carbon dioxide, biomass or as undegraded substrate, every reactor will clogg within months or years (clogging is more extensively discussed in Chapter 15, this volume).

6 Monitoring of Mixed Cultures in Waste Air Purification Reactors

As detailed insight into the biocenosis of biological waste gas reactors is lacking, the biology is most often seen as a "black box". As mentioned above, changing ratios of individual organisms may result in fluctuating efficiencies of the purification process and additionally every purification plant erected may contain a different biology ("accidental biology") with different unpredictable properties. Therefore, start-up times, stability, and effectiveness of these systems may vary strongly, often in an almost chaotic way. A solution of this problem could be the application of a "designed oligospecies culture system" (DOC system). In order to develop and use such systems in industrial applications and to make them economically feasible, one has to know exactly about the nature of all kinds of organisms in the biofiltration apparatus, their detection, survival behavior, and long-term efficiency.

For the elucidation and monitoring culture dependent and independent methods are used. The most often applied culture dependent method is plating a sample of the culture on an agar surface. Set growth conditions (mainly substrate availability and temperature) will determine which organisms will be cultured. However, it seems to be impossible to culture all viable organisms of a natural mixed biocenosis (AMANN et al., 1995).

Two culture-independent methods are most important. The FAME (fatty acid methyl esters) analysis makes use of the fact that the fatty acid composition of the cell membrane of each species is very distinct, related to its qualitative and quantitative composition. By a simple chemical procedure and a subsequent GC analysis accurate, reproducible knowledge about the biocenosis can be gained. Drawbacks result from the fact that the amount of fatty acids cannot be correlated with the number of organisms, since the fatty acid content differs from species to species. Additionally, growth conditions of the organisms may lead to changes to the fatty acid composition. FAME analysis was successfully applied for the monitoring of mixed cultures from sediments (PETERSEN and KLUG, 1994; KIEFT et al., 1997; SUNDH et al., 1997). DE CASTRO et al. (1997) investigated the biology of biofilters. Distinguishing between "real" biocenosis FAME and FAME originating from the biofilter matrix was hampered by the high background interference. When monitoring the biocenosis of biological trickling filters, this problem will not arise due to the inert character of packing. PLAGGEMEIER (1999) successfully applied the FAME analysis to compare mixed cultures in technically identical, but differently inoculated biotrickling filter units.

Up to now the most promising development for the monitoring of mixed bacterial cultures seems to be molecular genetic probing. The determination of the base sequences of ribosomal genes was the starting point for this development. The composition of these macromolecular assemblies is unusual: They consist of a small and a large subunit, which in turn are composed of so-called ribosomal RNA (rRNA) and a large number of small proteins. Together with the high copy number (10^3–10^5 per cell, due to physiological activity) and the simultaneous presence of multiple copies of the corresponding genes (rDNA) easy detection is possible accompanied by low detection levels. Comparison of entries in sequence databases – especially of the 16S rRNA of the small ribosomal subunit – led to the assembly of phylogenetic trees into which every newly isolated DNA sequence can be integrated.

In principal, every sequence can be used to construct nucleic acid probes of different specificity. The degree of detection specificity can be varied from an extremely high specific mode yielding only a single species to a very unspecific mode, detecting phylogenetical subunits. In a top-to-bottom approach with nested rRNA probes the position of unknown microorganisms in the phylogenetic tree is reached approximative in a hierarchical test procedure: First probes for the highest taxonomic level of the domains Archaea, Bacteria and Eukarya are used, followed by type probes specific for intermediate levels, e.g., as the orders Methanobacteriales and Methanococcales of the methanogens (RASKIN et al., 1994), and finally probes for single genera, species, and even subspecies.

The opposite approach allows for the examination for an ecosystem composition with respect to its phylogenetic subgroups or species. Addition of the subgroups' numerical share was compared with an all-embracing probe – leading to the detection of new subclasses of bacteria never known before.

With probes directed against species-specific genes, only one species, e.g., *Enterobacter agglomerans* (SELENSKA and KLINGMÜLLER, 1991), could be detected; the existence of a known gene or mRNA could unequivocally be demonstrated, and horizontal transfer of genes via plasmids, transposons, or uptake of free DNA was shown. This can be of special interest for research in biological waste air cleaning devices, permitting the detection of bacteria or their genes, which basically stand for the efficiency of the purification procedure.

The protocols for detection of specific microorganisms in soils should be easily adapted for detection requirements in biofilters and biotrickling filters as should be protocols derived from the application in sewage waters (e.g., GAYER-HERKERT and KUTZNER, 1991) for bioscrubbers and biomembrane reactors. SAKANO and KERKHOF (1998) used 16S rRNA probes as well as ammonia monooxygenase DNA probes for the description of changes, of the biocenosis in biofilter units. Concentration variations of members of the α-, β- and γ-subclasses of the proteobacteria during operation of a trickle bed reactor degrading Solvesso100 were demonstrated by HEKMAT et al. (1998) by the use of rRNA probes.

7 Emission of Germs from Biological Waste Air Filtration Plants

Biological waste air filtration is based on the degradative action of many different microorganisms. They exist in a complex technical ecosystem exhibiting a wealth of mutually dependent relationships (Sect. 6). The fact that huge amounts of air pass through this system causes the inevitable loss of a varying amount of microorganisms picked up by the air stream and due to their small size transported over more or less long distances. They may or may not cause harm to living beings around the treatment plants.

Therefore, it is relevant to know about the microorganisms present in the plant and about their tendency to escape from the system, their viability in air, and their potential to cause damage to living beings. In general, there are no specialized regimes already applied which could influence or reduce the germ content of biological waste air purification plants or of their clean gas. There may be a possible influence on the bioflora acting primarily on the compounds to be degraded, but the secondary flora at latest cannot be steered successfully, concerning the type and amount of individual microorganisms.

Therefore, one has to face the problem of germ emissions from plants of this type in a 2-fold way. First, one has to collect information about the germs in the system (Sect. 6) and, second, about the species in the clean gas.

Several aerobiological methods are available to measure germs in the air. Besides "Anderson" impactors frequently also impingement-based devices are used. Whereas the first fractionate individual cells and cell aggregates according to their size, the latter solely washes out the biological and non-biological particles into a collecting liquid, which is afterwards plated on agar media in order to count the viable cells. Gelatine-based systems are also applied where particles are sorbed on a gel matrix, which afterwards is solubilized in a liquid medium and cultured with nutrients (LUNDHOLM, 1982).

In general, therefore, culturable units are to be detected which may belong to pathogenic groups of microorganisms. In contrast to older investigations, which reported a decrease of the ratio of germs in the incoming air and those of the clean gas, the authors preliminary investigations indicated a substantial increase. Measurements at pilot plant and full technical scale biotrickling filters indicated, that the concentration of germs was in the range of 10^3–10^4 m^{-3} clean gas, sometimes exceeding these values. Normal levels are found to be about $5 \cdot 10^2$ m^{-3} (DIGIORGIO et al., 1996).

This means that more germs leave the system than enter it. The biofiltration devices are, therefore, presumably germ emission sources and no germ filters, although it should be admitted that more corresponding research has to be done. The sampling efficiency, e.g., of different devices and enumeration media are still debated controversely (see, e.g., STEWART et al., 1995; HUGENHOLTZ et al., 1995; STRAJA and LEONHARD, 1996; PARAT et al., 1999). Furthermore, the survival of germs outdoors is questionable, which is especially true for pathogenic strains, if they would exist in biofiltration plants (HANDLEY and WEBSTER, 1995; HATCH and DIMMICK, 1966; SHAFFER and LIGHTHART, 1997; HEIDELBERG et al., 1997).

However, the mere number of emitted germs does not tell us how dangerous they may be for living organisms. Due to a lack of sound data in the biofiltration area, one may use similar data from biological waste treatment plants, although this could mean an overestimation of fungi. In general, most of the environmentally important germs should not be pathogenic, as pathogenicity is a complex trait affording a lot of specialized capabilities of the respective strains (VANALPHEN et al., 1995).

For comparison, values for the occurrence of airborne bacteria and pathogen indicators during land application of sewage sludge are given (PILLAI et al., 1996). Accordingly, there was no indication for survival of *Salmonella* sp. or other fecal coliforms and coliphages. Where there was significant agitation of the material, however, pathogenic clostridia could be detected!

In addition, not only living microorganisms can be detected in air, but also their debris, e.g., endotoxins (see, e.g., TOLVANEN et al., 1998). In general, there is a good correlation between particle counts and the number of airborne bacteria.

The whole subject needs intensified research, as the methods are sophisticated enough to detect germs by physical devices and analyze them by culturing or non-culturing (PENA et al., 1999; MACNAUGHTON et al., 1999; HERNANDEZ et al., 1999) techniques or to detect MVOC (microbially produced volatile organic compounds; SENKPIEL, 1999). The data base should be enlarged in order to detract more accurate conclusions concerning the germ emission problem.

8 References

ABUMAIZAR, R. J., SMITH, E. H., KOCHER, W. (1997), Analytical model of dual-media biofilter for removal of organic air pollutants, *J. Environ. Eng. – ASCE* **123**, 606–614.

ALONSO, C., ZHU, Y., SUIDAN, M. T., KIM, B. R., KIM, B. J. (1999), Mathematical model for the biodegradation of VOCs in trickle bed biofilters, *Water Sci. Technol.* **39**, 139–146.

AMANN, R. I., LUDWIG, W., SCHLEIFER, K.-H. (1995), Phylogenetic identification and *in situ* detection of individual microbial cells without cultivation, *Microbiol. Rev.* **59**, 143–169.

ARINO, S., MARCHAL, R., VANDECASTEELE, J.-P. (1996), Identification and production of a rhamnolipidic biosurfactant by a *Pseudomonas* species, *Appl. Microbiol. Biotechnol.* **45**, 162–168.

ARNOLD, M., REITTU, A., VON WRIGHT, A., MARTIKAINEN, P. J., SUIHKO, M.-L. (1997), Bacterial degradation of styrene in waste gases using a peat filter, *Appl. Microbiol. Biotechnol.* **48**, 738–744.

AURIA, R., AYCAGUER, A. C., DEVINNY, J. S. (1998), Influence of water content on degradation rates for ethanol in biofiltration, *J. Air Waste Managem. Ass.* **48**, 65–70.

BALTZIS, B. C., WOJDYLA, S. M., ZAROOK, S. M. (1997), Modeling biofiltration of VOC mixtures under steady-state conditions, *J. Environ. Eng. – ASCE* **123**, 599–605.

BARTON, J. W., HARTZ, S. M., KLASSON, K. T., DAVISON, B. H. (1998), Microbial removal of alkanes from dilute gaseous waste streams: Mathematical modeling of advanced bioreactor systems, *J. Chem. Technol. Biotechnol.* **72**, 93–98.

BERTONI, G., BOLOGNESE, F., GALLI, E., BARBIERI, P. (1996), Cloning of the genes for and characterization of the early stages of toluene and *o*-xylene catabolism in *Pseudomonas stutzeri* OX1, *Appl. Environ. Microbiol.* **62**, 3704–3711.

BIELEFELDT, A. R., STENSEL, H. D. (1999), Modeling competitive inhibition effects during biodegradation of BTEX mixtures, *Water Res.* **33**, 707–714.

CAMEOTRA, S. S., SINGH, H. D., HAZARIKA, A. K., BARUAH, J. N. (1983), Mode of uptake of insoluble solid substrates by microorganisms. II: Uptake of solid *n*-alkanes by yeast and bacterial species, *Biotechnol. Bioeng.* **25**, 2945–2956.

CAMEOTRA, S. S., SINGH, H. D., BARUAH, J. N. (1984), Demonstration of extracellular alkane solubilizing factor produced by *Endomycopsis lipolytica* YM, *Biotechnol. Bioeng.* **26**, 554–556.

CESÁRIO, M. T., TURTOI, M., SEWALT, S. F. M., BEEFTINK, H. H., TRAMPER, J. (1996), Enhancement of the gas-to-water ethene transfer coefficient by a dispersed water-immiscible solvent: Effect of the cells, *Appl. Microbiol. Biotechnol.* **46**, 497–502.

CHERRY, R. S., THOMPSON, D. N. (1997), Shift from growth to nutrient-limited maintenance kinetics during biofilter acclimation, *Biotechnol. Bioeng.* **56**, 330–339.

CHO, K.-S., HIRAI, M., SHODA, M. (1992), Enhanced removal efficiency of malodorous gases in a pilot-scale peat biofilter inoculated with *Thiobacillus thioparus* DW44, *J. Ferment. Bioeng.* **73**, 46–50.

CHOU, M. S., HSIAO, C. C. (1998), Treatment of styrene-contaminated airstream in biotrickling filter packed with slags, *J. Environ. Eng. – ASCE* **124**, 844–850.

CHOU, M. S., HUANG, J. J. (1997), Treatment of methyl ethylketone in air stream by biotrickling filters, *J. Environ. Eng. – ASCE* **123**, 569–576.

CHOU, M. S., HUANG, Y. S. (1999), Treatment of ethyl ether in air stream by a biotrickling filter packed with slags, *J. Air Waste Managem. Ass.* **49**, 533–543.

CHOU, M. S., LU, S. L. (1998), Treatment of 1,3-butadiene in an air stream by a biotrickling filter and a biofilter, *J. Air Waste Managem. Ass.* **48**, 711–720.

CHUNG, Y. C., HUANG, C. P., TSENG, C. P. (1996a), Biodegradation of hydrogen sulfide by a laboratory-scale immobilized *Pseudomonas putida* CH11 biofilter, *Biotechnol. Prog.* **12**, 773–778.

CHUNG, Y. C., HUANG, C. P., TSENG, C. P. (1996b), Microbial oxidation of hydrogen sulfide with biofilter, *J. Environ. Sci. Health Part A – Environ. Sci. Eng. Toxic Hazard. Subst. Control.* **31**, 1263–1278.

CHUNG, Y. C., HUANG, C. P., TSENG, C. P. (1996c), Operation optimization of *Thiobacillus thioparus* CH11 biofilter for hydrogen sulfide removal, *J. Biotechnol.* **52**, 31–38.

CHUNG, Y. C., HUANG, C. P., PAN, J. R., TSENG, C. P. (1998), Comparison of autotrophic and mixotrophic biofilters for H_2S removal, *J. Environ Eng. – ASCE* **124**, 362–367.

CIOCI, F., LAVECCHIA, R., FERRANTI, M. M. (1997), High-performance microbial removal of ethanol from contaminated air, *Biotechnol. Tech.* **11**, 893–898.

COOPER, D. G., MACDONALD, C. R., DUFF, S. J. B., KOSARIC, N. (1981), Enhanced production of surfactin from *Bacillus subtilis* by continuous product removal and metal cation additions, *Appl. Environ. Microbiol.* **42**, 408–412.

CORKERY, D. M., O'CONNOR, K. E., BUCKLEY, C. M., DOBSON, A. D. W. (1994), Ethyl benzene degradation by *Pseudomonas fluorescens* strain CA-4, *FEMS Microbiol. Lett.* **124**, 23–27.

COSTERTON, J. W., LEWANDOWSKI, Z., DE BEER, D., CALDWELL, D., KORBER, D., JAMES, G. (1994), Biofilms, the customized microniche, *J. Bacteriol.* **176**, 2137–2142.

COX, H. H. J., DESHUSSES, M. A. (1998), Biological waste air treatment in biotrickling filters, *Curr. Opin. Biotechnol.* **9**, 256–262.

COX, H. H. J., DESHUSSES, M. A. (1999), Biomass control in waste air biotrickling filters by protozoan predation, *Biotechnol. Bioeng.* **62**, 216–224.

COX, H. H. J., HOUTMAN, J. H. M., DODDEMA, H. J., HARDER, W. (1993), Growth of the black yeast *Exophiala jeanselmei* on styrene and styrene-related compounds, *Appl. Microbiol. Biotechnol.* **39**, 372–376.

COX, H. H. J., FABER, B. W., VAN HEININGEN, W. N. M., RADHOE, H., DODDEMA, H. J., HARDER, W. (1996a), Styrene metabolism in *Exophiala jeanselmei* and involvement of a cytochrome P-450-dependent styrene monooxygenase, *Appl. Environ. Microbiol.* **62**, 1471–1474.

COX, H. H. J., MAGIELSEN, F. J., DODDEMA, H. J., HARDER, W. (1996b), Influence of the water content and water activity on styrene degradation by *Exophiala jeanselmei* in biofilters, *Appl. Microbiol. Biotechnol.* **45**, 851–856.

COX, H. H. J., MOERMAN, R. E., VAN BAALEN, S., VAN HEININGEN, W. N. M., DODDEMA, H. J., HARDER, W. (1997), Performance of a styrene-degrading biofilter containing the yeast *Exophiala jeanselmei*, *Biotechnol. Bioeng.* **53**, 259–266.

COX, C. D., WOO, H. J., ROBINSON, K. G. (1998), Cometabolic biodegradation of trichloroethylene (TCE) in the gas phase, *Water Sci. Technol.* **37**, 97–104.

DABROCK, B., RIEDEL, J., BERTRAM, J., GOTTSCHALK, G. (1992), Isopropylbenzene (Cumene) – A new substrate for the isolation of trichloroethene-degrading bacteria, *Arch. Microbiol.* **158**, 9–13.

DE BEER, D., STOODLEY, P., ROE, F., LEWANDOWSKI, Z. (1994), Effects of biofilm structures on oxygen distribution and mass transport, *Biotechnol. Bioeng.* **43**, 1131–1138.

DE CASTRO, A., ALLEN, D. G. G., FULTHORPE, R. R. (1997), Characterization of the microbial population during biofiltration and the influence of the inoculum source, in: Air Waste Management Ass. (Eds.) *Proc. Air Waste Management Ass. 90th Ann. Meeting*, Pittsburgh.

DE HEYDER, B., OVERMEIRE, A., VAN LANGENHOVE, H., VERSTRAETE, W. (1994), Ethene removal from a synthetic waste gas using a dry biobed, *Biotechnol. Bioeng.* **44**, 642–648.

DE HEYDER, B., VAN ELST, T., VAN LANGENHOVE, H., VERSTRAETE, W. (1997), Enhancement of ethene removal from waste gas by stimulating nitrification, *Biodegradation* **8**, 21–30.

DE HOLLANDER, G. R., OVERCAMP, T. J., GRADY, C. P. L. (1998), Performance of a suspended-growth bioscrubber for the control of methanol, *J. Air Waste Managem. Ass.* **48**, 872–876.

DESHUSSES, M. A. (1997), Biological waste air treatment in biofilters, *Curr. Opin. Biotechnol.* **8**, 335–339.

DESHUSSES, M., JOHNSON, C. T., LESON, G. (1999), Biofiltration of high loads of ethyl acetate in the presence of toluene, *J. Air Waste Managem. Ass.* **49**, 973–979.

DEVINNY, J. S., DESHUSSES, M. A., WEBSTER, T. S. (1999), *Biofiltration for Air Pollution Control*. Boca Raton, FL: Lewis Publishers.

DIGIORGIO, C., KREMPFF, A., GUIRAUD, H., BINDER, P., TIRET, C., DUMENIL, G. (1996), Atmospheric pollution by airborne microorganisms in the city of Marseilles, *Atmospher. Environ.* **30**, 155–160.

DIKS, R. M. M., OTTENGRAF, S. P. P., VRIJLAND, S. (1994), The existence of a biological equilibrium in a trickling filter for waste gas purification, *Biotechnol. Bioeng.* **44**, 1279–1287.

ENGESSER, K.-H. (1992), Mikrobiologische Aspekte der Biologischen Abluftreinigung, in: *Biotechniques for Air Pollution Abatement and Odor Control Policies, Studies in Environmental Science 51* (DRAGT, A. J., VAN HAM, J., Eds.), pp. 33–40. Amsterdam: Elsevier.

ENGESSER, K.-H., FISCHER, P. (1991), Degradation of haloaromatic compounds, in: *Biodegradation Natural and Synthetic Materials* Vol. 2 (BETTS, W. B., Ed.), pp. 15–54. Berlin: Springer-Verlag.

ENGESSER, K.-H., SCHULTE, P. (1989), Degradation of 2-bromo-, 2-chloro- and 2-fluorobenzoate by *Pseudomonas putida* CLB250, *FEMS Microbiol. Lett.* **60**, 143–148.

ENGESSER, K.-H., PIEPER, D. H., ROJO, F., TIMMIS, K. N., KNACKMUSS, H.-J. (1989), Simultaneous degradation of chloro- and methylaromatics via *ortho*-pathway by genetically engineered bacteria and natural soil isolates, in: *Recent Advances in Microbial Ecology* (HATTORI, T., ISHIDA, Y., MARUYAMA, Y., MORITA, R. Y., UCHIDA, A., Eds.), pp. 622–626. Kyoto: Japan Scientific Societies Press.

ENGESSER, K.-H., AULING, G., BUSSE, J., KNACKMUSS,

H.-J. (1990), 3-Fluorobenzoate enriched bacterial strain FLB 300 degrades benzoate and all three isomeric monofluorobenzoates, *Arch. Microbiol.* **153**, 193–199.

ENGESSER, K.-H., REISER, M., PLAGGEMEIER, T., LÄMMERZAHL, O. (1996), Why introduce biofiltration in industrial practice? in: *Wider Application and Diffusion of Bioremediation Technologies* Vol. 1, Chapter A1.1 (OECD, Ed.), pp. 115–122. Paris: OECD.

ENGESSER, K.-H., PLAGGEMEIER, T., KRÜGER, J. H. (1997), Microorganisms as demons for work: general remarks and some applications in biological waste air cleaning, in: *Biological Waste Gas Cleaning* (PRINS, W. L., VAN HAM, J., Eds.), pp. 5–18. Düsseldorf: VDI Verlag.

ERGAS, S. J., KINNEY, K., FULLER, M. E., SCOW, K. M. (1994), Characterization of a compost biofiltration system degrading dichloromethane, *Biotechnol. Bioeng.* **44**, 1048–1054.

ERGAS, S. J., SHUMWAY, L., FITCH, M. W., NEEMANN, J. J. (1999), Membrane process for biological treatment of contaminated gas streams, *Biotechnol. Bioeng.* **63**, 431–441.

EWERS, J., FREIERSCHRODER, D., KNACKMUSS, H. J. (1990), Selection of Trichloroethene (TCE) degrading bacteria that resist inactivation by TCE, *Arch. Microbiol.* **154**, 410–413.

FITCH, M. W., SPEITEL, G. E., GEORGIOU, G. (1996), Degradation of trichloroethylene by methanol-grown cultures of *Methylosinus trichosporium* OB3b PP358, *Appl. Environ. Microbiol.* **62**, 1124–1128.

FLETCHER, M. (1996), Bacterial attachment in aquatic environments: a diversity of surfaces and adhesion strategies, in: *Bacterial Adhesion: Molecular and Ecological Diversity* (FLETCHER, M., Ed.), pp. 1–24. New York: Wiley-Liss.

FORTIN, N. Y., DESHUSSES, M. A. (1999), Treatment of methyl *tert*-butyl ether vapors in biotrickling filters. 2. Analysis of the rate-limiting step and behavior under transient conditions, *Environ. Sci. Technol.* **33**, 2987–2991.

FURUKAWA, K., HIROSE, J., HAYASHIDA, S., NAKAMURA, K. (1994), Efficient degradation of trichloroethylene by a hybrid aromatic ring dioxygenase, *J. Bacteriol.* **176**, 2121–2123.

GAI, S. (1997), Abbau von Toluol und *m*-Kresol im Biofilmrieselbettreaktor, in: *Biological Waste Gas Cleaning* (PRINS, W. L., VAN HAM, J., Eds.), pp. 157–164. Düsseldorf: VDI-Verlag.

GAYER-HERKERT, G., KUTZNER, H. J. (1991), DNA-Sonden für den Nachweis von in Belebtschlamm eingeimpften Bakterien, *gwf Wasser Abwasser* **132**, 247–248.

GERRITSE, J., RENARD, V., VISSER, J., GOTTSCHAL, J. C. (1995), Complete degradation of tetrachloroethene by combining anaerobic dechlorinating and aerobic methanotrophic enrichment cultures, *Appl. Microbiol. Biotechnol.* **43**, 920–928.

GOMA, G., RIBOT, D. (1978), Hydrocarbon fermentation: Kinetics of microbial cell growth, *Biotechnol. Bioeng.* **20**, 1723–1734.

GUTNICK, D. L., ROSENBERG, E. (1977), Oil tankers and pollution: A microbiological approach, *Annu. Rev. Microbiol.* **31**, 379–396.

HALFMEIER, H., SCHÄFER-TREFFENFELDT, W., REUSS, M. (1993), Potential of *Thiobacillus ferrooxidans* for waste gas purification. Part 1. Kinetics of continuous ferrous iron oxidation, *Appl. Microbiol. Biotechnol.* **40**, 416–420.

HANDLEY, B. A., WEBSTER, A. J. F. (1995), Some factors affecting the airborne survival of bacteria outdoors, *J. Appl. Bacteriol.* **79**, 368–378.

HARTIKAINEN, T., RUUSKANEN, J., VANHATALO, M., MARTIKAINEN, P. J. (1996), Removal of ammonia from air by a peat biofilter, *Environ. Technol.* **17**, 45–53.

HARTMANS, S. (1995), Microbial degradation of styrene, *Biotransformations: Microb. Degrad. Health Risk Comp.* **32**, 227–238.

HATCH, M. T., DIMMICK, R. L. (1966), Physiological responses of airborne bacteria to shifts in relative humidity, *Bacteriol. Rev.* **30**, 597–603.

HEIDELBERG, J. F., SHAHAMAT, M., LEVIN, M., RAHMAN, I., STELMA, G. et al. (1997), Effect of aerosolization on culturability and viability of gram-negative bacteria, *Appl. Environ. Microbiol.* **63**, 3585–3588.

HEINZE, U., FRIEDRICH, C. G. (1997), Respiratory activity of biofilms: Measurement and its significance for the elimination of *n*-butanol from waste gas, *Appl. Microbiol. Biotechnol.* **48**, 411–416.

HEKMAT, D., LINN, A., STEPHAN, M., VORTMEYER, D. (1997), Biodegradation dynamics of aromatic compounds from waste air in a trickle-bed reactor, *Appl. Microbiol. Biotechnol.* **48**, 129–134.

HEKMAT, D., AMANN, R., LINN, A., STEPHAN, M., STOFFELS, M., VORTMEYER, D. (1998), Biofilm population dynamics in a trickle-bed bioreactor, *Water Sci. Technol.* **37**, 167–170.

HERNANDEZ, A., MARTINEZ, J. L., MELLADO, R. P. (1999), Detection of fungal spores from contaminated surfaces by the polymerase chain reaction, *World J. Microbiol. Biotechnol.* **15**, 39–42.

HINZ, M., SATTLER, F., GEHRKE, T., BOCK, E. (1994), Entfernung von Stickstoffmonoxid durch den Einsatz von Mikroorganismen – Entwicklung eines Membrantaschenreaktors, in: *VDI Bericht 1104 Biologische Abgasreinigung* (KrdL im VDI und DIN, Ed.), pp. 113–123. Düsseldorf: VDI-Verlag.

HOMMEL, R. K. (1990), Formation and physiological role of biosurfactants produced by hydrocarbon-utilzing microorganisms: Biosurfactants in hydrocarbon utilization, *Biodegradation* **1**, 107–119.

HOPKINS, G. D., SEMPRINI, L., MCCARTY, P. L. (1993), Microcosm and *in situ* field studies of enhanced biotransformation of trichloroethylene by phenol-utilizing microorganisms, *Appl. Environ. Microbiol.* **59**, 2277–2285.

HUGENHOLTZ, P., CUNNINGHAM, M. A., HENDRIKZ, J. K., FUERST, J. A. (1995), Desiccation resistance of bacteria isolated from an air-handling system biofilm determined using a simple quantitative membrane filter method, *Lett. Appl. Microbiol.* **21**, 41–46.

ITOH, N., YOSHIDA, K., OKADA, K. (1996), Isolation and identification of styrene-degrading *Corynebacterium* strains, and their styrene metabolism, *Biosci. Biotechnol. Biochem.* **60**, 1826–1830.

JOHNSON, G. R., OLSEN, R. H. (1995), Nucleotide sequence analysis of genes encoding a toluene benzene-2-monooxygenase from *Pseudomonas* sp. strain JS150, *Appl. Environ. Microbiol.* **61**, 3336–3346.

JONES, W. L., MIRPURI, R. G., VILLAVERDE, S., LEWANDOWSKI, Z., CUNNINGHAM, A. B. (1997), The effect of bacterial injury on toluene degradation and respiration rates in vapor phase bioreactors, *Water Sci. Technol.* **36**, 85–92.

JUTRAS, E. M., SMART, C. M., RUPERT, R., PEPPER, I. L., MILLER, R. M. (1997), Field-scale biofiltration of gasoline vapors extracted from beneath a leaking underground storage tank, *Biodegradation* **8**, 31–42.

KAR, S., SWAMINATHAN, T., BARADARAJAN, A. (1997), Biodegradation of phenol and cresol isomer mixtures by *Arthrobacter, World J. Microbiol. Biotechnol.* **13**, 659–663.

KEENER, W. K., WATWOOD, M. E., APEL, W. A. (1998), Activity-dependent fluorescent labeling of bacteria that degrade toluene via toluene 2,3-dioxygenase, *Appl. Microbiol. Biotechnol.* **49**, 455–462.

KENNES, C., COX, H. H. J., DODDEMA, H. J., HARDER, W. (1996), Design and performance of biofilters for the removal of alkylbenzene vapors, *J. Chem. Technol. Biotechnol.* **66**, 300–304.

KIARED, K., WU, G., BEERLI, M., ROTHENBUHLER, M., HEITZ, M. (1997), Application of biofiltration to the control of VOC emissions, *Environ. Technol.* **18**, 55–63.

KIEFT, T. L., WILCH, E., O'CONNOR, K., RINGELBERG, D. B., WHITE, D. C. (1997), Survival and phospholipid fatty acid profiles of surface and subsurface bacteria in natural sediment microcosms, *Appl. Environ. Microbiol.* **63**, 1531–1542.

KIM, Y.-H. (1999), Der mikrobielle Abbau von Etherverbindungen unter besonderer Berücksichtigung von Aralkyl- und Alkylethern, *Thesis*, University of Stuttgart, Germany.

KIRCHNER, K., HAUK, G., REHM, H.-J. (1987), Exhaust gas purification using immobilized monocultures (biocatalysts), *Appl. Microbiol. Biotechnol.* **26**, 579–587.

KLEINHEINZ, G. T., BAGLEY, S. T. (1998), Biofiltration for the removal and "detoxification" of a complex mixture of volatile organic compounds, *J. Ind. Microbiol. Biotechnol.* **20**, 101–108.

KLEINHEINZ, G. T., STJOHN, W. P. (1998), Method comparison for quantification of a complex mixture of petroleum VOCs in a biofiltration unit, *Environ. Toxicol. Risk Assessment:* 7th Volume **1333**, 262–271.

KLEINHEINZ, G. T., BAGLEY, S. T., STJOHN, W. P., RUGHANI, J. R., MCGINNIS, G. D. (1999), Characterization of alpha-pinene-degrading microorganisms and application to a bench-scale biofiltration system for VOC degradation, *Arch. Environ. Cont. Toxicol.* **37**, 151–157.

KRAAKMAN, N. J. R., VAN GROENESTIJN, J. W., KOERS, B., HESLINGA, D. C. (1997), Styrene removal using a new type of bioreactor with fungi, in: *Biological Waste Gas Cleaning* (PRINS, W. L., VAN HAM, J., Eds.), pp. 225–232. Düsseldorf: VDI-Verlag.

LANDA, A. S., SIPKEMA, E. M., WEIJMA, J., BEENACKERS, A. A. C. M., DOLFING, J., JANSSEN, D. B. (1994), Cometabolic degradation of trichloroethylene by *Pseudomonas cepacia* G4 in a chemostat with toluene as the primary substrate, *Appl. Environ. Microbiol.* **60**, 3368–3374.

LI, X. W., HOFF, S. J., BUNDY, D. S., HARMON, J., XIN, H., ZHU, J. (1996), Biofilter – A malodor control technology for livestock industry, *J. Environ. Sci. Health Part A – Environ. Sci. Eng. Toxic Hazard. Subst. Control* **31**, 2275–2285.

LU, C. S., LIN, M. R., CHU, C. H. (1999), Temperature effects of trickle-bed biofilter for treating BTEX vapors, *J. Environ. Eng. – ASCE* **125**, 775–779.

LUNDHOLM, I. M. (1982), Comparison of methods for quantitative determinations of airborne bacteria and evaluation of total viable counts, *Appl. Environ. Microbiol.* **44**, 179–183.

LYNCH, R. M., WOODLEY, J. M., LILLY, M. D. (1997), Process design for the oxidation of fluorobenzene to fluorocatechol by *Pseudomonas putida, J. Biotechnol.* **58**, 167–175.

MACNAUGHTON, S. J., CORMIER, M. R., JERKINS, T. L., DAVIS, G. A., WHITE, D. C. (1999), Quantitative sampling of indoor air biomass by signature lipid biomarker analysis, *J. Ind. Microbiol. Biotechnol.* **22**, 80–87.

MALHAUTIER, L., DEGORCEDUMAS, J. R., DEGRANGE, V., BARDIN, R., LECLOIREC, P. (1997), Serological determination of *Nitrobacter* species in a deodorizing granular activated carbon filter, *Environ. Technol.* **18**, 275–283.

MALLAKIN, A., WARD, O. P. (1996), Degradation of BTEX compounds in liquid media and in peat biofilters, *J. Ind. Microbiol.* **16**, 309–318.

MARCONI, A. M., BELTRAMETTI, F., BESTETTI, G., SOLINAS, F., RUZZI, M. et al. (1996), Cloning and characterization of styrene catabolism genes from *Pseudomonas fluorescens* ST, *Appl. Environ. Microbiol.* **62**, 121–127.

MARGARITIS, A., KENNEDY, K., ZAJIC, J. E., GERSON, D. F. (1979), Biosurfactant production by *Nocardia erythropolis, Dev. Ind. Microbiol.* **20**, 623–630.

MATTEAU, Y., RAMSAY, B. (1997), Active compost biofiltration of toluene, *Biodegradation* **8**, 135–141.

MORALES, M., REVAH, S., AURIA, R. (1998), Start-up and the effect of gaseous ammonia additions on a biofilter for the elimination of toluene vapors, *Biotechnol. Bioeng.* **60**, 483–491.

MORGENROTH, E., SCHROEDER, E. D., CHANG, D. P. Y., SCOW, K. M. (1996), Nutrient limitation in a compost biofilter degrading hexane, *J. Air Waste Managem. Ass.* **46**, 300–308.

MULLIGAN, C. N., CHOW, T. Y.-K., GIBBS, B. F. (1989), Enhanced biosurfactant production by a mutant *Bacillus subtilis* strain, *Appl. Microbiol. Biotechnol.* **31**, 486–489.

MURRAY, K., DUGLEBY, C. J., SALA-TREPAT, J. M. S., WILLIAMS, P. A. (1972), The metabolism of benzoate and methylbenzoates via the meta cleavage pathway by *Pseudomonas arvilla* mt-2 *Eur. J. Biochem.* **28**, 301–310.

NEU, T. R. (1996), Significance of bacterial surface-active compounds in interaction of bacteria with interfaces, *Microbiol. Rev.* **60**, 151–166.

NGUYEN, H. D., SATO, C., WU, J., DOUGLASS, R. W. (1997), Modeling biofiltration of gas streams containing TEX components, *J. Environ. Eng. – ASCE* **123**, 615–621.

O'CONNOR, K. E., DOBSON, A. D. W. (1996), Microbial degradation of alkenylbenzenes, *World J. Microbiol. Biotechnol.* **12**, 207–212.

O'CONNOR, K. E., BUCKLEY, C. M., HARTMANS, S., DOBSON, A. D. W. (1995), Possible regulatory role for non-aromatic carbon sources in styrene degradation by *Pseudomonas putida* CA-3, *Appl. Environ. Microbiol.* **61**, 544–548.

OH, Y. S., BARTHA, R. (1994), Design and performance of a trickling air biofilter for chlorobenzene and *o*-dichlorobenzene vapors, *Appl. Environ. Microbiol.* **60**, 2717–2722.

OH, Y. S., BARTHA, R. (1997a), Construction of a bacterial consortium for the biofiltration of benzene, toluene and xylene emissions, *World J. Microbiol. Biotechnol.* **13**, 627–632.

OH, Y. S., BARTHA, R. (1997b), Removal of nitrobenzene vapors by a trickling air biofilter, *J. Ind. Microbiol. Biotechnol.* **18**, 293–296.

OKKERSE, W. J. H., OTTENGRAF, S. P. P., OSINGA-KUIPERS, B., OKKERSE, M. (1999), Biomass accumulation and clogging in biotrickling filters for waste gas treatment. Evaluation of a dynamic model using dichloromethane as a model pollutant, *Biotechnol. Bioeng.* **63**, 418–430.

OLSEN, R. H., KUKOR, J. J., KAPHAMMER, B. (1994), A novel toluene-3-monooxygenase pathway cloned from *Pseudomonas pickettii* PKO1, *J. Bacteriol.* **176**, 3749–3756.

ORIGGI, G., COLOMBO, M., DEPALMA, F., RIVOLTA, M., ROSSI, P., ANDREONI, V. (1997), Bioventing of hydrocarbon-contaminated soil and biofiltration of the off-gas: Results of a field scale investigation, *J. Environ. Sci. Health Part A – Environ. Sci. Eng. Toxic Hazard. Subst. Control* **32**, 2289–2310.

PANKE, S., WITHOLT, B., SCHMID, A., WUBBOLTS, M. G. (1998), Towards a biocatalyst for (*S*)-styrene oxide production: Characterization of the styrene degradation pathway of *Pseudomonas* sp. strain VLB120, *Appl. Environ. Microbiol.* **64**, 2032–2043.

PARAT, S., PERDRIX, A., MANN, S., BACONNIER, P. (1999), Contribution of particle counting in assessment of exposure to airborne microorganisms, *Atmospher. Environ.* **33**, 951–959.

PEDERSEN, A. R., MØLLER, S., MOLIN, S., ARVIN, E. (1997), Activity of toluene-degrading *Pseudomonas putida* in the early growth phase of a biofilm for waste gas treatment, *Biotechnol. Bioeng.* **54**, 131–141.

PETERSEN, S. O., KLUG, M. J. (1994), Effects of sieving, storage, and incubation temperature on the phospholipid fatty acid profile of a soil microbial community, *Appl. Environ. Microbiol.* **60**, 2421–2430.

PENA, J., RICKE, S. C., SHERMER, C. L., GIBBS, T., PILLAI, S. D. (1999), A gene amplification–hybridization sensor based methodology to rapidly screen aerosol samples for specific bacterial gene sequences, *J. Environ. Sci. Health Part A – Toxic Hazard. Subst. Environ. Eng.* **34**, 529–556.

PHALE, P. S., SAVITHRI, H. S., RAO, N. A., VAIDYANATHAN, C. S. (1995), Production of biosurfactant "Biosur-Pm" by *Pseudomonas maltophila* CSV89: Characterization and role in hydrocarbon uptake, *Arch. Microbiol.* **163**, 424–431.

PIEPER, D. H., ENGESSER, K. H., KNACKMUSS, H.-J. (1989), Regulation of catabolic pathways of phenoxyacetic acids and phenols in *Alcaligenes eutrophus* JMP134, *Arch. Microbiol.* **151**, 365–371.

PILLAI, S. D., WIDMER, K. W., DOWD, S. E., RICKE, S. C. (1996), Occurrence of airborne bacteria and pathogen indicators during land application of sewage sludge, *Appl. Environ. Microbiol.* **62**, 296–299.

PLAGGEMEIER, T. (1999), Elimination der schwer wasserlöslichen Modellabluftinhaltsstoffe *n*-Hexan und Toluol im Biorieselbettverfahren, *Thesis*, University of Stuttgart, Germany.

PLAGGEMEIER, T., LÄMMERZAHL, O., ENGESSER, K.-H. (1997), Purification of *n*-hexane polluted air by use of a biological trickling filter: Monitoring of the primary-degrading microflora, in: *Biological*

Waste Gas Cleaning (PRINS, W. L., VAN HAM, J., Eds.), pp. 257–260. Düsseldorf: VDI-Verlag.

POL, A., VAN HAREN, F. J. J., OP DEN CAMP, H. J. M., VAN DER DRIFT, C. (1998), Styrene removal from waste gas with a bacterial biotrickling filter, *Biotechnol. Lett.* **20**, 407–410.

PRESSMAN, J. G., GEORGIOU, G., SPEITEL, G. E. (1999), Demonstration of efficient trichloroethylene biodegradation in a hollow-fiber membrane bioreactor, *Biotechnol. Bioeng.* **64**, 630.

QUINLAN, C., STREVETT, K., KETCHAM, M., GREGO, J. (1999), VOC elimination in a compost biofilter using a previously acclimated bacterial inoculum, *J. Air Waste Managem. Ass.* **49**, 544–553.

RAMSAY, B., MCCARTHY, J., GUERRA-SANTOS, L., KÄPPELI, O., FIECHTER, A. (1988), Biosurfactant production and diauxic growth of *Rhodococcus aurantiacus* when using *n*-alkanes as the carbon source, *Can. J. Microbiol.* **34**, 1209–1212.

RASMUSSEN, K., LEWANDOWSKI, Z. (1998), Microelectrode measurements of local mass transport rates in heterogeneous biofilms, *Biotechnol. Bioeng.* **59**, 302–309.

RASKIN, L., STROMLEY, J. M., RITTMANN, B. E., STAHL, D. A. (1994), Group-specific 16S rRNA hybridization probes to describe natural communities of methanogens, *Appl. Environ. Microbiol.* **60**, 1232–1240.

REIJ, M. W., HARTMANS, S. (1996), Propene removal from synthetic waste gas using a hollow fiber membrane bioreactor, *Appl. Microbiol. Biotechnol.* **45**, 730–736.

REIJ, M. W., KIEBOOM, J., DE BONT, J. A. M., HARTMANS, S. (1995), Continuous degradation of trichloroethylene by *Xanthobacter* sp. strain Pr2 during growth on propene, *Appl. Environ. Microbiol.* **61**, 2936–2942.

REIJ, M. W., HAMANN, E. K., HARTMANS, S. (1997), Biofiltration of air containing low concentrations of propene using a membrane bioreactor, *Biotechnol. Prog.* **13**, 380–386.

REIJ, M. W., KEURENTJES, J. T. F., HARTMANS, S. (1998), Membrane bioreactors for waste gas treatment, *J. Biotechnol.* **59**, 155–167.

REISER, M., FISCHER, K., ENGESSER, K.-H. (1994), Kombination aus Biowäscher- und Biomembranverfahren zur Reinigung von Abluft mit hydrophilen und hydrophoben Inhaltsstoffen, in: *VDI Bericht 1104 Biologische Abgasreinigung* (KrdL im VDI und DIN, Ed.), pp. 103–112. Düsseldorf: VDI-Verlag.

RIHN, M. J., ZHU, X. Q., SUIDAN, M. T., KIM, B. J., KIM, B. R. (1997), The effect of nitrate on VOC removal in trickle-bed biofilters, *Water Res.* **31**, 2997–3008.

ROCH, F., ALEXANDER, M. (1997), Inability of bacteria to degrade low concentrations of toluene in water, *Environ. Toxicol. Chem.* **16**, 1377–1383.

ROSENBERG, M., DOYLE, R. J. (1990), Microbial cell surface hydrophobicity: history, measurement, and significance, in: *Microbial Cell Surface Hydrophobicity* (DOYLE, R. J., ROSENBERG, M., Eds.), pp. 1–37, Washington, DC: ASM Press.

ROUHANA, N., HANDAGAMA, N., BIENKOWSKI, P. R. (1997), Development of a membrane-based vapor-phase bioreactor, *Appl. Biochem. Biotechnol.* **63-5**, 809–821.

SAEKI, H., AKIRA, M., FURUHASHI, K., AVERHOFF, B., GOTTSCHALK, G. (1999), Degradation of trichloroethene by a linear-plasmid-encoded alkene monooxygenase in *Rhodococcus corallinus* (*Nocardia corallina*) B-276, *Microbiology* (UK) **145**, 1721–1730.

SAKANO, Y., KERKHOF, L. (1998), Assessment of changes in microbial community structure during operation of an ammonia biofilter with molecular tools, *Appl. Environ. Microbiol.* **64**, 4877–4882.

SCHINDLER, I., FRIEDL, A. (1995), Degradation of toluene/heptane mixtures in a trickling-bed bioreactor, *Appl. Microbiol. Biotechnol.* **44**, 230–233.

SCHÖNDUVE, P., SARA, M., FRIEDL, A. (1996), Influence of physiologically relevant parameters on biomass formation in a trickle-bed bioreactor used for waste gas cleaning, *Appl. Microbiol. Biotechnol.* **45**, 286–292.

SENKPIEL, K. (1999), Gesundheitliche Gefahren durch biogene Luftschadstoffe, *Hyg. Mikrobiol.* **1999**, 45–49.

SELENSKA, S., KLINGMÜLLER, W. (1991), Direct detection of *nif*-gene sequence of *Enterobacter agglomerans* in soil, *FEMS Microbiol Lett.* **80**, 243–246.

SHAFFER, B. T., LIGHTHART, B. (1997), Survey of culturable airborne bacteria at four diverse locations in Oregon: Urban, rural, forest, and coastal, *Microb. Ecol.* **34**, 167–177.

SHAREEFDEEN, Z., BALTZIS, B. C., OH, Y.-S., BARTHA, R. (1993), Biofiltration of methanol vapor, *Biotechnol. Bioeng.* **41**, 512–524.

SHOJAOSADATI, S. A., ELYASI, S. (1999), Removal of hydrogen sulfide by the compost biofilter with sludge of leather industry, *Res. Conserv. Recycl.* **27**, 139–144.

SHURTLIFF, M. M., PARKIN, G. F., WEATHERS, L. J., GIBSON, D. T. (1996), Biotransformation of trichloroethylene by a phenol-induced mixed culture, *J. Environ. Eng. – ASCE* **122**, 581–589.

SLY, L. I., BRYANT, L. J., COX, J. M., ANDERSON, J. M. (1993), Development of a biofilter for the removal of methane from coal mine ventilation atmospheres, *Appl. Microbiol. Biotechnol.* **39**, 400–404.

SMET, E., VAN LANGENHOVE, H., VERSTRAETE, W. (1996), Long-term stability of a biofilter treating dimethyl sulfide, *Appl. Microbiol. Biotechnol.* **46**, 191–196.

SMET, E., VAN LANGENHOVE, H., VERSTRAETE, W.

(1997), Isobutyraldehyde as a competitor of the dimethyl sulfide degrading activity in biofilters, *Biodegradation* **8**, 53–59.

SORIAL, G. A., SMITH, F. L., SUIDAN, M. T., PANDIT, A., BISWAS, P., BRENNER, R. (1997), Evaluation of trickle bed air biofilter performance for BTEX removal, *J. Environ. Eng. – ASCE* **123**, 530–537.

SORIAL, G. A., SMITH, F. L., SUIDAN, M. T., PANDIT, A., BISWAS, P., BRENNER, R. C. (1998), Evaluation of trickle-bed air biofilter performance for styrene removal, *Water Res.* **32**, 1593–1603.

SPEITEL JR., G. E., MCLAY, D. S. (1993), Biofilm reactors for treatment of gas streams containing chlorinated solvents, *J. Environ. Eng.* **119**, 658–678.

STEWART, S. L., GRINSHPUN, S. A., WILLEKE, K., TERZIEVA, S., ULEVICIUS, V., DONNELLY, J. (1995), Effect of impact stress on microbial recovery on an agar surface, *Appl. Environ. Microbiol.* **61**, 1232–1239.

STOFFELS, M., AMANN, R., LUDWIG, W., HEKMAT, D., SCHLEIFER, K.-H. (1998), Bacterial community dynamics during start-up of a trickle-bed bioreactor degrading aromatic compounds, *Appl. Environ. Microbiol.* **64**, 930–939.

STRAJA, S., LEONARD, R. T. (1996), Statistical analysis of indoor bacterial air concentration and comparison of four RCS biotest samplers, *Environ. Int.* **22**, 389–404.

SUKESAN, S., WATWOOD, M. E. (1997), Continuous vapor-phase trichloroethylene biofiltration using hydrocarbon-enriched compost as filtration matrix, *Appl. Microbiol. Biotechnol.* **48**, 671–676.

SUKESAN, S., WATWOOD, M. E. (1998), Effects of hydrocarbon enrichment on trichloroethylene biodegradation and microbial populations in finished compost, *J. Appl. Microbiol.* **85**, 635–642.

SUNDH, I., NILSSON, M., BORGA, P. (1997), Variation in microbial community structure in two boreal peatlands as determined by analysis of phospholipid fatty acid profiles, *Appl. Environ. Microbiol.* **63**, 1476–1482.

TAHRAOUI, K., RHO, D. (1998), Biodegradation of BTX vapors in a compost medium biofilter, *Compost Sci. Util.* **6**, 13–21.

TAKAMI, W., HORINOUCHI, M., NOJIRI, H., YAMANE, H., OMORI, T. (1999), Evaluation of trichloroethylene degradation by *E. coli* transformed with dimethyl sulfide monooxygenase genes and/or cumene dioxygenase genes, *Biotechnol. Lett.* **21**, 259–264.

TALBOT, N. J. (1999), Coming up for air and sporulation, *Nature* **398**, 295–296.

TANJI, Y., KANAGAWA, T., MIKAMI, E. (1989), Removal of dimethyl sulfide, methyl mercaptan, and hydrogen sulfide by immobilized *Thiobacillus thioparus* TK-M, *J. Ferment. Bioeng.* **67**, 280–285.

TOLVANEN, O. K., HANNINEN, K. I., VEIJANEN, A., VILLBERG, K. (1998), Occupational hygiene in biowaste composting, *Waste Managem. Res.* **16**, 525–540.

UCHIYAMA, H., OGURI, K., NISHIBAYASHI, M., KOKUFUTA, E., YAGI, O. (1995), Trichloroethylene degradation by cells of a methane-utilizing bacterium, *Methylocystis* sp. M, immobilized in calcium-alginate, *J. Ferment. Bioeng.* **79**, 608–613.

UTKIN, I. B., YAKIMOV, M. M., MATVEEVA, L. N., KOZLYAK, E. I., ROGOZHIN, I. S. et al. (1991), Degradation of styrene and ethylbenzene by *Pseudomonas* species-Y2, *FEMS Microbiol. Lett.* **77**, 237–241.

VANALPHEN, L., JANSEN, H. M., DANKERT, J. (1995), Virulence factors in the colonization and persistence of bacteria in the airways, *Am. J. Resp. Crit. Care Med.* **151**, 2094–2100.

VAN LOOSDRECHT, M. C. M., HEIJNEN, J. J. (1996), *Biofilm Processes, in Immobilized Living Cell Systems: Modeling and Experimental Methods* (WILLAERT, R. G., BARON, G. V., DE BACKER, L., Eds.), pp. 256–271, Chichester: John Wiley & Sons.

VELASCO, A., ALONSO, S., GARCIA, J. L., PERERA, J., DIAZ, E. (1998), Genetic and functional analysis of the styrene catabolic cluster of *Pseudomonas* sp. strain Y2, *J. Bacteriol.* **180**, 1063–1071.

VILLAVERDE, S., FERNÁNDEZ-POLANCO, F. (1999), Spatial distribution of respiratory activity in *Pseudomonas putida* 54G biofilms degrading volatile organic compounds (VOC), *Appl. Microbiol. Biotechnol.* **51**, 382–387.

VLIEG, J. E. T. V., DEKONING, W., JANSSEN, D. B. (1996), Transformation kinetics of chlorinated ethenes by *Methylosinus trichosporium* OB3b and detection of unstable epoxides by on-line gas chromatography, *Appl. Environ. Microbiol.* **62**, 3304–3312.

VOGT SINGER, M. E., FINNERTY, R. W., TUNELID, A. (1990), Physical and chemical properties of a biosurfactant synthesized by *Rhodococcus* species H 13-A, *Can. J. Microbiol.* **36**, 746–750.

WANI, A. H., BRANION, R. M. R., LAU, A. K. (1997), Biofiltration: A promising and cost-effective control technology for odors, VOCs and air toxics, *J. Environ. Sci. Health Part A – Environ. Sci. Eng. Toxic Hazard. Subst. Control* **32**, 2027–2055.

WARHURST, A. M., FEWSON, C. A. (1994), Microbial metabolism and biotransformations of styrene, *J. Appl. Bacteriol.* **77**, 597–606.

WARHURST, A. M., CLARKE, K. F., HILL, R. A., HOLT, R. A., FEWSON, C. A. (1994), Production of catechols and muconic acids from various aromatics by the styrene-degrader *Rhodococcus rhodochrous* NCIMB 13259, *Biotechnol. Lett.* **16**, 513–516.

WEBER, F. J., HARTMANS, S. (1995), Use of activated carbon as a buffer in biofiltration of waste gases with fluctuating concentrations of toluene, *Appl.*

Microbiol. Biotechnol. **43**, 365–369.

WEBER, F. J., HAGE, K. C., DE BONT, J. A. M. (1995), Growth of thc fungus *Cladosporium sphaerospermum* with toluene as the sole carbon and energy source, *Appl. Environ. Microbiol.* **61**, 3562–3566.

WECKHUYSEN, B., VRIENS, L., VERACHTERT, H. (1993), The effect of nutrient supplementation on the biofiltration removal of butanal in contaminated air, *Appl. Microbiol. Biotechnol.* **39**, 395–399.

WECKHUYSEN, B., VRIENS, L., VERACHTERT, H. (1994), Biotreatment of ammonia- and butanal-containing waste gases, *Appl. Microbiol. Biotechnol.* **42**, 147–152.

WHITED, G. M., GIBSON, D. T. (1991), Toluene-4-monooxygenase, a three-component enzyme system that catalyzes the oxidation of toluene to para-cresol in *Pseudomonas mendocina* KR1, *J. Bacteriol.* **173**, 3010–3016.

WILCOX, D. W., AUTENRIETH, R. L., BONNER, J. S. (1995), Propane-induced biodegradation of vapor phase trichloroethylene, *Biotechnol. Bioeng.* **46**, 333–342.

WU, G., CHABOT, J. C., CARON, J. J., HEITZ, M. (1998), Biological elimination of volatile organic compounds from waste gases in a biofilter, *Water Air Soil Pollut.* **101**, 69–78.

WU, G., CONTI, B., LEROUX, A., BRZEZINSKI, R., VIEL, G., HEITZ, M. (1999), A high performance biofilter for VOC emission control, *J. Air Waste Managem. Ass.* **49**, 185–192.

WÜBKER, S. M., FRIEDRICH, C. G. (1996), Reduction of biomass in a bioscrubber for waste gas treatment by limited supply of phosphate and potassium ions, *Appl. Microbiol. Biotechnol.* **46**, 475–480.

YANI, M., HIRAI, M., SHODA, M. (1998), Ammonia gas removal characteristics using biofilter with activated carbon fiber as a carrier, *Environ. Technol.* **19**, 709–715.

YEOM, S. H., YOO, Y. J. (1999), Removal of benzene in a hybrid bioreactor, *Proc. Biochem.* **34**, 281–288.

ZHANG, Y., MILLER, R. M. (1995), Effect of rhamnolipid (biosurfactant) structure an solubilization and biodegradation of *n*-alkanes, *Appl. Environ. Microbiol.* **61**, 2247–2251.

ZHANG, L., KUNIYOSHI, I., HIRAI, M., SHODA, M. (1991), Oxidation of dimethyl sulfide by *Pseudomonas acidovorans* DMR-11 isolated from peat biofilter, *Biotechnol. Lett.* **13**, 223–228.

ZHANG, L., HIRAI, M., SHODA, M. (1992), Removal characteristics of dimethyl sulfide by a mixture of *Hyphomicrobium* sp. I55 and *Pseudomonas acidovorans* DMR-11, *J. Ferment. Bioeng.* **74**, 174–178.

ZHANG, T. C., FU, Y.-C., BISHOP, P. L. (1995), Competition for substrate and space in biofilms, *Water Environ. Res.* **67**, 992–1003.

ZHOU, Q., HUANG, Y. L., TSENG, D. H., SHIM, H., YANG, S. T. (1998), A trickling fibrous-bed bioreactor for biofiltration of benzene in air, *J. Chem. Technol. Biotechnol.* **73**, 359–368.

ZHU, X. Q., ALONSO, C., SUIDAN, M. T., CAO, H. W., KIM, B. J., KIM, B. R. (1998), The effect of liquid phase on VOC removal in trickle-bed biofilters, *Water Sci. Technol.* **38**, 315–322.

ZUBER, L., DUNN, I. J., DESHUSSES, M. A. (1997), Comparative scale-up and cost estimation of a biological trickling filter and a three-phase airlift bioreactor for the removal of methylene chloride from polluted air, *J. Air Waste Managem. Ass.* **47**, 969–975.

Processes

13 Bioscrubbers

EGBERT SCHIPPERT
HORST CHMIEL
Saarbrücken, Germany

1 Introduction

Bioscrubbers provide a suitable tool for eliminating small concentrations of water-soluble organic impurities from waste air. The concept for bioscrubbers is based on the fact that microorganisms suspended in water can only extract dissolved substances from water. Like the oxygen required for aerobic degradation, the impurities to be degraded must overcome the phase boundary to become dissolved in the liquid phase.

With the exception of non-degradable or not further degradable substances, the final products generated by this degradation process, which supplies the cells with energy and elements for their anabolism, are CO_2, H_2O, and possibly N_2. The following parameters are considered to be most important for these processes (STOCKHAMMER et al., 1992):

- mass transfer area between gas and liquid,
- residence time,
- type and concentration of pollutants,
- type and concentration of microorganisms,
- adaptation by microorganisms,
- available oxygen,
- temperature,
- pH value.

The concept is implemented, in technical terms, by combining a scrubber column with an aerobic wastewater reactor, as shown in Fig. 1 (CHMIEL, 1991).

The flow of the scrubbing liquid inoculated with microorganisms from the wastewater reactor is countercurrent to the waste air in the scrubber. In this manner impurities are transferred to the scrubbing liquid which is then fed into the aerated wastewater reactor where the dissolved substances are degraded biologically. Both stages must be coordinated in order to achieve an steady state between pollutant absorption and microbial pollutant degradation in continuous operation.

Bioscrubbers are particularly suitable for waste air cleaning if a biological wastewater treatment plant already exists and if the components to be removed from the waste air

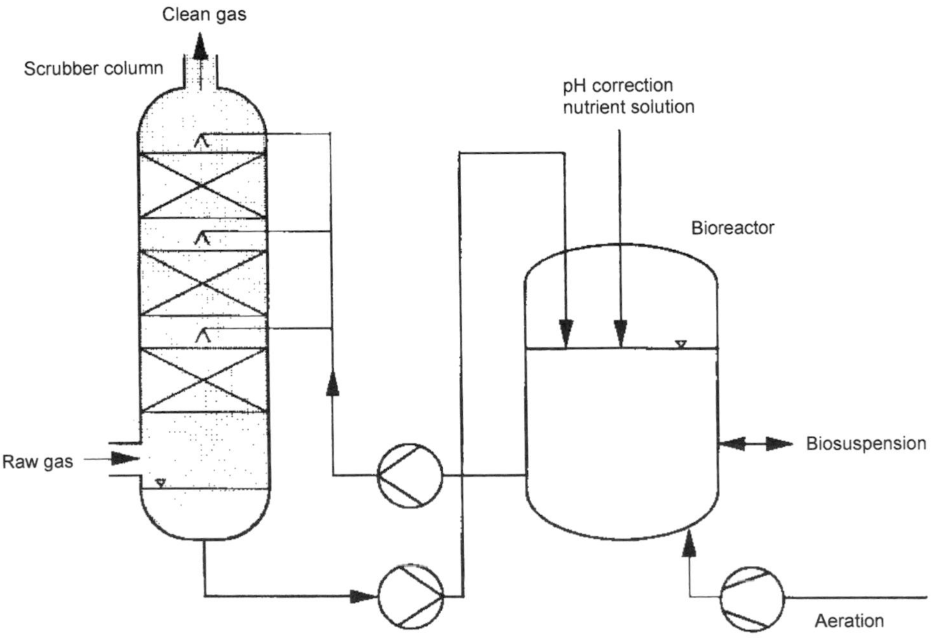

Fig. 1. Flow diagram of a bioscrubber for waste air cleaning.

stream can be easily dissolved in water and are biologically degradable.

Principally, any type of column can be used as a scrubber, e.g., sieve plate columns, bubble-cap columns, or packed columns. However – in contrast to the theoretical concept – it should be noted that after a short period of operation of bioscrubbers a biofilm develops on the fixed internal parts of the column. This can lead to partial blocking and to channeling, if the space for flow is too small, as is usually the case in packed columns.

In the planning and construction phases of gas scrubbers, appropriate measures should be implemented to prevent any considerable quantity of biomass from accumulating on the internal parts, so that gas and liquid can freely pass the scrubber. This can be achieved by using internal parts with an open structure and by selecting trickling densities high enough that the developing shear force prevents a biofilm from forming on the internal parts. This trickling density should be $>20\ m^3\ m^{-2}\ h^{-1}$.

2 Design of the Activated Sludge Tank

The main design criterium for these activated sludge tanks is the establishment of a steady state between the imported and the degraded substances at the residual concentration of the substances to be degraded in the washing liquid. Intermediate degradation products should not accumulate in the system.

Biological regeneration, i.e., degradation of waste air components which have been absorbed into the scrubbing liquid and leave the scrubber with it, occurs in separate activated sludge tanks with forced ventilation to supply the microorganisms with the oxygen needed for biodegradation.

Although the absorption stage and biodegradation stage are separated physically, there is significant feedback between these two technical stages caused by the scrubbing liquid which is continuously circulated: the higher the residual loading of organic substances in the circulated water, i.e., the less effective biological regeneration, the lower the achieveable degree of absorption efficiency; on the other hand, the higher the residual load of the circulated water, the higher the rate of biodegradation. However, this is only valid for a relatively low limiting concentration.

Fig. 2 shows the most important operational and design parameters for bioscrubbers (SCHIPPERT and FRÖHLICH, 1993). It is particularly evident that this process is only suitable for relatively low concentrations of raw gas.

Fig. 3 is a schematic representation of the technical and microbiological parameters to be considered for design and operation of bioscrubbers (VDI Guidelines, 1996). In the following the essential aspects of these parameters are discussed in detail.

The kinetics of biodegradation is crucial for the dimensional design of the activated sludge tank, i.e., the substrate freight imported for degradation via the scrubber and the steady-state residual concentration required for the specified degree of absorption efficiency. This can often be described by so-called "Monod kinetics": For relatively low steady-state concentrations the biodegradation rate is almost proportional to this concentration (first order kinetics). Above this limiting concentration, a maximum degradation rate, which is not dependent on the substrate concentration, is ultimately achieved (zero order kinetics). Fig. 4 shows, for some solvents, the measured biodegradation rates which have been fitted by Monod kinetics. It can be seen that major variations occur in the increase of the degradation rate related to an increasing substrate concentration and the maximum degradation rate depending on the substrate to be degraded. Bioscrubbers operate in both, first or zero order, kinetics while multi-stage activated sludge tanks may operate in both orders. The size required for the activated sludge tank can be calculated by simple division: Substrate freight transported into scrubber [kg h^{-1}] : degradation rate at a defined steady-state residual concentration [kg $m^3\ h^{-1}$].

The dimensions of activated sludge tank(s) must be large enough for total biodegradation (i.e., not only the degradation of intermediate products but to the formation of carbon dioxide) of the maximum solvent load transported

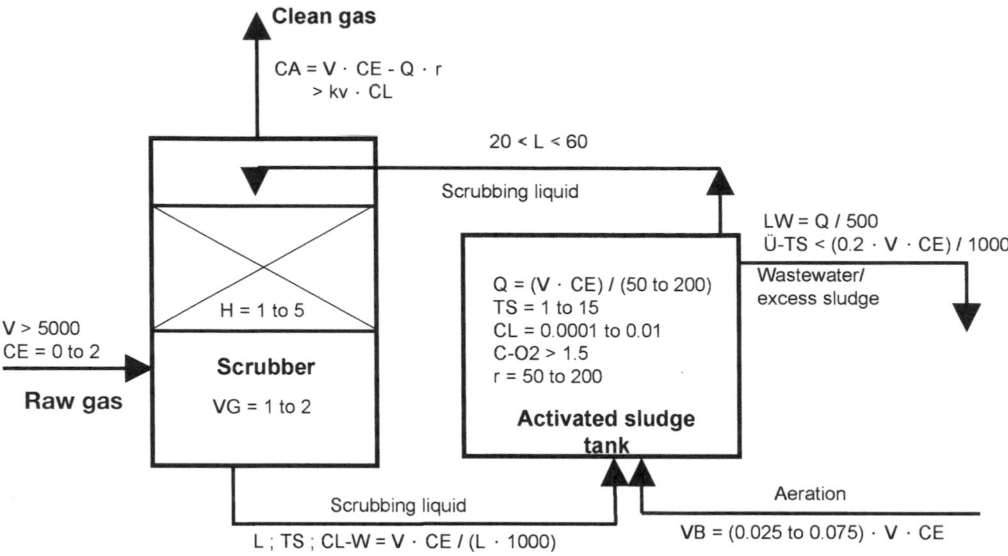

Fig. 2. Operational and design parameters for bioscrubbers. V: waste air quantity [$m^3 h^{-1}$], CE: crude gas concentration [$g m^{-3}$], CA: clean gas concentration [$g m^{-3}$], CL-W: solvent concentration in scrubber effluent [$g L^{-1}$], CL: steady-state residual solvent concentration in water [$g L^{-1}$], Q: size of activated sludge tank [m^3], L: specific scrubbing liquid circulation quantity [$(m^3 h^{-1}) m^{-2}$], r: biodegradation rate [$g (m^3 \cdot h)^{-1}$], Kv: distribution coefficient air–water [$(g m^{-3}) (g L^{-1})^{-1}$], TS: biomass concentration [$g L^{-1}$], Ü-TS: biological excess sludge quantity [$kg h^{-1}$], LW: wastewater quantity [$m^3 h^{-1}$], C-O_2: O_2 concentration in activated sludge tank [$mg L^{-1}$], VG: gas velocity in the adsorber [$m s^{-1}$], H: packing height [m], VB: aeration rate [$m^3 h^{-1}$].

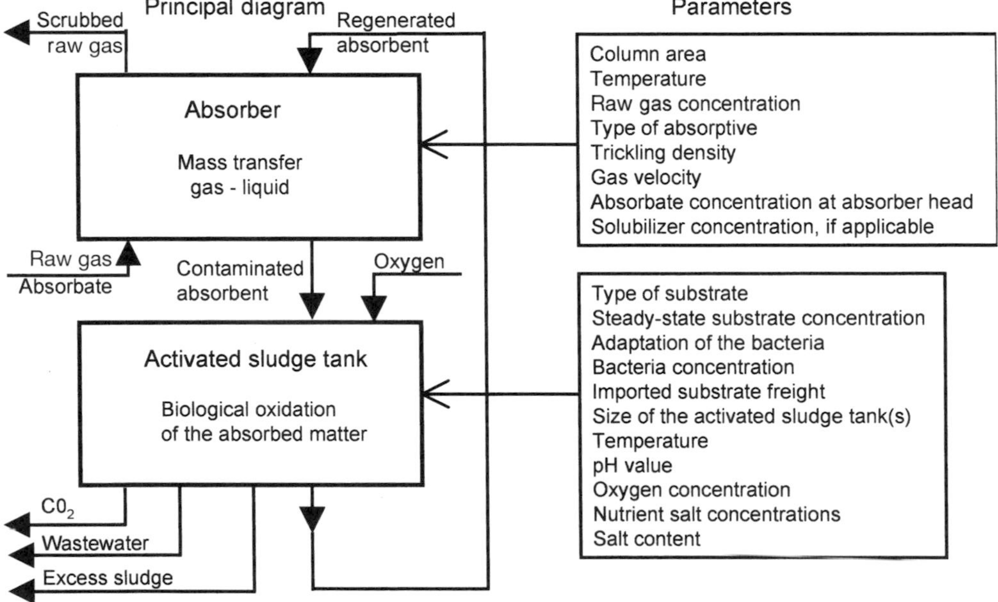

Fig. 3. Important parameters for the design of bioscrubbers.

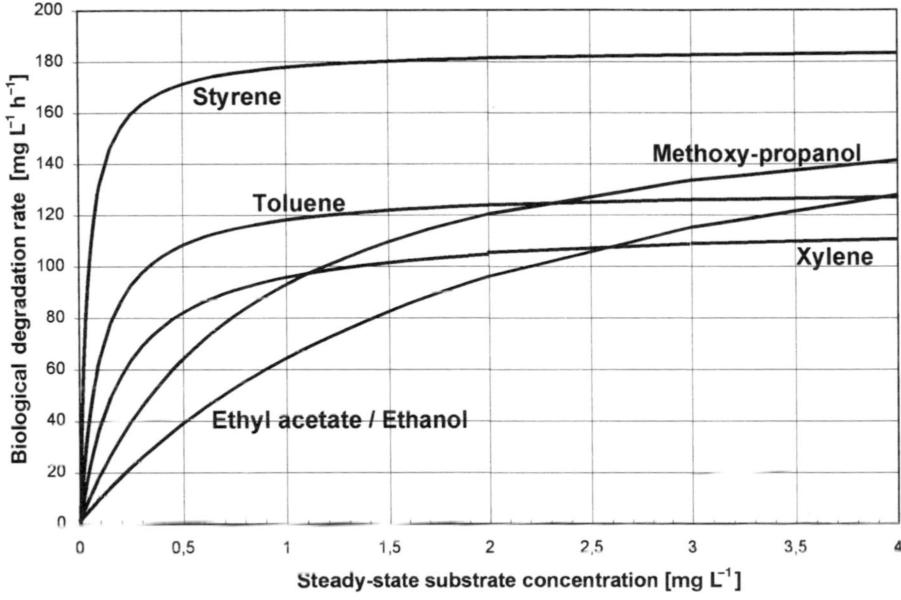

Fig. 4. Degradation rate of solvents in bioscrubbers as a function of the steady-state substrate concentration.

from the scrubber at the steady-state residual substrate concentration required for the desired scrubber efficiency in the activated sludge tank. To calculate the maximum pollutant load, the average load can be taken over a period of time similar to the residence time of water in the activated sludge tank (residence time = volume of activated sludge : circulating pump output).

In general, the required steady-state residual concentration is not known, as it depends on the design of the absorption column and particularly on the specified clean gas value. Of course, the substrate freight to be degraded also depends on the scrubber efficiency which increases with the absorption efficiency. The effects of feedback and operations in which multi-stage activated sludge tanks are advantageous will be dealt with later in detail. Moreover, the design of activated sludge tanks in bioscrubbing plants can be similar to that used in aerobic biological wastewater treatment, but with some essential differences:

- In bioscrubbers, the residence time of water is 1–2 orders of magnitude higher (20–40 d), the ratio of wastewater discharge to degradation of pollutant load is accordingly lower.
- In the waste air quantities of nitrogen and phosphorus usually are not sufficient and must be added. This provides, within certain limits, the opportunity of restricting activated sludge growth by reduced addition of nutrient salts.
- The quantity of excess sludge (to be disposed of) is considerably smaller than, e.g., in municipal sewage treatment plants. In relation to the degraded substrate quantity it can be only 10% of the quantities generated in municipal sewage treatment plants.
- Clarification from suspended matter of the relatively low quantities of wastewater generated is usually not possible in gravitational separators, because very fine biomass particles are formed due to the strong shear forces active in the plant and the very high sludge age of the biocenosis. Although these fine particles can be removed from the wastewater by microfiltration or ultrafiltration, it is usually not necessary to implement such expensive separation processes (see next point), unless it is intended to increase the biomass concentration.

- The excess sludge does not contain any toxic substances, particularly no heavy metals.
- A biological waste air treatment plant is not a wastewater treatment plant, but rather a waste air treatment plant producing small quantities of biologically pretreated wastewater. Authorities issuing permits often miss this point.

3 Design of the Absorption Column

Scrubber cross-sections, packed column height and scrubbing liquid quantities are determined during the design process for absorption columns. Waste air quantities, raw gas concentrations, the defined clean gas concentrations, residual loading in the scrubbing liquid which is discharged at the head of the column, and the distribution coefficient for the absorbate between air and water are used as a basis for design.

The column can be calculated and designed according to the HTU-NTU model by COLBURN (1939) which is based on the simplified two-film model of mass transfer in scrubber packed columns. The model concepts and calculations underlying these models are shown in Fig. 5 and 6.

The results of these basic equations, particularly for the design of bioscrubbers, will be illustrated in more detail in the following section. The influence of the so-called air/scrubbing liquid distribution coefficient will be dealt with here: This coefficient is higher, the poorer the solubility and the greater the volatility (vapor pressure) of the absorbate. With higher values of this distribution coefficient, the quantity of the scrubbing liquid which is required to obtain a given degree of absorption efficiency increases. The distribution coefficients for some solvents are specified in Tab. 1. For the absorption of ethanol using water as the scrubbing liquid, e.g., only 0.43 $m^3\ h^{-1}$ are required as a minimum quantity of scrubbing liquid for 1,000 $m^3\ h^{-1}$, whereas minimum requirements for ethyl acetate are 2.8 $m^3\ h^{-1}$, and at least 270 $m^3\ h^{-1}$ for toluene. Evidently, there are limits to using water as a scrubbing liquid, particularly when the waste air constit-

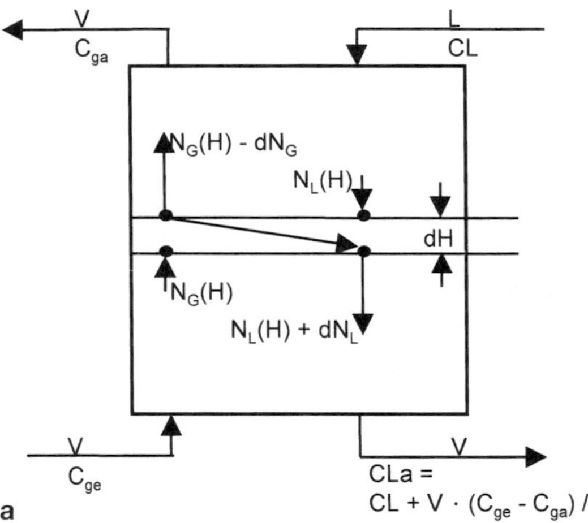

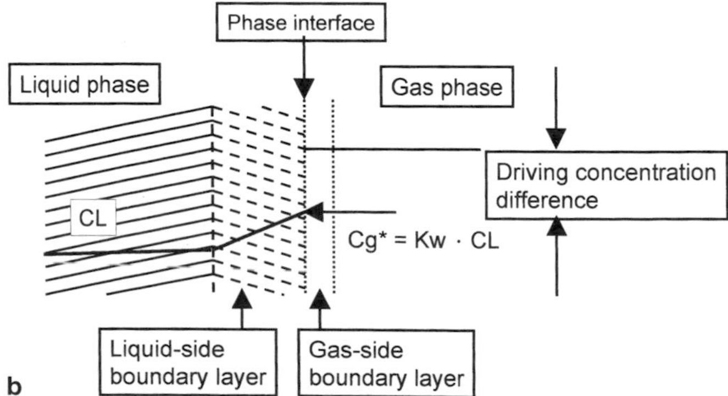

Fig. 5. (**a**) Mass balance in the column, (**b**) two-film model for mass transfer.

$$Cgl(H)_{(G)} = K_W \cdot C(H)_{(L)}$$ Gas - liquid equilibrium

$$dN_G = K_G \cdot F \cdot \left[C(H)_{(G)} - Cgl(H)_{(G)} \right] \cdot dH$$

$$dC_G = \frac{dN_G}{V}$$ Mass transfer, Fick's law

$$\left[Cgl(H)_{(L)} - C_{LE} \right] \cdot L = \left[C(H)_{(G)} - Cga \right] \cdot V$$ Mass balance between absorber effluent and one level in the column height H

Summarized and integrated via the column

$$\frac{K_g \cdot F}{V} \int_0^H 1\, dH = \int_{C_{GE}}^{C_{GA}} \frac{1}{C_G - K_W \cdot \left[(C_G - C_{GA}) \cdot \frac{V}{L} + C_{LE} \right]} dC_G$$

$$H \cdot \frac{K_g \cdot F}{V} = \frac{1}{1 - K_W \cdot \frac{V}{L}} \cdot \ln \left[\frac{C_{GE} - K_W \cdot C_{LE}}{C_{GA} - K_W \cdot C_{LE}} \cdot \left(1 - K_W \cdot \frac{V}{L} \right) + K_W \cdot \frac{V}{L} \right]$$

or acc. to Colburn

$$H = HTU \cdot NTU$$

with

Height of a transfer unit — No. of transfer units

$$HTU = \frac{V}{K_g \cdot F} \quad \text{und} \quad NTU = \frac{1}{1 - K_W \cdot \frac{V}{L}} \cdot \ln \left[\frac{C_{GE} - K_W \cdot C_{LE}}{C_{GA} - K_W \cdot C_{LE}} \cdot \left(1 - K_W \cdot \frac{V}{L} \right) + K_W \cdot \frac{V}{L} \right]$$

$Cgl(H)_G$	Gas equilibrium concentration
K_W	Distribution coefficient air/water
H	Column height
$C(H)_L$	Concentration in water in height H
K_g	Mass transfer coefficient
F	Column cross-section area
N_G	Mole flow of absorptive
$C(H)_G$	Concentration in gas in height H
V	Waste air quantity
L	Scrubbing liquid quantity
C_{ge} C_{ga}	Crude gas + clean gas concentration

Simplified assumptions:
- Straight equilibrium line
- No absorptive concentration
- Isotherm absorption
- Density water = 1
- Steady-state
- Mass transfer coefficient constant
- Equimolecular diffusion

Fig. 6. Basic equation for column design.

Tab. 1. Bioscrubber Plant with and without an Equalizing Pre-Filter with Countercurrent Flow

	K_v [(g m^{-3}) (g L^{-1})] or Minimum Quantity of Scrubbing Liquid (m^3 h^{-1}) per 1,000 m^3 h^{-1} Waste Air	Practical Minimum Water Quantity (m^3 h^{-1}) per 1,000 m^3 h^{-1} Waste Air
Formaldehyde	0.012	0.5
Phenol(at) pH=8	0.015	0.5
Butyl glycol	0.013	0.5
Ethanol	0.43	1
Butanol	0.55	1
Cyclohexane	1.0	2
Isobutanol	1.5	3
Acetone	1.67	3
Ethyl acetate	2.8	6
Butyl acetate	5	10
Methyl isobutylketone	5.3	10
Biphenyl	11	20
Naphthalene	17	35
Diethyl ether	35	70
Diisopropyl ether	89	180
Styrene	110	220
1,3,5-Trimethylbenzol	240	480
Toluene	270	540
Xylene	280	560

uents to be absorbed have distribution coefficients of >5. Then the required quantities of scrubbing liquid and, therefore, the energy needed to pump the circulating water are too high. More information on the successful use of bioscrubbers in industry can be found in the Guidelines by the Association of German Engineers (VDI Guidelines, 1996).

4 Examples of Conventional Bioscrubbers in Industrial Applications

Stockhammer (STOCKHAMMER et al., 1992) has described two examples of industrial appli-

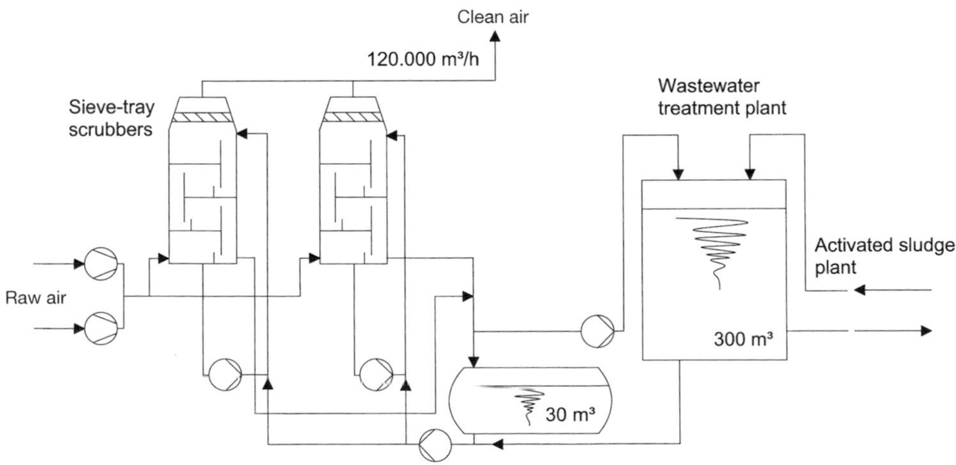

Fig. 7. Flow diagram of bioscrubber for odor mitigation in a production plant.

cation of conventional bioscrubbers. Due to the reasons mentioned earlier a sieve plate column was used.

The first case deals with the treatment of 100,000 m^3 h^{-1} of indoor air in a production plant which smelled unpleasantly of sulfur compounds. It was possible to deodorize the indoor air by integrating an existing on-site wastewater treatment plant.

Fig. 7 shows the flow diagram of this plant. Mass transfer took place in two scrubbers in parallel operation, each fitted with three sieve trays. The scrubbing liquid was continuously cycled via an aerated wastewater reactor which was supplied with fresh biosuspension by a second circulation process.

The possibility of combining the bioreactor with an existing wastewater treatment plant provided a sound argument in favor of the use of bioscrubbers.

The second example, described by STOCKHAMMER of the successful elimination of pollutants from the waste air of a plastics production plant is also based on the above mentioned fundamental concept.

Within capacity expansion measures attempts were made to draw up a new concept for waste air cleaning in this production plant. In this case, the objective was to treat the waste gases which contained much higher concentrations of pollutants than the example previously described. The volumetric flow of the waste air was roughly 10,000 m^3 h^{-1}, the pollutants being mainly monomers from the production of plastics. Due to the batch operation of the plant, pollutant concentration greatly fluctuated and distinct peaks in concentration occurred.

In order to adhere to the limit values as defined in the Technical Directive for Air Pollution Control (TA Luft, 1986), efficient removal of at least 80% had to be achieved for these pollutant concentrations.

The first results with the testing plant (Fig. 8) can be summarized as follows:

- The 80% degree of efficiency, as defined in the Technical Directive for Air Pollution Control, was achieved and exceeded for two-stage operation of the scrubbing column with 3 sieve trays per stage.

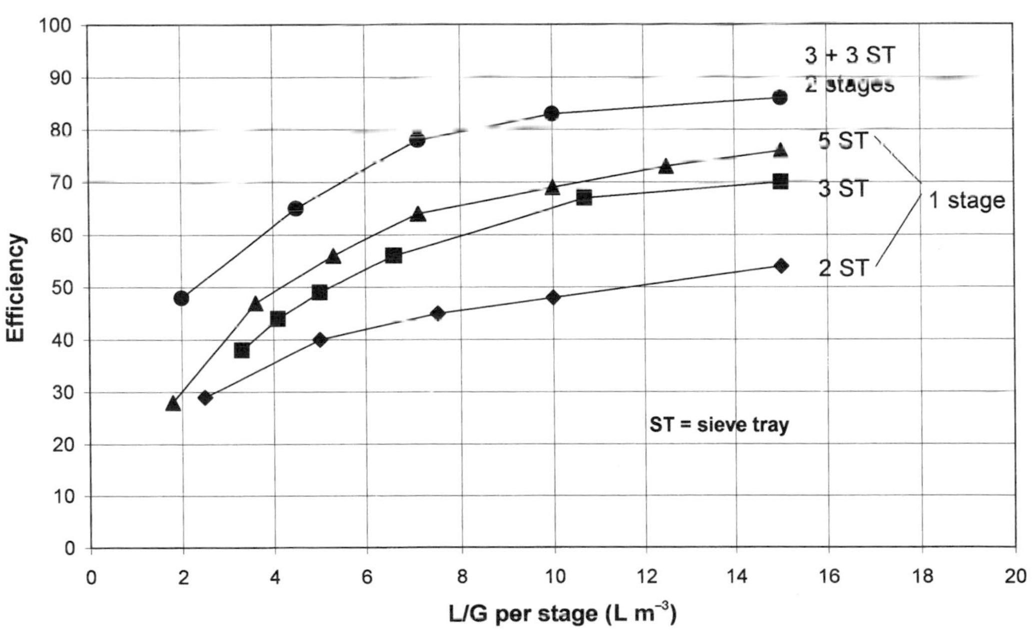

Fig. 8. Efficiency as a function of the liquid–gas ratio and number of sieve trays.

- Microbiological regeneration of the scrubbing liquid in the regeneration tank was rapid and sufficient (residence time required <15 min).
- The required efficiency could be maintained on the long term once the wastewater treatment plant was integrated and transfer rates were between 5 and 10% h^{-1}.

In view of the present results, it can be stated that the limit values as in the Technical Directive for Air Pollution Control could be maintained, i.e., the waste air problem in question could be solved technically by using a bioscrubber.

Extensive experience with the concept to integrate the on-site wastewater treatment plant has been gained from the long-term and successful use of industrial bioscrubbers for odor removal and the operation of the testing plant:

- It was evident that the degradation capacity of the "fresh" degrading microorganisms from the wastewater treatment plant was always adequate in the bioscrubber. This can be justified by the fact that the corresponding pollutants were also present in the wastewater so that adaptation of the microorganisms from the wastewater treatment plant was sufficient.
- After a standstill, start up of bioscrubbers could be carried out within a short time. Thus, standstill periods in production can be easily bridged.
- By varying the transfer rate, changes in operating conditions and possible malfunctions could be reacted to quickly.
- Major fluctuations in the pollutant concentrations of the waste air did not cause problems, because a part of the biological pollutant degradation was shifted to the wastewater treatment plant.
- Autonomous operation, i.e., completely independent of the wastewater treatment plant, was possible for a period of several days. This meant that any minor malfunction occurring was dealt with. However, when breakdowns of a lengthy duration took place, production had to be stopped as the treatment plant also treated the wastewater from production.

All these points show that biological waste air cleaning using a bioscrubber – under suitable conditions – can provide an interesting, promising, and advantageous alternative to rival processes, even for difficult waste air problems. In this case, the existence of an on-site wastewater treatment plant is of decisive importance.

5 Special Types of Bioscrubbers

One particular problem in the above mentioned industrial sector are the strongly fluctuating pollutant concentrations in waste air.

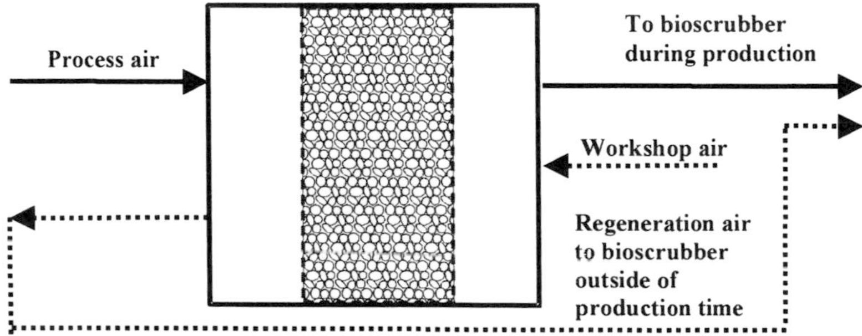

Fig. 9. Equalizing filter with countercurrent flow.

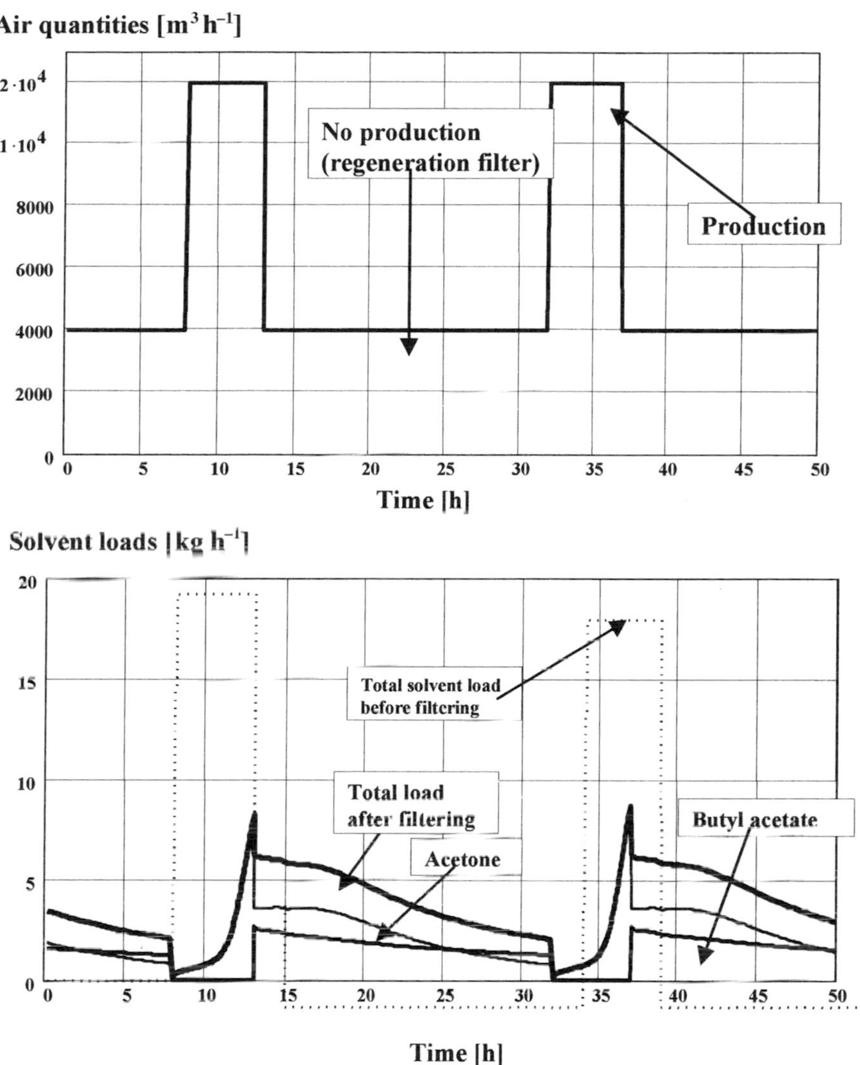

Fig. 10. Simulation calculation for equalizing filter with countercurrent flow.

To combat this relatively frequent problem, a process variant was developed to level out the concentration peaks by adsorption (CHMIEL and SCHIPPERT, 1999).

In industrial applications, waste air emissions from production are frequently not generated continuously, but rather at repetitive intervals. Typical examples are:

- Emissions are generated only during a working shift of 8 h per day only on work days.
- Emissions are generated during batch processes and take the form of a bell-shaped curve, e.g., during drying or hardening processes.

If these emissions are to be reduced, bioscrubbers must be designed to accommodate maximum pollutant load and maximum air quantities, although much lower levels of emission or no emissions at all are generated for most of the day.

Therefore, a simple process, which prevents strong fluctuations in the concentration and/or the loading of pollutants to be treated in the waste air plant would be beneficial.

This can be achieved by using an equalizing activated carbon filter with countercurrent flow (Fig. 9). Fig. 10 is a graphical representation of the results of a simulation calculation for this type of equalizing filter with countercurrent flow for operation upstream of the bioscrubber. The following process parameters were used for this example:

Process air quantity:	12,000 $m^3\ h^{-1}$ for 5 h d^{-1}
Process air load:	butyl acetate: 0.7 g m^{-3}, acetone: 0.9 g m^{-3} for 5 h d^{-1}, relative humidity: 40%, temperature: 20 °C min., 30 °C max.
Clean gas concentration:	< 100 mg m^{-3} total solvents
Regeneration air quantity:	4,000 $m^3\ h^{-1}$

Essential results of this simulation calculation are:

- The maximum solvent load to be degraded (average time 1 h) is reduced to less than one third.
- Butyl acetate, having a high air–water distribution coefficient of 8 (mg m^{-3} per g L^{-1}) and thus being unfavorable for the bioscrubber process with water/activated sludge, is present in the bioscrubber only during periods of no production (i.e., in the desorption phase). Therefore, the scrubbing waste quantity can be related to the air quantity in the desorption phase and amounts to only one third of the quantity required for processes without an equalizing filter with countercurrent flow.

Tab. 2 shows the major differences in the most important process parameters (the safety margins common to these types of plants have been taken into account) for a bioscrubber plant with and without an equalizing pre-filter with countercurrent flow. It is evident that the investment and operating costs for bioscrubbers can be reduced considerably by using equalizing activated carbon filters with countercurrent flow.

As mentioned before, good solubility is essential for effective absorption of pollutants from waste air. The so-called "biosolve" process was developed for cases where the solubility coefficient of air–water is very high, e.g., toluene (SCHIPPERT and FRÖHLICH, 1993).

However, even for substances with poor water solubility the absorption capacity of the scrubbing medium can be drastically increased by adding a high-boiling oil to the water/activated sludge mixture, in the form of a coarse dispersion with a proportion of roughly 10–30% in the bioscrubber system (SCHIPPERT, 1993). The reason for this increase in absorption capacity is that the solubility of the substance to be absorbed is 2- to 4-fold higher in order of magnitude in the high boiler than in water (Tab. 3). Therefore, the circulating absorption liquid, which consists of the activated

Tab. 2. Distribution Coefficients for Silicone Oil–Water

	Bioscrubber *without* Equalizing Filter	Bioscrubber *with* Equalizing Filter with Countercurrent Flow
Scrubber diameter [mm]	3,000	2,000
Scrubbing liquid circulation quantity [$m^3\ h^{-1}$]	250	100
Activated sludge tank volume [m^3]	240	80
Maximum oxygenation capacity [kg $O_2\ h^{-1}$]	36	12

Tab. 3. Scrubbing Liquid Quantities for the Absorption of Organic Substances and Air–Water Distribution Coefficients

Solvent	m [(g L^{-1}) (g L^{-1})]
Ethyl acetate	0.8
Methyl isobutyl ketone	4.1
Diethyl ether	3.9
Dichloromethane	12
Toluene	270
Xylene	280
Styrene	370
Cyclohexane	2,100
n-Hexane	22,000

sludge/high boiler dispersion, can absorb 100–1,000 times higher quantities of substance depending on the proportion of the high boiler in the mixture. As the Fig. 11 shows, the high-boiling oil acts as a type of solubilizer between air and water/activated sludge in the biosolve process.

The substances absorbed in the scrubbing liquid are distributed according to the respective oil–water distribution coefficient to the organic and the aqueous phase. The concentration in the organic (nonpolar) high-boiling phase is up to 100–1,000 times higher. This distribution equilibrium is disturbed by the biological degradation of the absorbed substance through the activated sludge in the aqueous phase with the result that this substance diffuses to the aqueous phase from the high-boiler. In this way, the high-boiler is regenerated indirectly and can then again absorb substances from the waste air.

Of course, not any high-boiling oil can be used in biological systems. Some of the chemical, physical, and biological properties of the methyl silicone oil used are: intrinsic vapor pressure <1 mg m^{-3}, water solubility <1 mg L^{-1} viscosity <100 cST, no biological degradability, odorless, no toxic or inhibiting effect on microorganisms (HEKMAT and VORTMEYER, 1999). Fig. 12 shows that it is possible to improve the removal efficiency for, e.g., toluene and xylene from practically zero to >90%.

Several large-scale plants for air quantities of between 10,000 and 150,000 m^3 h^{-1} containing odorous substances and solvents with poor water solubility have been constructed and operate according to the biosolve process. Fig. 13 shows the flow diagram of a biosolve plant for the treatment of waste air from a conditioning plant for paint sludge.

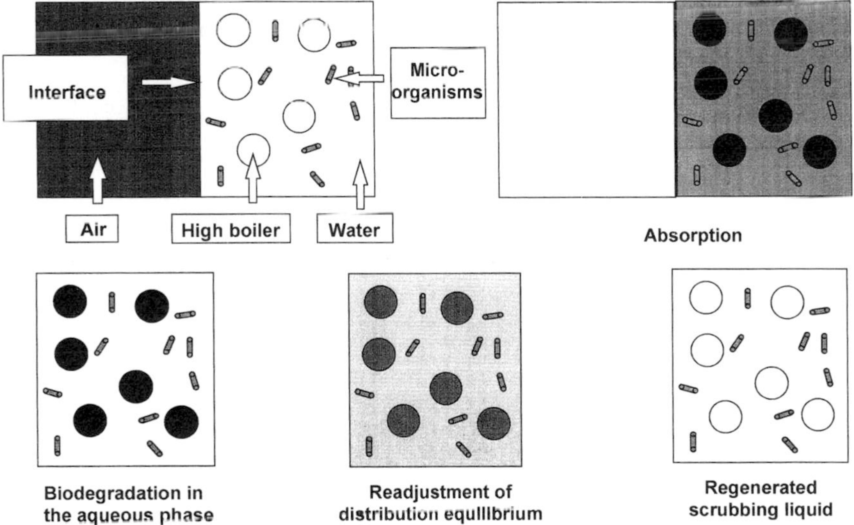

Fig. 11. Principle of the biosolve process.

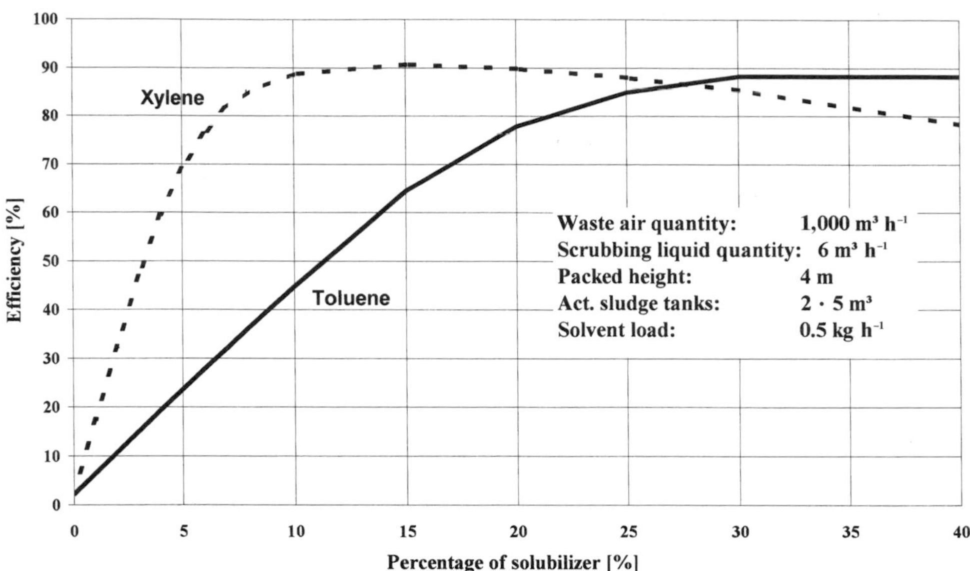

Fig. 12. Dependence of efficiency on the percentage of solubilizer.

As can be seen in Fig. 13b, the technical requirements for the biosolve process with wastewater treatment, activated sludge separation, and silicone oil recirculation are considerably greater than the bioscrubber process using only water and activated sludge.

The scrubber can be designed according to the principles shown in Fig. 6. Only the air–water distribution coefficient K_w is to be replaced by the term $K_w \cdot (1 + m \cdot R)^{-1}$ in which case R is the weight percentage of the high-boiler in the scrubbing liquid. A multi-stage design of the activated sludge plant is often an advantage, because the biology of the biosolve process often has to be operated in the kinetic range of the first order with regard to substrate concentration. In contrast to this, biodegradation usually involves zero order kinetics for bioscrubber operating on the water–activated sludge principle.

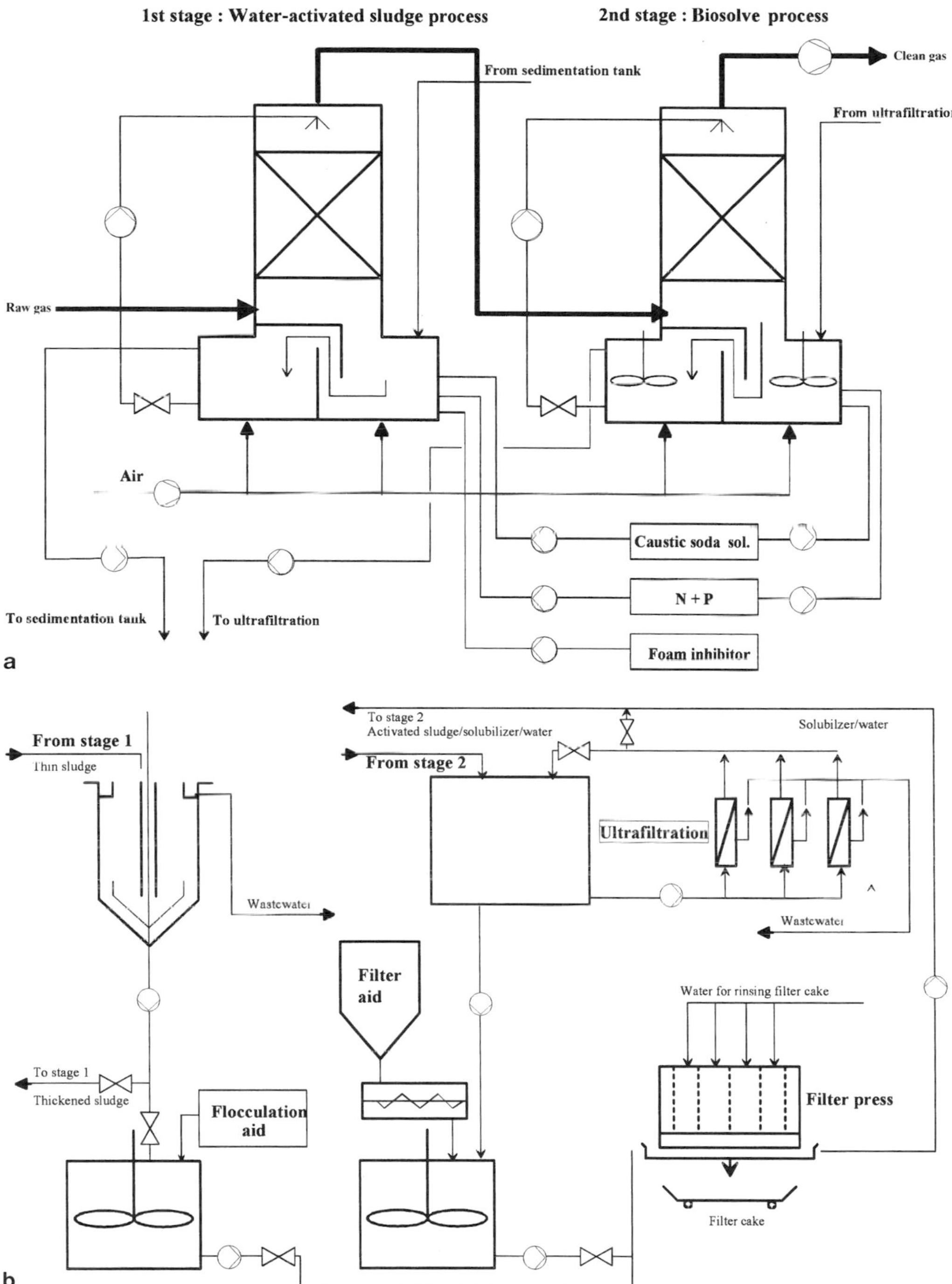

Fig. 13. (**a**) Two-stage biological waste air treatment plant for the elimination of solvents with poor and good water solubility, (**b**) treatment of wastewater and excess sludge from a two-stage biological waste air treatment plant.

6 References

CHMIEL, H. (1991), *Bioprozeßtechnik 2*. Stuttgart: Gustav Fischer Verlag.

CHMIEL, H., SCHIPPERT, E. (1999), Verfahren zur Vergleichsmäßigung von Schadstoffkonzentrationen in Abgasströmen, *German Patent Appl.* 19936965.8.

COLBURN, A. P. (1939), The simplified calculation of diffusional processes, *Chem. Eng.* **35**, 211–236.

HEKMAT, D., VORTMEYER, D. (1999), Biologischer Abbau von schwer wasserlöslichen, flüchtigen, aromatischen Verbindungen aus Abluft, *Chem. Ing. Tech.* **71**, 1290–1294.

KELLNER, C., FLAUGER, M. (1998), Reduction of VOCs in exhaust gas of coating machines with a bioscrubber, *Proc. Air & Waste Management Association's Annual Meeting & Exhibion*, 7 pp 98-WAA.02P. Pittsburgh, PA: Air & Waste Management Assoc.

NISHIMURA, S., YODA, M (1997), Removal of hydrogen sulfide from an anaerobic biogas using a bioscrubber, *J. Water Sci. Technol.* **36**, 349–356.

SCHIPPERT, E., (1993), Abscheidung von Lösemitteln im Biowäscher, *VDI Berichte 1034*, Fortschritte bei der thermischen, katalytischen, sorptiven und biologischen Abgasreinigung, pp. 577–609. Düsseldorf: VDI Verlag.

SCHIPPERT, E., FRÖHLICH, S. (1993), Biowäscher und das Biosolv-Verfahren im Abbau industrieller Schadstoffe, (WURSTER, B., Ed.), pp. 1–30. Köln: Verlag TÜV Rheinland.

STOCKHAMMER, S. et al. (1992), Einsatz von Biowäschern in der chemischen Produktion, *Chem. Ing. Tech.* **1**, 148–155.

TA Luft (Technische Anleitung zur Reinhaltung der Luft) (1986), *Gemeinsames Ministerialblatt G 3191 A*, Vol. 37, No. 7, pp. 93–144. Bergisch-Gladbach: Heider-Verlag.

VDI-Guidelines (Verein Deutscher Ingenieure Richtlinien 3478) (1996), Biologische Abgasreinigung, Biowäscher und Rieselbettreaktoren, in: *VDI/DIN-Handbuch Reinhaltung der Luft*, Vol. 6. Berlin: Beuth Verlag.

WHALEY, M., MONROIG, P., GUEZ, L., EXPOSITO, A. (1998), Allergan – pharmaceutical bioscrubber system case study, *Proc. Air & Waste Management Association's Annual Meeting & Exhibition*, 8P 97-RA114A.05. Pittsburgh, PA: Air & Waste Management Assoc.

Wuebker, S.-M., Friedrich, C. G. (1996), Reduction of biomass in a bioscrubber for waste gas treatment by limited supply of phosphate and potassium ions, *Appl. Microbiol. Biotechnol.* **46**, 475–480.

14 Biofilters

KLAUS FISCHER

Stuttgart, Germany

1 Introduction

As for all biological procedures, a main advantage of the biofilter technique is the genuine degradation of the pollutants. Therefore, pollutants are removed and not shifted from one environmental section to the other. However, biological purification of waste gases can only be applied where air contaminating substances are really degradable. Degradation is performed by microorganisms which are settled on a solid surface.

First indications of the application of biofilters can already be found in the 1920s. In 1923, e.g., BACH described the possibilities of waste gas cleaning at wastewater treatment plants as follows (BACH, 1923): Odor protection in wastewater treatment plants becomes necessary, if prevention of hydrogen sulfide leakage into the air is not successful, which results in bad odors. The rooms from which hydrogen sulfide leaks out into the air can be covered with gauze mats, grating engage, etc., and raise material which eliminates hydrogen sulfide by chemical bonding or by biochemical or catalytic disintegration.

It is not known whether biological waste gas filters were in fact applied at that time. In 1957, R. D. POMEROY received Patent No. 2793096 for a process called *"Deodorizing gas streams by the use of microbiological growth"*. In 1953, a single filter bed using soil as filter material was built in a Californian wastewater treatment plant.

Soil was also used as filter material for the biofilter in Genf-Villette (1964). On the other hand, compost from refuse was already used as filter material in the composting plant at Duisburg (1966). At this time, investigations on the removal of hydrogen sulfide and mercaptans using soil were carried out in the United States by CARLSON and LEISER (1966). Several soil filters were built at a wastewater treatment plant near Seattle. With these filters, it was proved that odor removal is also based on biological degradation, and not only on adsorption.

Applications of the biofilter technology in agriculture were also known at that time. In 1967, a biofilter was constructed for purification of waste gas from manure drying. For this purpose, DRATWA (1967) built a biofilter with a filter material height of only 8 cm. The first doctoral thesis on biofilters including extensive basic investigations was published in Germany by HELMER in 1972.

A lot of biofilters were built in Central Europe in the 1980s. At the end of the decade about 500 plants were in operation in Germany.

2 Fundamentals of the Biofilter Technique

For all waste gas purification procedures, there are specific fields where they can be applied preferentially. Depending on the kind of waste gas components, the amount of waste gas, temperature, and further limiting conditions the most suitable procedure has to be found. In the following, the fundamentals for the application of the biofilter technique are described.

2.1 Preconditions for the Biofilter Technique

Several important conditions must be met when applying the biofilter technique:

- water-soluble waste gas components,
- biologically degradable waste gas components,
- waste gas temperatures between 5 °C and 60 °C,
- no toxic compounds in the waste gas,
- only little dust and fat in the waste gas.

Since microorganisms are always surrounded by a water film and adjust their metabolic and degradation activities accordingly, waste gas components must at least have a certain water solubility. Therefore, in practical application non-degradable gas compounds, e.g., halogenated hydrocarbons cannot be eliminated. However, it has to be pointed out that some inorganic gases such as hydrogen sulfide (H_2S) or ammonia (NH_3) can also be oxidized biologically. The final products of this degradation process (elementary sulfur, sulfate, and nitrate) concentrate and, as a result, modify the pH value of the filter material. In the case of these inorganic gases, the great advantage of biological procedures to produce almost no waste does not exist.

The presence of toxic substances in waste air causes similar problems. It is known, e.g., that higher SO_2 and NO_X concentrations strongly damage the microorganisms. Treatment of combustion air with biofilters is difficult, although very often highly loaded odorous gases are produced in dryers directly heated using agricultural products (turnips, hay).

2.2 Physical Fundamentals

In order to remove the waste gas components from the gas stream they must first be brought into contact with the filter material. The waste gas or "raw gas" is a homogenous mixture of the main air components and small pollutant amounts. In the biofilter this waste gas mixture is divided into countless tiny partial flows, and it streams through small channels, pits, and spaces of the filter material (Fig. 1). Larger flow channels can also develop.

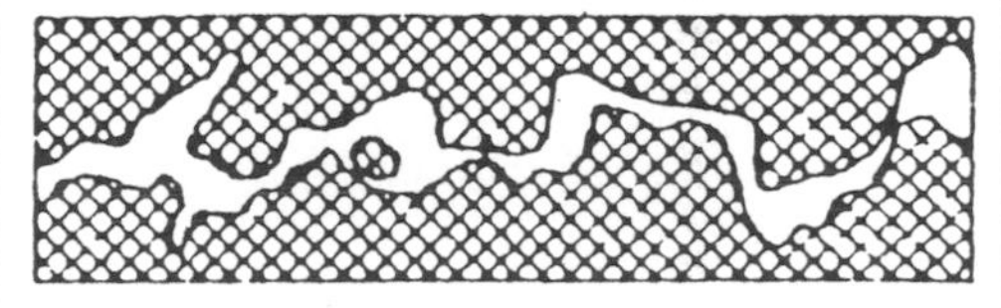

Fig. 1. Flow channels in the filter material.

Theoretical and experimental investigations show that the flow velocity is directly proportional to the pressure loss. This means that a laminary (uniform) flow must exist. This was confirmed by numerous measurements, but only to a certain extent. In the case of higher loads and/or flow velocities, this linear relationship is not given (SABO, 1990). Obviously, turbulent fluctuations exist then.

Together with the air the pollutant molecules flow through the pores as described above. Here, numerous opportunities to contact with the pore walls or the water film allow for bonding with the filter material. As these bondings are a mixture of adsorption and absorption they are summarized with the general term "sorption". In this context it is important to describe the adverse process of desorption. In this process gaseous molecules of the filter material, which can be degradation products (such as CO_2) or pollutant molecules transferred to the gas phase during high fluctuations of concentration, leave the surface and enter the gas phase (Tab. 1).

Accordingly, the decisive steps for bioscrubbers and biofilters are: absorption in the water phase and diffusion into the water-sided boundary layer. Furthermore, the degradation of pollutants by the microorganisms of the biofilm is very important. In the long run, the required concentration gradient can only be achieved by this combination of processes.

2.3 Biological Fundamentals

Bacteria, fungi, and actinomycetes are found as degrading microorganisms in the filter material. Easily degradable and slightly water-sol-

Tab. 1. Water Solubility of Organic Waste Gas Components at Equilibrium Concentrations of 1000 mg m^{-3} Air

Substance	Solubility in mg L^{-1} Water
Ethanol	2300
Butanol	1800
Toluene	32
Xylene	32
Dichloromethane	79
Hexane	1

uble waste gas components are decomposed mainly by bacteria. Fungi possibly play a role in the decomposition of aromatic hydrocarbons. Optimal conditions are important: moisture, sufficient nutrient supply, a favorable pH value (5.5–8), and acceptable temperatures (from 5 °C to ca. 60 °C, wih an optimum at ca. 20–40 °C) are the main parameters.

When a biofilter is put into operation, the microorganisms need a certain time for adaptation, which depends on the waste gas components and the filter material. It can take a few days, but also several months until the filter is fully in service. In order to accelerate the adaptation process already adapted microorganisms can be added to the filter material. Biofilters usually do not require inoculation. Usually favorable conditions for growing of the biocenose exist in the filter material.

3 Description of the Biofilter Procedure

The mode of operation of the biofilter can be described as follows: The air containing pollutants is pressed through the filter material where the microorganisms grow. At the filter material the pollutants are sorbed and removed from the waste air. Material is constantly regenerated by the degradation process of the microorganisms and is available again for the sorption of new waste gas components.

The construction of a biofilter is basically shown in Fig. 2. The heart of the biofilter is the filter material. However, a continuous process in the filter is only guaranteed if the waste gas shows specific qualities. Therefore, pretreatment steps such as dust cleaning, warming, cooling, and – depending on the application – humification of waste gas are also necessary.

In the pipelines and pretreatment stages a certain pressure is required in order to transport the waste gas up to the filter. The loss of pressure in the biofilter is approximately 200–2000 Pa m^{-1} filter height. It depends on the kind of filter material and its state (FISCHER, 1990).

3.1 Filter Material and Filter Material Disposal

Favorable filter materials to be used in biofilters should have the following qualities (VDI 3477, 1991):

- a regular and even structure (even penetration by waste gases, small pressure drop),
- sufficient voids (small pressure loss, small specific filter resistance, low energy demand, good drainage and oxygenation; the void fraction can amount to 20%–80% of the filter material, depending on its type and condition),
- a large surface area of the carrier material (for microorganisms and for adsorption),

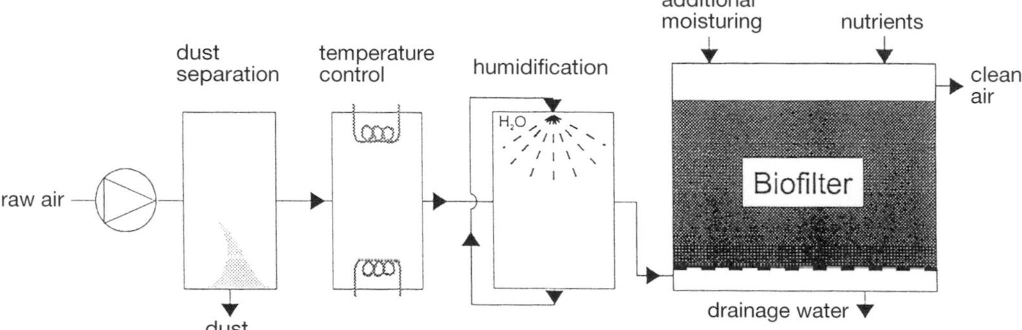

Fig. 2. Construction of a biological filter system.

- good water retention capacity (constant moisture level),
- no extreme pH fluctuations of the filter material (good buffering of pH fluctuations in the waste gas),
- a low rate of decomposition of filter material (structure stability, regular flow, and constant pressure loss results in longevity and little maintenance activities),
- negligible specific odor (short reach of the odor, no unpleasant odor quality),
- good nutrient supply (N, P, K, and micronutrients),
- a favorable price.

One material can hardly show all qualities required here. Therefore, mixtures of different materials are mostly used. At present, the following filter materials are known:

- bark products, chopped bark, and bark compost,
- wood products, chopped wood, and torn root wood,
- compost of biowaste and green waste,
- heather, brushwood,
- fiber materials, fibrous peat, and coconut fibers.

In addition, other compounds with inert materials such as lava, porous clay, polystyrene, or other fiber materials such as peat replacement products are used.

The mass density of the filter material can reach 1000 kg m^3 depending on its type and conditions.

Depending on the application and the material composition the filter material has a service lifetime of approximately 2–5 years. The biological degradation processes cause slow mineralization of the filter material so that not only the waste gas components, but also the filter material itself are removed by the microorganisms. By this, the material becomes more and more fine-grained and usually the pressure loss of the biological filter system also increases.

If the pressure loss of the filter becomes too high or its efficiency decreases too much, the filter material must be changed. Sometimes it is sufficient to shift the filter material and sieve the fine part. New material can be mixed and the new mixture can be directly introduced again, with the advantage that most of the adapted microorganisms are maintained so that the filter gets back to full service very quickly.

Disposal of garbage filter material depends on its prior use. If the waste gas is expected to contain hardly degradable compounds, potential contamination must be clarified with chemical analyses. Depending on the pollutant concentrations the filter material must be properly disposed in a landfill or combusted in an incinerating plant. However, in most cases used filter material is free from pollutants and can be used in gardening.

3.2 Filter Moisture

Sufficient moisture of the filter material is one of the most important preconditions to operate the biofilter successfully. Depending on the kind of filter material the optimal water content is kept at 40–60%. Moisture conditions should be uniform throughout the entire filter. Only in the case of homogenous circumstances a long service lifetime with high efficiency is guaranteed. If material dries up in a part of the filter the flow velocity will increase there. Due to the higher rate material will dry up even quicker so that crack initiation will start as a result and large amounts of waste gas may pass the filter uncleaned.

Maintaining regular moisture conditions in the entire filter is often the most difficult task for the operator. In the early days of the biofilter technique filters were mostly sprinkled from the top, e.g., with sprinklers used in agriculture. However, this way of moistening often results in a very moist upper layer while the lower part of the filter dries up more and more.

Homogenous filter moisture conditions can only be guaranteed over a relatively long time when the raw air is saturated with water. In the case of waste gas from composting plants, e.g., this is often the normal situation. On the other hand, with industrial waste gas prehumidification will usually be necessary. In general, packed column or spray towers are used for prehumidification. Prehumidification using scrubbers is also used simultaneously for dust cleaning and for buffering temperature fluctuations.

To check the filter moisture is relatively expensive. Measuring instruments for the continuous moisture measurement of the material are on the market. They are rather costly and not yet proven for all filter materials. Tests can of course also be done by sampling the filter material from different parts and different depths of the biofilter. The sampled material has to be dried in the laboratory at 105 °C and the water content is determined from the difference of weight.

Even if premoistening works well, an additional irrigation system should be installed. The upper parts of open single-level filters in particular are warmed up by solar radiation and dry up quickly.

Infiltrating water emerging from the biofilter or seeping water is mostly heavily contaminated with highly concentrated water-soluble organic substances (e.g., waste gas components, decomposition products, humic substances), inorganic nutrient salts, sludge particles, etc. It should not be used for remoisturing the filter material because this may cause high additional odor loads. Infiltrating water is usually treated in a wastewater treatment plant.

3.3 Nutrient Supply

The nutrient content of the filter material usually is sufficient to supply the contained microorganisms. Periodical addition of dissolved phosphorus and nitrogen salts is recommended in the case of predominantly inert filter material or specific waste gas components. Especially in the case of high concentrations of easily degradable waste gas components a periodical nutrient supply is necessary. Normally, the components C, N, and P can be applied in the general ratio of 100:5:1. The required amounts of nitrogen and phosphorus can be calculated from the carbon content of the pollutants to be removed.

3.4 Measuring and Control Technology

Measurements of the volume flow of the waste gas are important and can, e.g., be done between premoistening and the biofilter. Anemometers or orifice meters are used for measuring the volume flow.

Analysis of the waste gas composition is considerably more difficult. The composition of raw air mostly can only be determined by expensive gas chromatographic analyses in the laboratory. This should always be done before planning and dimensioning waste gas purification plants.

At present continuous analysis is only possible of a total parameter "organic carbon" with the aid of a flame ionization detector (FID). Analytical techniques for single organic components are under development.

Continuous analysis of the filter moisture would be favorable to check the filter material. However, at present this is only possible in part (see Sect. 3.2). Furthermore, regular analyses of nutrient content and pH value should be carried out.

To characterize the mechanical qualities of the filter material, the following laboratory analyses are useful:

- analysis of porosity,
- analysis of bulk density,
- analysis of water holding capacity.

Sieve analyses for the detection of size grading are of rather little value: The results of determinations with dry material actually do not correspond to the real operating conditions. Larger aggregates form in the case of usual material moisture conditions and, therefore, sieve analyses of the filter material at real operating conditions are only possible with very few materials.

4 Construction and Procedure Variants

Compared to other procedures, a relatively large filter volume is often necessary for waste gas cleaning with biofilters. The reason is a limited degradation efficiency of the microorganisms. High space requirements have often prevented the utilization of biofilters in industry, although, depending on the waste gas composition, biological procedures would be possible. A number of process variants which facilitate the application of biofilters have been developed recently; they require less space or are transportable:

- open single-level filters,
- closed single-level filters,
- container filters and high performance filters,
- varieties as tower and horizontal filters.

The different characteristics of these filter types and their advantages and disadvantages are described in the following.

4.1 Open Single-Level Filters

At present, the most frequently used variant of a biofilter is the single level filter. The first known biofilters, e.g., the filter at Genf-Villette (composting from refuse) built in 1964 or the filter of the composting plant at Duisburg (1966) were single-level filters.

The filter heights are 1.0–3.0 m. The scheme of such a filter is shown in Fig. 3. The raw air enters the filter material from below. Different air distribution systems are used, such as perforated or slit elements of special concrete, epoxy or wood, grates, drainage pipes, aeration stones, etc. Furthermore, a single-level filter must show a system for the diversion of surface water and of surplus moistening water.

For maintenance of the filter a passable filter bed is favorable. The borders of the filtering facility have to be constructed in a way that waste gas breakthroughs do not occur. Sprinklers from agriculture can be used for moistening the filter material. However, it is considerably better for a good operation of the filter, if the raw air is already saturated with water. Single-level filters of this type are dimensioned up to sizes of approximatey 2,000 m^2. They are suitable for the treatment of waste gas volumes of up to 100,000 $m^3\ h^{-1}$. At large filter areas, a subdivision into individual filter bed segments should always be planned so that maintenance operations can be carried out without having to switch off the whole plant.

4.2 Closed Single-Level Filters

The "classical" open biofilters have proven to be useful in many places, but despite higher costs more and more biofilters are built as closed plants due to technical and often also legal reasons:

- Closed biofilters are nearly independent of the climate. The filter material remains homogenous for a long time and, therefore, higher loads are possible. Nutrients are not rinsed by rainwater.
- The clean gas of closed biofilters can be collected in a chimney and, therefore, it can be controlled very easily.

As with the open single-level filters, the waste gas has to be moistening in a pretreatment step. In closed filters the direction of the waste gas flow can be directed from bottom to top or from top to bottom, while for moistenring the filter material a downstream is favorable. The

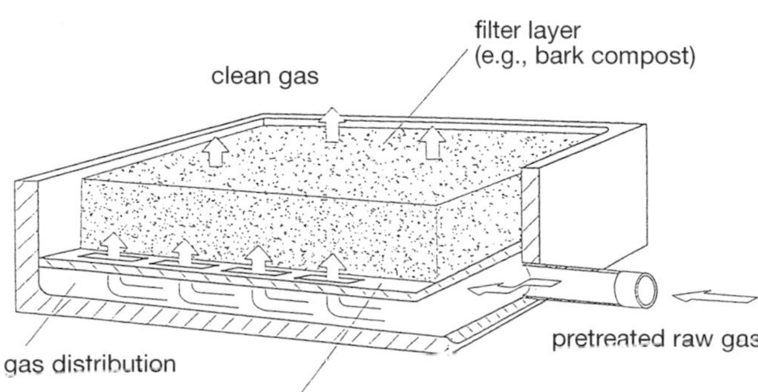

Fig. 3. Single-level filter.

moistening equipment can be integrated into the ceiling or the wall constructions. Using this moisturing technique fertilizer solutions can also be added simultaneously.

Although the filter material in closed biofilters has a longer service lifetime, facilities for the replacement of material must be planned, e.g., removable roof surfaces or movable side constructions. Furthermore, entry hatches and lockable hand holes are necessary in order to permit sufficient control of the filter.

4.3 Container Filter and High Performance Filter

Principally, container filters are small and transportable single-level filters. Where necessary they can be operated one on another (multi-level filter), in parallel, or in a row. For moistening of raw air the required blowers as well as analytical measurements can also be integrated into the containers. Problems concerning gas breakthroughs at the borders are prevented by baffles. The applied filter volume loads for container filters are in the range of about 50–150 m^3 m^{-3} h^{-1}. Sometimes the direction of waste gas in these filters is from top to bottom with the advantage of a better distribution of moistening water. In most cases the filter bulk in the container is 1.5–1.8 m high (Fig. 4).

High performance filters are characterized by filter volume loads exceeding 150 m^3 m^{-3} h^{-1}. The higher degradation efficiency is caused by process optimization and finer filter materials.

A homogenous flow can be achieved over a long service lifetime. The available filter volume of high performance filters is used at almost 100% compared to the "classical" biofilter with only around 50% (BARDTKE et al., 1992).

A large surface available as both a sorption area and a surface area for microorganisms is achieved with specific fine-grained filter materials. Using these filter materials the density of microorganisms increases about a factor of 10–100. As a result, the degradation efficiency achieved per m^3 of filter volume increases correspondingly.

4.4 Varieties: Tower and Horizontal Filters

Tower filters were developed at the beginning of the 1980s (KNEER, 1981). Using filter materials of up to 6 m height, a considerable pressure loss is recognized. Material can be delivered mechanically in the lower part of the filter and – possibly after moistening – released again above. The filter material can also be exchanged relatively easily and quickly in this way. However, experience of the last years show that the development of a stratification of different microorganisms in the filter bed is favorable for many waste gas compositions. Such a stratification would be possible in tower filters, however, it fails because of the constant mechanical settlement of material.

The direction of flow in *horizontal filters* is different from "classical" biofilters with vertical flow. For horizontal filters it must be guaranteed that no gas breakthroughs in the filter bed are caused by material settlement. This is possible with a corresponding filter material and by building filters as slim high towers. Filters can also be in the form of an annulus coil. In this case raw air is led via the central waste gas pressure tube. In addition the moistening

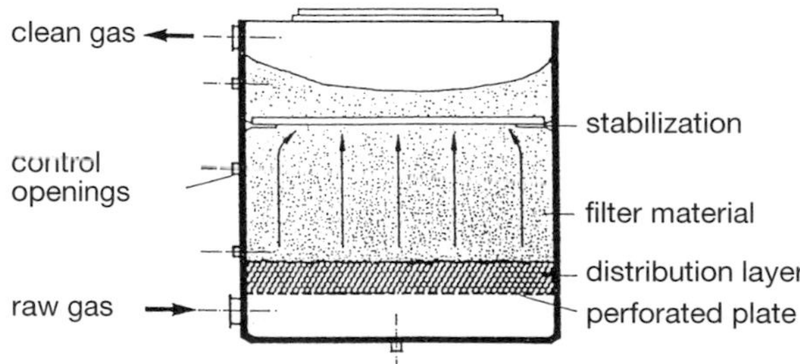

Fig. 4. Container filter.

technique can also be integrated here. The horizontal filter is of advantage especially in the case of little available space. Several horizontal filters are successfully operated on the technical scale.

5 Dimensioning

Dimensioning of the biofilter mainly depends on the kind of waste gas, the gas flow, and the filter material applied. Fine filter materials show a higher degradation efficiency due to a higher density of microorganisms per unit volume. On the other hand, the pressure loss increases with fine materials. In order to achieve a regular flow through the filter and to prevent air breakthroughs a minimum height of the filter bed of 1.0 m is generally recommended.

The VDI Guideline 3477 (1991) mentions three loading paramcters which can also be used for the calculation of filter dimensions:

- The *filter area load* is the simplest loading parameter. It is defined as the waste gas flow related to the filter area ($m^3\ m^{-2}\ h^{-1}$). This loading parameter has been applied very often for dimensioning single-level filters. From the experience with existing plants it is known that single level filters for odor treatment can be loaded with approximately 60–120 $m^3\ m^{-2}\ h^{-1}$.
- The *filter volume load* is defined as the waste gas volume flow related to the filter volume ($m^3\ m^{-3}\ h^{-1}$). Normally, the filter volume loads are in the range of approximately 50–100 $m^3\ m^{-3}\ h^{-1}$.

For biofilters used for cleaning of industrial waste gas the best dimensioning parameter will be the specific filter load:

- The *specific filter load* is calculated from the mass of the waste gas components passing through the filter per filter volume and unit of time. The parameter is given in g $m^{-3}\ h^{-1}$. If the filter is used for deodorizing the specific filter load can also be expressed in OU $m^{-3}\ h^{-1}$. Applied to industrial waste gas, the degradation efficiency of microorganisms for individual waste gas components must be known. Depending on the substance a range of 10–100 g $m^{-3}\ h^{-1}$ was found. From this and the desired and/or required efficiency of the biofilter the dimensions of the biofilter can be calculated.

From the dimensioning parameters presented above it becomes clear that for planning a biological filter system one has to determine first which volume flows and which waste gas components should be cleaned. For industrial waste gas with gas components for which little experience is available, the application of a pilot plant is always recommended. Based on such operating data correct dimensioning and exact cost calculations can be made.

6 Costs of the Procedure

Above all, investment costs for the application of biological filter systems are influenced by the size and kind of biofilter. For large single-level filters (approximately 1000 m^2) the specific investment costs are about 300 EUR per m^2. Costs increase to approximately 600 EUR per m^2 for small single-level filters (EITNER, 1990). For container filters and multilevel filters the specific investment costs often exceed 1,000 EUR m^{-3}. Additional costs may arise for waste gas collection in the firm and primary treatment (e.g., dust cleaning).

The operating costs of a biological filter system consist of costs for energy, personnel, maintenance, and filter materials; the costs for water consumption can in general be neglected:

- *Energy costs* depend on the filter material used. With compost, e.g., the basic data for calculation are 1.8–2.5 kWh 1,000 per m^3.
- The average *requirement of personnel* is 0.8–1.0 h per m^2 of filter area per year.
- *Maintenance costs* consist of construction engineering (approximately 1% of

investment costs) and machine engineering (approximately 7.5% of investment costs).

- *Filter material costs* depend on the material and the service lifetime of the filter. In general, the service lifetime is between 2–4 years. Material costs are between 20–50 EUR per m^3. In addition, considerable transport costs of up to 100% of the pure costs of materials may occur according to the distance and the bulk density.
- *Costs for material disposal* must also be included. Depending on the waste gas composition and filter material used it can be a fertilizer (compost) or, in the worst case, a toxic waste.
- *Water consumption* depends on the temperature and humidity of the waste gas. For calculation the following data are recommended: 1.0–1.5 m^{-3} water per m^2 filter area and year.

The operating costs of small biofilters are approximately 0.5 EUR per 1,000 m^3 of treated air. For large biofilters operating costs decrease to approximately 0.1 EUR per 1,000 m^3 of waste gas.

7 New Developments of the Biofilter Technique

7.1 Low-Cost Systems

There is a development of more and more effective filters with high degradation capacity at one hand and a recent contrary trend at the other: new low-cost systems using large areas or cleaning waste gas less efficiently to be applied for filters with very low loads. Two developments are to be found at present:

- *Biofilters without prehumidification:* The moistening system is integrated into the filter material.
- *Passive biofilters:* The waste gas is under little pressure and flows passively through the filter material.

In *biofilters without prehumidification* the prescrubber which often causes high investment and operation costs is replaced by moistening systems included in the filter. At present, this is only possible in the case of small loads. The moistening techniques used are known from agriculture and irrigation of dry areas. In such filters some kind of droplet irrigation is used instead of water spray techniques. They are, e.g., applied for waste gas cleaning of a wastewater treatment plant near a hotel in Egypt (MANGELSEN, 1999).

In *passive biofilters* the waste gas is not pressed or sucked through the filter material with the aid of a blower. Instead, the pressure of the gases or the natural air pressure variations are used to achieve gas replacement or a weak gas flow. Passive biofilters are already successfully applied

- to treat biogas from landfills with the aid of a "window" in the landfill surface,
- to minimize gas escapes in the area of the biogas collecting system,
- to eliminate putrefaction waste gases at tomb chambers,
- to cover compost windrows at simple open compost plants.

The principle is demonstrated with the example of biogas cleaning using a "window" in a landfill. Biofilters can be applied to clean biogases from landfills in different ways. The gas can be collected using, e.g., a gas fountain and is subsequently treated in a biofilter. Another possibility used in several landfills is as follows: An area of at least 1,000 m^2 of the landfill is, instead of soil, covered with a slack shift of filter material of maximally 0.5 m height. Since the covering of the remaining landfill surface is almost gasproof, it is assumed that the biogases from the landfill will flow into the biofilter where they are cleaned. Experience with this system is good up to now. For odor compounds efficiencies of 95% were found, although the biofilter areas were only moistened by natural rainfalls, even in summer (FISCHER, 1997).

7.2 Rotor Biofilter

In a rotor biofilter the filter material is located in a rotating filter drum which is stored at rolling. The filter drum can be turned continuously or step-by-step. As a result, the filter material is aerated constantly, and it can be moistened simultaneously. The gas flow is directed from the outside to the inside. As a result, a large air distribution area is available compared to a single-level filter. Further advantages are:

- reduction of breakthroughs at the borders,
- avoidance of waste gas breakthroughs,
- regular-depth moistening by direct currents of air and water,
- less pressure losses and, therefore, decreasing energy costs,
- easy replacement of material.

Such a pilot system was proven to be successful in the foundry of a car factory and in a food production plant (SABO et al., 1994)

7.3 Biofiltex

The Biofiltex procedure represents a link between biofilter and trickling filter systems. A textile tissue made of synthetic fibers serves as a surface area for microorganisms. The fibers are moistened step-by-step with a nutrient solution. For achieving a short adaptation time, the reactor is inoculated with adapted microorganisms. Advantages of the procedure are:

- high degradation efficiency in a very small filter volume,
- little pressure loss,
- easy cleaning.

A pilot filter was used with great success for waste gas treatment in a foundry (SCHNEIDER et al., 1997).

7.4 Combinations of Techniques

Combinations of biofilters with other waste gas purification procedures are used in the case of very high quality requirements. The following combinations are realized in full-scale plants:

- combinations of biofilters with bioscrubbers,
- combinations of biofilters with adsorption plants.

If a bioscrubber and a biofilter are arranged in series, the first stage should be the bioscrubber followed by the biofilter. By this, slightly soluble and easily removable substances can be removed with the bioscrubber. Simultaneously, it is used for prehumidification of waste gas.

When an active carbon adsorber is combined with a biofilter the same succession is recommended. By this, high loading peaks, e.g., can be buffered by the adsorber. Here the biofilter serves as a post-cleaning stage. Furthermore, this combination can be used to regenerate the activated carbon filter at times of low loads. The desorbed waste gas from the regeneration process is directly led to the biofilter where it is biologically eliminated.

Other possible combinations of biofilters with ultraviolet radiation treatment, plasma techniques, and ionization plants are being tested at present. It is not forseeable whether these combinations prove really effective in the long run. Using combinations the costs for waste gas treatment always increase so that one of the essential advantages of the biofilter technique is lost: good efficiency in connection with relatively low investment and operating costs.

8 References

BACH, H. (1923), Schwefel in Abwasser, *Gesundheits-Ing.* **38**, 370–376.

BARDTKE, D., FISCHER, K., SABO, F. (1992), Entwicklung und Erprobung von Hochleistungsbiofiltern. *Abschlußbericht 89/005/3*, Projekt Europäisches Forschungszentrum für Maßnahmen zur Luftreinhaltung (PEF). Karlsruhe, Germany.

CARLSON, D. A., LEISER, C. P. (1966) Soil Beds for the Control of Sewage Odours, *J. Water Pollut. Control Fed.* **38**, 829 ff.

DRATWA, H. (1967), Die Geschmacksbeseitigung in

der Landwirtschaft unter besonderer Berücksichtigung der Kottrocknung von Ablegebetrieben, *Dtsch. Geflügelwirtsch.* **19**, 276–277.

EITNER, D. (1990), Biofilter in der Praxis, in: *Biologische Abluftreinigung* (FISCHER, K., Ed.), pp. 55–73. Ehningen: expert-Verlag.

FISCHER, K. (1990), Biofilter: Aufbau, Verfahrensvarianten, Dimensionierung, in: *Biologische Abluftreinigung* (FISCHER, K., Ed.), pp. 35–54. Ehningen: expert-Verlag.

FISCHER K. (1997), Passive Biofilter zur Behandlung von Deponiegasemissionen – 5 Jahre Betriebserfahrungen, *Stuttgarter Berichte zur Abfallwirtschaft* Vol. 67.

HELMER, R. (1972), Sorption und mikrobieller Abbau in Bodenfiltern bei der Desodorierung von Luftströmen, *Stuttgarter Berichte zur Siedlungswasserwirtschaft* Vol. 49.

KNEER, F. X. (1981), Neuartige Filtereinrichtung Bio-Filter. Bundesforschungsauftrag PTB 8117, *Abschlußbericht*.

MANGELSEN, M. (1999), Entwicklung und Erprobung eines Biofilters zur Abluftdesodorierung im mediterranen Raum, *Thesis*, University of Stuttgart, Germany.

POMEROY, R. D. (1957), Deodorizing gas streams by the use of microbiological growth, *U.S. Patent* 2793 096.

SABO, F. (1990), Verfahrenstechnische Grundlagen der biologischen Abluftreinigung, in: *Biologische Abluftreinigung* (FISCHER, K., Ed.), pp. 13–34. Ehningen: expert-Verlag.

SABO, F., HANDTE, J., FISCHER, K. (1994), Entwicklung und Erprobung des Rotor-Biofilters, *Wasser Luft Boden* **11–12**, 64–65.

SCHNEIDER, T., EISENRING, R., SABO, F. SCHIRZ, S. (1997), Entwicklung und Erprobung eines neuartigen Biotropfkörperreaktors mit einem auf Textilien basierenden Filtermaterial, *Proc. Int. Symp. Biological Waste Gas Cleaning*, Maastricht, NL, 28–29 April 1997, pp. 321–328. Düsseldorf: VDI-Verlag.

VDI 3477 (1991), Biologische Abgas-/Abluftreinigung, Biofilter, VDI-Richtlinie, December 1991.

15 Treatment of Waste Gas Pollutants in Trickling Filters

THORSTEN PLAGGEMEIER
Stuttgart, Germany

OLIVER LÄMMERZAHL
Köngen, Germany

List of Symbols

a^L	specific wetted area [$m^2\ m^{-3}$]
h	reactor height [m]
$D_{S,eff}$	effective diffusion coefficient in the biofilm [$m\ s^{-1}$]
m_S	rate of substrate degradation due to maintenance [$g\ (h\ g)^{-1}$]
r_S	substrate elimination rate [$g\ (h\ m^3)^{-1}$]
r'_S	volumetric reaction rate in the biofilm [$g\ (h\ m^3)^{-1}$]
S	substrate concentration in batch experiments [$g\ m^{-3}$]
w^G	superficial gas flow rate [$m\ h^{-1}$]
w^L	superficial liquid flow rate [$m\ h^{-1}$] (x coordinate [m])
X	concentration of suspended biomass [$g\ m^{-3}$]
$Y_{X/S}$	true yield coefficient on substrate [–]
β^{OG}	overall gas phase related mass transfer coefficient [$m\ s^{-1}$]
δ	thickness of the active biofilm [m]
η_r	biofilm effectivity factor [–]
K	gas–liquid distribution coefficient [–]
$\mu(S)$	microbial growth rate [$1\ h^{-1}$]
μ_{max}	maximum microbial growth rate [$1\ h^{-1}$]
ρ^G	substrate concentration in the gas phase [$g\ m^{-3}$]
ρ^L	substrate concentration in the liquid phase [$g\ m^{-3}$]
ρ^S	substrate concentration in/near the biofilm [$g\ m^{-3}$]

1 Introduction

Biofilters and bioscrubbers are well-tried and reliable methods to degrade waste gas pollutants such as organic solvents and odorous compounds. A third "biological" way is the application of biotrickling filters. Biotrickling filters are reactors that degrade a substance rather than filtering one out. "Trickle bed bioreactors" or "fixed-film bioscrubbers" are other terms that describe biotrickling filters.

Biotrickling filtration is a classical technique described by the patent *Procedure for the Purification of Air and Oxygen Containing Gas Mixtures* (in German) of PRÜß and BLUNK dating back to 1941. The patent described many design features of modern biological trickling filters for waste gas purification. However, this technology was nearly overlooked until the late 80s. Since then, biotrickling filters have developed from a "very promising technique" (OTTENGRAF and DIKS, 1992) to a standard method for the purification of wastegas streams of 100,000 $m^3\ h^{-1}$ or more (KELLNER and VITZTHUM, 1997). HEKMAT and VORTMEYER (1994) define trickling filters as "the state of the art in biological waste gas purification".

Trickling filters are in some aspects the combination of biofilters and bioscrubbers. Similar to biofilters, the biocatalyst is immobilized on a packing material. In the biofilter usually an organic material is used, whereas in trickling filters inert materials form the carrier for the microorganisms. While the air flows through the packing, the air pollutants are absorbed in the liquid phase covering the biofilm and are subsequently degraded by the immobilized microorganisms. In contrast to the biofilter apparatus, the incoming air does not necessarily need to be humidified, but the packing will be wetted continuously or intermittently by a recirculated medium. Trickling filters require a minimum of structural expenditure such as preconditioning because the pollutant is simultaneously absorbed and degraded in one unit. The possibility to monitor and control more parameters than during standard biofiltration makes trickling filtration the best choice to treat "difficult" waste gas problems, like the purification of air contaminated with compounds giving rise to inorganic salts (like halogenated or sulfur containing compounds) (DIKS et al., 1994). Trickling filters offer research opportunities because the reactors containing the inert carriers and the media are clearly defined. This feature allows one to erect and operate a "glassy" system and eases modeling of biological trickling filters compared to modeling of biofilters.

2 Description of the Principle – Advantages and Disadvantages

A biological trickling filter for the treatment of waste gas consists of a combined absorber–reactor column (Fig. 1). The packing used in this unit is made from inert material (e.g., PP-Raschig rings) and serves as carrier for biofilm forming microorganisms. The waste air flows cocurrently (counter- or cross-flow operation may also be utilized) with a recirculated liquid through the packing. While passing through the column, compounds from the waste gas diffuse into the biofilm where they are degraded by microbial activity to carbon dioxide, water, heat, and cell substance. Compounds that are not completely degraded will form intermediates (i.e., partially oxidized compounds like organic acids), and compounds containing halogens will give rise to inorganic salts. In contrast to a biofilter, in which these substances would accumulate and eventually limit the function of the filter, a trickling filter washes

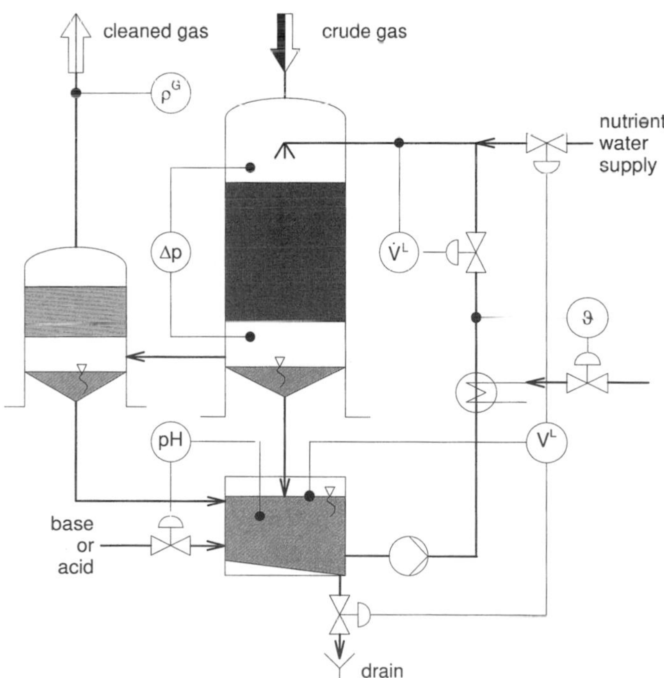

Fig. 1. Schematic drawing of a biological trickling filter. Parameters measured are the pH and the temperature of the liquid medium, its volume and volumetric flow rate, the pressure drop over the packing, and the cleaned gas humidity when leaving the droplet separator. The shown trickling filter is operated in cocurrent mode.

Fig. 2. Photograph of a technical scale biological trickling filter. This combined bioscrubber – biotrickling filter is used for the purification of 20,000 $m^3 h^{-1}$ of acetone and ethanol polluted waste air.

out the inorganic salts in the circulated medium. This waste stream may then be treated separately. On the other hand, the medium supplies nutrients like nitrogen salts, sulfate, phosphate, and trace elements to the biofilm. By modifying the composition of the medium the pH may be adjusted to a desired value, or a nutrient can be omitted to limit the growth of the microorganisms (see Sect. 4.2). Surplus biomass being rinsed down by the circulated medium can be separated from the medium, and needed nutrients can be added together with fresh water to replace the water loss by evaporation. The circulation and thereby the wetting of the filter packing can be carried out continuously or intermittently. The latter operation mode is assumed to ease the transfer of sparingly soluble compounds to the biofilm by keeping the water layer covering the microorganisms as thin as possible. Partial drying of the biofilm and thereby water stress will also lower the growth yield and delay clogging of the column (see Sect. 4.2). Another way to enhance water stress can be obtained by decreasing the water activity of the medium by salt addition. These points reflect the operational opportunities offered by a separate recirculated medium.

Homogeneous wetting of the packing is an important factor for achieving a high overall elimination capacity of the reactor. Usually, multi-nozzle systems are used. The task can be facilitated by additionally wetting the inlet air by a prescrubber unit (see Fig. 2).

The choice of the correct packing material for trickling filters is critical. Firstly, the material must be easily colonized by a biofilm and must be homogeneously wetted by the medium. Secondly, one must decide the geometry of the packing. A packing with a high specific area (in a range of 300–400 $m^2 m^{-3}$) will allow the formation of a large catalytically active biofilm, but the excess biomass will lead to comparatively early clogging of the column. Clogging results in a pressure drop which leads to biofilter failure. Choosing a packing with a low specific area (e.g., structured packing with a large void space) offers longer stable operation of the unit at a lower pressure drop, but since the catalytically active surface is low, the elimination capacity reached with this equipment may be insufficient.

The goal is to bring as much biocatalyst in contact with the waste air and to keep the biomass yield as low as possible without affecting the reaction rate of the pollutant to allow for stable long-time operation.

3 Construction and Operation Guidelines

3.1 General

Information on the types and characteristics of the air pollutants, such as their concentrations and toxic potentials on microorganisms, is an important consideration when designing trickling filters. Dust, resins, oil droplets, and strongly alkalizing or acidifying inorganic compounds should be removed before entering the reactor. The gas temperature should be adapted to biologically acceptable values (in a range of 10–40 °C), and the gas humidity should be taken into account to design a proper wetting of the packing material. The filter packing must comply with the requirements set out above, however, it is advisable to realize a turbulent current to enhance the gas–liquid transfer. The reactor itself may be designed following the principles used for absorption columns from corrosion resistant materials. The distribution of the medium over the cross-section of the reactor should be uniform. Monitoring of the medium pH and conductivity parameters is a basic requirement. Main nutrients (N, S, and P salts) should be measured and supplemented as needed. More detailed information on construction and design principles is given in VDI-Richtlinie No. 3478 (1996).

Operating a pilot plant for a specific duration at the original location is the most expensive, but also the best way to build a reliable biological waste gas purification plant. An on-site pilot plant will face the same operational tasks as a later main plant, as there may be shifts in the daily and weekly load, the composition of the crude gas, and other parameters influencing the success of the waste air treatment.

3.2 Counter- or Cocurrent Operation

From the point of chemical engineering, the air should be forced through the column countercurrently to obtain the maximum absorption rate (OTTENGRAF, 1986). This operation mode may lead to back-stripping (OCKELOEN et al., 1996): the mixing of input untreated gas with exiting spent wetting. The exiting medium will wash out some of the pollutant from the untreated gas according to its mass transfer coefficient. A strict theoretical approach predicts that there will be no or negligible regeneration of the liquid medium since there is no aeration tank. The medium loaded with pollutant is recirculated to the reactor head, where it is in contact with the purified air leaving the column allowing pollutant to be stripped out again according to its mass transfer coefficient. The overall elimination efficiency will, therefore, be lowered. However, in practice a trickling filter may be operated countercurrently with good success, since the contact time of gas and liquid medium is too short to allow for significant washing or stripping effects. Some regeneration of the liquid medium takes place in the medium tanks and pipelines. Nevertheless, it is advisable to employ cocurrent operation to reduce the liquid holdup and to improve the liquid distribution in the column, a critical parameter for obtaining high elimination capacities (LÄMMERZAHL et al., 1997). Microorganisms confronted with the highest pollutant concentrations will receive fresh medium and thereby a maximum of nutrients. Most experiments and technical plants are currently designed for cocurrent operation.

Another way is to wet the packing in a crosscurrent mode. This employs forcing the air through the packing horizontally while wetting the packing from above. By applying this operation mode, all zones of the filter – the ones being initially confronted with the crude gas as well as the rear zones – will receive "fresh" medium containing all the nutrients. This can be advantageous if the biocenosis in the filter is functionally heterogenous and unequally distributed in the direction of gas flow. The wetting of a labscale reactor for the purification of motor vehicle emissions from road tunnels containing carbon monoxide and different nitrogen oxides works crosscurrently with success (ROBRA et al., 1997). This method suffers from the circumstance that a homogeneous gas distribution is difficult to maintain because of design difficulties. Distribution of the liquid medium over the packing is achieved by using multi-nozzle systems mounted distal to the packing. The waste air will preferentially flow through this unobstructed space, and not through the packing.

4 Clogging of the Filter Packing

4.1 Definition of the Problem

Microorganisms use substrate for metabolism, thereby converting it to carbon dioxide, water, heat, and cell substance. Depending on the substrate, inorganic salts may be formed. Incomplete degradation can lead to the accumulation of intermediate compounds. During trickling filter operation, soluble compounds are rinsed out by the liquid medium. Insoluble compounds such as the microorganisms are retained in the filters. Of course, it is necessary that a biofilm remains on the packing. Stability of the biofilm is required for catalytic activity of the filter unit. Over time, shedding of the biofilm will lead to clogging. SMITH et al. (1996) state "For stable long-term continuous operation of highly loaded trickle bed air biofilters, the prevention of plugging due to accumulating biomass is essential for avoiding biofilter failure (...)". Clogging leads to a decrease in the elimination capacity by depriving the oxidatively active biofilm of final electron acceptors and by a decline of the bioflm surface being available for mass transfer. Further clogging will be manifested by an increase in the pressure drop. Besides really becoming overgrown (Fig. 3), dislodgement of biomass due to low adhesion properties of the biofilm can also lead to the clogging of trickling filters. Dislodged biomass will often not leave the column.

In a bioscrubber, it is theoretically possible to operate the reactor in a state of equilibrium between microbial growth and deterioration, so that no net growth occurs. In other words, the degradation of the pollutant will be completely used for maintenance metabolism. This idea can also be adapted to biotrickling filters. DIKS et al. (1994) postulate the existence of this equilibrium for a dichloromethane biotrickling filter system. Their assumptions are based on the carbon balance of the system: the amounts of input reduced carbon and oxidized carbon leaving the reactor are in equilibrium, i.e., the dichloromethane reduction equals the carbon dioxide production. The growth yield in this system approaches zero. This may be due to extensive grazing of predatory organisms (like protozoa) or stress (possibly due to the accumulation of chloride ions). Unfortunately, these results do not represent the majority of trickling filters: most laboratory and technical plants are sooner or later confronted with the clogging problem.

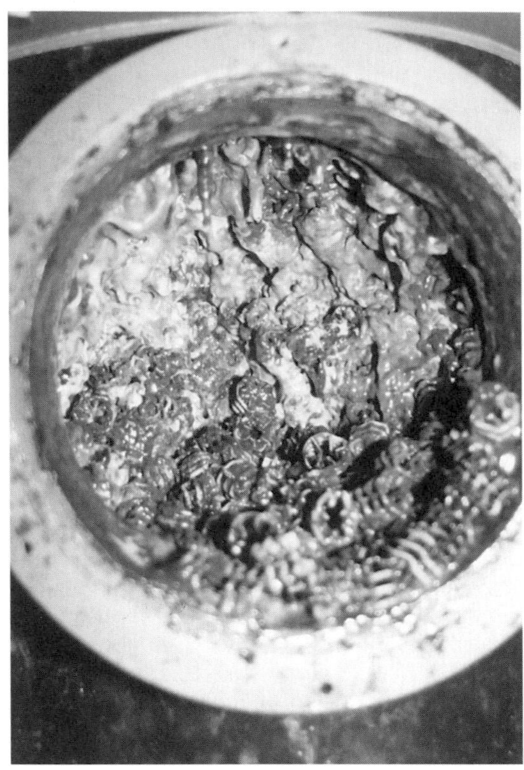

Fig. 3. Clogged packing of a semi-technical trickling filter for the purification of toluene containing waste air after 18 months of continuous operation. Fillings: PP NORPAC-similar 1″.

4.2 Strategies to Avoid Clogging

Different strategies to control biomass formation have been evaluated. They follow two main strategies: either to prevent extensive growth of biofilm, or to remove surplus biomass. The most common method is hindering microbial growth by modifications of the circulated liquid medium. This is accomplished by:

(1) modifying the medium so that a nutrient is limiting or using nitrate instead of ammonia as nitrogen source for anabolism, or
(2) adding inert salts that increase the ion strength and lower the water activity.

According to the literature, both approaches are successful.

Nutrient limitation will limit the surplus growth of the microorganisms (SCHÖNDUVE et al., 1996; WÜBKER and FRIEDRICH, 1996; WÜBKER et al., 1997), however, the absence of important nutrients can also lead to a significant decrease in the overall elimination capacity of the plant (WECKHUYSEN et al., 1993; WEBER and HARTMANS, 1996). Another contrary effect is that some microorganisms form storage compounds (like extracellular slimes) from the supplied carbon source when growth is limited. Slime accumulation also hastens clogging. SMITH et al. (1996) decreased biomass by supplying nitrate instead of ammonia as nitrogen source. Not only the biomass formation was lowered, but the removal efficiency of the filter was improved. Increased efficiency is due to nitrate reaching deeper layers of the biofilm and thereby being a more efficient electron acceptor. Addition of inert salts leads to a lower water activity in the filter. This additional stress to microorganisms leads to a more stable metabolism in the system. This approach, applicable to large scale plants, is promising, since it does not necessarily affect the degradation rate and requires a minimum in additional expenditure. SCHÖNDUVE et al. (1996) limited growth in an ethyl acetates–toluene trickling filter system by addition of sodium chloride to a final concentration of 0.4 mol L^{-1}. WEBER and HARTMANS (1994) also used sodium chloride, but at 0.2 mol L^{-1}, and achieved the best results when working with a fungal culture system instead of a bacterial one, possibly because of a higher maintenance requirement of the fungi. VAN LITH et al. (1994) limited surplus growth by applying sodium chloride in systems for the purification of ethanol, methylacrylate and dichloromethane polluted air. In their experiments, the maximum specific growth rate (μ_{max}) reached zero at sodium chloride concentrations of 0.2–0.8 mmol L^{-1} depending on the experimental conditions. The pollutant degradation at that salt concentration amounted to 50% of the maximal degradation rate.

Removal of excess biomass by mechanical measures is the most rigorous way to avoid clogging. In labscale reactors, biomass removal is accomplished by periodically moving the filter bed (WÜBKER et al., 1997), backwashing with medium fluidization (SMITH et al., 1996, 1998), or by sodium hydroxide wash (WEBER and HARTMANS, 1996). In all cases, the biotrickle filter must be taken off-line and biocatalysis capacity is compromised until regrowth of the biofilm is achieved. Removal of the filter bed is not scaleable to large plants because hundreds of m^3 of fillings cannot easily be moved, nor be backwashed (since the constructions would not stand the flooding with water and the resulting mass). However, the removal of excess biomass can be eased by design modifications of the trickling filter unit, as presented, e.g., by BRONNENMEIER et al. (1994).

In industrial practice, biological waste gas purification plants are usually started-up by the inoculation with sewage sludge often containing only few bacteria or fungi capable of degrading the air pollutant of interest, but many predatory protozoa. This results in delay. Sometimes it prevents functional starting-up of the plant. This circumstance leads to the propagation of inoculation with monocultures or "DOC" (designed oligospecies culture) systems. Usually, predating organisms are underrepresented or omitted in these systems. The most "natural" way to prevent clogging is, therefore, probably the one mentioned by COX and DESHUSSES (1999). They increased the stability of a trickling filter unit during a 77 d experiment by adding of protozoa which grazed on the degraders sufficiently to keep them at a certain level. Compared to a control filter, the biomass formation was reduced by approx. 15%. Future research will show if this method is adaptable to large-scale plants and long-term operation.

5 Theoretical Fundamentals and Modeling of Biological Trickling Filters

The majority of theoretical fundamentals are covered in Chapter 12, this volume. Modeling is important for the scaling-up of experimental results obtained with labscale biofiltration units to technical scale plants. Application of an adequate model, and fitting of the experimental data, can help in defining important parameters, e.g., the rate limiting step.

Biological trickling filters are especially attractive for modeling because many parameters like the packing surface are known, can be measured – like the amount and composition of the biomass on the packing and in the liquid phase – or can be varied – like the wetting. However, many variables remain. Most modeling approaches are derived from chemical engineering. In biological fixed-bed reactors the catalytically active biomass is unevenly distributed in the system – in terms of mass as well as composition – (Fig. 4). The wetting of the packing as well as the flow of the gas through the column are not homogeneous, and the specific surface area of the biofilm will change during process time due to "filling-in-the-cracks" (SMITH et al., 1998). Therefore, average biofilm thicknesses, pore volumes, or particle sizes have to be estimated in the calculation of important parameters. In most cases, the reactor is modeled as a whole, and experimental macrokinetic data usually fit quite well to the postulated models of the respective research groups.

Fig. 4. Unevenly distributed biomass in a semi-technical trickling filter for the purification of toluene containing waste air after 6 months of continuous operation. Fillings: reticulated PU foam, 10 pores per inch, $40 \cdot 40 \cdot 40\ mm^3$.

An expression for the outlet concentration in the gas phase and hence the removal efficiency after solving the differential mass balances for both the gas and liquid phases (DIKS et al., 1991; HARTMANS and TRAMPER, 1991; GOSSEN, 1991) is

$$-w^{G} \cdot \left[\frac{d\rho^{G}}{}\right] = \pm \beta^{OG} \cdot a^{L} \cdot (\rho^{G} - K \cdot \rho^{L}) \quad (1)$$

In Eq. (1) + refers to cocurrent and – to countercurrent operation with respect to the relative flow direction of both phases. For the liquid phase it follows:

$$-w^{L} \cdot \left[\frac{d\rho^{L}}{}\right] = \beta^{OG} \cdot a^{L} \cdot (\rho^{G} - K \cdot \rho^{L}) - \delta \cdot a^{L} \cdot \eta_{r} \cdot r'_{S} \quad (2)$$

In Eqs. (1, 2) the term $\beta^{OG}\ a^{L}$ represents the quality of the gas–liquid mass transfer, while $\delta\ a^{L}\ \eta\ r'_{S}$ equals the elimination capacity per unit of volume of the filter bed.

The boundary conditions are related to the recirculation of the liquid phase and the inlet concentration of the gas phase.

In general, it is of interest to determine the limiting step of the pollutant degradation, which can be of microbiological (reaction limitation) or physical nature (mass transfer from the gas to the liquid phase or diffusion limitation inside the liquid phase and the biofilm). There is no consensus about the nature of the mass transfer itself. Most groups proceed on the assumption that the biofilm is homogeneously wetted and, therefore, a mass transfer gas–liquid phase is required. RUTENFRANZ et al. (1994) discussed the possibility that the pollutant may directly transfer from the gas phase to the biofilm which is only wetted at discrete time periods. Based on that assumption, they formulated a "surface moistening model".

The mass transport and degradation of a specific substrate inside a flat biofilm at steady state conditions can be described by the following differential Eq. (3) (DIKS et al., 1994):

$$D_{S,\mathrm{eff}} \cdot \frac{d^2 \rho_S}{} = -r_S \tag{3}$$

In this equation, $D_{S,\mathrm{eff}}$ is the effective diffusion coefficient of the substrate, and r_S is the local substrate elimination rate in the biofilm, that can be related to microbial growth, as well as to maintenance m_S.

$$-r_S = \frac{\mu(S) \cdot X}{Y_{X/S}} + m_S \cdot X \tag{4}$$

The term $m_S\,X$ reflects the flow of energy required for basic metabolism. The microbial growth $\mu(S)$ is described by the Monod equation in which μ_{max} is the maximum growth rate and K_S the Monod constant, for the substrate concerned:

$$\mu\,(S) = \mu_{max} \cdot \frac{S}{K_S + S} \tag{5}$$

In a biological system, the pollutant degradation kinetics are of great importance. Obtaining microkinetic data is extremely difficult, since parameters influencing these kinetics will change over the course of time because of biomass growth (at least until a theoretical steady state is reached) and changes in biomass composition (since trickling filters are septicly driven mixed culture systems) (PLAGGEMEIER et al., 1997). Chemical parameters, like pH, nutrient concentrations, and salinity will also influence the degradation kinetics, but can easily be measured. Compared to biofilters, biofilms in trickling filters often reach thicknesses of several millimeters. Two scenarios are possible:

(1) the biofilm is completely or
(2) only partially penetrated by the substrate of interest.

In the latter case, deeper parts of the biofilm are completely inactive with respect to substrate degradation, which is often the rule as shown by microelectrode measurements of DE BEER et al. (1990). Their experiments were carried out for the investigation of the nitrification performance of a biofilm consortium, and they used ammonium and nitrate sensitive electrodes.

Strict application of Monod kinetics would lead to the assumption that at least in deeper parts of the biofilm the pollutant degradation is of first order and, therefore, the substrate concentration would never reach zero. Even when assuming zero order kinetics, a fully penetrated biofilm is possible if the degradation is reaction limited rather than diffusion limited. If diffusion limitation occurs, the biofilm will not be fully penetrated. The latter case was observed, e.g. by by OTTENGRAF (1981) in biofilter experiments dealing with toluene as a pollutant. A good overview of biofilm kinetics is given by VAN DEN HEUVEL (1992). The fact that biofilms are not flat homogeneous layers that fully cover a carrier, but are mountain-like shapes with liquid (COSTERTON et al., 1994) or possibly even gas filled channels must always be considered in theoretical approaches. Biofilm modeling fundamentals were developed, e.g., by BENEFELD and MOLZ (1985) with consideration of diffusional transport of oxygen and nutrients and biofilm growth and decay. RITTMANN and MCCARTY (1980) modeled a steady state biofilm. Mixed culture biofilms were modeled by WANNER and REICHERT (1996) with some consideration of the porosity of the film and other factors that had been simplified in former models.

Therefore, modeling of a filter biofilm is not precise. One will have to make many assumptions (e.g., zero order kinetics, a flat biofilm) and hence many details are overlooked. Nevertheless, when fitting experimental data to the model, important information can be derived, as shown by DIKS and OTTENGRAF (1994). A couple of examples for trickling filter modeling are given in the following paragraph.

HEKMAT and VORTMEYER (1995) investigated the biodegradation of ethanol and polyalkylated benzenes in a labscale trickling filter. They correlated the obtained data with results derived from a mathematical model based on stationary differential mass balances. The degradation process for ethanol is oxygen limited, while in the case of polyalkylated benzenes, it is substrate limited. The model results and the experimental data correlated well. A biotrickling filter for the removal of monochlorobenzene was investigated by MPANIAS and BALTZIS (1998) as well as a detailed mathematical model, accounting for mass transfer and kinetic effects based on substrate and oxygen availability. Like other groups, they stated the critical effect of oxygen on the performance of the device. From our own results, we can underline that the terminal electron acceptor may often be the limiting factor even in biological waste air purification reactors.

6 Trickling Filtration for Waste Gas Treatment in Practice

The degradation of many compounds of environmental interest was shown to be successful in trickling filters on a labscale as well as on a semi-technical scale. However, nowadays most research focuses on trickling filters, while at present trickling filters are applied in industrial practice to a lesser extent than biofilters. This is mainly due to higher costs for the construction of such plants, and in addition, trickling filters do not only offer the possibility to monitor more parameters than biofilters, this is also necessary to maintain the function of the filter. In research, these monitoring opportunities are a fundamental requirement for the elucidation of the governing processes in the filters, but in practice they increase personnel costs. Nevertheless, the authors are confident that trickling filter technology will prove to be the most reliable biological waste gas purification technology, and the only one really capable of dealing with difficult waste gas problems in long operational terms.

An example of a large scale biological trickling filter is a plant for the purification of toluene containing waste gas streams in a range of 150,000 $m^3 h^{-1}$ at average concentrations of 450 mg m^{-3} released during the preparation of rotogravure inks (KELLNER and VITZTHUM, 1997). This trickling filter consists of 8 containers housing approx. 640 m^3 of reticulated polyurethane foam, offering both a large void space and a sufficient surface for biofilm colonization. Another trickling filter by the same manufacturer was erected for the purification of acetone and ethanol containing waste gas streams of 20,000 $m^3 h^{-1}$. The apparatus consists of a bioscrubber unit followed by a trickling filter (Fig. 2).

Tab. 1 summarizes efforts of purifying waste gas streams containing toluene by trickling filtration. In most experiments, high elimination capacities can be obtained at very high specific loads, thereby lowering the efficiency, or high efficiencies can be reached at very low specific loads (and hence low elimination capacities). Where available, data have been selected for efficiencies ranging from 50–80%.

7 Conclusions

Compared to biofilters, biotrickling filters offer the possibility to monitor more parameters during biological purification of waste gas. This feature will help research and application to learn more about the process itself.

Biotrickling filters are the best choice to deal with substrates giving rise to inorganic salts such as halogenated hydrocarbons, because these salts are rinsed by a recirculated liquid wetting medium which may be collected and treated separately. Since an inert packing is used, the biomass can be removed quantitatively if necessary to elucidate the composition of the biocatalyst and its mass. Biomass removal can be required if the reactor column cloggs by the accumulation of excess biomass. Strategies to avoid clogging include the application of stress or constructional modifications. Because the method is more expensive than biofiltration, biotrickling filtration is applicable to the purification of waste air streams difficult to treat like halogenated organic compounds. Future biological waste gas purification systems will most likely combine the most promising aspects of biofiltration and biotrickling filtration.

Tab. 1. Toluene Degradation in Biological Trickling Filters

Reference	Reactor Specifications	"Best" Elimination Capacity [g C_{org} m^{-3} h^{-1}]	Obtained at an Efficiency of [%]
WOLFF (1992)	168 L[b]	62[c]	65[c]
SCHINDLER (1994)	100 L[a], Dinpac 1	30[c]	40[c]
PEDERSEN and ARVIN (1995)	4.5 L[b], steel pall rings	32[b]	54[b]
PEDERSEN et al. (1997)	19.3 L[a], polyvinyl difluoride	5.5[b]	67[b]
WEBER and HARTMANS (1994)	70 L[a], pall rings (50 mm)	32[a]	47[b]
WEBER and HARTMANS (1996)	106 L[b], pall rings (50 mm)	35[a]	54[b]
WÜBKER et al. (1997)	3.7 L[b], polyamide beads (8 mm)	7.3[b]	56–72[a]
GAI (1997)	21 L[b], HIFLOW fillings	48[c]	46[c]
Unpublished data by the authors	150 L, PP NORPAC-similar 1″	21	70

[a] Data directly given by cited author.
[b] Data calculated by the authors from data given by cited author.
[c] Data derived by the authors from graphs given by cited author, therefore, provisory some variance.

8 References

BENEFELD, L., MOLZ, F. (1985), Mathematical simulation of a biofilm process, *Biotechnol. Bioeng.* **27**, 921–931.

BRONNENMEIER, R., FITZ, P., TAUTZ, H. (1994) Reinigung von Lackiererei-Abluft mit einem Gitterträger-Biofilter, in: *VDI-Bericht 1104 Biologische Abgasreinigung* (KRdL im VDI und DIN, Ed.), pp. 203–215. Düsseldorf: VDI-Verlag.

COSTERTON, J. W., LEWANDOWSKI, Z., DEBEER, D., CALDWELL, D., KORBER, D., JAMES, G. (1994), Biofilms, the customized microniche, *J. Bacteriol.* **176**, 2137–2142.

COX, H. H. J., DESHUSSES, M. A. (1999), Biomass control in waste air biotrickling filters by protozoan predation, *Biotechnol. Bioeng.* **62**, 216–224.

DE BEER, C., VAN DEN HEUVEL, J. C., SWEERTS, J. P. R. A. (1990), Microelectrode studies in immobilized biological systems, in: *Physiology of Immobilized Cells* (DE BONT, J. A. M., VISSER, J., MATTIASSON, B., TRAMPER, J., Eds.), pp. 613–624. Amsterdam: Elsevier Science Publishers.

DIKS, R. M. M., OTTENGRAF, S. P. P. (1991), Verification studies of a simplified model for the removal of dichloromethane from waste gases using a biological trickling filter; Part I, *Bioproc. Eng.* **6**, 93–99.

DIKS, R. M. M., OTTENGRAF, S. P. P. (1994), Technology of trickling filters, in: *VDI-Bericht 1104 Biologische Abgasreinigung* (KRdL im VDI und DIN, Ed.), pp. 19–37. Düsseldorf: VDI-Verlag.

DIKS, R. M. M., OTTENGRAF, S. P. P., VRIJLAND, S. (1994), The existence of a biological equilibrium in a trickling filter for waste gas purification, *Biotechnol. Bioeng.* **44**, 1279–1287.

GAI, S. (1997), Abbau von Toluol und *m*-Kresol im Biofilmrieselbettreaktor, in: *Biological Waste Gas Cleaning* (PRINS, W. L., VAN HAM, J., Eds.), pp. 157–163. Düsseldorf: VDI Verlag.

GOSSEN, C. A. (1991), Abgasreinigung mit fixierten Bakterien im Rieselbett, *Thesis*, University of Technology, Munich, Germany.

HARTMANS, S., TRAMPER, J. (1991), Dichloromethane removal from waste gases with a trickle-bed bioreactor, *Bioproc. Eng.* **6**, 83–92.

HEKMAT, D., VORTMEYER, D. (1994), Modeling of biodegradation processes in trickle bed bioreactors, *Chem. Eng. Sci.* **49**, 4327–4345.

HEKMAT, D., VORTMEYER, D. (1995), Modeling of Biodegradation processes in trickle-bed bioreactors, *Chem. Eng. Sci.* **49**, 4327–4345.

KELLNER, C., VITZTHUM, E. (1997), Biologische Ablufttreinigung mit dem Biotropfkörperverfahren: Umsetzung vom Pilotversuch bis zur Großanlage mit toluolbeladener Abluft bei einem Druckfarbenhersteller, in: *Biological Waste Gas Cleaning* (PRINS, W. L., VAN HAM, J., Eds.), pp. 61–66. Düsseldorf: VDI Verlag.

LÄMMERZAHL, O., PLAGGEMEIER, T., ENGESSER, K.-H. (1997), Vergleich der Auswirkungen verschiedener Einflußgrößen auf die Leistung eines Biotricklingfilters bei der Reinigung 1-butanol- bzw. styrolhaltiger Abluftströme, in: *Biological Waste Gas Cleaning* (PRINS, W. L., VAN HAM, J., Eds.), pp. 193–196. Düsseldorf: VDI Verlag.

MPANIAS, C. J., BALTZIS, B. C. (1998) An experimental and modeling study on the removal of monochlorobenzene vapor in biotrickling filters, *Biotechnol. Bioeng.* **59**, 328–343.

OCKELOEN, H. F., OVERKAMP, T. J., GRADY JR., C. P. L. (1996), Engineering model for fixed-film bioscrubbers, *J. Environ. Eng. ASCE* **122**, 191–197.

OTTENGRAF, S. P. P. (1986), Chapter 12 exhaust gas purification, in: *Biotechnology* 1st Edn., Vol. 8 (REHM, H., REED, G., Eds.), pp. 425–452. Weinheim: VCH.

OTTENGRAF, S. P. P., DIKS, R. M. M. (1992), Process technology of biotechniques, in: *Biotechniques for Air Pollution Abatement and Odor Control Policies* (DRAGT, A. J., VAN HAM, J., Eds.), pp. 17–31. Amsterdam: Elsevier Science Publishers.

PEDERSEN, A. R., ARVIN, E. (1995), Removal of toluene in waste gases using a biological trickling filter, *Biodegradation* **6**, 109–118.

PEDERSEN, A. R., MØLLER, S., MOLIN, S., ARVIN, E. (1997), Activity of toluene-degrading *Pseudomonas putida* in the early growth phase of a biofilm for waste gas treatment, *Biotechnol. Bioeng.* **54**, 131–141.

PLAGGEMEIER, T., LÄMMERZAHL, O., ENGESSER, K.-H. (1997), Purification of *n*-hexane polluted air by use of a biological trickling filter: Monitoring of the primary-degrading microflora, in: *Biological Waste Gas Cleaning* (PRINS, W. L., VAN HAM, J., Eds.), pp. 257–260. Düsseldorf: VDI Verlag.

PRÜß, M., BLUNK, H. (1941), Verfahren zur Reinigung von luft- oder sauerstoffhaltigen Gasgemischen. *Patentschrift Nr. 710954*, Berlin: Reichspatentamt.

RITTMANN, B. E., MCCARTY, P. L. (1980), Model of Steady-state biofilm kinetics, *Biotechnol. Bioeng.* **22**, 2343–2357.

ROBRA, K.-H., WELLACHER, M., KIRCHMEIER, F., LEISTENTRITT, R., PUCHER, K. (1997), Abluftreinigung für Straßentunnel und Tiefgaragen durch carboxidotrophe Mischpopulationen, in: *Biological Waste Gas Cleaning* (PRINS, W. L., VAN HAM, J., Eds.), pp. 131–140. Düsseldorf: VDI Verlag.

RUTENFRANZ, C., HOLLEY, W., MENNER, M. (1994), Einfluß des Wasser-Luft-Verteilungsgleichgewichts von Schadstoffen auf das Einsatzgebiet von Biofilter, Tropfkörper oder Biowäscher, in: *VDI-Bericht 1104 Biologische Abgasreinigung* (KRdL im VDI und DIN, Ed.), pp. 57–68. Düsseldorf: VDI Verlag.

SCHINDLER, I., FRIEDL, A., SCHMIDT, A. (1994), Abbaubarkeit von Ethylacetat, Toluol und Heptan in Tropfkörperbioreaktoren, in: *VDI-Bericht 1104 Biologische Abgasreinigung* (KRdL im VDI und DIN, Ed.), pp. 135–147. Düsseldorf: VDI Verlag.

SCHÖNDUVE, P., SÁRA, M., FRIEDL, A. (1996), Influence of physiologically relevant parameters on biomass formation in a trickle-bed bioreactor used for waste gas cleaning, *Appl. Microbiol. Biotechnol.* **45**, 286–292.

SMITH, F. L., SORIAL, G. A., SUIDAN, M. T., BREEN, A. W., BISWAS, P., BRENNER, R. C. (1996), Development of two biomass control strategies for extended, stable operation of highly efficient biofilters with high toluene loadings, *Environ. Sci. Technol.* **30**, 1744–1751.

SMITH, F. L., SORIAL, G. A., SUIDAN, M. T., PANDIT, A., BISWAS, P., BRENNER, R. C. (1998), Evaluation of trickle bed air biofilter performance as a function of inlet VOC concentration and loading, and biomass control, *J. Air Waste Manag. Assoc.* **48**, 627–636.

VAN DEN HEUVEL, J. C. (1992), Mass transfer in and around biofilms, in: *Biofilms – Science and Technology* (MELO, L. F., BOTT, T. R., FLETCHER, M., CAPDEVILLE, B., Eds.), pp. 239–250. Dordrecht: Kluwer Academic Publishers.

VAN LITH, C. P. M., OTTENGRAF, S. P. P., DIKS, R. M. M. (1994), The control of a biotrickling filter, in: *VDI-Bericht 1104 Biologische Abgasreinigung* (KRdL im VDI und DIN, Ed.), pp. 169–180. Düsseldorf: VDI Verlag.

VDI-Richtlinie 3478 (1996), Biologische Abgasreinigung Biowäscher und Rieselbettreaktoren/ Biological waste gas purification bioscrubbers and trickle bed reactors, in: *VDI/DIN-Handbuch Reinhaltung der Luft* Vol. 6 (KRdL im VDI und DIN, Ed.). Düsseldorf: VDI Verlag.

WANNER, O., REICHERT, P. (1996), Mathematical Modeling of mixed-culture biofilms, *Biotechnol. Bioeng.* **49**, 172–184.

WEBER, F. J., HARTMANS, S. (1994), Toluene degradation in a trickle bed reactor – Prevention of clogging, in: *VDI-Bericht 1104 Biologische Abgasreinigung* (KRdL im VDI und DIN, Ed.), pp. 203–215. Düsseldorf: VDI Verlag.

WEBER, F. J., HARTMANS, S. (1996), Prevention of clogging in a biological trickle-bed reactor removing toluene from contaminated air, *Biotechnol. Bioeng.* **50**, 91–97.

WECKHUYSEN, B., VRIENS, L., VERACHTERT, H. (1993), The effect of nutrient supplementation on the biofiltration removal of butanal in contaminated air, *Appl. Microbiol. Biotechnol.* **39**, 395–399.

WOLFF, F. (1992), Biologische Abluftreinigung mit einem intermittierend befeuchteten Tropfkörper, in: *Biotechniques for Air Pollution Abatement and Odour Control Policies* (DRAGT, A. J., VAN HAM, J., Eds.), pp. 49–62. Amsterdam: Elsevier Science Publishers.

WÜBKER, S.-M., FRIEDRICH, C. G. (1996), Reduction of biomass in a bioscrubber for waste gas treatment by limited supply of phosphate and potassium ions, *Appl. Microbiol. Biotechnol.* **46**, 475–480.

WÜBKER, S.-M., LAURENZIS, A., WERNER, U., FRIEDRICH, C. (1997), Controlled biomass formation and kinetics of toluene degradation in a bioscrubber and in a reactor with a periodically moved bed, *Biotechnol. Bioeng.* **55**, 686–692.

16 Membrane Processes and Alternative Techniques

MARTIN REISER
Stuttgart, Germany

1 Introduction

Biological waste gas purification using the established biofilter, bioscrubber and trickle bed reactor procedures, offers solutions for a great number of problems in the field of reduction of VOC and odor emissions. Unfortunately, for a far greater number of emission problems, there is no biological waste gas purification procedure.

Attempts to find new niches for the application of microorganisms in waste gas purification repeatedly focus on the development of new variations on the established techniques.

The starting points for these variations lie often in the attempt to extend the limits of biological waste gas purification. Examples of such limits are:

- In conventional biofilters the filter area load is often in the range between 50 and 150 $m^3 m^{-2} h^{-1}$ (VDI Guideline 3477, 1991). An increase of this load is possible, e.g., by optimization of the gas flow in the filter bed. This also permits the application of biofilters at restrained spatial conditions.
- The industrial application of trickle bed reactors is often prevented by the occurrence of clogging. There is, therefore, no sufficient guarantee of reliability. One possible solution to the clogging problem is the employment of new textile materials, and different modes of operation.
- Owing to the necessary transfer of the pollutant from air into the washing water, the bioscrubber is only suitable for water-soluble components with a Henry coefficient <15 Pa $m^3 mol^{-1}$ (WITTORF, 1996). With the addition of solubilizers, an implementation is also possible for hydrophobic components.
- Owing to the ejection of microorganisms from biological waste gas purification plants, these procedures cannot be used in a closed air circuit. In addition to other advantages, a separation of biofilm and gas through a membrane causes effective avoidance of the emission of germs.

2 Variations on the Established Procedures

2.1 Biofilters with New Technologies

The development of new biofilter techniques was and is influenced, whether a new field of application is available or not. The original open single-level filters are still in use in sewage treatment plants and composting plants. Only the demand for biofilters in the field of industrial plants with less available space led to the construction of closed container systems.

New developments require in addition more complex procedures, which can only be justified, if they result in significant economical benefit.

One new concept requires that biofilters must be larger than it would normally be the case, because of inhomogeneous flow through the filter material. Thus, in the "rotor biofilter", material is continuously moved preventing the formation of channels. The most recent experimental plant with this technique makes an additional continuous discharge of material and also makes humidification possible, by immersion of the barrel into water (Fig. 1).

In the experimental plant shown, the use of an inert filter material and continuous humidification is sufficient for this to be called a trickle bed reactor. Consequently, depending on the problem, it is possible with this design to unite the advantages of both techniques:

- mode of operation as optimal flowed biofilter, e.g., of hardly water-soluble waste gas components,
- mode of operation as trickle bed reactor, e.g., if it is necessary to control the pH value.

The data to date justifies the relatively high design costs (HANDTE, 1999).

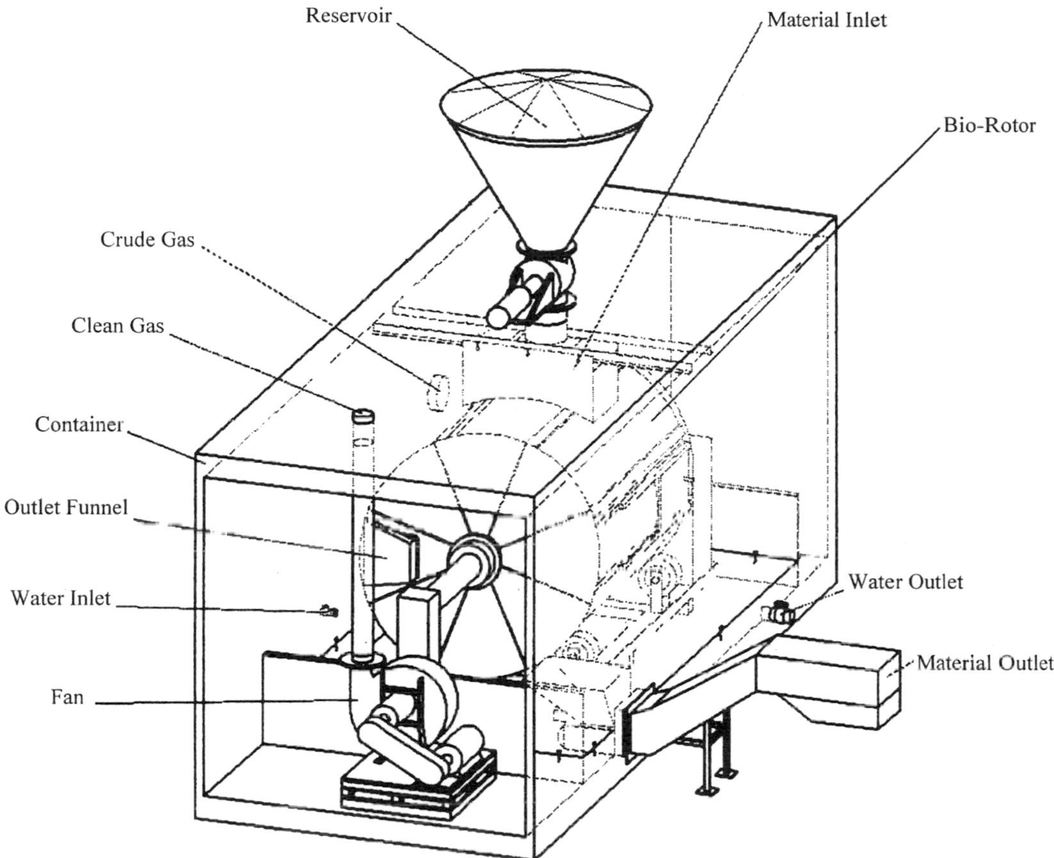

Fig. 1. Schematic drawing of a rotating biofilter (Fa. Handte Umwelttechnik, Tuttlingen, Germany).

2.2 Variations on Trickle Bed Reactors

Trickle bed reactors are given the greatest attention in the field of research and development. These systems are especially suitable for basic investigations owing to the many possibilities for control and regulation. Also at large plants, the procedure offers good solutions for specific problems (KELLNER, 1999).

In general, the aim of new developments is to prevent clogging. This is made possible by a special mode of operation and by using new support materials.

One of the latest developments, which are already field-tested in larger plants, is the Biotex technique (SABO et al., 1999). The packing used here is a textile material (Fig. 2).

Strong growth of biomass on this material is also possible under certain conditions. But this, however, results not in clogging, but in a slight increase of the pressure drop. Because of the properties of the inert textile material, a problem-free clean-up is also possible here. During further studies, it should be examined whether automatized cleaning is possible, without manual intervention. This is a necessary precondition for the market launch of this technique.

2.3 Specific Kinds of Bioscrubbers

Conventional bioscrubbers consist of an absorber and a tank for the biological regeneration of the scrubbing fluid. These two appara-

Fig. 2. Cleaning of the textile filter material in the Biotex Procedure (Fa. Reinluft Umwelttechnik, Stuttgart, Germany; photo: Fraunhofer Association, Stuttgart, Germany).

tus are, to a large extent, already optimized on account of their usage in other fields. For the combination of both apparatus in the bioscrubber, the potential for technical innovations is also limited. In the case of this system, innovations pursued in the event that an unusual problem cannot be solved by using a bioscrubber.

For the use of scrubber technology in the case of poorly water-soluble waste gas components, the Biosolv technique was developed. In this technique solubilizers – especially silicone oil – are added to the scrubbing fluid. The waste gas components are then first dissolved in the silicone oil stage. In a second step, the microorganisms located at the boundary between water and silicone oil regenerate the hydrophobic phase and degrade the compounds in the watery phase (SCHIPPERT, 1999).

A new modified bioscrubber, which is to be located procedurally at the boundary to the trickle bed reactor, is called BioWave. With this technique a reverse jet scrubber is used for biological waste gas treatment. In a froth zone the transfer from pollutants and particulates from the gas phase to the liquid phase takes place. A biomass slurry is pumped around as a scrubbing liquid. Such a bioscrubber is already used to clean off-gases from the wood panels industry. It has been found that BioWave is used best for gas streams with high temperature, a high particulate content and volatile organic compounds (VOC) (VAN LITH, 1999).

3 Membrane Procedures in Biological Waste Gas Purification

3.1 The Use of Membranes in Waste Gas Purification

The innovation potential, which the use of membranes brings into many fields, can be very high. This technique is not established in the area of waste gas purification in the way that it is with waste water treatment. It is only in the area of gas permeation and steam permeation, (e.g., in the recovery of hydrocarbons in petrol tank exhaust fumes), that the procedures are already state of the art (CEN et al., 1993).

For biological waste gas purification, the application of membranes is limited, up to now, to small-scale experimental plants. In this case, pore membranes are mainly used. A great advantage of such microporous systems is that they are already used in other fields with great success and that a great variety of membrane and module types are commercially available at relatively low prices.

The use of dense membranes brings further advantages, especially in the case of compounds with poor water solubility. Because certain modules (e.g., those with membranes made of silicone rubber) are not generally available, the technical realization of these requires greater expenditure.

3.2 Combination of Membranes and Microorganisms

Studies about membrane reactors for biological waste gas purification are so far concerned mainly with unusual problems. An example of this is the elimination of substances that lead to critical products during their degradation or elimination of hydrophobic compounds, which cannot be removed satisfactorily with the established techniques.

The membrane processes can be divided into two groups, depending on whether microporous membranes or dense solution–diffusion membranes are applied.

3.2.1 Membrane Reactors with Microporous Membranes

Membranes with a pore diameter <0.2 µm work here as mechanical separation between the exhaust air and the microorganisms in their aqueous environment. The driving force for the mass transfer is a pressure gradient from feed to permeate and the solubility of the exhaust air compounds in water or in the biofilm. The membrane itself is not highly involved in the mass transfer mechanism. The pores only prevent the contact between gaseous phase and liquid.

At the permeate site the membrane is the built-up surface for the biofilm. The compact structure of the membrane modules permits a very large specific exchange surface (within the range of 1,500 and 8,000 m^2 m^{-3}), particularly when using hollow fibers.

Membrane bioreactors with microporous membranes are, therefore, also suitable for components with which no sufficient efficiency can be obtained due to the bad mass transfer, e.g., in a conventionally dimensioned bioscrubber. However, with this procedure also hydrophilic substances are preferred.

An example of a practical application is the removal of substances such as dichloromethane or propene where very good results are reported (HARTMANS et al., 1992; REIJ, 1997). Another example is an attempt to purify tunnel exhaust gases, loaded with nitrogen monoxide, with a hydrophobic microporous membrane showing interesting results at a laboratory scale study with stack modules (HINZ et al., 1994). By the use of hollow fibers an application of this technique in a full-scale plant appears possible.

The development of a BAF (biological air filter) for air quality control during a manned space mission simulation is probably the most spectacular application (HENSSEN et al., 1997).

Hydrophobic microporous membranes were also used here. The schematic structure of this reactor is shown in Fig. 3.

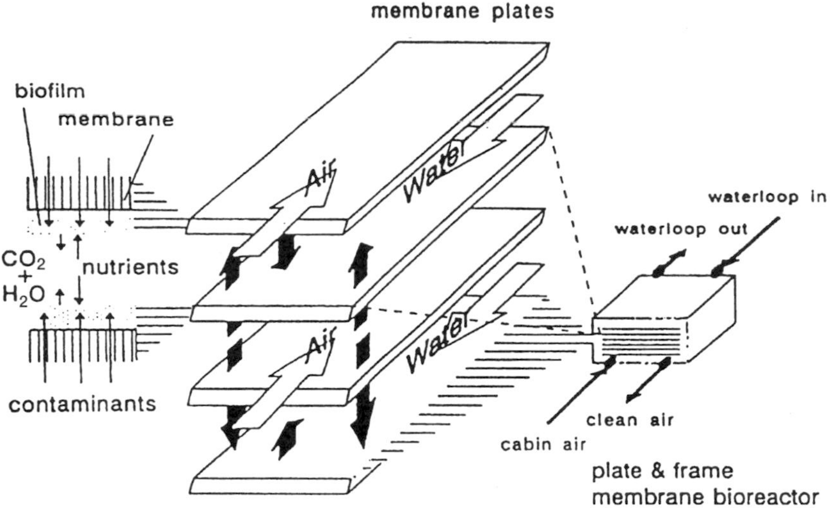

Fig. 3. Schematic drawing of a biological air filter (BAF) with a hydrophobic porous membrane for treatment of indoor atmospheres (HENSSEN et al., 1997).

3.2.2 Membrane Reactors with Dense Membranes

Membranes with a pore diameter of <2 nm are called, by definition, "dense membranes". Here the openings on the membrane surface are smaller than the space requirement of the membrane-forming polymer molecules. The mass transfer here differs substantially from the transfer in microporous membranes. It is usually described with the solution–diffusion model:

Components, which are suitable for it, are solved in a first step in the membrane and, provided that a concentration gradient exists, in a second step they are transported by diffusion to the permeate side.

In the case of biological waste gas purification, the concentration gradient is upheld by the microorganisms, which are degrading the volatiles dissolved in the membrane in the permeate side biofilm. For the case of tube membranes made of polydimethylsilicone (PDMS) the principle is shown in Fig. 4.

When using hydrophobic membranes, this principle of mass transfer results in a preference for substances that have poor water solubility. This is explained partly by the fact that the absorption is better because of the sub-

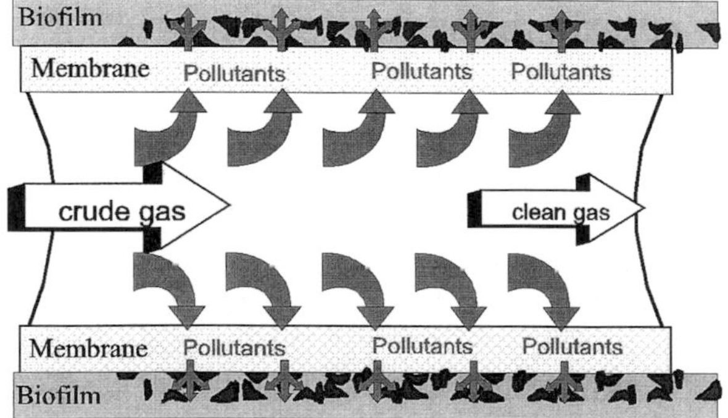

Fig. 4. Schematic picture of the mass transfer by using dense membranes in biological waste gas purification.

stantially smaller Henry coefficient. For example, the Henry coefficient for toluene in the sytem air–silicone rubber is approximately 250 times smaller than in water. Further it is to be assumed that the molecules arrive directly from the membrane to the microorganisms, without a genuine solution step being necessary in the water (REISER, 1999). Thus the substrate concentrations available to the bacteria are substantially larger. Therefore, according to Michaelis–Menten kinetics, an increase in the reduction rate results.

An example of the application of this technique with dense membranes is the removal of sulfur compounds from burn exhaust gases. With this procedure hollow fibers with a polyvinylalcohol membrane were used (KRAUTWURST, 1991). Experiments for the elimination of VOC with silicone rubber membranes also had promising results (REISER et al., 1997).

There have already been results in realistic conditions: Conventional silicone tubes have been used in a membrane bioreactor for removing a mixture of solvent steams in varying concentrations. The main component of this steam was toluene, and the legal threshold value had been maintained (Fig. 5).

The use of hollow fibers with a polyetherimid-PDMS composite membrane (manufacturer: GKSS Research Center, Geesthacht, Germany) made still further improvements possible: Good efficiencies could still be obtained with specific loads with toluene between 400 and 800 g m^{-3} h^{-1}.

3.2.3 Benefits of Membrane Procedures in Biological Waste Gas Purification

The application of the membrane technique in biological waste gas purification brings some advantages. In particular with the purification of waste air with hydrophobic compounds very good results are to be obtained due to the special type of mass transfer. There are also additional advantages offered by this technique as a result of the variation in the design from established processes (Tab. 1).

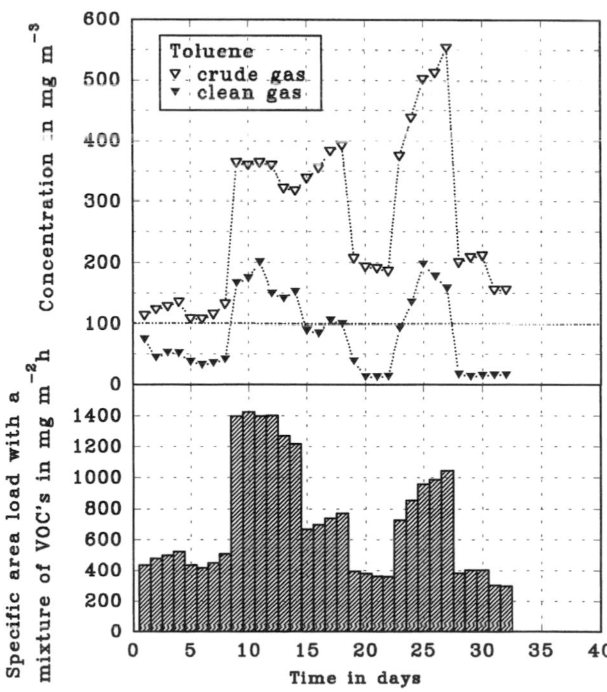

Fig. 5. Cleaning performance of a semi-technical membrane bioreactor with exhaust air containing toluene under industrial conditions The emission threshold value of toluene is 100 mg m^{-3}.

Tab. 1. Process Engineering Caused Benefits of Membrane Procedures Compared with the Remaining Techniques of Biological Waste Gas Purification

	Membrane Processes	Bioscrubber	Biofilter	Trickle Bed Reactor
Humidification of the crude gas	not necessary	not necessary	absolutely necessary	usefull
Blockage by closing up (clogging)	not possible	possible	theoretically possible	a well known fact
Long term operation with monocultures	possible (no contact of bacteria and crude gas)	not possible	not possible	not possible
Emission of germs	not possible	present	present	present

4 Conclusions

Procedures that are genuine alternatives to biofilter, bioscrubber and trickle bed reactor are few. This is primarily a result of the selection procedure: If a problem is not to be solved by the classical techniques, biological waste gas purification itself is normally not taken into account any more.

The techniques shown above, however, explain how basic innovative approaches can extend the application possibilities of biological waste gas purification.

The use of membranes opens completely new aspects, which cannot be offered by any of the established techniques. Thus, the separation from air and microorganisms leads to the fact that biological waste gas purification can also be used in closed cycles.

The problem of germ emission, which is discussed critically with some applications, does not appear in the membrane procedures. Finally, the use of dense hydrophobic membranes shows how compounds with poor water solubility can be more effectively eliminated than is possible with conventional procedures.

For the introduction to the market of the membrane techniques, further optimization of the industrially manufactured membrane modules is necessary. If suitable membrane modules are available at appropriate prices, as they are for modules in wastewater treatment, for example, the membrane technology can bring important new procedures to biological waste gas purification technology.

5 References

CEN, Y., MECKL, K., LICHTENTHALER, R.N. (1993), Nichtporöse Membranen – von der Grundlagenforschung zur technischen Anwendung, in: *Proc. Aachener Membrane Kolloquium*, pp. 117–164. Düsseldorf: GVC-VDI – Gesellschaft Verfahrenstechnik und Chemieingeieurwesen.

HANDTE, J. (1999), Rotorfilter als Tropfkörper, *Workshop* Biologische Abluftreinigung (Fraunhofer-Institutszentrum Stuttgart, 20.–21.September).

HARTMANS, S., LEENEN, E. J. T. M., VOSKULLEN, G. T. H. (1992), Membrane bioreactor with porous hydrophobic membranes for waste gas treatment, in: *Biotechniques for Air Pollution Abatement and Odor Control Policies* (DRAGT, A. J., VAN HAM, J., Eds.), pp. 103–106. Amsterdam: Elsevier Science Publishers.

HENSSEN, M. J. C., VAN DER WAARDE, J. J., KEUNING, S., PAUL, P. G., BREUKERS, R. J. L. H., BINOT, R. A. (1997), Application of closed biofiltration systems for treatment of indoor atmospheres, in: *Biological Waste Gas Cleaning* (PRINS, W. L., VAN HAM, J., Eds.), pp. 365-372. Düsseldorf: VDI-Verlag.

HINZ, M., SATTLER, F., GEHRKE, T., BOCK, E. (1994), Entfernung von Stickstoffmonoxid durch den Einsatz von Mikroorganismen – Entwicklung eines Membrantaschenreaktors, in: *VDI-Bericht 1104* Biologische Abgasreinigung (KRdL im VDI und DIN, Ed.), pp. 113–123. Düsseldorf: VDI-Verlag.

KELLNER, C. (1999), Abluftreinigung mit Rieselbettreaktoren, in: *VDI-Bericht 1478* Fortschritte in der Luftreinhaltetechnik (KRdL im VDI und DIN, Ed.), pp. 397–406. Düsseldorf: VDI-Verlag.

KRAUTWURST, B. (1991), Beiträge zur Permeation

von Gasen durch Membranen in reaktive Aufnehmerflüssigkeit – Umgekehrte Pervaporation. *Thesis*, University of Erlangen, Germany.

REIJ, M. W. (1997), Membrane bioreactor for waste gas treatment, *Thesis*, University of Wageningen, The Netherlands.

REISER, M. (1999), Reinigung von Abluft mit schlecht wasserlöslichen Inhaltsstoffen im Biomembranreaktor, *Stuttgarter Berichte zur Abfallwirtschaft* 73. Bielefeld: Erich Schmidt Verlag.

REISER, M., FISCHER, K. BARDTKE, D. (1997), Der Biomembranreaktor – vom Labor zur industriellen Anwendung, in: *Biological Waste Gas Cleaning* (PRINS, W. L., VAN HAM, J., Eds.), pp. 181–188. Düsseldorf: VDI-Verlag.

SABO, F., SCHNEIDER, T., BERNECKER, C. (1999), Textiles Trägermaterial für die biologische Abluftreinigung – Ergebnisse eines Langzeitversuches, in: *VDI-Bericht 1478* Fortschritte in der Luftreinhaltetechnik (KRdL im VDI und DIN, Ed.), pp. 407–416. Düsseldorf: VDI-Verlag.

SCHIPPERT, E. (1999), Biowäscher in der biologischen Abluftreinigung – Stand der Technik und neuere Entwicklungen, *Workshop* Biologische Abluftreinigung (Fraunhofer-Institutszentrum Stuttgart, 20.–21.September).

VAN LITH, C. P. M. (1999), Biofilter und seine Kombination mit anderen Verfahren – Fallbeispiele, *Workshop* Biologische Abluftreinigung (Fraunhofer-Institutszentrum Stuttgart, 20.–21.September).

VDI-Richtlinie 3477 (1991), Biologische Abgas/Abluftreinigung Biofilter/Biological Waste Gas/-Waste Air Purification Biofilters, in: *VDI/DIN-Handbuch Reinhaltung der Luft*, Vol. 6 (KRdL im VDI und DIN, Ed.). Düsseldorf: VDI-Verlag.

WITTORF, F. (1996), Abluftreinigung mit Biofiltern – Stand der Technik und Perspektiven, in: *Praxis der biotechnologischen Abluftreinigung* (MARGESIN, R., SCHNEIDER, M., SCHINNER, F., Ed.), pp. 55–76, Heidelberg: Springer-Verlag.

Applications

17 Commercial Applications of Biological Waste Gas Purification

DEREK E. CHITWOOD
Riverside, CA, USA

JOSEPH S. DEVINNY
Los Angeles, CA, USA

List of Abbreviations

BTEX	benzene, toluene, and xylene
EBRT	empty bed residence time
PLC	programmable logic controller
PVC	polyvinyl chloride
TPH	total petroleum hydrocarbon
VOC	volatile organic compounds

1 Background

1.1 Needs

Concerns over industrial releases of odors, toxic compounds, and smog forming chemicals have grown steadily. Increased public awareness of health impacts, declining tolerance for industrial offensiveness, expanding industrial activity, and greater prosperity have led to insistence that air discharges be cleaned up even when dischargers complain that costs are high. Regulations have forced some companies to move to less impacted areas and have made others close down. All are interested in seeking to meet regulatory requirements at the lowest possible cost.

Contaminants have commonly been removed by thermal treatment or activated carbon adsorption. Thermal treatment, either direct flaring or catalytic oxidation at lower temperature, is effective when contaminant concentrations are high enough to provide a significant portion of the energy required. However, when concentrations are low and additional fuel is required, a great deal of energy is wasted because the entire airflow must be heated in order to oxidize a small amount of contaminant. Activated carbon acts by surface adsorption, and can produce very clean effluent air. However, the amount of contaminant adsorbed per unit weight of activated carbon is a function of the airborne concentration. When concentrations are low, the adsorptivity of activated carbon is often reduced. The amount of activated carbon to be regenerated or replaced for each kilogram of contaminant removed becomes large, and the treatment is again impractical.

Biological treatment fills the need for an economical means of treating low concentration contaminants. It has been most widely used in odor control, and has gained greatest acceptance in Europe where odor regulations are strict and fuel costs are high. In recent years, however, it has been increasingly used for the control of toxic compounds and those that contribute to the formation of smog.

1.2 Biological Treatment

The concept of biological air treatment at first seems contradictory: Organisms suspended in the air cannot be biologically active. Air can be treated, however, if it is brought in close contact with a water phase that contains active organisms. Contaminants dissolve in the water where microbial degradation keeps concentrations low, driving further dissolution.

The needs for interphase contaminant transfer and for maintaining the water phase microbial ecosystem dictate the basic form of biological air treatment systems. A reactor must provide a large water surface area for efficient phase transfer and support for a substantial biomass. There are three general plans for biological treatment systems: biofilters, biotrickling filters, and bioscrubbers (OTTENGRAF, 1986). In biofilters, the porous medium is kept damp by maintaining the humidity of the incoming air and by occasional sprinkling (Fig. 1). In biotrickling filters, water flows steadily over the porous bed and is continuously collected and recirculated. In a bioscrubber, phase transfer occurs in one container as the water is sprayed through the air or trickled over a packed bed, then the water is transferred to a liquid phase bioreactor where the contaminant is degraded. Bioscrubbers are less commonly used, and will not be further discussed.

There are some secondary considerations for design. An important operating cost is the power for pumping air. The system must be designed so that the pressure required to drive the air through the system, called the "head loss", is small. It is also necessary to have a means for removing biomass so that it will not clog the system. In biotrickling filters, the flowing water may carry away biomass as rapidly as it grows, maintaining a steady state microbial ecosystem. However, heavily loaded biotrickling filters often have problems with clogging. Lightly or moderately loaded biofilters may reach steady state through the establishment of a culture of predatory microbes that consume the decomposers, but again heavy contaminant loads can cause rapid growth and pore clogging. Research on means to control or remove biomass remains active.

Water is readily available in biotrickling filters, but biofilters must be provided with humidification or irrigation systems to maintain the water content of the biofilm. All of the systems require a source of nutrients such as nitrates and phosphates.

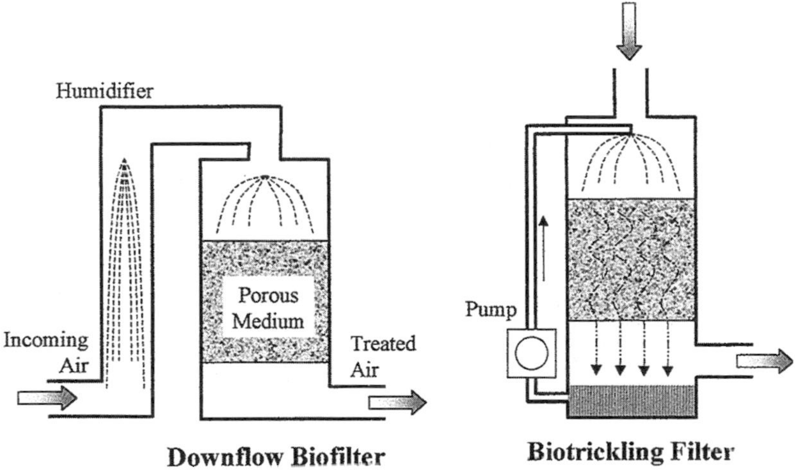

Fig. 1. Two common forms of biological air treatment.

Each biological system must have an initial seed culture of microorganisms. In most cases, a diverse mixed culture comes with the medium or is applied separately. It is presumed that those species capable of degrading the contaminant will grow rapidly, dominate the system and create a dense culture ideal for efficient treatment.

A major design characteristic of a biological treatment system is the detention time. This is the average time that a parcel of air is held in the system. For a given flow, the minimum detention time determines the size of the reactor, and so is a prime factor in determining the costs of the system. Because true porosity is often hard to measure, empty bed residence time (EBRT), calculated using the total volume of the bed, is often used to describe systems.

1.3 Biofilters

Biofilters are the most common form of biological air treatment system. The earliest and simplest were made by digging trenches, installing perforated pipe to serve as an air distribution system, and refilling the trenches with permeable soil. The Bohn Biofilter Corporation has installed many successful biofilters that are only slightly more complex, consisting of soil placed over a distribution system and contained within concrete traffic barriers (BOHN and BOHN, 1998). Soil biofilters are inexpensive. Soil is cheap, and the construction methods are very simple. Soil does not decompose significantly, and so does not compact. It is easily rewetted if it is inadvertently allowed to dry. However, the pores are small, so that head losses are high when flow rates are high, and high contaminant loads will cause clogging. These problems are avoided by making soil biofilters large, typically with detention times of a few minutes.

More complex biofilters are a pit or box filled with sieved compost or wood chips (Fig. 2). Once again the air is introduced through a distribution system at the bottom, typically perforated plastic pipes embedded in a layer of gravel or a layer of perforated concrete blocks. The compost or wood chips are placed above this, commonly in a layer about 1 m thick. A sprinkler system is provided to keep the medium moist, and drainage is provided to remove collected water in the event of rain or overwatering. While these biofilters are somewhat more elaborate than soil biofilters, they can still be made cheaply. Because the particle size can be controlled, the organic media provide large pores to reduce head loss. They also provide nutrients as they slowly decompose. The diverse and active culture of microorganisms on compost allows it to start working immediately. The reduced head loss and ideal microbial environment mean that biofilters can be made smaller.

Compost does have disadvantages, however. As it decomposes, small particles are generated, and the material softens and compacts.

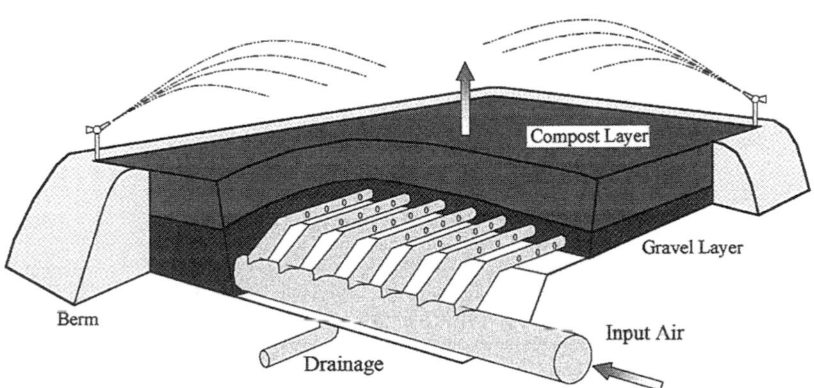

Fig. 2. Schematic of open bed compost biofilter.

Compaction along the vertical axis tends to close the pores, increasing head loss. Horizontal shrinkage creates fissures in the medium and pulls it away from the walls of the vessel. Air is diverted from flowing through the medium to flowing through open channels, and treatment fails.

Compost also shrinks if it dries. Ideally, humidification and irrigation systems should prevent drying, but these occasionally fail. Once the compost has dried, it may become hydrophobic, making rewetting difficult.

The need to keep biofilters moist is a significant design consideration. Large amounts of air are passed through biofilters, so that air with a relative humidity even slightly lower than saturation will cause a slow but steady drying that eventually disrupts operation. Relative humidity is a function of air temperature, and even modest warming will also cause drying (VAN LITH et al., 1997). Such warming can occur because of the pressure increase caused by a blower or because of heat generated in the biofilter by microbial oxidation of contaminants. To prevent problems, incoming air is often passed through a humidifier, typically a packed trickling bed. However, any warming of the air that occurs downstream (which is always the case for metabolic warming) again can reduce relative humidity.

The water content of the porous medium can also be supplemented by direct irrigation, usually through sprinkling the top of the bed. This provides a direct control, but it is sometimes difficult to ensure homogeneity. It is surprisingly difficult to arrange sprinklers to water uniformly, and partial clogging of the sprinkler nozzles commonly disrupts the distribution. Fine mist generators are better because the mist droplets move with the air. Even when water is sprinkled uniformly on the surface of the bed, downward flow tends to "finger", concentrating in streams in some areas and bypassing others. Biofilters are now often operated with both humidifiers and irrigation systems, so that each system can compensate for the shortcomings of the other.

Porous medium biofilters may be operated with either upward or downward airflow. (In a few special designs using other supports such as rotating disks, the air flows horizontally.) Construction costs are lower for upward flow because the biofilter can be uncovered. The bottom of the container that supports the medium also serves to contain the air, so no additional structure is needed. If downward flow is chosen, a cover is necessary to contain the air. Because biofilters are often large, this can add substantially to costs. However, there is a second constraint. Irrigation water must be added from the top so that it will drain downwards to wet the interior of the biofilter. Drying is most severe where the air first enters the medium. In an upflow biofilter, this means that the bottom of the medium is most prone to drying, and that irrigation water reaches this critical zone only after trickling down through 1 m (typically) of porous medium. Trickling is often non-uniform, so maintaining a homogeneous water distribution is difficult. The irrigation water also occupies pore space and flows counter current to the air. In some systems, the resulting backpressure can significantly decrease the air flow during irrigation. If the air and water are added at the top, the incoming air immediately contacts a layer of wet medium, raising its humidity and helping to maintain dampness throughout the bed and little or no added head loss is caused during irrigation.

The difficulties with compost and the desire to produce higher specific activities have lead to the use of mixed or inorganic media. Monsanto (MCGRATH and NIEUWLAND, 1998) used a proprietary mixture of foam plastic beads and organic matter. Several biofilters have been built using lava rock, which is a gravel with large pores. Biotrickling filters are often made with plastic shapes like those used in cooling towers, and biofilters have been filled with large-pore polyurethane foam. The inorganic materials solve the problems of compaction and can provide a highly uniform large pore medium for low head loss. The formed media include far less "dead space" within the particles, and may be much lighter. However, inorganic media provide no nutrients or seed culture. While they are easy to rewet if they are inadvertently dried, their low water holding capacity means that they will dry more easily. In all, the inorganic media hold the potential for high degradation rates and correspondingly smaller biofilters, but the systems must be managed more carefully. An appropriate seed culture must be found and applied uniformly to the medium. Nutrients must be added regularly, and the water content must be carefully controlled.

The most advanced biofilters are thus contained systems with permanent inorganic media and highly capable control systems. Sensors measure moisture in the air and on the medium, input and output contaminant concentrations, leachate pH, head loss, air flow rates, and other parameters (DEVINNY, 1998). A programmable logic controller may monitor biofilter operation, adjust operating parameters, and shut down the system in response to anomalous conditions. These most advanced systems hold the promise for high treatment rates and smaller, better controlled biofilters, but there are substantial increases in costs for equipment, maintenance, and operation. The choice between cheap low-tech systems and more expensive advanced biofilters is still a difficult one.

1.4 Biotrickling Filters

Providing a continuous flow of water over the surface of the porous medium precludes drying and provides a means for precise control of medium pH and nutrient content. As the water is pumped back to the top of the reactor, acids, bases, or nutrients can be added. The amount of water present in the reactor at any time is substantially larger than in a biofilter, allowing larger amounts of soluble contaminants to dissolve. These factors mean that biotrickling filters are often capable of higher specific rates of treatment: more contaminant is degraded per unit volume and the reactors can be smaller. However, the greater mechanical complexity requires more capital investment and maintenance. Biotrickling filters are more likely to be applied where contaminant concentrations are at high levels that would cause clogging in biofilters.

1.5 Applications for Biological Systems

Biological treatment is certainly not appropriate for all applications. Most obviously, microorganism able to transform the contaminant to a harmless product must be available. Commonly, organic contaminants are converted to carbon dioxide and water, but other processes are possible. Many biological systems are used to convert hydrogen sulfide, a strongly odorous compound, to highly soluble sulfate ions that are captured in the water. Mineral dusts cannot be treated, however, and will accumulate to eventually clog the biofilter. Some chlorinated hydrocarbons degrade only very slowly, and biological systems have not been successful in practical application to these vapors.

Contaminants that are very poorly soluble in water are more difficult to treat. A low water–air partition coefficient means concentrations in the water phase, and thus biodegradation rates, will be low (HODGE and DEVINNY, 1995). This restriction applies only to very low solubilities: benzene and toluene, e.g., are commonly thought of as not very soluble in water, but they are readily treated by biofiltration. However, some compounds like isopentane and chlorinated solvents, which combine low solubility and some resistance to biodegradation, are poor candidates for biological treatment.

Biological systems are generally larger than alternatives, and cannot be used where space is limited. Simple compost biofilters are restricted to layers of medium about 1 m thick because greater depths will cause compaction. A high volume biofilter must, therefore, have a large footprint, or bear the structural costs of constructing multiple layers.

Biological systems will be most successful for low concentrations of readily degradable contaminant in large volumes of air.

2 Applications

There are now many hundreds of biofilters in operation worldwide. The following examples are chosen to represent the range of complexity, from the simplest to the most elaborate (DRAGT and VAN HAM, 1991).

2.1 Soil Bed Biofilters (BOHN and BOHN, 1998)

Sunshine Plastics of Montebello, California, was required to control VOC emissions or face pollution penalties. The facilities emit 170 m^3 min^{-1} of waste air containing a mixture of propanol, ethanol and acetone at a total VOC concentration between 100 and 1,000 ppm. The air permit granted by the South Coast Air Quality Management District required a minimum 70% reduction in VOC emissions from the facilities.

A 486 m^3 soil bed biofilter was installed for VOC control by Bohn Biofilter Corporation of Tucson, Arizona (DEVINNY et al., 1999). The biofilter is 90 cm deep and provides an EBRT of 2.5 min. The biofilter was constructed above ground, adjacent to the facility in the parking lot. Concrete traffic dividers were used for the walls of the biofilter. The inlet air to the biofilter is humidified by fogger nozzles in the main air pipe and additional water is added to the bed through sprinklers controlled by timers. The schedule of plant operation is variable. The biofilter operates only while the presses are working, and the blower is on about 80 h per week. Despite the use of a soil bed, the head loss across the bed is relatively low (5 cm) because of the long air residence time. The inlet and outlet concentrations are continuously monitored. Results indicate 95% destruction of VOC in the biofilter, indicating much better removal than required. The biofilter costs approximately $ 78,000. Overall, it is typical of the simple soil bed that is cheap and effective, but large.

2.2 Open Compost Biofilter for Treating Odors from a Livestock Facility (NICOLAI and JANNI, 1998)

Odors from livestock facilities are a significant problem for some communities. Sources of odors include buildings, manure storage, and manure land applications. A set of simple and low cost biofilters was installed at a pig farm to control odors from sow gestation and farrowing (birthing) barns. The odors are caused chiefly by hydrogen sulfide and ammonia.

Readily available equipment and materials were used in the construction of the biofilter. The dimensions of the biofilter were based on available space and acceptable pressure drop across the medium. Standard agricultural ventilation fans were used for the air blowers. They are not designed to operate above a static pressure greater than 62 Pa, and tests determined that a bed depth of 28 cm caused no more than 50 Pa of pressure loss across the medium at the maximum flow rate. Ventilation rates vary throughout the year, depending on the temperature in the barns. The total for the three biofilters ranged from 641 m^3 min^{-1} in winter to 4,634 m^3 min^{-1} in summer. The biofilters were sized to have only a 5 s average EBRT but varied from 18 s in winter to 3 s in summer.

Odors originate from manure that collects in a pit beneath the barn floor. Odorous air is drawn out of the pit by fans located below the floor level. In the summer months, additional wall fans are activated to increase the air replacement rate in the building. Air is transferred through plywood ducts leading to air plenums below the medium. The plenums are constructed of shipping pallets covered with a plastic mesh to prevent the medium from dropping through the pallet openings and clogging the plenum. The medium used is an equal weight of compost (unspecified) and chipped brush. The total surface area of the biofilters was 189 m^2.

The biofilters are effective at removing odors, hydrogen sulfide and ammonia. Odors were reduced by an average of 82% during the first 10 months of operation. Hydrogen sulfide concentrations were decreased by approximately 80% and ammonia concentrations were lowered by 53%. Despite the high flow rates and extremely short detention time in the summer months, the effluent hydrogen sulfide concentration was consistently less than 100 ppb.

The use of available materials made the system very inexpensive. The biofilter cost less than $ 10,000, or approximately $ 0.13 per m^3 s^{-1} of air treated. Operational costs, including rodent control, are estimated to be approximately $ 400 per year.

2.3 Open Bed Compost Biofilter for Wastewater Plant Odor Control (CHITWOOD, 1999)

The Ojai Valley Sanitary District in California operates a 7.9 m^3 min^{-1} wastewater treatment plant that includes a 280 m^2 below grade, open bed biofilter. The biofilter was installed to control odors, chiefly hydrogen sulfide. Builders were particularly concerned about potential complaints from users of an adjacent bicycle path.

The biofilter uses wood chips from lumber waste as the medium. They are strips from 3–30 cm long. The biofilter is designed to treat 225 m^3 min^{-1} of waste air removed from the plant's headworks, grit chamber, and grit classifier. Air is driven by a centrifugal blower designed to deliver 225 m^3 min^{-1} at a pressure of 1,250 Pa. Air is humidified by passing it through a spray chamber and is then distributed through fourteen 25 cm diameter schedule 80 PVC laterals beneath the wood chips (Fig. 2). Each lateral is 15 m long and has a pair of 1.6 cm diameter holes drilled every 15 cm along its length. Pairs of holes were drilled at 90° from each other and the pipe was laid so that each hole faces 45° from the center bottom. The air is thus directed outwards and downwards from the pipe. The laterals are 1.2 m apart and are covered with 15 cm of 2 cm diameter acid resistant, smooth river rock. The depth of the medium above the rock is 0.9 m. 15 cm of chipped bark were added for beautification. Six sprinklers controlled by a timer provide irrigation of the biofilter once daily. No nutrients were added because it was presumed that the organic medium would provide necessary nutrients and the biofilter was not inoculated. Results indicate greater than 99% reduction in hydrogen sulfide with an average inlet concentration of 4.5 ppm. A 70% reduction of total VOC was observed with an average inlet concentration of 5 ppm measured as methane equivalent.

2.4 Inorganic Biofilter for Odor Control at a Wastewater Treatment Facility (DECHANT et al., 1999)

The Cedar Rapids Water Pollution Control Facility in Iowa is designed to treat 2.45 m^3 s^{-1} of municipal and industrial wastewater. The treatment facility serves approximately 145,000 people as well as wet corn milling, pulp paper, and other industries. In the early to mid 1990s the organic loading rate of the facility was substantially increased. Concurrent with this change, odor complaints increased markedly. Studies indicated that the wastewater trickling filters were the major source of odors with a typical hydrogen sulfide concentration of between 50 ppm and 350 ppm. The primary clarifiers and air floatation thickeners also produced odors. In an effort to reduce complaints, several air pollution control technologies were evaluated.

Biofiltration was chosen because the combined capital cost and operating cost (present worth) were significantly less than for chemical scrubbers. Lava rock was chosen for the biofilter medium for its resistance to low pH conditions. Operating costs are expected to be less than those for a compost biofilter because the lava rock is presumed to be permanent. The system includes two biofilters in parallel. Each is 11 m wide and 21 m long. The depth of the medium is 1.8 m. They are designed to treat 4,238 m^3 min^{-1} of waste air from the trickling filter, primary clarifiers and air floatation thickeners. The resulting design EBRT is 15 s.

The biofilters are housed in a concrete structure coated with a polyvinyl chloride liner. The medium is supported by fiberglass reinforced plastic. Air is not humidified before entering the biofilters but the medium is irrigated with 3.78 m^3 min^{-1} of water from the secondary effluent for 10 min every hour. This also serves to wash out excess acid and maintain a bed pH of between 2 and 2. 5 and to add necessary nutrients for the microbial growth. Since the beginning of operation, the average inlet concentrations have been about 200 ppm of hydrogen sulfide with removal efficiencies consistently between 90% and 95%. Measurements show an almost 50% removal of VOC. In the first 6 months of operation there have been 90% fewer odor complaints.

2.5 Biofilter Treating Gasoline Vapor at a Soil Vapor Extraction Site (Wright et al., 1997)

Weathered gasoline vapors from a soil vapor extraction system were treated using a biofilter system in Hayward, California. The system comprised four small biofilters initially operated in parallel and later operated as two sets of in-series biofilter systems. Each of the biofilters was 1.2 m long by 1.2 m wide with a bed depth of 95 cm. The medium was a mixture of compost from sludge solids and wood products mixed with an equal portion of perlite used as a bulking agent. Crushed oyster shells were used as a buffer. The EBRT was approximately 2 min during the parallel stage of the remediation and 1 min when operated in series. The relative humidity of the inlet air was increased by passing the air through a humidification chamber fitted with fogger nozzles. Excess water was recycled. As is typical of a soil vapor extraction system the inlet concentration was highly variable. Initially the inlet concentration was 2.7 g m^{-3} as total petroleum hydrocarbon (TPH), and decreased to 0.9 g m^{-3} by day 22 of the project. The final inlet concentration during the course of the study was 0.4 g m^{-3}.

Bed drying was a notable problem. However, after operators thoroughly mixed the bed, rewetted the medium and initiated regular direct irrigation, biofilter performance increased. The authors suggest that an improved inoculation would have helped to shorten the adaptation time and that humidification of the air should be supplemented with periodic direct application of water to the bed by either soaker hoses or a sprinkler system.

The sustained removal efficiency was 90% of the total petroleum hydrocarbons at an EBRT of 1.8 min and an inlet concentration of 0.4 g m^{-3} was observed. Removal efficiencies for the BTEX compounds were even better. However, some compounds (believed to be methyl substituted alkanes and cycloalkanes) where more poorly removed, probably because of their more recalcitrant nature.

2.6 Biofilter Treating VOC Emissions from an Optical Lens Manufacturer (Standefer et al., 1999)

A major optical lens manufacturer in Massachusetts emits a proprietary mixture of VOC containing alcohols and ketones from its coating process. The facility releases 127.5 m^3 min^{-1} of waste air with a peak concentration of approximately 0.2 g m^{-3} (1.55 kg h^{-1}). The purpose of VOC reduction was to assure compliance with the Massachusetts Department of Environmental Protection requirements and to maintain their minor source designation under the Clean Air Act. To meet these requirements, a 90% removal efficiency goal was set.

A biofilter was chosen over other technologies because of its relatively low capital and operating costs and restrictions on water consumption and wastewater disposal. A 3.4 m wide by 13.4 m long (45 m^2) biofilter was designed, constructed and installed by PPC Biofilter. To prevent drying of the medium bed, the relative humidity of the air is raised from 6% to >95% by a 2.8 m^3 counter current packed tower filled with 5 cm pall rings. The biofilter is a downflow, induced draft (negative pressure) system, with a proprietary medium 1.5 m in depth allowing a 27s EBRT at an approach velocity of 6 cm s^{-1}. PPC Biofilter research shows that typically 50–70% of overall removal occurs in the first 35–45 cm of a biofilter bed. Because oxidation of the VOC compounds is an exothermic reaction, down flow allows application of irrigation water where the highest consumption of water occurs.

A programmable logic controller (PLC) managed supplemental water addition. Water was added when the weight of the medium as measured by load cells was below a set level. The PLC activates a solenoid valve that allows irrigation of the bed by 27 fine mist nozzles. The same PLC allows all pertinent data to be logged via a personal computer.

The biofilter has met the removal efficiency goals. Independent results demonstrate that the biofilter has removed 93% of the alcohols and 82% of the ketones with a total VOC removal of more than 91% in the first two years of operation. The hazardous air pollutant of concern has had an average removal efficiency of 97%.

2.7 Advanced Biofilter for Controlling Styrene Emissions (PUNTI, personal communication; THISSEN, 1997)

The Synergy Biofilter system, manufactured by Otto Umwelttechnik in Germany and supplied by Biorem Technologies, Inc. in the United States, is an example of a technologically advanced system combining a biofilter and an adsorber. The system treats styrene emissions from a manufacturer in Germany. During working hours, the biofilter treats 84,000 m^3 h^{-1} of air containing up to 160 mg m^3 of styrene and reduces concentrations to less than 30 mg m^3, a removal efficiency of 81%. Considerable amounts of acetone are also present. The system includes humidification of the air stream and direct irrigation of the biofilter beds with find mist (Fig. 3). The microbial support medium are pellets of carbon coated concrete, which are adsorbent and essentially permanent.

An adsorber was installed downstream of the biofilter bed. This eased requirements for treatment efficiency in the biofilter, allowing use of a smaller bed. The surface load is 371 m^3 m^{-3} h^{-1}, for an empty bed detention time of only about 10 s. During non-working hours, a smaller flow of air is heated and passed through the adsorber to remove accumulated styrene, then returned to the biofilter. Some makeup air from the facility is continuously added to cool the air and to remove residual styrene vapors from the factory work area. This approach also supplies small amounts of styrene to the biomass in the bed to reduce losses of activity during non-working hours.

The system includes monitors for input and output concentrations, temperature, pH, airflow, and several other system parameters. The sprinkler system is activated automatically on a predetermined schedule.

3 References

BOHN, H.-L., BOHN, K.-H. (1998), Accurate monitoring of open biofilters, *Proc. 1998 Conf. Biofiltration (an Air Pollution Control Technology)*, The Reynolds Group, Tustin, CA, October 22–23, 1998. pp. 9–14.

CHITWOOD, D. E. (1999), Two-stage biofiltration for treatment of POTW off-gases, *Thesis*, University of Southern California, Los Angeles, CA.

DECHANT, D., BALL, P., HATCH, C. (1999), Full-scale validation of emerging bioscrubber technology for odor control, in: *Proc. Water Environ. Fed. 72nd Ann. Conf. Exposition*, New Orleans.

DEVINNY, J. S. (1998), Monitoring biofilters used for air pollution control. Practice Periodical of Hazardous, Toxic, and Radioactive Waste Management, *Am. Soc. Civil Eng.* **2**, 78–85.

DEVINNY, J. S., DESHUSSES, M. A., WEBSTER, T. S. (1999), *Biofiltration for Air Pollution Control.*

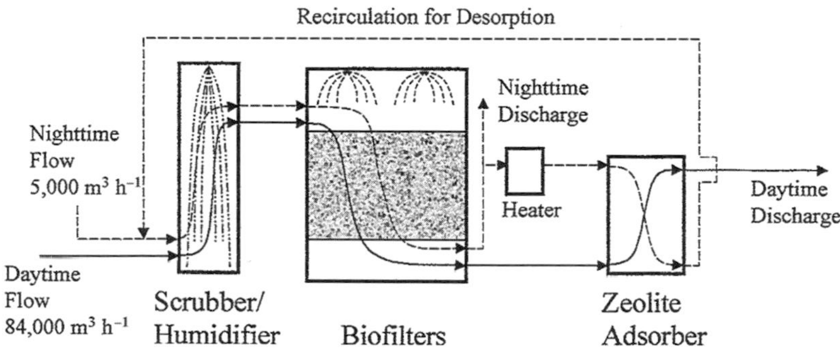

Fig. 3. Biofilter system with adsorber.

Boca Raton, FL: CRC Lewis Publishers.

DRAGT, A. J., VAN HAM, J. (Eds.) (1991), Biotechniques for air pollution abatement and odor control policies, *Proc. Int. Symp.,* Maastricht, The Netherlands, 27–29 October. Amsterdam: Elsevier.

HODGE, D. S., DEVINNY, J. S. (1995), Modeling removal of air contaminants by biofiltration, *J. Environ. Eng.* **121**, 21–32.

MCGRATH, M. S., NIEUWLAND, J. C. (1998), Case study: effectively treating high levels of VOCs using biofiltration, in: *Proc. 1998 Conf. Biofiltration (an Air Pollution Control Technology)*, The Reynolds Group, Tustin, CA.

NICOLAI, R., JANNI, K. (1998), Biofiltration – adaptation to livestock facilities, in: *Proc. 1998 Conf. Biofiltration (an Air Pollution Control Technology)*, The Reynolds Group, Tustin, CA.

OTTENGRAF, S. P. P. (1986), Exhaust gas purification, in: *Biotechnology 1st Edn.*, Vol. 8 (REHM, H.-J., REED, G., Eds.), pp. 425–452. Weinheim: VCH.

STANDEFER, S., WILLINGHAM, R., DAHLSTROM, R. (1999), Commercial biofilter applied to an optic lens manufacturer to abate VOCs, in: *Proc. 92nd Ann. Meeting Exhibition Air Waste Management Ass.*, St. Louis, M.

THISSEN, I. N. (1997), Biological treatment of exhaust air. Practical experience with combined processes for treatment of organic solvents, *Fachmagazin für Kreislaufwirtschaft, Abwassertechnik und Luftreinigung*.

VAN LITH, C., LESON, G., MICHELSEN, R. (1997), Evaluating design options for biofilters, *J. Air Waste Managem. Ass.* **47**, 37.

WRIGHT, W. F., SCHROEDER, E. D., CHANG, P. Y., ROMSTAD, K. (1997), Performance of a pilot-scale compost biofilter treating gasoline vapor, *J. Environ. Eng.* **123**, 547–555.

III Preparation of Drinking Water

18 Potable Water Treatment

ROLF GIMBEL
HANS-JOACHIM MÄLZER
Mülheim, Germany

1 Introduction

Water is an elementary and essential component of all processes of life on our blue planet. In spite of its relatively basic chemical structure water has a lot of exceptional properties establishing its central role in nature. Water is used as a reaction and transport medium in the range from microscopic to macroscopic dimensions, as a carrier of energy or as food.

Especially as drinking water for human use, water is called the "number one food", which cannot be substituted by any other substance. Due to this special utilization the demands on drinking water quality must be correspondingly high. Besides temperature and possibly necessary pressure on the distribution system these demands are particularly on kind and quantity, i.e., concentration of water contents. Principally an enormous amount of water on earth exists as shown in Fig. 1. The predominating part of the whole global water occurrence exists as seawater or polar and glacier ice which only can be used as drinking water with unusually high expenditures such as desalination of seawater or transport of icebergs.

The relatively small amount of the so called "fresh water" (about 0.3%) remains in a natural water cycle driven by solar radiation and is used by humans, animals, and plants in most diverse ways. As a result, many different substances enter the water cycle. These substances can be desired, harmless, disturbing, or of potential concern to human health if the water is used as drinking water. Disturbing or unhealthy contaminants in the water can possibly be eliminated by a self-cleaning mechanism of the natural water cycle so that the water could be used as drinking water without any special treatment. This effect can be achieved by drinking water obtained from groundwater with sufficiently protected aquifers. In many cases water which is taken from the natural water cycle for the use as drinking water has to be treated with suitable processes to eliminate disturbing and harmful contaminants. These contaminants can be of natural origin, but increasingly they are of anthropogenic origin such as wastewater inlets. This fact usually leads to special efforts in drinking water treatment processes for the elimination of contaminants which cannot be eliminated via so called "natural treatment processes". For this reason protection of water resources becomes more and more important. Generally, sustainable management of our aquatic ecological system is demanded and, furthermore, we have to make sure that future generations will be able to obtain drinking water from the resource of fresh water.

The sustainable use of the resource water requires that drinking water is not wasted and also that water is used sensibly and carefully for domestic, small and large scale industrial

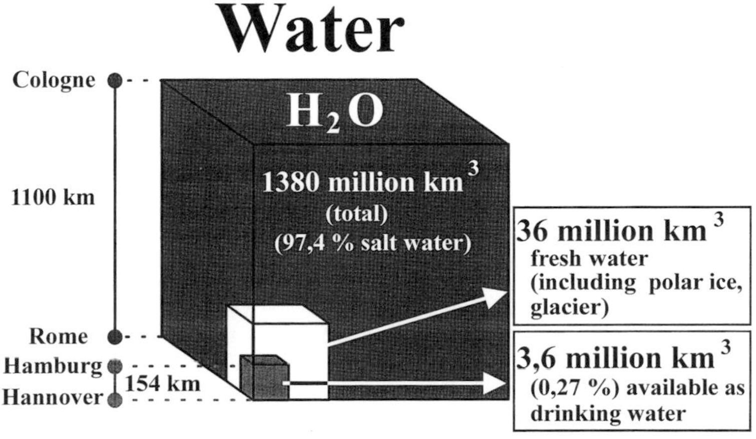

Fig. 1. Worldwide amounts of water.

purposes. Regional features, such as the natural water quantity and quality, play major roles and must be considered. Nevertheless, the consumption behavior in households and small scale industries of fundamentally water-poorer regions (e.g., South Africa, Spain, Australia), which is shown in Tab. 1, indicates that these regions do not necessarily have the lowest per capita consumption.

Furthermore, it has to be considered that water of lower quality, e.g., rain water which is collected from roof drains should be used for limited purposes in households only. This is because such low quality waters usually do not follow the strict quality standards for physicochemical and microbiological parameters applied for drinking water.

2 Demands on Drinking Water Quality and Treatment

Drinking water is a water which must be suitable as food for direct consumption by humans. Furthermore, various additional domestic utilizations of water require a quality corresponding to the quality of drinking water. This applies with priority to water, which is used for the preparation of meals and beverages as well as for the cleaning of dishes and kitchenware, etc. Also, water which is used for body hygiene and for laundry should possess the quality of drinking water, if possible.

Tab. 1. Public Water Supply (Households and Small Scale Industries) in Different Industrial Nations (BGW-Bundesverband der deutschen Gas- und Wasserwirtschaft, 1996)

Nation	Daily per Head Consumption	Percentage Amount of Groundwater
Australia	316 L	no data
Austria	170 L	99%
Belgium	120 L	no data
Denmark	155 L	95%
France	157 L	63%
Germany	132 L	73%
Hungary	121 L	91%
Italy	251 L	88%
Luxembourg	178 L	no data
Norway	180 L	15%
South Africa	276 L	no data
Spain	210 L	30%
Sweden	203 L	24%
Switzerland	242 L	no data

In a global view, the demands on potable water quality are controlled by a large number of national (e.g., German Drinking Water Regulation: Verordnung über Trinkwasser und über Wasser für Lebensmittelbetriebe, 1990), German Industrial Standard DIN 2000 (DIN 2000 – Zentrale Trinkwasserversorgung, 1973), multinational (e.g., European Union Guideline: Council Directive 80/778/EEC, 1980; Council Directive 98/33/EC, 1998), and international (e.g., WHO Guideline: Guidelines for Drinking Water Quality: Health Criteria and Other Supporting Information, 1996) rules or guidelines. National regulations can exhibit certain differences among each other. In the same way as multinational and international regulations they are subject to a constant revision due to new scientific findings regarding health-relevant substances in water. However, some guiding principles of the German Industrial Standard DIN 2000 (DIN 2000 – Zentrale Trinkwasserversorgung, 1973) can be used exemplarily as a general orientation regarding the demands on water quality:

- The demands on potable water quality must be oriented towards the properties of a groundwater of perfect condition, which is taken from the natural water cycle from a sufficient depth and after the passage of sufficiently filtering aquifer layers.
- Drinking water must be free of pathogenic germs and must not be harmful to health.

Taking into account the general rule that in nature a concentration of exactly zero is nearly impossible and that such a value cannot be controlled, this statement means that especially a microbiological quality is demanded so that diseases are not caused and spreading of epidemics is avoided. Therefore, pathogenic virus like, e.g., poliomyelitis and hepatitis virus causing polio and jaundice, pathogenic bacteria like, e.g., *Salmonella* and *Shigella* causing typhus and dysentery or parasites like, e.g.,

Cryptosporidium and *Giardia* must not be contained in the drinking water or have to be kept below a threshold value corresponding to an acceptable health risk. Furthermore, potable water must be constituted in such a way that it does not cause acute or chronic health damage after lifelong consumption and use.

- Drinking water should be poor in germs.

However, it is not demanded that drinking water should be free of germs or should be sterile.

- Drinking water should be appetizing and tempting for consumption. It should be free of color and odor and should be clear, cold, and of good taste.

The meaning of this point is further extended to raw water and the way of water catchment, which should not be disgusting in any way.

- The content of dissolved substances should stay within certain limits.

Here, the major water constituents are mentioned, but the limits are not defined.

- Drinking water and the materials in contact with it should be matched to avoid corrosion.

Especially the optimum materials should be chosen to avoid corrosion, but the water must also fit to some quality demands and should not be, e.g., very aggressive to calcium carbonate.

- Drinking water should be provided in sufficient amounts and with sufficient pressure.

This demand should assure that, e.g., washing, cleaning, and food preparation is also done with drinking water and that a high hygienic standard can be achieved and kept. A low or changing pressure in the distribution net should be avoided to prevent resuction and contamination of the water in the pipe.

The limits mentioned in the DIN 2000 are further defined in the German Drinking Water Regulation (Verordnung über Trinkwasser und über Wasser für Lebensmittelbetriebe, Trinkwasserverordnung – TrinkwV, 1990). At present this regulation is in accordance with the Drinking Water Guideline of the European Community published in 1980 (Council Directive 80/778/EEC, 1980).

An overview of the German regulation is given in Tab. 2. Limits are fixed for microbiological, chemical, and physical parameters. The microorganisms listed under the microbiological demands are indicator organisms, which must not be pathogens, but which indicate a possible pollution of the drinking water. Additionally, guiding values are given in this decree for colony counts and for sulfite reducing clostridia. The chemical-physical demands define the limits due to toxicological and due to other but non toxicological reasons. The reasons for these limits are aesthetic (e.g., for color, turbidity, odor, Fe, Mn, Al), but the excess of these limits may also indicate pollution of the raw water or insufficient treatment or disturbance of the treatment process.

The pH value is of special interest to avoid corrosion in the pipes and must be kept between 6.5 and 9.5. The pH must be adjusted not to dissolve calcium carbonate, if metal or cement materials are used. The limits of chlo-

Tab. 2. Drinking Water Quality Demands in Germany on the Basis of the Regulation from 1990 (Verordnung über Trinkwasser und über Wasser für Lebensmittelbetriebe, Trinkwasserverordnung – TrinkwV, 1990) which is in Accordance with the EC Drinking Water Guideline from 1980 (Council Directive 80/778/EEC, 1980)

Drinking water decree/according to EC drinking water guidelines)
– Microbiological demands
Escherichia coli (0 in 100 mL)
Coliform germs (0 in 100 mL)
Fecal *Streptococcus* (0 in 100 mL)
– Chemical-physical demands
Limits for toxicological reasons:
As, Pb, Cd, CN^-, F^-, Ni, No_3, NO_2^-, Hg
Polycyclic aromatic hydrocarbons
Organic chlorinated hydrocarbons
Pesticides
Sb, Se
Limits without toxicological reasons:
Sensoric parameters:
Color, turbidity, odor
Physico-chemical parameters:
Temperature, pH, conductivity, oxygen demand
Chemical parameters:
Al, NH_4^+, Ba, B, Ca, Cl, Fe, K, N, Mg, Mn, Na, phenol, PO_4^{3-}, Ag, SO_4^{2-}, hydrocarbons, chloroform extractable substances, surface active agents

ride and sulfate may be also seen in this context.

Turbidity is a parameter of high importance, because turbidity may indicate pollution of the drinking water by microorganisms, which may be attached to particulate substances. The turbidity limit is fixed to 1.5 FNU in the German Drinking Water Regulation (Verordnung über Trinkwasser und über Wasser für Lebensmittelbetriebe, Trinkwasserverordnung – TrinkwV, 1990), but experience has shown that a lower turbidity should be maintained at the effluent of the water treatment plant, especially if surface water or groundwater influenced by surface water is treated. For such conditions an additional recommendation is given by the Drinking Water Commission (Anforderung an die Aufbereitung von Oberflächenwässern zu Trinkwasser im Hinblick auf die Eliminierung von Parasiten, 1997) in Germany, in which the turbidity value of a drinking water is suggested with a maximum of 0.2 FNU. In an international report under the participation of numerous members of the International Water Services Association (IWSA) it is stated, that in most cases a turbidity <0.1 FNU should be aimed at for potable water (Gimbel et al., 1997). Furthermore, relatively high turbidity will result in a high consumption of disinfectants and in a high production of disinfection by-products.

Recently, a new drinking water guideline (Council Directive 98/33/EC, 1998) of the EC was published. As in other member states of the EC, the German Drinking Water Regulation (Verordnung über Trinkwasser und über Wasser für Lebensmittelbetriebe, Trinkwasserverordnung – TrinkwV, 1990) is reworked at the moment in order to adapt it to the new EC Guideline.

The ideal state of a groundwater, as it is represented for instance in the German Industrial Standard DIN 2000 (DIN 2000 – Zentrale Trinkwasserversorgung, 1973), obtained from a sufficiently protected aquifer can be found only rarely in nature and becomes even rarer by the general increase of contamination of our environment. But, also water with nearly no anthropogenic contamination often needs to be treated before it can be used as drinking water with the required quality and sufficient security. Examples are calcium carbonate dissolving groundwater and spring water, Fe and Mn containing groundwater, and in particular suspended solids of inorganic and organic species in surface water (e.g., rivers, lakes, and reservoirs). The treatment problems occurring with those waters can usually be solved with nature-near conventional treatment processes like groundwater enrichment, river bank filtration, aeration, sedimentation, slow sand filtration, and rapid sand filtration. Biological processes often play a major role with these treatment processes.

However, special demands on potable water treatment are required by the increase of raw water contamination of anthropogenic origin. This increased contamination can be detected regarding dissolved inorganic and organic water contaminants as well as microbiological parameters. Examples for these are:

- Effects from untreated or not sufficiently treated sewage which is discharged into surface waters by municipalities, industries, and intensive livestock farming. Here, much too high concentrations of undissolved water contaminants (e.g., bacteria, viruses, parasites) and dissolved water contaminants (e.g., heavy metals, nutrients for eutrophication like phosphate, non-natural organic contaminants, xenobiotics) can occur. In this case, occurring shock loads can represent a special problem for potable water treatment.
- Algae and algae-borne substances, which can appear with unnaturally high concentrations due to eutrophication processes especially in standing or very slow floating surface waters.
- Increasing hardness of groundwaters (calcium, magnesium) due to acidic precipitation released by waste gas in the atmosphere.
- Pesticides and fertilizer residues in groundwater and surface waters due to agricultural use.

To treat such raw water loads safely in water treatment plants, even under extreme circumstances like shock loads and varying water demands, it usually requires physicochemical treatment technologies to guarantee a water quality which corresponds to the strict quality requirements of drinking water. Such physicochemical treatment technologies may be precipitation, flocculation, flotation, oxidation with ozone, adsorption on activated carbon, membrane processes, and ion exchange.

As it is schematically shown in Fig. 2, each raw water consists of a mixture of pure water and different, partially unwanted substances. Usually, energy and more or less large amounts of additives need to be added during potable water treatment, which should be understood to include possible necessary disinfection. The result is a qualitative and quantitative change of the substances in the water, so that only those substances should remain in the product "drinking water" which are desired or at least not disturbing.

At first it has to be pointed out that disturbing substances can enter the drinking water during the treatment process because of impurities in the additives themselves. These substances can be called by-products in a broader sense. This can be of importance, e.g., using iron or aluminium salts as coagulants and flocculants, because they are partially obtained as residuals of industrial processes and used in the potable water treatment process after suitable preparation steps. Potentially related problems are to be considered in general by a water works operator. Thus, article 10 (Quality assurance of treatment, equipment and materials) of the European Union Drinking Water Guideline (Council Directive 98/33/EC, 1998) contains following statement:

"Member States shall take all measures necessary to ensure that no substances or materials for new installations used in the preparation or distribution of water intended for human consumption or impurities associated with such substances or materials for new installations remain in water intended for human consumption in concentrations higher than is necessary for the purpose of their use and do not, either directly or indirectly, reduce the protection of human health provided for in this directive ...".

Similar requirements were embodied in national regulations in former times. For example, the German Drinking Water Regulation of 1990 contains a comprehensive regulation concerning species and maximum permissible amount of additives used in water treatment.

New substances can be created by chemical or biochemical conversions depending upon the treatment process. These newly created substances can be called by-products in a strict sense. Typical examples for these are organic chlorine compounds (in particular trihalomethanes, THM), which result during oxidation and disinfection with chlorine. Bromate is a possible by-product of the oxidation with ozone. Tolerable concentrations of such by-products are controlled, e.g., in the Drinking Water Guideline of the EC and in the German Drinking Water Regulation (Verordnung über Trinkwasser und über Wasser für Lebensmittelbetriebe, Trinkwasserverordnung – TrinkwV 1990). Not controlled are the tolerable concentrations of bioavailable organic compounds,

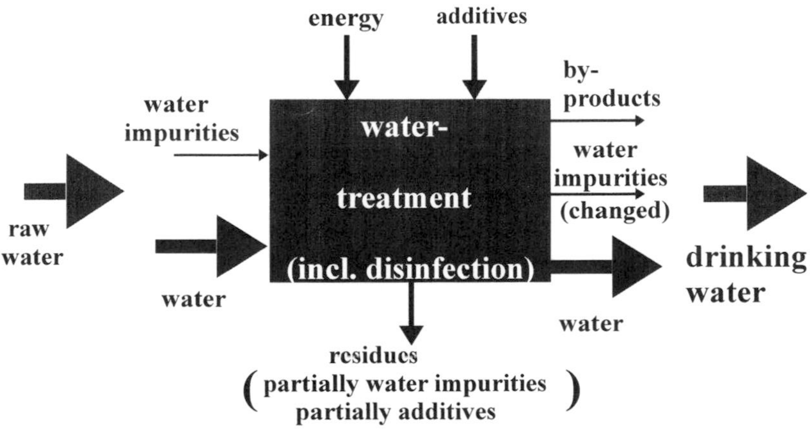

Fig. 2. Fluxes of substances during potable water treatment.

for instance (Council Directive 98/33/EC 1998). Such compounds can result from relatively bioresistant compounds during the treatment with ozone or chlorine and they can lead to an undesired bacterial aftergrowth in the drinking water in the distribution system.

Residues which are accumulated during the water treatment consist at least of substances removed out of the raw water and in many cases by the far predominant part of which additives consist. Especially the residues resulting from the treatment of surface water with flocculation and filtration (mainly sludge) lead to increasing problems in several countries. Thus, e.g., the aim of modern water treatment in Germany, which should be poor of residues and disturbing substances, is that the residues possibly consist only of substances which had to be removed from raw water. Furthermore, there should possibly be no substances in the drinking water from contaminated additives or by-products resulting from the treatment process.

Further to the minimization of residues and by-products, the safe retention or inactivation of pathogens is at present subject of intensive considerations and examinations of consisting treatment facilities. Special cause for this are the epidemic diseases which have recently occurred in several countries, which were initiated in many cases by *Cryptosporidium* and *Giardia* in drinking water.

Especially with the last examples it becomes clear that in principle water treatment has to prefer the elimination of turbidity causing substances (including pathogens) rather than the inactivation of pathogens (e.g., by chlorine addition). In particular, parasites like *Cryptosporidium* and *Giardia* cannot be inactivated by the usual chlorine addition, although this can be entirely possible in practice for the so-called indicating parameters like *E. coli* number of germs and coliform germs. Therefore, an examination for these indicating parameters can only pretend to indicate a well disinfected potable water.

Beyond that one has to consider the possibly enormous formation of health precarious by-products, if substantial chlorine amounts were used for the oxidation or disinfection of undesired water contaminants. However, one should not refuse the use of chlorine or compounds of chlorine (e.g., chlorine dioxide, chloramine, hypochlorite), if a hygienically safe drinking water supply can be guaranteed only in this way. In such cases, one may consider the formation of disinfection by-products as secondary.

Generally, however, one should always try to keep the expenditures of potable water treatment as low as possible, which can be achieved by optimal protection of raw water. The highest safety of the consumer that he will be supplied at any time with perfect drinking water can be guaranteed in this way.

3 Types of Raw Water and Treatment Processes

3.1 Groundwater Treatment

Due to many different local circumstances when groundwater is formed, very different types of groundwater which usually require very diverse processes of treatment for perfect drinking water are received. Here, the following spheres of influence are important (Fig. 3):

- Load of precipitation in the atmosphere (e.g., CO_2, SO_x, NO_x, aerosols).
- Physical, chemical, and microbiological processes in the layer of humus and in underlaying unsaturated region of percolating water. Hence, especially relatively high concentrations of CO_2 can be originated in water which may lead subsequently to a high hardness of water (contents of Ca^{2+} and Mg^{2+} ions) in the lower aquifer.
- Physicochemical and microbiological processes when the water flows through the lower aquifer in saturated ranges, including mixing processes with water of other origins and compositions. The important factors are summarized schematically in Fig. 3, whereas the present local geogeneous conditions can be exceptionally different.
- Additional load of the groundwater by direct anthropogenic influences (e.g., sewage infiltration, percolating water from not sufficiently sealed waste disposals, inappropriate use of organic solvents, effects due to intensive

agricultural use of pesticides and fertilizers, use of herbicides).

In general, the following processes to treat groundwater are mainly used as single steps or, if necessary, in combination with others.

• Aeration to reduce the concentrations of CO_2 (eventually CH_4, H_2S) and to raise the oxygen concentration.

An excess of dissolved gases like carbon dioxide may occur due to previous biological activity. If the groundwater contains high amounts of biodegradable organic substances the oxygen concentration in the groundwater may be about zero and methane and in some cases hydrogen sulfide may occur in the groundwater due to microbiological activity and must be reduced by stripping. Filtration over a calcareous material will be also possible to reduce an excess of carbon dioxide. If high amounts of oxygen are needed, oxygen may be directly added instead of aeration.

• Oxidation (eventually flocculation and sedimentation or flotation) and filtration to reduce iron and manganese concentration. Under anaerobic conditions the content of iron and manganese in the groundwater (as Fe^{2+} or Mn^{2+}) may amount to about 50 mg L^{-1} depending on the specific conditions, including the minerals of the aquifer.

Iron and manganese may be contained in quite low concentrations in the groundwater, if aerobic conditions predominate, because they will be almost fixed in the aquifer as oxides, oxide hydrates, or carbonates. If strongly anaerobic conditions predominate, iron and manganese may also be contained only in low concentrations, because they may be fixed as sulfides in the aquifer. The oxidation can be supported by addition of chemical oxidants and the filtration may be supported by a previous flocculation/sedimentation or flotation step.

• Biodegradation and filtration to reduce DOC (dissolved organic carbon) and ammonium concentration.

If the biodegradable DOC is high, it should be reduced in the treatment process to avoid regermination of the drinking water in the distribution system and to avoid an excess concentration of disinfection by-products. These by-products are especially forced, if one tries

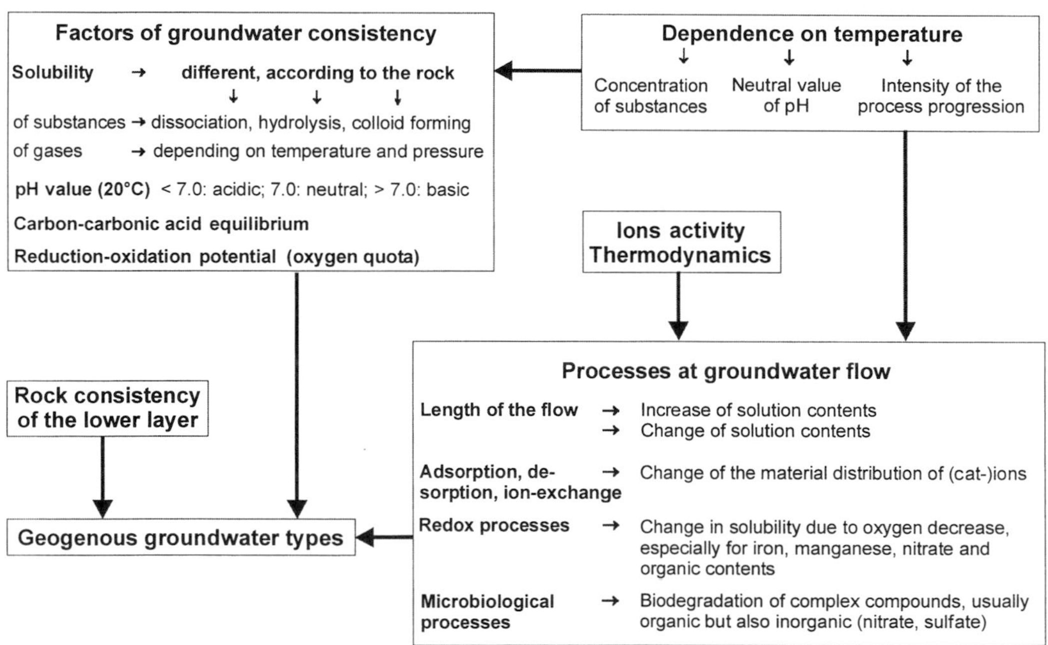

Fig. 3. Factors and processes of geogenous groundwater consistency (HÖLTING, 1991).

to avoid regermination by adding relatively high amounts of chlorine (eventually as post-chlorination of the distribution system). If the water contains ammonium due to anaerobic conditions in the aquifer, a nitrification should be carried out by biologically active filters. Breakpoint chlorination to oxidize ammonium should be avoided, again due to the formation of by-products. To some extent the elimination of DOC in a flocculation/sedimentation or flotation step will also be possible.

- Denitrification and filtration to reduce nitrate concentration.

Nitrate occurring in groundwater mainly originates from fertilizers used in agriculture, and there are many cases where it has to be reduced to a limit according to the drinking water regulations. The German decree, e.g., limits the NO_3 concentration to 50 mg L^{-1}. If it is not possible to meet the threshold value by operating the abstraction wells in an appropriate manner (if there are wells with quite low nitrate contents) biodegradation processes under anaerobic conditions are used and in some cases even membrane or ion exchange processes.

- Treatment with activated carbon to reduce dissolved natural organic matter (NOM).

Depending on the side-specific conditions the groundwater may contain relatively high NOM values. If there is a need to add chlorine to the water, NOM may form high concentrations of undesired halogenated by-products. Besides flocculation or eventually membrane filtration (especially nanofiltration) the use of powdered activated carbon (in many cases in combination with a flocculation step) or filtration through granular activated carbon is common.

- Activated carbon filtration to reduce organic (micro)pollutants.

Organic (micro)pollutants may be contained in the groundwater due to human activities. In many cases such pollutants, which may occur in the concentration range of only some ng L^{-1} or some µg L^{-1}, can be classified as xenobiotics, i.e., they are not of natural origin. Pesticides may occur in agriculturally used areas, organic pollutants like, e.g., chlorinated or unchlorinated hydrocarbons occur from industrial or urban sources of pollution.

- Softening to reduce the hardness.

Depending on the composition of the aquifer the groundwater may contain too much hardness, especially if the aquifer material contains calcareous components and if strong biodegradation processes (production of CO_2) have taken place previously. In most cases softening is done by decarbonization (e.g., precipitation of calcium carbonate) with addition of lye (lime or caustic soda). In recent times membrane filtration as a so called nanofiltration (NF) is increasingly used. Besides effective softening also very good eliminations of sulfate and organic matter (NOM and xenobiotics) can be achieved using nanofiltration.

- Dosage of Na(OH) or $Ca(OH)_2$ to adjust the pH value.

Due to the raw water quality or to the treatment process the water may (still) have the tendency to dissolve or to precipitate calcium carbonate. Both effects are undesired and may damage the water distribution system. Anyway, the pH must be kept between 6.5 and 9.5 if, e.g., the EC Guideline should apply.

3.2 Treatment of River Water

In Germany, e.g., the use of river water for drinking water supply was already quite common in the last century, especially in big cities and regions of high population densities. At the end of the last century, due to insufficient water treatment, several cholera and typhus epidemics broke out in some of these cities and regions. Today, German water works treating river water use bank filtration or groundwater recharge, slow sand filtration respectively, which are very safe and efficient processes for the elimination of pathogenic microorganisms. In the last decades people became more and more conscious of the possible disadvantages of using groundwater, because the water level may be lowered if too much water is taken from the aquifer for too long times. In parallel the river water quality in Germany could be improved significantly in the last years by operating wastewater treatment plants. The use of river water became more common for drinking water treatment, especially in regions of high population densities and low groundwater resources.

Yet, in most cases river water is not used directly, but bank filtration or groundwater recharge is carried out as shown in Fig. 4. In bank filtration the abstraction wells are usually located in a row parallel to the river. The river water infiltrates into the aquifer and is taken out after a subsoil passage. In groundwater recharge the water is taken directly from the river. After pretreatment it is infiltrated again into the aquifer. The infiltration can be done by wells or by basins which act like slow sand filters. By re-infiltration the aquifer is recharged and the abstraction well is protected against possible shock loads of pollutants in the river. The subsoil passage of the water acts like a natural purification step. The main processes in the subsoil are:

- diffusion and dispersion,
- mixing,
- sorption,
- ion exchange,
- solution and precipitation,
- hydrolyzation,
- biodegradation,
- equalization of temperature,
- filtration,
- elimination of microorganisms in general,
- elimination of pathogenic microorganisms.

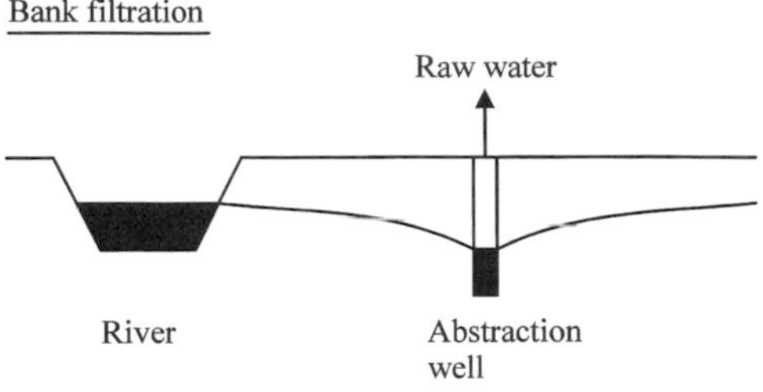

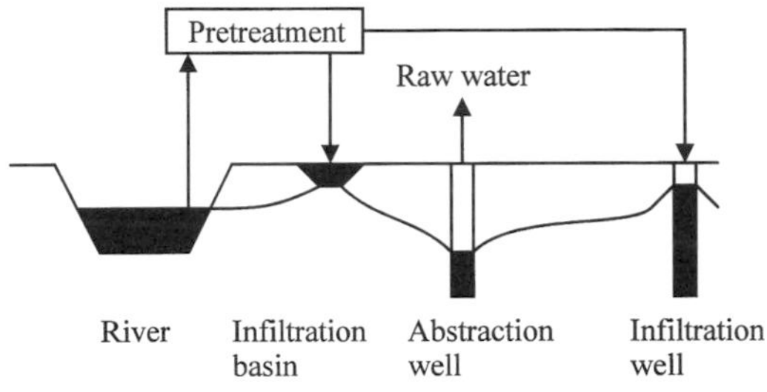

Fig. 4. Bank filtration and groundwater recharge.

During subsoil passage concentrations of pollutants in the river water may be generally reduced by diffusion, dispersion, and by mixing with unpolluted groundwater in the aquifer or in the abstraction well. Furthermore, inorganic pollutants may be reduced by sorption, precipitation, and ion exchange. Organic pollutants may be reduced by sorption or biodegradation. Changes of the temperature of the river water are equalized and particulate substances are eliminated by filtration. Especially the concentrations of pathogenic microorganisms are effectively reduced by the subsoil passage.

In the treatment of river water turbidities must be eliminated. These turbidities may consist of inorganic and organic colloidal or particulate matter which may be partially biodegradable. Especially in winter when the water temperature may be quite low nitrification processes become very slow and the river water often contains ammonium, which must be oxidized in the treatment process. In summer the oxygen concentration in the river water may decrease due to the lower oxygen solubility at higher temperatures. Anaerobic zones may occur in the aquifer due to a high oxygen consumption for nitrification and the biodegradation of dissolved and particular organic matter causing denitrification and a solution of iron and manganese, if they are contained in the aquifer. Due to the biodegradation processes in the aquifer the water may contain excess carbon dioxide and decarbonization may be necessary. As the river is exposed to potential pollution by wastewater compounds, by accidental spills or runoff, organic micropollutants may be contained in the water or their sudden occurrence must be expected. There are many possibilities to cover all the needed treatment steps and their combination depending on the compositions and the variation of the river water compounds. In Fig. 5 some examples for bank filtration post-treatment and groundwater recharge pre- and post-treatment are given.

Bank filtration	Groundwater recharge (Mülheim Process)	
Bank filtration	River	
Abstraction well	Preozonation	
Ozonation	Flocculation	
DM-Filtration	Sedimentation	
GAC-Filtration	Main ozonation	
Disinfection	DM-Filtration	
pH-Adjustment	GAC-Filtration	
	Infiltration	Ground-water recharge
	Abstraction well	
	Disinfection	
	pH-Adjustment	

Fig. 5. Examples of process schemes for bank filtration and groundwater recharge treatment.

In the bank filtration process the river bed acts like a self cleaning crossflow filter and turbidities are reduced very effectively. Subsoil passage acts like a first treatment step and parasites (e.g., *Cryptosporidium* and *Giardia*), bacteria, viruses, and biodegradable substances are reduced during bank filtration until the water is taken out at the abstraction well. The degree of elimination strongly depends on the characteristics of the aquifer and the hydraulic conditions including the water level in the river and in the aquifer and the operating conditions of the abstraction pump. In the pre-treatment process, which may start with an ozonation step, organic compounds are oxidized, but they are not mineralized, just the biodegradability is improved. In the following dual media filtration step iron, manganese and residual turbidities are reduced, ammonium is oxidized, and mainly due to biodegradation dissolved organic carbon is also reduced to some extent. In the granular activated carbon (GAC) filter organic substances are further reduced by adsorption and by biodegradation. After disinfection and pH adjustment the water is distributed as drinking water.

In the groundwater recharge process, shown as an example in Fig. 5, the water is taken from the river and pre-ozonated to improve the following flocculation of turbidities. After sedimentation a main ozonation is carried out to improve the biodegradability of organic substances. Both ozonation steps also cause the inactivation of microorganisms (especially pathogens) as well as the biodegradation of substances causing taste and odor. In dual media filtration ammonium is nitrified and dissolved organic carbon is degraded to some extent. The following GAC filters act as biologically active adsorption filters. They reduce biodegradable organic compounds and eliminate organic micropollutants such as pesticides. The pretreated water is infiltrated by infiltration basins and wells. During the subsoil passage an equalization of temperature is achieved, additional purification can be observed, and especially DOC is reduced again.

3.3 Treatment of Water from Lakes and Reservoirs

The water quality of lakes and reservoirs is strongly influenced by the seasonal changes concerning, e.g., temperature, light intensity, and rainfall. The self cleaning processes in lakes and reservoirs are restricted to biodegradation, agglomeration, sometimes precipitation, sedimentation, and to the exchange of volatile compounds with the atmosphere. Yet, depositions at the bottom of lakes and reservoirs may be remobilized under special conditions. Generally, lakes and reservoirs act as a sink for pollutants and nutrients which are accumulated. The accumulation of nutrients leads to eutrophication and deterioration of the water quality. Therefore, the treatment of water from lakes and reservoirs must be adapted to the seasonal changes and to the degree of trophication. According to the degree of trophication algae will grow and the reduction of algae is one of the most important issues in the treatment of water from reservoirs or lakes. In most cases the concentration of phosphorous compounds is responsible for the degree of trophication.

For example, according to the concentration of total phosphorus in German lakes and reservoirs the degree of trophication can be classified (SCHWOERBEL, 1993):

- ultra-oligotrophic: (about 4 mg P m^{-3})
- oligotrophic: (about 10 mg P m^{-3})
- mesotrophic: (10–35 mg P m^{-3})
- eutrophic: (35–100 mg P m^{-3})
- hypertrophic: (>100 mg P m^{-3})

The phosphorous compounds are almost all of anthropogenic origin:

- wastewater: 55%
- agriculture: 26%
- detergents: 17%
- natural sources: 2%

Lakes and reservoirs may be protected against pollution and nutrients by strict pollution control, but shock loads of turbidities are always possible and occur due to heavy rainfalls.

In summer the production of algae will take place in the upper part of a lake or a reservoir. Due to photoassimilation the CO_2 concentration is reduced and the pH and the oxygen concentration rise. If the algae die later in the year they settle at the bottom and will be biodegraded. In these zones anaerobic conditions may occur, and dissolved iron and manganese concentrations may rise. Even denitrification, desulfurication and methane formation may be observed. The higher the degree of eutrophication the higher is the algal growth and the higher developed are the anaerobic processes close to the bottom. Pesticides and algae-borne substances causing taste and odor may also occur in the raw water and must be reduced.

Fig. 6 shows possible process schemes for the treatment of water from lakes and reservoirs according to their degree of trophication. These schemes are only proposals and may be varied according to the specific treatment problem. In general, one has to consider, that besides the elimination of algae the elimination of microorganisms like pathogenic bacteria, viruses, and parasites is important.

For instance, most of the lakes in Germany are oligotrophic or oligotrophic-mesotrophic. In the oligotrophic case, addition of potassium permanganate may be necessary to oxidize manganese. Special attention should be paid to the flocculation step, which should be optimized with regard to the elimination of turbidity, algae, and dissolved organic carbon. Especially if organic colloids and small algae should be eliminated the adjustment of the pH, a flash mixing and a multi-tank aggregation step may be necessary. The addition of an anionic flocculant aid like polyacrylate may be advised. After filtration over a dual media filter (DM filter) the water will be disinfected and stabilized by pH adjustment.

If an oligotrophic-mesotrophic water is treated addition of powdered activated carbon (PAC) may be necessary to reduce algae-borne substances causing taste and odor. The addition of PAC may also be necessary, if pesticides seasonally occur in the water. The powdered activated carbon is eliminated by the following flocculation and DM filtration step, which should be optimized for this special purpose. A secondary flocculation may be necessary to reduce residual turbidities before marble filtra-

Oligothrophic	Oligotrophic-mesotrophic	Mesotrophic-eutrophic
$KMnO_4$	PAC	Microstraining
Flocculation	$KMnO_4$	$KMnO_4$ or O_3
DM Filtration	Flocculation	Flocculation
Disinfection	DM Filtration	DM Filtration
pH Adjustment	Sec. Flocculation	Sec. Flocculation
	Marble Filtration	Marble Filtration
	Disinfection	Ozonation
	pH Adjustment	GAC-Filtration
		Disinfection
		pH Adjustment

Fig. 6. Examples of process schemes for lake and reservoir water treatment.

tion with final demanganization and a deacidification.

In the case of a mesotrophic-eutrophic water microstraining as the first step may reduce the concentration of algae considerably. Pre-ozonation may be necessary to inactivate mobile algae and other microorganisms and to improve the elimination of algae and turbidity by flocculation/filtration. High ozone dosages in the pre-ozonation step should be avoided because ozone may break up the algal cells and may oxidize the manganese to permanganate. After flocculation, DM filtration, secondary flocculation, and marble filtration a main ozonation will oxidize taste and odor causing substances and will improve the biodegradability of dissolved organics. The granular activated carbon (GAC) filtration will simultaneously act as an adsorption filter and as a biodegradation filter. Finally, the water is disinfected and stabilized.

4 Processes of Water Treatment

Today, common processes of water treatment can be related, depending on the degree of dispersion of the water contaminants, to solid–liquid separation, or to the elimination and chemical conversion of dissolved water contaminants (organic and inorganic), respectively. An overview of the diverse treatment possibilities matching many different treatment goals is given in Fig. 7. This overview is based on a proposal of HABERER (1987) and is represented in an updated and expanded form.

As can be seen in Fig. 7, treatment goals which can be related to solid–liquid separation as well as to separation and conversion of dissolved water contaminants, can be achieved by diverse processes (e.g., flocculation, rapid filtration, slow sand filtration, membrane filtration). Special potable water treatment processes can often include further process steps. For example, softening by the so called decarbonization process consists first of a chemical flocculation step, a superimposed or following sedimentation step, and usually a subsequent filtration step.

In the following, the main drinking water treatment processes will be discussed briefly. The processes of biological filtration, slow sand filtration, artificial recharge of groundwater, and bank filtration are not further discussed because these processes are described in more detail in chapters 20–22, this volume.

4.1 Flocculation and Precipitation

Fine suspended particulate water contaminants (e.g. algae, viruses, bacteria, protozoa, iron hydroxide complexes, calcite particles) can be present especially in surface water, but also in corresponding pre-treated groundwater (e.g., after decarbonization), in such forms that they only are separated insufficiently with technical sensible residence times in sedimentation or flotation plants. Eventually, such substances cannot also be separated sufficiently in filtration facilities. Using flocculation such fine suspended and colloidal particles can be brought together into better removable aggregates. As shown in Fig. 8, flocs can be separated in different subsequent processes, whereas the choice of the separation process depends on the concentration of flocs in the influent and on the requested clean water quality. Furthermore, dissolved and disturbing substances can be removed from the water by deposition into or accumulation onto the flocs, respectively, due to co-precipitation and sorption processes. Floc formation is usually driven by added flocculation chemicals. The type of floc formation has to be synchronized with the subsequent floc separation process.

The most important definitions used for flocculation and precipitation are explained as follows:

- Flocculation: Production of visible and removable flocs.
- Coagulation: Mechanism which supports floc formation. Herein, the repulsive forces between the particles are decreased and their ability for agglomeration is increased.
- Bridging: Mechanism which supports floc formation. Herein, the particles are connected to each other by long-chain molecules (polymers).

	disinfection	deacidification	iron removal	manganese removal	turbidity removal	NH_4^+ oxidation	odor removal	taste removal	softening	desalination	NO_3^- removal	P removal (raw water)	removal of inorganic trace compounds and radionucleides	removal of polar organic compounds	removal of non-polar organic compounds
aeration		A	A_1	A_1		A_1B_1	A		B_2C_1						
precipitation									A_2		$(A)_2$		A		
flocculation					A_1							A_1	B	A	
sedimentation					A_2				C_2						
flotation					A_2'										
rapid filtration			A_2	A_2	A_3	A_2			A_3			A_2			
chemical filtration		B										B			
slow sand filtration	(A)					B_2								B	A
dry filtration			B	B		C									
BAC filtration						D								C_2	B_2
adsorption							B	B					C	D	C
ion exchange									B_1	A	B		D	E	
reverse osmosis										B	C			(F)	(D)
thermical treatment										C					
dosing											A_1				
NaOH, $Ca(OH)_2$		C							A_1						
O_2			A_1'	A_1'		A_1'									
O_3	B_1			A_1''										B_1 C_1 D_1	B_1 C_1
$KMnO_4$				A_1'''											
Cl_2	B_2C			A_1''''											
ClO_2	D														
phosphate, silicate		(D)							(D)						

A, B, C: different processes
A_1, A_2: subsequent treatment steps within a process
A_1', A_1'': alternative treatment steps

Fig. 7. Process alternatives in potable water treatment – overview matrix.

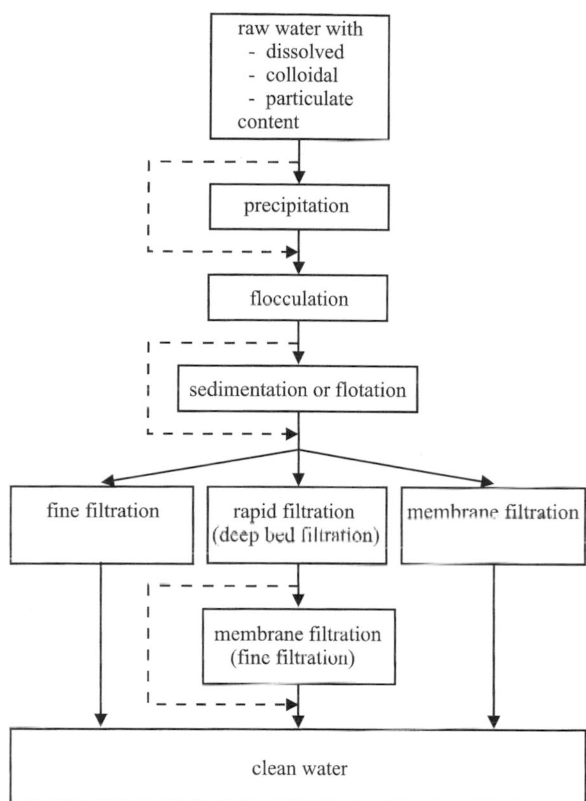

Fig. 8. Typical possibilities for the position of precipitation and flocculation in potable water treatment processes.

- Flocculant: Substances which are used to format flocs. These substances are mainly iron or aluminium(III) salts.
- Coagulant aid: Substances which improve the floc formation (organic polymers).
- Destabilization: Primary process during flocculation. Destabilization transforms the particles into a coagulable condition by coagulation or bridging.
- Transport processes: Secondary process during flocculation. These processes bring the particles into contact and lead to the formation of macro-flocs. Transport processes forced by diffusion due to thermal (Brownian) motion are called perikinetic. Transport processes forced by flow motion (velocity gradients) are called orthokinetic.
- Precipitation: Separation of compounds of low solubility from solutions. The precipitation reaction is a chemical reaction which always happens, if the solubility product of one of the involved substances is exceeded.
  Example: ($CaCO_3$ precipitation)
  Because metal salts are used among others as flocculants they can precipitate as oxide hydrates when overdosed. These hydroxides of low solubility do not precipitate as compact crystals like salts, but in form of voluminous flocs.
- Co-precipitation: Inclusion and carrying of colloidal particles and dissolved substances into the voluminous, gelatinous precipitates (oxide hydrates) of the metal salts which were used as flocculant.

Fig. 9 gives an overview of the possible processes occurring during flocculation and precipitation.

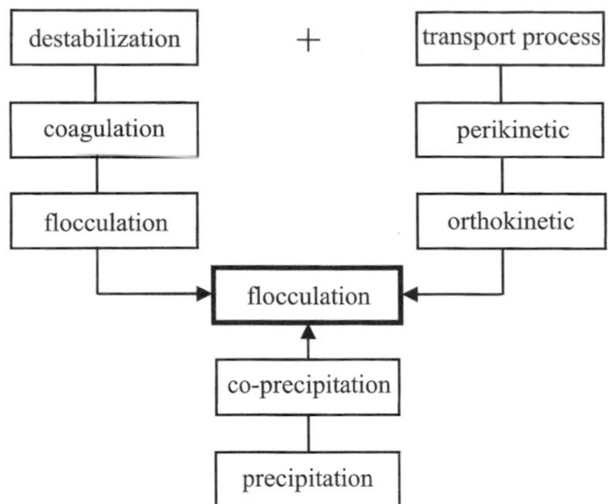

Fig. 9. Schematic presentation of the processes during flocculation and precipitation.

Addition of flocculants is characteristic for the operation of flocculation processes. Trivalent metal salts are of importance here.

$Al_2(SO_4)_3 \cdot 18H_2O$	aluminium(III) sulfate
$AlCl_3 \cdot 6H_2O$	aluminium(III) chloride
$Fe_2(SO_4)_3 \cdot 9H_2O$	iron(III) sulfate
$FeCl_3 \cdot 6H_2O$	iron(III) chloride

Depending on the pH value of the aqueous solution different hydrolysis products originate from Fe(III) or Al(III) salts. These products either initialize the two destabilization processes coagulation and/or bridging or the so called co-precipitation. For the formation of bigger flocs transport mechanisms are required which bring the primary particles or smaller flocs into contact. These transport mechanisms limit the velocity at which the flocculation process takes place. For very small particles (particles <1 μm) perikinetic transport predominates, whereas for particles >1 μm orthokinetic transport predominates.

Both transport processes can be described by the methods mainly based on the work of SMOLUCHOWSKI (1916). For the very important case of orthokinetic flocculation in practice the change in number of particles (N/N_0) as a function of flocculation time t, the average velocity gradient $\bar{G}$, the so called collision efficiency α and the floc volume fraction ϕ, can be described as follows:

$$N/N_0 = f(\alpha \cdot \bar{G} \cdot \phi \cdot t)$$

The average velocity gradient can be calculated from the energy input P as follows:

$$\bar{G} = \left[\frac{P}{\mu \cdot V}\right]^{\frac{1}{2}}$$

where μ is the dynamic viscosity and V is the volume of the reactor.

Both equations show that an increased energy input leads to an increase of the floc formation rate. However, the maximum possible floc size is limited by shear forces of which flocs break up, especially for bigger floc sizes, can occur.

The basic principle of a classical flocculation plant is shown in Fig. 10. At this point it has to be mentioned that there is a wide variety of different designs for flocculation units used in practice. In many cases the units chemical dosing, destabilization, floc formation and floc separation are integrated altogether into one facility. Fundamentally, a possibly fast and homogeneous mixing of the flocculant is to be carried out. Today, mixing times in the range of a few seconds are possible in modern facilities.

Before flocculation plants are designed one should execute generally orientated flocculation trials on a laboratory scale (eventually only as jar tests) (Technische Regel: Arbeitsblatt W 218, Flockung in der Wasseraufbereitung – Flockungstestverfahren, 1998). Herewith, approximately optimal chemical flocculation conditions (pH value, type and amount of flocculant dosage) can be derived. With regard to an optimal design of flocculation plants further statements require specific experience or the operation of a semi-commercial pilot plant (flow rate at least 1 $m^3 h^{-1}$).

4.2 Sedimentation and Flotation

Sedimentation and flotation are two processes of solid–liquid separation mainly used in technical plants as coarse cleaning stages and usually followed by a filtration stage. Both processes have the effect of the earth's gravitational pull in common. In technical plants this leads to usable settling processes for particles (single particles, particle agglomerates, prior

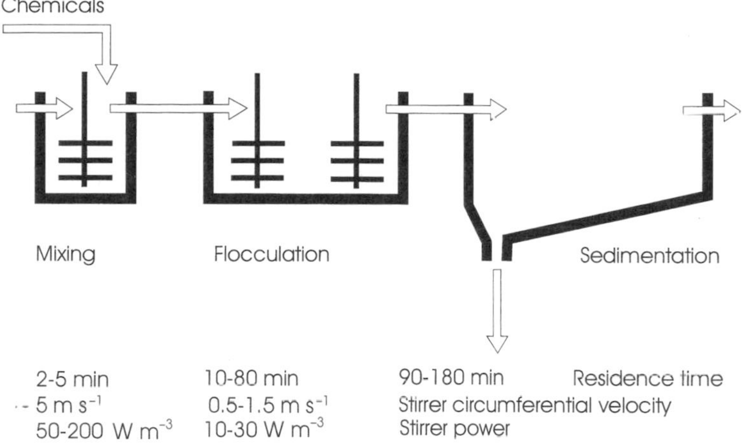

Fig. 10. Principal scheme of a classical flocculation plant followed by a sedimentation tank (SONTHEIMER, 1978).

built flocs, etc.) larger than about 10 μm with a density larger than the density of the surrounding water. Particles with a density smaller than the density of the surrounding water (e.g., single particles, aggregates, especially flocs with attached air bubbles) move upwards due to acting hydrostatic uplift forces. The sedimentation or flotation velocity v can be estimated for smaller particles on the basis of a simple balance of the gravity force, hydrostatic uplift force, and hydrodynamic resistance whereas the flow around the particles is assumed to be strictly laminar.

$$v = \frac{g}{18} \cdot \frac{\Delta\rho}{\eta} \cdot d^2$$

where

d = particulare diameter
g = the gravitational acceleration constant,
$\Delta\rho$ = the difference of solid density (including possibly attached air bubbles) and the water density and
η = the dynamical viscosity of the water.

Especially sedimentation processes play a major role in nature, e.g., in lakes and dam reservoirs. In some reservoirs and treatment plants which are eventually charged with high turbidity loads especially after storms (e.g., monsoon rainfalls), storage basins are integrated prior the actual plant to use the effects of the very simple sedimentation process.

For instance, in technical settling tanks a single particle does not settle down to the ground vertically, because the motion due to sedimentation is superimposed by the horizontal water flow into the direction of the tank outlet. The idealized relations for a settling tank are exemplarily represented in Fig. 11. The particle will only reach the tank bottom if its settling velocity v is larger than the quotient of the flow rate $\dot{V}$ and the sedimentation tank surface A. This proportion is called surface loading. In the ideal case it corresponds to the settling velocity of those particles which can be only just kept in the tank. As can be seen in Fig. 11 all particles with settling velocities larger than $\dot{V}/A$ will be kept completely. Particles with smaller settling velocities cannot be removed completely.

In the neighborhood of the tank bottom the solid concentration will increase in time so far that the particles will impede each other (the so-called impeded settling, Fig. 11), followed by a gradual compression of the accumulated solids. This process is used for the thickening of the sludge.

Particles which are found in water treatment able to settle down do not behave like single particles. Generally, there is a whole spectrum

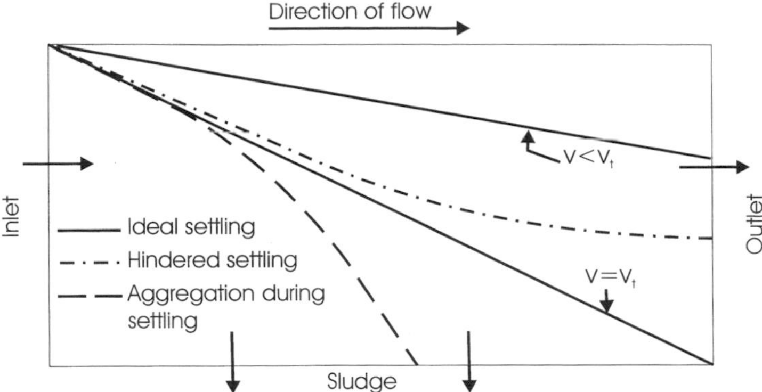

Fig. 11. Vertical cross section of a horizontal flow sedimentation tank (vt: settling velocity of a particle reaching the bottom of the tank just at the outlet).

of different sized particles with different settling velocities. Different settling velocities, influences of the wind, and gradients of temperature lead to further contacts between the particles whereby larger and faster settling particles can be formed. Due to the extranious influences (especially wind) so-called secondary flows can arise, whereby flocs with settling velocities $>\dot{V}/A$ can possibly be carried out of the tank.

For the increase of the surface loading $\dot{V}/A$ two ways are mainly committed in sedimentation:

- suppression of secondary flow by special constructive measures (e.g., baffles),
- increase of the sedimentation area at the same base of the settling tank by installation of angular positioned lamellas, pipes, etc.

Modern sedimentation plants, at which particle aggregation is completed before sedimentation, have a residence time of the water of about 5–30 min. Older circular or rectangular tanks with integrated aggregation are designed for residence times of about 3 h.

In flotation, air bubbles can be attached onto particulate water contaminants (e.g., single particles, particle agglomerates, hydroxide flocs, etc.) whereby the whole formation is raised in the water system to the water surface where it is concentrated and removed by suitable removal devices. Faster separations can be achieved here compared to sedimentation which allows for relatively small bases and, therefore, small building volumes of flotation tanks. Furthermore, flotation is generally more flexible than sedimentation due to the controllable creation of gas bubbles. For instance, special advantages of flotation can be shown, if algae are separated (especially if they occur only relatively temporary during so-called algal blooms) as well as flocs which were produced in water, e.g., with high loads of humic substances. Recent results from Scandinavia also show that, especially for the separation of parasites (*Cryptosporidium* and *Giardia*) with optimized flotation, exceptionally high separation efficiencies could be achieved (Kiuru, 1997).

The most important prerequisites for good working flotation are possibly small densities of the solids or flocs, respectively, and in particular good adhesion of air bubbles on the solids, which is usually given for prior built hydroxide flocs. Besides this, smaller floc sizes are of advantage in flotation rather than in sedimentation. In potable water treatment the so-called dissolved air flotation is predominantly used. The basis for this is given with the Henry law:

$$c_i = K_{Hi} \cdot p_i$$

where

c_i = the concentration of the dissolved gaseous component i in the water,
K_{Hi} = the Henry constant of the gaseous component i,

p_i = the partial pressure of the gaseous component i in the atmosphere above the water (usually air with about 80% nitrogen and about 20% oxygen).

Fig. 12 exemplarily shows a scheme of a flotation plant with a prior connected flocculation plant for potable water treatment. Here, dissolved air flotation is driven in the so-called recycle process, in which part of the draining clear water of the flotation is supersaturated with air and finally led back into the flotation tank and relaxed again.

Usually, the floating velocity of the particles attached onto air bubbles is higher than the settling velocity of the particles, if corresponding sedimentation plants are used. Therefore, the surface loadings are 3–5 times larger than in sedimentation plants. Furthermore, flotation plants deliver sludge which shows a higher solids content than sludge from sedimentation tanks.

Besides the influences given by the pre-flocculated raw water (e.g., temperature, type and amount of particulate water contaminants), the effect of the flotation process itself can be influenced by many adjustable values in contrast to sedimentation. These values are in particular

- saturation pressure (usually about 4–10 bar),
- content of recycled flow (usually about 4–12%),
- type of relaxation,
- surface loading (usually about 4–12 m h^{-1}),
- type and operating velocity of the removal device.

Similar to the flocculation process the operation of pilot plants is required for an optimal design of plants.

4.3 Rapid Filtration

Rapid filtration can be regarded as the central processing unit in potable water treatment. At present it is used in nearly all large scale water works in Germany, provided that water treatment is necessary at all. Most rapid filters used for the removal of turbidity consist of granular filter bed materials in the particle size range of millimeters (e.g., sand, pumice stones, filter coke) and a filter bed height of about 0.5–2.5 m (Fig. 13). The sensible use of rapid filtration is given if the volume concentration of solids contained in the raw water

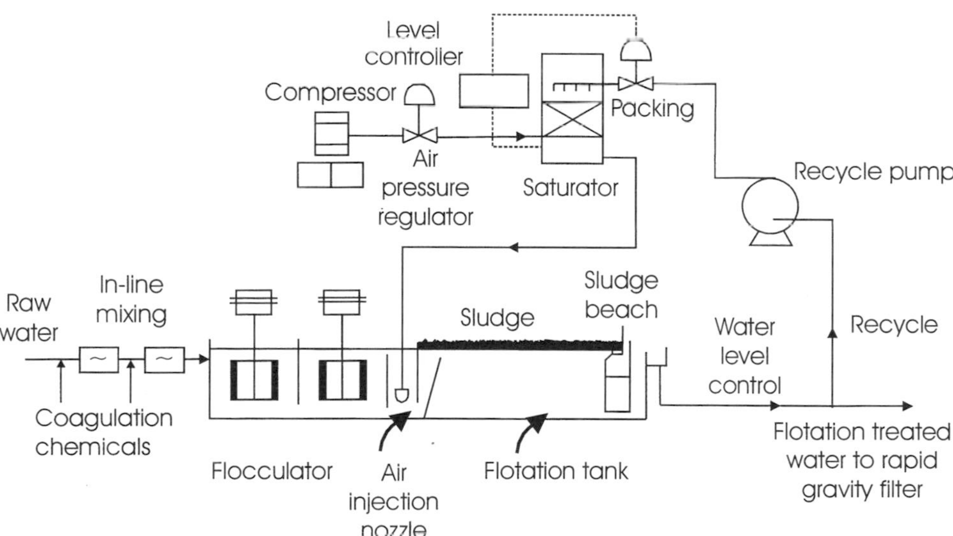

Fig. 12. Process scheme of a flocculation/flotation plant for potable water treatment.

does not exceed about 0.05%. With a correct design for a special treatment problem the filter should act in a way that, after a certain working period which is related to its turbidity storage capacity, a tolerable decrease in quality of the filtrate is given as well as an acceptable increase of the pressure drop. After this period, which is also called filter running time, regeneration of the filter is carried out. During this regeneration the filter is temporarily loaded with an upwards directed flow of such a high velocity that the filter material rises and the particles which were retained in the filter bed before, are flushed out. When the flow is stopped and the filter material is settled down, a new filter period can be started.

The retention of particles in deep bed filters requires two subsequent steps (GIMBEL, 1989):

- transport of the particles within the filter bed to the surface of the filter material,
- attachment of the particles after contact.

Both steps are the result of different forces acting on the particle which were based on flow resistance, inertia forces, the earth's gravitational pull and physicochemical interactions between the surfaces of the particles and the filter grain. Because of the special conditions in water treatment practice, exceptionally complicated circumstances are given. Therefore, it is recommended to run pilot trials for rapid filter design (Technische Regeln: Arbeitsblatt W 211-Filtration in der Wasseraufbereitung, Part 2: Planung und Betrieb von Filteranlagen, 1987).

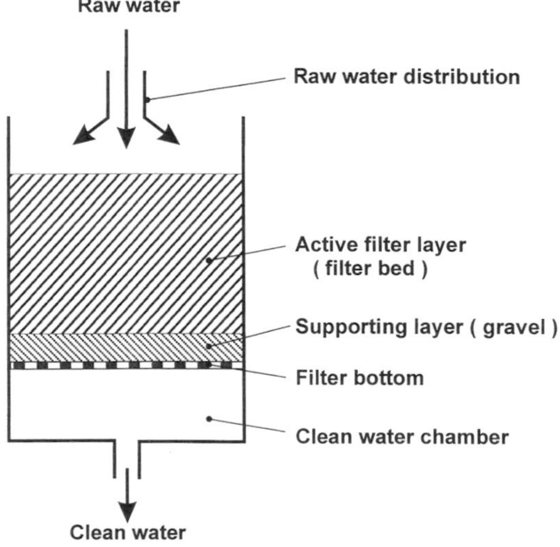

Fig. 13. Schematic design of a rapid filter.

If fine granular filter materials are used, only short filter layers are required to receive a good cleaning performance. However, this is combined with the disadvantage of relatively short filter running times, because the filter will be exhausted soon and loose its turbidity storage capacity. The degree of irregularity of the filter material is defined as the proportion of the mesh sizes of two test screens with mass passages of 60% or 10%, respectively (d60/d10). This proportion should be usually <1.5.

A classic rapid filter schematically shown as a so-called mono-layer filter in Fig. 13, consists of a raw water distribution system, an effective filter layer which is considered to be homogenous, and a filter bottom which carries the filter material and which is responsible for a uniform transfer of the filtered water over the cross sectional area of the filter into the clean water chamber. The filter bottom is made of steel or reinforced concrete and is either equipped with slotted filter nozzles or composed of special porous components. To secure perfect functioning of the filter bottom, especially if fine filter materials are used, one or more so-called supporting layers are used between the effective filter layer and the filter bottom. The supporting layer consists of a granular material with a grading 2–4 times coarser than the layer above.

Usually, the flow in rapid filters is directed downwards. The particle size distribution of the used filter material should be possibly narrow, because finer particles will be concentrated on the top of the filter layer after the filter was backwashed. As a result of this, rapid filtration can turn into cake filtration, which means that the deposition of particles cannot occur throughout the entire depth of the filter bed and that the pressure drop will strongly increase. To avoid this and to use the greatest possible depth of the filter for the deposition of particles, dual media filters (Fig. 14) are used in many cases. In such filters the upper layer consists of relatively coarse material of low density, whereas the lower layer shows a fine grading of high density. If the grain sizes and densities of the filter materials in both layers are matched correctly, the arrangement of the layers will be the same even after backwashing. The big pore size of the upper layer permits a high turbidity storage capacity, whereas the particle retention is not very effective. The lower layer acts in the opposite way. For this reason, dual media filters can achieve much longer filter running times than single media filters with comparable operating conditions. In some cases, triple media filters could be used in practice successfully. Two typical examples for the design of multi-media filters are shown in Tab. 3.

To avoid the above mentioned disadvantageous properties of single media filters they could be flown through directed upwards, whereas the usual sizing effect of the filter material during backwashing is of advantage for using the greatest possible depth of the filter for particle retention. Such filters tend to lift their upper layers after a certain turbidity load which will lead to a spontaneous turbidity breakthrough.

Fig. 14 shows the design of rapid filters built as closed filters, which are also called pressure filters. Additional realizations used in practice are shown in Fig. 15. Closed rapid filters are usually built as cylindrical containers made of steel. They can achieve filter areas of about 30 m^2 as standing models. Lying models allow much higher filter areas but often have relatively small filter layers. The usual pressure drops vary between 0.2 and 2 bar at filtration velocities of 5–30 m h^{-1}. The filtration velocity is defined as the proportion of the water volumetric flow rate and cross-sectional area of the filter.

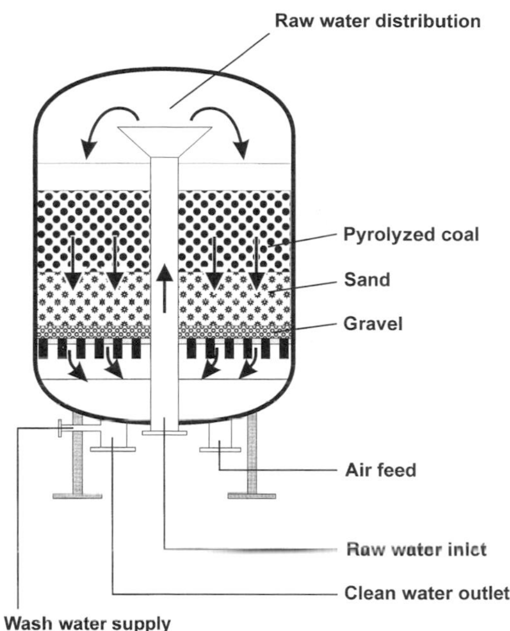

Fig. 14. Design of a dual media filter as a closed filter.

Large water treatment plants often use open rapid filters. The pressure drop arising within the filter bed has to be compensated for by corresponding submerging of the filter layer. The height of the submerging usually varies between 0.3 and 3 m columns of water, which correspondents to a pressure drop of about 0.03–0.3 bar. Usual filtration velocities are in the range of about 3–15 m h^{-1}. Open filters are mainly designed as rectangular tanks made of reinforced concrete. Filter areas of about 150 m^2 can be achieved.

Tab. 3. Typical Designs of Multi-Media Filters

Material	Filter Bed Height [m]	Loose Weight [kg m^{-3}]	Grain Size [mm]
Filter coke, pumice stones, etc.	0.5–1.2	500–700	1.7–2.5
sand	0.6–1.5	1,000	0.8–1.2
Activated carbon	0.3–0.6	250–350	3–5
filter coke, etc.	0.6 1.2	500–750	1.5–2.5
sand	0.5–0.8	1,000	0.6–0.8

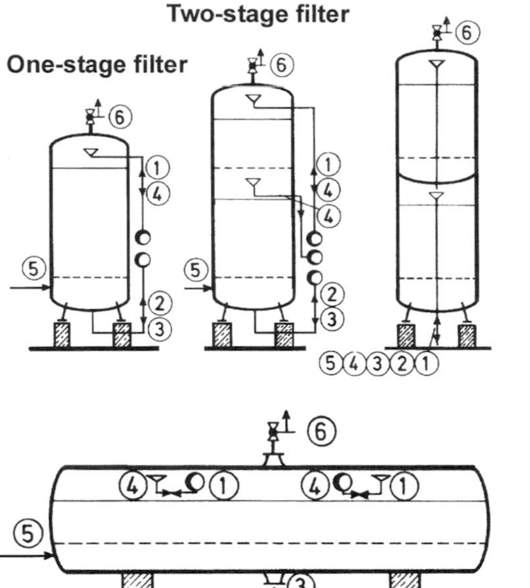

Fig. 15. Different types of closed rapid filters (ROENNEFAHRT, 1983).

The different applications of rapid filters in water treatment can be summarized in the following groups:

- Separation of single particles:

Such tasks can be found, especially while removing turbidity from surface waters which were treated to become drinking water. Here, it deals with inorganic compounds (e.g., clay minerals) as well as organic compounds like plant cells, algae, bacteria, etc. A further and frequently used application is the separation of $CaCO_3$ particles after decarbonization. In all these cases biological processes are mainly of subordinate importance.

- Separation of particle agglomerates and hydroxide flocs:

If the turbidity causing solids can only be filtered poorly as single particles. Agglomeration mostly is carried out ahead of rapid filtration or the particles are embedded, due to a flocculation process, into iron or aluminum hydroxide flocs, in which also disturbing dissolved substances are embedded. These applications are often for drinking water treatment from surface water. The turbidity removal can also strongly be influenced by biological effects within the filter bed.

- Iron and manganese removal:

The main application for rapid filters in the treatment of groundwater is the removal of bivalent iron and manganese ions. They have to be transferred into a form able to be separated by chemical or biochemical oxidation before or during the actual filtration process. Besides chemical-catalytic effects biological effects also are of importance at this common use of rapid filters.

In many cases, single media filters are sufficient for turbidity removal, if the turbidity of raw water is relatively small. For higher turbidity loads one should use multi-media filters. The importance of multi-media filters has strongly increased in recent years, because due to economical reasons the process of the so-called floc filtration (or direct filtration) is used frequently more. Here, the water is directly fed after a flocculation step onto the rapid filter without any preceeding steps like sedimentation or flotation.

Pre-treatment of water is essential, especially for direct filtration. Type and amount of the used flocculants are of importance as well as the circumstances when dosing it (residence time until the filter, energy input in the flocculation unit). Furthermore, dosed coagulant and filter aids, which nowadays are increasingly made of synthetically, soluble polymers (usually based on polyacrylamide), can clearly enhance the performance of the flocs with regard to the filtration. Similar effects can be achieved with oxidative treatment of the raw water, e.g., with ozone or chlorine. Their influences are also of interest when using contact filtration (or direct filtration), where the flocculant is directly added prior to filtration so that floc formation and retention takes place within the filter bed. Generally, the pre-treatment of raw water, which mainly depends on experience, strongly influences the operation performance of a rapid filter.

Usually, the space between the filter grains is completely filled with deposits of turbidity and water. In contrast to this air flows in parallel flow with water through the filter media in the so-called trickling filtration (Fig. 16). This special procedure can be used in water treatment, if high aerobic degradation or oxidative processes take place within the filter bed, so that additional air becomes necessary to supply oxygen. This is true for the treatment of water, e.g., with high loads of ammonium and iron. Trickling filtration is mainly used as a pre-treatment step of a common rapid filtration, because turbidity removal in the dry filter can cause instabilities.

For the operation of rapid filters it has been proven useful to keep the filtration velocity constant during the filter run. This requires a suitable regulation, because the flow resistance in the filter bed increases with filter running time due to the retained deposits of particles. A constant filtration velocity can be achieved using a throttle at the filter outlet which correspondingly decreases, if the pressure drop increases, or using a correspondingly increasing submerging, if open rapid filters are used. In some cases one prefers the use of a variable filtration velocity, where this velocity decreases to half of initial rate during the filter run. Depending on the raw water quality, a better average filtrate quality can be achieved in this way than with a constant filtration velocity at a similar filtration unit because the probability of detachment and washout of deposits caused by flow forces is strongly decreased due to decreasing velocity.

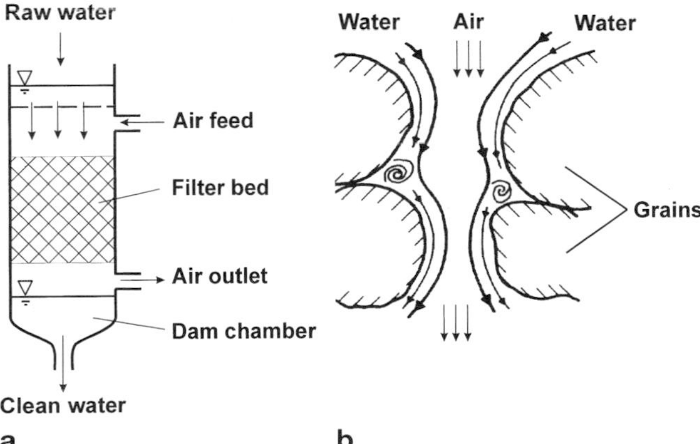

Fig. 16. Schematic representation of a trickling filter; (**a**) full view, (**b**) flow in the filter layer.

Filter running times of rapid filters usually range between 10 and 100 h. As is qualitatively shown in Fig. 17 the filtrate quality (expressed as concentration c) as well as the pressure drop of the filter bed (Δp_V) will change during the filter running time. For continuous turbidity control only turbidimetry with a scattered light angle of 90° is used in practice as a measure of the turbidity value. Currently particle analyzers are also used. These analyzers measure light scattering and extinction of single particles and deliver information about size and quantity of the turbidity causing particles.

As is shown in Fig. 17 one can distinguish between three characteristic periods of a filter run. The turbidity of the filter effluent decreases until time t_1, which means that the quality of the filtrate is improved. This period is called ripening, and it can be especially pronounced if chemical-catalytic or biological effects are important for the removal of turbidity, e.g., at deferrization and demanganization. Between t_1 and t_2 the turbidity of the effluent remains nearly constant. At time t_2 the breakthrough period of the filter starts, which continues as long as the maximum permissible turbidity concentration is reached and the filter has to be backwashed. Fig. 17 also shows the increase of the pressure drop of the filter bed. The pressure drop increases with the filter running time and reaches a maximum permissible value at time t_4. One of the fundamental tasks for optimization in rapid filtration is to unite times t_3 and t_4, if possible, because these times are the limiting criteria for the filter running time. In practice, due to security reasons, t_3 is often chosen to be 10–20% larger than t_4, and the maximum pressure drop is taken as the primary criterion to stop a filter run.

The rapid filter has to be regenerated after the filter run. Usually, this happens by backwashing in which the filter is loaded with an upwards directed strong flow, so that the deposits of particles which were attached onto the filter grains are detached and flushed out.

In the case of mono-layer filters backwashing is often executed only with water, whereas the flow rate should be so high that the filter bed will expand by at least 10% and the fixed bed will be transmuted into a fluidized bed in which the filter grains will carry out irregular motions. The rising, turbid backwashing water is taken out of the filter over suitable flumes, funnels, or lids. Today, pure water backwashing is mainly spread in the USA. It has the disadvantage that bonds of filter material in the upper filter region may not completely break apart. The so built filter grain clusters can concentrate in the lower regions of the filter bed in time and can considerably decrease the filter performance. In many cases additional washing of the surfaces is carried out with strong jets of water to avoid such problems. In Europe, the most preferred possibility to cope with these problems and to achieve intensive filter grain cleaning is the additional use of air as rinsing medium. In a first step the just submerged filter layer is raised only by air, which chops up eventually appearing filter grain aggregates. Subsequently washing with air and water and then only with water is often carried out in many cases.

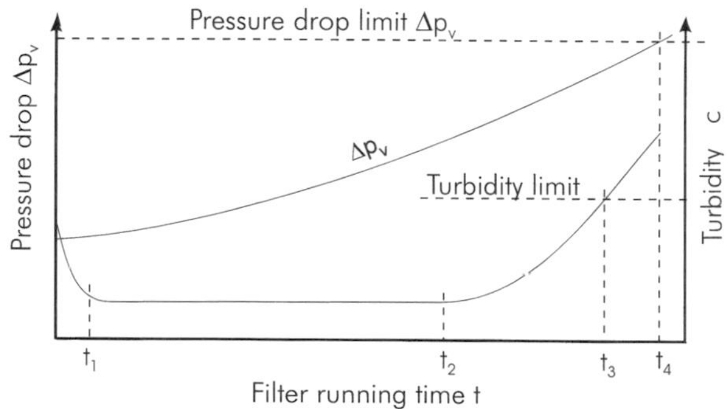

Fig. 17. Change of pressure drop and turbidity concentration in the effluent vs. filter running time.

An analogous step is carried out with multi-layer filters, but in most cases with omission of simultaneous air–water backwashing, because the risk that filter material is flushed out together with backwashing water is very high. As the last step in multi-layer filter backwashing a so-called separation washing with water is executed, where decomposition of the different filter materials is achieved so that the initial succession of formations of the filter is restored. This washing occurs with relatively high filtration velocities, whereas the expansion of the filter bed can increase to up to 40%.

Usually filter backwashing requires about 10–20 min at all. The rinsing air rate as well as the backwashing water rate vary between 40 and 100 m h^{-1} with filter bed expansions between 10% and 50% depending on grain size and density. In a well designed filter backwashing program the water consumption per washing should not be higher than 2–5 m^3 per m^2 filter area. Backwashing which is too intensive can be of disadvantage, if the turbidity removal of the filter strongly depends on biological or chemical-catalytic processes, because then a relatively long ripening time is again needed.

4.4 Adsorption

Adsorption processes are used for the elimination of undesired organic compounds by accumulation on the large inner surface of an adsorbent (usually activated carbon). The specific surface of activated carbon may be as high as 1,000 m^{-2} g. The accumulation of the organic molecules on the surface of the adsorbent is caused by their attachment due to physical-chemical interactions. The adsorption process with activated carbon plays an important role in water treatment. The removal depends of the polarity of the organic compound and decreases with increasing polarity. Because of the almost non-polar characteristic of the activated carbon surface and the more or less non-polar characteristic of many organic compounds as, e.g., chlorinated or non-chlorinated hydrocarbons and pesticides, the organic compounds have the tendency to leave the water, which is a polar solvent, and to accumulate on the non-polar activated carbon surface. For the description and understanding of the adsorption mechanisms the knowledge of the special characteristics of the activated carbon and detailed information about the adsorption equilibrium and the adsorption kinetics are necessary SONTHEIMER et al., 1988).

Activated carbon can be made of different natural raw products containing carbon. The activated carbon (and, therefore, also the raw product) must meet certain demands concerning purity as specified, e.g., by the European Committee for Standardization (CEN) for the concentration limits of toxic compounds which can by extracted by water. Wood, peat, anthracite, hard coal and brown coal are usually used as raw products but activated carbon may even be made from coconut shells or pit stones. In activated carbon production the raw product is carbonized in a first step to transform the carbon containing material to pure carbon. In a second step the carbon is (usually thermically) activated to produce fractures, cracks, and finally the large inner surface.

The adsorption equilibrium is defined by the equilibrium between the concentrations of the dissolved organic compounds in the water and the adsorbed compounds on the activated carbon. The adsorption equilibrium will be reached after a sufficient time. If more than one organic compound is dissolved in the water, the adsorption equilibrium is influenced by all compounds present which compete for attachment to the surface. In case of such competition the adsorption equilibria are lower than the equilibria which are reached, if only one single compound is adsorbed.

In water treatment processes the empirically determined Freundlich equation is usually used to describe the adsorption equilibrium between an organic compound dissolved in water and adsorbed on activated carbon. The terms K_F and n are constants, q is the quantity of organic compound per unit of adsorbant and c is the equilibrium concentration remaining in solution

$$q = K_F \cdot c^n.$$

The Freundlich equation (also called Freundlich isotherm) is in most cases suitable for the description of measured adsorption equilibria. If the measured data of q and c are plotted on a double logarithmic scale the Freundlich equation can be easily described by a straight line.

In case of a competitive multi-compound adsorption the equilibria of the different compounds are influenced by each other. A qualitative example for the competitive adsorption of two compounds A and B is shown in Fig. 18. A is much better adsorbable than B. The isotherm of the compound A which was measured in the presence of different concentrations of compound B is shown in Fig. 18a. The initial concentration of compound A (C_{A0}) was kept constant and the initial concentration of compound B (C_B) was varied. With increasing initial concentration of compound B the amount of the adsorbed compound B decreases. Fig. 18b shows the isotherm of compound B at a constant initial concentration C_{B0} and variable initial concentrations of compound A (C_A). Comparing Figs. 18a and 18b it becomes obvious that the equilibria of the less adsorbable compound B are much more influenced by the presence of the better adsorbable compound A than vice versa. There exist different theories to calculate multi-compound adsorption equilibria which are described in literature (e.g., SONTHEIMER et al., 1988).

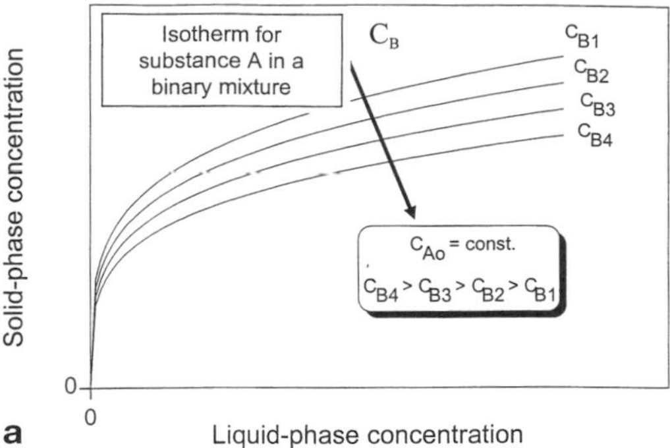

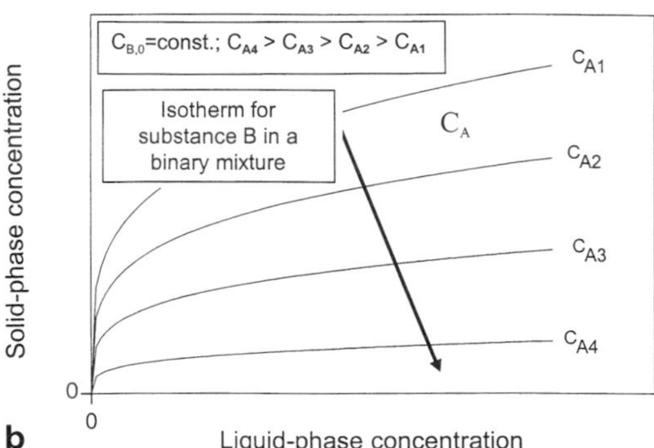

Fig. 18 a, b. Isotherms for simultaneous adsorption of two compounds. Compound A is better adsorbable than compound B.

Besides the adsorption equilibrium knowledge of the adsorption kinetics is necessary for a complete description of the adsorption process, especially if granular activated carbon filters are used. In a first step the adsorption kinetics describe the mass transfer of the organic compounds from the solution to the outer surface of the activated carbon grain. This process can be described by the film diffusion model. Furthermore, the sorbate is transferred from the grain surface through macro- and mesopores into the microporous system inside the grain. This process can be described by the so-called pore diffusion or surface diffusion models (ZIMMER and SONTHEIMER, 1989).

Activated carbon is used in water treatment as powdered and as granular activated carbon. The particle diameter of powdered activated carbon is usually between 20 and 80 μm, which results in rather fast adsorption kinetics. The diameter of granular activated carbon particles is usually between 1 and 5 mm, which results in slower kinetics and thus in a large influence of the kinetics on the adsorption process if granular activated carbon filters are used. Different processes using powdered or granular activated carbon are presented in the following paragraphs and the advantages and disadvantages will be discussed briefly.

Powdered Activated Carbon

Powdered activated carbon (PAC) can be added as suspension at different positions of a treatment process. Altogether three technical steps will be necessary:

- addition to the water and mixing,
- adsorption of undesired compounds, stirring for 20–30 min to achieve almost equilibrium,
- removal of the PAC, e.g., by sedimentation, flotation, or filtration.

It can be assumed that the equilibrium will be almost reached after a contact time of about 20–30 min, if the turbulence by stirring is high enough and the particle diameter is low enough (20–40 μm).

For example, the position of the PAC dosage may precede a flocculation/precipitation step. The carbon sludge may be removed together with the flocculation/precipitation sludge in a sedimentation step. Because of their low sedimentation velocity not all carbon particles will be removed by sedimentation, and for the almost complete removal of the added PAC an additional rapid filtration will be necessary.

The use of PAC will be of high economic efficiency, if the addition is only necessary sometimes, e.g., in case of a periodical occurrence of pesticides in the raw water. A disadvantage is a remaining concentration of the undesired compounds which are not adsorbed according to the equilibrium. A further disadvantage must be seen in the fact that the reactivation of used PAC is not economically possible.

Granular Activated Carbon

Granular activated carbon (GAC) is used as a filter material in layers which are passed by flowing water. While water passes the GAC filter the undesired compounds adsorb on the GAC and are removed. As the adsorption capacity of the GAC is slowly exhausted the concentration of the undesired compound increases after a certain operation time. Fig. 19 shows schematically the operation of a GAC filter which is charged by a water flow rate $\dot{V}$ containing an adsorbable compound with the concentration c_0. The breakthrough curve c/c_0 as a function of the filter running time t_F is also shown. Because of the adsorption kinetics a so-called adsorption front builds up in the GAC filter. The shape of the adsorption front depends on the adsorption equilibrium and the adsorption kinetics. Dependent on the equilibrium, the kinetics, and the filter velocity v_F the adsorption front moves through the GAC layer and after a certain filter running time finally reaches the end of the filter layer. At this time the breakthrough of the filter starts and the compound can be measured in the filter effluent. The effluent concentration increases with further running time and finally reaches the influent concentration. At this time the total breakthrough of the filter is achieved.

Fig. 19 shows an important difference between the use of PAC and GAC. After the use of PAC the water still contains the organic compound according to the equilibrium concentration, whereas the compound is not contained any more in the effluent water of a GAC filter during the first period of the filter running time. Furthermore, the adsorption capacity of the activated carbon can be used

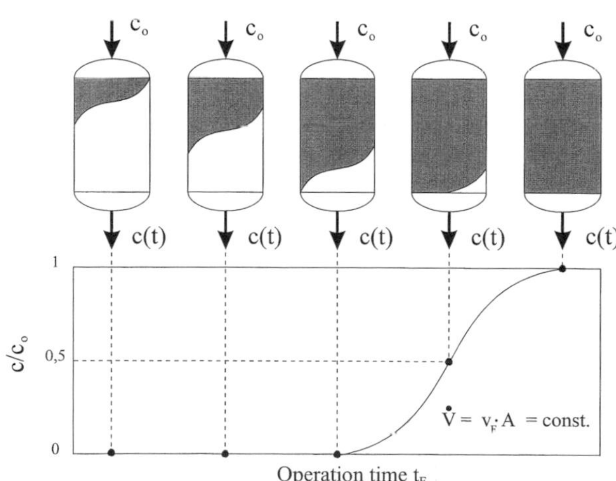

Fig. 19. Operation of a GAC filter.

much better in a GAC filter than by the addition of the powdered carbon. If changes of the raw water concentrations occur the effluent of the GAC filter will not be effected so much than the effluent of an adsorption step using PAC. Furthermore, reactivation of the exhausted activated carbon by a thermal process is possible using GAC, but not using PAC.

A disadvantage of the GAC filter must be seen in the decrease of the adsorption capacity by the so-called carbon fouling effect. This effect was first observed during the investigation of the adsorption behavior of 1,1,1-trichloroethane and is caused by pre-adsorption of natural organic matter (NOM) on the activated carbon. The amount of pre-adsorbed NOM increases with an increasing filter running time, and the adsorption capacity of the carbon decreases with an increasing amount of pre-adsorbed NOM.

The calculation of the breakthrough curves will not be explained in detail in this chapter. Altogether there are three equations necessary for the calculation of the time dependent effluent concentration. These equations describe the inner mass transfer in the carbon grain, the adsorption equilibrium and the mass balance within a differential filter element. The development of these equations, the initial and boundary conditions and the techniques for solving the system of these combined equations are explained, e.g., by SONTHEIMER et al. (1989).

GAC filters are usually operated with filter layers of 1.5–3 m height and filter velocities of 10–25 m h^{-1}. The filters can be open concrete basins or closed steel vessels. The inner side of the steel vessels may be covered by rubber to protect the steel from corrosion. Fig. 20 shows an example for a steel vessel as pressure filter. The GAC layer may be carried by a layer of gravel on a nozzle bottom. The filter is operated in the downflow mode until the adsorption capacity of the filter is exhausted or the effluent concentration of a certain compound exceeds a limit. If suspended solids are contained in the influent (which is not desired) they may be removed by the GAC filter and the pressure drop of the filter may increase. In this case the GAC filter must be backwashed and the adsorption front in the filter layer may be destroyed, if the GAC grains in the filter are mixed up too much during backwashing. A homogenous distribution of the adsorbed compounds in the filter may be the result of backwashing, and this may lead to shorter breakthrough times.

Another type of GAC filter is shown in Fig. 21. In this double layer filter two different filter materials are laying on separate gravel layers and on separate filter bottoms inside of one common filter vessel. In the upper layer pre-activated carbon, e.g., for the removal of suspended solids, may be contained. The lower layer contains GAC for the removal of organic compounds. Such filters are used if the influent contains, e.g., suspended solids, iron, manganese and adsorbable organic compounds. The filter layers can be backwashed separately. Removal of the suspended solids ahead of the GAC filter layer avoids frequent backwashing and the destruction of the adsorption front in the GAC filter layer.

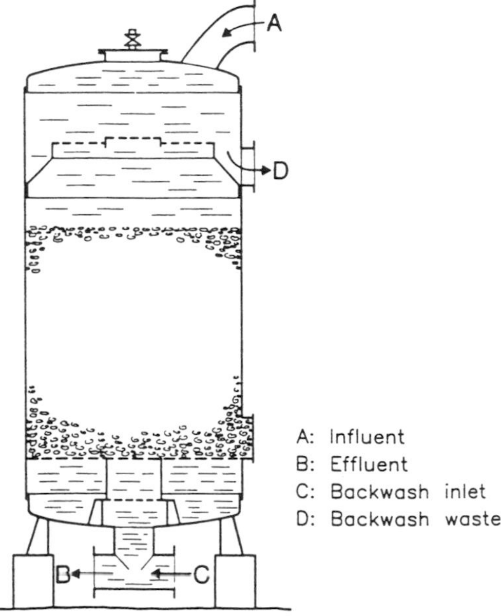

Fig. 20. Steel vessel as pressure GAC filter (SONTHEIMER et al., 1988).

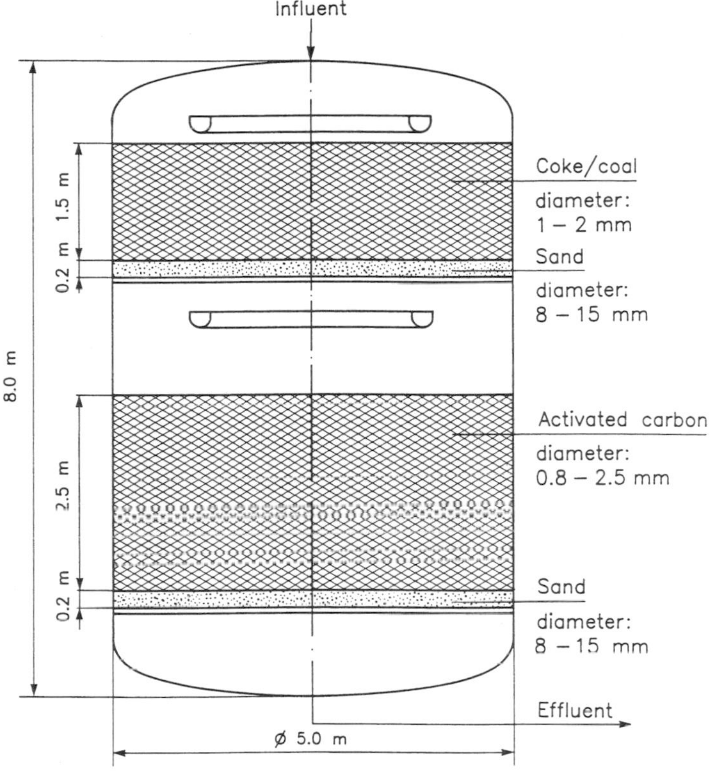

Fig. 21. Double layer filter (SONTHEIMER et al., 1988).

4.5 Gas Exchange

The exchange of gaseous substances with water is a frequently used process in potable water treatment. Typical examples are:

- enrichment of the water with oxygen to permit the oxidation of dissolved iron and manganese, e.g.,
- the input of ozone as oxidant,
- degasification of carbon dioxide for the deacidification of water,
- stripping of gases like hydrogen sulfide and methane,
- stripping of volatile organic compounds like halogenated hydrocarbons or organic solvents like tri- or perchloroethene.

The definition of a heterogenic equilibrium in a two-phase system (e.g., air and water) can be described by the Henry law (see Sect. 4.2). This equilibrium, which depends greatly on temperature and which can theoretically only be reached after infinite time, is called state of saturation. Depending on the gaseous substances dissolved in the water and depending on the "atmosphere" (mainly air) in which the water comes into contact with, the system has the tendency to reach a new equilibrium. Because one does not want to wait too long for this, technical gas exchangers are designed on the one hand with large exchange areas permiting a good gas transfer between the liquid and the gaseous phase, and on the other hand for adjusting good flow conditions so that a possibly good mass transfer, especially in the liquid phase (e.g., by turbulent flow), is achieved.

There are many technical facilities or devices which have been proven in potable water treatment in spite of their partially total different designs and modes of operation. A systematic distinction is possible with the following criteria, whereas the gaseous phase is always air:

distribution of the two phases air and water
- water drops distributed into air
- air bubbles distributed into water
- falling film

flow of both phases in relation to each other
- parallel flow
- reversed flow
- cross flow

direction of flow of the water
- upwards directed flow
- downwards directed flow
- horizontal flow

charge, discharge and feedback of air
- forced aeration
- operation without aeration (vacuum degasification)
- self-suction of air by the principle of the water jet air pump

number of stages of the facility or device
- single stage
- multi stage

process pressure
- pressure below atmospheric
- atmospheric

In spraying a large mass transfer area is created by distributing possibly small water drops into the air. Therefore, water is sprayed through divers constructions into reaction chambers or eventually into the environment. In bubble aeration air is sucked through hollow shafts of centrifugal aerators or rotating wings and then distributed as possibly small bubbles into the water of the reaction chamber. In a jet apparatus, which works with the principle of the water jet air pump, air is sucked from the outside and intensively mixed with water in a nozzle so that the air is distributed with very fine bubbles to the water.

In the so called flat path aeration, air is brought from below into the water which floats horizontally over perforated plates (eventually porous material like frits). This principle is especially used in deacidification and for the input of ozone.

In spraying processes the suction effect of falling water drops onto the surrounding air is used. To enlarge the mass transfer area in trickling processes usually several stages (cascades) are utilized in appropriate facilities. To improve mixing of the two phases units like perforated plates, pipe grids or baffles of different designs are built into the flow direction of the water.

The mass transfer area in devices can be enlarged by originating falling films. Special formed surface elements or filling materials are embedded into these devices. The water flows parallel or reversed to the air flow. Typical designs of the devices are the corrugated plate column, the profile bloc column or the package column and the pipe grid cascade.

4.6 Oxidation

Oxidation is a process which occurs in the natural environment and which is also used in water treatment. By addition of oxidants like oxygen (O_2), hydrogen peroxide (H_2O_2), ozone (O_3), potassium permanganate ($KMnO_4$), chlorine (Cl_2), chlorine dioxide (ClO_2), and hypochlorite (ClO^-) the structure of dissolved organic and inorganic water compounds, colloids and microbiological compounds may be changed. The oxidation reactions may be also caused by UV radiation and the formation of OH radicals (OH).

If the oxidation process is used for the inactivation of microorganisms such as, e.g., bacteria and viruses, the process is defined as disinfection (see Sect. 5). Some oxidation reactions may be accelerated by microorganisms such as, e.g., in iron and maganese removal by biologically active filters. Some oxidants may be used in combinations leading to the so-called advanced oxidation processes (AOP) such as, e.g., the combination of H_2O_2 and O_3 or H_2O_2 and UV.

In the past chlorine was frequently used as oxidant to perform a breakpoint chlorination but the undesired formation of halogenated organic compounds may lead to health concerns and, therefore, today the breakpoint chlorination does not meet any more the standards for drinking water treatment in many countries. In Germany, e.g., the use of chlorine in water treatment is restricted and chlorine is allowed to be added only for the purpose of disinfection, but not for general oxidation.

In oxidation processes electrons (e^-) are transferred from one molecule to another. The compound which is oxidized releases one electron as, e.g.

$$Fe^{2+} \cdot Fe^{3+} + e^-$$

The inversion of this process is reduction. In aqueous solution free electrons (e^-) cannot exist durably. Therefore, the electrons released by one process must be taken up by another. The oxidizing agent (oxidant) takes up the electrons and is reduced itself by this process. Oxidants always attempt to take up free electrons and act as an electron acceptor.

The most important oxidants in water treatment and the chemical oxidation reactions are summarized in Tab. 4. It becomes obvious that the reactions are strongly dependent from the pH because the H^+ or OH^- ions are always involved in the reactions. The normal potential E_o which is also listed in Tab. 4.2 can be used for the evaluation of the oxidizing power of an oxidant. According to the normal potential the OH radical is the most and oxygen the least powerful oxidant. Besides the power of the oxidant the kinetics of the oxidation reactions are of general importance and may even be crucial in technical oxidation processes (BUXTON et al., 1988; NETA, 1988).

The most important natural oxidant is oxygen. Oxygen is frequently used in water treatment and can be easily added to the water by aerators (see Sect. 4.5), e.g., to oxidize Fe^{2+} and Mn^{2+} in anaerobic groundwaters. In the aeration process one must take care that the water is not contaminated by pollutants in the air. The use of pure oxygen may be of advantage, if the required oxygen concentration in a treatment process is higher than the oxygen concentration that can be achieved by aeration.

Since several decades ozone is successfully used in water treatment because it is a quite powerful oxidant. Ozone may be produced from pure oxygen or from de-humidified air by the so-called silent electrical discharge process. The produced mixture of ozone and oxygen (and nitrogen) may be added to the water by gas exchangers (see Sect. 4.5). Ozone may be used for the oxidation of inorganic compounds (e.g., Fe^{2+}, Mn^{2+}, but not NH^{4+}) as well as for the oxidation of organic compounds (especially compounds which affect taste, odor, or the color of the water), which become easier to remove or lose their undesired properties. In all reactions with ozone the pH as well as the concentrations of HCO_3^- and CO_3^{2-} play an important role. Furthermore, the inactivation of viruses and bacteria is an additional positive effect of ozonation.

With the use of ozone in water treatment undesired products may be formed. In the presence of bromide ozonation may produce bromate which may react with other organic

Tab. 4. Reactions of Most Important Oxidants in Water Treatment and their Normal Potential E_o

Oxidant	Reaction	E_o (Volt) at 25 °C
OH radical (e.g., produced by UV radiation)	$OH^\circ + H^+ + e^- \rightarrow H_2O$	2.3
Ozone	$O_3 + 2H^+ + 2e^- \rightarrow O_2 + H_2O$	2.1
Hydrogen peroxide	$H_2O_2 + 2H^+ + 2e^- \rightarrow 2H_2O$	1.8
Chlorine dioxide	$ClO_2 + 2H_2O + 5e^- \rightarrow Cl^- + 4OH^-$	1.7
Permanganate	$MnO_4^- + 4H^+ + 3e^- \rightarrow MnO_2 + 2H_2O$	1.7
Hypochlorous acid	$HOCl + H^+ + 2e^- \rightarrow Cl^- + H_2O$	1.5
Chlorine	$Cl_2 + 2e^- \rightarrow 2Cl^-$	1.4
Hypochlorite	$ClO^- + H_2O + 2e^- \rightarrow Cl + 2OH^-$	0.9
Oxygen	$O_2 + 2H_2O + 4e^- \rightarrow 4OH$	0.4

compounds producing bromoorganic substances. Furthermore, biodegradable (assimilable) organic carbon may be produced by the ozonation which may cause problems concerning regermination of the drinking water in the distribution system. Finally it must be mentioned that ozone is very toxic and, therefore, special attention must be paid to safety precautions which are necessary for the production and use of ozone.

Permanganate is used as a dark purple colored solution to oxidize, e.g., Fe^{2+} or Mn^{2+} or to combat the growth of algae in open water basins. Permanganate is relatively easy to handle and is frequently used in special cases, if short time sanitation of treatment plants and distribution systems is necessary.

Since the beginning of the 20th century the use of chlorine (gas) and chlorine containing chemicals (hypochlorite, chloroamine, and chlorine dioxide) as oxidants has widely spread. Because of the low price of chlorine and its compounds and the high disinfection efficiency these substances are frequently used.

Chlorine gas, which is produced in the chlorine-alkali electrolysis process, can be dosed from gas bottles or vessels to the water. Because of the toxicity special safety precautions must be considered. Solutions of hypochlorite are easier to handle and are easily dosed especially to smaller water flows in the desired and suitable concentrations. In the presence of ammonium (NH^{4+}) chloroamines are formed by the addition of chlorine or hypochlorite to the water. Chloroamines can also be used as oxidants or as disinfectants. The efficiency of chloroamines is lower than the efficiency of chlorine or hypochlorite, but lasts much longer which can be advantageous in special cases. They are frequently used in the USA, but their use is not allowed, e.g., in Germany. The reaction of chlorine and hypochlorite with natural organic matter forms undesired disinfection by-products which are sometimes reasons for hygienic concerns. Especially for trihalomethanes, which may be formed by these reactions, concentration limits are fixed in most of the drinking water quality regulations.

Much less disinfection by-products (trihalomethanes and chloroamines) are formed, if chlorine dioxide (ClO_2) is used. Chlorine dioxide is very unstable and, therefore, must be produced at the site where it should be added. The amount of chlorine dioxide which can be added is restricted by the formation of the toxic chlorite (ClO_2^-). Furthermore, addition of chlorine dioxide may cause problems concerning taste and odor of the drinking water.

4.7 Chemical Softening (Decarbonization)

Washing machines and dishwashers are frequently used in highly industrialized countries. Complexing agents in washing powder and ion exchangers integrated in the machines have come more and more in use to reduce the hardness of the water and to save detergents. But the complexing agents in washing powders and the salts necessary for regeneration of the ion exchangers pollute the environment and at least the water. Therefore, in developed countries the demand for soft water arose. In Germany, e.g., more than 50 technical treatment plants for softening are operated in water works at present.

Softening can be done by chemical means by the addition of milk of lime or lime water or by the addition of caustic soda: The following reactions take place where always calcium carbonate is precipitated.

$$Ca(HCO_3)_2 + Ca(OH)_2 \rightarrow 2\,CaCO_3\downarrow + 2\,H_2O$$
$$CO_2 + Ca(OH)_2 \rightarrow CaCO_3\downarrow + H_2O$$

or

$$Ca(HCO_3)_2 + NaOH \rightarrow CaCO_3\downarrow + Na(HCO_3) + H_2O$$
$$Ca^{2+} + CO_2 + 2\,NaOH \rightarrow CaCO_3\downarrow + 2\,Na^+ + H_2O$$

Besides the precipitation of calcium the carbon dioxide and the carbon hydroxide is also reduced. Therefore, these processes are also frequently called "decarbonization". If caustic soda is used for the process, the elimination of non-carbonate hardness is also possible. The precipitation of magnesium is usually not carried out because magnesium carbonate is quite soluble and the insoluble magnesium hydroxide only precipitates at pH values >10.5.

Furthermore, it will be possible to precipitate calcium carbonate by reduction of the carbon dioxide concentration by physical means as by aeration or by vacuum.

$$Ca(HCO_3)_2 \rightarrow CaCO_3\downarrow + CO_2\uparrow + H_2O$$

Chemical softening is quite frequently used. The treatment scheme is shown in Fig. 22. The softening reactor may be of the slow decarbonization or the rapid decarbonization type – the difference will be explained later. $Ca(OH)_2$ is added as milk of lime or as lime water. Additionally or alternatively caustic soda can be added. The calcium carbonate precipitates as a sludge or as pellets. As the effluent of the softening reactor still contains particular calcium carbonate a following filtration and the addition of flocculant or flocculant aid will be necessary. To adjust the pH value addition of carbon dioxide or an aeration may be necessary. A divided treatment is possible to reduce the size of the softening plant. Only one stream is softened to a very high degree and is mixed with the untreated water. Besides precipitation of calcium carbonate the precipitation of iron and manganese will be favored at high pH values and heavy metals will also be eliminated to some degree.

As already mentioned, two different types of decarbonization reactors are used: the rapid and the slow decarbonization reactors. The slow decarbonization process can be performed in normal flocculation/sedimentation tanks. Milk of lime and raw water are mixed in the central mixing zone. The precipitants are separated by sedimentation in the outer zone. A recirculation of the precipitated sludge and the addition of flocculants (iron salts) favor the formation of larger flocs with a higher sedimentation velocity. The process is called slow decarbonization, not because of the slow reaction kinetics, but because of the slow water velocity in the reactor. The slow velocity is necessary for the sedimentation of the precipitated hydroxides.

The rapid decarbonization works with a fluidized bed. Milk of lime or caustic soda is added together with the raw water at the bottom of the reactor. Fine sand with a grain size of 0.2–0.6 mm is inside the reactor. The sand bed is fluidized due to the upstreaming water

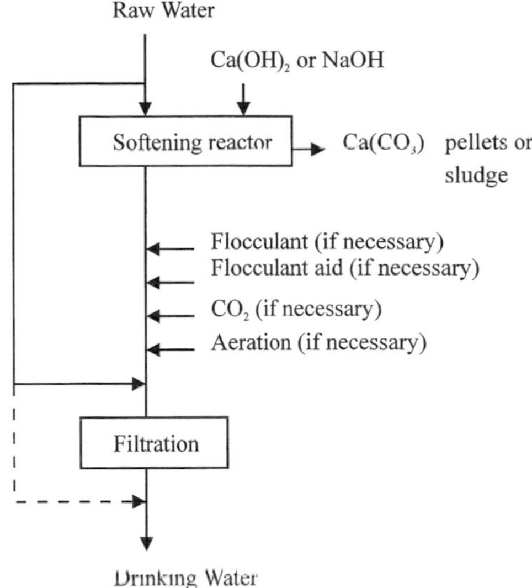

Fig. 22. General scheme of the softening process.

with a velocity of about 50 to 120 m h^{-1}. The calcium carbonate precipitates at the surface of the sand forming a coating of calcium carbonate on the sand grains. Due to further precipitation the size of the grains called pellets rises. If the grain size exceeds a limit of the diameter, the pellets are taken out and replaced by new sand. Rapid decarbonization may be disturbed by turbidities, high molecular organic substances, phosphate, and iron which may hamper the growth of the pellets and may produce suspended precipitate of calcium carbonate causing a high effluent turbidity. In these cases the use of slow decarbonization will be advised.

New slow decarbonization systems have been developed to speed up the decarbonization process as shown in Fig. 23. In the influent of the plant the water to be treated and lime water are mixed. With the use of several flocculation tanks, a controlled energy input and the addition of a flocculant aid flocs of high density are produced, which settle in a sedimentation tank with thickener and scraper. A lamella separator is additionally installed to improve the sedimentation. Excess sludge is partially recirculated. To avoid the sedimentation of the recirculated sludge in the flocculation chambers the sludge can be milled. With the improvement of the flocculation and the sedimentation the calcium carbonate flocs settle much faster than in the normal slow decarbonization. Therefore, this process is called “rapid slow decarbonization”.

During the last years people became more and more conscious of the importance of the type and quality of the milk of lime. The turbidity in the effluent of the decarbonization reactors stems not only from precipitates of the raw water, but also contains residuals of the milk of lime. If the particle size of the calcium hydroxide particles is too large they will not dissolve completely during the decarbonization process and will act as a crystallization germ for the precipitating calcium carbonate. These precipitants will neither attach to the pellets nor sediment in the available treatment time and will be taken out by the effluent water stream. Therefore, the efficiency of the decarbonization strongly depends upon the particle size of the milk of lime and their dissolving kinetics. The smaller the particle size and the faster the dissolving kinetics the lower the effluent turbidity and the lower the amount of milk of lime needed for softening. To produce a highly efficient milk of lime the following conditions must be kept:

- use of pure CaO of high reactivity,
- controlled and optimized temperature in the exothermic mixing process of CaO and water,
- use of water free of carbon dioxide, hydrogen carbonate, and sulfate.

The use of lime water for decarbonization will also be generally of advantage because calcium hydroxide has been already dissolved and will react very fast. Yet there is a big disadvantage in the use of lime water because of the large size of the lime water preparation units. Because of the low solubility of calcium hydroxide in water, about up to 20% of the treated water will be needed for lime water preparation. Therefore, the use of lime water was only possible in small water works but new processes for the preparation of lime water have been developed.

4.8 Deferrization and Demanganization

Iron and manganese concentrations in the drinking water are undesired because they will produce precipitates of oxide hydrates in contact with air. Deferrization and demanganization are the processes frequently necessary to

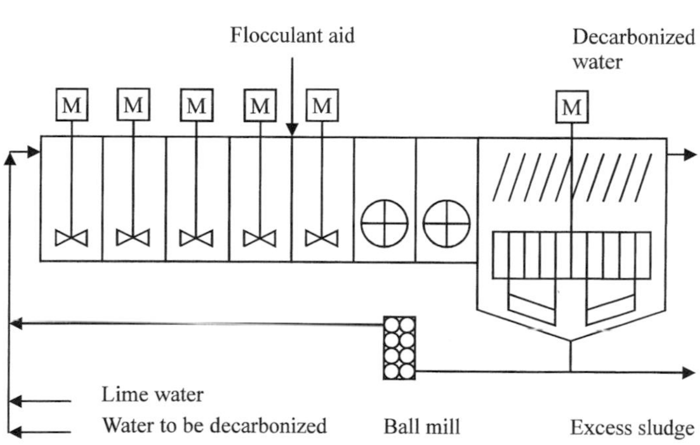

Fig. 23. “Rapid slow decarbonization” process.

reduce the concentration in the groundwater, which may be about 1–10 sometimes up to 50 mg L^{-1}, to the limits demanded by the drinking water decree (e. g., in Germany 0.2 mg Fe L^{-1} and 0.05 mg Mn L^{-1}). According to the chemical reactions the oxidation reactions are mainly dependent on the redox potential, the pH and the temperature as shown in Tab. 5.

The next higher redox potential will only be achieved, if the concentration of the reduced species of the actual step is about zero. Therefore, the iron oxidation will be hampered by sulfide and methane, but not by ammonium or nitrite and the oxidation of manganese will only be possible, if all previous oxidation reactions have been finished. The deferrization and the demanganization processes are usually carried out in filters. The chemical reactions are very slow, but the kinetics are speeded up by microbiological activity in the filters due to enzymes and the metabolisms of specialized iron and manganese oxidizing bacteria which will populate the filters in most cases. Furthermore, autocatalytical effects of the precipitated iron and manganese oxide hydrates speed up the kinetics of the reactions. Often both effects take place in the filters. Because of the complexity of the different mechanisms and reactions the layout of deferrization and demanganization filters is quite difficult and preliminary filter experiments should be done to determine the optimum pH, filter velocity, filter bed depths and filter material. The filter bed depths can be found between 1 and 3 m and filter velocities of about 10–20 m h^{-1} and sometimes up to 40 m h^{-1} can be observed. For biological deferrization the pH should be between 5.5 and 7.8 and for the demanganization higher than 6.5. The autocatalytic processes will need pH values clearly higher than 6.0 for deferrization and clearly higher than 7.0 for demanganization. The backwashing of the filters should be done softly. In autocatalytic processes high turbidities in the filter influents should be avoided because they may block the catalytic surfaces. In biological processes the backwashing should not be done with disinfected water to avoid damage to the microorganisms.

Tab. 5. Redox Reactions and Redox Potential

Reaction	Redox Potential in mV (pH about 7, 25 °C)
Sulfide oxidation to SO_4^{2-}	> -150
Methane oxidation	0
Fe(II) oxidation to Fe(III)	>150
NH_4^+ or NO_2^- oxidation	>500
Mn(II) oxidation to Mn(III)	>650

With the addition of chemical oxidants like potassium permanganate an oxidation process at lower redox potentials as shown in Tab. 5 is possible.

4.9 Membrane Filtration

Since many years desalination by reverse osmosis can be considered as a standard process for drinking water production from brackish water or sea water. For the treatment of inland water the use of membrane processes like ultra-, micro-, and nanofiltration have become more and more common in the last years.

Fig. 24 shows the possible use of membrane processes depending on the compounds which should be removed from the raw water. If surface water is considered, removal of turbidity and a final disinfection will always be necessary. The use of membrane processes in surface water treatment has become a point of increasing interest in the last years due to the high removal efficiencies for parasites and for viruses. For the removal of these particulate substances microfiltration (MF) and ultrafiltration (UF) may be used, but the almost complete removal of both, parasites and viruses, is only possible by UF. If the removal of organic compounds from the water is required, a combination of powdered activated carbon with MF or UF may be possible. If in addition inorganic compounds shall be removed membrane processes like nanofiltration (NF) or reverse osmosis (RO) will be necessary. NF will be especially suitable if beside the natural organic matter (NOM) sulfate and hardness should be removed from the raw water. By the use of RO the removal of undesired organic compounds and desalination will be possible.

Fig. 25 schematically shows the structures of the membranes of the different membrane

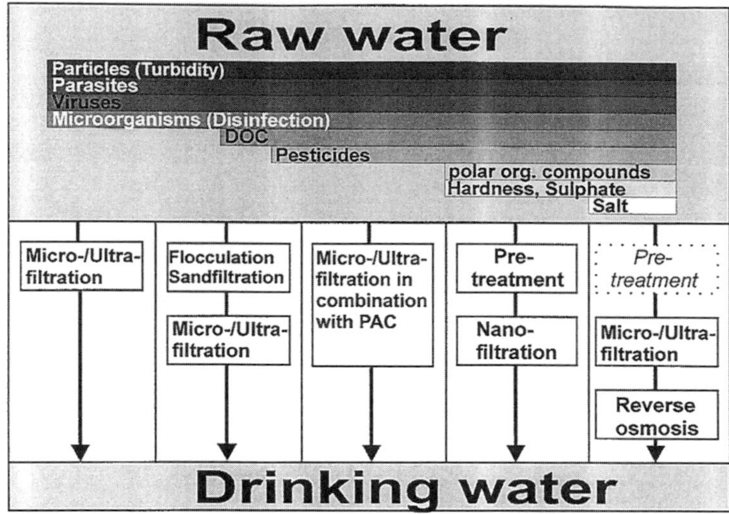

Fig. 24. Selective removal of surface water compounds by membrane processes.

processes. Up to now almost membranes made of organic materials are used in drinking water treatment. The RO membrane is characterized as a dense membrane without microscopic pores. The frequently used solution–diffusion model describes the mass transport through the membrane under the assumption that the permeating compounds are at first solved in the membrane (distribution equilibrium) and then move through the membrane which can be described by diffusive transport mechanisms.

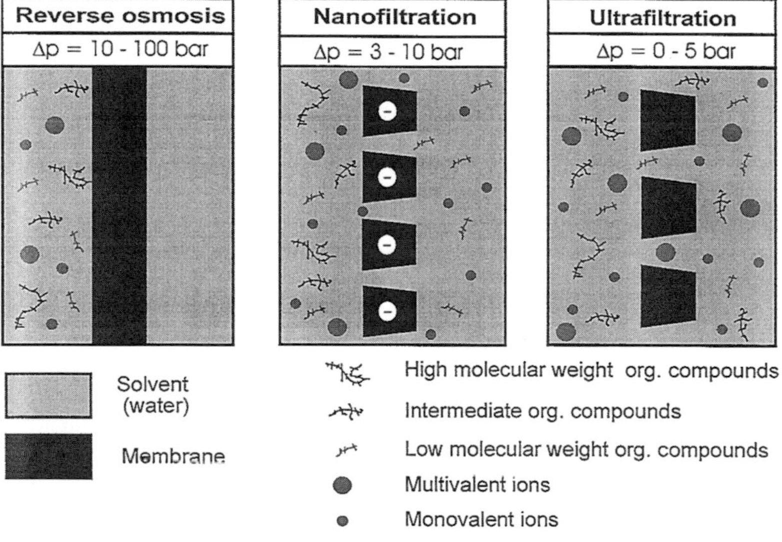

Fig. 25. Schematic structures of membranes.

The nanofiltration membrane can be characterized as a microporous membrane (pore diameter about 1 nm) with a charged surface, caused by carboxylic and sulfonic groups. Because of the usually negative charge of the membrane surface especially anions are removed very well. UF and MF membranes are real pore membranes and the removal can be explained approximately by sieving effects.

The principle of a membrane filtration process is shown in Fig. 26. The module containing the membrane is fed with raw water by a pump. The water flows tangentially over the membrane inside of the module (cross flow). The pressure inside of the module can be adjusted by a valve in the concentrate effluent pipe of the module. Because of the transmembrane pressure difference between the concentrate side and the permeate side of the membrane the water is pressed through the membrane. RO and NF membranes are operated without recirculation, but several membrane modules are subsequently connected in the concentrate effluent line. The UF can be operated in the recirculation mode (cross flow, valve V2 is closed) as well as in the dead end mode (valve V1 is closed).

The following parameters are important for the characterization of a membrane process:

Rejection R

$$R = 1 - \frac{c_P}{c_F}$$

c_P = concentration of a compound in the permeate,
c_F = concentration of a compound in the feed.

Recovery ϕ

$$\phi = \frac{\dot{V}_P}{\dot{V}_F}$$

$\dot{V}_P$ = flow of the permeate,
$\dot{V}_F$ = flow of the feed.

Using membrane processes to treat inland water a recovery as high as possible should be attempted, because the permeate flow becomes as small as possible under this condition. But this will cause a strong rise of concentration of the water compounds along the membrane surface on the concentrate side as it is shown

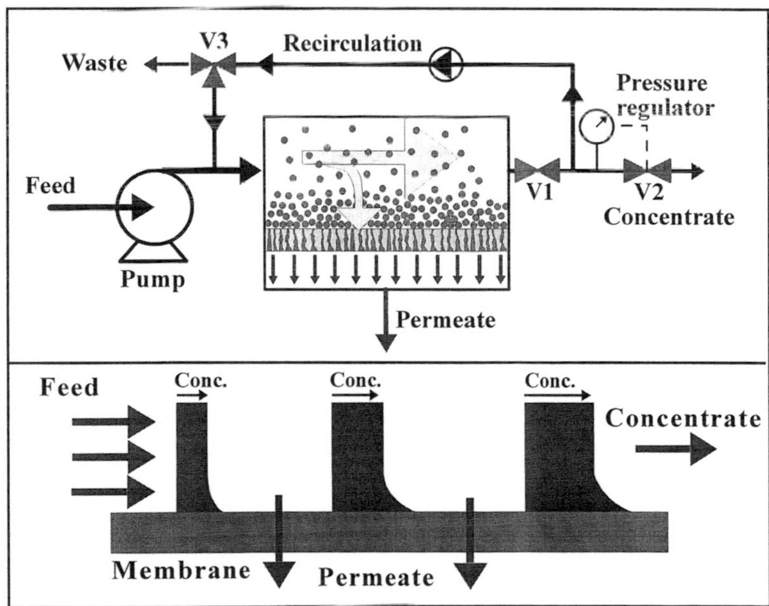

Fig. 26. Principle of a membrane filtration process.

in the bottom part of Fig. 26. Furthermore, this can result in the formation of a covering layer on the membrane surface. A decrease of the rise of concentration can be achieved by enhancing the cross flow velocity and by reducing the water flux through the membrane. In RO and NF processes the solubility of dissolved salts can be exceeded by the rise of concentration, and covering layers may grow up on the membrane which are very dense and hardly to remove. This effect is called scaling and should be avoided under any circumstances. Therefore, the recovery of these processes is only about 75–85%. In MF and UF processes covering layers can be reduced by regular flushing procedures.

In UF and MF processes in drinking water treatment capillary membranes are used in most cases because they can be backwashed and because the ratio of the total membrane surface to the total module volume is very high. A UF or MF module consists of many thousands of capillaries which are embedded in epoxy at both ends.

Fig. 27 shows an example for the construction of a commonly used UF module with the effluent for the filtrate in the pressure vessel. These modules have a total surface of up to 75 m^2 per module and they are vertically positioned in a treatment plant. If the capillaries are integrated in module elements with a central permeate collection pipe, as shown in Fig. 28, and such elements are positioned in horizontal pressure pipes, total surfaces of more than 100 m^2 per pressure pipe can be achieved.

Fig. 29 schematically shows the construction of a spiral wound reverse osmosis module element as it is used for RO or NF processes. Two membrane sheets which are divided by the so-called permeate spacer (connected hydraulically to the collection pipe) are glued at the central pipe. The two membrane sheets are al-

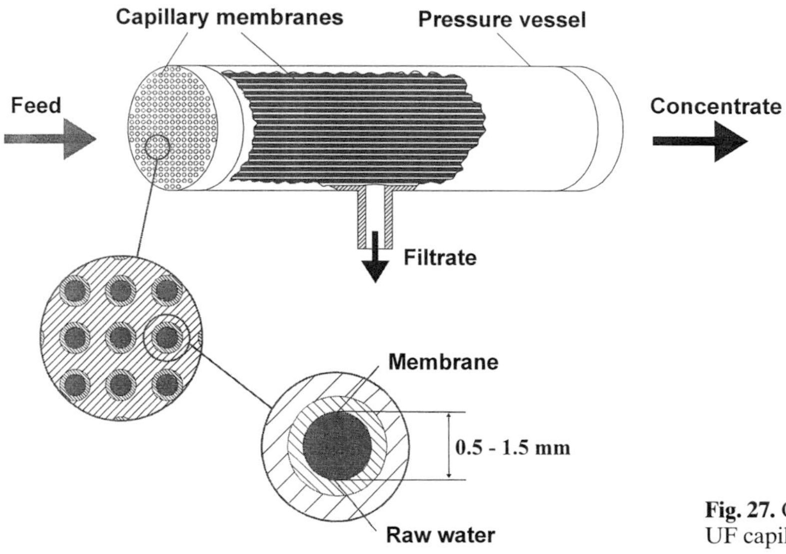

Fig. 27. Construction scheme of a UF capillary module.

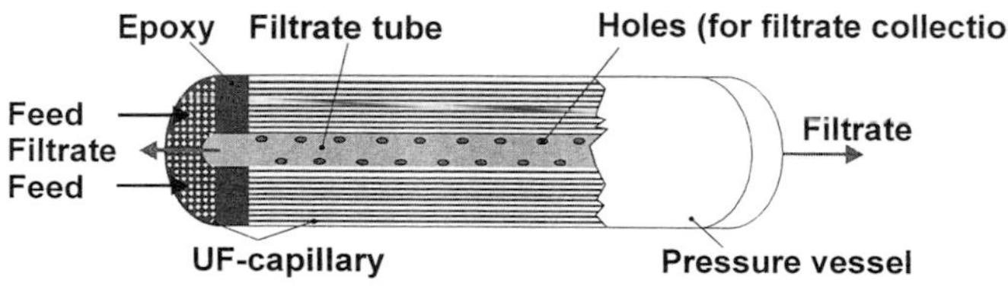

Fig. 28. Construction scheme of a capillary module with central permeate collection pipe.

so glued at their edges. Then a fabric, a so-called feed spacer, is laid on the membrane bag and finally the membrane bag and the feed spacer are rolled around the permeate collection pipe. The resulting flow conditions within such a spiral wound element can be compared to a cross flow arrangement. The feed flows through the feed channel between the membrane bags in axial direction. Parts of the water permeate through the membrane and flow through the spiral wounded permeate channel to the permeate collection pipe, by which it is removed from the membrane element.

The MF and UF processes are used in drinking water treatment, especially for the advanced removal of particles from surface waters or groundwater and spring waters influenced by surface water. Plants with a treatment capacity of some thousand m^3 per hour are built today.

The NF process is especially suitable for drinking water treatment, if the hardness is to be reduced and in addition organic compounds (NOM, but also non-polar and especially polar trace compounds) is to be removed significantly. The simultaneous removal of several undesired compounds by just one treatment step is a great advantage of the NF process, especially from an economic point of view.

5 Disinfection

Spreading of epidemics by infected drinking water until today is a potential danger. The first possibility to avoid the occurrence of pathogenic microorganisms in drinking water should be the use of unpolluted and well protected groundwater, which should be taken out of an aquifer with fine pores and good filtration characteristics. Such waters are not always available in the necessary quality and amount. In such cases other waters must be used and treated for the drinking water supply. The disinfection should finally guarantee the hygienic security of the drinking water supply, and it is the last step of the multi-barrier system (pollution control, water treatment, disinfection) for a safe drinking water supply.

By the disinfection process pathogenic germs like viruses, bacteria, and parasites should be killed or inactivated and the amount of microorganisms is generally reduced. Drinking water must not be sterile, but it is not allowed that it contains pathogenic microorganisms causing disease. The disinfection efficiency is determined by the

- kind of disinfectant,
- resistance of the microorganisms to the disinfectant,

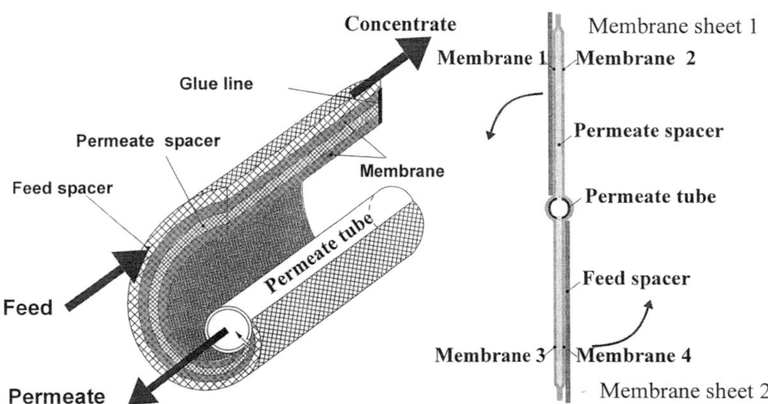

Fig. 29. Construction scheme of a spiral wound RO membrane element.

- kind of the disinfection process,
- amount of added disinfectant,
- contact time,
- quality parameters (e.g., temperature, pH, turbidity, number and kind of aggregation of the microorganisms, DOC, ammonium).

The German drinking water regulations (Verordnung über Trinkwasser und über Wasser für Lebensmittelbetriebe, Trinkwasserverordnung – TrinkwV, 1990), e.g., allow for chemical disinfection only the use of the following substances:

- chlorine,
- sodium, calcium or magnesium hydrochlorite (and chlorite of lime),
- chlorine dioxide,
- ozone.

The following limits are given by the German drinking water regulations (Verordnung über Trinkwasser und über Wasser für Lebensmittelbetriebe, Trinkwasserverordnung – TrinkwV, 1990) for the use of chemical disinfectants (Tab. 6).

If chlorine or hypochlorites are used hypochloric acid will be formed in the reaction with water. The hypochloric acid is the effective disinfectant, but it dissociates at pH values >6 to protons and hypochlorite, which has a much lower disinfection efficiency than the hypochloric acid. At a pH of about 7.5 nearly equal amounts of hypochlorite and hypochloric acid can be found. Hypochloric acid reacts with ammonium to chloroamines, which have a low disinfection efficiency and cause taste and odor. The direct use of chloroamines for disinfection, e.g., is not allowed in Germany. If the water contains higher amounts of humic acids and/or bromide the formation of trihalomethanes (THM) is possible. Such substances are supposed to be cancerogenic and their concentration is, therefore, limited (e.g., in Germany to 10 μg L^{-1}; Verordnung über Trinkwasser und über Wasser für Lebensmittelbetriebe, Trinkwasserverordnung – TrinkwV, 1990). The THM formation will be enforced by higher temperatures and pH values. The maximum and minimum concentration of the disinfectant after disinfection are not given to ensure a protection of the distribution net, but to ensure sufficient addition of the disinfectant.

Chlorine dioxide has become more and more important for disinfection in the last years. Chlorine dioxide does not form trihalomethanes and does not react with ammonium. Yet it oxidizes dissolved organic carbon compounds and is reduced to chlorite. The chlorite concentration, e.g., is limited by the German drinking water regulations to a concentration of 0.2 mg L^{-1}. If the concentration of dissolved organic carbon exceeds 2 mg L^{-1} it can be expected that the maximum allowed concentration of 0.4 mg L^{-1} chlorine dioxide must be exceeded to achieve a residual minimum concentration of 0.05 mg L^{-1} and that the maximum concentration of chlorite will also be exceeded.

Tab. 6. Limits for Disinfectants and Disinfection By-Products According to the German Drinking Water Regulations (Verordnung über Trinkwasser und über Wasser für Lebensmittelbetriebe, Trinkwasserverordnung – TrinkwV, 1990)

Limit	Chlorine, Hypochlorites, (Free Chlorine)	Chlorine Dioxide	Ozone
Maximum addition (mg L^{-1})	1.2	0.4	10
Maximum concentration after disinfection (mg L^{-1})	0.3	0.2	0.05
Minimum concentration after disinfection (mg L^{-1})	0.1	0.05	–
Disinfection by-product	THM	Chlorite	THM
Limit for disinfection by-product (mg L^{-1})	0.01	0.2	0.01

THM: chloroform + monobromedichloromethane + dibromemonochloromethane + bromoforme

Ozone has a much higher disinfection efficiency than hypochloric acid or chlorine dioxide. It reacts quite fast and a long time disinfection and protection of the distribution net is not possible using ozone for disinfection. Ozonation improves the biodegradability of organic substances and, therefore, the tendency for a regermination after ozonation is quite high. Ozonation should not be used for the final disinfection, but it is frequently used in surface water treatment in combination with a following biodegradation step like, e.g., a granular activated carbon filtration. If the water contains bromide it is possible, that bromomethanes and bromate is formed by the ozonation, which may restrict the use of ozone.

For physical disinfection the following procedures are possible:

- thermal disinfection ($T > 70\,°C$),
- UV light,
- ultrafiltration (see Sect. 4.9).

Because of the high energy consumption the thermal disinfection is not used in water works. Nevertheless, the thermal disinfection is a useful mean to disinfect the warm water supply and distribution systems and to avoid the growth of thermophilic microorganisms, especially such as *Legionella*.

The use of UV light for disinfection is quite a new process. The UV light damages the DNA of the microorganisms and prevents reproduction. UV disinfection does not produce detectable disinfection by-products and does not improve the biodegradability of the DOC. If high nitrate concentrations occur in the water the formation of nitrite may be possible, which is limited (e.g., in Germany to 0.1 mg L^{-1}). The following conditions should be kept for UV disinfection:

- UV extinction of the water at 254 nm ($<8\ m^{-1}$),
- low turbidity (<0.3 FNU),
- low concentrations of iron (<0.03 mg L^{-1}) and manganese (<0.02 mg L^{-1})
- minimum intensity of 400 J m^{-2},
- emission of the UV lamp at about 254 nm (to avoid photo reactions at smaller wavelengths, e.g., $NO_3^- \rightarrow NO_2^-$).

Acknowledgement

The authours would like to acknowledge the assistance of Dipl.-Ing. André Lerch in preparation and translation of the manuscript.

6 References

Anforderung an die Aufbereitung von Oberflächenwässern zu Trinkwasser im Hinblick auf die Eliminierung von Parasiten (1997), *Bundesgesundheitsblatt* 12/97.

BGW-Bundesverband der deutschen Gas- und Wasserwirtschaft (Ed.) (1996), *Entwicklung der öffentlichen Wasserversorgung 1990–1995*. Bonn: Wirtschafts- und Verlagsgesellschaft Gas und Wasser.

Buxton, G. V. et al. (1988), Critical review of rate constants for reactions of hydrated electrons and hydroxyl radicals in aqueous solution, *J. Phys. Chem. Ref. Data* **17**, 513–759.

Council Directive 80/778/EEC of 15 July, 1980 relating to the quality of water intended for human consumption (30. 8. 1980), *Official Journal of the European Communities* OJ L229, p. 11.

Council Directive 98/33/EC of 3 November, 1998 on the quality of water intended for human consumption (5. 12. 98), *Official Journal of the European Communities* OJ L330, p. 32.

DIN 2000 – Zentrale Trinkwasserversorgung (1973), Fachnormenausschuß Wasser im Deutschen Institut für Normung e.V. Berlin, Köln: Beuth-Verlag.

Gimbel, R. (1989), Theoretical approach to deep bed filtration, in: *Water, Wastewater, and Sludge Filtration* (Vigneswaran, S., Ben Aim, R., Eds.), pp. 17–56. Boca Raton, FL: CRC Press.

Gimbel, R. et al. (1997), Removal of microorganisms by clarification and filtration processes, in: *International Report* 6, pp. 1–6, IWSA Congress 1997. Oxford: Blackwell Science.

Guidelines for Drinking Water Quality: Health Criteria and Other Supporting Information (1996) 2nd Edn., Vol. 2, World Health Organization Geneva. Mastercom/Wiener Verlag.

Haberer, K. (1987), Technologische Alternativen in der Wasseraufbereitung, in: *DVGW-Fortbildungskurse Wasserversorgungstechnik für Ingenieure und Naturwissenschaftler*, Kurs 6: Wasseraufbereitungstechnik für Ingenieure 3rd Edn., pp. 28.1-28.4, DVGW-Schriftenreihe Wasser Nr. 206. Bonn: ZfGW-Verlag.

Hölting, B. (1991), Geogene Grundwasserbeschaffenheit, in: *Handbuch Bodenschutz*, 6. Lfg. 1/91 (Rosenkranz, D., Einsele, G., Harress, H.-M.,

Eds.). Berlin: Erich Schmidt Verlag.

KIURU, H. (1997), Removal of microorganisms by clarification and filtration processes, Special contribution, in: *International Report, IWSA Congress*. Oxford: Blackwell Science.

NETA, P. et al. (1988), Rate constants for reactions of inorganic radicals in aqueous solution, *J. Phys. Chem. Ref. Data* **17**, 1027–1100.

ROENNEFAHRT, K. W. (1983), Filtration, in: *DVGW-Schriftenreihe Wasser* Nr. 206, 2nd End., pp. 6.1–6.34. Frankfurt: ZfGW-Verlag.

SCHWOERBEL, J. (1993), *Einführung in die Limnologie*. Stuttgart, Jena: Gustav Fischer Verlag.

SMOLUCHOWSKI, M. V. (1916), Versuch einer mathematischen Theorie der Koagulationskinetik kolloider Lösungen, *Z. phys. Chem.* **92**, 126.

SONTHEIMER, H. (1978), Flocculation in water treatment, in: *The Scientific Basis of Flocculation* (IVES, K. J., Ed.), Series E, Applied Science No. 27. Sijthoff & Noordhoff, The Netherlands.

SONTHEIMER, H., FRICK, B., FETTIG, J., HÖRNER, H., HUBELE, C., ZIMMER, G. (1988), *Activated Carbon for Water Treatment* 2nd Edn. DVGW-Forschungsstelle am Engler-Bunte-Institut der Universität Karlsruhe, Germany.

Technische Regeln: Arbeitsblatt W 218, Flockung in der Wasseraufbereitung-Flockungstestverfahren (November 1998), *DVGW Regelwerk*. Bonn: Wirtschafts- und Verlagsgesellschaft Gas und Wasser mbH.

Technische Regel: Arbeitsblatt W 211-Filtration in der Wasseraufbereitung, Part 2: Planung und Betrieb von Filteranlagen (September 1987), *DVGW Regelwerk*. Frankfurt: ZfGW-Verlag.

Verordnung über Trinkwasser und ¸ber Wasser für Lebensmittelbetriebe (Trinkwasserverordnung – TrinkwV) (12. Dezember 1990), BGBI. I, p. 2613.

ZIMMER, G., SONTHEIMER, H. (1989), Beschreibung der Adsorption von organischen Spurenstoffen in Aktivkohlefiltern, *Vom Wasser* **72**, 1–19.

Further Reading

Degrèmont-Water Treatment Handbook 6. Edn., Vols. 1 and 2 (1991). Paris: Lavoisier Publishing.

GROMBACH, P., HABERER, K., TRUEB, E. U. (1985), *Handbuch der Wasserversorgungstechnik*. München Wien: R. Oldenbourg Verlag.

PORTER, M. C. (Ed.) (1990), *Handbook of Industrial Membrane Technology*. Park Ridge: Noyes Publications.

IVES, K. J. (1975), *The Scientific Basis of Filtration*. Nordhoff-Leyden, The Netherlands.

Membranfiltration in der Trinkwasseraufbereitung (1999), Dokumentation zum BMBF-Statusseminar in Roetgen vom 10. November 1999, Berichte aus dem IWW Rheinisch-Westfälisches Institut für Wasserforschung gemeinnützige GmbH Vol. 28.

MUTSCHMANN, J., STIMMELMAYR, F. (1995), *Taschenbuch der Wasserversorgung*, 11th Edn. Stuttgart: Franckh-Kosmos.

Ullmann's Encyclopedia of Industrial Chemistry (1996), Vol. A28 Water. Weinheim: VCH.

Wasserchemie für Ingenieure (1993), *Lehr- und Handbuch Wasserversorgung* Vol. 5. München: R. Oldenbourg Verlag.

American Water Works Association (1990), *Water Quality and Treatment – A Handbook of Community Water Supplies* 4th Edn. New York: McGraw-Hill.

19 Hygienic Aspects of Drinking Water

DIRK SCHOENEN

Bonn, Germany

1 Introduction

It has been known for a very long time that drinking water affects human health, as was already described in antiquity. Although there was a lack of chemical and microbiological methods of investigation, the taste and smell of water was checked with great care. In addition, it was already recommended that water had to be boiled before drinking in order to preclude a risk to health.

In Central Europe, there have only been occasional hygiene-based approaches to improving the quality of water up to the 18th century. Since the number of people using the water remained small owing to the small output of the water supply systems, little importance was attached to the cases of illness because of the generally unfavorable epidemiological situation of earlier centuries. The individual wells in the cities did not necessarily supply unhygienic water.

However, new water supply technologies led to a fundamental change at the middle of the 19th century. Since this time, it was also possible to build large water works and establish extensive municipal networks, so that the inhabitants of the rapidly growing cities were supplied from one water source. Diseases broke out in large populations which had not been observed up to that time. The Hamburg cholera epidemic in 1892 is a well-known example of this (KOCH, 1893).

As shown in Tab. 1, only cholera and typhus were known waterborne diseases in the first decades of the 19th century. In the meantime, a large number of disease agents that can be transmitted via drinking water has been identified (Tab. 1, THOFERN, 1990)). The great importance of epidemics related to drinking water with large numbers of cases of infection in various countries is shown by a compilation made by STÜRCHLER and published in 1990 (Tab. 2).

Tab. 3 lists diseases which can be transmitted via water. In an analysis of the diseases which can be spread via drinking water, a fundamental distinction must be made between agents that are already present in untreated water in amounts to entail a risk to health, and agents which only proliferate in the network of distribution to such an extent that they can endanger health. The causative agents of disease which may already be present in untreated water in numbers representing a health risk and which are distributed with the drinking water will be described below.

For a long time, interest in drinking water hygiene was concentrated almost exclusively on chemical constituents of the water. There has been a major change owing to the observation of the spread of parasites with water in industrialized countries. Tab. 4 lists incidents caused by parasites in water. At present, the greatest epidemiological importance is attached to Cryptosporidia, not least owing to the epidemic in Milwaukee in 1993 with more than 400,000 cases of infection in a population of 1.6 million (MAC KENZIE et al., 1999). In developing countries, how-ever, the classical causative agents of bacterial and viral disease continue to play a major role, as the cause of disease outbreaks due to drinking water.

2 Diseases Caused by *Cryptosporidium parvum*, *Giardia lamblia*, and *Toxoplasma gondii*

Cryptosporidium and *Giardia* are protozoal intestinal parasites which are widespread throughout the world. They can lead to severe diarrhea and non-specific general symptoms in humans. The persistent forms of *Cryptosporidium* and *Giardia* (oocysts or cysts) are transmitted orally via the feces and can also spread with the water. In order to cause an outbreak of the disease, only a few parasites have to be ingested. Even a single parasite may lead to an infection. Severe illnesses occur especially in individuals with impaired immunity, e.g., AIDS patients. This also applies to infants with a immune system that has not yet fully developed. In immunocompetent individuals, the symptoms of disease are very much less pronounced and subside after 6 weeks at the latest, even without specific treatment. The persistent forms of the parasites are not only ex-

Tab. 1. Waterborne Disease Outbreaks Associated with Drinking Water in the 19th and 20th Century (THOFERN, 1990)

Place	Year	Disease	Number of Cases	Number of Deaths
London	1854	cholera	–	–
Switzerland	1867	cholera	–	–
Halle	1871	typhoid fever	282	11
Lausen (Swizerland)	1872	typhoid fever	130	–
Stuttgart	1872	typhoid fever	180	14
Zürich	1884	typhoid fever	1,621	148
Hamburg	1885/88	typhoid fever	15,804	1,214
Chemnitz	1888	typhoid fever	2,516	–
Berlin	1889	typhoid fever	632	–
Altona	1891	typhoid fever	685	47
Hamburg	1892	cholera	16,956	8,605
Nietleben (Irrena.)	1893	cholera	122	52
München (Kaserne)	1893	typhoid fever	331	–
Paderborn	1893	typhoid fever	150	–
Paris	1894	typhoid fever	419	–
Beuthen	1897	typhoid fever	474	–
Maidstone (GB)	1897	typhoid fever	1,748	–
Paderborn	1898	typhoid fever	234	32
Gräfrath (Solingen)	1898/99	typhoid fever	155	–
Bochum	1900	typhoid fever	ca. 500	43
Gelsenkirchen	1901	typhoid fever	3,200	350
Detmold	1904	typhoid fever	780	54
Gräfrath (Solingen)	1904	typhoid fever	118	11
Greiz	1908	typhoid fever	140	–
St. Petersburg	1908	cholera	9,000	4,000
Altwasser	1909	typhoid fever	622	32
Reutlingen	1909	typhoid fever	290	–
Jena	1915	typhoid fever	537	60
Pforzheim	1919	typhoid fever	ca. 4,000	400
Alfeld	1923/24	typhoid fever	over 1,100	100
Hannover	1926	typhoid fever	ca. 2,500	260
Rostow/Don	1926	gastroenteritis	ca. 16,000	–
Rostow/Don	1926	typhoid fever	2,935	–
Lyon	1928	typhoid fever	3,000	300
Yugoslavia	1938	weil disease	390	8
Philadelphia	1944	hepatitis A	344	–
Westerode	1945/46	typhoid fever	ca. 400	26
Neu-Ötting	1946	typhoid fever	400	–
Klafeld	1946	typhoid fever	325	10
Greifswald	1947	typhoid fever	–	–
Neu-Ötting	1948	typhoid fever	ca. 600	96
Waldbröl	1949	typhoid fever	127	11
Drolshagen	1951	typhoid fever	51	–
Thereker Mühle	1953	typhoid fever	ca. 50	–
Drolshagen	1955	typhoid fever	92	–
New Delhi	1955	hepatitis A	28,745	–
Hagen	1956	typhoid /paratyphoid fever	500	–
California	1965	salmonellosis	16,000	3
East-Slovakia	1967	tularemia	228	–
Worchester	1969–71	hepatitis A	1,174	–
Heidenau (Pima)	1971	shigellosis	482	–
Worbis	1972	shigellosis	ca. 1,400	–
Dingelstedt (Thür.)	1972	hepatitis A	ca. 40	–

Tab. 1. Continued

Place	Year	Disease	Number of Cases	Number of Deaths
Ismaning	1978	shigellosis	2,400	–
Jena	1980	typhoid fever	63	–
Halle (Beesen)	1981	rota viral enteritis	–	–
Oberes Vogtland	1983	*Aeromonas* enteritis	–	–
Bristol (GB)	1985	giardiasis	108	–
Carrolton (USA)	1987	cryptosporidiosis	ca. 13,000	–
Ayrshire (GB)	1988	cryptosporidiosis	27	–
Oxfordshire, Swindon (GB)	1989	cryptosporidiosis	ca. 5,000	–

– = no statement

Tab. 2. Waterborne Disease Outbreaks Associated with Drinking Water 1980–1988 with More than 100 Cases (STÜRCHLER, 1990)

Year	Country	Source/Agent	Number of Cases
1980	Sweden	*Campylobacter*	2,000
1981	Algeria	Hepatitis NANB	790
1981	USA	Rotavirus	1,500
1982	Taiwan	Poliovirus	1,000
1983	USA	*Giardia lamblia*	780
1984	India	*Shigella dysenteriae*	78,000
1985	USA	*Campylobacter*	150
1985	Sudan	*Vibrio cholerae*	1,175
1985	USA	*Giardia lamblia*	700
1985	Botswana	Hepatitis NANB	270
1986	Thailand	*Shigella dysenteriae*	10,090
1986	Somalia	Hepatitis NANB	2,000
1987	Thailand	*Vibrio cholerae*	910
1987	USA	*Cryptosporidium*	13,000
1987	USA	Norwalkvirus	350
1988	China	*Vibrio cholerae*	3,950
1988	Angola	*Vibrio cholerae*	32,500

creted by infected persons, but also by individuals without detectable symptoms of disease. The parasites also lead to diarrhea in animals. Especially young animals such as calves and lambs are affected. Up to 40% of calves with diarrhea are infected with *Cryptosporidium*. Farm animals, pets, and wild animals contribute to spreading the parasites.

Toxoplasma infections are also widespread throughout the world, but generally do not lead to conspicuous disease. In primarily healthy persons, only mild general symptoms and lymph node swelling occur – if at all. In AIDS, encephalitis may occur which not uncommonly is the immediate cause of death. However, severe lesions may also occur in neonates, if the mothers had first contact with the infectious agent during pregnancy and the infectious agent was transmitted to the fetus. Only one epidemic was observed up to now in which the infectious agent was transmitted via a central water supply (see Tab. 4, ISAAC-RENTON et al., 1988). There were already cases in the past of people who became ill after drinking surface water.

Tab. 3. Agents Causative of Diseases Spread via Drinking Water and Already Present in Untreated Water in Amounts to Entail a Health Risk (SCHOENEN, 1997)

Bacteria	Viruses	Protozoa
Vibrio cholerae	Poliovirus	*Entamoeba histolytica*
Salmonella typhi	Hepatitis A und E virus	*Giardia lamblia*
Salmonella paratyphi	Enterovirus	*Cryptosporidium parvum*
Salmonella spp.	Rotavirus	*Toxoplasma gondii*
Shigella spp.	Adenovirus	
Yersinia enteritidis	Norwalkvirus	
Campylobacter spp.	Coxsackievirus	
E. coli (EHEC)		

Agents causative of diseases spread via drinking water and growing in the distribution system:
Pseudomonas aeruginosa
Legionella pneumophila
Mycobacteria avium complex
Aeromonas hydrophila
Acinetobacter spp.
Yersinia enterocolitica
Acantamoeba spp.
Naegleria fowleri

3 Presence, Transmission, and Dissemination of Parasites in Water

In some cases, *Cryptosporidium* and *Giardia* are excreted by humans and animals in large amounts with the feces. Up to 10^{10} oocysts per day can be excreted by a calf. The parasites can pass directly into all surface waters and into the groundwater with the feces of humans and animals via sewage, liquid manure, or dung. The causative agent of toxoplasmosis can be spread in various ways. Raw meat and contact with cats is of upmost importance. Cats and other animals of feline group of species excrete the infectious agent with the feces. The infectious agents which are passed with the feces can also be transmitted via water.

The number of parasites in water varies within wide limits. A definitive correlation with the classical fecal indicators such as *E. coli* and coliform bacteria cannot be demonstrated. It is also not expected to be found as in other epidemic diseases. The parasite may occur in high (and, exceptionally extremely high) concentrations of up to several thousands per liter in any water contaminated by feces. When parasites are detected, the water always shows fecal contamination which can be analyzed by means of conventional bacteriological methods. However, the parasites can also be temporarily absent when infectious agents are not excreted for a time either by humans or by farm animals, domestic pets, or wild animals in the catchment area of the water or when they are not temporarily washed into the water. Any water with only occasional fecal contamination must thus be regarded as potentially contaminated by parasites, irrespective of the individual parasitological findings. Even streams and lakes in uninhabited areas without sewage inflow are affected. Contamination of karst water is a particular problem. In karst systems, the water quality may change very quickly and contaminated surface water can be transported over large distances. Only well-protected groundwater can be regarded as free of infectious agents, including parasites.

Tab. 4. Listing of Outbreaks that are Caused by Parasites (*Giardia lamblia, Cryptosporidium parvum*, and *Toxoplasma gondii*) via Drinking Water

Agents	Place	Population	Number of Cases	Origin of the Water	Treatment ot the Water	Cause
Giardia of 1965/66	Aspen, CO (USA)	1094 skiers	123	mountain stream, 3 well	chlorination	contamination, 2 wells by wastewater
Giardia 1969–73	Leningrad (USSR)	1419 travellers	23%	n.s.	n.s.	n.s.
Giardia 1973	Leningrad (USSR)	members of NASA	81%	n.s.	n.s.	n.s.
Giardia 1974/75	Roma NY (USA)	n.s.	proved: 350 calculated: 5,300	surface water	chlorination	temporary contamination
Giardia 1976	Camas, WA (USA)	6,000	600	mountain stream	filtration, disinfection	n.s.
Giardia 1977	Berlin, NH (USA)	1,500	750	n.s.	filtration	n.s.
Giardia 1979	Bradford, PA (USA)	n.s.	3,500	n.s.	disinfection	n.s.
Giardia 1982	Mjövik (S)	600	(*Giardia*) 56 (Gastro-enterities) 454	groundwater	filtration to pH regulation	backwater of wastewater and conversion into the well
Cryptosporidium 1984	Bruan Station, TX (USA)	5,900	2,006	groundwater	chlorination	contamination of the well
Giardia 1985	Bristol (GB)	n.s.	108	n.s.	n.s.	contamination during repair of a water pipe
Cryptosporidium 1986	Sheffield, S. Yorks (GB)	n.s.	84	surface water	n.s.	drain of faeces of cows during a storm water runoff
Cryptosporidium 1987	Carroll County, GA (USA)	32,400	12,960	surface water	traditional treatment	faulty treatment
Cryptosporidium 1988	Ayrshire (GB)	24,000	27	n.s.	n.s.	seep from liquid manure of cows and contamination of drinking water over a layed up pipe
Cryptosporidium 1989	Swindon/Oxfordshire (GB)	741,092	516	surface water	traditional treatment	feedback of backwash water into the raw water
Cryptosporidium 1989–90	Humberside (GB)	n.s.	n.s.	n.s.	n.s.	n.s.
Cryptosporidium 1990	Loch Lomond (GB)	n.s.	147	surface water	n.s.	n.s.

Cryptosporidium 1990–91	Isle of Thanet (GB)	177,300	47	surface water	traditional treatment	treatment not adapted to load
Cryptosporidium 1991	Pennsylvania, PA (USA)	n.s.	551	groundwater	chlorination	faulty treatment
Cryptosporidium 1991	South London (GB)	n.s.	44	n.s.	traditional treatment	n.s.
Cryptosporidium 1992	Jackson County, OR (USA)	160,000	43	spring and surface water	chlorination	faulty treatment
Cryptosporidium 1992	Bradford (GB)	50,000	125	surface water	slow sand filtration and chlorination	intense rainfall and one filter did not yet show full work after routine maintenance
Cryptosporidium and 1992/1993	Warrington (GB)	38,200	47	groundwater from sandstone	natural underground filtration and chlorination	intensive rainfall no treatment
Cryptosporidium 1993	Washington, DC (USA)	600,000	n.s.	n.s.	filtration, chlorination	insufficient filtration
Cryptosporidium 1993	Milwaukee, WI (USA)	1,600,000	403,000	surface water	traditional treatment	faulty treatment
Cryptosporidium *Campylobacter* 1993	Northumberland private Wasser-vers. (GB)	200	43	several sources	terminal filter and UV disinfection	intensive rainfall and brown coat on the quartz lamp of UV-reactors
Toxoplasma 1994–95	Great Victoria area British Columbia (CAN)	219,000	proved: 100 calculated: 7,718	surface water	chloramine	no filtration

n.s.: no statement.

4 Measures to Eliminate Causative Agents of Disease (Including Parasites) from the Water

There are various means of avoiding dissemination of infectious agents. On the one hand, these measures comprise water protection. On the other hand, pathogens can be removed by water processing and by disinfection. In particular, water processing in the form of filtration and disinfection, especially with chemical agents, shows various degrees of effectiveness in this respect. The persistent forms of *Cryptosporidium, Giardia*, and *Toxoplasma* show high resistance to environmental influences. Even after being in the water for a long time, they remain infectious. Due to their high resistance, the persistent forms of parasites cannot be killed adequately by disinfection of drinking water with chlorine, chlorine dioxide, or UV radiation. Conventional disinfection cannot preclude the risk of infection by parasites in water. This explains the epidemics spread by water supplies in which fecal contamination could no longer be detected in disinfected water, but infectious parasites were still present. The parasites can be killed reliably only by heating. In laboratory investigations, a relatively high killing rate could be achieved with ozone at high concentrations and long periods of exposure (FAUST and ALY, 1999; WHITE, 1999).

In processing of drinking water by filtration with or without flocculation, the parasites behave like particles or other microorganisms including bacterial and viral pathogens. The parasites can be reliably removed from the water only by protection of untreated water or in combination with sufficient filtration. The efficiency of processing must be adapted to the quality of untreated water. A breakthrough of the filters has to be avoided, even if several unfavorable factors such as high contamination of untreated water and high throughput of water coincide.

The great importance of inadequate processing by filtration and the dissemination of parasites with drinking water are the causes of the epidemics listed in Tab. 4. A total of 26 instances are listed. In eight cases, the authors do not specify the cause (Tab. 5). In eight further cases, there is no processing or it is not adequate to eliminate parasites.

There are fundamental differences which are also important in terms of infectious disease hygiene between conventional disinfection of drinking water on the one hand and processing of drinking water by filtration on the other. Using disinfection the microorganisms are killed, but their remnants remain in the water. Using filtration, the microorganisms including the causative agent of disease are removed from the water as a whole. However, it is more important from the point of view of infectious disease hygiene to note that not all causative agents of disease are eliminated equally well by filtration and disinfection, as shown in Tab. 6. A distinction must be made not only between the various groups of microorganisms (bacteria, viruses, parasites), but also as to whether the causative agents of disease are present singly and freely suspended in water which is typical for laboratory investigations, or whether they are located in particles of fecal origin as they occur in their natural environment. Only the freely suspended bacteria

Tab. 5. Reasons for the Outbreaks Caused by Parasites Listed in Tab. 4 (SCHOENEN et al., 1997)

Number of Outbreaks	
Giardiasis	9
Cryptosporidiosis	16
Toxoplasmosis	1
total	26
Cause:	
no statement	8
contamination of the wells with wastewater or surface water	3
intensive rainfalls	4
absence of treatment	1
faulty filtration	6
filtration backwash water into raw water	1
contamination of drinking water already flown into distribution system	
– on repair of pipe	1
– over a layed up pipe	1

Tab. 6. Elimination of Microorganisms Including the Causative Agents of Diseases from Water Depending on:
- procedure, – filtration, chemical, or physical disinfection
- group of microorganisms, – bacteria, viruses and parasites
- whether the microorganisms are single and freely suspended in water, as they are typically examined in laboratories or whether they are located in particles of fecal origin, as they occur in the natural environment

Procedure	Microorganisms Inclusive Pathogenic Agents					
	Single, Freely Suspended			In Particles of Fecal Origin		
	Bacteria	Viruses	Parasites	Bacteria	Viruses	Parasites
Filtration*	+	+	+	+	+	+
Disinfection						
Chlorine, chlorine dioxide	+	+	–	–	–	–
Ozone	+	+	(+)∅	–	–	–
UV light	+	+	(–)°	–	–	–
Thermal (>90 °C)	+	+	+	+	+	+

\+ Microorganisms including causative agents of diseases can be eliminated reliably.
– Elimination of microorganisms including causative agents of diseases not guaranteed.
* Filtration is hygienically sufficient, if the colony count in the filtrate is $< 100\ mL^{-1}$ and *E. coli* as well as coliform organisms are not detectable in 100 mL.
∅ Ozone has shown a quite good effect against parasites in laboratory examinations when the concentration was high and the contact time was long. Experiences in practice is still missing for a final judgement.
° UV light may possibly kill parasites as well as bacteria and viruses as early investigations show (SCHOENEN et al., 1997).

and viral agents, but not the parasites, can be reliably killed or inactivated by means of conventional drinking water disinfection with chlorine, chlorine dioxide, or UV radiation. (Early investigations show that parasites may possibly be killed by UV radiation as are bacteria and viruses.) In the same way, bacteria and viral pathogens which are attached to particles of fecal origin are well protected from the action of the disinfectant. On the other hand, parasites as well as bacterial and viral pathogens bound to particles can be reliably filtered. Optimal filtration can also remove single, freely suspended bacterial and viral pathogens from the water. The elimination of causative agents of disease in filtration corresponds to the purification of water in fine-pore groundwater aquifers. After filtration and before any disinfection the number of colonies may not exceed $100\ mL^{-1}$, and the fecal indicators *E. coli* and coliform bacteria may not be detectable in 100 mL. Only thermal disinfection (<90 °C) can kill or inactivate bacterial, viral, and parasitic causative agents of disease with equal reliability, irrespective of whether they are present singly in free suspension or whether they are present in fecal particles. The requirement of boiling water, if safe drinking water cannot be made available in any other way, therefore, leads to reliable protection from bacterial, viral, and parasitic pathogens.

In connection with preparation of drinking water, a high infectious disease hygienc quality has been ascribed again and again to groundwater without any further stipulations. Depending on the geological formation and its origin the water is just subject to fecal contamination as is surface water. Only under very specific conditions it is safe in terms of infectious disease hygiene. Only the fact that groundwater is used to supply drinking water thus does not mean that the water is free of causative agents of disease, including parasites. Qualitatively safe drinking water always shows a colony count of <100 mL and is always free of *E. coli* and coliform bacteria in 100 mL.

If it is processing and not disinfection with chlorine, chlorine dioxide, or UV radiation that makes a crucial contribution to avoid transmission of parasites with water, this must be appropriately taken into consideration in testing water. The classical bacteriological appraisal parameters such as colony counts, *E. coli* and coliform bacteria react sensitively to disinfection when these agents are not attached to particles. In order to rule out the health risk from persistent forms of parasites in water, the water must already be free of detectable fecal contamination before infection. This stipulation is not new. The classical bacteriological parameters for assessing drinking water (colony count, *E. coli* and coliform bacteria) were always applied before disinfection before introduction of the first German Regulations on Drinking Water in 1975 (Anonymous, 1975). In contrast, in the USA a distinction between the microbiological requirements after processing depending on whether this was before disinfection or only after disinfection was never made. However, if the water is not to be tested before, but after disinfection, as e.g. outlined in the EU Directive on Drinking Water, spores of *Clostridium perfringens* can be used for monitoring. This stipulation is based on the fact that the spores largely behave like parasites do in processing and disinfection. Above all, the spores (like the parasites) are highly resistant to disinfectants.

It has been repeatedly suggested in the past that turbidity of the water should be used as a parameter for assessing the infectious disease hygiene when monitoring drinking water. Special attention was given to the suggestion in connection with the risk from parasites.

In the USA, turbidity has been a parameter for infectious disease hygiene evaluation since 1974. It is undisputed that there is an especially high risk for transmission of causative agents of disease (including parasites) when the water has a high content of turbid substances (USEPA, 1989, 1994). However, up to now it has not been possible to derive a sufficiently substantiated threshold value for turbidity. The measurement of turbidity cannot replace infectious disease hygiene monitoring of the drinking water on the basis of conventional microbiological investigation parameters. It must be taken into consideration that the change of turbidity can only be measured with an increase of the colony counts from 0 to 100,000 mL^{-1} or more. However, only a maximum colony count of 100 mL^{-1} is tolerated in drinking water.

5 Conclusions

Even though all questions connected with the transmission of parasites by water have not been clarified, some important conclusions on the operation and surveillance of water supplies can be inferred:

- According to the present state of knowledge, well-protected groundwater resources which do not show any signs of fecal contamination on the long-term and regular testing can be regarded as safe with respect to transmission of the causative agents of disease, including the persistent forms of parasites.
- Even if only occasionally they show fecal contamination, all surface waters and groundwaters must be regarded as potentially contaminated with causative agents of disease, including parasites.
- Parasites as well as bacterial and viral pathogens bound to particles must be removed from the water in processing by means of filtration, since they are not killed by disinfection with chlorine, chlorine dioxide, and UV radiation conventionally used in processing drinking water. The required elimination must then be maintained without interruption. Even if several unfavorable factors coincide, e.g., severe contamination of untreated water and high water throughput, a passage of the pathogens must be precluded. The processing must be adapted to the most unfavorable untreated water quality anticipated. The return of the filter backwash water into the untreated water automatically leads to a high contamination of the untreated water and thus to an immediate risk due to the accumulation of pathogens. Even today the following specifications for filtration which were already formulated in 1896 by LÖFFLER are still valid without reservations:
  "To provide satisfactory results for a sand filter the following conditions must be fulfilled:
  1. a good untreated water with as low contamination as possible,

2. a slow rate of filtration,
3. constant activity of the filter,
4. the aliquots of water obtained at the beginning of each filtration period must be discarded."

- The backwash water of the filters must be discarded or subjected to further purification before it can be returned to the untreated water.
- Investigations under practical conditions are required to show whether disinfection with ozone at high concentrations and long times of action leads to adequate killing of parasites.
- According to what is known up to now, satisfactory drinking water quality in terms of infectious disease hygiene can be assumed, if the water does not show any fecal contamination before disinfection, like a well-protected groundwater. Disinfection should be used to avoid a residual risk from bacterial and viral pathogens, but not to eliminate demonstrable fecal contamination.
- All presently known epidemic pathogens which can be transmitted via water, including the persistent forms of the parasites *Cryptosporidia* and *Giardia* can be reliably killed by heating. In case of emergency when satisfactory drinking water cannot be made available in any other way boiling water prior to its use provides reliable protection.

New observations and further results of investigation will close the still existing gaps in knowledge and permit even greater safety of supply. If necessary, additional stipulations must be made with regard to processing and monitoring. The information now available no longer leaves any doubt that protection from parasites can only be ensured by comprehensive protection of untreated water and careful processing, but not by conventional disinfection of drinking water with chlorine or chlorine dioxide.

6 References

Anonymous (1975), Verordnung über Trinkwasser und über Brauchwasser für Lebensmittelbetriebe (Trinkwasser-Verordnung, *Bundesgesetzblatt* Teil 1, 453–461.

Faust, S. D., Aly, O. M. (1999), *Chemistry of Water Treatment* 2nd Edn. Boca Raton, FL: Lewis Publishers.

Isaac-Renton, J., Bowie, W. R., King, A., Irwin, G. S., Ong, C. S. et al. (1988), Detection of *Toxoplasma gondii* oocysts in drinking water, *Appl. Environ. Microbiol.* **64**, 2278–2280.

Koch, R. (1893), Wasserfiltration und Cholera, *Zbl. Hyg.* **14**, 393–426.

Löffler, F., Oesten, G., Sendtner, R. (1896), Wasserversorgung, Wasseruntersuchung und Wasserbeurteilung, in: *Handbuch der Hygiene* Vol. 1 (Weyl, T., Ed.). Jena: Gustav Fischer Verlag.

Mac Kenzie, W. R., Hoxie, N. J., Proctor, M. E., Gradus, M. S., Blair, K. A. et al. (1994), A massive outbreak in Milwaukee of *Cryptosporidium* infection transmitted through the public water supply, *New Engl. J. Med.* **331**, 161–167.

Schoenen, D. (1997), Möglichkeiten und Grenzen der Trinkwasserdesinfektion unter besonderer Berücksichtigung der historischen Entwicklung, *Gas- Wasserfach gwf (Wasser/Abwasser)* **138**, 61–74.

Schoenen, D., Botzenhart, K., Exner, M., Feuerpfeil, I., Hoyer, O. et al. (1997), Vermeidung einer Übertragung von Cryptosporidien und Giardien mit dem Wasser, *Bundesgesundhbl.* **40**, 466–475.

Stürchler, D. (1990), Epidemien. Auch heute aktuell, *Therapiewoche Schweiz*, 468–470.

Thofern, E. (1990), *Die Entwicklung der Wasserversorgung und der Trinkwasserhygiene in europäischen Städten vom 16. Jahrhundert bis heute, unter besonderer Berücksichtigung der Bochumer Verhältnisse*. Bochum: Wasserbeschaffung mittlere Ruhr GmbH.

USEPA (1989), *Surface Water Treatment Rule (SWRT) Drinking Water, National Primary Drinking Water Regulations; Filtration, Disinfection; Turbidity, Giardia lamblia, Viruses, Legionella, and Heterotrophic Bacteria*. Final Rule. Fed. Reg. 54:124:27486, June.

USEPA (1994), *National Primary Drinking Water Regulations; Enhanced Surface Water Treatment Requirements*. Proposed Rule. Fed. Reg. 59:145: 38832, July.

White, G. C. (1999), *Handbook of Chlorination and Alternative Disinfectants.* New York, Chichester: John Wiley & Sons.

20 Artificial Groundwater Recharge and Bank Filtration

Gudrun Preuß
Ulrich Schulte-Ebbert
Schwerte, Germany

List of Abbreviations

AOC	assimilable organic carbon
AOX	adsorbable organic halogens
BOD	biological oxygen demand
COD	chemical oxygen demand
DOC	dissolved organic carbon
E_h	redox potential
NOM	natural organic matter
NTA	nitrilotriacetic acid
PAH	polycyclic aromatic hydrocarbons
POC	particulate organic carbon
TOC	total organic carbon

1 Introduction

Today the use of artificial recharge of groundwater is spread world wide. Artificial recharge of groundwater is used for a broad range of applications (JOHNSOHN and FINLAYSON, 1989; KIVIMÄKI and SUOKKO, 1996; FRYCKLUND, 1998). These are, e.g., drinking water production, storage of fresh water, improvement of raw water quality, aquifer recovery, formation of barriers against saltwater intrusions, preservation of natural wetlands, infiltration of storm water runoff or disposal of treated sewage effluents. In all these applications water is infiltrated through infiltration systems into the underground and finally reaches the groundwater. The many different technically approved infiltration devices can be divided into surface recharge, subsurface recharge and bank filtration or induced recharge (Fig. 1). The different methods of recharge are explained in the following sections.

In all surface or subsurface infiltration techniques and the technique of bank filtration a granular deep bed filter acts as cleaning media. In this filter the water is cleaned to a certain extent before it is infiltrated into the aquifer. The filter consists of layers of sand or gravel and in the case of bank filtration of fine river or lake sediments. The underground passage through the gravel and sand of the aquifer also contributes to the purification of the infiltrated water. 4 major mechanisms are responsible for the purification of the infiltrated water:

(1) mechanical straining,
(2) sedimentation,
(3) sorption and
(4) biodegradation.

The different infiltration systems show a distinct demand upon the quality of the infiltrated water. Surface infiltration techniques are usually better suited to cope with suspended material in the infiltration water. These sys-

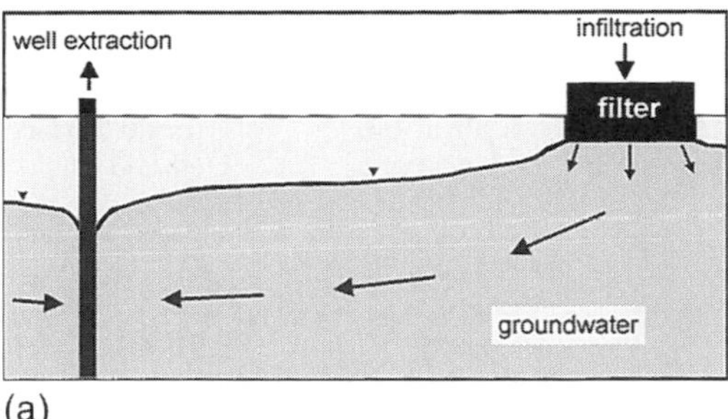

(a)

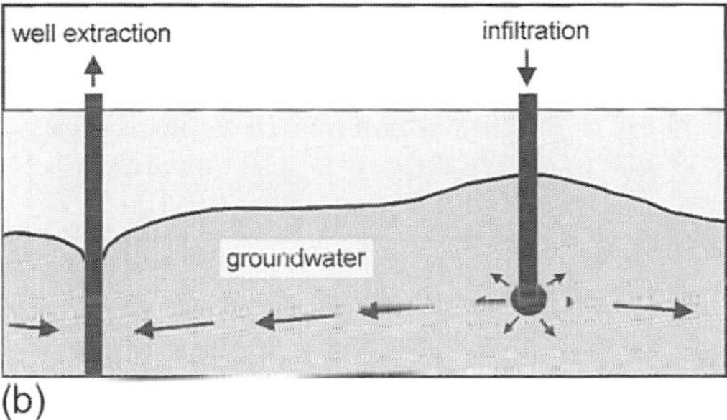

(b)

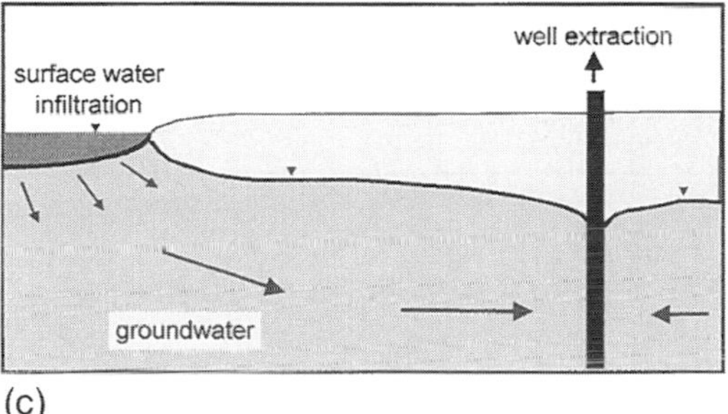

(c)

Fig. 1a–c. Functional schemes of the major infiltration techniques for artificial groundwater recharge, (**a**) surface infiltration, (**b**) subsurface infiltration, (**c**) bank filtration.

tems can be relatively easy cleaned when the filter surface is clogged by particulate matter. These systems also show a considerably higher degree of purification due to biochemical/bacterial activities.

The highest demands upon the quality of the infiltrated water and the quality of the purification process are set by systems used for the artificial recharge of groundwater for drinking water production. In these systems the water filtration processes which claim the common factor of biological/biochemical activity are of particular importance. Due to this and the limited space of this short introduction to artificial groundwater recharge the authors will mainly concentrate on infiltration techniques for drinking water production and the relevant purification mechanisms.

Artificial recharge of groundwater has been used for the drinking water supply in several European countries and in the USA since the 19th century. For groundwater recharge for drinking water production surface water is mainly infiltrated by slow sand filtration or bank filtration. At the beginning the most important reason for the application of this technique was the elimination of pathogens. Later the elimination of chemical compounds gained more importance. Due to the increasing need for drinking water, the overexploitation and contamination of natural groundwater and surface water, slow sand filtration and bank filtration have become important means of solving water supply problems all over the world. Especially in the last years slow sand filtration and bank filtration are receiving increased attention because of their potential for removing cysts of protozoan pathogens, which are resistant against common disinfection methods and, therefore, a possible risk for drinking water.

2 Infiltration Techniques

Artificial recharge of groundwater can be done by different infiltration techniques. These techniques can be subdivided into surface infiltration, subsurface infiltration and bank filtration. The type of infiltration technique that should be chosen to solve a special infiltration problem depends on the hydrological and geological boundary conditions, the available ground space, the composition of the infiltrated water and the degree of purification to be achieved.

In Tab. 1 the major boundary conditions are compiled for some of the basic infiltration techniques.

Tab. 1. Comparison of Different Infiltration Techniques (SCHÖTTLER, 1993; ORL-ETHZ, 1970; STUYFZAND, 1997; BERGER, 1997)

Infiltration Technique	Raw Water Quality (Pretreatment)	Purification Potential	Infiltration Capacity $[m^3 (m^2 d)^{-1}]$	Ground Surface Demand $[m^2 (m^3 d)^{-1}]$
Surface infiltration				
Slow sand filtration	not very high to high/pretreatment	high	1.0–4.0 10 exceptional	0.25–1.0
Flooding of grass- and woodland	not very high	high	0.2–1.0	1.0–5.0
Trench infiltration	not very high/pre-treatment	medium – high	0.5–3.0	0.33–2.0
Trough-trench-systems	not very high	high	depending on constructive conditions	depending on constructive conditions
Irrigation	not very high	high	0.01–1.0	1.0–100
Subsurface infiltration				
Infiltration wells	very high/pre-treatment	medium	depending on geological conditions	insignificant
Horizontal infiltration galleries	very high/pre-treatment	medium	depending on geological conditions	insignificant
Seepage trenches	very high/pre-treatment	medium	10–200	insignificant
Bank filtration				
River/lake/pond bank filtration	depending on purification aim	medium	depending on geological and hydraulic conditions	insignificant

2.1 Surface Infiltration

In cases where a well permeable aquifer lies relatively near the surface and where the necessary space is available it is preferable to use the technique of surface infiltration. The useful infiltration techniques can be divided into three main categories (Sects. 2.1.1–2.1.3).

2.1.1 Recharge through Overgrown Soil

This category contains techniques like surface irrigation, periodically flooding of grassland and/or woodland and the infiltration by trenches. All these techniques use the purification potential of the activated soil biocenosis. Fig. 2 gives a schematic view of a system designed for periodically flooding of grassland.

Small trenches are used to divert the raw water on the grassland. Parcels are divided by small walls. The infiltration water percolates through the soil and the unsaturated zone of the underground and reaches the groundwater.

Periodical flooding of woodland follows the same scheme. The infiltration by trenches is also very similar. Here a number of trenches in the grassland are periodically flooded. Normally these systems are operated in a manner that periods of infiltration alternate with peri-

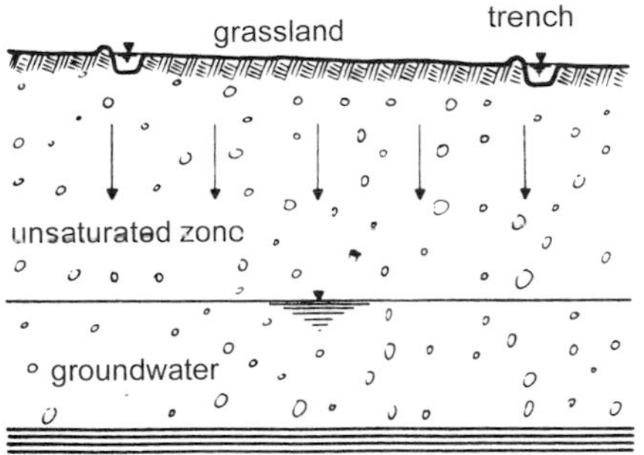

Fig. 2. Schematic diagram of infiltration by flooding of grassland, after ORL-ETHZ (1970).

ods in which the land surface can fall dry. This is necessary to allow a recreation of the soil organisms.

2.1.2 Infiltration Basins (Slow Sand Filtration)

Open infiltration basins are the most common devices for artificial recharge of groundwater. Fig. 3 shows a schematic cross section of an open infiltration basin.

The infiltration water flows over cascades. This leads to an enrichment of oxygen in the water. The oxygen serves as electron acceptor for the biologically induced degradation processes in the filter/soil and on the following underground passage of the infiltrated water.

The infiltration basin is filled with a layer of filter sand (0.5–1 m). The water passes through the filter sand, percolates through the unsaturated zone under the sand filter and reaches the groundwater. Due to the relatively slow filter velocities of 1–4 m d^{-1} this infiltration through sand filters is called slow sand filtration.

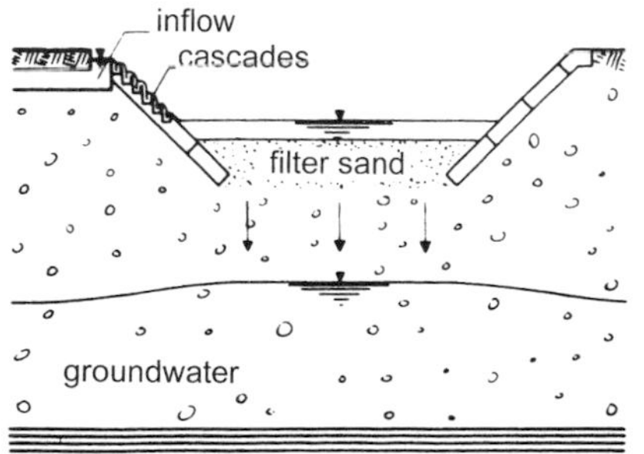

Fig. 3. Schematic diagram of infiltration by open basins (slow sand filtration), after ORL-ETHZ (1970).

The infiltration basins often have an oblong rectangular form and a typical surface of some 100–10,000 m^2. The slopes of the basins are commonly reinforced by concrete but can also be planted like dikes or embankments. Normally the basins are operated in an intermittent manner. Periods of infiltration alternate with periods in which the surface of the sand filter can dry. This is necessary to minimize filter clogging in the upper zone of the sand filter where most of the suspended material is held back.

A good overview of the technique of slow sand filtration can be found in GRAHAM and COLLINS (1996) and GRAHAM (1988).

2.1.3 Trough-Trench Systems

Trough-trench systems are mainly used for non-central storm water infiltration in urban areas. The system consists of a flat trough in the land surface (Fig. 4). The trough is filled with soil and covered with grass. A trench filled with porous material (e.g., gravel, broken volcanic rocks) is positioned under the trough and is used as interim storage room for the water that percolates through the soil layer of the trough. The trench stands in hydraulic contact with the surrounding sediments and water can infiltrate into the underground.

The system combines the high purification capability of the activated soil biocenosis of the trough, the relatively low demand upon the raw water quality and the ability to store large amounts of the relatively unpredictable storm water runoff, which would normally not infiltrate fast enough (REMMLER and HÜTTER, 1997). Shallow infiltration troughs fit well into urban areas and – besides the use as infiltration device – can be used as garden areas.

A good selection of information about trough-trench systems and other techniques to infiltrate storm water runoff is given in SIEKER and VERWORN (1996).

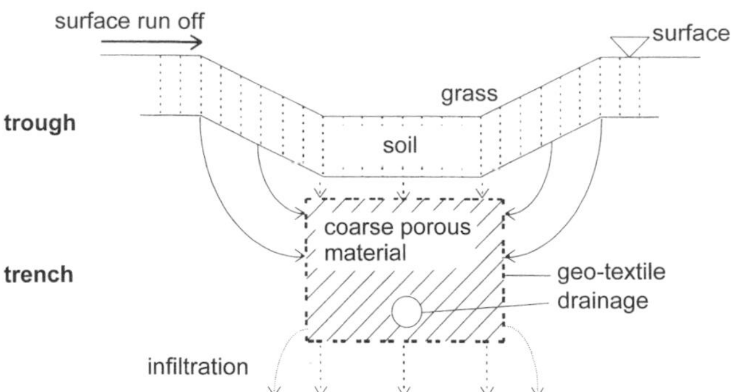

Fig. 4. Schematic cross-section through a trough-trench system (SIEKER, 1998).

2.2 Subsurface Infiltration

In cases where thick non-permeable layers at the surface of the aquifer hinder the installation of surface infiltration systems, the use of subsurface infiltration systems is appropriate. This may also be true in cases where the vast amount of land needed for surface infiltration systems is not available or too expensive. Subsurface infiltration systems are not easy to clean and consequently have a high demand upon the water quality with respect to suspended matter, growth of bacteria in the infiltration systems, degassing, redox reactions and reactions between infiltration water and aquifer material.

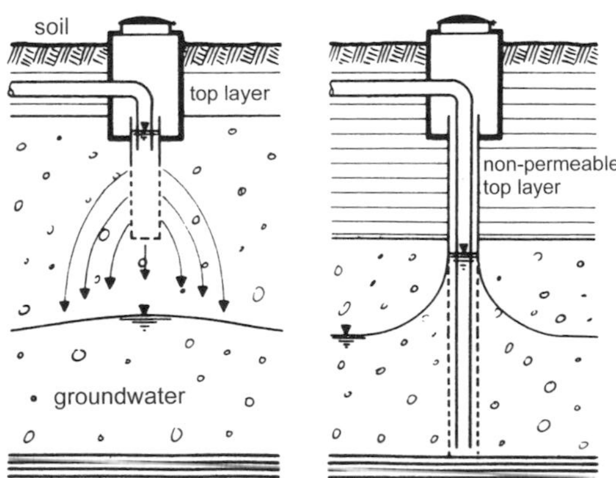

Fig. 5. Schematic cross section of infiltration wells. Dug infiltration well on the left and injection well on the right, after ORL-ETHZ (1970).

2.2.1 Infiltration Wells

Infiltration wells can be divided into injection wells and dug infiltration wells. Dug infiltration wells consist of a relatively large gauged shaft filled with gravel or coarse sand. The shaft penetrates the low permeable top soil layer and the infiltration water is recharged through the gravel or sand core filling (Fig. 5).

Injection wells are the most widely used subsurface infiltration technique. The construction of an injection well is very similar to the construction of a vertical discharge well. Injection wells are usually drilled. The well design consists of a well tubing with perforated or slotted casing at least in the lower part of the well. The infiltration water is recharged through the surrounding gravel pack into the aquifer (Fig. 5).

An example of the infiltration of water by deep wells in a dune area in The Netherlands is given by STAKELBEEK et al. (1997).

2.2.2 Horizontal Infiltration Galleries

Horizontal infiltration galleries use the same infiltration principle as injection wells. They consist of a number of horizontally oriented well tubings with perforated or slotted casings. The water is infiltrated through these casings and the surrounding gravel pack. Horizontal infiltration galleries are usually built in an open construction method.

2.2.3 Seepage-Trenches

A seepage-trench is an infiltration system consisting of a deep trench filled with coarse-grained sand onto which water is fed through a pipe system. Fig. 6 shows a schematic cross section of a seepage trench. The trench is usually 0.4–1.0 m wide, 5–10 m deep and several 10 m long. The sand used to fill the trench has a grading of 0.6–1.5 mm. The trench is covered with removable cover sheets. These sheets can be removed to clean the surface of the sand filter. The upper portion of the trench is protected against the surrounding soil and sediments by concrete plates. The lower portion stands hydraulically in contact with the aquifer. This and the depth of the trench make high filter velocities possible.

2.3 Bank Filtration

During bank filtration water seeps from a surface water into the underground (Fig. 1c). The reason for this is a potential drop between the head of the surface water and the groundwater potential. This can be due to high water and with special respect to artificial groundwater recharge due to damming up or the draw down of the groundwater level by withdrawal of groundwater. The possible amount of bank filtrate that can be infiltrated highly depends on the permeability of the pond, lake or river sediments, the aquifer sediments and the potential difference between the head of the surface water and the groundwater potential.

The infiltration of surface water into the aquifer is a filtration process in which complex physical, chemical and biological factors have a combined effect on the purification of the infiltrated water.

KOERSELMANN (1997) gave an example of the purification efficiency of pond infiltration and KUHLMANN and SCHÖTTLER (1993) described the behavior of NTA during bank filtration.

3 Biology of Filter Layers

In contrast to the wealth of the biological processes of infiltration systems used for artificial groundwater recharge, the literature on the biology and ecology of bank filtration layers and slow sand filter beds is rather sparse. The common organisms of the interstitial sand community consist mainly of bacteria, protozoa and metazoa (Fig. 7). Various algae species occur especially in the flooding layer on the filter beds. Slow sand filters as well as other infiltration layers can be regarded as ecological systems because of the various

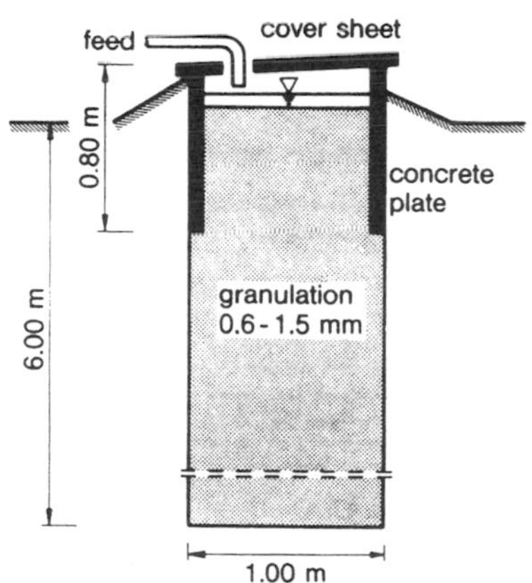

Fig. 6. Cross section of a seepage-trench (SCHMIDT and MEYER, 1989).

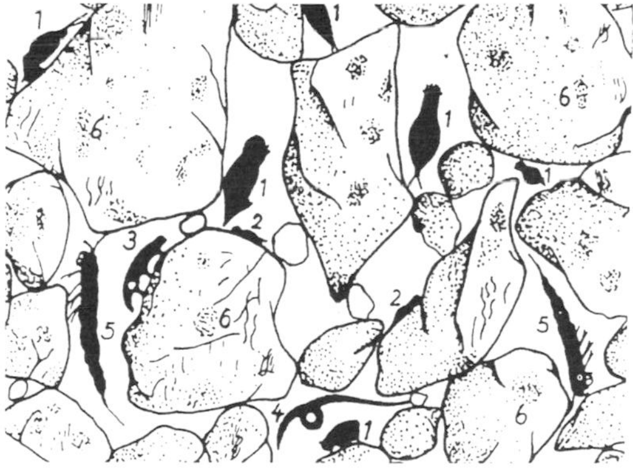

Fig. 7. Consortium of various organisms in the interstices of sandy filter layers (UHLMANN, 1975); (1) rotatoria, (2) ciliates, (3) tardigrada, (4) nematodes, (5) arthropods, (6) microbes.

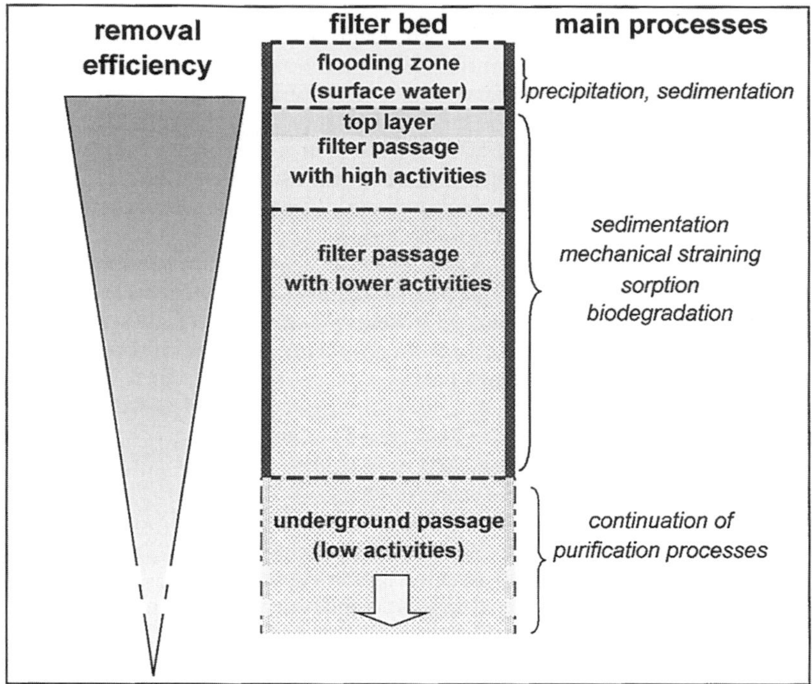

Fig. 8. Purification process during infiltration of water.

interactions between the different groups of organisms inhabiting the aquatic and solid phases (SCHMIDT, 1988). The distribution of the various groups of organisms depends on the filter depth, on seasonal effects and on the chemical quality and flow velocity of the infiltrating water. The percentage composition of the carbon weight from organic matter and microorganisms usually shows the following sequence: non-living material > bacteria > algae > protozoa (DUNCAN, 1988). These organisms and additionally also metazoa guarantee the regeneration of the filter layers as well as the adaptation capacity to changing operating or environmental conditions (Fig. 8). The biological actions and interactions enable artificial groundwater recharge to be an effective and sustainable technique for water management.

3.1 Algae

Algae can be found especially in the flooding zone and the first centimeters of the filter layer. They need sun light for their existence and are not able to grow in deeper filter layers. They assimilate organic and inorganic compounds and offer biological surfaces for adsorption, i.e., for heavy metals. Their oxygen production can stabilize the aerobic conditions during the infiltration process. Aerobic conditions are necessary for the microbial mineralization of organic compounds. Depending on the actual algae activities, the produced oxygen can persist even during the night and catalyze the microbial degradation processes (NAKAMOTO et al., 1996). On the other side extensive growth of algae can lead to a reduction of the flow velocity by clogging. Also odor and taste problems can occur by the production of undesirable algal metabolites. These problems can be avoided by intermittent operation techniques or by covering the filter beds.

3.2 Bacteria

The bacterial consortium of the filter layers contains allochthonous and autochthonous microorganisms. The first are transported by the infiltrating water into the deeper layers, but are not able to grow at the special environmental conditions. Also pathogens are allochthonous microbes and are normally eliminated by physical and biological mechanisms. The second are part of the natural community in the underground or filter bed. They are well adapted to the specific conditions of this habitats. Autochthonous bacteria show physiological and growth activities under underground conditions characterized by rather low temperatures and low nutrient concentrations.

The number of microorganisms depends on the depth of the filter layer and the nutrient conditions of the infiltrating water. Fig. 9 demonstrates the distribution of different bacterial groups in a slow sand filter bed. The numbers indicate the removal of allochthonous microbes especially, coliforms and *E. coli*, which are indicators for fecal contamination. Also mesocarbophilic bacteria, adapted to rather high nutrient concentrations decrease with the depth of the filter layer. However, the number of oligocarbophilic bacteria, which are adapted to rather low nutrient concentrations shows no significant decrease with increasing filter depth. This indicates that oligocarbophilic bacteria must be regarded as a part of the natural microbial community of the sand filter, guaranteeing the biodegradation processes during infiltration.

Many investigations during the last decades confirmed that more than 90% of the autochthonous microbes in the filter body or underground were attached at the solid matter. The microbial communities in the underground or filter beds must be regarded as very diverse, with a high physiological capacity (GHIORSE and WILSON, 1988; KÖLBEL-BOELKE et al., 1988; MARSHALL, 1988; PREUß and NEHRKORN, 1996, ESCHWEILER et al., 1998). Such microbial diversity is a necessary prerequisite for the biodegradation of a wide spectrum of different organic substances transported in the infiltrating water.

The structure of the microbial communities strongly changes during the infiltration processes. Investigations in a groundwater catchment area in Germany demonstrated, that the

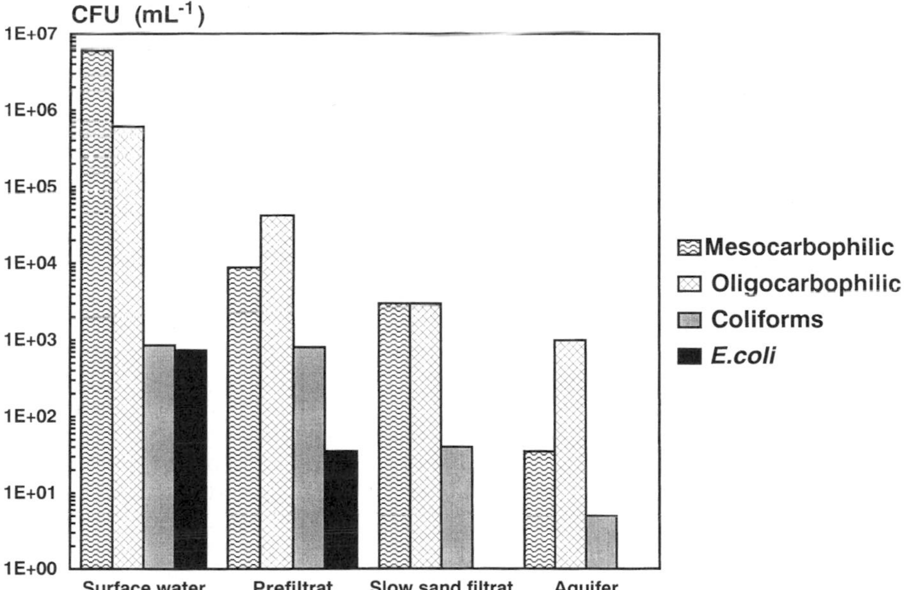

Fig. 9. Counts of bacteria in a slow sand filter (PREUß and NEHRKORN, 1996).

succession of different physiological bacterial groups like denitrifying or sulfate reducing bacteria corresponded well with the change of redox zones during bank filtration. Also during slow sand filtration the composition of the microbial communities completely changes, indicating a groundwater specific microbiota at the end of the infiltration processes and underground passage of water (PREUß and NEHRKORN, 1996).

3.3 Protozoa

Protozoa like flagellates, ciliates, and amoebas are known as predators (NOVARINO et al., 1997). They are regarded as one important factor for the elimination of pathogens or allochthonous bacteria during artificial groundwater recharge. Experiments with model systems containing sand with and without protozoa have demonstrated a significant increase of bacteria removal in the presence of ciliates (LLOYD, 1996). Most of protozoan organisms live in the upper horizon of the infiltration zones. Cleaning procedures for slow sand filters can destroy the function of this important top layer. To stabilize the biological removal of undesired bacteria, these procedures should be conducted during warmer weather, when the regrowth of the protozoan population could be faster.

3.4 Metazoa

Animals like worms, nematodes, annelids, or arthropods are known as natural inhabitants of the interstitial phase of sandy aquifers and sandy biologically active filters. Their ecological role includes the predation of bacteria and protozoa as well as the cutting and reduction of particulate detritus. Among other things they are characterized by their small size, which ranges from 0.01–5 mm. With their activities they increase the permeability of the filter layers. Normally they are not pathogenic but autochthonous organisms in the subsurface environment adapted to the specific groundwater conditions. Allochthonous metazoa, i.e., some rotatoria are reduced up to 99% after slow sand filtration. The number of nematodes and sometimes copepods can be higher in the water after slow sand filtration (SCHELLART, 1988).

4 Redox Succession during Infiltration

Microbial degradation activities can cause several changes in the environmental conditions of the filter layers. Heterotrophic microorganisms obtain energy by using organic carbon as electron donor and inorganic compounds as electron acceptors. The complete aerobic degradation processes result in the mineralization of organic matter and the production of CO_2. The pH value decreases during these processes and the oxygen concentration decreases during infiltration. During slow sand filtration the aerobic processes are dominant also in deeper filter layers because of the technical aeration of the infiltrating water and the additional oxygen production by algae. In the presence of ammonia nitrifying bacteria can grow producing nitrate (Fig. 10).

During bank filtration anaerobic conditions can often be observed. Depending on the input of biological degradable organic matter the successive use of alternative and less efficient terminal electron acceptors can lead to the development of different redox zones (Fig. 11). Distinct anaerobic types of bacteria reduce nitrate, iron, manganese, sulfate and CO_2. During these anaerobic degradation processes the E_h value decreases. At low E_h levels microbial methanogenesis is possible. Depending on the thickness of the surface water sediments, the TOC content of these sediments and the infiltrating water, and the hydrochemical, biological and hydraulic boundary conditions of the aquifer, the succession of the redox zones can occur on a flow distance of a few centimeters up to several 10 m.

Because of the inefficiency of the anaerobic energy metabolism, anaerobic processes are associated with the formation of reduced compounds such as ethanol, butyrate or methanol and H_2S. In some cases the changes of redox conditions and the production of anaerobic

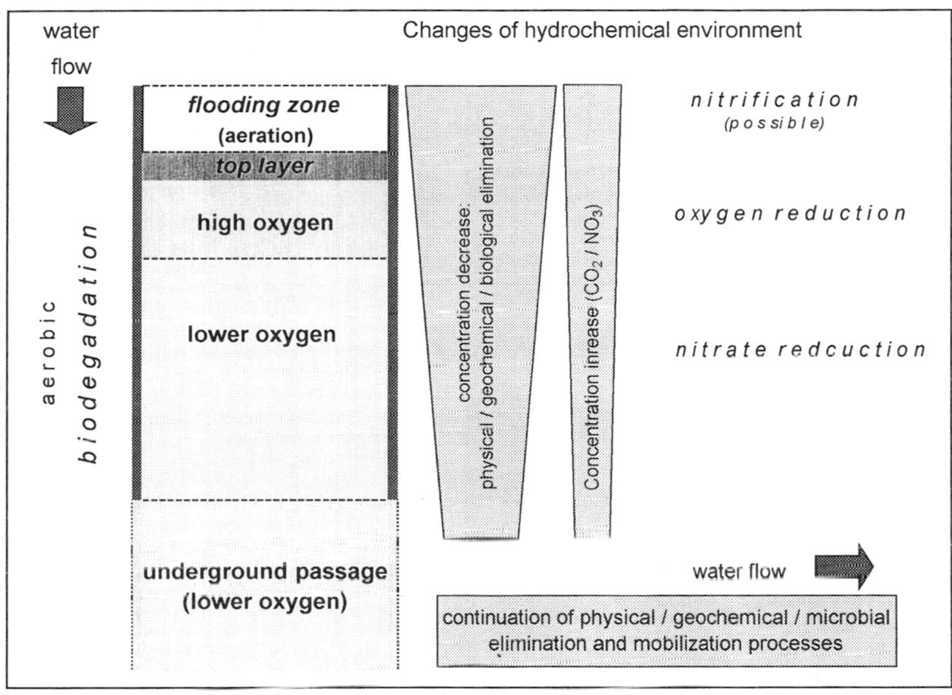

Fig. 10. Changing environmental conditions during slow sand filtration.

metabolites in groundwater habitats indicate organic contamination of the infiltrating water (Chapelle, 1993).

5 Purification Mechanisms

During artificial groundwater recharge the infiltrated surface water is freed of chemical and biological contaminants to gain a clean and healthy groundwater usable for the drinking water supply. The main processes during the purification of infiltrating water are:

(1) mechanical straining
(2) sedimentation
(3) sorption
(4) biodegradation

Physical-chemical particle removal mechanisms like sedimentation, mechanical straining and sorption are significant factors during all infiltration techniques as well as during bank filtration (Hofmann and Schöttler, 1998). Evidence of biologically mediated particle removal for particles <2 µm could also be observed (Weber-Shirk and Dick, 1997). In water filtration through sands, the adsorptive electrostatic processes of attraction and neutralization have been considered to be important mechanisms responsible for the removal of bacteria and particulate contaminants during artificial groundwater recharge. Biological processes can remove both particle bound and soluble compounds more efficiently after adsorption.

Dissolved and also some smaller particle bound compounds are removed mainly by the degradation activity of microorganisms. As mentioned above the microbial degradation of organic matter used as electron donor depends on the successive use of the inorganic electron acceptors oxygen, iron, manganese, nitrate, and sulfate. These microbial activities addi-

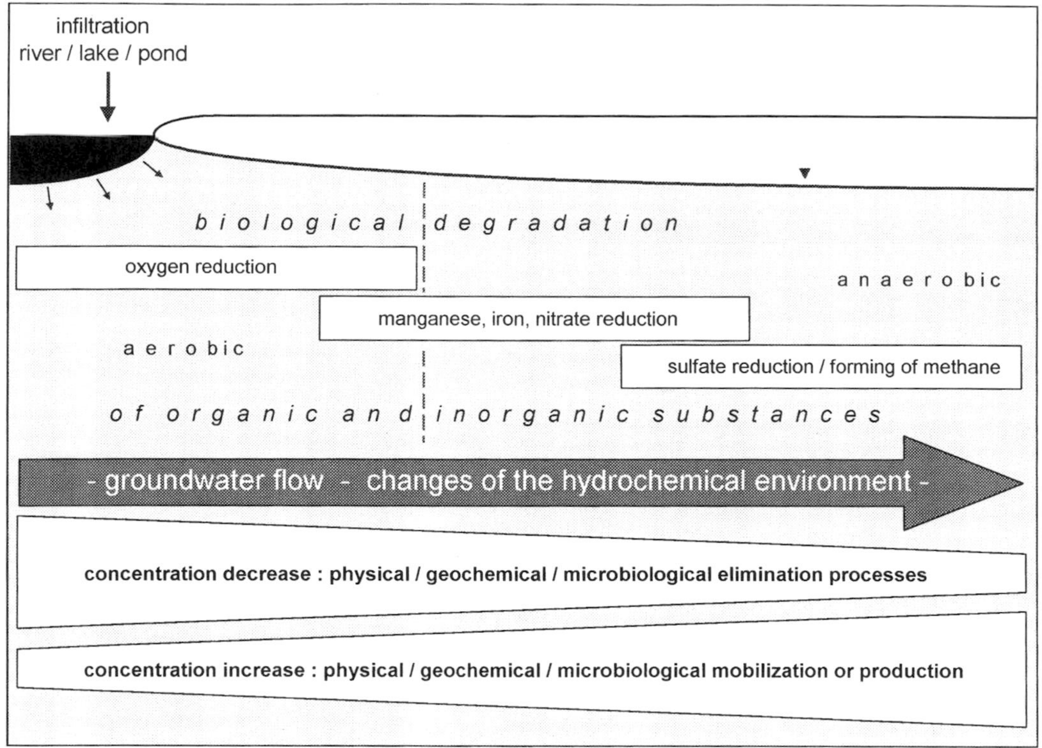

Fig. 11. Changing environmental conditions during bank filtration.

tionally influence the transport, adsorption and mobilization processes by changing the redox and pH conditions. The effect of the biological and physico-chemical processes during the infiltration of water depends on the depth of the filter layer (Fig. 8). A lot of investigations confirm that biological activities decrease with increasing depth of the filter layer (Ellis and Aydin, 1995; Preuß and Nehrkorn, 1996; Lehtola et al., 1996) but still go on in the deeper underground layers. According to that, the deeper filter layers and also the additional underground passage are important for further purification and especially for the stabilization of the bacteriological and hygienic conditions in the infiltrate.

The mechanisms of the removal of microorganisms by infiltration processes are biological, physical and physico-chemical mechanisms. The removal of microorganisms mainly depends on the properties of the filter layer (i.e., grain size), hydraulic conditions, effects of chemical compounds in the water (inhibitors, nutrients), the autochthonous microflora (antibiotics produced by algae or fungi) and the activity of predators (protozoa, metazoa) and parasites (bacteriophages).

6 Removal Capacity

Various hydrogeochemical, physical, and biological properties of the actual filter system effect the purification processes during artificial groundwater recharge (Fig. 12).

Filtration and sorption processes depend on the morphological structure of the filter layer like pore structure, grain size, and surface clogging, i.e., at the top layer. The surface structure of the grains is not constant but influenced by

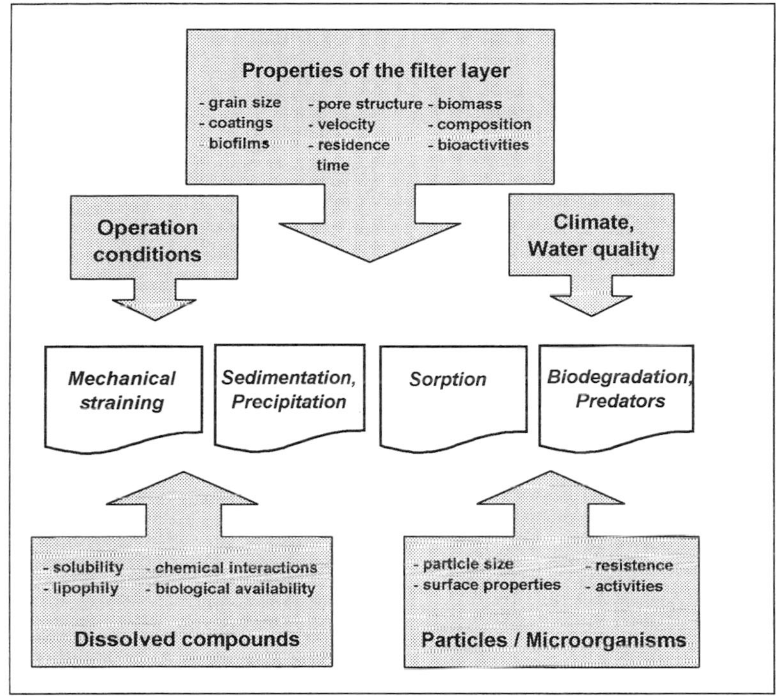

Fig. 12. Effects on purification efficiency of artificial groundwater recharge.

organic and inorganic coatings and the development of a biofilm. The surface conditions and especially the biofilm strongly influence the sorption and degradation processes in the filter layer (WHEELER at al., 1988).

The transport but also the sedimentation of chemical compounds, particles and microbes are determined additionally by hydrogeochemical properties of the layer, i.e., mainly infiltration velocity and residence time, alteration and direction of the flow velocity and the effect of dead-end pores. These properties are determinated by the natural hydrogeological aquifer conditions but also by the actual operation conditions of the water work.

The quality, quantity and physiology of the involved microbiota determine the biological degradation processes. They can be effected by seasonal factors like temperature (WELTE and MONTIEL, 1996; SEGER and ROTHMAN, 1996).

Besides these filter properties, the behavior of contaminants during infiltration and underground passage is determined also by the physical-chemical properties of the substance including water solubility, lipophility or volatility. Possible interactions with other compounds play an important role for precipitation, transport or elimination of dissolved substances. The possible chemical reactions that influence the behavior of compounds are complex formation, oxidation and precipitation, e.g., calcium carbonate precipitation. Also the concentration of the substances and the biological availability under the actual environmental conditions play an important role during the purification processes.

The behavior of particulate substances is effected by the morphological and chemical properties of the particles like size, weight, sedimentation properties, mechanical, physical, chemical and biological stability, and the surface characteristics. The behavior of microorganisms during infiltration depends on the physiological and morphological properties of the infiltrated microorganisms and their growth potential in the subsurface. The morphological surface properties of the microbes like charge, capsules, surface structure, appendages, mobility, resting forms (spores), morphological shape are important due to the removal of microbes during infiltration.

6.1 Removal of Organic Matter

Slow sand filtration and bank filtration have consistently demonstrated their effectivity in treating water from high quality water sources. Many authors described the removal of organic matter after monitoring various sum parameters. After slow sand filtration the turbidity usually decreases 50% or more. The observed removal of organic compounds (DOC, TOC, NOM, COD or AOC) during slow sand filtration or bank filtration varies from 10–70%, depending on the monitored parameter, the operating conditions and the quality of the infiltrated raw water (SCHELLART, 1988; COLLINS et al., 1992; LEHTOLA et al., 1996; MIETTINEN et al., 1996; HAMBSCH and WERNER, 1996; ODEGAARD, 1996; WRICKE et al., 1996). High levels of biomass were not always correlated with high levels of NOM removal. Organic precursor removal in slow sand filters must be regarded as a function of both the microbiological maturity and adsorption capacity of the top layer and of the filter bed (COLLINS et al., 1996).

In laboratory scale slow sand filters the removal of BOD_5, was up to 93% running with wastewater and depending on the flow velocity. Gravel prefiltration could optimize the removal of suspended solids (ELLIES, 1987). A rapid decrease of POC (particulate organic carbon) could be demonstrated after 3 cm of infiltration (ELLIS and AYDIN, 1995).

6.2 Removal of Xenobiotics and Trace Elements

Several investigations have been done to assess the removal of xenobiotics like pesticides during artificial groundwater recharge. Sufficient elimination potential up to 100% can be stated for hydrophobic pesticides like DDT or heptachlor with a high sorption tendency with inorganic and organic filter materials during artificial groundwater recharge. Some studies show a mean retention during the infiltration process of approximately 10% for atrazine or simazine and up to 100% for diuron and other substances with higher water solubility (HEINZ et al., 1995).

AOX, chlorobenzenes and chlorophenols are removed after slow sand filtration conditions after 110 d with rates between 94% and 99% (ZACHARIAS et al., 1995). Long-time investigations at the river Rhine demonstrated the efficiency of bank filtration in regard to various organic and inorganic compounds. Although many trace metals and organic pollutants were detected in the river, concentrations in the bank filtrate were in general low with problematic levels only for some pesticides (STUYFZAND and KOOIMAN, 1996).

Heavy metals seem to be removed during artificial groundwater recharge depending on the actual concentrations in the raw water. High removal of iron and manganese has been observed in the surface layer of slow sand filtration beds (LEHTOLA et al., 1996; MÄLKKI, 1996).

Cyanobacteria and their toxins can adversely affect the resulting water quality especially in summer during algae bloom. Filtration through soil and sandy sediment apparently results in a rather efficient removal of cyanobacterial toxins and cells, except in very massive bloom situations. The removal of cyanobacterial microcystins during filtration must be regarded as a result of adsorption and biodegradation processes (LATHI et al., 1996).

6.3 Removal of Microorganisms and Pathogens

The conventional purpose using slow sand filtration or bank filtration for artificial groundwater recharge was to eliminate pathogenic organisms in the resulting drinking water. Most of the investigations show, that especially slow sand filtration is an efficient technique for the removal of microorganisms (POSTILLION et al., 1989; TANNER and ONGERTH, 1990; LEHTOLA et al., 1996; PREUß and NEHRKORN, 1996; PETERS et al., 1997). Fig. 13 shows the decrease of various bacteria during artificial groundwater recharge including gravel

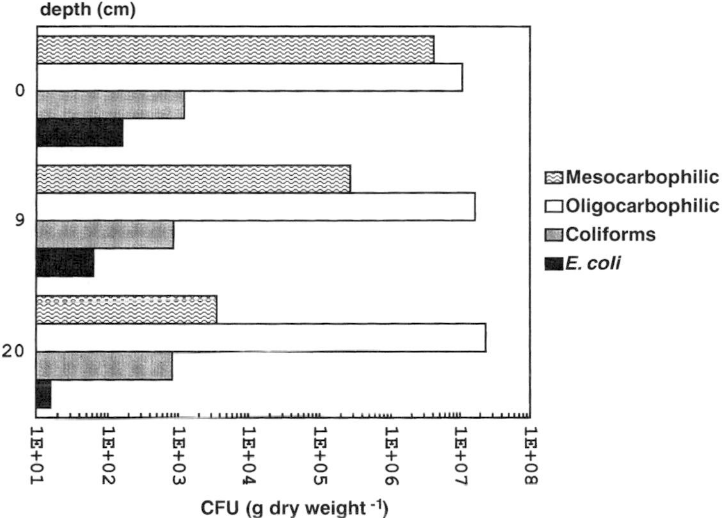

Fig. 13. Decrease of bacteria counts during artificial groundwater recharge.

prefiltration, slow sand filtration and underground passage.

The data in Fig. 13 and Tab. 2 indicate that hygienic critical microbes like coliforms and especially *E. coli*, determined as indicators for fecal contamination, can be removed during artificial groundwater recharge depending on the quality of the surface water. Also, after bank filtration the microbial biomass can be reduced to 10–20% of those in the corresponding surface water (MIETTINEN et al., 1996). The approximate reduction capacity for various microbes during slow sand filtration is shown in Tab. 2.

Usually, most of the removal of bacteria during bank or slow sand filtration occurs within the uppermost infiltration layers where microfauna and flora are most abundant. In this top layer both predation and adsorption of bacteria are very effective. Adsorption to biological and non-biological surfaces and mechanical straining also play a role in all depths of the filter. Investigations with model systems have confirmed, that the transport of bacteria

Tab. 2. Removal Capacity of Slow Sand Filtration for Several Organisms

	Removal (log units)	References
Standard plate counts	1–3	RACHWAL et al. (1996), TANNER and ONGERTH (1990), PREUß and NEHRKORN (1996), WHO (1977)
Coliforms	2–3	RACHWAL et al. (1996), PREUß and NEHRKORN (1996), LELAND and DAMEWOOD (1990), WICHMANN (1980)
E. coli	2–3	RACHWAL et al. (1996), TANNER and ONGERTH (1990), GILBERT et al. (1976), WHO (1977)
Streptococci	3	GILBERT et al. (1976)
(Oo)cysts (Protozoa)	2–4	RACHWAL et al. (1996), SCHULER et al. (1988), FOGEL et al. (1993)
Viruses	1–3	RACHWAL et al. (1996), DIZER et al. (1983), POWELSEN et al. (1993)

increased with increasing size of the sand grain (ELLIES and AYDIN, 1995).

The removal of protozoan cysts from pathogens like *Cryptosporidium* sp. or *Giardia* sp. by filtration through sand is generally assumed to be very efficient. In this sense, slow sand filtration may adequately compensate for the comparatively low efficiency of disinfection against encysted protozoa. It has been observed by several authors, that the removal of cysts and oocysts during infiltration of water was adversely effected when biological activity in the filter layers was reduced, i.e., by cold operating conditions (FOGEL et al., 1993).

Investigations of the behavior of enteric viruses suggest that the elimination of virus indicators such as poliovirus and retrovirus may be at least as efficient as the removal of fecal bacteria through slow sand filtration. Results from laboratory adsorption experiments confirm the importance of biomass in enhancing the removal of viruses from the aqueous phase (WHEELER et al., 1988; STUYFZAND, 1989).

7 Conclusions

Bank filtration and slow sand filtration are some of the oldest treatment technologies for the public drinking water supply. Because of the simplicity of this technology it is attractive for use in small or even large communities in industrial countries as well as in developing countries. Results of NELLOR et al. (1989) and CHAPMAN and RUSH (1990) indicate that artificial groundwater recharge normally has no measurable impact on groundwater quality or the health of people. Depending on the actual operating conditions and the quality of the surface water pathogens and toxic xenobiotics can be removed efficiently during the treatment steps. In addition some investigations confirm a significant decrease of mutagenicity formation potential during bank filtration (STUYFZAND, 1989; MIETTINEN et al., 1996). Tab. 3 summarizes the removal efficiency of artificial groundwater recharge investigated in a groundwater catchment area in Germany.

A description or prognosis of the actual biological reactions during artificial groundwater recharge is very difficult, because the purification process is a combination of various single processes depending on the river water quality or meteorological conditions. The biological water purification processes during infiltration and underground passage depend on special boundary conditions. Besides general water quality requirements the following parameters are important:

(1) oxygen,
(2) degradable DOC (concentration and composition),
(3) inorganic nutrients (concentration and composition),
(4) temperature,
(5) structure and physiological status of the biocenosis,
(6) filter material (composition, pore size etc.),
(7) infiltration velocity, infiltration amount,
(8) light, pH value, CO_2.

Consequently, the following factors are to be considered with regard to the optimization of biological purification processes:

(1) high quality of the used surface water,
(2) pretreatment techniques if appropriate,
(3) high oxygen content,
(4) low algae mass growth,
(5) low and continuous operating rate of the filter,
(6) a long underground passage after slow sand filtration.

The biological and chemical actions and interactions, the relatively unsophisticated operation of the infiltration devices and the natural purification mechanisms enable artificial groundwater recharge to be an efficient and sustainable technique for water management.

Tab. 3. Assessment of the Efficiency and Purification Mechanisms during Artificial Groundwater Recharge (after SCHULTE-EBBERT and SCHÖTTLER, 1995)

Substance Groups	Substance	Concentration Decrease		Purification Processes				Microbiological Degradation	
		Slow Sand Filtration Aerobic	Bank Filtration Anaerobic	Mechanical Straining	Sorption	Sedimentation Precipitation	Microbiological Degradation	Aerobic	Anaerobic
Microorganisms	Enterobacteria	xxxx	xxxx	xx	xx	–	–		
Heavy metals	Fe, Mn	xxxx	mobilization	–	–	xxxx	–		
	Ni	xx	xx	–	xxx	x	–		
	Cu, Cd, Pb, Zn and other	xx	xx	–	xx	xx	–		
PAH	diverse	xxx	xxx	xx	xx	–	x	x	x
Triazines	diverse	xx	xxx	–	x	–	xxx	xx	xxx
Biogenic substances	diverse	xx	x	–	x	–	xxx	xx	x
Volatile hydrocarbons	diverse	–	xxx	–	x	–	xxx	–	xxx

–: not or only marginally, x: low, xx: important, xxx: very important, xxxx: dominant.

8 References

BERGER, H. (1997), Bau und Inbetriebnahme neuer Infiltrationsanlagen im Wasserwerk Wiesbaden-Schierstein, in: 2. Deutsch-Niederländischer Workshop Künstliche Grundwasseranreicherung – Artificial Groundwater Recharge, (DVGW), *DVGW-Schriftenreihe Wasser Nr. 90*, pp. 237–267. Bonn, Germany.

CHAPELLE, F. H. (1993), *Groundwater Microbiology and Geochemistry*. New York: John Wiley & Sons.

CHAPMAN, P. A., RUSH, B. (1990), Efficiency of sand filtration for removing *Cryptosporidium* oocysts from water, *J. Med. Microbiol.* **32**, 243–245.

COLLINS, M. R., EIGHMY, T. T., FENSTERMACHER JR., J. M., SPANOS, S. K. (1992), Removing natural organic matter by conventional slow sand filtration, *J. AWWA* **84**, 80–90.

COLLINS, M. R., EIGHMY, T. T., FENSTERMACHER JR., J. M., SPANOS, S. K. (1996), Using granular media amendments to enhance NOM removal, *J. AWWA* **88**, 48–61.

DIZER, H., SEIDEL, K., LOPEZ, J. M., FILIP, Z., MILDE, G. (1983), Verhalten enterotropher Viren im Grundwasser unter Modellbedingungen, in: Mikroorganismen und Viren in Grundwasserleitern (DVGW), *DVGW-Schriftenreihe Wasser Nr. 35*, pp. 89–99. Bonn, Germany.

DUNCAN, A. (1988), The ecology of slow sand filters, in: *Slow Sand Filtration: Recent Developments in Water Treatment Technology* (GRAHAM, N. J. D., Ed.), pp. 163–180. Chichester: Ellis Horwood.

ELLIS, K.V. (1987), Slow sand filtration as a technique for the tertiary treatment of municipal sewages, *Water Res.* **2**, 403–410.

ELLIS, K. V., AYDIN, M. E. (1995), Penetration of solids and biological activity into slow sand filters, *Water Res.* **29**, 1333–1341.

ESCHWEILER, B., KILB, B., KUHLMANN, G., ZIEMANN, E. (1998), DNA analysis to study the microbial diversity in recharge groundwater, in: *Artificial Recharge of Groundwater* (PETERS, J. H., Ed.), *3rd Int. Symp. Artificial Recharge of Groundwater – TISAR 98*, Rotterdam, pp. 129–134. Brookfield: A. A. Balkema.

FOGEL, D., ISAAC-RENTON, J., GUASPARINI, R., MOOREHEAD, W., ONGERTH, J. (1993), Removing of *Giardia* and *Cryptosporidium* by slow sand filtration, *J. AWWA* **85**, 77–84.

FRYCKLUND, C. (1998), Artificial recharge of groundwater for public water supply. Potential and limitations in boreal conditions, *Thesis*, Royal Institute of Technology, Stockholm.

GHIORSE, W. C., WILSON, J. T. (1988), Microbial ecology of the terrestrial subsurface, *Adv. Appl. Microbiol.* **33**,107–172.

GILBERT, R. G., GERBA, C. P., RICE, R. C., BOUWER, H., WALLIS, C., MELNICK, J. L. (1976), Virus and bacteria removal from wastewater by land treatment, *Appl. Environ. Microbiol.* **32**, 333–338.

GRAHAM, N. (1988), *Slow Sand Filtration: Recent Developments in Water Treatment Technology*. Chichester: Ellis Horwood.

GRAHAM, N., COLLINS, R. (1996), *Advances in Slow Sand and Alternative Biological Filtration*, Chichester, New York: John Wiley & Sons.

HAMBSCH, B., WERNER, P. (1996), The removal of regrowth enhancing organic matter by slow sand filtration, in: *Advances in Slow Sand and Alternative Biological Filtration* (GRAHAM, N., COLLINS, R., Eds.), pp. 21–28. Chichester, New York: John Wiley & Sons.

HEINZ, I., FLESSAU, A., ZULLEI SEIBERT, N., KUHLMANN, B., SCHULTE-EBBERT, U. et al. (1995), Economic efficiency calculations in conjunction with the drinking water directive (Directive 80/778/EEC), Part III: The parameter for pesticides and related products, *Final Report for the European Commission – DGXI*. Contact No. B4 – 3040/94/000223/Mar B1, Dortmund, Germany.

HOFMANN, T., SCHÖTTLER, U. (1998), Behavior of suspended and colloidal particles during artificial groundwater recharge, *Progr. Coll. Polymer Sci.* **111**, 184–188.

JOHNSON, A. I., FINLAYSON, D. J. (1989), *Artificial Recharge of Groundwater*. Chichester, New York: American Society of Civil Engineers.

KIVIMÄKI, A. L., SUOKO, T. (1996), Artificial recharge of groundwater. *Proc. Int. Symp.* Nordic Hydrological Programme NHP, Report No. 38. Helsinki Finnish Environment Institute.

KOERSELMANN, W. (1997), Reducing the trophic status of artificial recharge areas: enhanced water treatment or management?, in: 2. Deutsch-Niederländischer Workshop Künstliche Grundwasseranreicherung – Artificial Groundwater Recharge, (DVWG), *DVGW-Schriftenreihe Wasser Nr. 90*, pp. 5–16. Bonn, Germany.

KÖLBEL-BOELKE, J., ANDERS, E. M., NEHRKORN, A. (1988), Microbial communities in the saturated groundwater environment II: Diversity of bacterial communities and *in vitro* activities in a Pleistocene sandy aquifer, *Microb. Ecol.* **16**, 31–48.

KUHLMANN, B., SCHÖTTLER, U. (1992), Behavior and effect of NTA during anaerobic bank filtration, *Chemosphere* **24**, 1217– 1224.

LATHI, K., KILPONEN, J., KIVIMÄKI, A.L., ERKOMAA, K., SIVONEN, K. (1996), Removal of cyanobacteria and their hepatotoxins in soil and sediment columns, in: *Artificial Recharge of Groundwater, Proc. Int. Symp.* (KIVIMÄKI, A.L., SUOKO, T., Eds.), Nordic Hydrological Programme NHP Report No. 38, pp. 209–214. Helsinki: Finnish Environment Institute.

LEHTOLA, M., MIETTINEN, I., VARTIAINEN, T., MARTI-

KAINEN, P., PARTANEN, H. (1996), Changes in microbiology and water chemistry during slow sand filtration, in: *Artificial Recharge of Groundwater, Proc. Int. Symp.* (KIVIMÄKI, A. L., SUOKO, T., Eds.), Nordic Hydrological Programme NHP Report No. 38, pp. 197–202. Helsinki: Finnish Environment Institute.

LELAND, D. E., DAMEWOOD, M. (1990), Slow sand filtration in small systems in Oregon, *J. AWWA* **82**, 50–59.

LLOYD, B. J. (1996), The significance of protozoal predation and adsorption for the removal of bacteria by slow sand filtration, in: *Advances in Slow Sand Filtration* (GRAHAM, N., COLLINS, R., Eds.), pp. 95–106. Chichester, New York: John Wiley & Sons.

MÄLKKI, E. (1996), Decrease of organic load from the surface water by biological treatment, in: *Artificial Recharge of Groundwater, Proc. Int. Symp.* (KIVIMÄKI, A. L., SUOKO, T., Eds.), Nordic Hydrological Programme NHP Report No. 38, pp. 209 214. Helsinki: Finnish Environment Institute.

MARSHALL, K. C. (1988), Adhesion and growth of bacteria at surfaces in oligotrophic habitats, *Can. J. Microbiol.* **34**, 503–5060.

MIETTINEN, I., VARTIAINEN, T., MARTIKAINEN, P. (1996), Changes in water microbiology and chemistry during bank filtration of humus-rich lake water, in: *Artificial Recharge of Groundwater, Proc. Int. Symp.* (KIVIMÄKI, A. L., SUOKO, T., Eds.), Nordic Hydrological Programme NHP Report No. 38, pp. 203–208. Helsinki: Finnish Environment Institute.

NAKAMOTO, N., YAMAMOTO, M., SAKAI, M., NOZAKI, K., IWASE, N. et al. (1996), Role of filamentous diatom as an automatic purifier in a slow sand filter, in: *Advances in Slow Sand and Alternative Biological Filtration* (GRAHAM, N., COLLINS, R., Eds.), pp. 139–148. Chichester, New York: John Wiley & Sons.

NELLOR, M. H., MIELE, R. P., BAIRD, R. (1989), Health effects of groundwater recharge, in: *Artificial Recharge of Groundwater* (JOHNSON, A. I., FINLAYSON, D. J., Eds.), pp. 225–234. New York: American Society of Civil Engineers.

NOVARINO, G., WARREN, A., BUTLER, H., LAMBOURNE, G., BOXSHALL, A. et al. (1997), Protistan communities in aquifers: a review, *FEMS Microbiol. Rev.* **20**, 261–276.

ODEGAARD, H. (1996), The development of an ozonation/biofiltration process for the removal of humic substances, in: *Advances in Slow Sand and Alternative Biological Filtration* (GRAHAM, N., COLLINS, R., Eds.), pp. 39–50. Chichester, New York: John Wiley & Sons.

ORL-ETHZ (1970), *Richtlinie für die künstliche Anreicherung (Neubildung) von Grundwasser,* Blatt 516023.

PETERS, J. H., SCHIJVEN, J. F., HOOGENBOEZEM, W., VAN BAAR, M. J. C., BERGSMA, J. et al. (1997), Fate of pathogens and concequences for the design of artificial recharge systems, 2. Deutsch-Niederländischer Workshop Künstliche Grundwasseranreicherung – Artificial Groundwater Recharge (DVGW), *DVGW-Schriftenreihe Nr. 90*, pp. 163–180. Bonn, Germany.

POSTILLION, F. G., EXPOSITO, D. M., RUSIN, P. A., SINCLAIR, N. A., GERBA, C. P. (1989), Bacterial fingerprinting to trace source of coliform bacteria during artificial recharge, in: *Artificial Recharge of Groundwater* (JOHNSON, A. I., FINLAYSON, D. J., Eds.), pp. 220–224. New York: American Society of Civil Engineers.

POWELSON, D. K., GERBA, C. P., YAHYA, M. T. (1993), Virus transport and removal in wastewater during aquifer recharge, *Water Res.* **27**, 583–590.

PREUß, G., NEHRKORN, A. (1996), Succession of microbial communities during bank filtration and artificial groundwater recharge, in: *Artificial Recharge of Groundwater, Proc. Int. Symp.* (KIVIMÄKI, A.L., SUOKO, T., Eds.), Nordic Hydrological Programme NHP Report No. 38, 209–214. Helsinki: Finnish Environment Institute.

RACHWAL, A. L., BAUER, M. J., CHIPPS, M. J., COLBOURNE, J. S., FOSTER, D. M. (1996), Comparison between slow sand and high rate biofiltration, in: *Advances in Slow Sand and Alternative Biological Filtration* (GRAHAM, N., COLLINS, R., Eds.), pp. 3–10. Chichester, New York: John Wiley & Sons.

REMMLER, F., HÜTTER, U. (1997), Storm water management in urban areas: Risk and case studies, in: *Proc. XXVII IAH Congr. Groundwater in the Urban Environment*, Nottingham, UK, September (CHILTON, J., Ed.), pp. 1–27. Rotterdam, Brookfield: Balkema.

SCHELLART, J. A. (1988), Benefits of covered slow sand filtration, in: *Slow Sand Filtration: Recent Developments in Water Treatment Technology* (GRAHAM, N. J. D., Ed.), pp. 253–264. Chichester: Ellis Horwood.

SCHMIDT, C. (1988), Development of a slow sand filter model as a bioassay, in: *Slow Sand Filtration: Recent Developments in Water Treatment Technology* (GRAHAM, N. J., Ed.), pp. 191–205. Chichester: Ellis Horwood.

SCHMIDT, W. D., MEYER, R. (1989), Status and experiences made with the artificial recharge of groundwater in the federal republic of Germany, in: *Artificial Recharge of Groundwater* (JOHNSON, A. I., FINLAYSON, D. J., Eds.), pp. 528–537. New York: American Society of Civil Engineers.

SCHÖTTLER, U. (1993), Techniken zur ökologisch vertretbaren Grundwasserbewirtschaftung, in: DVGW-LAWA Kolloquium Ökologie und Wassergewinnung (DVWG), *DVGW-Schriftenreihe Wasser Nr. 78*, pp. 87–110. Bonn, Germany.

SCHULER, P. F., GHOSH, M. M., BOUTROS, S. N. (1988), Comparing the removal of *Giardia* and *Cryptosporidium* using slow and diatomaceous earth filtration, in: *Am. Water Works Ass., Annual Conf.*, Orlando, FL, June 88, pp.789–805. Denver: American Water Works Association (AWWA).

SCHULTE-EBBERT, U., SCHÖTTLER, U. (1995), Systemanalyse des Untersuchungsgebietes "Insel Hengsen", in: *Schadstoffe im Grundwasser* Vol 3: *Verhalten von Schadstoffen im Untergrund bei der Infiltration von Oberflächenwasser am Beispiel des Untersuchungsgebietes "Insel Hengsen" im Ruhrtal bei Schwerte* (SCHÖTTLER, U., SCHULTE-EBBERT, U., Eds.), pp. 475 - 513. Weinheim: VCH.

SEGER, A., ROTHMAN, M. (1996), Slow sand filtration with and without ozonation in nordic climate, in: *Advances in Slow Sand Filtration* (GRAHAM, N., COLLINS, R., Eds.), pp. 95–106. Chichester, New York: John Wiley & Sons.

SIEKER, F., VERWORN, H. (Eds.) (1996), *Proceedings of the Seventh International. Conference on Urban Storm Drainage*, Vols. I–III, University of Hannover, Germany, September 9–13, 1996.

SIEKER, F. (Ed.) (1998), *Naturnahe Regenwasserbewirtschaftung, Stadtökologie* Vol. 1. Berlin.

STAKELBEEK, A., ROOSMA, E., HOLZHAUS, P. M. (1997), Deep well infiltration in the North-Holland dune area, in: 2. Deutsch-Niederländischer Workshop Künstliche Grundwasseranreicherung – Artificial Groundwater Recharge, (DVWG), *DVGW-Schriftenreihe Wasser Nr. 90*, pp. 269–282. Bonn, Germany.

STUYFZAND, P. J. (1989), Quality changes of river Rhine and Meuse water upon basin recharge in Netherlands' coastal dunes: 30 years of experience, in: *Artificial Recharge of Groundwater* (JOHNSON, A. I., FINLAYSON, D. J., Eds.), pp. 235–247. Chichester, New York: American Society of Civil Engineers.

STUYFZAND, P. J. (1997), Fate of pollutants during artificial recharge and bank filtration in the Netherlands, in: 2. Deutsch-Niederländischer Workshop Künstliche Grundwasseranreicherung – Artificial Groundwater Recharge, (DVWG), *DVGW-Schriftenreihe Wasser Nr. 90*, pp. 131–146. Bonn, Germany.

STUYFZAND, P. J., KOOIMAN, J. W. (1996), Elimination of pollutants during artificial recharge and bank filtration: a comparison, in: *Artificial Recharge of Groundwater, Proc. Int. Symp.* (KIVIMÄKI, A. L., SUOKO, T., Eds.), Nordic Hydrological Programme NHP Report No. 38, pp. 223–232. Helsinki: Finnish Environment Institute.

TANNER, S. A., ONGERTH, J. E. (1990), Evaluating the performance of slow sand filters in northern Idaho, *J. AWWA* **82**, 51–61.

UHLMANN, D. (1975), *Hydrobiologie. Ein Grundriß für Ingenieure und Naturwissenschaftler.* Stuttgart: Gustav Fischer Verlag.

WEBER-SHIRK, M. L., DICK, R. I. (1997), Physical-chemical mechanisms in slow sand filters, *J. AWWA* **89**, 87–100.

WELTE, B., MONTIEL, A. (1996), Removal of BDOC by slow sand filtration: Comparison with granular activated carbon and effect of temperature, in: *Advances in Slow Sand Filtration* (GRAHAM, N., COLLINS, R., Eds.), pp. 95–106. Chichester, New York: John Wiley & Sons.

WHEELER, D., BETRAM, J., LLOYD, B. J. (1988), The removal of viruses by filtration through sand, in: *Slow Sand Filtration: Recent Developments in Water Treatment Technology* (GRAHAM, N. J. D., Ed.), pp. 207–229. Chichester: Ellis Horwood.

WHO (World Health Organization) (1977), Slow sand filtration for community supply in developing countries. A selected and annotated bibliography, *Bulletin No. 9*, pp. 3–50. Leidschendam, The Netherlands.

WICHMANN, K. (1980), Versickerungsversuche mit gereinigtem Abwasser zur Grundwasseranreicherung auf den Nordseeinseln, *Wasser Boden* **4**, 164–169.

WRICKE, B., PETZOLDT, H., HEISER, H., BORMANN, K. (1996), NOM removal of biofiltration after ozonation – results of a pilot plant test, in: *Advances in Slow Sand Filtration* (GRAHAM, N., COLLINS, R., Eds.), pp. 51–60. Chichester, New York: John Wiley & Sons.

ZACHARIAS, B., LANG, E., HANERT, H. H. (1995), Biodegradation of chlorinated aromatic hydrocarbons in slow sand filter simulating conditions in contaminated soil – pilot study for *in situ* cleaning of industrial site, *Water Res.* **29**, 1663–1671.

21 Biofilms in Biofiltration

HANS-CURT FLEMMING

Duisburg, Germany

1 Introduction

Biofiltration represents one of the most important steps in drinking water treatment. Biofiltration enhances biological substrate removal by providing a surface for the attachment and growth of indigenous water microorganisms; this is common for all kinds of biofilters used in drinking water treatment such as slow and rapid sand filters and granular activated carbon (GAC) filters. All of these processes depend on biofilms which have to mature on the filter surfaces. This process is known as "ripening" of the filter material (GALVIN, 1992). The interactions between design, operation, and substrate availability in biofilters control the composition and activity of the microorganisms in the filters. Details of types of biofilters and processes are described in Chapter 22, this volume.

In slow sand filters, a "Schmutzdecke" develops which holds strong biological activity and high numbers of surface-bound microorganisms, with higher colonization of larger grains (ESCH and NEHRKORN, 1988). The population declines sharply with depth (DUNCAN, 1988). Degradation of organic materials begins even in the supernatant water and continues in the "Schmutzdecke" on the upper sand layer and in the lower layers. Since the filters are normally in the open air, on account of their size, there is occasional massive growth of cyanobacteria and algae. It is not only bacteria living in sand filters, but also protozoa and metazoa. Thus, a biocoenosis is created which in its multiplicity produces the desired effects: mechanical removal of particulate substances aided by extracellular polymeric substances (EPS) from the microorganisms coating the sand surface with a biofilm, degradation of dissolved organic substances by the microbial population, and predation of bacteria by protozoa and higher organisms (SCHWEISFURTH, 1986). Pathogenic bacteria such as *Escherichia coli*, which serve as indicators for the hygienic-microbiological quality assessment of water, have not been reported in biofilter biofilms, but this may change when more adequate detection methods are applied. In plants which practice wet harrowing as a cleaning technique, "Schmutzdecke"-like high population and biomass down to the depth of 30 cm was observed. The reduction of tetrazolium chloride as a measure of respiratory electron transport and the ability to mineralize ^{14}C-benzoate was significantly higher indicating a better use of the entire filter bed. In such a filter, high levels of acclimated and metabolically active biomass was detected (EIGHMY et al., 1992). The performance of biofilters for removing particulate and dissolved matter from the water phase includes not only organic compounds, but also, e.g., ammonia, nitrate, iron, and manganese. In assessing the performance of activated carbon filters, the contribution of the biofilm has been acknowledged for some time (SCHWEISFURTH, 1986).

This chapter does not focus on particular biofiltration processes, but on the key feature of biofilters, i.e., that the organisms are organized in biofilms – regardless of the filter material and the type of process. They are immobilized in a matrix of extracellular polymeric substances (EPS). Biofilms are an ubiquitous form of life in all environments in contact with water and can exist even under extreme conditions (FLEMMING, 1991). They provide the "self-cleaning" potential of nature in soils and sediments, and there is virtually no surface material which cannot be colonized by microorganisms (COSTERTON et al., 1987). In technical publications on biofilters, biofilms are usually treated as a "black box" without any further analysis of structure, population, and functional interrelations within the biological matrix. For the better understanding of microbial factors affecting biological filtration processes, some features of the biofilm form of life are presented in this chapter. Fig. 1 shows an SEM micrograph of a biofilm on a sand filter.

The structure of a mature biofilm in a drinking water biofilter is complex and characterized by strong heterogeneity. It is maintained by the EPS which keep the microorganisms together and attach them to the filter material surface. EPS can represent the major part of the organic biomass. Also, they sorb metal ions such as Fe^{2+}, Fe^{3+}, Mn^{2+}, Ca^{2+}, Mg^{2+}, Al^{3+}, etc. (GALVIN, 1992; FLEMMING et al., 1996) as well as organic and in particular humic substances which contribute to the brown or black color of the biofilm.

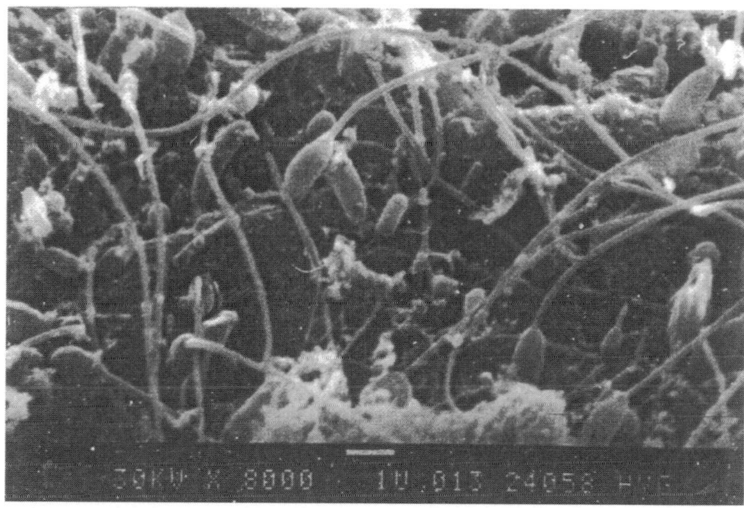

Fig. 1. Scanning electron micrograph (SEM) of a biofilm in a drinking water sand filter (courtesy of Dr. G.-J. TUSCHEWITZKI).

2 Extracellular Polymeric Substances (EPS)

EPS form the geometry and morphology of the biofilm and, thus, also determine their physico-chemical properties such as diffusion resistance and gradients of oxygen and substrates. Thus, they are of great importance for the function of biofilms. The production of EPS is a general feature of microorganisms in most environments. Microbial EPS are biosynthetic polymers (biopolymers). GEESEY (1982) defined EPS as "extracellular polymeric substances of biological origin that participate in the formation of microbial aggregates". Under natural conditions, EPS production seems to be an important feature of survival. As most environmental bacteria occur in microbial aggregates such as biofilms, the structural and functional integrity is essentially based on the presence of an EPS matrix. EPS consist of varying proportions of carbohydrates, proteins, nucleic acids, lipids/phospholipids, and humic substances. Thus, EPS contain a huge variety of chemical structures. In many earlier studies, different types of species-specific and non-specific polysaccharides were considered to be the main constituents of EPS, so that isolation and purification procedures were often focused on the carbohydrate fraction of the EPS. However, more extensive analyses of EPS showed that proteins frequently were abundant in EPS from pure cultures of gram-negative and gram-positive bacteria (e.g., BROWN and LESTER, 1980; PLATT et al., 1985; ARVANITI et al., 1994; JAHN and NIELSEN, 1995) as well as from biofilms (e.g., BROWN and LESTER, 1980; JAHN and NIELSEN, 1995; NIELSEN et al., 1997). A number of publications have shown that proteins were even predominant over polysaccharides and represented the largest fraction in the EPS of biofilms and activated sludges from wastewater systems (RUDD et al., 1983; NIELSEN et al., 1997; BURA et al., 1998; DIGNAC et al., 1998; JORAND et al., 1998). Among the nucleic acids, DNA has regularly been found in the EPS from wastewater biofilms (BROWN and LESTER, 1980; URBAIN et al., 1993; NIELSEN et al., 1997; BURA et al., 1998), but also in the extracellular material from pure cultures (PLATT et al., 1985; ARVANITI et al., 1994; JAHN and NIELSEN, 1995; WATANABE et al., 1998). In addition, accumulation of humic substances in the EPS matrix of wastewater biofilms and activated sludges seems to be common (NIELSEN et al., 1997; JAHN and NIELSEN, 1998).

The presence of EPS is considered to be the basic prerequisite for the formation as well as the maintenance of the integrity of biofilms. EPS are responsible for the mechanical stability of these microbial aggregates which is a result of intermolecular interactions between many different macromolecules; multivalent cations often promote and enforce the interactions between EPS. The result of these interactions is the formation of a three-dimensional, gel-like matrix surrounding the EPS-producing cells. At first glimpse, the EPS matrix looks simply random and without any perceivable order. From the point of view of microbial ecology, however, it has an important function, i.e., to create the gel matrix in which the organisms can be fixed next to each other with a long retention time. This function should be achieved with the lowest expense of energy and nutrients. The fact that only 1–2% of organic matter are required to bind 98–99% of water and to form a stable gel demonstrates the success of the strategy.

After discovering the universal presence of EPS-containing structures, the formation of a gel-like network has been regarded to be the most important general function permitting microorganisms to live in aggregated communities. This primary function of EPS seems to provide an important advantage of survival for the immobilized microorganisms, which may be the basis of various other functions of EPS. More recent studies suggest that lectin-like proteins also contribute to the formation of the three-dimensional network of the biofilm matrix by cross-linking polysaccharides directly or indirectly through multivalent cation bridges (HIGGINS and NOVAK, 1997). However, the main function of extracellular proteins in biofilms is still seen in their role as enzymes performing the digestion of exogenous macromolecules in the microenvironment of the immobilized cells. Thus, they provide low-molecular-weight nutrients which can readily be taken up and metabolized by the cells. This is a very rapid process because of the short retention time of water in a biofilter. The role of other EPS components as structural elements of the biofilm matrix remains to be established. However, it is expected that EPS components such as nucleic acids, lipids, or humic substances significantly influence the rheological properties and thus the stability of biofilms, as can be deduced from basic laboratory studies on the properties of polymer mixtures.

3 Life in Biofilms

The vast majority of microorganisms on earth live in biofilms (COSTERTON et al., 1987). An interesting aspect of this form of life is that sessile bacteria are inherently different from their planktonic counterparts. In the sessile state, bacteria express different genes, they grow at different rates and produce EPS in large amounts. Genetic responses of bacteria at solid surfaces have been demonstrated using reporter gene techniques (GOODMAN and MARSHALL, 1995). In *Pseudomonas aeruginosa,* attachment to surfaces resulted in increased expression of genes involved in the biosynthesis of alginate, the major extracellular polysaccharide (DAVIES et al., 1993; HOYLE et al., 1993). However, the factors regulating the expression of biofilm phenotypes are still poorly understood. Decreased growth rates and/or EPS production have been associated with increased resistance to antimicrobial agents (GILBERT and BROWN, 1995). Another protective effect is credited to enhanced water retention by the EPS matrix preventing or delaying desiccation and increasing survival of EPS-producing micoorganisms (ROBERSON and FIRESTONE, 1992; OPHIR and GUTNICK, 1994). The EPS matrix is supposed to play a significant role in the nutrition of immobilized bacteria by adsorbing organic nutrients and metal ions from the surrounding medium. This represents an important mechanism for the acquisition of carbon and energy sources in oligotrophic environments – one reason why biofiltration even works with water-containing low concentrations of organic carbon. Extracellular enzymes can degrade organic matter secreted or entrapped from the environment and concentrated within the EPS matrix. The formation of complexes of extracellular enzymes such as lipases with alginates and concomitant stabilization of the enzymes has been demonstrated (WINGENDER et al., 1990). Interaction of secreted enzymes with polysaccharides of the gel matrix may prevent loss of digestive enzymes into the bulk fluid leading to a localized reservoir of high catabolic enzyme activity in close proximity to the immobilized cells within biofilms and allowing the cells to maintain a certain level of control over enzymes in their microenvironment.

Microelectrode studies and the use of fluorescently labeled rRNA-targeted oligonucleotide probes have shown that an important characteristic of microbial aggregation is the heterogenous spatial arrangement of microorganisms giving rise to gradients of solutes such as protons, oxygen, ammonia, nitrite, nitrate, sulfide, organic substrates, and microbial products within the dimensions of biofilms (KÖHL and JÜRGENSEN, 1992; DE BEER et al., 1993; COSTERTON et al., 1994; WIMPENNY and KINNIMENT, 1995; SCHRAMM et al., 1997). In aerobic environments oxygen is consumed in outer layers of biofilms through respiration, so that in deeper regions fermentative species or anaerobic organisms such as sulfate reducers or methanogens can develop and be active. Thus, aerobic and anaerobic habitats allow the coexistence of different physiological groups of microorganisms in close proximity, which may lead to metabolic versatility of biofilm communities. Transport processes may be irregular due to the structural heterogeneity of biofilms. COSTERTON et al. (1994) have shown that water channels appeared to transport oxygen into the depth of biofilms, but diffusion limitation and oxygen utilization still resulted in very low oxygen levels at the centers of cellular microcolonies in the biofilm. Thus, in heterogenous biofilm systems, convective flow through channels and pores combined with molecular diffusion into and out of constituent microcolonies determine solute transport, whereas in more confluent microbial biofilms, the penetration of solutes is dominated by molecular diffusion. Observations by transmission electron microscopy of steady state biofilms of *P. aeruginosa* or dental plaque organisms suggested that nutrient diffusion limitations in lower parts of biofilms may lead to starvation, cell death, and lysis of microorganisms (WIMPENNY and KINNIMENT, 1995). In general, lysis of bacteria and generation of gas by anaerobes in deeper zones of biofilms can lead to destabilization and sloughing of biofilm fragments. DE BEER et al. (1993) determined vertical microprofiles of ammonium, oxygen, nitrate, and pH within nitrifying aggregates (diameter approximately 100–120 µm) grown in a fluidized-bed reactor. Fig. 2 shows the distribution of oxygen concentration within a biofilm (DE BEER et al., 1993).

Both ammonium consumption and nitrate production occurred in the outer parts of the aggregates, even under conditions when ammonium and oxygen penetrated the whole aggregate, suggesting that nitrifying bacteria were localized in distinct areas within the aggregates. In trickling filter biofilms, oxygen consumption and nitrite/nitrate production were restricted to the upper 50–100 µm of the

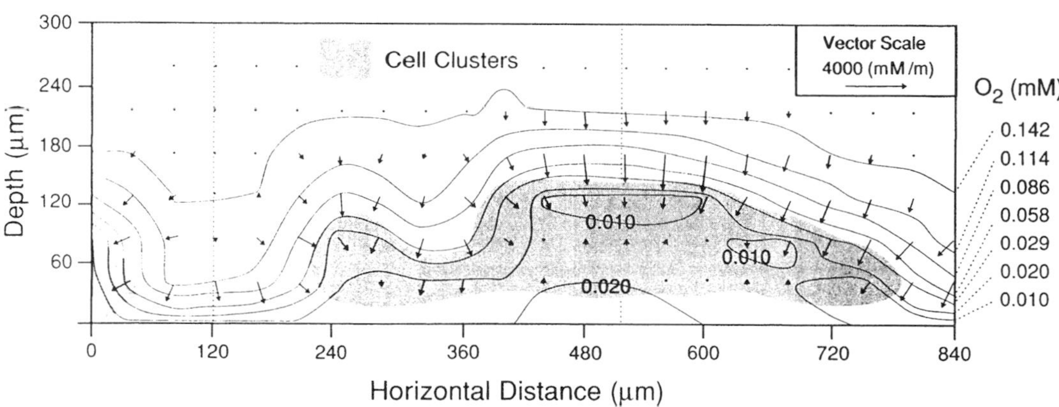

Fig. 2. Oxygen concentration profiles in a biofilm (after DE BEER et al., 1993).

biofilm (SCHRAMM et al., 1997). Ammonia and nitrite oxidizers were found in a dense layer of cells and cell clusters in the upper part of the biofilm in close vicinity to each other, favoring the sequential two-step metabolism from ammonia to nitrate.

Limited penetration into microbial aggregates is suspected to be one of several resistance mechanisms of biofilm organisms against the action of biocides and antibiotics (GILBERT and BROWN, 1995). Whether or not the EPS matrix constitutes a penetration barrier and protective structure greatly depends on the nature of the antimicrobial agent. In mucoid strains of *P. aeruginosa*, slime formation afforded protection against chlorine, but not against hydrogen peroxide (WINGENDER et al., 1999). Chlorine penetration into biofilms was found to be a function of simultaneous diffusion into the matrix and reaction with EPS resulting in the neutralization of the biocide (DE BEER et al., 1994). Generally, it is assumed that mass transfer of antimicrobial agents to microorganisms within aggregates may be restricted by boundary layers surrounding biofilm surfaces, by the EPS matrix acting as diffusion barrier, and by the interaction with matrix components. However, this does not always apply to small molecules and ions, which can rapidly penetrate the matrix (DE BEER et al., 1994). It is commonly assumed that mass transport in biofilms was only due to diffusion, however, convectional transport within biofilms was demonstrated by LEWANDOWSKI et al. (1994). Although flow velocity in the channels was about three orders of magnitude lower than in the adjacent water phase, mass transport was much higher than by diffusion only.

Horizontal gene transfer may be facilitated due to the close proximity of aggregated cells and the accumulation of DNA in the EPS matrix. Several mechanisms of gene transfer are likely to occur in the natural environment: conjugation, transduction, and natural transformation (LORENZ and WACKERNAGEL, 1994). Little is known about the exchange of genetic information within biofilms, although a few studies have demonstrated gene exchange in pure culture biofilms (ANGLES et al., 1993; LISLE and ROSE, 1995) and plasmid transfer in river epilithon (BALE et al., 1988). These observations suggest that gene transfer can occur in microbial aggregates, possibly contributing to phenotypic changes of constituent microorganisms.

Another form of information transfer resides in the phenomenon of quorum sensing which is a signaling mechanism in response to population density (SWIFT et al., 1996). This kind of cell-to-cell communication is mediated by low-molecular weight diffusible signal molecules (autoinducers), which are chemically described as N-acyl-L-homoserine lactones (AHL). Extracellular accumulation of AHL above a critical threshold level results in transcriptional activation of a range of different genes with concomitant expression of new phenotypes. This may be one reason why populations with high cell densities characteristic of microbial aggregates display biological properties which are different from populations with lower cell densities typical of microbial suspensions in aqueous environments. AHL allow bacteria to sense when cell densities in their surroundings reach the minimal level for a coordinative population response to be initiated. AHL control gene expression of diverse processes. They are involved in the regulation of bioluminescence, swarming, conjugation, and the production of extracellular products such as antibiotics, surfactants, and enzymes (SWIFT et al., 1996). Cell-to-cell communication via AHL seems to be of fundamental importance for aggregated bacteria to adapt dynamically in response to prevailing environmental conditions. Studies on the role of AHL in biofilter biofilms are to be awaited and certainly will further illuminate the interactions of the organisms.

4 Biofilter Microbial Communities – an Example

The distribution of biomass in a filter is typically not continuous, but decreases gradually from the influent to the effluent. MOLL and SUMMERS (1999) have shown this using phospholipids as a marker for viable biomass. Phospholipids are suitable to assess living biomass, because this class of biomolecules decays very rapidly after cell death and, thus, is representative of intact cells (TUNLID and WHITE, 1989; FINDLAY and DOBBS, 1993). In Fig. 3, the viable biomass profile in a sand filter prior to backwashing is shown. Both attached and planktonic biomass followed the same trend with filter depth. The suspended biomass made up 40% of the total biomass at the top and medium filter positions but only 10% of the total biomass at the bottom of the filter.

Ozonation increases the bioavailability of organic matter in water, which is generally known (LECHEVALLIER, 1991). This could also be shown for attached biomass in biofilters. If the water was ozonated, the reaction products with natural organic matter (NOM) led to a 2- to 3-fold increase in the viable biomass at the top of the filters compared to filters fed with non-ozonated water. Biomass decreased quickly with increasing depth in filters treating both ozonated and non-ozonated water as the partially oxidized NOM became more bioavailable. Biofiltration following ozonation is recommended in drinking water treatment scenarios to decrease microbial regrowth in distribution systems, for removal of by-products of disinfection by ozonation. It is also beneficial for the removal of precursors of chlorine disinfection by-products.

For obvious reasons, backwashing with disinfectants, in particular with chlorine, and operation at 5 °C were detrimental to biomass growth and substrate removal (MOLL and SUMMERS, 1999). This indicates that longer contact times are needed, if chlorinated water is used for backwashing, or in winter in order to achieve similar performance to the warmer seasons. Varying the pH of the source water did not significantly affect biomass concentrations or removal of DOC or disinfectant by-product precursors. Therefore, steps taken to enhance NOM removal before filtration, such as enhanced coagulation, which decreases the pH value, or softening, which increases the pH value, will have little impact on biofilter performance.

Some observations of the study of MOLL and SUMMERS are summarized here: The greatest impact on biofilter performance and microbial community structure was given by the contact time, preozonation, and the presence of disinfectants. The populations in a biofilter have been analyzed based on phospholipid fatty acid (PLFA) analysis, a method which is

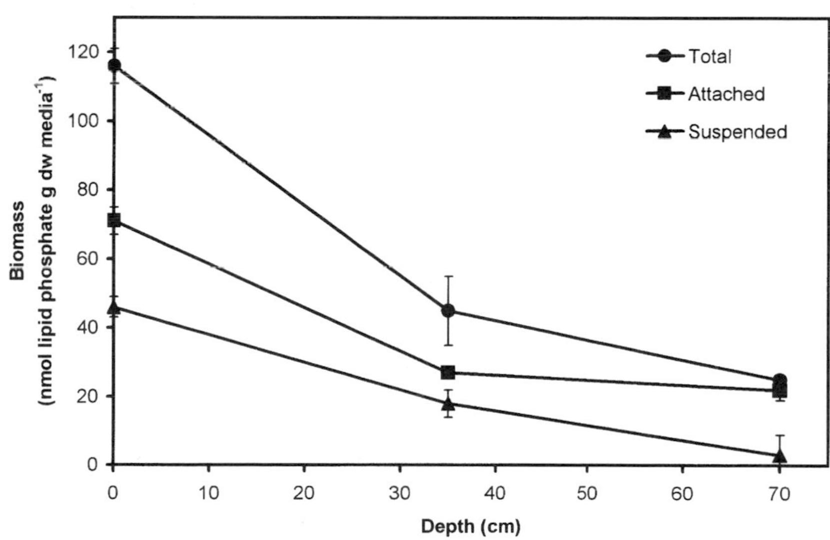

Fig. 3. Biomass profile of a biofilter treating conventionally treated water from Ohio River (after MOLL and SUMMERS, 1999).

well suited to characterize summarily large microbial populations (WHITE, 1993). Gram-negative bacterial fatty acid markers (monoenoic acids) were predominant in all biofilters (63–65 mole %). Eukaryotic markers (polyenoic acids) were next most abundant (8.2–9.3 mole %), followed by fatty acids common to gram-positive bacteria (terminally branched saturated acids, 2.0–5.5 mole %) and, not generally expected in a fully aerated system, sulfate-reducing bacteria (SRB, branched monoenoic acids, 1.5–3.1 mole %). The presence of strictly anaerobic SRB in biofilters can be explained by the structure of biofilms which allows for anaerobic zones below aerobic colonies (see Fig. 2). This is common when the consumption of oxygen is faster than the diffusion through the biofilm matrix (DE BEER et al., 1994). Microeukaryotic markers, particularly for protozoa, were depleted in the filter treating non-ozonated water and at the bottom of the filter treating ozonated water, indicating that a lower prokaryotic biomass may cause a decrease in the relative abundance of protozoan predators. Biofilters operated at acidic pH values were depleted from gram-positive bacteria at the top of the filters. These were replaced by markers for gram-negative bacteria. The filters treating water of neutral pH value had a greater relative abundance of markers for gram-positive bacteria than the filters treating waters with high and low pH values. The tops of the filters treating alkaline water were enriched with markers for eukaryotes. Shifts in the PLFA profiles indicated a gradient in the growth rate of the microbial communities colonizing the filters from log phase growth at alkaline pH value towards stationary growth at acidic pH value. PLFA profiles of the filters operated at different temperatures indicated an increasing gradient in markers for gram-negative bacteria and microeukaryotes as the biofilter operation temperature decreased. These fatty acids are replaced by general fatty acids (normal saturates), and markers for gram-positive bacteria and sulfate-reducing bacteria in the community from the filter operated at 35 °C. These changes in membrane fatty acid composition may be important for the maintenance of membrane fluidity at different temperatures (WHITE, 1993).

Further investigations on the utilization of different carbon sources as determined by the BIOLOG system revealed that carboxylic acids and amino acids were better utilized by the communities treating neutral or acidic water compared to the filter treating alkaline water, regardless of the operation temperature. Carbohydrates were better utilized by the microorganisms inhabiting the filter operated at pH 7.5. Fig. 4 shows the percentage of compounds used in each of the substrate classes represented by the higher number of compounds: carbohydrates (28 compounds), carboxylic acids (24 compounds), and amino acids (20 compounds) (MOLL and SUMMERS, 1999).

This study indicates that increasing the ozone dose prior to biofiltration may select for microorganisms better adapted to quickly degrade assimilable organic carbon. Likewise, increasing the temperature and decreasing the influent DOC concentration may select for microbial populations adapted for utilization of precursors to TOX (total organic halides), THM (trihalomethanes), HAA6 [haloacetic acids (six)], and chlorine demand (MOLL and SUMMERS, 1999).

5 Future Outlook

Biofiltration works very successfully (Chapter 22, this volume), although the processes within the biofilm are still not yet fully understood. However, a deeper knowledge of the mechanisms with which the biofilm organisms metabolize their substrates and the conditions under which they perform best are still worth further research. The role of EPS in biofilm stability is currently under investigation. This may shed light on detachment during backwashing and sloughing events. Operating parameters such as filtration rate and backwash conditions may affect the biofilm present in biofilters. However, the effects of specific backwashing strategies on substrate removal and biofilm activity and amount are not fully known, but under investigation (BROWER et al., 1996; COLTON et al., 1996). Biofilm research is about to reveal the three-dimensional structure of microbial consortia and how they are

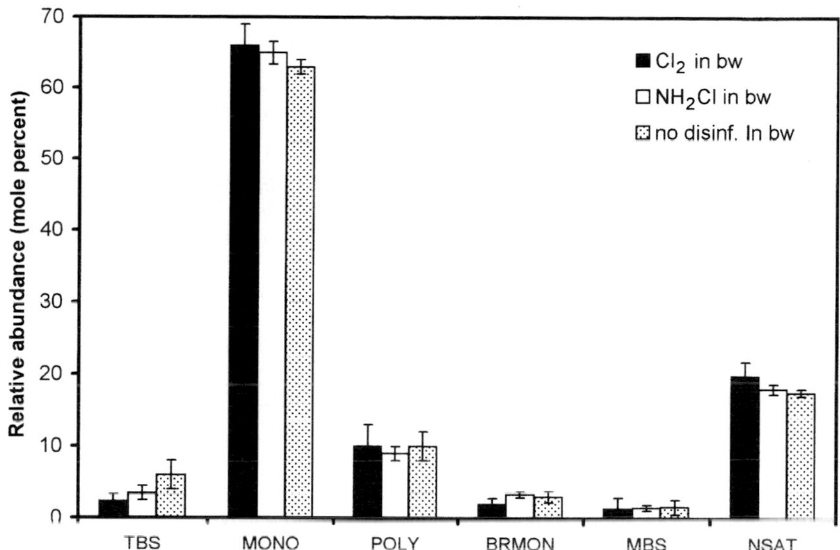

Fig. 4. Impact of ozonation on the utilization of BIOLOG substrates by pilot-scale biofilter microbial communities (after MOLL and SUMMERS, 1999).

regulated. The most intriguing feature in this context is the still open question of cell–cell communication in response to the conditions operating a biofilter. Thus, although biofiltration is a very old technique for drinking water treatment, further improvement can be expected in terms of performance of the filtration reactors.

6 References

ANGLES, M. L., MARSHALL, K. C., GOODMAN, A. E. (1993), Plasmid transfer between marine bacteria in the aqueous phase and biofilms in reactor microcosms, *Appl. Environ. Microbiol.* **59**, 843–850.

ARVANITI, A., KARAMANOS, N. K., DIMITRACOPOULOS, G., ANASTASSIOU, E. D. (1994), Isolation and characterization of a novel 20 kDa sulfated polysaccharide from the extracellular slime layer of Staphylococcus epidermidis, *Arch. Biochem. Biophys.* **308**, 432–438.

BALE, M. J., FRY, J. C., DAY, M. J. (1988), Transfer and occurrence of large mercury resistance plasmids in river epilithon, *Appl. Environ. Microbiol.* **54**, 972–978.

BROWER, J. B., BARFORD, C. C., HAO, O. J. (1996), Biological fixed-film systems, *Water Environ. Res.* **68**, 469–479.

BROWN, M. J., LESTER, J. N. (1980), Comparison of bacterial extracellular polymer extraction methods, *Appl. Environ. Microbiol.* **40**, 179–185.

BURA, R., CHEUNG, M., LIAO, B., FINLAYSON, J., LEE, B. C. et al. (1998), Composition of extracellular polymeric substances in the activated sludge floc matrix, *Water Sci. Technol.* **37**, 325–333.

COLTON, J. F., HILLIS, P., FITZPATRICK, C. S. B (1996), Filter backwash and start-up strategies for enhanced particulate removal, *Water Res.* **30**, 2502–2507.

COSTERTON, J. W., CHENG, K.-J., GEESEY, G. G., LADD, T. I., NICKEL, J. C. et al. (1987), Bacterial biofilms in nature and disease, *Annu. Rev. Microbiol.* **41**, 435–464.

COSTERTON, J. W., LEWANDOWSKI, Z., DE BEER, D., CALDWELL, D., KORBER, D., JAMES, G. (1994), Biofilms, the customized microniche, *J. Bacteriol.* **176**, 2137–2142.

DAVIES, D. G., CHAKRABARTY, A. M., GEESEY, G. G. (1993), Exopolysaccharide production in biofilms: substratum activation of alginate gene expression by *Pseudomonas aeruginosa, Appl. Environ. Microbiol.* **59**, 1181–1186.

DE BEER, D., VAN DEN HEUVEL, J. C., OTTENGRAF, S. P. P. (1993), Microelectrode measurements of the activity distribution in nitrifying bacterial aggregates. *Appl. Environ. Microbiol.* **59**, 573–579.

DE BEER, D., SRINIVASAN, R., STEWART, P. S. (1994), Direct measurement of chlorine penetration into biofilms during disinfection, *Appl. Environ. Microbiol.* **60**, 4339–4344.

DIGNAC, M.-F., URBAIN, V., RYBACKI, D., BRUCHET, A., SNIDARO, D., SCRIBE, P. (1998), Chemical description of extracellular polymers: implication on activated sludge floc structure, *Water Sci. Technol.* **38**, 45–53.

DUNCAN, A. (1988): The ecology of slow sand filters, in: Slow Sand Filtration: *Recent Developments in Water Treatment Technology* (GRAHAM, N. J. D., Ed.), pp. 163–180. Chichester: Ellis Horwood.

EIGHMY, T. T., COLLINS, M. R., SPANOS, S. K., FENSTERMACHER, J. (1992), Microbial populations, activities and carbon metabolism in slow sand filters, *Water Res.* **26**, 1319–1328.

ESCH, P., NEHRKORN, A. (1988), Scanning electron microscopic examination of the microbial population in slow sand filters, *Zbl. Bakt. Hyg. B* **85**, 569–579.

FINDLAY, R. AN DOBBS, F. C. (1993), Quantitative description of microbial communities using lipid analysis, in: *Handbook of Methods in Aquatic Microbial Ecology* (KEMP, P. F., SHERR, B. F., SHERR, E. B., COLE, J. C., Eds.), pp. 271–284. Boca Raton, FL: Lewis Publishers.

FLEMMING, H.-C. (1991), Biofilms as a particular form of microbial life, in: *Biofouling and Biocorrosion in Industrial Water Systems* (FLEMMING, H.-C., GEESEY, G. G., Eds.), pp. 3–9. Heidelberg: Springer-Verlag.

FLEMMING, H.-C., J. SCHMITT, MARSHALL, K. C. (1996), Sorption properties of biofilms, in: *Environmental Behaviour of Sediments* (CALMANO, W., FÖRSTNER, U., Eds.), pp. 115–157. Chelsea, MI: Lewis Publishers.

Galvìn, R. M. (1992), Ripening of silica sand used for filtration, *Water Res.* **26**, 683–688.

GEESEY, G. G. (1982), Microbial exopolymers: ecological and economic considerations, ASM News **48**, 9–14.

GILBERT, P., BROWN, M. R. W. (1995), Mechanisms of the protection of bacterial biofilms from antimicrobial agents, in: *Microbial Biofilms* (LAPPIN-SCOTT, H. M., COSTERTON, J. W., Eds.), pp. 118–130. Cambridge, MA: Cambridge University Press.

GOODMAN, A. E., MARSHALL, K. C. (1995), Genetic responses of bacteria at surfaces, in: *Microbial Biofilms* (LAPPIN-SCOTT, H. M., COSTERTON, J. W. Eds.), pp. 80–98. Cambridge, MA: Cambridge University Press.

HIGGINS, M. J., NOVAK, J. T. (1997), Characterization of exocellular protein and its role in bioflocculation, *J. Environ. Eng.* 123, 479–485.

HOYLE, B. D., WILLIAMS, L. J., COSTERTON, J. W. (1993), Production of mucoid exopolysaccharide during development of Pseudomonas aeruginosa biofilms, *Appl. Environ. Microbiol.* **61**, 777–780.

JAHN, A., NIELSEN, P. H. (1995), Extraction of extracellular polymeric substances (EPS) from biofilms using a cation exchange resin, *Water Sci. Technol.* **32**, 157–164.

JAHN, A., NIELSEN, P. H. (1998), Cell biomass and exopolymer composition in sewer biofilms, *Water Sci. Techol.* **37**, 17–24.

JORAND, F., ZARTARIAN, F., THOMAS, F., BLOCK, J. C., BOTTERO, J. Y. et al. (1995), Chemical and structural (2D) linkage between bacteria within activated sludge flocs, *Water Res.* **29**, 1639–1647.

KÖHL, M., JÜRGENSEN, B. B. (1992), Microsensor measurements of sulfate reduction and sulfide oxidation in compact microbial communities of aerobic biofilms, *Appl. Environ. Microbiol.* **58**, 1164–1174.

LECHEVALLIER, M. (1991), Biocides and the current status of biofouling control in water systems, in: *Biofouling and Biocorrosion in Industrial Water Systems* (FLEMMING, H. C., GEESEY, G. G., Eds.), pp. 113–132. Heidelberg: Springer-Verlag.

LEWANDOWSKI, Z., STOODLEY, P., ALTOBELLI, S., FUKUSHIMA, E. (1994), Hydrodynamics and kinetics in biofilm systems – recent advances and new problems, *Water Sci. Technol.* **29**, 223–229.

LISLE, J. T., ROSE, J. B. (1995), Gene exchange in drinking water and biofilms by natural transformation, *Water Sci. Techol.* **31**, 41–46.

LORENZ, M. G., WACKERNAGEL, W. (1994), Bacterial gene transfer by natural genetic transformation in the environment, *Microbiol. Rev.* **58**, 563–602.

MOLL, D. M., SUMMERS, R. S. (1999), Assessment of drinking water filter microbial communities using taxonomic and metabolic profiles, *Water Sci. Technol.* **39**, 83–89.

NIELSEN, P. H., JAHN, A., PALMGREN, R. (1997), Conceptual model for production and composition of exopolymers in biofilms, *Water Sci. Techol.* **36**, 11–19.

OPHIR, T., GUTNICK, D. L. (1994), A role for exopolysaccharides in the protection of microorganisms from desiccation, *Appl. Environ. Microbiol.* **60**, 740–745.

PLATT, R. M., GEESEY, G. G., DAVIS, J. D., WHITE, D. C. (1985), Isolation and partial chemical analysis of firmly bound exopolysaccharide from adherent cells of a freshwater bacterium, *Can. J. Microbiol.* **31**, 657–680.

ROBERSON, E. B., FIRESTONE, M. K. (1992), Relationship between desiccation and exopolysaccharide production in a soil *Pseudomonas* sp., *Appl. Environ. Microbiol.* **58**, 1284–1291.

RUDD, T., STERRITT, R. M., LESTER, J. N. (1983), Extraction of extracellular polymers from activated sludge, *Biotechnol. Lett.* **5**, 327–332.

SCHRAMM, A., LARSEN, L. H., REVSBACH, N. P., AMANN, R. I. (1997), Structure and function of a nitrifying biofilm as determined by microelectrodes and fluorescent oligonucleotide probes, *Water Sci. Technol.* **36**, 263–270.

SCHWEISFURTH, R. (1986), Biofiltration in drinking water treatment, in: *Biotechnology* 1st Edn. Vol. 8 (REHM, H.-J., REED, G., Eds.), pp 399–423. Weinheim: VCH.

SWIFT, S., THROUP, J. P., WILLIAMS, P., SALMOND, G. P. C., STEWART, G. S. A. B. (1996), Quorum sensing: a population-density component in the determination of bacterial phenotype, *Trends Biochem. Sci.* **21**, 214–219.

TUNLID, A., WHITE, D. C. (1990), Use of lipid biomarkers in environmental samples, in: *Analytical Microbiology Methods* (FOX, A., MORGAN, S. L., LARSSON, L., ODHAM, G., Eds.), pp. 259–274. New York, London: Plenum Press.

URBAIN, V., BLOCK, J. C., MANEM, J. (1993), Bioflocculation in activated sludge: an analytical approach, *Water Res.* **27**, 829–838.

WATANABE, M., SASAKI, K., NAKASHIMADA, Y., KAKIZONO, T., NOPARATNARAPORN, N., NISHIO, N. (1998), Growth and flocculation of a marine photosynthetic bacterium *Rhodovulum* sp., *Appl. Microbiol. Biotechnol.* **50**, 682–691.

WHITE, D. C. (1993), In situ measurement of microbial biomass, community structure, and nutritional status, *Phil. Trans. R. Soc. Lond. A* **344**, 59–67.

WIMPENNY, J. W. T., KINNIMENT, S. L. (1995), Biochemical reactions and the establishment of gradients within biofilms, in: *Microbial Biofilms* (LAPPIN-SCOTT, H. M., COSTERTON, J. W., Eds.), pp. 99–117. Cambridge, MA: Cambridge University Press.

WINGENDER, J. (1990), Interactions of alginate with exoenzymes, in: *Pseudomonas Infection and Alginates. Biochemistry, Genetics and Pathology* (GACESA, P., RUSSELL, N. J., Eds.), pp. 160–180. London: Chapman and Hall.

WINGENDER, J., GROBE, S., FIEDLER, S., FLEMMING, H.-C. (1999), The effect of extracellular polysaccharides on the resistance of *Pseudomonas aeruginosa* to chlorine and hydrogen peroxide, in: *Biofilms in Aquatic Systems* (KEEVIL, W., GODFREE, A. F. HOLT, D. M., DOW, C. S., Eds.), pp. 93–100. Cambridge: Royal Society of Chemistry.

22 Biofiltration Processes for Organic Matter Removal

WOLFGANG UHL
Mülheim, Germany

1 General and Historical Aspects

In this chapter only bioprocesses in deep bed rapid filters and activated carbon adsorbers for drinking water are understood and discussed as biofiltration processes for organic matter removal. Actually, processes like slow sand filtration or ground filtration are also biofiltration processes, but would require special and detailed discussion.

In Europe the application of biofiltration processes for the removal of natural organic matter has a long tradition which goes back to the 1970s. It is very often applied in Germany, France, and The Netherlands. In North America this process has gained special attention since the late 1980s.

A typical process applying biofiltration for natural organic matter removal is the so-called Mülheim process which is depicted in Fig. 1 and was described by SONTHEIMER et al. (1978). It applies preozonation, precipitation/flocculation and sedimentation, main ozonation, dual media rapid filtration, and activated carbon rapid filtration. After this treatment the water is subjected to further ground filtration and is safety-chlorinated before distribution. Organic substances of higher molecular weight are partly oxidized by ozonation and their biodegradability is increased. Biodegradation then is carried out by bacteria which are sessile in deep-bed rapid filters and granular activated carbon rapid filters.

An overview of different treatment plants applying biofiltration for organic matter removal in drinking water treatment is given by RITTMANN and HUCK (1989). However, this is the state of the late 1980s.

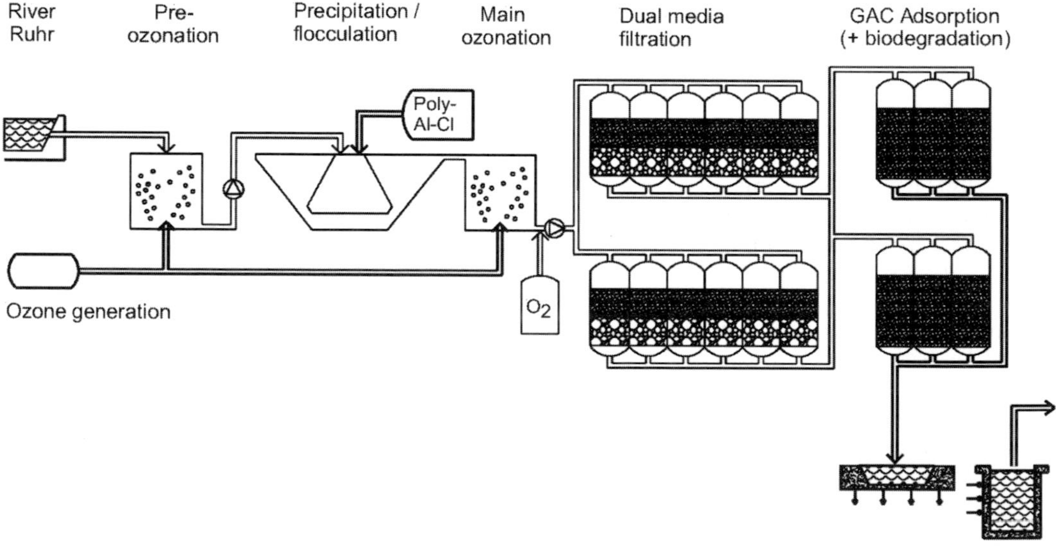

Fig. 1. The Mülheim process: a typical process where biofiltration is taken advantage of in surface water treatment.

2 Methods of Determination of Biodegradable Organic Matter (BOM)

2.1 General

During the 1980s and 1990s several methods for the determination of biodegradable organic matter were developed. They were targeted to answer two questions. First, consumers and legislation in many European countries claimed that drinking water should be microbiologically safe, but at the same time as natural as possible (e. g., Anonymous, 1973). Thus, in order to assure that drinking water could be distributed without safety chlorination water works and researchers looked for methods to determine concentrations of organic matter able to support bacterial regrowth during distribution.

Secondly, dissolved organic matter (DOM) can cause taste and odor problems and serve as precursors for trihalomethanes (THM) which may be formed during chlorination (see Chapter 18, this volume). Since it became more and more evident that biological processes could be taken advantage of in drinking water treatment for the removal of dissolved organic matter, methods were required to determine the concentration of DOM which could be removed by biofiltration.

For single substrates biomass produced and substrate consumed are clearly interrelated for stoichiometric and energetic reasons. However, in drinking water treatment complex mixtures of mostly unknown composition and at relatively low concentration serve as substrates. Therefore, depending on the question to be answered (possible regrowth or removal of organic carbon), the one or the other method is more suitable.

2.2 AOC Method

The AOC method (AOC: assimilable organic carbon) was originally developed by VAN DER KOOIJ et al. (1982a) in The Netherlands. It uses the metabolic diversity of three bacterial strains to degrade a very broad spectrum of organic substrates usually occurring in drinking water. All three can use nitrate and ammonia as nitrogen sources. Among other substances *Pseudomonas fluorescens* P17 can utilize carboxylic and amino acids, alcohols, and carbohydrates for growth (VAN DER KOOIJ et al., 1982b). Growth of *Spirillum* sp. NOX is – besides carboxylic acids – supported by, e.g., oxalate, formate, and other substances (VAN DER KOOIJ and HIJNEN, 1984) typically formed by ozonation of drinking water and not utilized by *P. fluorescens* P17. If starch-based coagulants or filter aids are applied in water treatment a third strain, *Flavobacterium* sp. S12, should be used in the assay (VAN DER KOOIJ and HIJNEN, 1981).

The original AOC method is depicted in Fig. 2. The sample is usually taken in duplicate in carbon-free, sterile borosilicate Erlenmeyer flasks, cleaned with chromium sulfuric acid or muffled at 500 °C. After pasteurization at 60 °C for 30 min and cooling to ambient temperature the samples are inoculated with *P. fluorescens* P17 and *Spirillium* sp. NOX in a way that the initial concentration in the samples is about 500 CFU mL^{-1} (usually, *Flavobacterium* sp. is only used when starch or maltose-like compounds are to be expected). The samples are incubated at 15 °C and proliferation of the inoculated species is monitored by plating serial dilutions daily (nutrient-poor agar, spread plate technique).

P. fluorescens P17 and *Spirillium* sp. NOX can be distinguished on the plates by the morphology of colonies. AOC is calculated from the maximum concentration of culturable bacteria of the two strains. As both strains can use acetate as carbon source, acetate is used as surrogate. AOC is calculated as the sum of the product of the viable counts for each strain and the inverse of the yield of the respective strain on acetate. It is expressed as µg acetate–C equivalents per L, i.e.,

$$\mathrm{AOC} = \frac{N_{\mathrm{P17}}}{Y_{\mathrm{P17}}} + \frac{N_{\mathrm{NOX}}}{Y_{\mathrm{NOX}}} \tag{1}$$

N_{P17} and N_{NOX} are the bacterial concentrations and Y_{P17} and Y_{NOX} are the yield-coefficients for P17 and NOX, respectively.

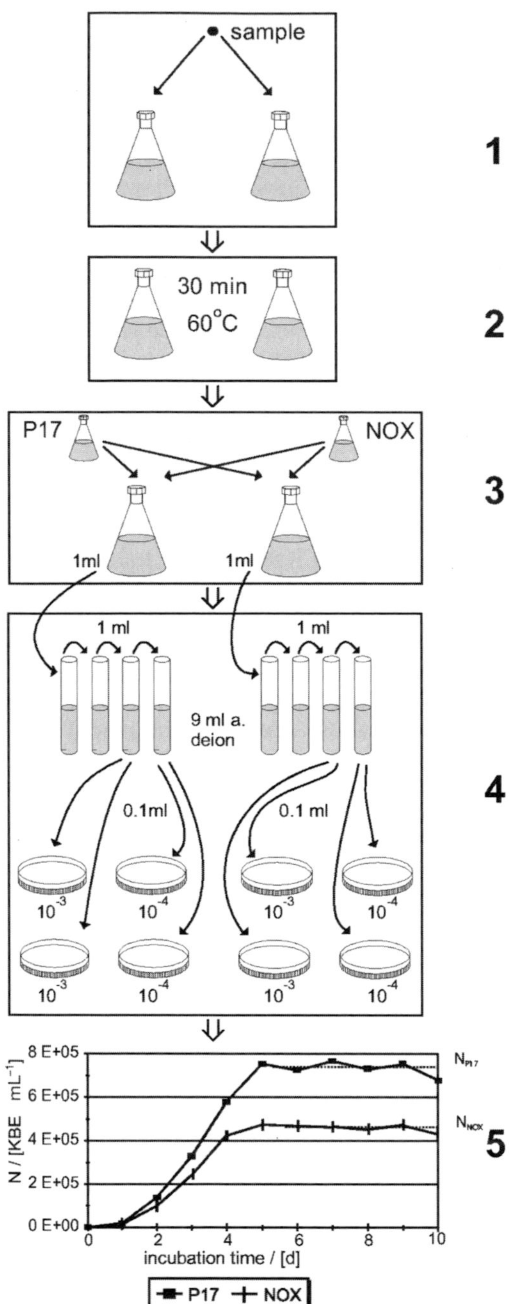

Fig. 2. Procedure for AOC determination according to the method of VAN DER KOOIJ. (1) Sampling in sterile flasks, (2) pasteurization, (3) inoculation and incubation (4) sampling, and determination of culturable bacteria concentration, (5) calculation of N_{P17} and N_{NOX} from growth curve.

Values for yield coefficients determined by VAN DER KOOIJ and HIJNEN (1985) are $4.1 \cdot 10^6$ CFU μg^{-1} acetate–C for P17 and $1.2 \cdot 10^7$ CFU μg^{-1} acetate–C for NOX. For reasons of quality assurance yield coefficients should be determined repeatedly in each laboratory. Typical AOC concentrations for drinking water are in the range from 5–200 µg acetate–C equivalents per L. The precision of the method is reflected by the coefficient of variation, which was calculated to about 10% in the author's laboratory (from 360 duplicate determinations), which is in accordance with the experiences of others.

Some modifications of the original method are applied by different researchers. These may lead to different results. Thus, special attention has to be taken to the variant chosen when comparing the results from different laboratories. In North America the procedure according to the proposed standard method is often applied (Anonymous, 1995). This involves incubation in 40 mL vials instead of 1 L Erlenmeyer flasks. As it was shown by KAPLAN and BOTT (1988) for natural organic matter the surface-to-volume ratio of the incubation vessels considerably influences the viable count of bacteria, but not for acetate. This results in the calculation of much higher AOC concentrations when applying smaller vessels compared to the original method (about 3-fold with 40 mL vials).

Also a nutrient salts medium is sometimes added which may increase the yield of the test strains for organic matter and thus result in higher AOC concentrations.

2.3 BRP Method

Another method to determine regrowth potential of bacteria in drinking water was developed by WERNER (1985) and WERNER and HAMBSCH (1986), the BRP method (biological regrowth potential). It is not so much aimed at utilizable organic carbon concentrations, but more on the question how the bacteria originally present in the water sample can multiply. The method is depicted in Fig. 3. The sample is sterilized by filtration with 0.2 µm pore size filters, and a nutrient salts medium is added.

Bacterial proliferation is then monitored as turbidity at 12° forward scattering. From the

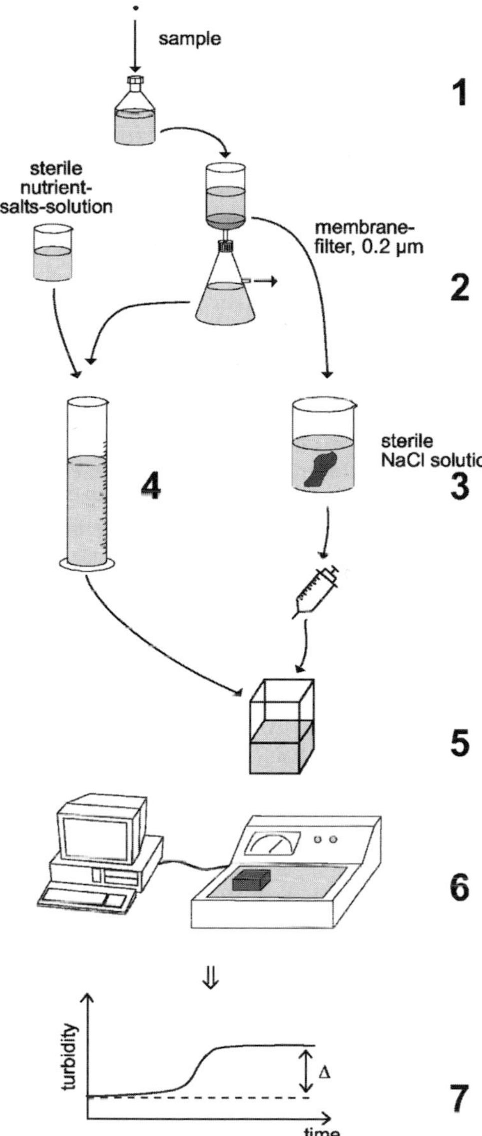

Fig. 3. Determination of bacterial regrowth potential according to the method of WERNER. (1) Sampling, (2) sterile filtration, (3) resuspension, (4) addition of nutrient salts solution, (5) inoculation, (6) turbidity measurement, (7) calculation of regrowth factor from difference in turbidity.

curves of turbidity over time the growth rate is calculated by fitting a Monod type growth function. Usually a regrowth factor is given as a result, i.e., the turbidity reached after a plateau is divided by the turbidity in the beginning. Sometimes a concentration of assimilated organic carbon may be given which is calculated from a regression of regrowth factors for water samples spiked with acetate.

It was often claimed that an advantage of the method is the use of a mixture of autochthonous bacteria which may be better adapted to the substrates than the specific strains applied in the AOC method. However, the amount of bacteria available from the sample may often be too small. Then bacteria from other environments (mostly from GAC filter effluents of drinking water treatment plants) may be collected. Disadvantages are that the inoculum needed is relatively large in comparison to the AOC method (about $5 \cdot 10^4$ mL^{-1} total count in comparison to about $5 \cdot 10^2$. Thus a large increase in bacterial concentration has to be found to be reflected in the regrowth factor. Furthermore, the risk of contamination of the sample with biodegradable carbon is higher during sterile filtration compared to pasteurization, and the equipment used is relatively expensive.

2.4 BDOC Methods

In contrast to the AOC and BRP methods the BDOC method (BDOC: biodegradable dissolved organic carbon) aims at the determination of organic carbon concentrations that can be removed from water. Thus, it is not the amount of biomass formed in the test which is determined, but the decrease in dissolved organic carbon concentration.

In the procedure for the original BDOC method developed by SERVAIS et al. (1987, 1989) (Fig. 4) the sample is filter-sterilized and then reinoculated with part of the sample that was filtered with filters of 2 µm pore size for the removal of particles and of protozoa. Incubation at 20 °C in the dark is carried out for a period of up to 30 d. Carbon utilization is monitored by regularly measuring dissolved organic carbon (DOC) concentration. BDOC is calculated from the difference of the initial concentration and the plateau at minimum. (DOC may increase slightly after a certain time as a result of bacterial lysis.)

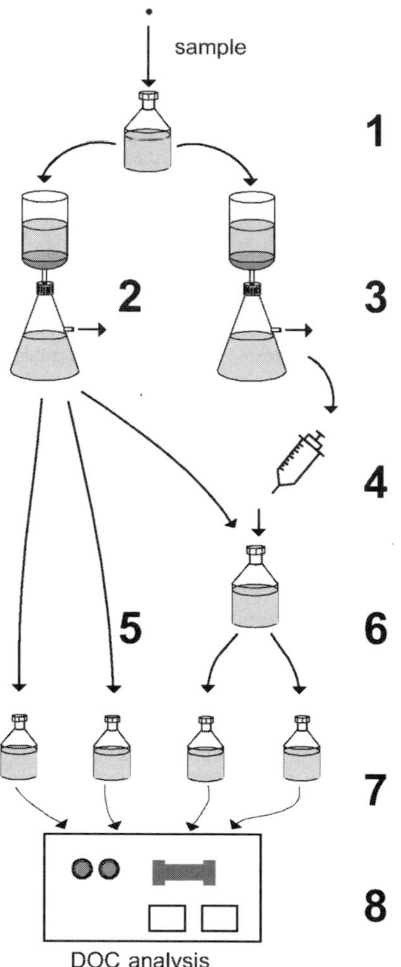

Fig. 4. Procedure of BDOC determination according to the method of SERVAIS. (1) Sampling, (2) sterile filtration, (3) filtration for removal of particles and protozoa, (4) inoculation, (5)-(8) sampling and DOC analysis.

As this method requires relatively long incubation times a BDOC method with sessile bacteria, i.e., an increased bacterial inoculum, was developed by JORET et al. (1988). The inoculum usually consists of sand from rapid filters from drinking water treatment which was washed until no more DOC washout could be observed. ALLGEIER et al. (1996) investigated this method thoroughly. With the sand inoculum DOC degradation was completed after approx. 7 d. The ideal sand–liquid ratio was found at approx. 20% (v/v) and the headspace–liquid ratio at approx. 80%.

In another modification of the BDOC method developed by FRIAS et al. (1992) the sample is circulated through a bed of sintered glass beads with sessile bacteria. It is actually mostly equivalent to the methods developed by HUBELE (1985) and by MOGREN et al. (1990), which used anthracite or sand, respectively, as carrier.

The above mentioned methods are discontinuous and require several days or several weeks until the results are available, so that continuous methods were developed by LUCENA et al. (1990) and RIBAS et al. (1991). In these, water taken directly from the sampling point or a reservoir is pumped continuously through a bed with sand and sessile bacteria. With a residence time of the sample of about 1 h BDOC is calculated from the difference of inlet and outlet concentrations.

2.5 Results of Comparative Investigations

Comparisons of the AOC and BRP methods were carried out by HUCK et al. (1990) and ANDERSON et al. (1990) who investigated the correlation between AOC concentration and regrowth factor. They found a significant correlation in some cases. However, in other cases the results were contradictory.

The AOC method and the original BDOC method according to SERVAIS et al. (1987) were compared by PRÉVOST et al. (1989, 1992). In direct comparison AOC concentrations were much lower than BDOC concentrations. VOLK et al. (1992) compared AOC concentrations with the BDOC method of JORET et al. (1988) and found a linear relationship ($r=0.769$; $n=31$) between AOC and BDOC. By regression a factor of 90 µg AOC per mg BDOC was calculated.

BDOC determinations with suspended and sessile bacteria inocula of different origin were compared in a round robin test by BLOCK et al. (1992). They always found that BDOC was lower when a suspended inoculum was used. However, as in all assays, the incubation time was only 3 weeks, and it might be possible that degradation was not already finished in the assays with suspended inoculum.

3 Effects of Some Treatment Processes on BOM

3.1 Ground Filtration, River Bank Filtration, Slow Sand Filtration

As expected, a decrease in bacterial regrowth potential can be observed when the water is treated by ground filtration or river bank filtration, due to biological processes in the subsurface. WERNER and HAMBSCH (1988) and HAMBSCH (1992) investigated bacterial proliferation in river water and in river water treated by river bank filtration. They showed that the bacterial regrowth potential decreased along the flow path, while the main decrease was observed during the first few meters. Later the decrease was only marginal.

VAN DER KOOIJ (1987) showed that ground filtration resulted in very low AOC concentrations. However, problems may arise with clogging of the infiltration wells, if the water contains considerable amounts of AOC. HIJNEN and VAN DER KOOIJ (1992) investigated clogging of infiltration wells in dependence of the AOC concentration of raw waters. They found that the runtime between necessary regenerations was less than one year when AOC was $>10\ \mu g\ L^{-1}$.

3.2 Precipitation/Flocculation/ Coagulation

During precipitation and flocculation (see Chapter 18, this volume) not only fine suspended particles are removed, but also dissolved organic matter. This may be due to precipitation initiated by changes in pH, but also to adsorption on flocs and particles formed in the treatment stage. Part of this organic matter is biodegradable, resulting in a reduction of AOC or regrowth potential. VAN DER KOOIJ (1990), e.g., showed that coagulation decreases AOC for more than 50%. However, as shown in the same reference, when applying starch-based coagulants a considerable increase in AOC can be observed.

HUCK et al. (1990) showed that polyaluminum chloride may have an inhibitory effect on the growth of *P. fluorescens* P17 in the AOC assay. Thus, special care has to be taken when investigating AOC in samples from precipitation/flocculation. According to the authors' own experience required incubation times may be much higher until the bacteria grow in the samples.

3.3 Ozonation

Ozonation is an oxidation process (see Chapter 18, this volume). The reaction of ozone with natural organic matter can take place via the molecular ozone pathway and the free radical pathway (HOIGNÉ and BADER, 1976). Both pathways result in the formation of higher oxidized organic compounds and lower molecular weight and better biodegradability (see e.g., KUO et al., 1977; GILBERT, 1988; TAKAHASHI et al., 1995).

HAMBSCH and WERNER (1993) studied the effect of ozonation on bank filtrate with a DOC concentration of 2.8 mg L^{-1}. The effect of the ozone dose on the regrowth factor of BRP is shown in Fig. 5. It can be seen that the regrowth factor, which is a measure for the biomass formed during incubation, is approximately linear with the ozone dose. It increases from about 4 without addition of ozone to about 20 at an ozone dose of 1.9 mg O_3 mg^{-1} DOC.

For AOC an obviously linear relationship with the ozone dose was also found. VAN DER KOOIJ et al. (1989) investigated the effect of ozone on river water treated by coagulation, sedimentation, and rapid sand filtration (Fig. 6). At low zone doses (i.e., up to 1.5 mg O_3 mg^{-1} DOC) they found a linear relationship between ozone dose and AOC formation. For higher ozone doses AOC formation leveled off and obviously led to a plateau. This may be the result of the effect that at further addition of ozone part of the low-molecular weight compounds are oxidized to carbon dioxide and, in

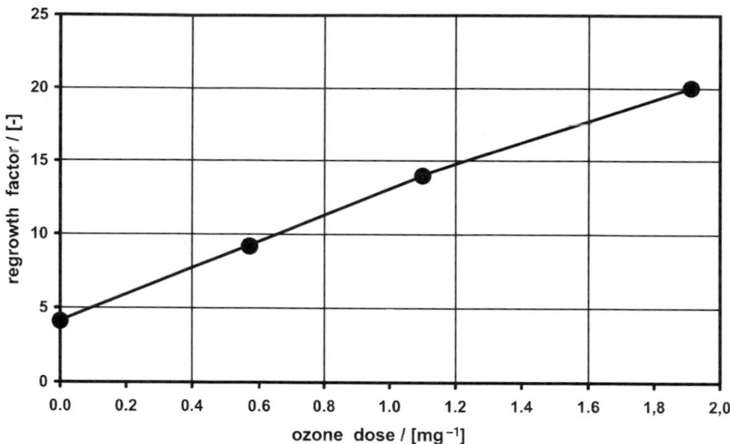

Fig. 5. Effect of the ozone dose on the regrowth factor of bank filtrate (adapted from HAMBSCH and WERNER, 1993).

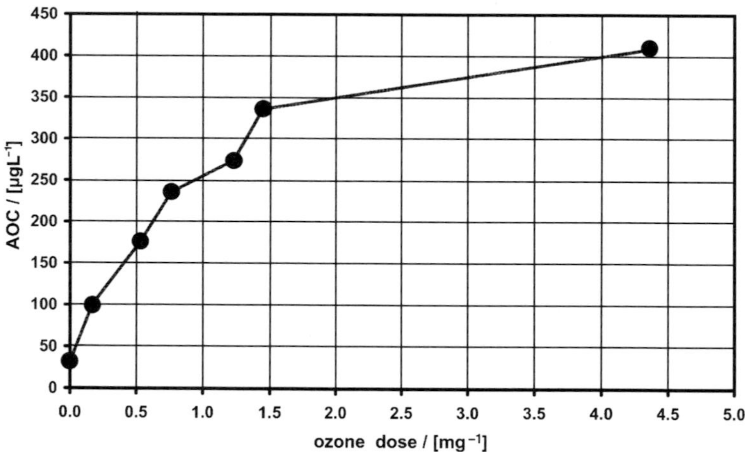

Fig. 6. Effect of the ozone dose on AOC (adapted from VAN DER KOOIJ et al., 1989).

parallel, newly formed from remaining higher-molecular weight compounds.

When applying ozonation in drinking water treatment special care has to be taken that the following treatment steps are well designed for the removal of easily biodegradable compounds. Otherwise, the composition of the water after treatment could eventually support bacterial regrowth to a higher extent than before.

3.4 Chlorination

Chlorination, usually applied for disinfection purposes, is an oxidation process like ozonation and results in an increase in biodegradability of organic matter. HAMBSCH and WERNER (1993) investigated the effect of chlorination on the bacterial regrowth potential of a fulvic acids solution. It was shown that an increasing chlorine dose (molar ratios of Cl:C of 0.1 and 1.0 were investigated) also increased the portion of DOC oxidized during regrowth from approximately 2% in the unchlorinated solution to about 6% at a Cl : C-ratio of 1.0. Moreover, it was shown that chlorination not only produced low-molecular weight compounds which could be degraded by bacteria. The low molecular weight compounds were removed after chlorination using gel chromatography, and the regrowth experiment was redone. This revealed that even high-molecular weight compounds were better susceptible to utilization by bacteria after chlorination.

The effect of increased bioavailability of organic matter by chlorination was also shown by VAN DER KOOIJ (1987). This effect is particularly important for the distribution of drinking water. First, if free chlorine is not maintained until the endpoints of the distribution system, considerable regrowth can occur. Secondly, bacteria located in biofilms on the walls of the distribution system are able to protect themselves by slimes and capsules. High concentrations of easily biodegradable substances can support their proliferation and release into the bulk liquid.

3.5 Typical AOC Concentrations over a Treatment Train

Fig. 7 summarizes the effects discussed above. It shows the AOC concentrations after different treatment steps of a waterworks treating river water by slow sand/ground filtration, ozonation, rapid dual-media filtration, and granular activated carbon adsorption. Finally, chlorine is dosed for post-disinfection.

It can clearly be seen from Fig. 7 that AOC is greatly reduced by ground filtration. AOC is increased by approximately 30 µg L^{-1}. using ozonation. In rapid filtration and the following granular activated carbon filtration AOC is further decreased by biological processes of bacteria which are mainly sessile on the filter media. Finally, AOC is again increased by chlorination.

4 Characterization of Biofilters for BOM Removal

4.1 Filter Types and Media

As already mentioned, in this chapter only filtration through granular media in open or closed rapid filters is understood and discussed as biofiltration. Actually, slow sand filtration is also a biofiltration process. However, conditions and mechanisms are partly different and not as well controlled as in deep bed filters. Ground filtration is an extreme variant of bio-

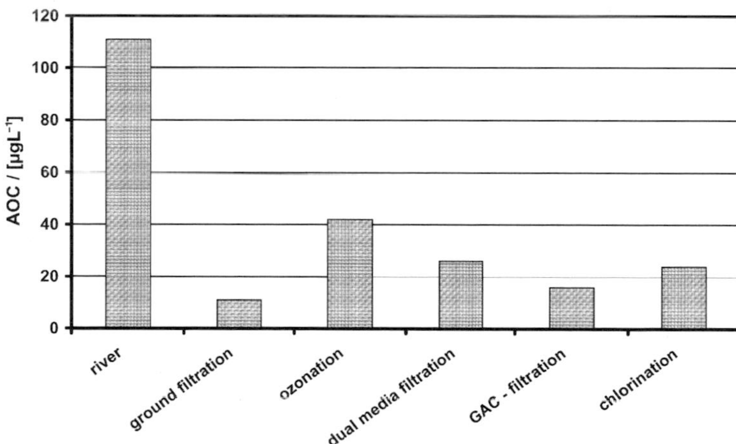

Fig. 7. AOC concentrations after different treatment steps of a waterworks operated according to the Mülheim process.

filtration.

Typical types of deep bed and granular activated carbon filters are discussed in detail in Chapter 15, this volume. For their performance as biofilters not so much the construction is essential but the carrier material for sessile microorganisms and the operating conditions.

Typical carrier materials for sessile microorganisms mainly responsible for BOM removal in drinking water treatment are gravel, sand, and anthracite as non-porous materials, and granular activated carbon (GAC) as a porous material. GAC may be chemically reactive, e.g., by reducing disinfectans such as chlorine or ozone already in the top part of the filters. Other, chemically inert porous carrier materials are pumice and sintered glass. However, these are relatively rarely used, probably because of minor mechanical stability (which may cause problems after several backwashing cycles) and cost.

Trickling filters, in which air flows through the filter media in parallel with water, are of minor importance for BOM removal in drinking water treatment. They are, therefore, also excluded from further discussion in this chapter. Their application is mainly restricted to processes for deferrization and demanganization (see Chapter 18, this volume).

4.2 Suspended Bacteria

Concentrations and species of suspended bacteria in biofilters are influenced by processes of deposition and attachment from the liquid phase to the carrier media, but also by detachment of bacteria from the media into the liquid. Together with growth and mortality of bacteria in the filters this may result in changes of species composition predominant in the influent and the effluent, respectively.

WILCOX et al. (1983) investigated the influents and effluents of biological activated carbon filters which were operated with ozonated and non-ozonated water over a period of one year. As the influents of the filters with ozonated water showed lower bacterial concentrations, the effluents mainly showed similar concentrations. Mostly they were higher in the effluents compared to the influents. Similar experiences were reported by SERVAIS et al. (1991, 1994) who monitored bacterial concentrations of GAC filters for about one year. During the first 150 d after startup often much higher concentrations in the effluent were observed. Then obviously a quasi-steady state was reached and effluent concentrations remained approximately 60% higher than in the influent.

On the other hand DONLAN and YOHE (1983) and BREWER and CARMICHAEL (1979) observed lower bacterial concentrations in the effluents than in the influents of GAC filters. DEN BLANKEN (1982) found that suspended bacterial concentrations were decreased by approximately 70% at empty bed contact times of 15 min. MORISSETTE et al. (1995) investigated dual-media filters with sand and activated carbon. In these, heterotrophic plate count bacteria were decreased by about 16% compared to single-layer sand filters. There they were diminished by about 43%.

WERNER (1982) classified bacteria from influents and effluents of activated carbon filters. A difference between the distribution of bacterial species in influents and effluents, respectively, was found. The spectrum of utilizable substrates was higher in the influents. Obviously, the part of acetate and oxalate utilizing bacteria was higher in the effluent, independent of whether the influent was ozonated or not.

In summary, to the authors' own experience, in drinking water treatment a decrease in suspended bacteria concentration from the influent to the effluent of biofilters can usually be observed, if the influents are not directly dis-infected. Increases in bacterial concentrations are mainly restricted to initial phases of operation of such filters of 30–150 d. The reasons are not completely clear, but it is assumed that inorganic and organic nutrient sources for phosphate and/or nitrogen may be released from the solid media during early operation, thus increasing the yield of heterotrophic bacteria on organic substrates.

4.3 Sessile Bacteria

Results of investigations on sessile bacteria concentrations are hard to compare. This is due to the fact that methods to release sessile bacteria from the filter media, especially activated carbon, are much different. For cultural methods it is necessary to completely relieve the bacteria from the surface into the liquid, which is impossible. When microscopic methods are applied part of the bacteria are washed away during sample preparation. However, a general overview can be obtained from the literature.

In general, there is an agreement that filter materials from drinking water treatment biofilters are not completely covered by a closed biofilm. Scanning electron microscopic examinations by CAIRO et al. (1979), LECHEVALLIER et al. (1984), and LECHEVALLIER and MCFETERS (1990) showed that macropores, cracks, and crevices of activated carbon grains were occupied by bacteria, partly covered by or embedded into a matrix of extracellular polymeric substances (see also Chapter 21, this volume). This was also emphasized by TUSCHEWITZKI et al. (1983) and WERNER (1982). They found that in the upper layer of activated carbon filters fed with ozonated water the surface of the filter grains was occupied to about 50%. With increasing bed depth the degree of occupation decreased.

SERVAIS et al. (1994) determined bacterial occupation monitoring the formation of radioactive CO_2 during incubation of the media sample with radioactive glucose. By this they investigated the time course of bacterial colonization of activated carbon. During the first 100 d bacterial density increased linearly with time and then remained constant. Over a bed depth of 3 m a decrease of about 70% was observed.

Typical bacterial densities in biofilters for drinking water treatment range between 10^5 and 10^9 bacteria per mL bed volume. Compared to concentrations in the liquid phase they are ususally between 10 and 1,000 times higher. Generally much higher densities are observed in activated carbon filters compared to sand filters.

Analysis of bacterial genera found in biofilters for drinking water treatment revealed that they were typical autochthonous bacteria also found in raw water (e.g., MCELHANEY and MCKEON, 1978).

For example, PARSONS et al. (1980) found bacteria of the genera *Acinetobacter, Alcaligenes, Moraxella*, and *Pseudomonas*. This was also found by the author (Tab. 1).

4.4 Behavior of Pathogens in Biofilters

There is special concern that pathogens may survive in biofilters for drinking water treatment and possibly be a source of contamination. Systematic investigations showed that pathogens dosed into laboratory scale

Tab. 1. Identification of Heterotrophic Plate Count (HPC) Bacteria in Interstitial Water and on Different Support Material (+ + + very often identified, + + often identified, + rarely identified, – never identified) (from UHL and GIMBEL, 1996)

Genus	Late Summer		Spring	
	Influent and Effluent	On Carrier Material (Activated Carbon, Pumice)	Influent and Effluent	On Carrier Material (Activated Carbon, Pumice)
Flavobacterium spp.	+ + +	+ + +	+ + +	+ + +
Moraxella spp.	+ + +	+ + +	+ + +	+ + +
Micrococcus spp.	+ +	+ +	+	+ +
Alcaligenes spp.	+	–	–	–
Actinomyces spp.	–	+	–	–
Acinetobacter spp.	+	–	–	–
Bacillus spp.	–	+ + +	+	+ + +
Gram-positive rods (not further identified)	+	+	+	+

biofilters for drinking water treatment operated under typical conditions did not proliferate, but were diminished in concentration with time. In a thorough study ROLLINGER and DOTT (1987) showed that indicator organisms such as *E. coli, S. faecalis*, and *Klebsiella pneumoniae* were not able to survive in activated carbon biofilters operated with drinking water.

CAMPER et al. (1985) showed that pathogens added to a suspension of sterile activated carbon in sterile river water adsorbed to the carbon and stayed without decline over the period of investigation of 14 d. Their concentration quickly decreased when autochthonous bactera from river water were added. After addition of autochthonous bacteria the concentration of pathogens continuously declined from $7 \cdot 10^5$ per g activated carbon to zero after 7 d of further incubation.

These experiments, together with an interpretation of kinetic data, explain why pathogens usually do not survive in biofilters for drinking water treatment. They are accustomed to high concentrations of biodegradable organic matter. Thus, at high concentrations their growth rate is high, but at low concentrations their growth rate is very low, while typical autochthonous water bacteria reach their maximum growth rate already at very low concentrations. As a result pathogens are outcompeted at the low concentrations of biodegradable organic matter usually valid in drinking water treatment (i.e., far below 1 mg L^{-1} as carbon).

4.5 The Role of Protozoa and Metazoa in Biofilters

During the operation of biofilters protozoa and metazoa are deposited on the filter media and may proliferate. WERNER (1982) investigated the density of protozoa of activated carbon biofilters in drinking water treatment and found about 450 individuals per mL bed volume. HUSMANN (1982) investigated protozoa in such filters in detail. In the top of the filters he found about 100 individuals per mL bed volume, whereas after 6 m bed depth their density had decreased by a factor of approximately 50. He proved that their species were similar to those found in raw water.

Even though they did not explicitly investigate on protozoa, SERVAIS et al. (1994) concluded from their observations that protozoa considerably contributed on the flux of organic matter in the biofilters investigated. Net bacterial production in the filters was found to be about 20 times higher directly after startup of the filters compared to the periods of pseudo-steady state. This was supported by the fact that during the initial phase bacteria in the filter effluent were much bigger in size than later. For aquatic ecosystems it is known that protozoa preferably feed on bigger bacteria (GONZALES et al., 1990). This supports the hypothesis of considerable grazing during pseudo-steady state.

It is difficult to compare the biomass of protozoa and bacteria in biofilters. Calculations can only give a rough estimate. From the numbers cited above and the typical concentrations of sessile bacteria it can be calculated that bacterial biomass is expected to be in the same order or up to 100 times higher than the biomass of protozoa. However, conversion by protozoa can be considerable.

5 Effects of Process Parameters on Efficiency of Biofilters for BOM Removal

5.1 Filter Media

Filter media chosen for biofilters can have a high impact on the removal efficiency for biodegradable organic matter. Their surface characteristics are very important for the deposition and attachment of bacteria which in turn are responsible for the oxidation of BOM. When comparing adsorptive media such as GAC with inert media like sand it is important to make sure that the adsorption capacity is completely exhausted. Even if this is the case, biofilters with adsorptive media usually are superior.

LeChevallier et al. (1992), e.g., investigated the effect of different filter media on the removal of total organic carbon and AOC. Comparison of a multiple media filter (anthracite/sand/garnet) with a GAC filter revealed that GAC performed much better than the multiple media for AOC removal (55% vs. 28%) and for total organic carbon (TOC) removal (45% vs. 17%). A dual-media filter with GAC and sand performed only slightly worse than GAC filters. This may be due to the fact that the removal profile is relatively steep in filters with GAC. Therefore, the maximum removal efficiency already may be reached at low bed depths. In such cases increasing the depth will have only little effect on the removal of organic matter.

Fig. 8 shows the removal efficiency of biofilters with different media for removal of dissolved organic carbon (DOC). It can clearly be seen that GAC gives a much higher efficiency than inert materials. Pumice, although coarser, gives a slighly better performance than sintered glass. Sand, of the same grain size as pumice, gives the poorest performance.

Wang et al. (1995) clearly showed that higher removal of BOM in GAC filters compared to anthracite and sand media coincided with higher biomass concentrations on the carriery material. It is often assumed that this is the result of the very rough surface of activated carbon which protects sessile bacteria from shear forces. But, however, even when porous inert materials such as pumice are compared with GAC, this protection is available.

The reasons for higher biomass accumulation and BOM removal in GAC filters are most probably that attractive forces between GAC and bacteria are much stronger than between non-adsorptive media such as sintered glass or sand and bacteria. Daniels (1980) discusses in detail attractive and repulsive forces between bacteria and solid surfaces. Steady state can be described by adsorption isotherms such as the Langmuir isotherm which is based on a steady state between attachment and detachment. Consequently, Billen et al. (1992) determined rate constants for adsorption and desorption from the Langmuir isotherm. They found that the adsorption rate constant was about two times higher for activated carbon compared to sand, while the desorption rate constant was five times lower.

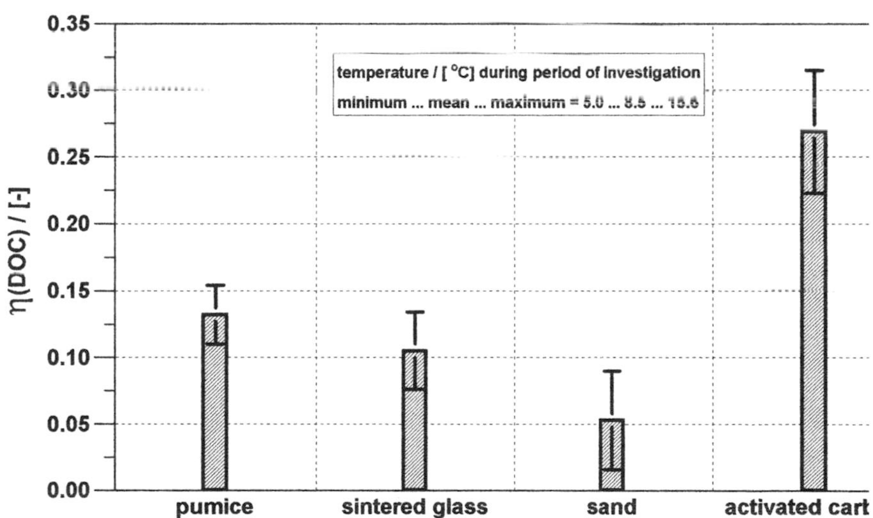

Fig. 8. Removal efficiency for dissolved organic carbon (DOC) in different media biofilters and 95% confidence intervals (superficial velocity: 6 m h^{-1}; bed depth: 2 m; temperature: 5–15 °C).

5.2 Flow Rate, Contact Time

Flow rate may influence on the performance of biofilters in two different ways. First, mass transfer of organic matter to sessile bacteria may be improved by increasing the flow rate and the amount of bacteria deposited on the material per unit of time. This should improve the removal efficiency for BOM at a fixed bed depth. Second, the contact time is reduced, which should impair efficiency.

In general, for a fixed bed depth, a decrease in BOM removal efficiency can be observed with increasing superficial flow rate, as shown in Fig. 9. Thus, the eventually promoting effect of increased mass transfer and bacterial deposition is overruled by the detrimental effect of decreasing contact time. Obviously, researchers agree, the contact time is decicious and not the flow rate. This is in coincidence with the fact that by empirical modeling a first order rate reaction can be found (see below). It indicates that mass transfer is of minor importance. A study carried out by URFER-FRUND (1998) proved this assumption .

5.3 Temperature

It is generally agreed upon that the performance of biofilters for BOM removal is impaired by decreasing temperatures. However, no information could be found in the literature how this effect can be quantified. In own investigations the author operated two biofilters with GAC as filter media in parallel over a period of one year. While one was kept at almost constant temperature the other was subjected to seasonal changes of temperature. An empirical model (see below) of a first order reaction with a temperature dependent rate constant was applied. For the temperature dependency an activation energy according to the Arrhenius law of 33 kJ mol^{-1} was found. This implies that the rate constant decreases by approximately 40% when temperature decreases from 20 °C to 10 °C.

5.4 Deposition of Flocs and Particles

It can be expected that flocs, e.g., from precipitation/flocculation in front of a biofilter,

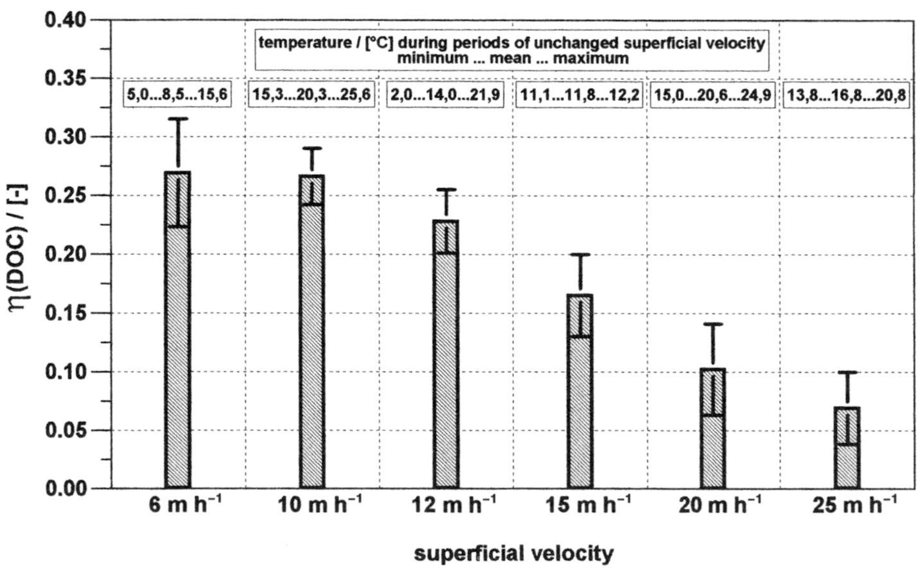

Fig. 9. Impact of flow rate on the relative removal efficiency in biofilters with GAC as filter media and 95% confidence intervals.

may cover sessile bacteria. This could lead to an increased mass resistance for transport of organic matter to the bacteria. Furthermore, polyaluminum chloride often used for flocculation can inhibit bacterial growth and thus the utilization of biodegradable organic matter (HUCK et al., 1991; ANDERSON et al. 1990). In a study of PRÉVOST et al. (1995) it was shown that the deposition of flocs in biofilters greatly impaired their performance. Backwashing led to short improvement.

Consequently, PALINSKI et al. (1998) developed and tested a dual media filter in which the top layer is made of so-called permeable synthetic collectors. These are highly effective in capturing flocs and particles. Thus, the second layer consisting of activated carbon is better protected from the flocs. By this, bioprocesses in the lower layer are less hindered, and a considerable improvement of BOM removal between two backwashings is obtained. However, this promising process is not out of pilot scale at the time this article is written. Difficulties are in cleaning of the permeable synthetic collectors.

5.5 Backwashing

Backwashing biofilters may influence their performance in two ways. On the one hand bacteria sessile on the filter media are relieved from flocs and particles which have a detrimental effect. On the other hand, due to the higher shear forces during backwashing, already sessile bacteria may be washed out of the biofilters, which results in minor performance.

Several researchers have investigated the effect of backwashing on removal of BOM. HASCOET et al. (1986) found that BDOC removal was impaired immediately after backwashing of the biofilters investigated (GAC, about one year in operation, $v = 6\text{–}9\ \text{m h}^{-1}$), but already after one hour the biofilters returned to their original efficiency. This is in coincidence with results of MILTNER et al. (1995) who investigated performance and biomass on the filter media. Bacterial density on the grains was not measurably influenced by backwashing. HOZALSKI and BOUWER (1998) carried out experiments under thoroughly controlled conditions and found that only about 20% of biomass were removed by backwashing. Therefore, backwashed biofilters maintained their treatment levels for BOM similar to that before backwashing.

In North America backwashing of biological filters often is performed with chlorinated water. Compared to non-chlorinated backwashing the efficiency of BOM removal was impaired by chlorinated backwash water. The biofilters recovered their original performance within 1–2 d (MILTNER et al., 1995). AHMAD et al. (1994) also observed that backwashing with chlorinated water had a detrimental effect on AOC removal efficiency.

6 Modeling of Biofilters for BOM Removal

6.1 Empirical Steady State Modeling

Attempts have been made to model removal of biodegradable organic matter by empirically testing the fit of first order rate reactions on data for biofilter performance. HUCK and ANDERSON (1992) and HUCK et al. (1994) suggested a first order rate reaction in BOM concentration.

According to UHL (2000) such an empirical approach is very well suited to describe the removal of dissolved organic carbon (DOC) and of AOC in biofilters. For process analysis the removal efficiency $\eta_z(x)$ for a parameter x at a given bed depth z is defined as

$$\eta_z(x) = \frac{x_{\text{inf}} - x_z}{x_{\text{inf}}}, \qquad (2)$$

where the index inf indicates the value of the parameter at the influent. For DOC, clearly only part of it, i.e., BDOC is biodegradable. Consequently, a maximum removal efficiency η_{max} can be expected and defined. If a first order reaction is assumed in the interstitial water of the bed, the removal efficiency as a function of bed depth can be calculated according to Eq. (3)

$$\frac{\eta_z}{\eta_{max}} = 1 - e^{-(k_{app} \cdot \varepsilon_{bed})\, EBCT_z} \tag{3}$$

In Eq. (3) k_{app} is the apparent rate constant for a first order reaction, ε_{bed} is the interstitial porosity and $EBCT_z$ is the empty bed contact time from influent to bed depth z.

Fig. 10 shows experimental data (45 data sets) obtained with a biofilter with GAC as carrier, obtained over a period of one year with 95% confidence intervals. The fit obtained by non-linear optimization for an empirical first order reaction model according to Eq. (3) with 95% confidence intervals is also shown. It can be concluded that the empirical model excellently fits the data.

AOC is generally assumed to be completely biodegradable. Therefore, it should be possible to completely remove AOC in biofilters. However, experimental data very often do not show this. Fig. 11 shows the same approach for AOC. The non-linear fit is not as good as for DOC. However, it has to be taken into account that AOC cannot be determined with the same precision as DOC, and the number of data sets was much smaller (15 compared to 45).

The observation that there seems to be a saturation function for the removal of biodegradable organic matter in biofilters was also observed by several other researchers, e.g., SERVAIS et al. (1991), WRICKE et al. (1996), and CARLSON et al. (1996).

6.2 Mechanistic Modeling Approaches

As shown above the behavior of biofilters can be described by empirical models. However, these do not reveal anything about the mechanisms involved. As a consequence, it is possible to describe measurements already made, but due to the lack of understanding the behavior cannot be predicted very well.

Consequently, several mechanistic models for the removal of biodegradable organic matter in drinking water treatment biofilters were developed. RITTMANN and MCCARTY (1980a) developed the so-called S_{min} model for completely mixed reactors. This model assumes that the attachment of bacteria from the liquid phase can be neglected. Thus, a minimum substrate concentration for a stable biofilm guaranteeing substrate removal must exist. The model was experimentally verified by RITT-

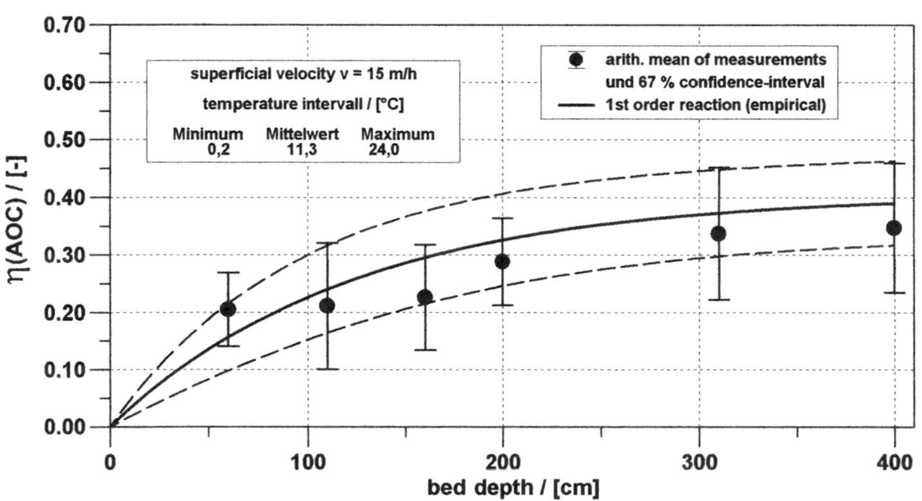

Fig. 10. Removal efficiency η(DOC) for dissolved organic carbon as a function of bed depth and empirical modeling of a first order reaction with 95% confidence intervals for non-linear fitting.

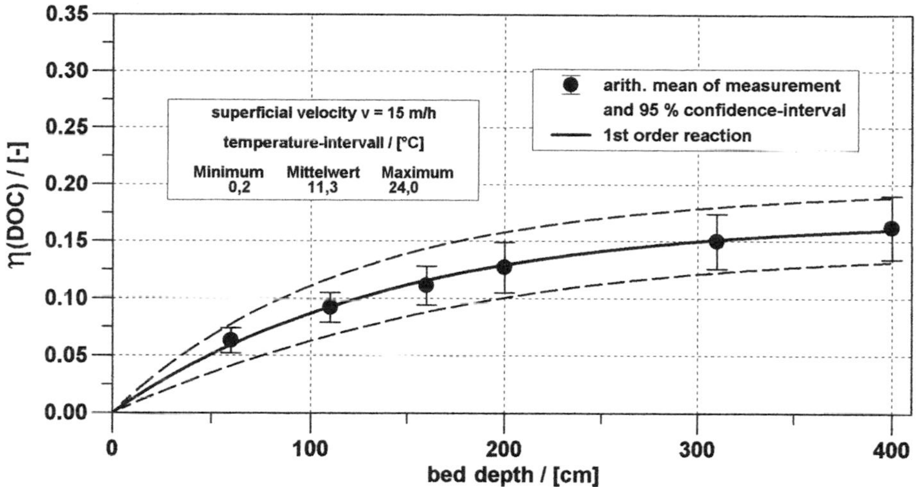

Fig. 11. Removal efficiency $\eta(\mathrm{AOC})$ for assimilable organic carbon as a function of bed depth and empirical modeling of a first oder reaction with 67% confidence intervals for non-linear fitting.

MANN and MCCARTY (1980b) with acetate as model substrate. The minimum substrate concentration was found at approximately 350 µg acetate carbon, while in biofilters for drinking water treatment (which are equivalent plug flow reactors) much lower AOC and BDOC concentrations are usually achieved.

ZHANG (1996) developed the $S_{\min}$ model in such a way that it became applicable to plug flow reactors. As in real situations for the mixture of unknown substrates parameters like yield coefficients, Monod constants and diffusion coefficients are unknown, ZHANG and HUCK (1996) applied the model to data from pilot and full scale biofilters for drinking water treatment and determined four unknown parameters by non-linear regression. The fitted prediction of the model for AOC effluent concentration of the biofilters followed the measured data, but absolute deviations were quite high in many cases.

The so-called Chabrol model was published by BILLEN et al. (1992). According to this model higher molecular weight organic substances in the water passing the biofilter are first hydrolized to monomers by exoenzymes. BDOC is treated as rapidly and slowly hydrolizable BDOC. The Chabrol model ob-viously was the first applied to drinking water treatment which took into account attachment and detachment of suspended bacteria as well as temperature dependency of biodegradation. The applicability of the model to drinking water treatment was investigated by MERLET et al. (1992) and by LAURENT et al. (1999) and proved quite well.

The Chabrol model requires a large number of parameters which were determined by the examination of bacteria in marine ecosystems. Furthermore, exoenzymatic activities for hydrolysis of high-molecular weight compounds has to be determinied by the use of radioactively labeled substances. Therefore, UHL (2000) developed another model only requiring data which can mainly be obtained by analysis of bioprocesses in drinking water treatment. In the following, this model will be discussed as an example.

Drinking water treatment biofilters are typical plug flow reactors. Therefore, a mass balance over an infinitely small volume element of the filter at bed depth z yields

$$\left(\frac{\partial S_z}{\partial t}\right) = r_{\mathrm{s,gr}} + r_{\mathrm{s,maint}} - \frac{Q}{A \cdot \varepsilon_{\mathrm{bed}}} \cdot \left(\frac{\partial S_z}{\partial z}\right) \tag{4}$$

where $r_{\mathrm{s,gr}}$ and $r_{\mathrm{s,maint}}$ are the rates for substrate consumption coupled to growth and maintenance of sessile bacteria, Q is the superficial velocity, A the cross sectional area of the filter. Eq. (4) describes the change of substrate concentration S as a function of time at a given bed depth z. Equivalent balances for suspended and attached bacteria concentrations X_{sus} and X_{att} yield

$$\left(\frac{\partial X_{sus,z}}{\partial t}\right) = -\frac{1}{\varepsilon_{bed}}\left(a_{bed}\cdot j_{x,att} - a_{bed}\cdot j_{x,det}\right) - \frac{Q}{A\cdot\varepsilon_{bed}}\left(\frac{\partial X_{sus}}{\partial z}\right) \quad (5)$$

and

$$\left(\frac{\partial X_{att,z}}{\partial t}\right) = \left(a_{bed}\cdot j_{x,att} - a_{bed}\cdot j_{x,det}\right) + \left(r_{X,gr} + r_{X,mort} + r_{X,graz}\right) \quad (6)$$

where $r_{X,gr}$, $r_{X,mort}$, and $r_{X,graz}$ are the rates for bacterial growth, mortality as a consequence of endogenous metabolism and of grazing. $j_{X,att}$ and $j_{X,det}$ are the fluxes of suspended bacteria attaching and detaching on the media, a_{bed} is the volume-specific surface area of the media. The growth rate is described by Monod type kinetics in the model, in which temperature dependency is taken into account by the Arrhenius equation. Mortality and maintenance are described as first order in attached bacteria concentration. Rates of attachment and detachment are, as in the Chabrol model, described by Langmuir type kinetics.

Grazing by protozoa is difficult to measure and describe. In the model a steady state situation between protozoa and attached bacteria is assumed. The rate of mortality as a consequence of grazing can be expressed in this way as a function of substrate and attached bacteria concentration. Attachment and detachment are, as in the Chabrol model, described by Langmuir type kinetics.

The model was yet solved for quasi-steady state situations. It is able to give quite good descriptions of the behavior of biofilters in drinking water treatment. Fig. 12 shows the model prediction of the influence of flow rate on DOC removal. This influence can most clearly be seen at bed depths of 100–200 cm. At low flow rates maximum removal efficiency is already obtained at these bed depths, whereas an increase in flow rate results in a stratifica-

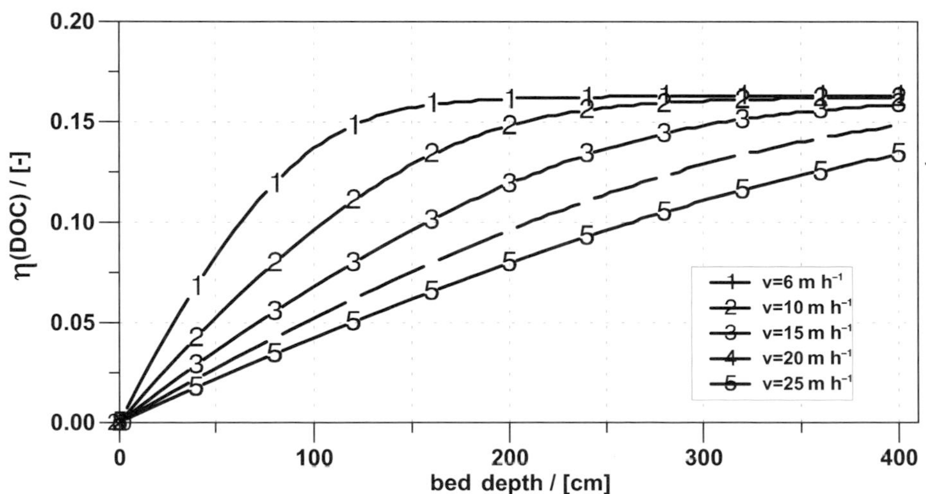

Fig. 12. Model prediction of the influence of flow rate on removal of dissolved organic carbon in biofilters ($DOC_{in} = 1.85$ mg L^{-1}; $\eta_{max} = 0.16$ mg L^{-1}; temperature = 11 °C).

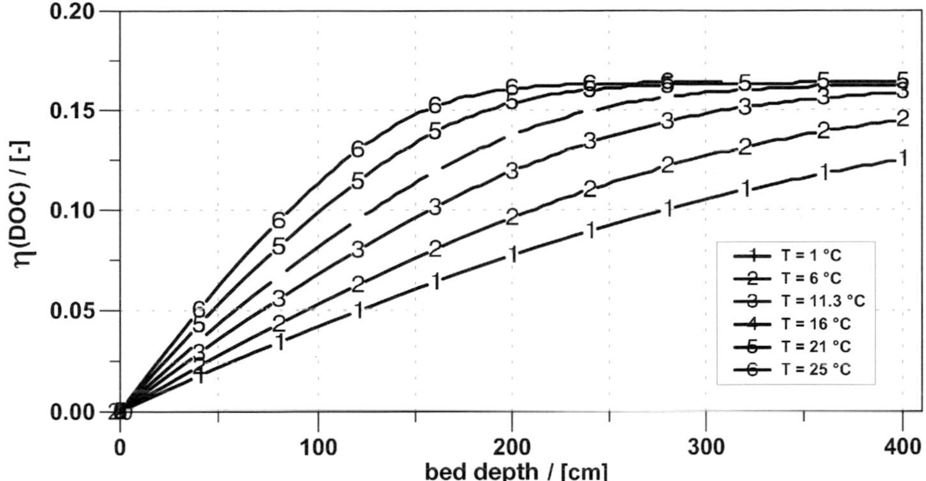

Fig. 13. Model prediction of the influence of temperature on removal of dissolved organic carbon in biofilters ($DOC_{in}=1.85$ mg L^{-1}; $\eta_{max}=0.16$ mg L^{-1}; $v=15$ m h^{-1}).

tion of DOC removal into deeper parts of the bed. Thus, at low flow rates deeper beds do not make sense, but are necessary at high flow rates to ensure sufficient removal of biodegradable matter.

Fig. 13 shows the influence of temperature at a flow rate of 15 m h^{-1}. For example, it can be depicted from the simulation that about two times the bed depth is needed at 1 °C to achieve the same DOC removal when compared to 16 °C.

Yet the different processes playing a role in biofilters for drinking water treatment are not completely elucidated and understood. Therefore, it cannot be concluded that one or another model is superior. But altogether they can help in the understanding of the mechanisms involved in and predicting the behavior of biofilters.

7 References

AHMAD, R., AMIRTHARAJAH, A., AL-SHAWWA, A., HUCK, P. M. (1994), Optimum backwashing strategies for biological filters, in: *Proc. AWWA Water Qual. Technol. Conf.* (Am. Water Works Ass., Ed.), pp. 817–832. Denver, CO: AWWA.

ALLGEIER, S. C., SUMMERS, S., JACANGELO, G., HATCHER, V. A., MOLL, D. M. H. S. M. et al. (1996), A simplified and rapid method for biodegradable dissolved organic carbon measurement, in: *Proc. AWWA Water Qual. Technol. Conf.* (Am. Water Works Ass., Ed.). Denver, CO: AWWA.

ANDERSON, W. B., HUCK, P. M., FEDORAK, P. M. (1990), Comparison of the van der Kooij and Werner Methods for the determination of AOC and bacterial regrowth potential in drinking water, in: *Proc. 1990 AWWA Water Qual. Technol. Conf.* (Am. Water Works Ass., Ed.), pp. 1157–1175. Denver, CO: AWWA.

Anonymous (1973), DIN 2000: Zentrale Trinkwasserversorgung – Leitsätze für Anforderungen an Trinkwasser; Planung, Bau und Betrieb der Anlagen, [Fachnormenausschuß Wasserwesen (FNW) im Deutschen Normenausschuß (DNA), Ed.] Berlin: Beuth-Verlag.

Anonymous (1995), Assimilable organic carbon (proposed), in: *Standard Methods for the Examination of Water and Wastewater* (Am. Public Health Org./Am. Water Works Ass./Water Env. Fed., Eds.), pp. 9–39 – 9–43. Washington: DC.

BILLEN, G., SERVAIS, P., BOUILLOT, P., VENTRESQUE,

C. (1992), Functioning of biological filters used in drinking water treatment – the Chabrol model, *J. Water Supp. Res. Technol. – AQUA* **41**, 231–241.

BLOCK, J. C., MATHIEU, L., SERVAIS, P., FONTVIELLE, D., WERNER, P. (1992), Indigenous bacterial inocula for measuring the biodegradable dissolved organic carbon (BDOC) in waters, *Water Res.* **26**, 481–486.

BREWER, W. S., CARMICHAEL, W. W. (1979), Microbiological characterization of granular activated carbon filter systems, *Am. Water Works Ass. J.* **71**, 738–740.

CAIRO, P. R., MCELHANEY, J., SUFFET, I. H. (1979), Pilot plant testing of activated carbon adsorption systems, *Am. Water Works Ass. J.* **71**, 660–673.

CAMPER, A. K., LECHEVALLIER, M. W., BROADAWAY, S. C., MCFETERS, G. A. (1985), Growth and persistence of pathogens on granular activated carbon filters, *Appl. Environ. Microbiol.* **50**, 1378–1382.

CARLSON, K. H., AMY, G. L., GARSIDE, J., BLAIS, G. (1996), Ozone induced biodegradation and removal of NOM and ozonation byproducts in biological filters, in: *Advances in Slow Sand and Alternative Biological Filtration* (GRAHAM, N., COLLINS, R., Eds.), pp. 61–69. Chichester: John Wiley & Sons.

DANIELS, S. L. (1980), Mechanisms involved in sorption of microorganisms to solid surface, in: *Adsorption of Microorganisms to Surfaces* (BITTON, G., MARSHALL, K. C., Eds.), pp. 7–58. New York : John Wiley & Sons.

DEN BLANKEN, J. G. (1982), Microbial activity in activated carbon filters, *J. Environ. Eng.* **108**, 405–425.

DONLAN, R. M., YOHE, T. L. (1983), Microbial population dynamics on granular activated carbon used in treating surface impounded groundwater, in: *Treatment of Water by Granular Avtivated Carbon: Adv. Chem., Ser. 202* (MCGUIRE, M. J., SUFFET, I. H., Eds.), pp. 337–354. Washington: American Chemical Society.

FRIAS, J., RIBAS, F., LUCENA, F. (1992), A method for the measurement of biodegradable organic carbon in waters, *Water Res.* **26**, 255–258.

GILBERT, E. (1988), Biodegradability of ozonation products as a function of COD and DOC elimination by the example of humic acids, *Water Res.* **22**, 123–126.

GONZALES, J. M., SHERR, E. B., SHERR, B. F. (1990), Size-selective grazing on bacteria by natural assemblages of estuaries flagellates and ciliates, *Appl. Environ. Microbiol.* **56**, 583–584.

HAMBSCH, B. (1992), Untersuchungen zu mikrobiellen Abbauvorgängen bei der Uferfiltration, *Thesis*. University of Karlsruhe, Germany.

HAMBSCH, B., WERNER, P. (1993), Control of bacterial regrowth in drinking water treatment plants and distribution systems, *Water Supply* **11**, 299–308.

HASCOET, M. C., SERVAIS, P., BILLEN, G. (1986), Use of biological analytical methods to optimize ozonation and GAC filtration in surface water treatment, in: *Proc. 1986 AWWA Ann. Conf.* (Am. Water Works Ass., Ed.), pp. 205–222. Denver, CO: AWWA.

HIJNEN, W. A. M., VAN DER KOOIJ, D. (1992), The effect of low concentrations of assimilable organic carbon (AOC) in water on biological clogging of sand beds, *Water Res.* **26**, 963–972.

HOIGNÉ, J., BADER, H. (1976), The role of hydroxyl radical reactions in ozonation processes in aqueous solutions, *Water Res.* **10**, 377.

HOZALSKI, R. M., BOUWER, E. J. (1998), Deposition and retention of bacteria in backwashed filters, *Am. Water Works Ass. J.* **90**, 71–85.

HUBELE, C. (1985), Adsorption und biologischer Abbau von Huminstoffen in Aktivkohlefiltern, *Thesis*. University of Karlsruhe Germany.

HUCK, P. M., ANDERSON, W. B. (1992), Quantitative relationships between the removal of NVOC, chlorine demand and AOX formation potential in biological drinking water treatment, *Vom Wasser* **78**, 281–303.

HUCK, P .M., FEDORAK, P. M., ANDERSON, W. B. (1990), Methods for determining assimilable organic carbon and some factors affecting the van der Kooij method, *Ozone Sci. Eng.* **12**, 377–392.

HUCK, P. M., FEDORAK, P. M., ANDERSON, W. B. (1991), Formation and removal of assimilable organic carbon during biological treatment, *Am. Water Works Ass. J.* **83**, 69–80.

HUCK, P. M., ZHANG, S., PRICE, M. L. (1994), BOM removal during biological treatment: a first-order model, *Am. Water Works Ass. J.* **86**, 61–71.

HUSMANN, S. (1982), Aktivkohlefilter als künstliche Biotope stygophiler und stygobionter Grundwassertiere, *Arch. Hydrobiol.* **95**, 139–155.

JORET, J. C., LEVI, Y., DUPIN, T., GILBERT, M. (1988), Rapid method for estimating bioeliminable organic carbon in water, in: *Proc. AWWA Ann. Conf.* (Am. Water Works Ass., Ed), pp. 1715–1725. Denver, CO: AWWA.

KAPLAN, L. A., BOTT, T. L. (1988), Measurement of assimilable organic carbon in water distribution systems by a simplified bioassay technique, in: *Proc. AWWA Water Qual. Technol. Conf.* (Am. Water Works Ass., Ed.), pp. 475–498. Denver, CO: AWWA.

KUO, P. P. K., CHIAN, S. K., CHANG, B. J. (1977), Identification of end products resulting from ozonation and chlorination of organic compounds commonly found in water, *Environ. Sci. Technol.* **11**, 1177–1181.

LAURENT, P., PRÉVOST, M., CIGANA, J., NIQUETTE, P., SERVAIS, P. (1999), Biodegradable organic matter removal in biological filters: evaluation of the chabrol model, *Water Res.* **33**, 1387–1398.

LeChevallier, M. W., McFeters, G. A. (1990), Microbiology of activated carbon, in: *Drinking Water Microbiology. Progress and Recent Developments* (McFeters, G. A., Ed.), pp. 104–119. New York: Springer-Verlag.

LeChevallier, M. W., Hassenhauer, T. S., Camper, A. L., McFeters, G. A. (1984), Disinfection of bacteria attached to granular activated carbon, *Appl. Environ. Microbiol.* **48**, 918–923.

LeChevallier, M. W., Becker, W. C., Schorr, P., Lee, R. G. (1992), Evaluating the performance of biologically active rapid filters, *Am. Water Works Ass.* J. **84**, 136–146.

Lucena, F., Frias, J., Ribas, F. (1990), A new dynamic approach to the determination of biodegradable dissolved organic carbon in water, *Environ. Sci. Technol.* **12**, 343–347.

McElhaney, J., McKeon, W. R. (1978), Enumeration and identification of bacteria in granular activated carbon columns, in: *Proc. AWWA Water Qual. Technol. Conf.* (Am. Water Works Ass., Ed.), pp. 1–22. Denver, CO: AWWA.

Merlet, N., Prévost, M., Merlet, Y., Coallier, J. (1992), Enlèvement de la matière organique dans les filtres CAB, *Rev. Sci Eau* **5**, 143–164.

Miltner, R. J., Summers, R. S., Wang, J. Z. (1995), Biofiltration performance: Part 2, Effect of backwashing, *Am. Water Works Ass. J.* **12**, 64–70.

Mogren, E., Scarpino, P. V., Summers, R. S. (1990), Measurement of biodegradable dissolved organic carbon in drinking water, in: *Proc. 1990 AWWA Ann. Conf.* (Am. Water Works Ass., Ed.), pp. 573–587. Denver, CO: AWWA.

Morissette, C., Prévost, M., Johnston-Main, K., Coallier, J., Millette, R. et al. (1995), Comparing several filter configurations for direct biological filtration at the city of Montreal treatment facility for the removal of biological organic matter (BOM), disinfection by-products (DBP) and ammonia, in: *Proc. 1995 AWWA Water Qual. Technol. Conf.* (Am. Water Works Ass., Ed.), pp. 1855–1880. Denver, CO: AWWA.

Palinski, A., Uhl, W., Gimbel, R. (1998), Optimierung der Mehrschichtfiltration durch Einsatz permeabler synthetischer Kollektoren, *Chem.-Ing Tech.* **70**, 983–987.

Parsons, F., Wood, P. R., DeMarco, J. (1980), Bacteria associated with granular activated carbon columns, in: *Proc. AWWA Water Qual. Technol. Conf.* (Am. Water Works Ass., Ed.), pp. 271–296. Denver, CO: AWWA.

Prévost, M., Duchesne, D., Coallier, J., Desjardins, R., Lafrance, P. (1989), Full-scale evaluation of biological activated carbon filtration for the treatment of drinking water, in: *Proc. AWWA Water Qual. Technol. Conf.* (Am. Water Works Ass., Ed.), pp. 147–165. Denver, CO: AWWA.

Prévost, M., Coallier, J., Mailly, J., Desjardins, R., Duchesne, D. (1992), Comparison of biodegradable organic carbon (BOC) techniques for process control, *J. Water Supp. Res. Technol. – AQUA* **41**, 141–150.

Prévost, M., Niquette, P., MacLean, R. G., Thibault, D., Desjardins, R. (1995), Factors affecting the performance stability of first stage sand-activated carbon filters for the removal of biodegradable organic matter and ammonia, in: *Proc. 1995 AWWA Water Qual. Technol. Conf.* (Am. Water Works Ass., Ed.), pp. 391–410. Denenver, CO: AWWA.

Ribas, F., Frias, J., Lucena, F. (1991), A new dynamic method for the rapid determination of the biodegradable dissolved organic carbon in drinking water, *J. Appl. Bacteriol.* **71**, 371–378.

Rittmann, B. E., McCarty, P. L. (1980a), Model of steady-state biofilm kinetics, *Biotechnol. Bioeng.* **22**, 2343–2357.

Rittmann, B. E., McCarty, P. L. (1980b), Evaluation of steady-state biofilm kinetics, *Biotechnol. Bioeng.* **22**, 2359–2373.

Rittmann, B. E., Huck, P. M. (1989), Biological treatment of public water supplies, *Crit. Rev. Env. Control* **19**, 119–184.

Rollinger, Y., Dott, W. (1987), Survival of selected bacterial species in sterilized activated carbon filters and biological activated carbon filters, *Appl. Environ. Microbiol.* **53**, 777–781.

Servais, P., Billen, G., Hascoet, M. C. (1987), Determination of the biodegradable fraction of dissolved organic matter in waters, *Water Res.* **21**, 445–450.

Servais, P., Anzil, A., Ventresque, C. (1989), Simple method for determination of biodegradable dissolved organic carbon in water, *Appl. Environ. Microbiol.* **55**, 2731–2734.

Servais, P., Billen, G., Ventresque, C., Bablon, G. P. (1991), Microbial activity in GAC filters at the Choisy-le-Roi treatment plant, *Am. Water Works Ass. J.* **83**, 62–68.

Servais, P., Billen, G., Bouillot, P. (1994), Biological colonization of granular activated carbon filters in drinking water treatment, *J. Environ. Eng.* **120**, 888–899.

Sontheimer, H., Heilker, E., Jekel, M. R., Nolte, H., Vollmer, F. H. (1978), The Mülheim process, *Am. Water Works Assoc. J.* **70**, 393–396.

Takahashi, N., Nakai, T., Satoh, Y., Katoh, Y. (1995), Ozonolysis of humic acid and its effect on decoloration and biodegradability, *Ozone Sci. Eng.* **17**, 511–525.

Tuschewitzki, G. J., Werner, P., Dott, W. (1983), Biologische Besiedelung und mineralische Ablagerungen auf Filtermaterialien zur Trinkwasseraufbereitung, *gwf Wasser/Abwasser* **124**, 521–526.

UHL, W. (2000), Einfluß von Schüttungsmaterial und Prozessparametern auf die Leistung von Bioreaktoren bei der Trinkwasseraufbereitung, in: *Berichte aus dem Rheinisch-Westfälischen Institut für Wasserforschung* (IWW, Ed.). Mülheim: Eigenverlag.

UHL, W., GIMBEL, R. (1996), Investigations on the performance of fast-rate biological filters in drinking water treatment, in: *Advances in Slow Sand and Alternative Biological Filtration* (GRAHAM, N., COLLINS, R., Eds.), pp. 189–199. Chichester: John Wiley & Sons.

URFER-FRUND, D. (1998), Effects of oxidants on drinking water biofilter, *Thesis*. University of Waterloo, ON, Canada.

VAN DER KOOIJ, D. (1987), The effect of treatment on assimilable organic carbon in drinking water, in: *Treatment of Drinking Water for Organic Contaminants* (Proc. 2nd Natl. Conf. on Drinking Water, Edmonton, Alberta) (HUCK, P. M., TOFT, P., Eds.), pp. 317–328. New York: Pergamon Press.

VAN DER KOOIJ, D. (1990), Assimilable organic carbon (AOC) in drinking water, in: *Drinking Water Microbiology. Progress and Recent Developments* (MCFETERS, G. A., Ed.), pp. 57–87. New York: Springer-Verlag.

VAN DER KOOIJ, D., HIJNEN, W. A. M. (1981), Utilization of low concentration of starch by a *Flavobacterium* species isolated from tap water, *Appl. Environ. Microbiol.* **41**, 216–221.

VAN DER KOOIJ, D., HIJNEN, W .A. M. (1984), Substrate utilization by an oxalate-consuming *Spirillum* species in relation to its growth in ozonated water, *Appl. Environ. Microbiol.* **47**, 551–559.

VAN DER KOOIJ, D., HIJNEN, W. A. M. (1985), Measuring the concentration of easily assimilable organic carbon in water treatment as a tool for limiting regrowth of bacteria in distribution systems, in: *Proc. 1985 AWWA Water Qual. Technol. Conf.* (Am. Water Works Ass., Ed.). Denver, CO: AWWA.

VAN DER KOOIJ, D., VISSER, A., HIJNEN, W. A. M. (1982a), Determining the concentration of easily assimilable organic carbon in drinking water, *Am. Water Works Ass. J.* **74**, 540–545.

VAN DER KOOIJ, D., VISSER, A., ORANJE, J. P. (1982b), Multiplication of fluorescent pseudomonads at low substrate concentrations in tap water, *Antonie van Leeuwenhoek* **48**, 229–243.

VAN DER KOOIJ, D., HIJNEN, W. A. M., KRUITHOF, J. C. (1989), The effects of ozonation, biological filtration and distribution on the concentration of easily assimilable organic carbon (AOC) in drinking water, *Ozone Sci. Eng.* **11**, 297–311.

VOLK, C., RENNER, C., JORET, J. C. (1992), La mesure du CODB: un index du potentiel de reviviscence bactérienne des eaux, *Rev. Sci. Eau* **5**, 189–205.

WANG, J. Z., SUMMERS, R. S., MILTNER, R. J. (1995), Biofiltration performance: Part 1, Relationship to biomass, *Am. Water Works Ass. J.* **87**, 55–63.

WERNER, P. (1982), Mikrobiologische Untersuchungen der Aktivkohlefiltration zur Trinkwasseraufbereitung, in: *Zur Optimierung der Ozonanwendung in der Wasseraufbereitung – Veröffentlichungen des Bereichs und des Lehrstuhls für Wasserchemie und der DVGW-Forschungsstelle am EBI* Vol. 19 (Engler-Bunte-Institut University of Karlsruhe, Ed.). Karlsruhe: Eigenverlag.

WERNER, P. (1985), Eine Methode zur Bestimmung der Verkeimungsneigung von Trinkwasser, *Vom Wasser* **65**, 257–270.

WERNER, P., HAMBSCH, B. (1986), Investigations on the growth of bacteria in drinking water, *Water Supply* **4**, 227.

WERNER, P., HAMBSCH, B. (1988), Messung der Wachstumsrate von Bakterien bei der Aufbereitung von Oberflächenwässern, *Vom Wasser* **70**, 93.

WILCOX, D. P., CHANG, E., DICKSON, K. L., JOHANSSON, K. R. (1983), Microbial growth associated with granular activated carbon in a pilot water treatment facility, *Appl. Environ. Microbiol.* **46**, 406–416.

WRICKE, B., PETZOLDT, H., HEISER, H., BORNMANN, K. (1996), NOM-removal by biofiltration after ozonation – results of a pilot plant test, in: *Advances in Slow Sand and Alternative Biological Filtration* (GRAHAM, N., COLLINS, R., Eds.), pp. 51–60. Chichester: John Wiley & Sons.

ZHANG, S. (1996), Modeling biological drinking water treatment processes, *Thesis*. University of Alberta, Edmonton, Canada.

ZHANG, S., HUCK, P. M. (1996), Removal of AOC in biological water treatment processes: a kinetic modeling approach, *Water Res.* **30**, 1195–1207.

23 Perspectives of Waste, Wastewater, Off-Gas and Drinking Water Management

CLAUDIA GALLERT
JOSEF WINTER
Karlsruhe Germany

1 Introduction

After the Second World War an expanding industry was essential to restore and improve the life standard and stimulate the economy in Europe. The tribute for this booming industrial development (the so-called "Wirtschaftswunder"), however, was severe air pollution and the release of huge masses of domestic and industrial waste and wastewater. Due to the fact that atmospheric pollution was rapidly diluted and dislocated and solid wastes could be deposited within defined, spacially limited sanitary landfill areas, in the beginning of the industrial boom these pollutions were not recognized as serious as was the deterioration of surface waters, e.g., rivers or lakes, by pollutants of wastewater. The extent of pollution often exceeded the natural self-purification capacity of aquatic ecosystems and severe environmental harm was visible for everybody. Epidemic mortality of whole populations of fish or other water organisms by depletion of oxygen or the presence of toxicants in the water demanded for counteractions by legislative state authorities. Technically compliable and environmentally acceptable atmospheric, aquatic, and terrestric pollution limits had to be defined, fixed by state laws and controlled by administrative offices. The standards for wastewater, waste, and off-gas treatment as well as for drinking water preparation were defined and progressively strengthened with improving technological treatment and purification possibilities.

2 Wastewater Handling

Domestic and industrial wastewater had to be purified to meet the quality standards which were in accordance with the actually valid boundary values, practically obtainable with standardized, widely experienced "state of the art" treatment technologies. For this purpose the wastewater had to be collected, transported into public or industrial sewer systems, and treated in sewage or industrial wastewater treatment plants to remove organic and inorganic pollutants, as required by environmental laws and enforced by state control agencies. The boundary limits in the purified wastewater for residual carbon (e.g., measured as biological oxygen demand, BOD, or chemical oxygen demand, COD), nitrogen (total nitrogen or ammonia nitrogen), and phosphorus (in particular soluble *ortho*-phosphate) to be met for disposal into surface waters became more stringent with time. Improvements and the development of new processes for wastewater purification were stimulated by this.

Due to the complexity of the pollutants in different wastewater types or even in a certain wastewater, combined multi-stage processes for physical, chemical, and biological removal of organic pollutants and of nitrogen and phosphorus were required. Generally, the process development was always ahead of the exact knowledge about biological processes. A lack of detailed knowledge on even major metabolic pathways or, in particular, on single reactions within the complex ecosystem "wastewater" always was the bottleneck for specific improvement of wastewater purification techniques and treatment efficiencies, favoring a trial-and-error philosophy by civil engineers. Another nuisance was an apparently deep gap in the common scientific articulation of civil engineers (practical process designers and executers) and life scientists (basic researchers). Even today tracing the bottleneck reactions still is one of the obstacles to improve wastewater treatment.

2.1 Domestic Wastewater

In the 1920s treatment of domestic wastewater started in big cities with the construction of sewer systems and large treatment units. Later, many small wastewater treatment plants, often judged as less efficient, were built in smaller settlements all over the country to serve single towns or villages in less densely populated parts of the country. With time the development was directed more and more towards a centralization of wastewater treatment with huge treatment units serving whole regions. They were supplied with wastewater from several settlements, often via pumping stations to transport the wastewater over long distances. This development was subsidized by the government and favored by the inspecting administrations, since the treatment efficiency in these plants was considered more reliable (or easier to control) than in a high number of locally scattered, smaller wastewater treatment plants. However, except for exploding costs for additional pumping stations and the construction of new central sewage treatment plants or the extention of the capacity of exisiting plants, another possible source of environmental pollution was or still is created: thousands of miles of main sewers getting leaky with age and causing wastewater to trickle into soil and groundwater.

Since all wastewater sources of settlements, including rainwater from roofs of buildings or streets are conducted abundantly into mixed-water sewer systems, the wastewater reaching the treatment plant during or after rainy weather is highly diluted. All wastewater treatment facilities must be designed to cope with such unfavorable conditions for the hydraulics and for chemical and biological reactions. On the one hand dilution of wastewater is contraproductive for efficient chemical or biological treatment, whereas on the other hand rainwater is periodically required to flush the channels free of sediments, due to the construction of the sewer systems with little slope. Only in some communities a dual channel system for sewage and for rainwater is available.

For testing alternative wastewater handling, in some new settlement areas the general strategy of collection of all wastewater types in sewer systems for treatment in a central plant has meanwhile been reversed. Less polluted rainwater, e.g., is collected in natural or artifical ground depressions and oozed into the underground with the top soil layers serving as a natural, biologically active filter. Only restricted pavement areas are allowed to favor the trickling of as much rainwater as possible.

Grey water from single households can be purified biologically in special soil filters, planted with *Phragmites australis*, *Typha angustifolia* or other plants which develop an aerenchym. After removal of most of the pollutants by biofiltration the purified wastewater is oozed into the underground. The natural self-purification capacity of the top soil layers for wastewater components is extended into deeper layers of the soil by improving the oxygen supply via the aerenchym of planted vegetation.

To reduce the amount of waste and wastewater, the separation of night soil and urine was intended in separation toilets. After utilization these toilets are flushed with very little water. Whereas the solids are separated and composted in-house, the concentrated mixture of urine, some suspensa, and the flushing water is guided through a sewer system to a nearby biogas plant for wastewater stabilization and biogas production.

Compact treatment units for human excrements with solids separation and long-term hydrolyzation, pre-anoxic zone fixed-bed denitrification, fixed-bed activated sludge treatment for carbon removal and nitrogen oxidation, followed by pasteurization for direct disposal (if disposal standards are less strict) or microfiltration and UV irradiation for re-utilization of the purified wastewater have been developed on a very small scale for railway waggons or small ships (total volume 600–900 L). On a larger scale these units are available for single buildings to serve up to 20 inhabitant equivalents (e.g., Fa. Protec GmbH, Luhe-Oberwildenau, Germany). Several of these units have been operated successfully for years. If recontamination of the purified and ultra-filtrated wastewater can be prevented it can be recycled for toilet flushing. If the grey water from showers and bathrooms of houses is also purified in these mini-sewage treatment plants, much more flushing water for toilets than necessary is generated and no drinking water at all has to be wasted for toilets. Only little maintenance, usually once a year, is required.

At present some pilot projects are under investigation to test "zero-emission" concepts. These are, however, only realistic for single houses with some acres of garden or for settlements with sufficient area for oozing water into the ground. A double piping system to supply kitchen, bathroom, and laundry with high-quality drinking water and toilets with less pure, pre-purified rain and grey water could help to safe drinking water resources. A pre-purified rainwater from the roofs of the houses might even be used for laundry and thus further reduce the drinking water demand.

In the future, drastically increasing costs for supply of drinking water and wastewater treatment might lead to a decrease of the drinking water consumption in single households. A separation of costs for drinking water supply and wastewater treatment might favor individual on-site wastewater treatment systems for single housholds or small communities. Rain and grey water purification for re-utilization within the household, e.g., for toilet flushing or watering the garden may still sound futuristic (and for some people not acceptable), but may become necessary in the future.

2.2 Industrial Wastewater

Wastewater treatment in industry focussed on two approaches: Some companies favored a central treatment plant for all production units, requiring the enrichment of an "omnipotent" population at sub-optimal loading. Other companies favored smaller treatment units for every single production process, requiring the enrichment of specialized bacteria in each plant, that could be operated at maximal loading. In any case wastewater purification required a technically sophisticated combination of mechanical, chemical, and biological processes essential to purify the multiple components containing residing waste fluids from production processes.

If the wastewater from certain production processes contains xenobiotic substances and these substances cannot be adsorbed, precipitated, or biologically degraded, environmental protection has to go one step further and intervene into the production process. The process should be altered to avoid non-degradable substances in the wastewater or, if this is not possible, the concentrations should be reduced to the minimum and single wastewater streams should be recycled internally in order to minimize environmental pollution.

2.3 Effluent Quality and Future Improvements

Every purifed wastewater always contains some residual pollutants consisting of a small proportion of BOD (biodegradable residing organic substances) and a higher proportion of non-degradable COD (organic substances which resist rapid biological degradation and require more drastic conditions for chemical oxidation) as well as a certain salt freight. The BOD and part of the COD of the purified wastewater are degraded in the receiving water and some of the salt components may be precipitated. However, even if the biological self-purification capacity of receiving waters was not exceeded, traces of non-degradable wastewater components, such as detergents, household chemicals, antibiotics, pharmaceuticals, pesticides and fungicides, are washed into the groundwater. Some of these substances have hormonal activity and damage the fauna and flora in the receiving lakes and rivers or influence on human health. If the water body for the preparation of drinking water contains such contaminants they must be separated, e.g. by adsorption onto charcoal. Alternatively, membrane technologies may serve to separate trace pollutants from the bulk mass of water.

Future wastewater handling must more and more start at its sources and get away from an almost exclusive end-of-pipe-treatment, as it has been practiced hitherto. Two main goals have to be envisaged:

- reduction of the total amount of wastewater by water saving and recycling procedures and
- reduction or avoidance of biologically undegradable chemicals.

The first maxim requires a strict house- or fabrique-intern water saving regime and recycling techniques and the second maxim a change of production processes or human habits.

A complete closure of water cycles has already been practised by some industries, e.g., the paper recycling industry. However, new problems came up with the recycling of production water. Massive germination of the water during interim storage and re-utilization required permanent application of biocides. Due to fatty acid formation from carbohydrates by the contaminating microorganisms only low-quality papers could be produced from the recycled cellulose material. If these papers were moistened, an unpleasant odor developed and its acidity favors rapid decay.

To promote environmentally sound production processes in industry the so-called "Öko-audit" system was introduced by control authorities in Germany. Input material as well as products and wastes are analyzed and improvements of working procedures or production processes proceeded by the management of the company. With their participation in the "Ökoaudit" evaluation a company aggrees to reduce pollution year by year by a certain percentage compared to the present state. To make the consumer aware of the environmentally friendly production process applied (that may verify a somewhat higher retail price) the partcipating companies are allowed to print a respective certificate on their products

3 Solid Waste Handling

Until only recently, solid domestic wastes and residues from industrial production have been collected and simply deposited in sanitary landfills located outside residential areas. Except for a certain degree of homogenization, no other pre-treatment was considered necessary. Only highly poisonous industrial wastes were deposited underground in abundant salt or ore mining caves or tunnels.

Most waste pretreatment or treatment procedures other than just deposition in sanitary landfills were developed during the last 2–4 decades. Recently, a new deposition guideline of the European Community (EU-Richtlinie 94/904/EG, Suggestion of the European Commission, 11th March 1997) has been introduced defining construction, operation and after-care requirements of the three exclusively allowed sanitary landfill classes of the future for either:

- dangerous industrial wastes,
- non-dangerous wastes such as domestic refuse, and
- inert mono-waste material without chemical and biological reactivity.

The EU Guideline prescribes a leakage proof construction of bottom and top seals and focuses on deposition techniques, whereas the German technical instruction for the handling of domestic wastes (TA Siedlungabfall) is much more stringent. Except for comparably detailed prescriptions for the construction of sanitary landfills the "TA Siedlungsabfall" defines a maximum organic dry matter content of 3 or 5% of the wastes for deposition in class I or class II sanitary landfills to ascertain an inert or quasi-inert behavior after deposition. Class III deponies are mono-deponies for certain inert, non-dangerous waste materials.

In the German federal waste recycling law (Kreislaufwirtschaftsgesetz-KrW/AbfG 1996) three top priorities were defined:

(1) Avoidance of wastes: to reduce the amount of waste material to a minimum,
(2) Recycling of wastes:
    - first, as secondary raw material (substance recycling) or
    - second, as a source of energy (energy recycling),
(3) Deposition of wastes.

Deposition of untreated wastes is restricted until 2005. Later on, only deposition of mechanically and biologically pretreated wastes with low residual respiratory activity or methane production capacity and a high lignin-to-carbohydrate coefficient (until 2025), or of fully inert wastes such as incineration slags or ashes (after 2025) will be allowed, according to the presently valid environmental law.

Whereas avoidance of wastes, especially of packaging wastes, still seems to have a real potential for improvement, recycling of waste material for other waste types has reached a high overall level.

Glass recycling, e.g., works at very high recovery rates, presumably due to the fact that glass products from recycled raw materials are of almost the original quality.

Paper recycling by standardized procedures is also well introduced and accepted. However, due to breakage of fibers with every recycling round paper recycling is more a process of downcycling. A certain percentage of fresh fiber material must be added to achieve a constant quality. Paper from recycled raw material has to compete with paper made from low-quality wood or waste wood, which is available in high quantities in the Northern, wood-producing countries of the world.

Whether plastic material should be recycled for the production of new plastic goods can be a matter of discussion. To achieve high qualities of recycled plastics, plastic wastes have to be cleaned from contaminants. Then the mixed plastic material must be separated into the different polymer fractions (which apparently does not work properly with the procedures available at present) for specific recycling of each polymer class. It might be more favorable to feed the mixed plastic wastes into energetic recycling by incineration and to produce new plastic material from the fossil fuel saved by the use of plastic wastes.

To reach the low carbon content of 5% or even 3% organic dry matter content for deposition the non-recyclable fractions of municipal or industrial wastes must be incinerated. Pyrolyisis alone would not suffice. Waste incineration leads to two main residual products: incineration gases and incineration slags. Both are highly polluted with toxic material, but concentrations of toxicants are not as high as in the fly ashes. Whereas purification of the off-gas from waste incineration can be considered a state-of-the-art process enforced by respective state laws, e.g., the 17th German Federal Ordinance on Protection from Immissions (17. BImSchV 1990), disposal or proper utilization of the ashes and slags is still a matter of controverse discussion. Incineration slags have been used to construct traffic noise protection dams along highways. Long-term reactions of the heavy metal oxides might, however, lead to re-mobilization and cause environmental harm.

After separation of the powderous fraction of the slags by sieving, the granular fraction was used as a raw material in the construction industry. However, there seem to be gaseous organic inclusions in the slags which slowly diffuse out of concrete walls of houses and cause harm to the inhabitants. In addition, some residual toxic organics remain in the ashes, even if the waste incineration efficiency was high.

In order to obtain less toxic incineration residues in the future, detoxification of slags and fly ashes is considered not only a desirable option, but a must – even if it seems too expensive at present. In Switzerland, a new technique was tested on the laboratory scale to separate toxic heavy metals from highly toxic filter ashes of waste incineration plants from non-toxic mineral products (*Chemische Rundschau* No. 15, 1998). The heavy metal ions react with hydrochloric acid to their chlorides which can be evaporated at 900 °C. This would avoid deposition of toxic fly ashes in mining shafts or solidification of the toxic material with concrete. To save energy, this treatment ideally should start with still hot ashes.

4 Soil Remediation

Due to an almost unlimited number of pollutants and due to different soil and underground structures, no general guideline for soil remediation is applicable. Since soil is an agglomerate of mineral compounds, including small-particulate matter such as clay, gravel, or stones, and – at least in the upper layers – of organic material (e.g., plant residues or organic fertilizers, humic substances, etc.) with changing adsorption capacities for toxicants at different moisture contents or water conductivities, the retention of hydrophilic, water-soluble or hydrophobic, water-insoluble contaminants varies. Particularly water-insoluble compounds finally tend to accumulate on top of the groundwater level.

Besides safeguarding of contaminated sites by, e.g., inertization and encapsulation, several methods for soil decontamination of highly polluted soils at former industrial sites have been practiced in the past, including *ex situ* soil treatment and *in situ* remediation. *Ex situ* treatment is better controllable, but expensive whereas *in situ* remediation is less controllable and less efficient with time, but cheaper.

In the case of locally restricted soil compartments which have been contaminated with highly toxic chemicals or metals (e.g., mercury) excavation and thermal or chemical treatment (incineration or chemical extraction of the pollutants) may be necessary to restrain spreading of the toxic contaminants or of toxic metabolites during sanitation. By excavation and, e.g., thermal soil treatment not only the original soil texture is destroyed, but also the redox state is varied towards oxidized sinter products. However, a high decontamination efficiency even of micropores can be obtained within a short treatment time. If the contaminants are highly volatile, on-site decontamination should be favored, since otherwise extensive precautions for transport in closed containers would be necessary, as would be the case for soil contaminated, e.g., with poisonous solvents or leaded anti-knock agents.

Except for those sites with highly toxic contaminants requiring *ex situ* treatment procedures, the more economic, but eventually less quantitative *in situ* procedures could be applied for a high number of contaminated sites. The procedures include soil stripping with solvents or water or gas venting, purification of the stripped liquid or ventilating gasses from the contaminants and reintroduction of the purified media, and in many cases *in situ* bioremediation. For *in situ* bioremediation up to now only a limited number of possible techniques have been developed for practical application. This is mainly due to the fact that the efficiency of biological in situ removal of contaminants is limited by numerous factors concerning the contaminants, e.g., low solubility, strong sorption onto the soil matrix, diffusion into macropores of soil and sediments, concerning the transport of nutrients and electron acceptors for the microbial activity, e.g., permeability and porosity of the soil, the ion exchange capacity, pH value, and redox potential.

The least invasive *in situ* remediation approach is the identification of intrinsic bioremediation, whereby it can be demonstrated that under suitable environmental conditions an indigenous microbial population exists and that degradation has already occurred and is continuing. In order to support or enhance this natural self-curing ability, bioaugmentation and biostimulation technologies are available and in some cases suited. In all these cases, degradation rates, degradation efficiency, and groundwater flow rates should be carefully monitored and the remediation area should be planned far enough downstream the groundwater to avoid transportation of contaminants or metabolites out of the remediation field. By construction of funnel-and-gate systems at the downstream end of the remediation field non-degraded contaminants and residual metabolites could be adsorbed to activated carbon or other suited materials and be prevented from migrating into non-contaminated areas.

5 Drinking Water Preparation

In some regions toxic substances trickled into the groundwater and must be separated during processing for drinking water preparation. This is a consequence of soil pollution by leachates from sanitary landfills, production residues or spillages from industry or over-fertilization or insecticide and pesticide application in agriculture. Separation, filtration and hygienization procedures have been developed and have reached high technological levels. Since contamination of groundwater is still increasing and many contaminants remain for decades, water purification procedures must have high priority at present and in the future, especially since the drinking water resources are limited.

Although techniques for complete purification of wastewater are principally available, the application of these multi-step procedures for drinking water preparation is not likely in the near future due to the very high water processing costs and the availability of less polluted water resources.

6 Off-Gas Purification

For removal of organic and inorganic pollutants from huge quantities of highly polluted waste gas streams of, e.g., the lime and cement industry, coal power plants, or waste incineration plants technical procedures are available for the separation of fly ashes, e.g., by gas cyclones, particle filtration or electrofiltration, of acid and alkaline gas impurities by washing procedures and of neutral trace gas components by adsorption/gas filtration. Off-gas from composting plants, pork and chicken breeding stables, etc. which is polluted with mainly volatile organic pollutants may be purified by gas washing and aerobic/anaerobic treatment of the washing water by using biofilters. Natural and synthetic filter materials have been applied as support materials for the development of active biofilms. To maintain a permanent high adsorption and degradation efficiency, the moisture content of the filter material must be kept at a certain minimum level and trace elements should eventually be added to supply the microorganisms forming the biofilm.

7 Future Strategies to Reduce Pollution and Conserve a Natural Environment

In industialized countries the main strategy for handling of domestic and industrial wastewater seems to be fixed for years or even decades due to high investments into sewer systems and to what is considered modern wastewater treatment. A high efficiency of carbon, nitrogen, and phosphate removal was intended in the past to avoid damage of the receiving ecosystems.

In Germany, centralized treatment with hundreds of miles of sewers and many pumping stations has almost been completed for domestic wastewater. "Spot solutions" for new settlement areas, for single houses, or very small settlements should be promoted to experience new small-scale process alternatives.

The real challenge for wastewater purification comes up in many developing countries. Central treatment units are unaffordable, and even if they would exisit the sewer systems would not be capable to handle the masses of rain water during the rainy season. This is why decentralized wastewater and waste treatment should be favored. Due to the still unreliable electric supply outside the mega-cities as a starting technology small-scale treatment systems of industrialized countries should be reduced to the basic components, requiring little or no electricity or skilled personnel for maintainence. Decentralized wastewater management should be favored, not only because a copy of our systems would not be affordable, but because the resources of wastewater could be better used. Domestic wastewater or wastes, if properly collected and treated, can be upgraded to yield valuable fertilizers and thus save money for mineral fertilizers. By decentralized treatment more farm land for nontoxic wastewater or waste compost is available and transport distances are short.

A process development that goes hand-in-hand with investigations on the respective microbiology is very important for the future development of wastewater treatment. Microbial reaction rates are higher in equatorial countries due to increased average annual temperatures.

Future microbial investigations for wastewater treatment should start with the complex ecophysiology and must finally trace and optimize single microbial bottle-neck reactions. Except for Anammox, the anaerobic ammonia oxidation process, microbiology always seemed to lag far behind technical verification.

Whereas in some branches of the food and feed industry starter cultures or even enzymes meanwhile are essential for production, the advantage of a broad application of starter cultures for wastewater treatment (bioaugmentation) in order to improve purification efficiencies or the degradation of trace compounds may still be doubted at present. Starter cultures containing genetically engineered specialists for certain xenobiotics which periodically appear in more than trace concentrations may, however, help to introduce or stabilize the establishment of the required metabolic capabilities. Starter cultures with an omnipotent population might be seeded only after complete process failure by toxicants in order to re-establish the microbial degradation potential more quickly in wastewater treatment plants receiving wastewater with little indigenous population.

A major problem at present and in the future is the handling of surplus sludge from wastewater treatment. Dewatering procedures must be improved and new and better sludge disintegration methods developed. Although in some cases the microbiological basis for the formation of bulking sludge is understood, reliable microbiological counteractions for avoidance of bulking are not yet available. For sludge disintegration enzyme engineering will hopefully create new, stable and powerful lytic enzymes in the future.

For water management in new settlement areas the development might go into the direction of dual water supply, on-spot treatment of slightly polluted wastewater, and oozing of purified wastewater in especially designed ecosystems. The concentrated wastewater streams should also be treated in close neighborhood to their generation. New settlement areas must be planned with little pavement (or existing settlement areas should be depaved) to retain most of the rain water for replenishing the groundwater body.

Industrial production processes with a better product-to-wastes relation have to be developed by applying new production processes or by more efficient utilization of the water, e.g., by internal water circles. Tailor-made treatment systems for every wastewater stream should be optimized with emphasis on procedures and on the microbiological capabilities, including the application of starter cultures (bioaugmentation).

The slogan "the waste of one company is the raw material of another company" should be verified countrywide and might be facilitated by respective data banks. Retail prices for all goods, including those imported from developing countries, should include the full, real or fictive costs for wastewater and residue treatment.

For solid wastes the potential to reduce the total amount in future must be fully used, in particular by the packaging industry. Improvement of distribution logistics may help to avoid one-way single-product packaging, packages for product arrangements, and a third packaging in addition for transport of larger package units.

Since incineration is the most expensive waste destruction system, it should be reserved only for those fractions which cannot be recycled or re-utilized. Biowaste composting and biowaste methanation are options for organic waste fractions with a high content of naturally occurring organics. Co-fermenatation of biowaste fractions with sewage sludge might also be taken into consideration, if excess digester volume is available. A combined mechanical and biological waste inertization could be an alternative to incineration, but cannot reach the low carbon content of the German TA Siedlungsabfall.

In developing countries direct re-utilization or product recycling seems to be more distributed than in highly industrialized countries due to a shortage of raw materials or due to restricted production or affordance. This is especially true for, e.g., plastic bottles or containers, which are often one-way articles in industrialized countries, but re-utilized several times in developing countries.

In industrialized countries drinking water management must in the future take care of trace pollutants with unknown effects to human health. New methods to analyze and separate residual agricultural or household chemicals or their metabolites must be developed.

Due to the high number of contaminated areas in almost every country and due to a restricted budget for soil remediation a ranking according to environmental risks should be performed. Then soil remediation techniques should be chosen which allow avoidance of further migration of the contaminants or their possibly toxic reaction products. Besides the common techniques for groundwater treatment (pump-and-treat) funnel-and-gate systems and reactive barriers increasingly have to be applied. For treatment of sites with low contaminant concentrations phytoremediation approaches for metals and organics, e.g., nitrocompounds and polycyclic aromatic hydrocarbons, have to be tested increasingly. Together with other near-natural processes and the monitored natural attenuation procedures sustainable strategies have to be developed to overcome the problems of contaminated sites. Furthermore, a variety of bacterial species and enzymes have been the target of genetic engineering to improve the performance of biodegradation, control degradation processes, and detection of chemical pollutants and their bioavailability. On the other hand avoidance of environmental contamination is the future challenge for which suitable and sustainable strategies can only be achieved by an interdisciplinary collaboration between all protagonists in research and industry. The wide-ranging experience accumulated with respect to the contamination of soils and groundwater must provide a special impetus for testing the environmental impact of new chemical products before they are introduced, thus preventing subsequent contamination. A benign "design chemistry" would, therefore, have to concentrate research on identifying forms of bonding which facilitate the development of biodegradable and environmentally sound substances in the circulation of chemical products.

The supply of good drinking water quality must especially be improved in developing countries in order to reduce mortality, particularly of children. Groundwater pumping through deep wells often exceeds the amount of newly formed groundwater. So wells are thrilled deeper and deeper. In costal regions this may cause salt water infiltration which contaminates the sweet water reserves.

Wastewater oozing and groundwater pumping often are in close spacial neighborhood, not maintaining a sufficient purification stretch for complete degradation and sufficient hygienization. Contamination of the well water with pathogenic micororgansims is favored by this mismanagement and the warm climate causes epidemies.

Off-gas purification by biological means expanded very much in the past. For biological off-gas purification existing gas ventilation, washing or filtration techniques and the respective technical equipment must be improved further.

Index

C

D

E

F

P

Q

R

S